Conversion Factors

Mass	$1\ kg = 1000\ g$
	$1\ g = 10^{-3}\ kg$
	$1\ u = 1.66 \times 10^{-27}\ kg \leftrightarrow 931.5\ MeV/c^2$
	$1\ slug = 14.6\ kg$
Length	$1\ m = 100\ cm = 1000\ mm$
	$1\ foot = 1\ ft = 0.305\ m$
	$1\ inch = 1\ in. = 2.54\ cm$
	$1\ mile = 1\ mi = 5280\ ft = 1609\ m$
	$1\ light\text{-}year\ (ly) = 9.46 \times 10^{15}\ m$
Time	$1\ min = 60\ s$
	$1\ h = 60\ min = 3600\ s$
	$1\ day = 24\ h = 8.64 \times 10^4\ s$
	$1\ yr = 365.24\ days = 3.16 \times 10^7\ s$
Force	$1\ N = 0.225\ lb$
	$1\ dyne = 10^{-5}\ N$
	$1\ lb = 4.45\ N = 16\ oz$
Energy	$1\ J = 10^7\ erg = 0.239\ cal$
	$1\ cal = 4.186\ J$
	$1\ kW \cdot h = 3.60 \times 10^6\ J$
	$1\ eV = 1.60 \times 10^{-19}\ J$
Power	$1\ W = 1\ J/s$
	$1\ hp = 745.7\ W$
Pressure	$1\ Pa = 1\ N/m^2$
	$1\ atmosphere = 1\ atm = 1.01 \times 10^5\ Pa = 14.7\ lb/in.^2$
	$1\ millimeter\ of\ Hg = 1\ mm\ Hg = 133\ Pa$
	$1\ bar = 10^5\ Pa$
	$1\ lb/in.^2 = 6.89 \times 10^3\ Pa$
Speed/velocity	$1\ m/s = 3.60\ km/h$
	$1\ mi/h = 0.447\ m/s$
	$100\ mi/h \approx 45\ m/s$
Temperature	$T(K) = T(°C) + 273.15$
	$T(°F) = \frac{9}{5}T(°C) + 32$
	$T(°C) = \frac{5}{9}[T(°F) - 32]$
Angle	$180° = \pi\ rad$

Useful Physical Data

Acceleration due to gravity near the Earth's surface	$g = 9.8 \text{ m/s}^2$
Atmospheric pressure	$1.01 \times 10^5 \text{ Pa}$
Density of water (4°C)	$1000 \text{ kg/m}^3 = 1 \text{ g/cm}^3$
Density of air (20°C, 1 atm)	1.20 kg/m^3
Absolute zero of temperature	$-273.15°C$
Speed of sound in air (20°C, 1 atm)	343 m/s
Radius of the Earth	$6.37 \times 10^3 \text{ km} = 6.37 \times 10^6 \text{ m}$
Mass of the Earth	$5.97 \times 10^{24} \text{ kg}$
Radius of the Moon	1740 km
Mass of the Moon	$7.35 \times 10^{22} \text{ kg}$
Average Earth–Moon distance	$3.84 \times 10^5 \text{ km}$
Radius of the Sun	$6.95 \times 10^5 \text{ km}$
Mass of the Sun	$1.99 \times 10^{30} \text{ kg}$
Average Earth–Sun distance	$1.50 \times 10^8 \text{ km}$

COLLEGE PHYSICS

Reasoning and Relationships

COLLEGE PHYSICS

Reasoning and Relationships

Nicholas J. Giordano
PURDUE UNIVERSITY

BROOKS/COLE
CENGAGE Learning™

Australia • Brazil • Japan • Korea • Mexico • Singapore • Spain • United Kingdom • United States

BROOKS/COLE
CENGAGE Learning™

College Physics: Reasoning and Relationships, First Edition

Nicholas J. Giordano

Editor-in-Chief: Michelle Julet

Publisher: Mary Finch

Managing Editor: Peggy Williams

Senior Development Editor: Susan Dust Pashos

Development Project Manager: Terri Mynatt

Development Editor: Ed Dodd

Development Editor Art Program: Alyssa White

Associate Development Editor: Brandi Kirksey

Editorial Assistant: Stephanie Beeck

Media Editors: Sam Subity/ Rebecca Berardy Schwartz

Marketing Manager: Nicole Mollica

Marketing Coordinator: Kevin Carroll

Marketing Communications Manager: Belinda Krohmer

Project Manager, Editorial Production: Teresa L. Trego

Creative Director: Rob Hugel

Art Director: John Walker

Print Buyer: Rebecca Cross

Permissions Editor: Mollika Basu

Production Service: Lachina Publishing Services

Text Designer: Brian Salisbury

Art Editor: Steve McEntee

Photo Researcher: Dena Digilio Betz

Copy Editor: Kathleen Lafferty

Illustrator: Dartmouth Publishing Inc., 2064 Design/Greg Gambino

Cover Designer: William Stanton

Cover Image: Ted Kinsman/Photo Researchers, Inc.

Compositor: Lachina Publishing Services

For product information and technology assistance, contact us at **Cengage Learning Customer & Sales Support, 1-800-354-9706.**

For permission to use material from this text or product, submit all requests online at **www.cengage.com/permissions.** Further permissions questions can be e-mailed to **permissionrequest@cengage.com.**

Library of Congress Control Number: 2008931521

ISBN-13: 978-0-534-42471-8
ISBN-10: 0-534-42471-6

Volume 1
ISBN-13: 978-0-534-46243-7
ISBN-10: 0-534-46243-X

Volume 2
ISBN-13: 978-0-534-46244-4
ISBN-10: 0-534-46244-8

Brooks/Cole
10 Davis Drive
Belmont, CA 94002-3098
USA

Cengage Learning is a leading provider of customized learning solutions with office locations around the globe, including Singapore, the United Kingdom, Australia, Mexico, Brazil, and Japan. Locate your local office at: **international.cengage.com/region**

Cengage Learning products are represented in Canada by Nelson Education, Ltd.

For your course and learning solutions, visit **www.cengage.com**

Purchase any of our products at your local college store or at our preferred online store **www.ichapters.com**

Printed in Canada
1 2 3 4 5 6 7 12 11 10 09

Brief Table of Contents

Table of Contents

(© Gandee Vasan/Stone/Getty Images)

© Cengage Learning/Charles D. Winters

© Charles Gupton/Stone/Getty Images

NASA/JPL-Caltech/University of Arizona/STScI

Chapter 27

Relativity 917

Chapter 28

Quantum Theory 954

Chapter 29

Atomic Theory 986

© Photo Researchers/Alamy

Preface

College Physics: Reasoning and Relationships is designed for the many students who take a college physics course. The majority of these students are not physics majors (and don't want to be) and their college physics course is the only physics class they will ever take. The topics covered in a typical college physics course have changed little in recent years, and even decades. Indeed, except for many of the applications, much of the physics covered here was well established more than a century ago. Although the basic material covered may not be changing much, the way it is taught should not necessarily stay the same.

GOALS OF THIS BOOK

Reasoning and Relationships

Students often view physics as merely a collection of loosely related equations. We who teach physics work hard to overcome this perception and help students understand how our subject is part of a broader science context. But what does "understanding" in this context really mean?

Many physics textbooks assume understanding will result if a solid problem-solving methodology is introduced early and followed strictly. Students in this model can be viewed as successful if they deal with a representative collection of quantitative problems. However, physics education research has shown that students can succeed in such narrow problem-solving tasks and at the same time have fundamentally flawed notions of the basic principles of physics. For these students, physics *is* simply a collection of equations and facts without a firm connection to the way the world works. Although students do need a solid problem-solving framework, I believe such a framework is only one component to learning physics. For real learning to occur, students must know how to reason and must see the relationships between the ideas of physics and their direct experiences. Until the reasoning is sound and the relationships are clear, fundamental learning will remain illusive.

The central theme of this book is to weave reasoning and relationships into the way we teach introductory physics. Three important results of this approach are the following:

1. Establishing the relationship between forces and motion.
2. A systematic approach to problem solving.
3. Reasoning and relationship problems.

1. Establishing the relationship between forces and motion.

All of Chapter 2 is devoted to Newton's laws of motion and what they tell us about the way force and motion are related. This is the central thread of mechanics. Armed with an understanding of the proper relationship between kinematics and forces, students can then reason about a variety of problems in mechanics such as "nonideal" cases in which the acceleration is not constant, as found for projectiles with air drag.

From Chapter 2, page 45

Newton's second law tells us that the acceleration of an object is given by $\vec{a} = (\sum \vec{F})/m$, where $\sum \vec{F}$ is the total force acting on the object. In the simplest situations, there may only be one or two forces acting on an object, and $\sum \vec{F}$ is then the sum of these few forces. In some cases, however, there may be a very large number of forces acting on an object. Multiple forces can make things appear to be very complicated, which is perhaps why the correct laws of motion—Newton's laws—were not discovered sooner.

Forces on a Swimming Bacterium ⊗

Figure 2.32 shows a photo of the single-celled bacterium *Escherichia coli*, usually referred to as *E. coli*. An individual *E. coli* propels itself by moving thin strands of protein that extend away from its body (rather like a tail) called flagella. Most *E. coli* possess several flagella as in the photo in Figure 2.32A, but to understand their function, we first consider the forces associated with a single flagellum as sketched in Figure 2.32B. A flagellum is fairly rigid, and because it has a spiral shape, one can think of it as a small propeller. An *E. coli* bacterium moves about by rotating this propeller, thereby exerting a force $\vec{F}_w$ on the nearby water. According to Newton's third law, the water exerts a force $\vec{F}_E$ of equal magnitude and opposite direction on the *E. coli*, as sketched in Figure 2.32B. One might be tempted to apply Newton's second law with the force $\vec{F}_E$ and conclude that the *E. coli* will move with an acceleration that is proportional to this force. However, this is incorrect because we have not included the forces from the water on the body of the *E. coli*. These forces are also indicated in Figure 2.32B; to properly describe the total force from the water, we must draw in many force vectors, pushing the *E. coli* in virtually all directions. At the molecular level, we can understand these forces as follows. We know that water is composed of molecules that are in constant motion, and these water molecules bombard the *E. coli* from all sides. Each time a water molecule collides with the *E. coli*, the molecule exerts a force on the bacterium, much like the collision of the baseball and bat in Figure 2.30. As we saw in that case, the two colliding objects both experience a recoil force, another example of action–reaction forces. So, in the present case, the *E. coli* and the water molecule exert forces on each other. An individual *E. coli* is not very large, but a water molecule is much smaller than the bacterium, and the force from one such collision will have only a small effect on the

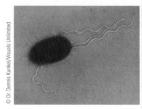
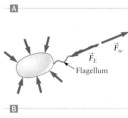

Figure 2.32 Ⓐ *E. coli* use the action–reaction principle to propel themselves. An individual *E. coli*

---

2. A systematic approach to problem solving.

Every worked example follows a five-step format. The first step is to "*Recognize the principle*" that is key to the problem. This step helps students see the "big picture" the problem illustrates. The other steps in the problem-solving process are "*Sketch the problem*," "*Identify the relationship*s," "*Solve*," and "*What does it mean?*" The last step emphasizes the key principles once more and often describes how the problem relates to the real world. Explicit problem-solving strategies are also given for major classes of quantitative problems, such as conservation of mechanical energy.

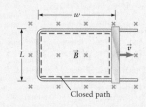

Figure 21.12 Example 21.3.

From Chapter 21, page 698

EXAMPLE 21.3 | Using Lenz's Law to Find the Direction of an Induced Current

Consider the sliding metal bar in Figure 21.12. The magnetic field is constant and is directed into the plane of the drawing. If the bar is sliding to the right, use Lenz's law to find the direction of the induced current.

RECOGNIZE THE PRINCIPLE

According to Lenz's law, the induced current produces a field that opposes the change in flux through the circuit loop.

SKETCH THE PROBLEM

Following our "Applying Lenz's Law" problem-solving strategy (step 2), the sketch in Figure 21.12 shows a dashed, rectangular path. We are interested in the current induced in this circuit, so we must consider the flux through this rectangle.

IDENTIFY THE RELATIONSHIPS

Step 3: The magnetic field is directed into the plane of this drawing, so the flux through the rectangular surface is directed into the plane. The area of this chosen surface is wL, and the magnetic flux through the surface is $\Phi_B = BwL$. Because the bar is sliding to the right, Φ_B is increasing and is downward.

SOLVE

The induced emf produces an induced current that opposes the downward increase in the flux, so the induced magnetic field must be directed *upward*. Applying right-hand rule number 1 (Chapter 20), this field direction is produced by a *counterclockwise* induced current.

What does it mean?

The flux through a given area may be "upward" or "downward," and its magnitude may be increasing or decreasing with time. The induced emf always *opposes any changes* in the flux.

---

1. **RECOGNIZE THE PRINCIPLE.** The induced emf always opposes changes in flux through the Lenz's law loop or path.

2. **SKETCH THE PROBLEM,** showing a closed path that runs along the perimeter of a surface crossed by magnetic field lines.

3. **IDENTIFY** if the magnetic flux through the surface is increasing or decreasing with time.

4. **SOLVE.** Treat the perimeter of the surface as a wire loop; suppose there is a current in this loop and determine the direction of the resulting magnetic field. Find the current direction for which this induced magnetic field opposes the change in magnetic flux. This current direction gives the sign (i.e., the "direction") of the induced emf.

5. Always *consider what your answer means* and check that it makes sense.

From Chapter 21, page 698

PROBLEM SOLVING | Dealing with Reasoning and Relationships Problems

1. **RECOGNIZE THE PRINCIPLE.** Determine the key physics ideas that are central to the problem and that connect the quantity you want to calculate with the quantities you know. In the examples found in this section, this physics involves motion with constant acceleration.

2. **SKETCH THE PROBLEM.** Make a drawing that shows all the given information and everything else that you know about the problem. For problems in mechanics, your drawing should include all the forces, velocities, and so forth.

3. **IDENTIFY THE RELATIONSHIPS.** Identify the important physics relationships; for problems concerning motion with constant acceleration, they are the relationships between position, velocity, and acceleration in Table 3.1. For many reasoning and relationships problems, values for some of the essential unknown quantities may not be given. You must then use common sense to make reasonable estimates for these quantities. Don't worry or spend time trying to obtain precise values of every quantity (such as the amount that the child's knees flex in Fig. 3.23). Accuracy to within a factor of 3 or even 10 is usually fine because the goal is to calculate the quantity of interest to within an order of magnitude (a factor of 10). Don't hesitate to use the Internet, the library, and (especially) your own intuition and experiences.

4. **SOLVE.** Since an exact mathematical solution is not required, cast the problem into one that is easy to solve mathematically. In the examples in this section, we were able to use the results for motion with constant acceleration.

5. Always *consider what your answer means* and check that it makes sense.

As is often the case, practice is a very useful teacher.

From Chapter 3, page 78

3. Reasoning and relationship problems. Many real-world applications require an estimation of certain key parameters. For example, the approximate force on a car bumper during a collision can be found by making a few simplifying assumptions about the collision and the way the bumper deforms, and estimating the mass of the car. Physicists find these "back-of-the-envelope" calculations very useful for gaining an intuitive understanding of a situation. The ability to deal with such problems requires a good understanding of the key relationships in the problem and how fundamental principles can be applied. Most textbooks completely ignore such problems, but I believe students can, with a careful amount of coaching and practice, learn to master the skills needed to be successful with these problems. This kind of creative problem solving is a valuable skill for students in all areas.

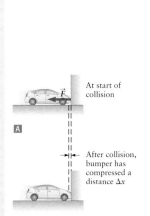

Figure 3.24 Example 3.6.
A When a car collides with a wall, the wall exerts a force F on the bumper. This force provides the acceleration that stops the car.
B During the collision and before the car comes to rest, the bumper deforms by an amount Δx. The car travels this distance while it comes to a complete stop.

At start of collision

After collision, bumper has compressed a distance Δx

From Chapter 3, page 78

EXAMPLE 3.6 | ® Cars and Bumpers and Walls

Consider a car of mass 1000 kg colliding with a rigid concrete wall at a speed of 2.5 m/s (about 5 mi/h). This impact is a fairly low-speed collision, and the bumpers on a modern car should be able to handle it without much damage to the car. Estimate the force exerted by the wall on the car's bumper.

RECOGNIZE THE PRINCIPLE

The motion we want to analyze starts when the car's bumper first touches the wall and ends when the car is stopped. To treat the problem approximately, we *assume* the force on the bumper is constant during the collision period, so the acceleration is also constant. We can then use our expressions from Table 3.1 to analyze the motion. Our strategy is to first find the car's acceleration and then use it to calculate the associated force exerted by the wall on the car from Newton's second law.

SKETCH THE PROBLEM

Figure 3.24 shows a sketch of the car along with the force exerted by the wall on the car. There are also two vertical forces—the force of gravity on the car and the normal force exerted by the road on the car—but we have not shown them because we are concerned here with the car's horizontal (x) motion, which we can treat using $\sum F = ma$ for the components of force and acceleration along x.

IDENTIFY THE RELATIONSHIPS

To find the car's acceleration, we need to estimate either the stopping time or the distance Δx traveled during this time. Let's take the latter approach. We are given the initial velocity ($v_0 = 2.5$ m/s) and the final velocity ($v = 0$). Both of these quantities are in Equation 3.4:

$$v^2 = v_0^2 + 2a(x - x_0) \tag{1}$$

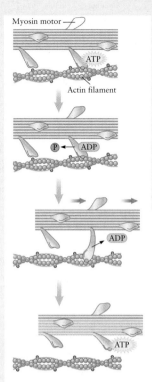

Figure 6.34 Some molecular motors move by "walking" along long strands of a protein called actin. These motors are the subject of much current research. We can use work–energy principles to understand their behavior.

From Chapter 6, page 195

Changing the Way Students View Physics

The *relationships* between physics and other areas of science are rapidly becoming stronger and are transforming the way all fields of science are understood and practiced. Examples of this transformation abound, particularly in the life sciences. Many students of college physics are engaged in majors relating to the life sciences, and the manner in which they need and will use physics differs from only a few years ago. For their benefit, and for the benefit of students in virtually all technical and even nontechnical disciplines, textbooks must place a greater emphasis on how to apply the *reasoning* of physics to real-world examples. Such examples come quite naturally from the life sciences, but many everyday objects are filled with good applications of fundamental physics principles as well. For instance, my discussion of molecular motors in the context of work and energy in Chapter 6 is unique, as are discussions of photosynthesis as a thermodynamic process in Chapter 16 and electricity in the atmosphere (lightning) in Chapter 19. Students must be made to see that physics is relevant to their daily lives and to the things they find interesting.

6.8 | ⓧ WORK, ENERGY, AND MOLECULAR MOTORS

In Section 6.7, we discussed how our ideas about work, energy, and power can be used to understand the behavior of motors and similar devices. The same ideas apply to all types of motors, including the molecular motors that transport materials within and between cells. Several different types of molecular motors have been discovered, one example of which is sketched in Figure 6.34. This motor is based on a molecule called myosin that moves along long filaments composed of actin molecules.

Myosin is believed to move by "walking" along a filament in steps, much like a person walks on two legs. This process takes place in muscles, where the filaments form bundles. A myosin molecule is also attached to its own muscle filament, so as it walks, the myosin drags one muscle filament relative to another. The operation of your muscles is produced by these molecular motors.

Calculating the Force Exerted by a Molecular Motor

The precise biochemical reactions involved in the myosin walking motion are not completely understood. However, we do know that each step has a length of approximately 5 nm (5×10^{-9} m) and that the energy for this motor comes from a chemical

Although much can be gained by bringing many new and current examples into the text, traditional physics examples such as inclined planes, pulleys, and resistors in series or parallel can still be useful pedagogical tools. A good example, however, must do more than just illustrate a particular principle of physics; students should also see clearly how the example can be expanded and generalized to other (and, I hope, interesting) situations. The block-and-tackle example in Chapter 3 is one such case, illustrating pulleys and tension forces in a traditional way but going on to describe how this device can amplify forces. This theme of force amplification is revisited in future chapters in discussions of torque and levers, work, hydraulics, and conservation of energy, and it is also applied to the mechanical function of the human ear. Returning to key themes throughout the text gives students a deeper understanding of fundamental physics principles and their relationship to real-world applications.

Encouraging student curiosity. Many important and fundamental ideas about the world are ignored in most textbooks. By devoting some time to these ideas, this book helps students see that physics can be extremely exciting and interesting. Such issues include the following. (1) Why is the inertial mass equal to the gravitational mass? (This question is mentioned in Chapter 3, revisited in the discussion of gravitation, and mentioned again in Chapter 27 on relativity.) (2) What is "action-at-a-distance," and how does it really work? (The concept of a *field* is mentioned in several places, including but not limited to the sections on gravitation and Coulomb's law.) (3) How do we know the structure of Earth's core? (4) What does color vision tell us about the nature of light? These issues and others like them are essentially unmentioned in current texts, yet they get to the heart of physics and can stimulate student curiosity.

Starting where students are and going farther than you imagined possible. Many students come to their college physics course with a common set of pre-Newtonian misconceptions about physics. I believe the best way to help students overcome these misconceptions is to address them directly and help students see where and how their pre-Newtonian ideas fail. For this reason, *College Physics: Reasoning and Relationships* devotes Chapter 2 to the fundamental relationships between force and motion as Newton's laws of motion are introduced. The key ideas are then reinforced in Chapter 3 with careful discussions of several applications of Newton's laws in one dimension. This approach allows us to get to the more interesting material faster, and, in my experience, the students are more prepared for it then.

Building on prior knowledge. A good way to learn is to build from what is already known and understood. (Learning scientists call this "scaffolding.") This book therefore revisits and builds on selected examples with a layered development, deepening and extending the analysis as new physical principles are introduced. In typical cases, a topic is revisited two or three times, both within a chapter and across several chapters. One example is the theme of amplifying forces, which begins in Chapter 3. This theme reappears in a number of additional topics, including the mechanics of the ear and the concepts of work and energy. Layered or scaffolded development of concepts, examples, and problem topics helps students see relationships between various physical principles.

From Chapter 3, page 74

Using Pulleys to Redirect a Force

Cables and ropes are an efficient way to transmit force from one place to another, but they have an important limitation: they can only "pull," and this force must be directed along the direction in which the cable lies. In many situations, we need to change the direction of a force, which can be accomplished by using an extremely useful mechanical device called a pulley. A simple pulley is shown in Figure 3.21A; it is just a wheel free to spin on an axle through its center, and it is arranged so that a rope or cable runs along its edge without slipping. For simplicity, we assume both the rope and the pulley are massless. Typically, a person pulls on one end of the rope so as to lift an object connected to the other end. The pulley simply changes the direction of the force associated with the tension in the rope as illustrated in Figure 3.21B, which shows the rope "straightened out" (i.e., with the pulley removed). In either case—with or without the pulley in place—the person exerts a force F on one end of the rope, and this force is equal to the tension. The tension is the same everywhere along this massless rope, so the other end of the rope exerts a force of magnitude T on the object. A comparison of the two arrangements in Figure 3.21—one with the pulley and one without—suggests the tension in the rope is the same in the two cases, and in both cases the rope transmits a force of magnitude $T = F$ from the person to the object.

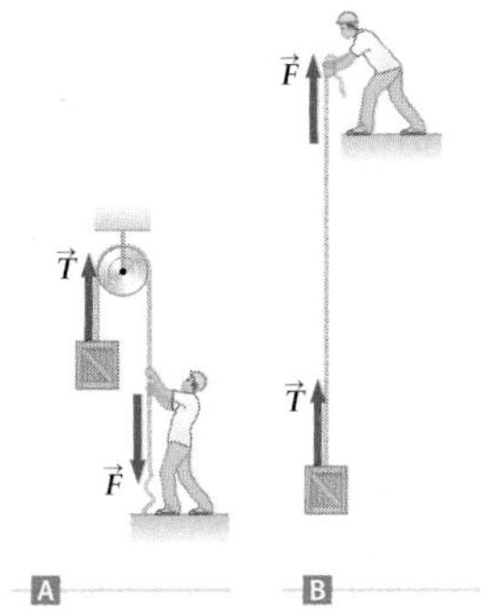

Figure 3.21 Ⓐ A simple pulley. If the person exerts a force F on a massless string, there is a tension T in the string, and this tension force can be used to lift an object. Ⓑ The string "straightened out." The pulley simply redirects the force.

Work, Energy, and Amplifying Forces

In Chapter 3, we encountered a device called the block and tackle and showed how it can amplify forces. Figure 6.8 shows a block and tackle that amplifies forces by a factor of two. We saw in Chapter 3 that if a person applies a force F to the rope, the tension in the rope is $T = F$. Because the pulley is suspended by two portions of the rope, the upward force on the pulley is $2T$ and the pulley exerts a total force $2T$ on the object to which it is connected, that is, the crate in Figure 6.8. Let's now consider how this process of force amplification affects the work done by the person. Suppose the person lifts his end of the rope through a distance L. That will raise the pulley by half that amount, that is, a distance of $L/2$. You can see why by noticing that when the pulley moves upward through a distance $L/2$, the sections of the rope on both sides become shorter by this amount, so the end of the rope held by the person must move a distance L.

When the pulley moves upward by a distance $L/2$, the crate is displaced by the same amount. The work done *by the pulley on the crate* is equal to the total force of the pulley on the crate ($2T$) multiplied by the displacement of the crate, which is $L/2$:

$$W_{\text{on crate}} = 2T(L/2) = TL$$

At the same time, the person does work on the end of the rope since he exerts a force $F = T$ and the displacement of the end of the rope is L. The work done *by the person on the rope* is equal to the force that he exerts on the rope (T) multiplied by the displacement of the rope, which is L, so

$$W_{\text{on rope}} = FL = TL$$

From Chapter 6, page 173

Thus, the work done on the rope is precisely equal to the work done on the crate.

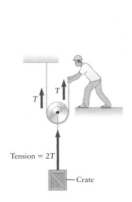

Figure 6.8 The person applies a force T to the rope. The block and tackle amplifies this force, and the total force applied to the crate by the rope is $2T$. However, when he moves the end of the rope a distance L, the crate moves a distance of only $L/2$.

Going the extra step: reasoning and relationship problems. Many interesting real-world physics problems cannot be solved exactly with the mathematics appropriate for a college physics course, but they can often be handled in an approximate way using the simple methods (based on algebra and trigonometry) developed in such a course. Professional physicists are familiar with these back-of-the-envelope calculations. For instance, we may want to know the approximate force on a skydiver's knees when she hits the ground. The precise value of the force depends on the details of the landing, but we are often interested in only an approximate (usually order-of-magnitude) answer. We call such problems *reasoning and relationship* problems because solving them requires us to identify the key physics relationships and quantities needed for the problem and that we also estimate values of some important quantities (such as the mass of the skydiver and how she flexes her knees) based on experience and common sense.

Reasoning and relationship problems provide physical insight and a chance to practice critical thinking (reasoning), and they can help students see more clearly the fundamental principles associated with a problem. A truly unique feature of this book is the inclusion of these problems in both the worked examples and the end-of-chapter problems. The ability to deal with this class of problems is an extremely useful skill for all students, in all fields.

Problem solving: a key component to understanding. Although reasoning and relationship problems are used to help students develop a broad understanding of physics, this book also contains a strong component of traditional quantitative problem solving. Quantitative problems are a component of virtually all college physics courses, and students can benefit by developing a systematic approach to such problems. *College Physics: Reasoning and Relationships* therefore places extra emphasis on step-by-step approaches students can use in a wide range of situations. This approach can be seen in the worked examples, which use a five-step solution process: (1) recognize the physics principles central to the problem, (2) draw a sketch showing the problem and all the given information, (3) identify the relationships between the known and unknown quantities, (4) solve for the desired quantity, and (5) ask what the answer means and if it makes sense. Explicit problem-solving strategies are also given for major classes of quantitative problems, such as applying the conservation of mechanical energy.

From Chapter 17, page 535

PROBLEM SOLVING | Calculating Forces with Coulomb's Law

1. **RECOGNIZE THE PRINCIPLE.** The electric force on a charged particle can be found using Coulomb's law together with the principle of superposition.

2. **SKETCH THE PROBLEM.** Construct a drawing (including a coordinate system) and show the location and charge for each object in the problem. Your drawing should also show the directions of all the electric forces—$\vec{F}_1$, $\vec{F}_2$, and so forth—on the particle(s) of interest.

3. **IDENTIFY THE RELATIONSHIPS.** Use Coulomb's law (Eq. 17.3 or 17.5) to find the magnitude of the forces $\vec{F}_1$, $\vec{F}_2$, ... acting on the particle(s)

4. **SOLVE.** The total force on a particle is the sum (the superposition) of all the individual forces from steps 2 and 3. Add these forces *as vectors* to get the total force. When adding these vectors, it is usually simplest to work in terms of the components of $\vec{F}_1$, $\vec{F}_2$, ... along the coordinate axes.

5. Always *consider what your answer means* and check that it makes sense.

PROBLEM SOLVING | Plan of Attack for Problems in Statics

1. **RECOGNIZE THE PRINCIPLE.** For an object to be in static equilibrium, the sum of all the forces on the object must be zero. This principle leads to Equation 4.2, which can be applied to calculate any unknown forces in the problem.

2. **SKETCH THE PROBLEM.** It is usually a good idea to show the given information in a picture, which should include a coordinate system. Figures 4.1 through 4.3 and the following examples provide guidance and advice on choosing coordinate axes.

3. **IDENTIFY THE RELATIONSHIPS.**
   - Find all the forces acting on the object that is (or should be) in equilibrium and construct a free-body diagram showing all the forces on the object.

   - Express all the forces on the object in terms of their components along x and y.
   - Apply the conditions[3] $\sum F_x$ and $\sum F_y = 0$.

4. **SOLVE.** Solve the equations resulting from step 3 for the unknown quantities. The number of equations must equal the number of unknown quantities.

5. Always *consider what your answer means* and check that it makes sense.

From Chapter 4, page 95

ORGANIZATION AND CONTENT

Translational motion. The organization of topics in this book follows largely traditional lines with one exception. Forces and Newton's laws of motion are introduced in Chapter 2 along with basic defining relationships from kinematics. In almost all other texts, kinematic equations are covered first in the absence of Newton's laws, which obscures the cause of motion. This text presents the central thread of all mechanics from the beginning, allowing students to see and appreciate the motivations for many kinematic relationships. Students can then address and overcome common misconceptions early in the course. They are also able to deal sooner with interesting and realistic problems that do not involve a constant acceleration. When forces and Newton's laws are introduced early, we can discuss issues such as air drag and terminal velocity at an earlier stage, which avoids giving students the impression that physics problems are limited to the mathematics of ideal cases. Chapters 2 and 3 are limited to one-dimensional problems for simplicity before moving on to two dimensions in Chapters 4 and 5. Major conservation principles (of energy and momentum) are introduced in Chapters 6 and 7 before moving on to rotational motion.

Rotational motion. In the same way that force is connected to acceleration in Chapters 2 and 3 while introducing the variables of translational motion, torque is connected to angular acceleration while the variables of rotational motion are introduced in Chapter 8. Parallel development of topics in translational and rotational motion is direct and deliberate. The central thread in Chapters 8 and 9 is once again Newton's laws of motion, this time in rotational form.

Fluids. Chapter 10 discusses fluids, including the principles of Pascal, Archimedes, and Bernoulli.

Waves. Chapters 11 through 13 cover harmonic motion, waves, and sound. Waves—moving disturbances that transport energy without transporting matter—provide a link to later topics in electromagnetism, light, and quantum physics.

Thermal physics. Chapters 14 through 16 on thermal physics have conservation of energy as their central thread. Thermodynamics is about the transfer of energy between systems of particles and tells how changes in the energy of a system can affect the system's properties.

Electricity and magnetism. Chapters 17 through 23 keep conservation of energy as an important thread, with additional development of concepts introduced earlier such as that of a field (action-at-a-distance). The topic of magnetism brings in the new concept of a velocity-dependent force. The importance of Maxwell's theory of electromagnetism is emphasized without undue mathematical details.

Light and optics. In Chapters 24 through 26, students can compare and contrast properties of light that depend on its wave nature with properties that require a particle (or ray) approach. Students can apply the principles of optics to model how the human eye works, including the mechanism of color vision.

Twentieth-century physics. Students are introduced to the modern concepts of relativity, quantum, atomic and nuclear physics in Chapters 27 through 31. Quantum physics reveals that matter, like light and other electromagnetic radiation, has both particle and wave properties.

Worked Examples

Worked examples are problems that are solved quantitatively within a chapter's main text. They are designed to teach sound problem-solving skills, and each involves a principle or result that has just been introduced. The worked examples also have other attributes:

- Extra emphasis is placed on step-by-step approaches that students can use in a wide range of situations. All worked examples use a five-step solution process: (1) *recognize* the physics principles central to the problem, (2) draw a *sketch* showing the problem and all the given information, (3) *identify the relationships* between the known and unknown quantities, (4) *solve* for the desired quantity, and (5) *ask what the answer means* and if it makes sense. Answers are boxed for clarity, and all examples emphasize the key final step of asking what the answer means, whether it makes sense, or what can be learned from it.
- Some worked examples are designated with the symbol Ⓡ as *reasoning and relationship problems,* designed for back-of-the-envelope solutions. A reasoning and relationship problem requires an approximate mathematical solution, a rough estimate of one or two key unknown quantities, or both. These examples and corresponding homework problems begin in Chapter 3, where Section 3.6 introduces and explains the notions of estimating and reasoning. Reasoning and relationship problems and examples are distributed throughout later chapters.
- Worked examples of special interest to life science students are designated with the symbol Ⓧ.

Problem-Solving Strategies

The five-step problem-solving method is adapted to suit broad classes of problems students will encounter, such as when applying Newton's second law, using the principles of conservation of momentum and energy, or finding the current in two branches of a DC circuit. For these classes of problems, a problem-solving strategy is highlighted within the chapter for special study emphasis.

Concept Checks

At various points in the chapter are conceptual questions, called "Concept Checks." These questions are designed to make the student reflect on a fundamental issue. They may involve interpreting the content of a graph or drawing a new graph to predict a relationship between quantities. Many Concept Check questions are in multiple-choice format to facilitate their use in audience response systems. Answers to Concept Checks are given at the end of the book. Full explanations of each answer are given in the *Instructor's Solutions Manual.*

From Chapter 6, page 187

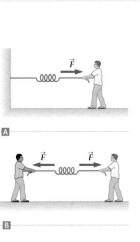

Figure 6.26 Concept Check 6.5.

CONCEPT CHECK 6.5 | Spring Forces and Newton's Third Law

Figure 6.26 shows two identical springs. In both cases, a person exerts a force of magnitude F on the right end of the spring. The left end of the spring in Figure 6.26A is attached to a wall, while the left end of the spring in Figure 6.26B is held by another person, who exerts a force of magnitude F to the left. Which statement is true?

   (1) The spring in Figure 6.26A is stretched half as much as the spring in Figure 6.26B.

   (2) The spring in Figure 6.26A is stretched twice as much as the spring in Figure 6.26B.

   (3) The two springs are stretched the same amount.

Insights

Each chapter contains several special marginal comments called "Insights" that add greater depth to a key idea or reinforce an important message. For instance, Insight 3.3 emphasizes the distinction between weight and mass, and Insight 16.1 explains why diesel engines are inherently more efficient than conventional gasoline internal combustion engines.

Diagrams with Additional Explanatory Labeling

Every college physics textbook contains line art with labeling. This book adds another layer of labeling that explains the phenomenon being illustrated, much as an instructor would explain a process or relationship in class. This additional labeling is set off in a different style.

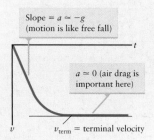

Figure 3.26 The skydiver's motion is initially like free fall; compare with Figure 3.15▲. Eventually, however, air drag becomes as large as the force of gravity and the skydiver reaches her terminal velocity v_{term}.

From Chapter 3, page 80

Figure 21.8 Different ways to produce an induced emf.

From Chapter 21, page 696

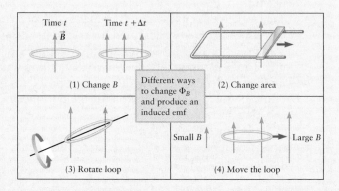

Chapter Summaries

To make the text more usable as a study tool, chapter summaries are presented in a modified "study card" format. Concepts are classified into two major groups:

    Key Concepts and Principles

    Applications

Each concept is described in its own panel, often with an explanatory diagram. This format helps students organize information for review and further study.

Relation between the electric field and the electric potential
Suppose the potential changes by an amount ΔV over a distance Δx. The component of the electric field along this direction is then

$$E = -\frac{\Delta V}{\Delta x} \qquad \textbf{(18.18)} \text{ (page 573)}$$

The electric field thus has units of volts per meter (V/m).

Capacitors
Two parallel metal plates form a *capacitor*. The *capacitance* C of this structure determines how easily charge can be stored on the plates. The charge on a capacitor is related to the magnitude of the *potential difference* between the plates by

$$\Delta V = \frac{Q}{C} \qquad \textbf{(18.30)} \text{ (page 581)}$$

This relation holds for any type of capacitor. For a parallel-plate capacitor, the capacitance is

$$C = \frac{\varepsilon_0 A}{d} \quad \text{(parallel-plate capacitor)} \qquad \textbf{(18.31)} \text{ (page 581)}$$

From Chapter 18, page 594

End-of-Chapter Questions

Approximately 20 questions at the end of each chapter ask students to reflect on and strengthen their understanding of conceptual issues. These questions are suitable for use in recitation sessions or other group work. Answers to questions designated SSM are provided in the *Student Companion & Problem-Solving Guide*, and all questions are answered in the *Instructor's Solutions Manual*.

13. Two workers are carrying a long, heavy steel beam (Fig. Q8.13). Which one is exerting a larger force on the object? How can you tell?

From Chapter 8, page 273   **Figure Q8.13**

End-of-Chapter Problems

Homework problems are designed to match the examples that are worked throughout the chapter. Most of these problems are grouped according to the matching chapter section. A final list of "Additional Problems" contains problems that bring together ideas from across the chapter or from multiple chapters. Unmarked problems are straightforward, and intermediate and challenging problems are indicated. Problems of special interest to life science students ⊗, reasoning and relationship problems ®, and problems whose solutions appear in the *Student Companion & Problem-Solving Guide* SSM are so indicated. Answers to odd-numbered problems appear at the end of the book.

6. ✳ A bar magnet is thrust into a current loop as sketched in Figure P21.6. Before the magnet reaches the center of the loop, what is the direction of the induced current as seen by the observer on the right, clockwise or counterclockwise?

Observer

From Chapter 21, page 718

Figure P21.6 Problems 6, 7, and 8.

ANCILLARIES

Using Technology to Enhance Learning

Enhanced WebAssign is the perfect solution to your homework management needs. Designed by physicists for physicists, this system is a reliable and user-friendly teaching companion. Enhanced WebAssign is available for *College Physics: Reasoning and Relationships,* giving you the freedom to assign

- Selected end-of-chapter problems, algorithmically driven where appropriate and containing an example of the student solution.

- Reasoning and relationship problems. Students are at risk of missing which crucial quantities must be estimated. A coached solution can help students learn how to attack these problems and arrive at a sensible answer.
- Concept Checks from the chapter, available in multiple-choice format for assignment.
- Algorithmically generated versions of the worked examples from the text. These can be assigned to students to help them prepare for homework problems.

Please visit **www.webassign.net/brookscole** to view an interactive demonstration of Enhanced WebAssign.

PowerLecture™ CD-ROM is an easy-to-use multimedia tool allowing instructors to assemble art with notes to create fluid lectures quickly. The CD-ROM includes prepared PowerPoint® lectures and digital art from the text as well as editable electronic files of the *Instructor's Solutions Manual* and the *Test Bank*. The CD also includes the ExamView® Computerized Test Bank, giving you the ability to build tests featuring an unlimited number of new questions or any of the existing questions from the preloaded *Test Bank*. Finally, the CD includes audience response system content specific to the textbook. Contact your local sales representative to find out about our audience response software and hardware.

Additional Instructor Resources

Instructor's Solutions Manual by Michael Meyer (Michigan Technological University), David Sokoloff (University of Oregon) and Raymond Hall (California State University, Fresno). This two-volume publication provides full explanations of Concept Check answers, answers to end-of-chapter questions, and complete solutions to end-of-chapter problems using the five-step problem-solving methodology developed in the text.

Test Bank by Ed Oberhofer (University of North Carolina at Charlotte and Lake-Sumter Community College) is available on the *PowerLecture*™ CD-ROM as editable electronic files or via the ExamView® test software. The file contains questions in multiple-choice format for all chapters of the text. Instructors may print and duplicate pages for distribution to students.

Student Resources

Student Companion & Problem-Solving Guide by Richard Grant (Roanoke College) will prove to be an essential study resource. For each chapter, it contains a summary of problem-solving techniques (following the text's methodology), a list of frequently-asked questions students often have when attempting homework assignments, selected solutions to end-of-chapter problems, solved Capstone Problems representing typical exam questions, and a set of MCAT review questions with explanation of strategies behind the answers.

Physics Laboratory Manual, third edition by David Loyd (Angelo State University) supplements the learning of basic physical principles while introducing laboratory procedures and equipment. Each chapter includes a prelaboratory assignment, objectives, an equipment list, the theory behind the experiment, experimental procedures, graphing exercises and questions. A laboratory report form is included with each experiment so that the student can record data, calculations, and experimental results. Students are encouraged to apply statistical analysis to their data. A complete *Instructor's Manual* is also available to facilitate use of this lab manual.

ACKNOWLEDGMENTS

Creating a new textbook is an enormous job requiring the assistance of many people. To all these people, I extend my sincere thanks.

Contributors

Raymond Hall of California State University, Fresno and Richard Grant of Roanoke College contributed many interesting and creative end-of-chapter questions and problems. Ray Hall's ideas also led to the design of the book's cover.

Accuracy Reviewers

David Bannon, *Oregon State University*

Ken Bolland, *The Ohio State University*

Stephane Coutu, *The Pennsylvania State University*

Stephen D. Druger, *University of Massachusetts—Lowell*

A. J. Haija, *Indiana University of Pennsylvania*

John Hopkins, *The Pennsylvania State University*

David Lind, *Florida State University*

Edwin Lo

Dan Mazilu, *Virginia Tech*

Tom Oder, *Youngstown State University*

Brad Orr, *University of Michigan*

Chun Fu Su, *Mississippi State University*

Manuscript Reviewers

A special thanks is due to Amy Pope of Clemson University for her thoughtful reading of the entire manuscript.

Jeffrey Adams, *Montana State University*

Anthony Aguirre, *University of California, Santa Cruz*

David Balogh, *Fresno City College*

David Bannon, *Oregon State University*

Phil Baringer, *University of Kansas*

Natalie Batalha, *San Jose State University*

Mark Blachly, *Arsenal Technical High School*

Gary Blanpied, *University of South Carolina*

Ken Bolland, *The Ohio State University*

Scott Bonham, *Western Kentucky University*

Marc Caffee, *Purdue University*

Lee Chow, *University of Central Florida*

Song Chung, *William Patterson University*

Alice Churukian, *Concordia College*

Thomas Colbert, *Augusta State University*

David Cole, *Northern Arizona University*

Sergio Conetti, *University of Virginia*

Gary Copeland, *Old Dominion University*

Doug Copely, *Sacramento City College*

Robert Corey, *South Dakota School of Mines & Technology*

Andrew Cornelius, *University of Nevada, Las Vegas*

Carl Covatto, *Arizona State University*

Nimbus Couzin, *Indiana University, Southeast*

Thomas Cravens, *University of Kansas*

Sridhara Dasu, *University of Wisconsin, Madison*

Timir Datta, *University of South Carolina*

Susan DiFranzo, *Hudson Valley Community College*

David Donnelly, *Texas State University*

Sandra Doty, *The Ohio State University*

Steve Ellis, *University of Kentucky*

Len Finegold, *Drexel University*

Carl Fredrickson, *University of Central Arkansas*

Joe Gallant, *Kent State University, Warren Campus*

Kent Gee, *Brigham Young University*

Bernard Gerstman, *Florida International University*

James Goff, *Pima Community College*

Richard Grant, *Roanoke College*

William Gregg, *Louisiana State University*

James Guinn, *Georgia Perimeter College, Clarkston*

Richard Heinz, *Indiana University, Bloomington*

John Hopkins, *Pennsylvania State University*

Karim Hossain, *Edinboro University of Pennsylvania*

Linda Jones, *College of Charleston*

Alex Kamenev, *University of Minnesota*

Daniel Kennefick, *University of Arkansas*

Aslam Khalil, *Portland State University*

Jeremy King, *Clemson University*

Randy Kobes, *University of Winnipeg*

Raman Kolluri, *Camden County College*

Ilkka Koskelo, *San Francisco State University*

Fred Kuttner, *University of California, Santa Cruz*

Richard Ledet, *University of Louisiana, Lafayette*

Alexander Lisyansky, *Queens College, City University of New York*

Carl Lundstedt, *University of Nebraska, Lincoln*

Donald Luttermoser, *East Tennessee State University*

Steven Matsik, *Georgia State University*

Sylvio May, *North Dakota State University*

Bill Mayes, *University of Houston*

Arthur McGum, *Western Michigan University*

Roger McNeil, *Louisiana State University*

Rahul Mehta, *University of Central Arkansas*

Charles Meitzler, *Sam Houston State University*

Michael Meyer, *Michigan Technological University*

Vesna Milosevic-Zdjelar, *University of Winnipeg*

John Milsom, *University of Arizona*

Wouter Montfrooij, *University of Missouri*

Ted Morishige, *University of Central Oklahoma*

Halina Opyrchal, *New Jersey Institute of Technology*

Michelle Ouellette, *California Polytechnic State University, San Louis Obispo*

Kenneth Park, *Baylor University*

Galen Pickett, *California State University, Long Beach*

Dinko Pocanic, *University of Virginia*

Amy Pope, *Clemson University*

Michael Pravica, *University of Nevada, Las Vegas*

Laura Pyrak-Nolte, *Purdue University*

Mark Riley, *Florida State University*

Mahdi Sanati, *Texas Tech University*

Cheryl Schaefer, *Missouri State University*

Alicia Serfaty de Markus, *Miami Dade College, Kendall Campus*

Marc Sher, *College of William & Mary*

Douglas Sherman, *San Jose State University*

Marllin Simon, *Auburn University*

Chandralekha Singh, *University of Pittsburgh*

David Sokoloff, *University of Oregon*

Noel Stanton, *Kansas State University*

Donna Stokes, *University of Houston*

Carey Stronach, *Virginia State University*

Chun Fu Su, *Mississippi State University*

Daniel Suson, *Texas A & M University, Kingsville*

Doug Tussey, *Pennsylvania State University*

John Allen Underwood, *Austin Community College, Rio Grande Campus*

James Wetzel, *Indiana University-Purdue University, Fort Wayne*

Lisa Will, *Arizona State University*

Gerald T. Woods, *University of South Florida*

Guoliang Yang, *Drexel University*

David Young, *Louisiana State University*

Michael Yurko, *Indiana University-Purdue University, Indianapolis*

Hao Zeng, *State University of New York at Buffalo*

Nouredine Zettili, *Jacksonville State University*

Focus Group Participants

Edward Adelson, *The Ohio State University*

Mark Boley, *Western Illinois University*

Abdelkrim Boukahil, *University of Wisconsin, Whitewater*

Larry Browning, *South Dakota State University*

Thomas Colbert, *Augusta State University*

Susan DiFranzo, *Hudson Valley Community College*

Hector Dimas, *Mercer County Community College*

David Donnelly, *Texas State University*

Taner Edis, *Truman State University*

Kevin Fairchild, *La Costa Canyon High School*

Joseph Finck, *Central Michigan University*

David Groh, *Gannon University*

Kathleen Harper, *The Ohio State University*

John Hill, *Iowa State University*

Karim Hossain, *Edinboro University of Pennsylvania*

Debora Katz, *United States Naval Academy*

Larry Kirkpatrick, *Montana State University*

Terence Kite, *Pepperdine University*

Lois Krause, *Clemson University*

Mani Manivannen, *Missouri State University*

Michael Meyer, *Michigan Technological University*

John Milsom, *University of Arizona*

M. Sultan Parvez, *Louisiana State University, Alexandria*

Amy Pope, *Clemson University*

Michael Pravica, *University of Nevada, Las Vegas*

Joseph Priest, *Miami University*

Shafiqur Rahman, *Allegheny College*

Kelly Roos, *Bradley University*
David Sokoloff, *University of Oregon*
Jian Q. Wang, *Binghamton University*
Lisa Will, *San Diego City College*

David Young, *Louisiana State University*
Michael Ziegler, *The Ohio State University*
Nouredine Zettili, *Jacksonville State University*

Class Testers

Abdelkrim Boukahil, *University of Wisconsin, Whitewater*
David Groh, *Gannon University*
Tom Moffet, *Purdue University*
Nouredine Zettili, *Jacksonville State University*

Extra Credit

Special thanks go to Lachina Publishing Services, especially Katherine Wilson, Jeanne Lewandowski, and Aaron Kantor, for keeping up with innumerable changes during the production stages of this book. Kathleen M. Lafferty diligently edited the copy. Steve McEntee expertly guided the art program, and Sarah Palmer coordinated the efforts of the art studio at Dartmouth Publishing Services. Greg Gambino created the "Stick Dude" art for worked examples and end-of-chapter problems that proved so popular with reviewers and class testers. Dena Digilio Betz went the extra mile to find photos from unusual sources. Ed Dodd skillfully guided the arrangement of pages during the hectic final phase of production.

I would like to thank all the team at Cengage Learning, including Michelle Julet, Mary Finch, Peggy Williams, John Walker, Teresa L. Trego, Sam Subity, Terri Mynatt, Alyssa White, Brandi Kirksey, Stefanie Beeck, and Rebecca Berardy Schwartz, for giving me the opportunity to undertake this project. I also want to give special thanks to Susan Pashos for her unwavering and tireless support, and to my wife for all her patience.

No textbook is perfect for every student or for every instructor. It is my hope that both students and instructors will find some useful, stimulating, and even exciting material in this book and that you will all enjoy physics as much as I do.

Nicholas J. Giordano

About the Author

Nicholas J. Giordano obtained his B.S. at Purdue University and his Ph.D. at Yale University. He has been on the faculty at Purdue since 1979, served as an Assistant Dean of Science from 2000 to 2003, and in 2004 was named the Hubert James Distinguished Professor of Physics. His research interests include the properties of nanoscale metal structures, nanofluidics, science education, and biophysics, along with musical acoustics and the physics of the piano. Dr. Giordano earned a Computational Science Education Award from the Department of Energy in 1997, and he was named Indiana Professor of the Year by the Carnegie Foundation for the Advancement of Teaching and the Council for the Advancement and Support of Education in 2004. His hobbies include distance running and restoring antique pianos, and he is an avid baseball fan.

COLLEGE PHYSICS

Reasoning and Relationships

Electric Forces and Fields

In this and the next two chapters, we focus on electric phenomena before moving on to describe magnetism in Chapter 20. Electric and magnetic forces both act on electric charges and currents. Electric charges and currents also act as *sources* of electric and magnetic fields. In fact, electricity and magnetism are really a unified phenomenon, often referred to as *electromagnetism,* and this unified picture leads to a theory of electromagnetic radiation. Light is a type of electromagnetic radiation, so electromagnetism also forms the basis for the theory of *optics,* which is discussed in Chapters 24 through 26. Our study of electricity and magnetism will thus take us quite a long way!

Benjamin Franklin's famous experiments with kites in thunderstorms showed that lightning involves the motion of electric charge. This motion and the associated electric and magnetic phenomena are the subject of Chapters 17 through 23. (© Gandee Vasan/Stone/Getty Images)

17.1 | EVIDENCE FOR ELECTRIC FORCES: THE OBSERVATIONAL FACTS

The discovery of electricity is generally credited to the Greeks and is thought to have occurred around 2500 years ago, which is approximately the era in which Aristotle lived. The Greeks observed electric charges and the forces between them in a variety of situations. Many of their observations made use of a material called amber, a plastic-like substance formed by allowing the sap from certain trees to dry and harden (Fig. 17.1A). The Greeks found that after amber is rubbed with a piece of animal fur, the amber can attract small pieces of dust. The same effect can be demonstrated with a piece of plastic and small bits of paper, although the Greeks did not have these materials to work with! This experiment (parts B and C of Fig. 17.1) is quite striking because neither the fur nor the plastic rod ordinarily attract paper. Somehow, the process of rubbing the two together "creates" an attractive force. Moreover, this force occurs even when the amber and the paper are *not* in contact, and there would be an attractive force even if the plastic rod and the paper were placed in a vacuum (although the Greeks did not try that).

The Greek word for amber is "elektron"; this is the origin of the terms *electricity* and *electron*. The Greeks did not discover the electron; even though the notion of a small charge-carrying particle was hypothesized by a number of early philosophers and scientists, there was no clear experimental demonstration of the existence of the electron until the late 1880s. The long history of scientific work on electricity is very interesting, but we will skip most of the details and simply state a few key results.

Basic properties of electric charge

- *There are two types of electric charge*, called **positive** and **negative**. The subatomic particle called a **proton** has a positive charge, and an **electron** has a negative charge.
- *Charge comes in quantized units. All* protons carry the *same* amount of charge $+e$, and *all* electrons carry a charge $-e$. We will discuss how charge is measured and the unit of electric charge below.
- *Like charges repel each other, unlike charges attract.* The electric force between two objects is repulsive if the objects carry "like" charge, that is, if both are positively charged or both are negatively charged (Fig. 17.2). The electric force is attractive if the two objects carry "unlike" charge. Here the terms *like* and *unlike* refer to the *signs* of the charges, not their magnitudes. So, the expression "like charges" means that the two charges are both positive or both negative. The expression "unlike charges" means that one charge is positive and the other is negative.
- *Charge is conserved.* The *total* charge on an object is the sum of all the individual charges (protons and electrons) carried by the object. The total charge can

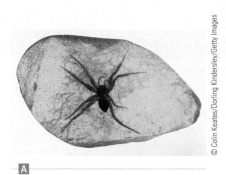
A

B

C

Figure 17.1 **A** Amber is the dried, hardened sap from certain trees. Small creatures or bits of plant material are often preserved inside. **B** When a plastic or amber rod is rubbed with fur, the rod acquires an electric charge. **C** A charged rod attracts small bits of paper and other objects.

be positive, negative, or zero. Charge can move from place to place, and from one object to another, but the total charge of the universe does not change.

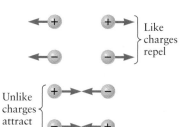

What Is Electric Charge?

You are likely familiar with the term *electric charge* or simply *charge*. Charge is a fundamental property of matter: the amount of charge that is "on" or "carried by" a particle determines how the particle reacts to electric and magnetic fields. In many ways, it is similar to the quantity we call *mass*; the mass of a particle determines how it reacts when a force acts on it. The mass of a particle is a measure of the amount of matter it carries, whereas the charge of a particle is a measure of the amount of "electric-ness" it carries. Charge and mass are both primary entries in our dictionary of physics terms, but it is not possible to give definitions for these fundamental entries in our dictionary.

In the SI system of units, charge is measured in **coulombs** (C) in honor of French physicist Charles de Coulomb (1736–1806). The charge on a single electron is[1]

$$\text{electron charge } = -e = -1.60 \times 10^{-19} \text{ C} \tag{17.1}$$

and the charge carried by a single proton is

$$\text{proton charge } = +e = +1.60 \times 10^{-19} \text{ C} \tag{17.2}$$

The symbol e is usually used to denote the *magnitude* of the charge on an electron or a proton. We will always take e to be a *positive* quantity ($e = +1.60 \times 10^{-19}$ C). We use the symbols q and Q to denote charge in general, such as the total charge on a bit of paper. When discussing a charged particle, it is common to say that the particle "has" a charge q or that it "carries" a charge q. These two expressions simply mean that the total charge of the particle is q.

Figure 17.2 The electric force between two like charges (charges with the same sign) is repulsive, whereas the force between unlike charges (those of opposite sign) is attractive.

17.2 | ELECTRIC FORCES AND COULOMB'S LAW

Imagine that you travel to a distant star where the laws of physics seem different from the ones that describe physics on Earth. While on this star, you discover a new type of force acting between particles; the force can be attractive or repulsive, and it is *extremely* large. In fact, the force between two objects each having a mass of only 1 g is large enough to hold an entire planet in orbit around this alien star! This force sounds very different from anything you have encountered before, but it *already exists* on Earth. It is the *electric force* between two charged objects.

Consider two charged objects that are so tiny that they can be modeled as point particles. If the charges carried by the two objects are q_1 and q_2 and they are separated by a distance r (Fig. 17.3), the electric force between the objects has a magnitude

$$F = k \frac{q_1 q_2}{r^2} \tag{17.3}$$

Equation 17.3 is called **Coulomb's law**. The constant k has the value

$$k = 8.99 \times 10^9 \text{ N} \cdot \text{m}^2/\text{C}^2 \tag{17.4}$$

Do not confuse the constant k in Equation 17.4 with the spring constant in Hooke's law. (Most physicists use the same symbol for both.) The direction of the electric force on each of the charges is along the line that connects the two charges and is illustrated in Figure 17.3 for the case of two like charges. As already mentioned, this force is repulsive for like charges, corresponding to a positive value of F in

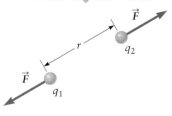

Like charges q_1 and q_2 repel.

COULOMB'S LAW:

$$|\vec{F}| = \frac{kq_1 q_2}{r^2} = \frac{q_1 q_2}{4\pi\varepsilon_0 r^2}$$

Figure 17.3 The electric force between two point charges q_1 and q_2 is given by Coulomb's law.

---

[1]Here we give the value of e to just three significant figures, but it is known to much greater precision than that. The latest experiments give $e = 1.60217653 \times 10^{-19}$ C.

Equation 17.3, whereas the force is attractive for unlike charges and F in Equation 17.3 is negative in that case.

Strictly speaking, the value of F in Equation 17.3 applies only for two point charges, but it is a good approximation whenever the sizes of the particles are much smaller than their separation r. Later in this chapter, we'll see how to build on Equation 17.3 to find the electric force in other situations, including nonpoint particles.

The mathematical form of Equation 17.3 is very similar to Newton's law of gravitation, with the constant k playing a role analogous to the gravitational constant G. Another way to write Coulomb's law is

$$F = \frac{q_1 q_2}{4\pi\varepsilon_0 r^2} \tag{17.5}$$

where ε_0 is yet another physical constant called the **_permittivity of free space_**, having the value

$$\varepsilon_0 = 8.85 \times 10^{-12} \frac{C^2}{N \cdot m^2} \tag{17.6}$$

The values of ε_0 and k are related by

$$\frac{1}{4\pi\varepsilon_0} = k$$

so these two forms of Coulomb's law, Equations 17.3 and 17.5, are completely equivalent. Why do we bother to write Coulomb's law in the slightly more complicated form in Equation 17.5? We do so because other equations of electromagnetism have simpler forms when written in terms of ε_0 instead of k. For calculating the electric force between charges, either form of Coulomb's law (Eq. 17.3 or 17.5) can be used.

Features of Coulomb's Law

Coulomb's law has several important properties.

1. We have already seen that the electric force is repulsive for like charges and attractive for unlike charges (Fig. 17.2). Mathematically, this property results from the product $q_1 q_2$ in the numerator in Equations 17.3 and 17.5. The factor $q_1 q_2$ is positive for like charges, so F is positive and the force tends to push the charges farther apart. For unlike charges the product $q_1 q_2$ is negative and the value of F in Equations 17.3 and 17.5 is also negative, and the particles are attracted to each other.

2. We have already noted that the form of Coulomb's law is very similar to Newton's universal law of gravitation. Both laws exhibit a $1/r^2$ dependence on the separation of the two particles. Therefore, a negative charge can move in a circular orbit around a positive charge, just like a planet orbiting the Sun, and that was an early model for the hydrogen atom. There is one very important difference, however: gravity is always an attractive force, whereas the electric force in Coulomb's law can be either attractive or repulsive.

3. The magnitude of F in Equations 17.3 and 17.5 is the magnitude of the force exerted on _each_ of the particles. That is, a force of magnitude F is exerted on charge q_1, and a force of equal magnitude and opposite direction is exerted on q_2. We should expect such a pair of forces, based on Newton's third law, the action–reaction principle.

Using Equation 17.3, let's now show that electric forces can be extremely large. Suppose you are given a box holding 1 g (0.001 kg) of pure electrons; that is, the box contains _only_ electrons. The mass of a single electron is $m_e = 9.11 \times 10^{-31}$ kg, so the total number of electrons N_e in the box is

$$N_e = \frac{1 \times 10^{-3}\ kg}{9.11 \times 10^{-31}\ kg/electron} = 1.1 \times 10^{27}$$

Each of these electrons carries a charge $-e$, so the total charge is

$$Q_{\text{total}} = N_e(-e) = (1.1 \times 10^{27})(-1.60 \times 10^{-19}\text{ C}) = -1.8 \times 10^8\text{ C}$$

Now suppose there are two of these boxes, each with charge Q_{total}, separated by a distance $r = 1$ m; for simplicity, we assume each box is small so that it can be modeled as a point particle. The magnitude of the total electric force is (Eq. 17.3)

$$F = \frac{kQ_{\text{total}}Q_{\text{total}}}{r^2} = \frac{(8.99 \times 10^9\text{ N}\cdot\text{m}^2/\text{C}^2)(-1.8 \times 10^8\text{ C})(-1.8 \times 10^8\text{ C})}{(1\text{ m})^2}$$

$$F = 3 \times 10^{26}\text{ N} \tag{17.7}$$

which is an *enormous* force. It is nearly a million times larger than the gravitational force exerted between the Sun and the Earth. All that from just two small containers of electrons!

The value found for F in Equation 17.7 is so large that you may be a little skeptical: if the electric force between two small pieces of matter is this large, there must be many staggering consequences. For example, the electric forces acting within ordinary matter, which contains both electrons and protons, must be huge. Why don't these forces dominate everyday life? The resolution of this apparent paradox is that it is essentially impossible to obtain a box containing only electrons. You probably know that a neutral atom contains equal numbers of electrons and protons. In fact, the term *neutral* means that the total charge is zero. If our two pointlike boxes had contained equal numbers of electrons and protons, their *total* charges would have been $Q_{\text{total}} = 0$ and the force in Equation 17.7 would be zero.

Ordinary matter always consists of equal, or nearly equal, numbers of electrons and protons. The total charge is then zero or very close to zero. At the atomic and molecular scale, however, it is common to have the positive and negative charges (nuclei and electrons) separated by a small distance. In this case, the electric force is not zero, and these electric forces are responsible for holding matter together. That is why solids can be quite strong.

EXAMPLE 17.1 | Using Coulomb's Law to Clean the Air

Coal-burning power plants produce large amounts of potential pollution in the form of small particles (soot). Modern smokestacks use devices called scrubbers to remove these particles from the smoke they emit. Scrubbers use a two-step process: electrons are first added to the soot particle, and an electric force then pulls the particle out of the smoke stream (Fig. 17.4). In this example, we analyze the force on a soot particle after electrons are added.

Consider a soot particle of mass $m_{\text{soot}} = 1.0$ nanogram ($= 1.0 \times 10^{-9}$ g $= 1.0 \times 10^{-12}$ kg, which corresponds to a diameter of a few microns) that has a charge q_{soot}. That is, some number of electrons have been added to give the particle this total charge. Suppose the collector has a total charge $q_{\text{collector}} = 1.0 \times 10^{-6}$ C and is small enough to be treated as a point charge at the rim of the smokestack. (a) If the separation between the collector and the soot particle is $r = 0.10$ m, what is the value of q_{soot} so that the electric force exerted on the particle is equal to its weight? (b) How many electrons must be added to the soot particle?

RECOGNIZE THE PRINCIPLE

The collector charge is assumed to be a point particle and the soot particle is also very small, so we can apply Coulomb's law for two point particles (Eq. 17.3). The electric force between the collector (charge $q_{\text{collector}}$) and the soot particle (charge q_{soot}) has a magnitude

$$F_{\text{elec}} = \frac{kq_{\text{collector}}q_{\text{soot}}}{r^2} \tag{1}$$

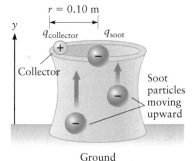

Figure 17.4 Example 17.1. A dust particle in a smokestack is attracted to the electric charge on the collector of a smokestack scrubber.

Figure 17.4 shows the soot particle and the collector charge. They are separated by a distance r, just like the charges in Figure 17.3.

IDENTIFY THE RELATIONSHIPS

We want the electric force on the particle in Equation (1) to be equal in magnitude to the particle's weight, so

$$F_{\text{elec}} = \frac{kq_{\text{collector}}q_{\text{soot}}}{r^2} = mg$$

SOLVE

(a) Solving for the charge q_{soot} leads to

$$q_{\text{soot}} = \frac{mgr^2}{kq_{\text{collector}}}$$

Inserting the given values of m and r, we find

$$q_{\text{soot}} = \frac{mgr^2}{kq_{\text{collector}}} = \frac{(1.0 \times 10^{-12} \text{ kg})(9.8 \text{ m/s}^2)(0.10 \text{ m})^2}{(8.99 \times 10^9 \text{ N} \cdot \text{m}^2/\text{C}^2)(1.0 \times 10^{-6} \text{ C})}$$

$$q_{\text{soot}} = \boxed{1.1 \times 10^{-17} \text{ C}}$$

(b) The charge on a single electron has a magnitude $e = 1.60 \times 10^{-19}$ C (Eq. 17.1), so our value of q_{soot} corresponds to

$$N = \frac{q_{\text{soot}}}{e} = \frac{1.1 \times 10^{-17} \text{ C}}{1.60 \times 10^{-19} \text{ C/electron}} = \boxed{69 \text{ electrons}}$$

which is the number of extra electrons added to the soot particle.

What have we learned?

It is amazing that only this small number of electrons is needed for the scrubber to successfully remove a soot particle from air. The total number of atoms in the particle is about 4×10^{14} (see Question 13 at the end of this chapter), so the fraction of "unbalanced" electrons on the particle is only $69/(4 \times 10^{14}) \approx 2 \times 10^{-13}$!

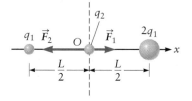

Figure 17.5 The total force on q_2 is equal to the sum of the forces from charges q_1 and $2q_1$. This is the principle of superposition at work.

Electric forces and the principle of superposition

Superposition of Electric Forces

So far, we have discussed the electric force between two point particles; let's now consider how to use Coulomb's law to deal with more complicated charge distributions. Suppose two particles of charge q_1 and $2q_1$ are separated by a distance L as shown in Figure 17.5. What is the force on a third charge q_2 placed midway between these two charges?

We can deal with this problem by first using Coulomb's law to find the force exerted by charge q_1 on q_2 and then using Coulomb's law a second time to calculate the force exerted by the charge $2q_1$ on q_2. The total force on q_2 is the sum of these two separate contributions, which is an example of the ***principle of superposition***. That is, the total force on q_2 is just the sum (the superposition) of the individual forces exerted on q_2 by all the other charges in the problem. Because force is a vector, we must be careful to add these forces as vectors.

From Figure 17.5, the separation between q_2 and the charge q_1 is $L/2$. Writing Coulomb's law for this pair of charges, we have

$$F_1 = \frac{kq_1q_2}{(L/2)^2} = \frac{4kq_1q_2}{L^2} \tag{17.8}$$

Here a positive value corresponds to a repulsive force (since like charges repel), and in the coordinate system in Figure 17.5 it corresponds to a force on q_2 in the $+x$ direction. Hence, F_1 in Equation 17.8 is the component of the force along the x axis, and the component along y is zero.

We can deal with the force from the charge $2q_1$ in a similar way. The separation of the charges is again $L/2$, so, applying Coulomb's law (Eq. 17.5), we find

$$F_2 = \frac{k(2q_1)q_2}{(L/2)^2} = \frac{8kq_1q_2}{L^2} \qquad (17.9)$$

From the geometry of Figure 17.5 we can see that this result is the x component of the force on q_2 because the electric force acts along the line that connects the two charges and here that line is along the x direction. For this reason, the y component of the force on q_2 is again zero. Equation 17.9 gives the magnitude of the force F_2. If they are "like" charges (i.e., if q_1 and q_2 are both positive or both negative so that the product q_1q_2 is positive), the force F_2 is in the $-x$ direction.

The total force on q_2 is the sum of Equations 17.8 and 17.9, but we must account for the direction (the sign) of each. Assuming all charges are of "like sign," the result is

$$F_{\text{total}} = +\frac{4kq_1q_2}{L^2} - \frac{8kq_1q_2}{L^2} = -\frac{4kq_1q_2}{L^2}$$

The negative sign here tells us that if the product $q_1q_2 > 0$ (consistent with our "like sign" assumption), the force exerted on q_2 is along the $-x$ direction. This example leads to the following strategy.

PROBLEM SOLVING | Calculating Forces with Coulomb's Law

1. **RECOGNIZE THE PRINCIPLE.** The electric force on a charged particle can be found using Coulomb's law together with the principle of superposition.

2. **SKETCH THE PROBLEM.** Construct a drawing (including a coordinate system) and show the location and charge for each object in the problem. Your drawing should also show the directions of all the electric forces—$\vec{F}_1$, $\vec{F}_2$, and so forth—on the particle(s) of interest.

3. **IDENTIFY THE RELATIONSHIPS.** Use Coulomb's law (Eq. 17.3 or 17.5) to find the magnitudes of the forces $\vec{F}_1$, $\vec{F}_2$, ... acting on the particle(s) of interest.

4. **SOLVE.** The total force on a particle is the sum (the superposition) of all the individual forces from steps 2 and 3. Add these forces *as vectors* to get the total force. When adding these vectors, it is usually simplest to work in terms of the components of $\vec{F}_1$, $\vec{F}_2$, ... along the coordinate axes.

5. Always *consider what your answer means* and check that it makes sense.

CONCEPT CHECK 17.1 | Superposition and the Direction of the Electric Force

An electron and a proton are each placed on the x axis, a distance L from the origin as shown in Figure 17.6. Another electron (electron 2 in Fig. 17.6) is placed to the right of the proton as shown. Is the *direction* of the electric force on electron 2 (a) in the $+x$ direction or (b) in the $-x$ direction, or (c) is the force zero?

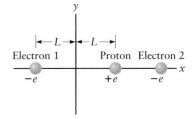

Figure 17.6 Concept Check 17.1.

EXAMPLE 17.2 | An Electric Dipole

Consider two point particles of charge $+q$ and $-q$ (with q positive) arranged as shown in Figure 17.7A. This arrangement of charge, called an *electric dipole*, is found in many molecules, so it is important in chemistry as well as physics. Suppose a third charge Q is placed on the x axis as shown in Figure 17.7A. Assuming Q is positive, what are the magnitude and the direction of the total electric force on Q?

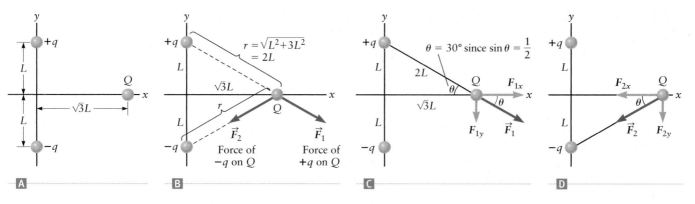

Figure 17.7 Example 17.2. Calculation of the force exerted by a dipole (charges $+q$ and $-q$) on a third charge Q.

RECOGNIZE THE PRINCIPLE

Our approach is based on Coulomb's law and the principle of superposition, following the general approach outlined in the "Calculating Forces with Coulomb's Law" problem-solving strategy.

SKETCH THE PROBLEM

We begin with the sketch in Figure 17.7A. This picture includes the coordinate axes and the positions of all the charges in the problem. We next draw in vectors $\vec{F}_1$ and $\vec{F}_2$ showing the forces on charge Q in Figure 17.7B. The force of charge $+q$ on charge Q is represented by $\vec{F}_1$. These charges are both positive, so the force is repulsive (away from $+q$) and $\vec{F}_1$ is directed to the lower right of the figure. The force exerted by $-q$ on Q is represented by $\vec{F}_2$. They are unlike charges, so this force is attractive (toward $-q$).

IDENTIFY THE RELATIONSHIPS

We apply Coulomb's law (Eq. 17.3) to find the magnitudes of $\vec{F}_1$ and $\vec{F}_2$. Using the Pythagorean theorem in Figure 17.7B, the distance from $+q$ to Q is $r = \sqrt{L^2 + 3L^2} = 2L$. Inserting into Equation 17.3 gives the magnitude of vector $\vec{F}_1$

$$F_1 = \frac{kqQ}{r^2} = \frac{kqQ}{(2L)^2} = \frac{kqQ}{4L^2}$$

The distance between $-q$ and Q is also r, and the magnitude of $\vec{F}_2$ is

$$F_2 = \frac{kqQ}{r^2} = \frac{kqQ}{4L^2} = F_1 \qquad (1)$$

SOLVE

We now add $\vec{F}_1$ and $\vec{F}_2$ *as vectors* to get the total force on Q. The components of these vectors are shown in parts C and D of Figure 17.7. For the components of $\vec{F}_1$, we have

$$F_{1x} = F_1 \cos \theta$$

and

$$F_{1y} = -F_1 \sin \theta \qquad (2)$$

where the angle θ is defined in Figure 17.7C. This angle is contained in a right triangle with $\sin \theta = \frac{1}{2}$, so $\theta = 30°$. Note that F_{1y} is negative; the negative sign in Equation (2) is due to the direction of $\vec{F}_1$ in parts B and C of Figure 17.7. We next write the components of $\vec{F}_2$. From Figure 17.7D, we get

$$F_{2x} = -F_2 \cos \theta$$

and

$$F_{2y} = -F_2 \sin \theta$$

We now sum the components of $\vec{F}_1$ and $\vec{F}_2$ to get the total force. For the x components, we find

$$F_{\text{total}, x} = F_{1x} + F_{2x} = F_1 \cos\theta - F_2 \cos\theta$$

Since $F_1 = F_2$ by Equation (1), we have

$$F_{\text{total}, x} = \boxed{0}$$

For the y components,

$$F_{\text{total}, y} = F_{1y} + F_{2y} = -F_1 \sin\theta - F_2 \sin\theta$$

$$F_{\text{total}, y} = -2F_1 \sin\theta = -\frac{kqQ}{2L^2} \sin\theta$$

We already found $\theta = 30°$, so

$$\sin\theta = \sin(30°) = \tfrac{1}{2}$$

Our final result is thus

$$F_{\text{total}, y} = -\frac{kqQ}{2L^2} \sin\theta = \boxed{-\frac{kqQ}{4L^2}}$$

What does it mean?
When Q is at this particular location, the force from the dipole on a positive charge Q is along the $-y$ direction. At other locations (e.g., away from the x and y axes), the force will have nonzero components along both x and y. Had the charge Q instead been $-Q$ (negative), the force would have the same magnitude as found here, but directed along $+y$.

CONCEPT CHECK 17.2 | Force on a Point Charge

Two particles of charge q and $-2q$ are located as shown in Figure 17.8. A third charge Q is placed on the x axis and it is found that the total electric force on Q is zero. In which region in Figure 17.8 is Q located?

(a) region A (b) region B (c) region C

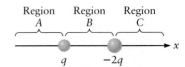

Figure 17.8 Concept Check 17.2.

CONCEPT CHECK 17.3 | Holding Together Two Charges

Two particles, both with charge $q = +1$ C, are located 1 m apart as sketched in Figure 17.9. If q_1 is held fixed in place, what is the magnitude of the force that must be exerted on q_2 to keep it from moving?

(a) 9×10^9 N   (b) 1 N   (c) zero   (d) 9×10^{18} N

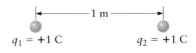

Figure 17.9 Concept Check 17.3.

17.3 | THE ELECTRIC FIELD

Coulomb's law gives the electric force between a pair of charges, but there is another way to describe electric forces. Imagine a single isolated point charge, that is, a point charge very far from any other charges as in Figure 17.10A. The presence of this charge produces an *electric field*, represented by the arrows in Figure 17.10A. This electric field is similar to the gravitational field near an isolated mass discussed in Chapter 5. A positive point charge produces field lines that emanate radially outward from the charge as sketched in Figure 17.10A. For a negative charge, the field lines are directed inward, toward the charge (Fig. 17.10B).

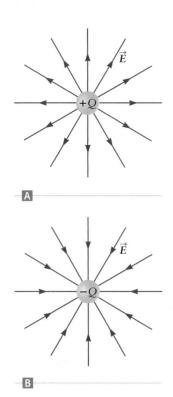

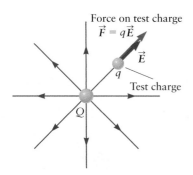

Figure 17.10 Electric field lines near a point charge placed at the origin. **A** If the charge is positive, the electric field is directed outward away from the charge, while **B** if the charge is negative, the field lines are directed inward toward the charge.

Electric field of a point charge

Force on test charge
$\vec{F} = q\vec{E}$

$\vec{E}$

q

Test charge

Q

Figure 17.11 The electric field at a particular point in space is related to the electric force on a test charge q placed at that location by $\vec{F} = q\vec{E}$.

The electric field is a *vector*—it has a magnitude and a direction—and is denoted by $\vec{E}$. The associated field lines are parallel to the direction of $\vec{E}$, whereas the density of field lines is proportional to the magnitude of $\vec{E}$. In parts A and B of Figure 17.10, the field lines are most dense in the vicinity of the charges; in both cases, the magnitude of $\vec{E}$ increases as one gets closer to the charge.

Using a Test Charge to Measure $\vec{E}$

Consider a particular point in space where the electric field is $\vec{E}$ (Fig. 17.11). This field might be produced by a point charge or by some other arrangement of charges. If we now place a charge q at this location, it experiences an electric force given by

$$\vec{F} = q\vec{E} \qquad (17.10)$$

The electric force is thus either parallel to $\vec{E}$ (if q is positive) or antiparallel (if q is negative). The charge q in Figure 17.11 (and Eq. 17.10) is called a **test charge**. By measuring the force on a test charge, we can infer the magnitude and direction of the electric field at the location of the test charge. Because force is measured in units of newtons (N) and charge in coulombs (C), the electric field (according to Eq. 17.10) has units of newtons/coulombs.

The electric field $\vec{E}$ is related to electric force by Equation 17.10, so we can use a Coulomb's law calculation of the force to find the electric field in many situations. For example, let's calculate the electric field a distance r from the charge Q in Figure 17.11. Here q is the test charge, and we want to calculate the electric field produced by Q. According to Coulomb's law, the magnitude of the electric force exerted on the test charge q is

$$F = \frac{kQq}{r^2}$$

Inserting this expression into our relation for the electric field (Eq. 17.10) gives

$$F = \frac{kQq}{r^2} = qE$$

which leads to

$$E = \frac{kQ}{r^2} \qquad (17.11)$$

This result is the magnitude of the electric field a distance r from a point charge Q. The direction of $\vec{E}$ is along the line that connects the charge producing the field to the point where the field is measured. The electric field is directed outward away from Q when Q is positive (as sketched in Fig. 17.11) and inward toward Q when Q is negative.

Electric Field Lines, the Inverse Square Law, and Action-at-a-Distance

According to Coulomb's law, the force between two point charges falls off as $1/r^2$, where r is the separation between the two charges. In a similar way, the electric field produced by a point charge (Eq. 17.11) also varies as $1/r^2$. The electric force thus obeys an **inverse square law**, just as we found for the gravitational force in Chapter 5 (Fig. 5.30). Roughly speaking, a charged particle "sets up" field lines in its neighborhood, and the density of field lines is proportional to the amount of charge on the particle. These electric field lines produce a force on a nearby test charge and on any other nearby charges, and the magnitude of that force is proportional to the density of field lines at the test charge. For a point charge Q as in Figure 17.12, the density of field lines—that is, the number of field lines per unit area of space—falls as one goes away from the charge. Because the field lines emanating from Q spread out in a three-dimensional space, the number of field lines per unit area falls as $1/r^2$ as shown by the geometry in Figure 17.12.

Another interesting aspect of the electric force is the question of **action-at-a-distance**. How do two point charges that interact through Coulomb's law "know" about each other? In other words, how is the electric force transmitted from one charge to another? In terms of the field line picture, every charge generates (or carries with it) an electric field through which the electric force is transmitted.

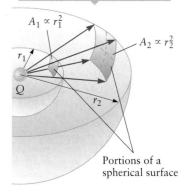

The field line area grows as r^2 because we live in a three-dimensional space.

Figure 17.12 As electric field lines emanate outward from a point charge, they intercept a larger and larger surface area. These surfaces are spherical, so their areas increase with radius as $A \propto r^2$.

EXAMPLE 17.3 | Electrons in a Television Tube

A conventional television picture tube called a cathode-ray tube (CRT) uses electrons to produce an image on its screen. A beam of electrons is produced at one end of the tube (Fig. 17.13) by a device called an "electron gun." The electrons have a very small velocity when they leave the vicinity of the gun and then move through a second region, where an electric field accelerates them to the final speed necessary for the CRT to operate properly.

The electric field in a typical TV tube has a value of 8.0×10^5 N/C. If the acceleration region has a length $L = 2.0$ cm, what is the speed of an electron when it exits this region?

RECOGNIZE THE PRINCIPLE

The electric field in the acceleration region produces a force on the electron according to Equation 17.10. If the force F and the electron's displacement L are parallel, the work done by the electric force on the electron is

$$W = FL$$

We can then use the work–energy theorem to calculate the final kinetic energy of the electron when it leaves the accelerator.

SKETCH THE PROBLEM

Figure 17.13 shows the directions of the electric field and the force. Since the charge on an electron is negative, the electric field must be directed to the left so as to give a force in the $+x$ direction (to the right).

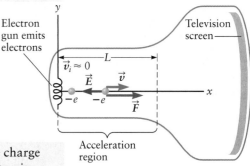

Figure 17.13 Example 17.3. A television picture tube. (Not to scale.)

IDENTIFY THE RELATIONSHIPS

The horizontal axis in Figure 17.13 is the $+x$ direction, and the electric field in the acceleration region is along $-x$ (to the left). We denote the x component of the electric field as E, so the value of E is negative. The charge on an electron is $-e$, so the x component of the force exerted on an electron is $F = -eE$. (Since the value of E is negative, F has a positive value; see Fig. 17.13.) This is the component of the force along x; the y component of the force in the acceleration region is zero. The work done on the electron by the electric force is

$$W = FL = -eEL$$

According to the work–energy theorem, this result equals the change in the electron's kinetic energy. Because the problem specified that the electron begins with a very small initial velocity, we assume the initial kinetic energy is approximately zero. If the electron's mass is m_e and its final speed (when it leaves the accelerator region) is v_f, we have

$$KE_f - KE_i = W = -eEL$$

$$KE_f = \tfrac{1}{2}m_e v_f^2 = -eEL \qquad (1)$$

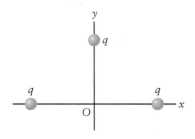

Figure 17.14 Concept Check 17.4.

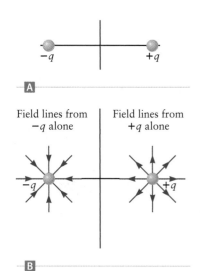

Figure 17.15 Example 17.4.

SOLVE

We can now solve for the final speed of the electron using Equation (1):

$$v_f^2 = -\frac{2eEL}{m_e}$$

$$v_f = \sqrt{-\frac{2eEL}{m_e}} \qquad (2)$$

The negative sign inside the square root might be a "sign" of trouble, but here it is correct. As we have already mentioned, the electric field is in the $-x$ direction and hence the value of E is negative. Inserting values of the various factors in Equation (2), including the electron mass from Appendix A, gives

$$v_f = \sqrt{-\frac{2eEL}{m_e}} = \sqrt{-\frac{2(1.60 \times 10^{-19} \text{ C})(-8.0 \times 10^5 \text{ N/C})(0.020 \text{ m})}{9.11 \times 10^{-31} \text{ kg}}}$$

$$v_f = \boxed{7.5 \times 10^7 \text{ m/s}}$$

What does it mean?

The electrons in a CRT have quite a high speed. In fact, our result for v_f is about 25% of the speed of light![2] In Section 17.5, we'll describe how the electric field in the acceleration region is generated.

CONCEPT CHECK 17.4 | Direction of an Electric Field

Three equal and positive charges q are located equidistant from the origin as shown in Figure 17.14. Is the direction of the electric field at the origin (a) along $+x$, (b) along $-x$, (c) along $+y$, or (d) along $-y$?

EXAMPLE 17.4 Electric Field of a Dipole

Consider an electric dipole consisting of two charges $+q$ and $-q$ as sketched in Figure 17.15A. They might be an electron and a proton or negative and positive ions that make up part of a molecule. Make a qualitative sketch of the electric field lines in the vicinity of this dipole.

RECOGNIZE THE PRINCIPLE

The electric field due to a single point charge is sketched in Figure 17.10. To find the electric field of a dipole, we use the superposition principle: the field of the dipole is the *sum* of the fields produced by the individual charges.

SKETCH THE PROBLEM AND SOLVE

Figure 17.15B shows the field lines from the charges $+q$ and $-q$ separately, and Figure 17.16 shows their sum. The fields from the two point charges must be added as *vectors*. This addition is shown carefully at point A, where the separate fields from $+q$ and $-q$ are drawn along with their sum.

What does it mean?

The dipole field in Figure 17.16 is an important field pattern that arises in many applications. Many molecules, including H_2O, act as electric dipoles, and the forces between such molecules involve the dipole electric field. We'll see a similar field pattern when we study the magnetic field produced by a magnetic dipole.

---

[2]The speed of our TV electrons is so high that a high-accuracy calculation of their motion requires the use of special relativity (Chapter 27). Including relativity would change our result here by about 6%.

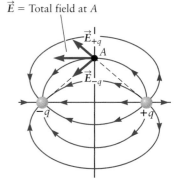

| EXAMPLE 17.5 | Electric Field from Two Point Charges |

Two point charges $Q_1 = 4.0\ \mu C$ and $Q_2 = -2.0\ \mu C$ are arranged as shown in Figure 17.17A. If $L = 3.0$ cm, what is the electric field at the origin?

SKETCH THE PROBLEM

Figure 17.17B shows the directions of the electric fields produced by the two charges along with the total field $\vec{E}_{\text{total}}$.

RECOGNIZE THE PRINCIPLE

The magnitude of the electric field from each charge can be calculated using Equation 17.11. The electric field at the origin is the vector sum of the field from each charge. We need to find both the magnitude and direction of $\vec{E}_{\text{total}}$.

IDENTIFY THE RELATIONSHIPS

The field from Q_1 has a magnitude (Eq. 17.11)

$$E_1 = \frac{kQ_1}{L^2} = \frac{(8.99 \times 10^9\ \text{N·m}^2/\text{C}^2)(4.0 \times 10^{-6}\ \text{C})}{(0.030\ \text{m})^2} = 4.0 \times 10^7\ \text{N/C}$$

The field from Q_2 has a magnitude

$$E_2 = \frac{kQ_2}{L^2} = \frac{(8.99 \times 10^9\ \text{N·m}^2/\text{C}^2)(2.0 \times 10^{-6}\ \text{C})}{(0.030\ \text{m})^2} = 2.0 \times 10^7\ \text{N/C}$$

SOLVE

Adding the electric fields as vectors is done graphically in Figure 17.17B. The magnitude of the total field E_{total} is

$$E_{\text{total}} = \sqrt{E_1^2 + E_2^2} = \sqrt{(4.0 \times 10^7\ \text{N/C})^2 + (2.0 \times 10^7\ \text{N/C})^2}$$

$$E_{\text{total}} = \boxed{4.5 \times 10^7\ \text{N/C}}$$

From the trigonometry in Figure 17.17B, the angle that $\vec{E}_{\text{total}}$ makes with the x axis is

$$\theta = \tan^{-1}\left(\frac{E_1}{E_2}\right) = \tan^{-1}\left(\frac{4.0 \times 10^7\ \text{N/C}}{2.0 \times 10^7\ \text{N/C}}\right) = \boxed{63°}$$

What does it mean?
When adding the fields from different charges, the fields must always be added as vectors.

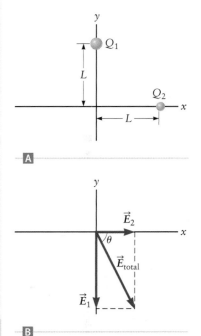

Figure 17.17 Example 17.5.

17.4 | CONDUCTORS, INSULATORS, AND THE MOTION OF ELECTRIC CHARGE

In Section 17.1, we described how the ancient Greeks used amber and fur to observe electric forces. To understand those experiments, we need to understand how charges can be transferred from one material to another.

The Structure of Matter from an "Electrical" Viewpoint

An atomic-scale picture of a *metal* such as copper is given in Figure 17.18A. Each Cu atom by itself is electrically *neutral*, with equal numbers of protons and electrons. When these neutral atoms come together to form a piece of copper metal, one

Figure 17.16 Example 17.4. The electric field of a dipole.

Insight 17.2
DRAWING ELECTRIC FIELD LINES
When you draw the electric field lines produced by a collection of point charges, the field lines must always begin and end on charges. As with the dipole in Figure 17.16, electric field lines begin at a positive charge and end at a negative charge.

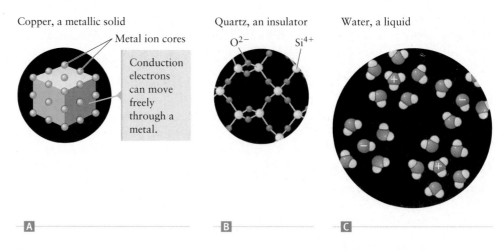

Copper, a metallic solid   Quartz, an insulator   Water, a liquid

Metal ion cores

Conduction electrons can move freely through a metal.

O^{2-} Si^{4+}

A   B   C

Figure 17.18 A In a metal such as copper, some of the electrons (called conduction electrons) are able to move freely through the entire sample. B In an insulator such as quartz, all the electrons are bound to ions and there are no conduction electrons. C Liquids and gases contain relatively small numbers of mobile charges (mainly ions).

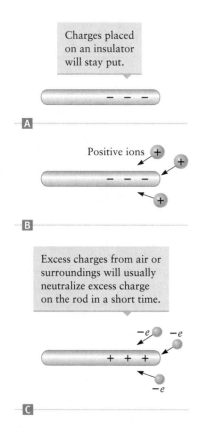

Charges placed on an insulator will stay put.

A

Positive ions

B

Excess charges from air or surroundings will usually neutralize excess charge on the rod in a short time.

$-e$ $-e$

$-e$

C

Figure 17.19 A When excess electrons are placed on an insulator, this charge generally stays where it is placed for a period of time (unlike in a metal, where the charge will move immediately). B and C Excess electrons or stray ions in the surrounding air are attracted to the insulator's surface, so the insulator's excess charge will eventually be neutralized.

or more electrons from each atom are able to "escape" from its "parent" atom and move freely through the entire piece of metal. These electrons are called *conduction electrons*. They leave behind positively charged ion cores that are bound in place and not mobile. In addition, a piece of metal can accept some additional electrons or let some of the conduction electrons be taken away, so the entire piece of metal can acquire a net negative or positive charge.

Amber, plastic, and quartz are *insulators*; electrons in these materials are *not* able to move freely through the material. A typical example is quartz, which is composed of SiO_2 molecules, each of which is a combination of one Si^{+4} and two O^{-2} ions. (See Fig. 17.18B.) These ions are bonded in place within the solid and are not able to move about. Unlike in metals, electrons cannot "escape" from these ions, and there are no conduction electrons available to carry charge through the solid. Extra electrons can be placed onto a piece of quartz, but they are not able to move about freely. Instead, an added electron would tend to stay in one spot (wherever it was initially placed), although the small amount of moisture that is often present will allow excess charge to gradually move about on the surface.

An atomic-scale picture of another important type of matter is shown in Figure 17.18C. A drop or cup of water consists of a collection of H_2O molecules, each of which has a total charge of zero and is thus electrically neutral. Each water molecule consists of two H atoms and one O atom; these atoms and their electrons are usually tightly bound to each other within a particular molecule. If all the atoms and electrons were bound in this way, there would be no "free" charges available to carry charge throughout the liquid, and water would behave as a perfect insulator. However, a few water molecules in any sample are always dissociated into free H^+ and OH^- ions. These ions can carry charge from place to place in much the same way that conduction electrons carry charge in a metal. (H^+ ions may also combine with neutral water molecules to form hydronium ions, H_3O^+, which along with OH^- ions can carry charge from place to place in a sample of water.) Most samples of water also contain impurities that contribute additional ions such as Na^+ or Cl^-, which act as mobile charge carriers.

The picture in a gas is similar to that found in a liquid; the constituent atoms and molecules are mostly neutral, but a few are present as ions able to carry charge from place to place. A few free electrons are usually present in a gas as well.

Placing Charge on an Insulator

To understand the electrical behavior of an object, we must understand what happens when charge is placed on or taken off the object. Let's begin with the case of an insulator such as quartz, and, for simplicity, assume there is no moisture on it and the air around it is very dry. If a few electrons are placed onto such an insulator, they will stay where they are put (Fig. 17.19A) because there are no free ions in an insulator and the insulator does not allow the extra electrons to move about.

In real life, excess charge will not stay on an insulator indefinitely. If an insulator contains some excess electrons, they will attract stray positive ions from the surrounding air; these stray ions will move onto the insulator and cause it to become neutral, or the electrons may combine with the ions to form neutral atoms (Fig. 17.19B). The amount of time needed to neutralize the charge depends on the amount of moisture in the air. It can be many minutes on a dry (usually winter) day or only a few seconds on a humid day. Any small amount of moisture on an insulator's surface forms a thin layer of water that enables charge to move from one spot to another.

Excess Charge Goes to the Surface of a Metal

Unlike the case of an insulator, electrons can move easily through a metal. So, if we place electrons on one spot of a piece of metal, there is no guarantee they will stay there, and we might imagine that they could go anywhere within the metal. In fact, any excess electrons on a piece of metal will all be distributed *on the surface* of the metal as sketched in Figure 17.20A. To understand this result, consider a piece of metal that is electrically isolated from the rest of the universe so that no new charge can jump onto or off the metal. If we wait long enough, all mobile charges (electrons) in the metal will come to rest and be in static equilibrium. Let's now imagine that some of these excess electrons are not on the surface but are in the interior instead. These electrons would each produce an electric field (according to Fig. 17.10), and the resulting electric field would cause other electrons in the metal to move. This result, however, contradicts our assumption that the system is in static equilibrium and means that the excess electrons cannot be inside the metal. Hence, for a metal in equilibrium, any excess electrons must be *at the surface* and *the electric field is zero inside a metal in equilibrium* (Fig. 17.20C).

This argument applies only to the *excess* charges placed on the metal. The interior of a neutral, uncharged metal contains equal numbers of positive and negative charges. These charges are present inside the metal at all times, and their electric fields cancel, leaving zero electric field inside the metal. Only the excess charge resides at the metal's surface.

Parts A and B of Figure 17.20 show how excess charge is distributed on a metal. In Figure 17.20A, some extra electrons have been added to the metal, so the surface has a net negative charge. In Figure 17.20B, we imagine that some electrons have been removed from the metal, so its net charge is positive. It is customary to draw this situation as if there are positive charges at the surface of the metal, even though that does *not* correspond to placing positive ions or protons onto the metal!

Charging an Object by Rubbing It

The Greeks discovered electricity when they rubbed a piece of amber with fur. The act of rubbing caused some charge to be transferred between the amber and the fur, and we now know that electrons moved from the fur to the amber. The amber thus acquired a net negative charge (an excess of electrons), whereas the fur was left with a net positive charge (an excess number of positive ions) as illustrated in Figure 17.21.

There is nothing unique about amber; a similar result is found if a rubber or plastic rod is rubbed with fur. Likewise, there is nothing special about fur. For example, a glass rod can be charged by rubbing it with a silk cloth, although in this case the glass acquires a net positive charge because electrons leave the glass and move to the silk. The mechanical process of rubbing is thus a general way to move electrons from one material to another. This is the basis of the Van de Graaff generator (Fig. 17.22), which is found in many science museums.

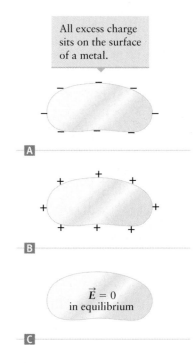

All excess charge sits on the surface of a metal.

A

B

$\vec{E} = 0$
in equilibrium

C

Figure 17.20 In static equilibrium, the electric field inside a metal must be zero. All excess charges then reside on the metal's surface. **A** If the excess charge is negative—that is, if there are extra electrons on the metal—these electrons will rest on the surface. **B** If the excess charge is positive, there is a deficit of electrons. This deficit of electrons will be at the surface, so the surface will have a net positive charge. **C** In all cases, the electric field is zero inside a metal in static equilibrium.

Rubbing an amber rod with fur transfers electrons from the fur to the amber.

A

Rubbing glass with silk transfers electrons from the glass to the silk.

B

Figure 17.21 ◢ When a piece of amber or a plastic rod is rubbed with fur, it becomes negatively charged. ◣ A glass rod rubbed with silk becomes positively charged.

Tony Freeman/Photo Edit

Figure 17.22 A Van de Graaff generator produces a large electric charge by rubbing an internal rubber belt. This belt transfers electric charge to the metal sphere at the top of the generator.

We mentioned in connection with Figure 17.1 that a charged object, such as a piece of amber, can be used to attract and move small pieces of paper and similar objects. Let's consider this attraction in a little more detail. Suppose we have a rubber rod and a small piece of paper, and that both are electrically neutral at the start of the experiment (Fig. 17.23). If the rubber rod is rubbed with fur, it will acquire a negative charge. If this charged rod is then brought near the paper, the paper is attracted to the rod. This effect is caused by electric forces between charges on the rod and the paper. The rod acquired its excess charge by rubbing, but the paper is electrically neutral (its net charge is zero), so why is an electric force exerted on the paper? Although the paper is electrically neutral, the presence of the rod nearby causes some movement of the charges in the paper. The rod is negatively charged, so electrons in the paper are repelled, and positive ions in the paper are attracted. Although these electrons and ions are not free to move very far, they can move a small amount while still being bound inside the paper. The electrons thus move a short distance away from the rod and the positive ions move toward the rod as sketched in Figure 17.23C, leaving a net positive charge on the portion of the paper nearest the rod and a net negative charge on the opposite side of the paper. The paper is then said to be *polarized*.

Because unlike charges attract, the positive side of the paper is attracted to the negatively charged rod. At the same time, the negative side of the paper is repelled from the rod. The positively charged side of the paper is closer to the rod, so because the electric force decreases as the distance between the charges increases, the magnitude of the attractive force is larger than the magnitude of the repulsive force. The net result is that the paper is *attracted* to the rod. In this way, there can be an electric force on an object even when the object is electrically neutral, provided the object is polarized.

The same principle is illustrated in Figure 17.24A, which shows water flowing from a container. The stream of water would ordinarily flow vertically downward due to the force of gravity, but a charged balloon placed near the stream exerts a force on the water and deflects the stream sideways. The electric field from the charged balloon polarizes the water molecules (Fig. 17.24B), leading to an attractive force as found with the paper in Figure 17.23.

CONCEPT CHECK 17.5 | Force on a Polarized Object

The negatively charged balloon in Figure 17.24A attracts the stream of water. Suppose a positively charged balloon was used instead. Which of the following statements would then be true?

(a) The water would be repelled from the positively charged balloon.
(b) The water would be attracted to the positively charged balloon.
(c) There would be no force on the water.

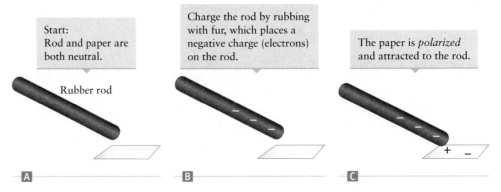

Start: Rod and paper are both neutral.

Rubber rod

Charge the rod by rubbing with fur, which places a negative charge (electrons) on the rod.

The paper is *polarized* and attracted to the rod.

A   **B**   **C**

Figure 17.23 ◢ A rubber rod that is initially neutral can ◣ be charged negatively by rubbing it with fur. ◔ This negatively charged rod can then polarize a nearby object such as a small piece of paper, resulting in electrical attraction between the rod and the paper.

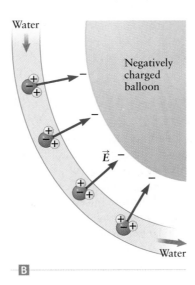

Water

Negatively charged balloon

$\vec{E}$

Water

A

B

Figure 17.24 **A** A stream of water is deflected by a nearby charged balloon. **B** The balloon polarizes the water, similar to the piece of paper in Figure 17.23C.

The Concept of Electrical "Ground"

Suppose a rubber rod is charged by rubbing so that it acquires some extra electrons and is then placed on a table. If we wait for some period of time (the time required depends on the amount of moisture in the room), we would find that the rod has lost its excess charge. If we watch very carefully with very sensitive electronic measuring equipment, we would find that the excess electrons originally on the rod move from the rod to the table and eventually into the ground below. The ground is generally quite moist, so it conducts charge well. Because the ground is literally everywhere, it provides a common path that excess charge can use to flow from one spot to another. In fact, the term *ground* is used generically to denote the path or destination of such excess charges, even if the true path does not involve any dirt. This notion of **electrical ground** plays an important role in many situations.

Charging by Induction

Suppose you are given a negatively charged rubber rod and are asked to transfer some of this charge to a piece of metal. You could accomplish this task by simply bringing the rubber rod into contact with the metal (Fig. 17.25). The negative charge on the rod is caused by an excess of electrons, and some of these electrons will move to the metal when it touches the rod; this process is just an example of charging by rubbing.

Now suppose you are given the same negatively charged rubber rod, but this time you are asked to give the metal a net *positive* charge. This task might seem to be impossible because the excess charges on the rod are negative, but an approach called **charging by induction** will accomplish it. This approach makes use of polarization and the properties of an electrical ground. The negatively charged rod is first brought near the metal (Fig. 17.26A), polarizing the metal. That is, some of the conduction electrons in the metal move to the opposite side—away from the rod—due to the repulsion of like charges, leaving the side of the metal near the rod with a net positive charge. A connection is then made from the piece of metal to electrical ground using a metal wire, so electrons are able to move even farther from the charged rod. In fact, some of the electrons will move off the original piece of metal and into the electrical ground region (Fig. 17.26B). The final step in the process is to remove the grounding wire, leaving the original piece of metal with a net positive charge (Fig. 17.26C). Notice that we have not put protons or any other

CHARGING AN OBJECT BY CONTACT
Before contact

Contact

After contact

Figure 17.25 If one charged object touches a second object, the second object will usually acquire some of the excess charge. Hence, the second object is charged "by contact."

Positive charge is
attracted to rod.

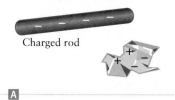

Charged rod

A

Negative charge
(electrons) will flow
to ground.

Ground

B

Final stage: Excess
charge is positive.

C

Figure 17.26 An object can be charged by the process of induction. **A** The object is first brought near a charged rod, polarizing the object. **B** The object is then connected to ground, and some electrons flow between it and ground. **C** The object is left with an excess charge when it is disconnected from ground.

positive charges on the metal. Instead, we have *removed* electrons. The final result is that the metal has a net positive charge.

17.5 | ELECTRIC FLUX AND GAUSS'S LAW

We can calculate the electric field due to a point charge using Equation 17.11. In principle, we can use this result to deal with any conceivable distribution of charges by treating the distribution as a collection of point charges and using superposition to find the total field. Fortunately, there is a simpler way to deal with complex charge distributions, based on *Gauss's law*. To understand Gauss's law, we must first define a quantity called the *electric flux*. In words, electric flux is equal to the number of electric field lines that pass through a particular surface multiplied by the area of the surface. Electric flux is denoted by the symbol Φ_E.

Some examples of electric flux calculations are given pictorially in Figure 17.27. For simplicity, these examples assume the electric field is constant in both magnitude and direction. In Figure 17.27A, $\vec{E}$ is perpendicular to a flat surface having a total area A. The electric flux through this surface is $\Phi_E = EA$. In this case, the flux is just the magnitude of the electric field multiplied by the area of the surface. Flux is a scalar, just an ordinary number; it is *not* a vector.

The value of Φ_E depends on how many field lines intersect the surface as illustrated in Figure 17.27B, where $\vec{E}$ is parallel to a surface. In this case, the field lines do not cross the surface, and $\Phi_E = 0$. If the field makes an angle θ with the surface (Fig. 17.27C), the flux is

$$\Phi_E = EA \cos \theta \qquad (17.12)$$

The three examples in Figure 17.27A–C all involve simple, flat surfaces. We'll often be interested in the flux through a *closed* surface such as a box or a sphere. The flux due to a constant field $\vec{E}$ through these closed surfaces is shown in Figure 17.27D and E. By convention, the flux through a surface is *positive* if the field is directed *out* of the region contained by the surface, whereas the flux is *negative* if $\vec{E}$ is directed *into* the region. For the cases with a constant field in parts D and E of Figure 17.27, the total flux through the entire closed surface is *zero* in each case. For the box in Figure 17.27D, there is a negative flux through the face of the box on the left and a positive flux through the face on the right. The total flux is the sum of these two contributions and is zero. Likewise, there is a negative flux through the left side of the spherical surface in Figure 17.27E and a positive flux through the right side, and the total flux is again zero.

Gauss's Law

Gauss's law asserts that the electric flux through any closed surface is proportional to the total charge q inside the surface, with

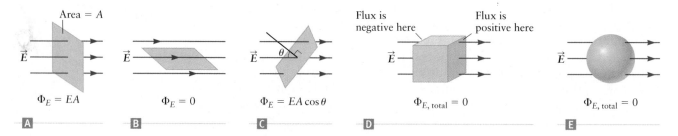

$$\Phi_E = EA \qquad \Phi_E = 0 \qquad \Phi_E = EA \cos\theta \qquad \Phi_{E,\text{total}} = 0 \qquad \Phi_{E,\text{total}} = 0$$

Figure 17.27 Finding the electric flux through a surface.

$$\Phi_E = \frac{q}{\varepsilon_0} \qquad\qquad \text{(17.13)} \quad \text{Gauss's law}$$

The constant of proportionality that relates flux and charge is ε_0, the same physical constant that enters Coulomb's law (Eq. 17.5).

Since electric flux depends on the magnitude and direction of the electric field, the left-hand side of Equation 17.13 depends on $\vec{E}$ while the right-hand side depends on the charge. We would like to use Gauss's law to calculate $\vec{E}$, but it is not immediately obvious how to do so. In particular, Φ_E is the total flux through a closed surface, so its value depends on the magnitude and direction of the electric field at *all* points on the surface. To see how to use Gauss's law, it is simplest to consider some examples.

Using Gauss's Law to Find $\vec{E}$ for a Point Charge

Let's first consider the familiar case of a single point charge. To apply Gauss's law, we must first choose the surface, called a **Gaussian surface**, that will be used in the flux calculation. Equation 17.13 holds for *any* closed surface, so our strategy is to choose a surface that will make the calculation of Φ_E as simple as possible. To this end, we choose a surface that *matches the symmetry of the problem.* For a point charge, we know that the electric field lines have a spherical symmetry (Fig. 17.28). This spherical symmetry means that the magnitude of the electric field depends only on the distance from the charge r and that $\vec{E}$ must be directed radially, either outward or inward (and not sideways). A surface that matches this symmetry is a sphere centered on the charge as sketched in Figure 17.28. Because of the symmetry, the magnitude of $\vec{E}$ is the same at all points on the sphere, and $\vec{E}$ is perpendicular to the sphere at all points where it intersects the surface.

A calculation of the flux through the sphere in Figure 17.28 is similar to the flux computation in Figure 17.27A. When $\vec{E}$ is perpendicular to a surface, the flux is equal to the magnitude of $\vec{E}$ multiplied by the area of the surface. So, for the flux in Figure 17.28, we have

$$\Phi_E = E A_{\text{sphere}}$$

where A_{sphere} is the area of our spherical Gaussian surface. If the radius of this sphere is r, then $A_{\text{sphere}} = 4\pi r^2$, and the flux is

$$\Phi_E = 4\pi r^2 E$$

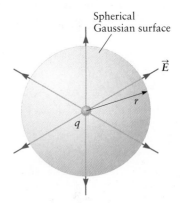

Spherical Gaussian surface

$\vec{E}$

r

q

According to Gauss's law, this flux is proportional to the total charge contained within the surface. Using Equation 17.13, we have

$$\Phi_E = 4\pi r^2 E = \frac{q}{\varepsilon_0}$$

We can now solve for E and find

$$E = \frac{q}{4\pi\varepsilon_0 r^2} \qquad\qquad \text{(17.14)}$$

Figure 17.28 To calculate the electric field near a point charge using Gauss's law, we choose a spherical Gaussian surface centered on the point charge.

which agrees with our previous result obtained using Coulomb's law (Eq. 17.11). The key to this application of Gauss's law was our choice of the Gaussian surface. This choice made the calculation of Φ_E straightforward because E has the same value over the entire surface and the electric field's direction is always perpendicular to the surface.

The result in Equation 17.14 shows that Coulomb's law is actually a special case of Gauss's law. That is, Gauss's law applied to the case of a point charge together with the relationship between the electric field and the electric force (Eq. 17.10) can be used to derive Coulomb's law. Hence, all the results we have found using Coulomb's law also follow from Gauss's law. In addition, Gauss's law can readily deal with many other situations that are mathematically very difficult or awkward to attack with Coulomb's law. We will spend the rest of this section working through a few such cases, using the general approach summarized below.

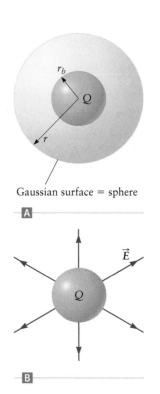

Gaussian surface = sphere

A

B

Figure 17.29 Example 17.6.

EXAMPLE 17.6 Electric Field from a Spherical Charge

Consider the uniform spherical ball of charge in Figure 17.29A, with total charge Q and radius r_b. Find the electric field at points outside the ball.

RECOGNIZE THE PRINCIPLE

We follow the strategy described in the "Applying Gauss's Law" problem-solving strategy. The key is to find a Gaussian surface that matches the symmetry of the charge distribution (step 1 in the strategy). The charge is spherically distributed in Figure 17.29A, so the electric field must also have spherical symmetry.

SKETCH THE PROBLEM

Step 2 in the strategy: The spherical symmetry means that $\vec{E}$ must be directed radially pointing either away from or toward the center of the ball of charge. If Q is positive, the field will be as sketched in Figure 17.29B.

IDENTIFY THE RELATIONSHIPS

Step 3: Choose a Gaussian surface. Because of the spherical symmetry, we can also say that the magnitude of $\vec{E}$ will depend only on the distance from the center of the ball of charge. We therefore pick a spherical Gaussian surface as sketched in Figure 17.29A. The electric field will have a constant magnitude at all points on this surface, and $\vec{E}$ will be perpendicular to the surface. This situation is similar to what we had in Figure 17.28 when we applied Gauss's law to the case of a point charge.

SOLVE

We next calculate the electric flux through our chosen Gaussian surface. The Gaussian surface is a sphere of radius r, so its surface area is $A_{\text{sphere}} = 4\pi r^2$. Denoting the electric field on the sphere by E and again noting that $\vec{E}$ is perpendicular to all points on the surface, the electric flux is

$$\Phi_E = EA_{\text{sphere}} = 4\pi r^2 E$$

Gauss's law relates this flux to the total charge contained within the spherical surface. We are interested in locations outside the ball ($r > r_b$), so the charge inside the Gaussian surface is the total charge of the ball Q. We thus have

$$\Phi_E = 4\pi r^2 E = \frac{Q}{\varepsilon_0}$$

which leads to

$$E = \boxed{\frac{Q}{4\pi\varepsilon_0 r^2}}$$

Since $1/(4\pi\varepsilon_0) = k$, we can also write this result as

$$E = \boxed{\frac{kQ}{r^2}}$$

What does it mean?

Our result for E is identical to the electric field of a point charge of magnitude Q located at the origin (Eq. 17.11). In fact, the electric field from *any* spherical distribution of charge is the *same* as the field from a point charge with the same total charge. Note that this result applies *only outside* the ball of charge.[3]

CONCEPT CHECK 17.6 | Finding the Electric Flux

Consider a point charge q located outside a closed surface as sketched in blue in Figure 17.30. The electric field lines from the charge penetrate the surface. What is the total electric flux through this surface?

(a) q/ε_0 (b) $-q/\varepsilon_0$ (c) 0

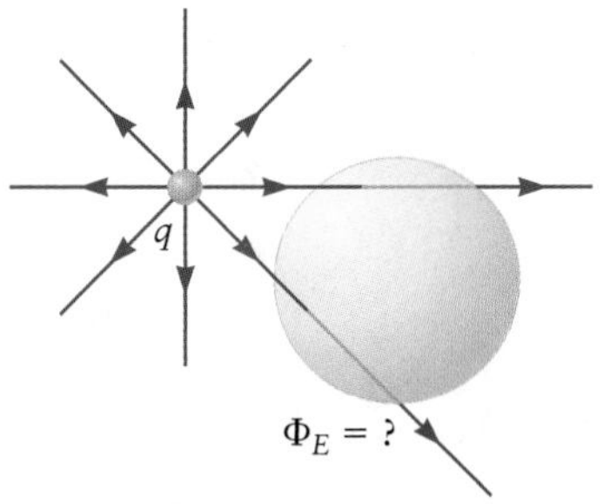

More Applications of Gauss's Law

Let's next discuss an application of Gauss's law that does not involve the spherical symmetry of a point charge or a ball of charge. Consider a very long, straight line of charge that lies along the x axis as sketched in Figure 17.31A. The line of charge has a total length L (where L is very large) and a total charge Q. How can we find the electric field produced by this charge distribution?

The first step in the "Applying Gauss's Law" problem-solving strategy is to recognize the symmetries of the charge distribution and of the electric field it produces so that you can choose a Gaussian surface. For step 2, the sketch in Figure 17.31B shows two observers who are studying the charged line and its electric field. These two observers are looking at the charged line from opposite directions. Because it is very long, the charged line will look the same to these observers, so the electric field must also look the same. Suppose (hypothetically) observer A finds that $\vec{E}$ points away from him and makes an angle θ with the x axis as sketched in Figure 17.31C. Since the observers are symmetrically arranged, observer B must find a similar result (Fig. 17.31D), with the electric field pointing away from him and also making an angle θ with the x axis. However, the two observers *must agree* on the magnitude and direction of $\vec{E}$, and they can only agree if $\theta = 90°$. The symmetry of

Figure 17.30 Concept Check 17.6. What is the total electric flux through this surface?

$\Phi_E = ?$

---

[3]We found a similar result in connection with gravitation in Chapter 5, where we noted that the gravitational force due to a spherically symmetric mass is the same as if all the mass were located at the center of the sphere.

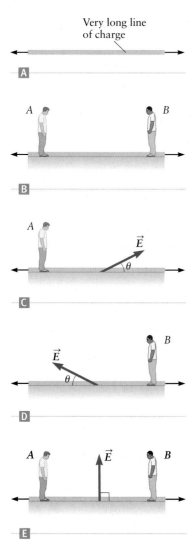

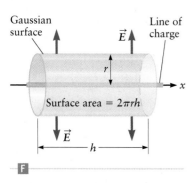

Figure 17.31 **A** Calculation of the electric field near a line of charge. **B–E** Symmetry requires that the electric field be perpendicular to the line. The magnitude of E depends only on the distance r from the line. **F** If we choose a cylindrical Gaussian surface, the electric field is constant in magnitude at the walls of the cylinder and $\vec{E}$ is parallel to the ends of the cylinder.

the line of charge means that $\vec{E}$ must be *perpendicular to the line*. The electric field could be directed away from the line or toward the line, but symmetry requires that it be perpendicular (Fig. 17.31E).

The next step is to choose a Gaussian surface that matches this symmetry. We wish (if possible) for $\vec{E}$ to be perpendicular to the surface, so we choose a cylinder centered on the line of charge (Fig. 17.31F). This cylinder can be divided into a curved section plus two flat, circular ends. The electric field is parallel to the circular ends, so the electric flux through the ends is zero. We want the magnitude of $\vec{E}$ to be the same at all points on the curved part of the cylinder. The magnitude of the field can only depend on the distance from the charged line, so E will be a constant on our Gaussian surface provided the charged line runs along the central axis of the cylinder.

Next (step 4), we calculate the flux through our chosen Gaussian surface. For a cylinder of radius r and length h, the area of the curved part of the cylinder is $A = 2\pi rh$. The flux through this curved surface is thus

$$\Phi_E = EA = 2\pi Erh$$

We now find the total charge inside the Gaussian surface and then apply Gauss's law. The total charge within the cylinder in Figure 17.31F is equal to the charge per unit length Q/L multiplied by the length of the cylinder h, so

$$q = \frac{Q}{L}h$$

Applying Gauss's law then gives

$$\Phi_E = 2\pi Erh = \frac{q}{\varepsilon_0} = \frac{Qh/L}{\varepsilon_0}$$

We can now solve for the magnitude of E and find

$$E = \frac{Q}{2\pi\varepsilon_0 Lr} \qquad (17.15)$$

CONCEPT CHECK 17.7 | **Electric Field from a Line of Charge**

If the charge Q on the line in Figure 17.31F is negative, is the direction of the electric field (a) perpendicular to the line and directed away from it, (b) perpendicular to the line and directed toward it, or (c) parallel to the line?

EXAMPLE 17.7 | **Field from a Flat Sheet of Charge**

Consider a large, flat sheet of charge as sketched in Figure 17.32A. If this sheet has a positive charge per unit area σ, find the electric field produced by the sheet.

RECOGNIZE THE PRINCIPLE

We again follow the steps outlined in the "Applying Gauss's Law" problem-solving strategy. We begin by considering the symmetry of the charge distribution and of the resulting electric field so that we can choose a Gaussian surface.

SKETCH THE PROBLEM

We again imagine how this charge distribution and field would appear to different observers who look at them from different directions (Fig. 17.32B). These observers must agree on the direction of $\vec{E}$, and as was the case with the charged line in Figure 17.31, the electric field must be perpendicular to the plane. If the charge on the plane is positive, the electric field points away from the plane.

IDENTIFY THE RELATIONSHIPS

Now that we have identified the direction of $\vec{E}$, we must choose a Gaussian surface. The cylinder's axis in Figure 17.32A is oriented perpendicular to the plane. The flux through the curved sidewall of the surface is zero because $\vec{E}$ is parallel to this part of the cylinder. The electric field $\vec{E}$ is perpendicular to the ends of the cylinder, so if the ends each have an area A, the flux is $\Phi_E = EA$ through each end of the cylinder. For a positively charged plane, the electric flux through each end of the cylinder is positive because the electric field lines pass outward through the cylinder. The total electric flux through the cylinder is thus

$$\Phi_E = 2EA$$

where the factor of two is needed to include the flux through both ends.

SOLVE

The total charge inside the Gaussian surface is equal to the charge per unit area on the sheet σ multiplied by the cross-sectional area of the cylinder A, so $q = \sigma A$. Inserting this into Gauss's law leads to

$$\Phi_E = \frac{q}{\varepsilon_0}$$

$$2EA = \frac{\sigma A}{\varepsilon_0}$$

Solving for the electric field, we get

$$\boxed{E = \frac{\sigma}{2\varepsilon_0}}$$

What does it mean?

The magnitude of the electric field is constant, independent of distance from the plane. For this reason, a large, flat sheet of charge is a very convenient way to produce a uniform electric field. You may think that a plane of charge is a rather artificial example, but such charged sheets are used often in practical devices such as the accelerator of the TV picture tube in Figure 17.13.

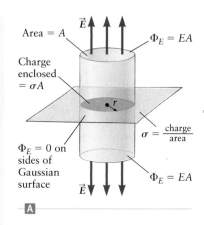

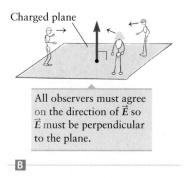

Figure 17.32 Example 17.7. **A** Calculation of the electric field near a charged plane. **B** Symmetry requires that the electric field be perpendicular to the plane.

Electric Field between Two Metal Plates

Figure 17.33 shows two parallel metal plates that carry charges $+Q$ and $-Q$. This arrangement, called a ***capacitor***, is used in many electric circuits. Let's calculate the electric field produced by this parallel-plate capacitor.

Recall from Section 17.4 that the excess charge on a metal in equilibrium is always located at the surface. For a single metal plate with a total charge of $+Q$ (Fig. 17.34), symmetry tells us that half the charge will reside on each side. (To be strictly correct, there would also be some charge on the edges, but if the plate is very thin, the amount of charge on the edges will be negligible.) When we consider two plates with opposite charges as in our capacitor (Fig. 17.33), the excess charges on the two plates will attract each other and draw all the excess charge to the inner sides of the two plates.

To find the electric field between the two capacitor plates in Figure 17.33, we use the result from Example 17.7 for the field from a single plane of charge. There we found

$$E = \frac{\sigma}{2\varepsilon_0} \quad \text{(magnitude of electric field from a single charged plane)} \quad (17.16)$$

where σ is the charge per unit area on the plane.

The capacitor in Figure 17.33 contains two plates; if each has an area A, they have a charge per unit area $\sigma_1 = +Q/A = +\sigma$ and $\sigma_2 = -Q/A = -\sigma$. According to the principle of superposition, the total field is the vector sum of the fields from each

Figure 17.33 Calculation of the electric field between two thin metal plates. This arrangement is called a parallel-plate capacitor.

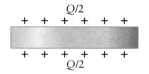

Figure 17.34 For a single thin metal plate, any excess charge is distributed evenly on the two sides.

plate. In the region between the plates, the fields from the two plates are parallel and add to give a total field

$$E = \frac{\sigma}{\varepsilon_0} \quad \text{(magnitude of electric field between two charged plates)} \quad \text{(17.17)}$$

The field between the two plates is thus twice the field from a single plate (Eq. 17.16). In terms of the charge Q and plate area A, the field is

$$E = \frac{Q}{\varepsilon_0 A} \quad \text{(17.18)}$$

We'll use this result in Chapters 18 and 19 as we analyze the properties of capacitors.

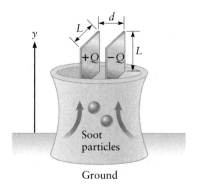

Figure 17.35 Example 17.8.

EXAMPLE 17.8 Electric Fields in a Smokestack Scrubber

In Example 17.1, we discussed a very simple model for a smokestack scrubber that removes undesirable particles from the air by first adding some excess electrons and then using electric forces to pull the particles out of the air. Consider again a soot particle of mass $m_{\text{soot}} = 1.0 \times 10^{-12}$ kg that travels upward in a smokestack. The scrubber in Example 17.1 added 69 electrons to the particle, and its charge was $q_{\text{soot}} = 1.1 \times 10^{-17}$ C. Assume the electric field in the scrubber is produced by two parallel, square plates of width $L = 1.0$ m and separation $d = 0.010$ m, with charges $\pm Q$. **(a)** What must be the value of the electric field between the plates so that force on the soot particle is equal to the weight of the particle? (A real scrubber would use a collection of many pairs of such plates in parallel.) **(b)** What charge Q on the scrubber's plates is required to produce the electric field in part (a)?

RECOGNIZE THE PRINCIPLE

We can find the magnitude of E by setting the electric force $q_{\text{soot}}E$ on a soot particle equal to its weight. We can then use our result for the electric field between two charged plates (Eq. 17.18) to relate the electric field to the charge $\pm Q$ on each plate.

SKETCH THE PROBLEM

The scrubber is shown in Figure 17.35. This arrangement is similar to the two metal plates in Figure 17.33, with $\vec{E}$ perpendicular to the plates.

(A) IDENTIFY THE RELATIONSHIPS

The electric force on a soot particle is

$$F_{\text{elec}} = q_{\text{soot}}E$$

where E is the magnitude of the field between the scrubber plates. This force is designed to equal the particle's weight, so we have

$$F_{\text{elec}} = q_{\text{soot}}E = m_{\text{soot}}g$$

SOLVE

Solving for E gives

$$E = \frac{m_{\text{soot}}g}{q_{\text{soot}}}$$

Inserting the mass and charge on the soot particle leads to

$$E = \frac{m_{\text{soot}}g}{q_{\text{soot}}} = \frac{(1.0 \times 10^{-12} \text{ kg})(9.8 \text{ m/s}^2)}{1.1 \times 10^{-17} \text{ C}} = \boxed{8.9 \times 10^5 \text{ N/C}}$$

This electric field is quite large, but it is still less than the electric fields at which air molecules are torn apart, an effect called dielectric breakdown (discussed in Chapter 18). Realistic scrubber designs do employ large electric fields.

(B) IDENTIFY THE RELATIONSHIPS

The electric field between two metal scrubber plates with charge $\pm Q$ is (Eq. 17.18)

$$E = \frac{Q}{\varepsilon_0 A}$$

For square plates with sides of length L, we have $A = L^2$ and

$$E = \frac{Q}{\varepsilon_0 L^2}$$

SOLVE

Solving for the charge Q, we find

$$Q = \varepsilon_0 E L^2$$

Inserting the given dimensions of our plates and the value of E from part (a) gives

$$Q = \varepsilon_0 E L^2 = [8.85 \times 10^{-12} \ C^2/(N \cdot m^2)](8.9 \times 10^5 \ N/C)(1.0 \ m)^2$$

$$Q = \boxed{7.9 \times 10^{-6} \ C}$$

What have we learned?
Our analysis of a scrubber has so far only considered the force and electric field that acts on the soot particle. We'll consider how to place the necessary charge Q on the plates when we discuss electric potential energy in Chapter 18.

17.6 | ⊗ APPLICATIONS: DNA FINGERPRINTING

Electric forces and fields are used in many applications, some of which are very old. The smokestack scrubber used for pollution control in Examples 17.1 and 17.8 was invented about 100 years ago. Let's now consider a relatively new application, the technique of DNA fingerprinting, that is widely used in biology. The DNA molecules to be studied are first broken into fragments using enzymes (Fig. 17.36A; a job for biochemists). The fragments have a range of sizes and form ions in solution with typical charges between $-100e$ and $+100e$. Each specific DNA molecule forms its own unique set of fragments. The DNA fragments are next placed at one end of a long tube containing a gel (Fig. 17.36B). An electric field is then turned on in the gel tube; this field can be produced using two metal plates as we analyzed in Section 17.5. Since the DNA fragments are ions, they experience an electric force that pushes them through the gel. In Figure 17.36B, we assume the ions are negatively charged, so the electric field is to the left and the force $\vec{F}_{elec} = q\vec{E}$ is to the right.

In addition to the electric force, there is also a drag force on the ions due to collisions with molecules in the gel. This drag force was described in Chapter 3; for a particle of radius r and velocity $\vec{v}$, the drag force is given by (Eq. 3.23)

$$\vec{F}_{drag} = -Cr\vec{v} \tag{17.19}$$

where C is a constant that depends on the properties of the gel. The DNA fragment ions move through the gel in response to the forces $\vec{F}_{elec}$ and $\vec{F}_{drag}$. At first, their velocity is very small, so the drag force is also small and the fragment ions accelerate. After a short time, their velocity increases to the point at which $\vec{F}_{drag}$ is equal in magnitude and opposite in direction relative to $\vec{F}_{elec}$. The total force is then zero,

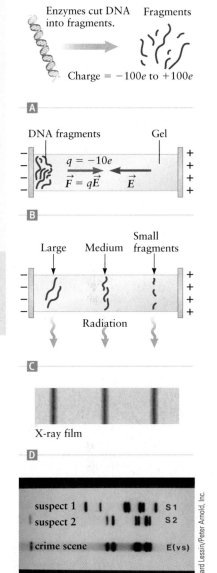

Figure 17.36 **A–D** DNA fingerprinting uses the electric force on fragments of DNA to separate the fragments according to their size and charge. **E** In a forensic analysis, the DNA fingerprints of suspects are compared to a "known" sample (i.e., one collected at the crime scene). Here it appears that suspect 2 is guilty.

and the ions move at a constant velocity through the gel in a process very similar to a skydiver moving through the air at her terminal velocity (Chapter 3). When an ion in the gel travels at its terminal velocity, we thus have $\vec{F}_{elec} + \vec{F}_{drag} = 0$. Inserting our results for the components of these forces along the gel tube gives

$$qE - Crv = 0$$

Solving for the ion's speed, we find

$$Crv = qE$$

$$v = \frac{qE}{Cr} \qquad (17.20)$$

The speed of the ion thus depends on its charge q and its size r, so different DNA fragments move through the gel at different speeds, with the smallest fragments moving fastest (assuming similar values of q). After some time, the fragments are distributed along the gel tube as sketched in Figure 17.36C, and the fragments are thus separated according to their size.

The final step in this process is to record the fragment locations, which is done by attaching radioactive isotopes (see Chapter 30) to the DNA before the fragments are made. These isotopes emit radiation that can be detected with photographic film or in other ways to record the position of the fragments within the gel tube (Fig. 17.36D).

The technique of separating ions according to their size by using electric forces to push the ions through a gel is called *electrophoresis*. This method has become an important tool in biology and forensics (Fig. 17.36E).

17.7 | "WHY IS CHARGE QUANTIZED?" AND OTHER DEEP QUESTIONS

In Section 17.1, we stated some "well-known" facts about electric charges. Let's now reexamine those facts in light of what we have learned about electric forces and fields.

How Are Protons Held Together?

We have mentioned many times that electric charge is carried by subatomic particles called protons and electrons. Indeed, we tend to take the existence of these particles for granted today, but what do they "look like" inside? Let's consider the proton first. We have seen that electric forces can be very large, and we know that like charges repel. If a proton is a small ball of positive charge as sketched in Figure 17.37A, the repulsive electric forces between different parts of a proton will tend to make these parts fly apart. How, then, is the proton held together?

Experiments have shown that a proton is actually an assembly of three particles called *quarks* as illustrated schematically in Figure 17.37B. The proton is composed of two particular types of quarks called "up" and "down." Each proton consists of one up quark of charge $-\frac{1}{3}e$, and two down quarks, each with charge $+\frac{2}{3}e$. The total charge of a proton is thus $-\frac{1}{3}e + \frac{2}{3}e + \frac{2}{3}e = +e$. That a proton is composed of three separate quarks does not really answer our initial question about how protons are held together because electric forces between the quarks will still tend to push them apart, especially since the down quarks are both positively charged. The large repulsive electric forces between quarks are overcome by an even more powerful attractive force called the *strong force*, which is a fundamental force of physics just as gravity and electromagnetism are fundamental forces. What's more, the strong force is so powerful that an individual quark is not able to escape from a proton under any circumstances. This is why we only see integral multiples of $\pm e$ in nature.

Asking simple questions about electric forces and Coulomb's law has thus led us to some deep questions about the fundamental structure of matter. The stability of

The Coulomb force tends to make a proton fly apart. There must be a strong attractive force to hold it together.

A

Down quark's charge $= +\frac{2}{3}e$ each

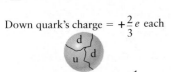

Up quark's charge $= -\frac{1}{3}e$

B

Figure 17.37 **A** A "ball" of electric charge will tend to fly apart due to the electric forces between different parts of the ball. If this ball is to be in mechanical equilibrium, there must be another force that overcomes the electric force and holds the ball together. **B** A proton is such a charged ball, made up of three quarks held together by the strong force.

matter relies on a force we had so far not even known to exist! We'll say more about the strong force and quarks in Chapters 30 and 31.

Electrons Are Point Charges

We have just described the internal structure of the proton in terms of quarks. You might now expect that the electron has a similar internal structure, but nature has another surprise for us. As far as physicists can tell, the electron is truly a *point* particle; the mass and charge of an electron are located at an infinitesimal "point" in space. Your "everyday" intuition will probably have difficulty accepting that a particle can be a true "point" in space. There is no way to explain it except to say that this is how nature seems to have designed the electron!

Conservation of Charge

Electric charge is conserved. That is, the total charge of the universe is a constant. This does *not* mean that the total number of protons or electrons in the universe is constant. Indeed, in Section 7.8 we mentioned that in certain situations a neutron can decay to form a proton, an electron, and an antineutrino (see also Chapter 31). The neutron is neutral, so the charge before this decay is zero. The total charge of all the decay products is also zero since the antineutrino is neutral and the proton and electron carry charges of $+e$ and $-e$, respectively. Hence, the net charge is the same before and after the decay, even though the number of charged particles changes.

The principle of **conservation of charge** is believed to be an exact law of nature. No violations of this principle have ever been observed, but there is no fundamental understanding of why. Again, all we can say is that this is the way the laws of physics seem to operate.

SUMMARY | Chapter 17

KEY CONCEPTS AND PRINCIPLES

Electric charge and Coulomb's law
There are two types of electric charge, positive and negative. Protons have a positive charge $+e$, and electrons carry a negative charge $-e$. Charge is measured in **coulombs** (C), with

$$e = 1.60 \times 10^{-19} \text{ C} \qquad \textbf{(17.1)} \text{ and } \textbf{(17.2)} \text{ (page 531)}$$

The electric force between two point charges q_1 and q_2 is given by **Coulomb's law.** If the charges are separated by a distance r, this force has a magnitude

$$F = k \frac{q_1 q_2}{r^2} = \frac{q_1 q_2}{4\pi\varepsilon_0 r^2} \qquad \textbf{(17.3)} \text{ and } \textbf{(17.5)} \text{ (pages 531 and 532)}$$

and is directed along the line joining the two charges.

The electric force is repulsive for like charges (charges with the same sign) and attractive for unlike charges (those of opposite sign). Here, k is a constant of nature with the value

$$k = 8.99 \times 10^9 \text{ N} \cdot \text{m}^2/\text{C}^2 \qquad \textbf{(17.4)} \text{ (page 531)}$$

and ε_0 is a universal constant called the permittivity of free space, with the value

$$\varepsilon_0 = 8.85 \times 10^{-12} \frac{\text{C}^2}{\text{N} \cdot \text{m}^2} \qquad \textbf{(17.6)} \text{ (page 532)}$$

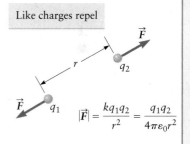
Like charges repel

$$|\vec{F}| = \frac{k q_1 q_2}{r^2} = \frac{q_1 q_2}{4\pi\varepsilon_0 r^2}$$

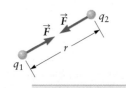
Unlike charges attract

(Continued)

Electric fields

The **electric field** $\vec{E}$ can be measured using a test charge q and is proportional to the electric force, with

$$\vec{F} = q\vec{E} \qquad \textbf{(17.10)} \text{ (page 538)}$$

The electric field due to a point charge q can be determined from Coulomb's law. Its magnitude is

$$E = \frac{kq}{r^2} \qquad \textbf{(17.11)} \text{ (page 538)}$$

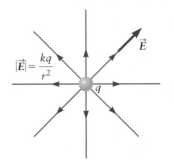

$$|\vec{E}| = \frac{kq}{r^2}$$

Electric flux and Gauss's law

When an electric field of magnitude E passes through a surface of area A while making an angle θ with the normal to the surface, the **electric flux** through the surface is

$$\Phi_E = EA\cos\theta \qquad \textbf{(17.12)} \text{ (page 546)}$$

Gauss's law relates the electric flux Φ_E through a closed surface to the amount of electric charge inside the surface:

$$\Phi_E = \frac{q}{\varepsilon_0} \qquad \textbf{(17.13)} \text{ (page 547)}$$

Gauss's law gives a powerful way to calculate the electric field in cases that are very symmetric. Coulomb's law can be derived from Gauss's law.

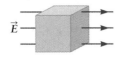

$$\Phi_E = EA\cos\theta$$

$$\Phi_{E,\ total} = 0$$

Conservation of charge

A fundamental conservation law of nature is that electric charge is conserved. Charge is also quantized; it comes in discrete amounts, such as electrons and protons, which have charges $-e$ (an electron) and $+e$ (a proton).

APPLICATIONS

Principle of superposition

Coulomb's law gives the electric field due to a single point charge (Eq. 17.11). For a collection of point charges, the principle of superposition states that the total electric field is the vector sum of the fields produced by the individual charges.

Electric field lines near a dipole

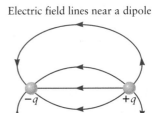

Conductors and insulators

In a metal, some of the electrons are free to move, whereas in an insulator, there are no mobile charges. For a metal in static equilibrium, the electric field inside is zero and all excess charges reside on the metal's surface. The electric field inside an insulator need not be zero.

DNA fingerprinting

The technique of **electrophoresis** uses the electric force to separate ions according to their size and charge. This is the basis of DNA fingerprinting.

1. When a glass rod is rubbed with silk, the rod becomes positively charged, but when a rubber rod is rubbed with fur, it becomes negatively charged. Suppose you have a charged object but don't know whether it carries a positive or a negative charge. Explain how you could use a glass rod and piece of silk to determine the sign of the charge on the unknown object.

2. Explain why two electric field lines cannot cross.

3. Determine the sign of each charge q_a, q_b, and q_c in Figure Q17.3.

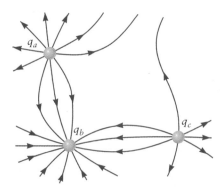

Figure Q17.3 Questions 3 and 4.

4. Consider the charges q_a, q_b, and q_c in Figure Q17.3. In terms of magnitude, which charge is the greatest? Which is the smallest? How do you know?

5. SSM Explain how two objects can be attracted due to an electric force, even when *one* object has *zero* net charge.

6. Explain how two objects can be attracted due to an electric force, even when *both* objects have *zero* net charge.

7. It is found that a charged point particle is repelled from a glass rod that has been charged by rubbing with silk. What is the sign of the charge on the particle?

8. The children in Figure Q17.8 are rubbing balloons on their hair and then placing the balloons on the wall and ceiling. If the rubbing process puts excess electrons on the balloon, how does the balloon then stay attached to the ceiling or wall?

9. The end of a charged rubber rod will attract small pellets of Styrofoam that, having made contact with the rod, will move violently away from it. Describe why that happens.

Figure Q17.8

10. Two point particles, each of charge Q, are located on the x axis at $x = \pm L$. Another charge q is now placed at the origin. (a) Show that the total force on q is zero. (b) Suppose q is now moved a very small distance away from the origin, along the x axis. What is the direction of the electric force on q? (c) Based on your result in part (b), explain why the origin is, or is not, a point of *stable* equilibrium for q.

11. Under normal atmospheric conditions, the surface of the Earth is negatively charged. What is the direction of the electric field near the Earth's surface?

12. When two objects (such as a glass rod and a silk cloth) are rubbed together, electrons can be transferred from one to the other. Can protons also be transferred? Explain why or why not.

13. In Example 17.1, we mentioned that there are approximately 4×10^{14} atoms in a dust particle of mass 1 ng. Estimate this number for yourself.

14. Explain why the child's hair in Figure 17.22 is "standing on end."

15. SSM Describe at least three ways in which Coulomb's law is similar to Newton's law of gravitation. Discuss at least two ways in which the two laws are different.

16. Explain why a test charge used to measure a field must be small in comparison with the field's source charge.

17. A charge is placed inside a partially inflated balloon. If the balloon is then further inflated to a larger volume and the associated surface area increases, does the electric flux though the balloon's surface change? Why or why not?

18. The electron was not discovered until well after the laws of electricity were determined. Protons were discovered even later, so scientists such as Coulomb and Gauss did not know that electric charge is quantized. How do you think that affected their way of thinking about electric phenomena?

19. An ion is released from rest in a region in which the electric field is nonzero. If the ion moves in a direction antiparallel (opposite) to the direction of the electric field, is the ion positively charged or negatively charged?

20. Figure Q17.20 shows the electric field lines outside several Gaussian surfaces, but does not show the electric charges inside. In which cases is the net charge inside positive, negative, or perhaps zero? How do you know?

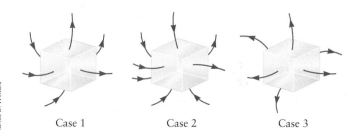

Case 1   Case 2   Case 3

Figure Q17.20

PROBLEMS

SSM = solution in Student Companion & Problem-Solving Guide
★ = intermediate ✪ = challenging
= life science application
R = reasoning and relationships problem

17.1 EVIDENCE FOR ELECTRIC FORCES: THE OBSERVATIONAL FACTS

1. What is the total charge of 1 mole of electrons?

2. You are given a container with 5.0 g of hydrogen atoms. What is the total charge on the electrons in the container? What is the total charge on the protons?

3. A piece of amber is charged by rubbing with a piece of fur. If the net excess charge on the fur is -8.5 nC (-8.5×10^{-9} C), how many electrons were added to the amber?

4. SSM ★ R Approximately how many electrons are in your body? What is the total charge of these electrons?

5. What is the net charge on an object that has an excess of 45 electrons?

6. ★ R Approximately how many electrons are in a penny? Assume the penny is made of pure copper.

7. The excess charge per unit area on a surface is $\sigma = -5.0 \times 10^{-3}$ C/m^2. How many excess electrons are there on a 1.0-mm^2 piece of the surface?

8. The charge per unit length on a glass rod is $\lambda = -7.5$ μC/m. If the rod is 2.5 m long, how many excess electrons are on the rod?

17.2 ELECTRIC FORCES AND COULOMB'S LAW

9. What is the magnitude of the electric force between two electrons separated by a distance of 0.10 nm (approximately the diameter of an atom)?

10. ★ Two point charges are separated by a distance r. If the separation is reduced by a factor of 1.5, by what factor does the electric force between them change?

11. What is the magnitude of the electric force between an electron and a proton in a hydrogen atom? Assume they are point charges separated by 0.050 nm.

12. Two particles with electric charges Q and $-3Q$ are separated by a distance of 1.2 m. (a) If $Q = 4.5$ C, what is the electric force between the two particles? (b) If $Q = -4.5$ C, how does the answer change?

13. ✪ Which graph in Figure P17.13, a, b, or c, best describes how the electric force between two point charges varies with their separation, r?

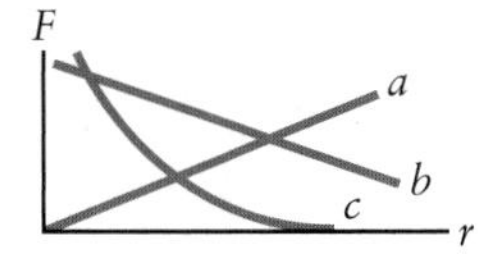

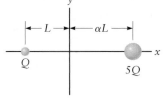

Figure P17.13

14. ★ A point charge with $q_1 = +2.5$ C is located at $x = -3.0$ m, $y = 0$, and a second point charge with $q_2 = +4.0$ C is at $x = +1.0$ m, $y = +2.0$ m. (a) What is the force exerted by q_1 on q_2? (b) What is the force exerted by q_2 on q_1?

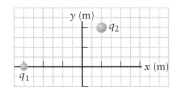

Figure P17.14

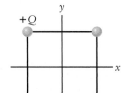

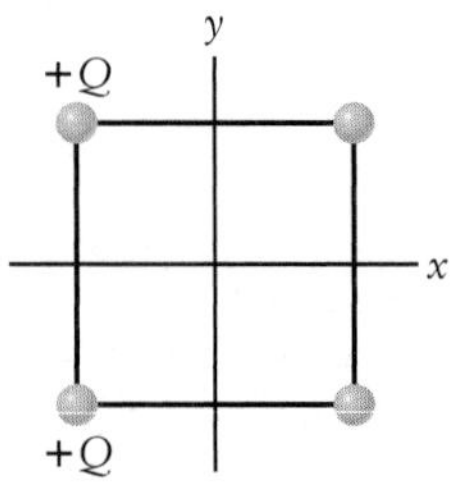

Figure P17.15

15. ★ R Four point particles are located at the corners of a square (Fig. P17.15). The two particles on the left each carry a charge $+Q$. If the electric field at the center of the square is directed to the right (along the $+x$ axis), what might the charges on the particles on the right be?

16. ✪ Two particles of charge Q and $5Q$ are located as shown in Figure P17.16. When a third charge is placed at the origin, it is found that the force on it is zero. Find α.

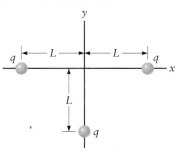

Figure P17.16

17. Two small, identical particles have charges $Q_1 = +3.0$ μC and $Q_2 = -5.0$ μC. If the electric force between the particles is 120 N, what is the distance between the particles?

18. ★ The particles in Problem 17 are conducting and are brought together so that they touch. Charge then moves between the two particles so as to make the excess charge on the two particles equal. If the particles are then separated by a distance of 5.3 mm, what is the magnitude of the electric force between them?

19. ✪ Three charges with $q = +7.5$ μC are located as shown in Figure P17.19, with $L = 25$ cm. (a) What are the magnitude and direction of the total electric force on the charge at the bottom? (b) What are the magnitude and direction of the total electric force on the charge at the right?

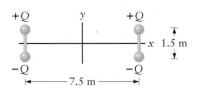

Figure P17.19 Problems 19, 20, and 47.

20. SSM ★ Three charges each with positive charge q are located as in Figure P17.19. What are the magnitude and direction of the force on an electron (charge $-e$) at the origin?

21. R You are given two boxes containing electrons, with 1.0 gram of electrons in each box. If these boxes are separated by 1.0 meter, estimate the gravitational force between them. Compare this force with the electric force found in Equation 17.7. Assume the boxes are both very small so that you can use the point-particle approximation.

22. ✪ Two electric dipoles with charges $\pm Q$, where $Q = 5.0$ C, are arranged as shown in Figure P17.22. The charges $\pm Q$ on the left are attached to each other by a rigid rod, and the two charges on the right are also rigidly attached to each other. Find the electric force between the two dipoles.

Figure P17.22

23. ✪ Particles of charge Q and $3Q$ are placed on the x axis at $x = -L$ and $x = +L$, respectively. A third particle of charge q is placed on the x axis, and it is found that the total electric force on this particle is zero. Where is the particle?

24. ✪ A point charge $q_1 = -1.5$ C is at the origin, and a second point charge $q_2 = +5.0$ C is at the point $x = 1.2$ m, $y = 2.5$ m. (a) Find the x and y coordinates of the position at which an electron would be in equilibrium. (b) Find the x and y coordinates of the position at which a proton would be in equilibrium. (c) Suppose the charges q_1 and q_2 were both doubled. How would your answers change? Why?

25. ★ Consider an electron and a proton separated by a distance of 1.0 nm. (a) What is the magnitude of the gravitational force between them? (b) What is the magnitude of the electric force between them? (c) How would the *ratio* of these gravitational and electric forces change if the distance were increased to 1.0 m?

26. ✪ You are on vacation in an alternate universe where nearly all the laws of physics are the same as in your home universe, but you notice that the charge on an electron in the alternate universe is $-0.9999e$ and the charge on a proton is $+e$. Consider now two spheres of copper, each with mass 1.0 kg. (a) What is the total charge on one of the spheres? (b) What is the electric force between the two spheres when they are separated by a distance of 0.50 m? Is this force attractive or repulsive? (c) Compare the magnitude of the electric force with the magnitude of the gravitational force between the two spheres.

27. ★ The mass of a typical car is $m = 1000$ kg, so its weight on the Earth's surface is $mg = 9800$ N. Suppose you have two containers, one with N electrons and another with N protons. These two containers are placed a distance 10 m apart, and the electric force between them is equal to the weight of the car. Find N.

28. ★ A helium nucleus contains two protons. What is the approximate magnitude of the electric force between these two protons? Is this force attractive or repulsive? The radius of this nucleus is about 1.0×10^{-15} m. Assume the protons are point particles.

29. SSM ★ An electric dipole is situated as shown in Figure P17.29, and the dipole charges are $\pm Q$, with $Q = 3.3\ \mu$C. A point particle with charge q is now placed at location A as shown in the figure. If the force on the particle is 2.5 N and is directed toward the origin, find q.

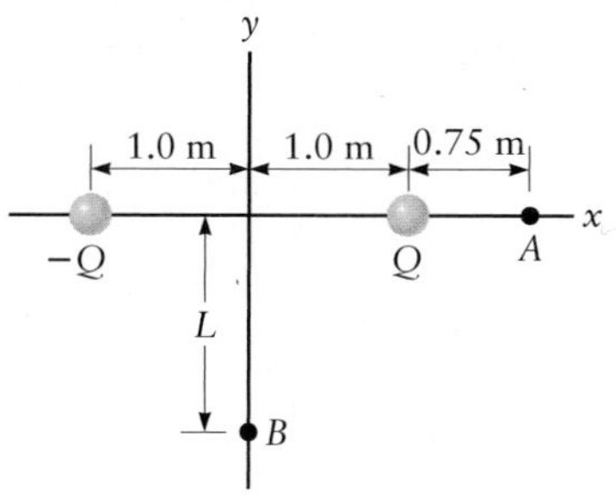

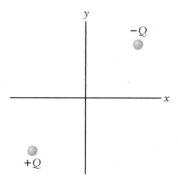

Figure P17.29 Problems 29 and 30.

30. ★ Consider again the dipole in Figure P17.29 with $Q = 3.3\ \mu$C. A point particle of charge $6.9\ \mu$C is placed at location B, and the magnitude of the electric force on this particle is found to be 0.035 N. Find L.

31. ★ Two point charges are a distance L apart, and the electric force exerted by one charge on the other is F_E. The charges are then moved farther apart, and the electric force between them decreases to $F_E/9$. What is the new separation between the charges?

17.3 THE ELECTRIC FIELD

32. A small piece of dust of mass $m = 1.0\ \mu$g travels through an electric air cleaner in which the electric field is 500 N/C. The electric force on the dust particle is equal to the weight of the particle. (a) What is the charge on the dust particle? (b) If this charge is provided by an excess of electrons, how many electrons does that correspond to?

33. What is the magnitude of the electric field at a distance of 1.5 m from a point charge with $Q = 3.5$ C?

34. Find the electric field a distance of 1.0 nm from an electron. Is this field directed toward or away from the electron?

35. ★ A point particle with charge $q = 4.5\ \mu$C is placed on the x axis at $x = -10$ cm. A second particle of charge Q is now placed on the x axis at $x = +25$ cm, and it is found that the electric field at the origin is zero. Find Q.

36. ✪ A point particle with charge $q = 4.5\ \mu$C is placed on the x axis at $x = -10$ cm and a second particle of charge $Q = 6.7\ \mu$C is placed on the x axis at $x = +25$ cm. (a) Determine the x and y components of the electric field due to this arrangement of charges at the point $(x, y) = (10, 10)$ (the units here are centimeters). (b) Determine the magnitude and direction of the electric field at this point.

37. ✪ A point charge with $q_1 = +2.5$ C is located at $x = -3.0$ m, $y = 0$, and a second point charge with $q_2 = +4.0$ C is at $x = +1.0$ m, $y = +2.0$ m. What are the components of the electric field at $x = +1.0$ m, $y = 0$?

38. A circular ring of radius 25 cm and total charge $500\ \mu$C is centered at the origin as shown in Figure P17.38. If the charge is distributed uniformly around the ring, what is the electric field at the origin?

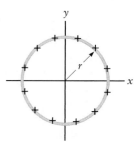

Figure P17.38

39. An electron is traveling through a region of space in which the electric field is along the $+x$ direction and has a magnitude of 2000 N/C. What is the acceleration of the electron? Give both the magnitude and direction of $\vec{a}$.

40. ✪ An interesting (but oversimplified) model of an atom pictures an electron "in orbit" around a proton. Suppose this electron is moving in a circular orbit of radius 0.10 nm (1.0×10^{-10}) and the force that makes this circular motion possible is the electric force exerted by the proton on the electron. Find the speed of the electron.

41. ★ Sketch the electric field lines near the electric dipole in Figure P17.41. Assume Q is positive.

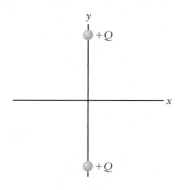

Figure P17.41

42. ★ Sketch the electric field lines near the two charges in Figure P17.42. Assume Q is positive.

Figure P17.42

43. ✪ A small, plastic sphere of mass $m = 100$ g is attached to a string as shown in Figure P17.43. There is an electric field of 100 N/C directed along the $+x$ direction. If the string makes an angle $\theta = 30°$ with the y axis, what is the charge on the sphere?

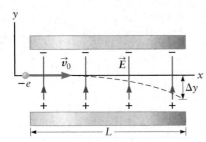

Figure P17.43

44. ✪ An electron is traveling through a region between two metal plates in which there is a constant electric field of magnitude E directed along the y direction as sketched in Figure P17.44. This region has a total length of L, and the electron has an initial velocity of v_0 along the x direction. (a) How long does it take the electron to travel the length of the plates? (b) What are the magnitude and direction of the electric force on the electron while it is between the plates? (c) What is the acceleration of the electron? (d) By what distance is the electron deflected (Δy) when it leaves the plates? (e) What is the final velocity of the electron? (Give the magnitude and direction or give the components of the velocity along x and y.)

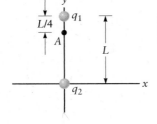

Figure P17.44

45. ☆ Two point particles with charges q_1 and q_2 are separated by a distance L as shown in Figure P17.45. The electric field is zero at point A, which is a distance $L/4$ from q_1. What is the ratio q_1/q_2?

Figure P17.45

46. ✪ Five point charges, all with $q = 7.5$ C, are spaced equally along a semicircle as shown in Figure P17.46. If the semicircle has a radius of 2.3 m, what are the magnitude and direction of the electric field at the origin?

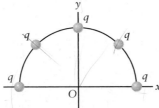

Figure P17.46

47. SSM ☆ Three point charges all with $q = -8.2$ C are located as shown in Figure P17.19 with $L = 4.5$ m. What are the magnitude and direction of the electric field (a) at the origin and (b) at the point $y = 6.8$ m on the y axis?

17.4 CONDUCTORS, INSULATORS, AND THE MOTION OF ELECTRIC CHARGE

48. Consider a metal sphere onto which a large number (suppose 10^6 or more) of extra electrons is placed (Fig. P17.48). (a) Explain how these electrons will

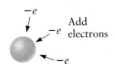

Figure P17.48
Problems 48 and 50.

be distributed over the interior and the surface of the sphere. (b) Draw the electric field lines inside and outside the sphere.

49. SSM One million electrons are placed on a solid chunk of metal shaped as shown in Figure P17.49. Draw the electric field lines inside and outside the metal.

Figure P17.49

50. Assume the sphere in Figure P17.48 is a piece of plastic (an insulator). How will that affect (change) the answer to Problem 48?

17.5 ELECTRIC FLUX AND GAUSS'S LAW

51. Consider an electron and the surface that encloses it in Figure P17.51. What is the electric flux through this surface?

Figure P17.51
Problems 51 and 52.

52. ☆ (a) Explain how the answer to Problem 51 would change if the surface were increased in size (expanded) by a factor of two. (b) How would the answer to Problem 51 change if three more electrons were placed inside the surface?

53. An electric dipole is enclosed in a spherical surface. What is the electric flux through the surface?

54. ☆ A cylinder has a volume V and a charge density ρ (Fig. P17.54). What is the electric flux through a sphere of radius R that encloses the cylinder?

Figure P17.54

55. ☆ Three point charges are located near a spherical Gaussian surface of radius R (Fig. P17.55). One charge ($+3Q$) is inside the sphere, and the others are a distance $R/3$ outside the surface. What is the electric flux through this surface?

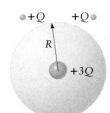

Figure P17.55

56. SSM ☆ Consider a very long, very thin plastic rod as sketched in Figure P17.56. Assume the rod has a length of $L = 1.0 \times 10^6$ m and a total charge of 1.0 C distributed evenly along the rod. Your job in this problem is to calculate the electric field a distance $r = 10$ cm from the rod. (a) What is the direction of the electric field? Give a qualitative answer. Is the field directed toward or away from the rod? (b) What Gaussian surface matches the symmetry of this problem? Can you pick a surface on which the electric field is constant and perpendicular to the surface or zero at all points on the surface? (c) What is the total charge inside your Gaussian surface? (d) Use Gauss's law to find the electric field.

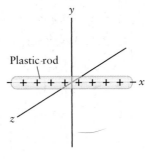

Figure P17.56

57. ✪ Use the step-by-step approach in Problem 56 to find the electric field produced by the hollow metal cylinder with charge per unit length $\lambda = 1.0 \times 10^{-6}$ C/m in Figure P17.57. Be sure to find the electric fields inside *and* outside the cylinder, assuming you are far from the ends and that the cylinder is extremely long.

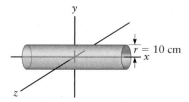

Figure P17.57

58. ✪ Consider three very large, parallel planes of charge that are equally spaced as shown in Figure P17.58. These planes are insulators, and each is charged uniformly with $+10\ \mu C$ on every square centimeter of area on the top surface of each plane. Use Gauss's law to determine the field at points A, B, C, and D.

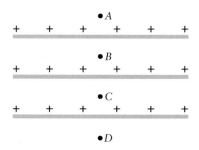

Figure P17.58 Problems 58 and 59.

59. ✪ Consider again the charged planes in Problem 58 (Fig. P17.58), but now assume the inner plane has a charge density of $-10\ \mu C/cm^2$ and the outer planes have charge densities $+10\ \mu C/cm^2$. Find the electric field at points A, B, C, and D.

60. ✪ A point charge q is located at the center of a cubical box with an edge length L. What is the electric flux through one face of the box?

61. ✪ Consider an arrangement of eight charges Q at the corners of a cube of size L. A spherical surface of radius $3L$ is arranged so as to completely contain these charges. What is the electric flux through this surface?

62. Consider the spherical Gaussian surface shown in Figure P17.62. It is situated near a point charge Q which is located a distance $2r$ from the center of the sphere and outside the Gaussian surface. Compute the net electric flux through this surface.

Figure P17.62

63. ✪ An insulating sphere has a total excess charge Q distributed uniformly throughout the sphere. Use Gauss's law to calculate the electric field inside the sphere at a distance r from the center of the sphere. Express your answer in terms of the radius of the sphere r_s and the total charge on the sphere Q.

64. ✪ Consider a solid metal sphere of radius R. An excess charge Q is placed on the sphere. (a) How will this excess charge be distributed on the sphere? (b) Make a qualitative sketch of the electric field lines outside the metal sphere. (c) Choose a Gaussian surface that matches the symmetry of the electric field. That is, choose a surface on which the electric field is a constant. (d) Apply Gauss's law to find the electric flux through your Gaussian surface. (e) Solve for the electric field a distance r from the center of the sphere, with $r > R$.

65. For the metal sphere in Problem 64, what is the electric field *inside* the sphere?

66. ✪ A spherical metal shell has a charge per unit area σ and a radius R. What is the magnitude of the electric field at a distance x from the surface of the sphere?

67. The electric flux through a large cardboard box is $500\ N \cdot m^2/C$. If the box is a cube whose edges are 1 m, what is the net charge inside the box?

68. ✪ Figure P17.68 shows a cylindrical capacitor; it consists of a solid metal rod of radius r_1 surrounded by a metal cylinder with inner radius r_2 and outer radius r_3. Suppose the capacitor has a length L (with L very large). Also suppose a charge $+Q$ is placed on the inner rod and a charge $-Q$ is placed on the outer cylinder. (a) Find the electric field inside the rod (i.e., for $r < r_1$). (b) Find the electric field in the space between the rod and the cylinder ($r_2 > r > r_1$). (c) Find the electric field outside the capacitor ($r > r_3$).

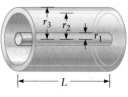

Figure P17.68

69. ✪ Consider a uniformly charged sphere, with radius R and charge per unit volume ρ. What is the magnitude of the electric field a distance $R/3$ from the center of the sphere?

17.6 APPLICATIONS: DNA FINGERPRINTING

70. ⊗ An electrophoresis experiment is performed on two fragments of DNA with radius $r_1 = 20$ nm and charge $q_1 = -5e$, and $r_2 = 30$ nm. The speed of the smaller fragment is two times greater than the speed of the larger fragment. What is the charge on the larger fragment?

71. ✪ ⊗ Consider two fragments of DNA of radius $r_1 = 30$ nm and $r_2 = 35$ nm when they are coiled up as they move through a gel. An electrophoresis analysis (Fig. 17.36) is carried out to separate these two fragments. Suppose the drag factor C in Equation 17.19 is $C = 5.0 \times 10^{-5}$ kg/(m · s), and assume each fragment has a charge $q = -10e$. (a) If the electric field has a magnitude of 1.0 N/C, what is the speed of each fragment in the gel? (b) How long will it take the smaller fragment to move a distance 1 cm (a typical distance used in DNA fingerprinting)? (c) After the time in part (b) has elapsed, what is the separation of the two different-size fragments in the electrophoresis tube?

72. SSM ✪ ⊗ A sample of DNA is prepared for fingerprinting. It contains three fragments of different sizes, all having charge $-8e$. After a certain amount of time, the fingerprint pattern in Figure P17.72 is found. What are the relative sizes of the three fragments? That is, assuming the fragments are spheres with radii r_A, r_B, and r_C, what are the ratios r_B/r_A and r_C/r_A?

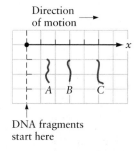

Figure P17.72

17.7 "WHY IS CHARGE QUANTIZED?" AND OTHER DEEP QUESTIONS

73. SSM A proton consists of three quarks of charge $+\frac{2}{3}e$, $+\frac{2}{3}e$, and $-\frac{1}{3}e$. The average spacing between the quarks is approximately 1.0×10^{-15} m. Assuming the quarks are arranged to form an equilateral triangle, find the magnitude of the total force on each quark due to the other two quarks. Do you think this arrangement of quarks is stable? Assume quarks are point charges.

74. Repeat Problem 73 for a neutron. A neutron consists of three quarks of charge $+\frac{2}{3}e$, $-\frac{1}{3}e$, and $-\frac{1}{3}e$.

75. ✪ A balloon of N_2 at atmospheric pressure and room temperature has a volume of 1.5 m³. What is the total charge of the electrons in the balloon?

76. Two infinite, parallel planes have excess charge densities σ and -2σ. If $\sigma > 0$, what are the magnitude and direction electric field between the planes?

77. ✪ Consider a half-infinite line of charge with charge per unit length λ. This line begins at the origin and then goes to infinity to the left (Fig. P17.77). Find the component of the electric field in the direction shown (the y direction, perpendicular to the line) at a distance r from the end of the line.

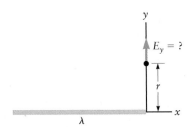

Figure P17.77

78. ✪ Two infinite, parallel, charged planes have charge per unit area $\sigma_1 = +3\sigma$ and $\sigma_2 = -\sigma$, respectively, with $\sigma > 0$ (Fig. P17.78). What are the magnitude and direction of the electric field at points A, B, and C?

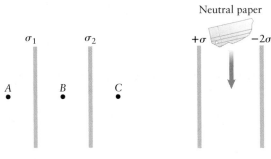

Figure P17.78   **Figure P17.79**

79. ✪ Two infinite, parallel, charged planes have charge per unit area $\sigma_1 = +\sigma$ and $\sigma_2 = -2\sigma$, respectively (Fig. P17.79). A small and electrically neutral bit of paper is dropped so as to fall between the plates. Will the paper be deflected by the electric field? If so, in what direction?

80. ✪ Two very large parallel plates have charge per unit area $+2.0$ μC/m² and -2.0 μC/m², respectively. A small grain of pollen of mass $m = 200$ mg and charge q hangs by a slender thread as shown in Figure P17.80. If the thread makes an angle $\theta = 30°$ with the vertical direction, what is q?

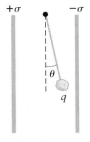

Figure P17.80

81. ✪ ✪ A water molecule (H_2O) is shaped as shown in Figure P17.81A. The hydrogen–oxygen bond length is about 9.6×10^{-11} m, and the bond angle is $\theta = 104°$. To a first approximation, the molecule is composed of three point ions (two H⁺ ions and one O⁻² ion). (a) What are the magnitude and direction of the electric field at the point midway between the O⁻² ion and the line connecting the two H⁺ ions? (b) We can model this system as the dipole in Figure P17.81B. If the electric field at the middle of this dipole is the same as your result in part (a), what is the

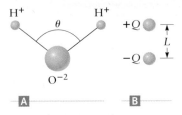

Figure P17.81

effective charge Q? (c) Use the effective dipole model to calculate the force between two water molecules that are separated by a distance of 0.30 nm (the typical separation in liquid water). Assume the dipoles of the two water molecules are oriented in opposite directions with their axes perpendicular to the line connecting the centers of the dipoles.

82. ✪ **Charged planet.** Lightning is striking somewhere on the planet at all times of the day. This and other meteorological phenomena produce a sustained vertical electric field—pointing inward toward the center of the Earth—that averages about 140 N/C near the surface. What is the average net charge per unit area on the Earth's surface?

83. ✪ A small drop of water, measuring 0.011 mm in diameter, hangs suspended above the ground due to the Earth's electric field (see Problem 82). How many extra electrons are on the drop of water?

84. ✪ Imagine that the Moon is locked into its orbit around the Earth by electrical forces and the gravitational force between the Earth and Moon is somehow "turned off." (a) What magnitude charge would need to be on each body? Approximate the Moon and the Earth as point particles and assume both have the same magnitude of charge. (b) How many moles of electrons would that be?

85. ✪ Repeat Problem 84. This time, assume the charges on the Moon and the Earth are not the same magnitude, but the ratio of the charges are the same as the ratio of their masses. Find (a) the charges on the Moon and the Earth and (b) the corresponding number of moles of electrons on each.

86. SSM ✪ An electron gun in the picture tube of a television set accelerates electrons from rest to a speed of 4.0×10^7 m/s along a distance of 1.0 cm. What is the magnitude of the uniform electric field used by the gun?

87. ✪ An electron with an initial velocity of 3.1×10^6 m/s enters a parallel-plate capacitor at an angle of 40° through a small hole in the bottom plate as shown in Figure P17.87. (a) If the plates are separated by a distance $d = 2.4$ cm, what minimum electric field must be present to prevent the electron from hitting the top plate? (b) Which plate has to have the positive charge? Assume there is a vacuum between the plates.

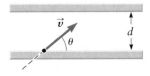

Figure P17.87 Problems 87 and 88.

88. ✪ Reconsider part (a) of Problem 87, replacing the electron with a proton and reversing the direction of the electric field. Compare your answer with the results obtained with an electron.

89. ✪ Consider a beam of protons traveling though a vacuum, where each proton has a velocity of 6.0×10^6 m/s. You need to design a means to change the direction of the proton beam by 90°. Consider the charged parallel plates separated by a distance

$d = 60$ cm as depicted in Figure P17.89. Small holes have been drilled into the bottom plate at right angles to one another to allow the beams to follow the trajectory shown. Calculate the electric field between the plates.

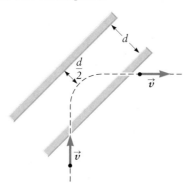

Figure P17.89

90. ✪ **Spray it to say it.** A particular inkjet printer uses electrostatic forces to direct droplets of ink to create letters, words, and images at a resolution of 300 dots per inch. Figure P17.90 shows a schematic of the process. Using ultrasound, 100,000 droplets of ink are produced per second out of the droplet gun in the inkjet's head, which moves across the page horizontally. Each drop has a diameter of 70 μm and moves at 20 m/s. Vertical displacement of the drops is achieved by placing a charge on each drop with an electron gun and then deflecting the drop through a uniform electric field provided by charged parallel plates. (a) Assume ink has a density equal to that of water (1000 kg/m³). What is the mass of each drop? (b) Each ink drop is sprayed with electrons until it has a charge of −650 nC. How many excess electrons does that correspond to? (c) A particular drop experiences an electric field of 100 N/C as it passes through the 0.50-mm-long parallel plates. How much is it displaced vertically as it leaves the plates? (d) What is the vertical component of velocity as it leaves the plates? (e) If the plates are not charged, the drop would land in the center of the paper, which is an additional 1.5 mm away from the plates. How far below this point does the drop strike the paper?

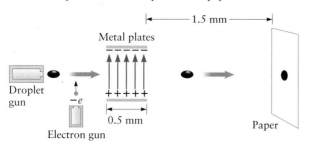

Figure P17.90

91. ✪ A charge of 1.8 nC is be placed in the center of three solids of equal volume as shown in Figure P17.91. Calculate the electric flux out of *one face* of the (a) tetrahedron if the charge is placed at its center. Repeat for (b) the cube and (c) the dodecahedron.

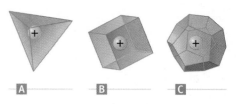

Figure P17.91

92. ✪ Eight point particles, each of charge Q, are located on the corners of a cube. What is the direction of the electric force on one of the particles from the others?

93. ✪ Two particles with the same charge are attached to strings that are 80 cm long and that hang from a ceiling as shown in Figure P17.93. If the angle between the strings is $\theta = 60°$ and the particles each have a mass of 250 g, what is the charge on each particle?

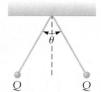

Figure P17.93

94. ✪ Consider three particles of charge Q, 2Q, and 4Q, where $Q = 2.4$ C, arranged as shown in Figure P17.94. If a particle of charge $q = -1.5$ C is placed at the origin, (a) what are the components of the electric force on this particle along both the x and y directions? (b) What are the components of the electric force on the particle of charge 4Q?

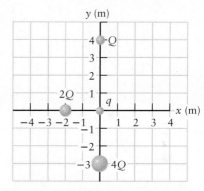

Figure P17.94

Chapter 18

Electric Potential

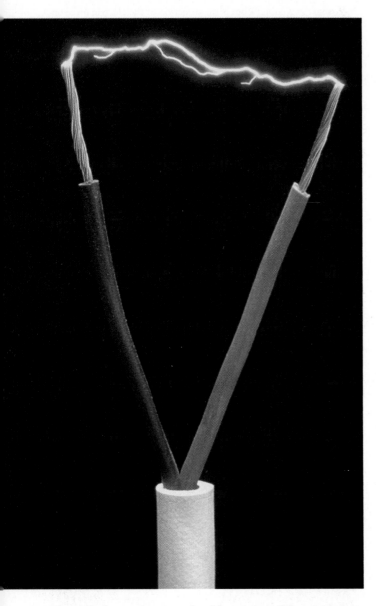

These sparks are one example of the motion of electric charges. This motion is determined by how the electric potential energy varies from place to place. (© Victor de Schwunberg/Photo Researchers, Inc.)

In Chapter 17, we discussed electric forces and electric fields, and learned how to calculate them using Coulomb's law and Gauss's law. From Chapter 6, we know that when a force acts on an object, the force may do mechanical work on the object. It should thus be no surprise that electric forces can do work on a charged object. Electrical work is related to *electric potential energy*, which is analogous in many ways to gravitational potential energy. Using electric potential energy, we will define a closely related quantity called the *electric potential*. These quantities and the underlying ideas are quite similar to mechanical potential energy and will lead us again to the principle of conservation of energy that we explored earlier in mechanical systems (Chapter 6). Extending these ideas to electric forces and systems will give us a more complete picture of electric phenomena and provide the basis for our work on electric circuits in Chapter 19.

Let's begin by reviewing the relationship between force and work. Figure 18.1A shows a region of space in which the electric field is constant, so $\vec{E}$ has the same magnitude and direction at all points. A point charge q in this region experiences an electric force

$$\vec{F} = q\vec{E} \tag{18.1}$$

If the charge q is positive, this force is parallel to $\vec{E}$ as shown in Figure 18.1A.

Suppose this charge moves a distance Δx, starting at point A and ending up at point B, and for simplicity we assume this displacement is parallel to the electric force $\vec{F}$. According to the definition of work in Chapter 6, the work done by the electric force on the charge is

$$W = F\,\Delta x \tag{18.2}$$

The electric force is conservative, so the work done on the charge is independent of the path it takes to go from A to B. We can now define the **electric potential energy**, which we denote by PE_{elec}. From Chapter 6, we know that the change in the potential energy associated with a particular conservative force is equal to $-W$, where W is the work done by that force (see Eq. 6.12). So, if the electric force does an amount of work W on a charged particle, the change in the electric potential energy is

$$\Delta PE_{elec} = -W \tag{18.3}$$

Combining this equation with Equations 18.1 and 18.2, the change in electric potential energy when the charged particle moves from A to B in Figure 18.1A is

$$\Delta PE_{elec} = -W = -F\,\Delta x = -qE\,\Delta x \tag{18.4}$$

Change in electric potential energy

Equation 18.4 gives the *change* in the potential energy as the charge moves through a displacement Δx, in a region where the electric field is parallel to the displacement. In Equation 18.4, the change in the potential energy depends on the starting and ending locations, but does *not* depend on the path taken. In Figure 18.1A, the displacement is along a line, but the charge may move from A to B along many other paths without affecting ΔPE_{elec}. All conservative forces, including the electric force, have this property: changes in the potential energy and the work done are independent of the path taken between two points. In the simple example shown in Figure 18.1A, the electric field is a constant, but the same results hold when $\vec{E}$ varies with position. In this more general case, the potential energy function will have a different mathematical form than the one in Equation 18.4, but the electric force is still conservative.

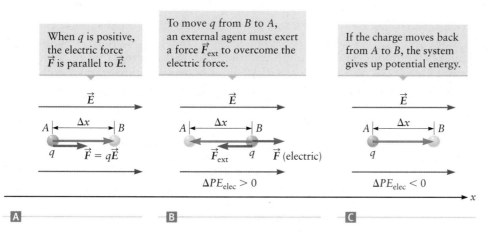

When q is positive, the electric force $\vec{F}$ is parallel to $\vec{E}$.

To move q from B to A, an external agent must exert a force $\vec{F}_{ext}$ to overcome the electric force.

If the charge moves back from A to B, the system gives up potential energy.

$\vec{E}$

$A \, \overset{\Delta x}{\longleftarrow} \, B$

$q \quad \vec{F} = q\vec{E}$

$\vec{E}$

$A \, \overset{\Delta x}{\longleftarrow} \, B$

$\vec{F}_{ext} \quad q \quad \vec{F}$ (electric)

$\Delta PE_{elec} > 0$

$\vec{E}$

$A \, \overset{\Delta x}{\longleftarrow} \, B$

q

$\Delta PE_{elec} < 0$

x

A   B   C

Figure 18.1 **A** The electric force on a charged particle can do work on the particle. **B** and **C** This work is connected with changes in the electric potential energy PE_{elec} of the particle. Here we assume a positively charged particle ($q > 0$).

If q_1 and q_2 are both positive, the electric force is repulsive and the work $W_E < 0$.

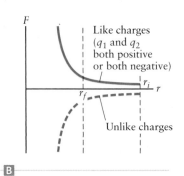

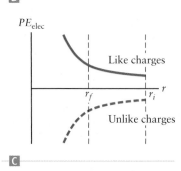

Figure 18.2 ◮ Two point charges exert an electric force on each other. If the separation between the two charges changes, this force does work. ◪ Plot of the force and ◪ electric potential energy as functions of the separation between q_1 and q_2. In B a negative value of F indicates an attractive force.

Electric potential energy of two point charges

Potential Energy Is Stored Energy

Electric potential energy is stored through the "potential" effect of the electric field on an electric charge. This effect is illustrated in Figure 18.1B, where the charge q is moved from point B to point A by a force $\vec{F}_{ext}$ directed to the left; this force is exerted by an external "agent," which might be your hand. The electric force on q is directed to the right (we assume q is positive), so the displacement is opposite to the electric force. The work done *by the electric field* on the particle is thus *negative*, and according to Equations 18.3 and 18.4 the change in the electric potential energy is *positive*. In words, we have stored a positive amount of energy in the system that is composed of the charge q and the electric field. That energy came from the positive amount of work done by the external agent (and the external force $\vec{F}_{ext}$) that moved the charge from B to A.

The entire process can be reversed if we allow the charge to move from A back to B again (Fig. 18.1C). Now the electric field does a positive amount of work on the particle because the electric force and the particle's displacement are parallel and the change in the potential energy is negative (Eq. 18.4). Hence, stored energy is taken out of the system, and this energy might show up as an increase in the kinetic energy of the particle when it reaches B.

Potential Energy of Two Point Charges

Let's now consider the electric potential energy for the slightly more complicated case of two charges q_1 and q_2 in Figure 18.2A. From Coulomb's law, we know that for two point charges separated by a distance r, the magnitude of the electric force exerted by q_1 on q_2 is

$$F = \frac{kq_1q_2}{r^2} \tag{18.5}$$

If we now assume they are like charges (i.e., both positive or both negative), the force is repulsive. The solid curve sketched in Figure 18.2B shows how F varies as a function of the separation r. If q_2 is brought closer to q_1, the electric force is directed opposite to the displacement (Fig. 18.2A), so the work done *by the electric field* on q_2 is negative. From our relation between work and potential energy (Eq. 18.3), we know that the electric potential energy *increases*, so $\Delta PE_{elec} > 0$. Mathematically, it can be shown that this electric potential energy is given by

$$PE_{elec} = \frac{kq_1q_2}{r} \tag{18.6}$$

We can also write PE_{elec} in terms of the permittivity of free space ε_0. From Chapter 17, $k = 1/(4\pi\varepsilon_0)$; using this expression in Equation 18.6 leads to

$$PE_{elec} = \frac{q_1q_2}{4\pi\varepsilon_0 r} \tag{18.7}$$

We can use either Equation 18.6 or 18.7 to calculate PE_{elec} for two point charges.

A plot of PE_{elec} is shown as the solid curve in Figure 18.2C. This is qualitatively similar to Coulomb's law for the electric force (Eq. 18.5), *except* that PE_{elec} varies as $1/r$, not $1/r^2$. If the charges are initially separated by a distance r_i and then brought together to a final separation r_f, the change in the potential energy is

$$\Delta PE_{elec} = PE_{elec,f} - PE_{elec,i} = \frac{kq_1q_2}{r_f} - \frac{kq_1q_2}{r_i} \tag{18.8}$$

We know that only changes in potential energy are important, so we could add a constant value to PE_{elec} in Equation 18.6 without changing ΔPE_{elec}. Conventional practice, however, is to use Equation 18.6 without any added constant, so PE_{elec} approaches zero when the two charges are very far apart (i.e., when r becomes infinitely large). The electric force also approaches zero in this limit.

The general behavior of the electric potential energy for two like charges is shown as the solid curve in Figure 18.2C. In this case, PE_{elec} is *positive* and becomes larger and larger as the charges are brought together. On the other hand, if the charges have opposite sign, the electric force is attractive; that is, it is negative as plotted by the dashed curve in Figure 18.2B. The potential energy is then also *negative* (according to Eq. 18.6) and is shown by the dashed curve in Figure 18.2C.

Insight 18.1
SIMILARITY BETWEEN ELECTRIC AND GRAVITATIONAL POTENTIAL ENERGY
The electric potential energy of two point charges (Eq. 18.6) falls to zero as the separation between the charges is made very large (infinite). This behavior is similar to that of the gravitational potential energy between two masses (Eq. 6.22), which also varies as $1/r$ and falls to zero when the two masses are very far apart.

EXAMPLE 18.1 Potential Energy of a Hydrogen Atom

A very simple model of a hydrogen atom is sketched in Figure 18.3, which shows an electron traveling in a circular orbit around proton. (We'll discuss more realistic models of an atom in Chapter 29.) If the distance between the electron and the proton is $r = 5.0 \times 10^{-11}$ m, what is the electric potential energy of this atom?

RECOGNIZE THE PRINCIPLE

We can treat the electron and proton as two point charges. We know their charges and the separation r is given, so we can calculate the potential energy using Equation 18.6.

SKETCH THE PROBLEM

Figure 18.3 shows the problem.

IDENTIFY THE RELATIONSHIPS

The electric potential energy is given by Equation 18.6, with $q_1 = +e$ (the proton) and $q_2 = -e$ (the electron). We have

$$PE_{elec} = \frac{kq_1q_2}{r} = \frac{k(+e)(-e)}{r} = -\frac{ke^2}{r}$$

This potential energy is negative because the electron and proton have opposite charges. This energy holds the atom together and is related to the ionization energy of the atom, the energy required to separate an electron from the rest of the atom. In this case, it is the energy needed to move the electron an infinite distance from the proton.

SOLVE

Inserting the given value of r, we find

$$PE_{elec} = -\frac{ke^2}{r} = -\frac{(8.99 \times 10^9 \ \text{N} \cdot \text{m}^2/\text{C}^2)(1.6 \times 10^{-19} \ \text{C})^2}{5.0 \times 10^{-11} \ \text{m}}$$

$$PE_{elec} = \boxed{-4.6 \times 10^{-18} \ \text{J}}$$

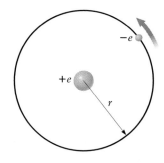

Figure 18.3 Example 18.1. A simplified model of an atom, with an electron in orbit around a proton.

What have we learned?
Our calculation has considered only the potential energy of the electron, but a better calculation of the ionization energy must also include the electron's kinetic energy (Section 29.1). Even so, let's compare our value of PE_{elec} with the measured ionization energy of hydrogen. It is conventional to quote positive values for the ionization energy, so our estimate is 4.6×10^{-18} J. The actual value for a hydrogen atom is approximately 2.2×10^{-18} J, so our simple calculation gives a value within about a factor of two of the correct answer.

Electric Potential Energy and Superposition

Our result for the electric potential energy of two point charges (Eq. 18.6) can be applied to other situations by using the principle of superposition. For example, if we have a collection of point charges, the total potential energy is the sum of the potential energies of each pair of charges. Note again that this electric potential energy arises from the interaction (the electric force) between two charges, so when we speak of potential energy we must keep in mind that it is the potential energy of a *system* of two particles.

The principle of superposition can be used to prove another important result. We know that the Coulomb's law force between two point charges is conservative, with a potential energy function given by Equation 18.6. More complicated charge distributions can always be treated as a collection of point charges arranged in some particular manner. We can therefore use superposition to conclude that electric forces between a collection of charges will always be conservative, no matter how complex the charge distribution or electric field may be.

Using the Principle of Conservation of Energy

In Chapter 6, we applied the principle of conservation of energy to solve various problems in mechanics, and we usually dealt with gravitational or elastic (spring) potential energy. The same approach can be applied to problems involving electric potential energy. The following problem-solving strategy is adapted from the one given in Chapter 6.

PROBLEM SOLVING | Applying the Principle of Conservation of Energy to Problems with Electric Potential Energy

1. **RECOGNIZE THE PRINCIPLE.** First determine the system whose energy is conserved. It might be two point charges, a collection of point charges, or a single charge in an electric field.

2. **SKETCH THE PROBLEM.** Draw a figure showing the initial and final states of all the charges in the system.

3. **IDENTIFY THE RELATIONSHIPS.** Identify the change in the electric potential energy. For a charge in a constant electric field, you can find ΔPE_{elec} using $\Delta PE_{\text{elec}} = -qE\,\Delta x$ (Eq. 18.4), whereas for two point charges we have $PE_{\text{elec}} = kq_1q_2/r$ (Eq. 18.6).

4. **SOLVE.** The next step depends on the problem.

 • If there are no external forces on the system, the sum of the electric potential energy plus kinetic energy is conserved and $KE_i + PE_i = KE_f + PE_f$.
 • If there is a nonzero external force, the work done by that force will equal the change in potential plus kinetic energy.
 • The electric force is conservative, so the work done by the electric force is independent of the path and equals $-\Delta PE_{\text{elec}}$.

5. Always *consider what your answer means* and check that it makes sense.

We'll apply this general approach in the next two examples.

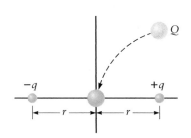

Figure 18.4 Example 18.2. A charge Q is taken from very far away and then moved to the origin. Like the gravitational force, the electric force is conservative; for both forces, the work done is independent of the path taken.

EXAMPLE 18.2 Moving Charges Around

Consider two point charges $+q$ and $-q$ separated by a distance $2r$ as shown in Figure 18.4. A third particle of charge Q is taken from rest very far away by an external force and brought to rest at the origin along the path shown in the figure. How much work is required to push or pull Q against the electric forces from the other charges?

RECOGNIZE THE PRINCIPLE

We start with step 1 in our strategy for applying the principle of conservation of energy to problems with electric potential energy. The charge Q moves in an electric field produced by the charges $\pm q$; these three charges are the system. The initial and final velocities of Q are zero, so its kinetic energy is zero at both the start and end. According to the principle of conservation of energy, the total work done on Q (the work done by the electric forces plus the work done by the external agent) must be zero. (The work done by one is the negative of the work done by the other.) So, the work done on Q by the external agent that moves it to the origin must equal ΔPE_{elec}, the change in its electric potential energy.

SKETCH THE PROBLEM

Step 2 in our strategy is the sketch in Figure 18.4, which shows one possible path the charge Q might take to get to the origin.

IDENTIFY THE RELATIONSHIPS

Step 3: We use the result for the change in the potential energy in Equation 18.8. We apply this relation twice, once for the pair of charges $+q$ and Q, and again for the pair $-q$ and Q. The charge Q begins very far from the others, so r_i is very large (infinite) in both cases, whereas the final location has $r_f = r$ for both pairs (since r is the final distance of charge Q from each of the other two charges).

SOLVE

Step 4: We solve for the change in the electric potential energy:

$$\Delta PE_{elec} = \frac{kqQ}{r} - \frac{kqQ}{r_i} + \frac{k(-q)Q}{r} - \frac{k(-q)Q}{r_i}$$

The terms involving r_i are each zero because r_i is infinite, whereas the other two terms (involving r) have opposite signs and cancel. Hence,

$$\Delta PE_{elec} = 0$$

The work done by the external agent that moves the charge to the origin is thus $\boxed{\text{zero}}$.

What does it mean?
The work required to move Q to a point midway between two particles of charge $+q$ and $-q$ is zero because the two particles have opposite charge, making their contributions to the total potential energy cancel. Note also that ΔPE_{elec} is independent of the path taken by Q.

EXAMPLE 18.3 | Moving Charges (Again)

A point charge q with mass m is initially at rest, a distance L from a second charge Q (Fig. 18.5). Both charges are positive, so they repel each other. Charge Q is held fixed at the origin, whereas q is released and travels to the right. What is the speed of charge q when it reaches a distance $2L$ from the origin?

RECOGNIZE THE PRINCIPLE

We again follow our strategy for applying the principle of conservation of energy to problems with electric potential energy. Our system is just the two charges Q and q. Because there are no external forces on the system, its energy must be conserved. We can thus write

$$KE_i + PE_i = KE_f + PE_f \qquad (1)$$

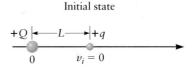

Figure 18.5 Example 18.3.

Because charge Q is held fixed, KE is the kinetic energy of charge q, and PE is the electric potential energy of the two charges.

SKETCH THE PROBLEM

Step 2: Figure 18.5 shows the problem. This sketch shows both the initial and final states of the system.

IDENTIFY THE RELATIONSHIPS

Step 3: We can find the change in the potential energy using $PE_{elec} = kqQ/r$ (Eq. 18.6). We can then calculate the change in the kinetic energy and, from that, the final speed.

Charge q is initially at rest, so the initial kinetic energy is $KE_i = 0$. The initial separation between the two charges is L, so the initial potential energy is (Eq. 18.6)

$$PE_i = PE_{\text{elec}, i} = \frac{kqQ}{L}$$

The final separation is $2L$, so

$$PE_f = PE_{\text{elec}, f} = \frac{kqQ}{2L}$$

Denoting the final speed by v_f, the final kinetic energy is

$$KE_f = \tfrac{1}{2}mv_f^2$$

SOLVE

Step 4: Inserting all this information in Equation (1) leads to

$$KE_i + PE_{\text{elec}, i} = 0 + \frac{kqQ}{L} = KE_f + PE_{\text{elec}, f} = \frac{1}{2}mv_f^2 + \frac{kqQ}{2L}$$

Solving for v_f gives

$$\frac{1}{2}mv_f^2 = \frac{kqQ}{L} - \frac{kqQ}{2L} = \frac{kqQ}{2L}$$

$$v_f^2 = \frac{2}{m}\left(\frac{kqQ}{2L}\right) = \frac{kqQ}{mL}$$

$$\boxed{v_f = \sqrt{\frac{kqQ}{mL}}}$$

What have we learned?
Electric forces are conservative, so we can use conservation of energy principles to analyze many situations involving electric forces and electric potential energy.

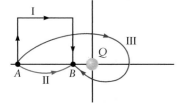

Figure 18.6 Concept Check 18.1.

CONCEPT CHECK 18.1 | Moving a Charge from Place to Place

A point charge Q is held fixed at the origin while an external force $\vec{F}_{\text{ext}}$ is used to move a second charge q from point A to point B along the three different paths shown in Figure 18.6. The charge q begins from rest and has zero velocity when it reaches B. Is the work done *by the external force* largest along (a) Path I, (b) Path II, or (c) Path III, or (d) is the work done by $\vec{F}_{\text{ext}}$ the same for all three paths?

18.2 | ELECTRIC POTENTIAL: VOLTAGE

Electric potential energy is a property of a *system* of charges or of a point charge in an electric field (where the field is created by other charges). In either case, this electric potential energy is not the property of a single charge alone. For example, in the case of two point charges, the potential energy is associated with the Coulomb's law force that acts between them. For this reason, the potential energy in Equation 18.6 depends on the values of both charges involved in the interaction. Now let's consider a slightly different way of viewing electric potential energy: as a property associated with just the electric field.

In Chapter 17, we introduced the notion of a *test charge* in our definition of the electric field. We can measure the electric field at a particular location in space by placing a test charge q at the location of interest (Fig. 18.7) and then measuring the force on the test charge. The electric force is related to the electric field by $\vec{F} = q\vec{E}$, so we can deduce the field from the measured force using (Eq. 17.10)

$$\vec{E} = \frac{\vec{F}}{q} \qquad (18.9)$$

If this field is produced by a second point charge Q, the magnitude of the electric field is

$$E = \frac{kQ}{r^2}$$

The value of this electric field does *not* depend on the test charge; the test charge merely gives us a convenient way to measure the field.

We now treat the electric potential energy in a similar manner. We place a test charge q at a particular location and measure the potential energy of this charge. We then define the *electric potential V* through the relation

$$V = \frac{PE_{\text{elec}}}{q} \qquad (18.10)$$

The electric potential is often referred to as simply "the potential." The SI unit of potential is the volt (V) in honor of Alessandro Volta, 1745–1827. According to Equation 18.10, the volt is related to other SI units by

$$\text{units of electric potential: } 1 \text{ V} = 1 \text{ J/C} = 1 \text{ N} \cdot \text{m/C} \qquad (18.11)$$

The volt is widely employed in the measurement of many electrical quantities. In particular, the electric field is usually measured in units of volts per meter. We thus have several different ways to express the units of the electric field. The ones most commonly used are

$$\text{units of electric field: } 1 \text{ V/m} = 1 \text{ N/C} \qquad (18.12)$$

Using an Electric Potential Difference to Accelerate Charged Particles

Many of the important concepts associated with electric potential are at work in a conventional (and old fashioned) television picture tube. A TV tube uses a beam of electrons to produce an image on a screen. As shown in Figure 18.8, this electron beam is produced by passing a stream of electrons through a region called the "accelerator," which consists of two parallel metal plates containing small holes that allow the electrons to enter and exit the central region. Some external electronic circuitry is attached to the metal plates, placing positive charge on one plate and negative charge on the other. We calculated the electric field between two parallel

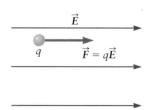

Figure 18.7 The electric field at a particular point in space can be determined by measuring the force on a test charge q placed at that point. In a similar way, changes in the potential energy of a test charge as it is moved from place to place can be used to measure changes in the electric potential.

Electric potential V

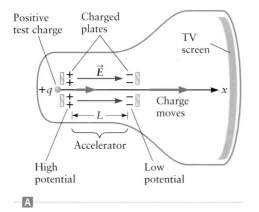

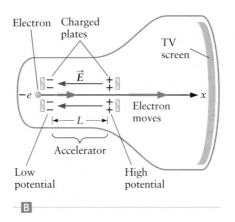

Figure 18.8 Electric field and potential inside a TV tube. **A** Hypothetical arrangement using the electric field to accelerate a positive test charge $+q$. **B** Realistic TV tube, with an electron moving through the accelerator from left to right.

charged plates in Chapter 17, where we found that for plates of a given charge and area this field is constant (Eq. 17.17).

Let's first analyze the case of a positive test charge moving through the accelerator shown in Figure 18.8A. Here the electric field is directed from left to right, which is the $+x$ direction in Figure 18.8A. Suppose the electric field has a magnitude E and the plates are separated by a distance L. We imagine that the test charge q is initially placed at the entry hole on the left and then moves toward the plate on the right. The magnitude of the electric force on the test charge is $F = qE$, and the charge moves a distance L from left to right. The work done on the test charge by the electric field is thus $W = FL = qEL$. From Equation 18.4, we then know that the corresponding change in the electric potential energy is

$$\Delta PE_{\text{elec}} = -qEL$$

From the definition of electric potential (Eq. 18.10), the change in the potential is

$$\Delta V = \frac{\Delta PE_{\text{elec}}}{q} = -EL \tag{18.13}$$

We say that there is a ***potential difference*** ΔV between the two plates. The electric field is directed to the right in Figure 18.8A, so the negative sign in Equation 18.13 means that the plate on the left is at a higher potential than the plate on the right. We say that the plate on the left is at "high" potential and the plate on the right is at "low" potential.

Now we replace the test charge by an electron and consider the motion of the electron as it travels through the accelerator. Since an electron is negatively charged, we must reverse the direction of the electric field; the electric field is now negative because it is along the $-x$ direction in Figure 18.8B. This change is needed because we want the electric force on the electron to push it from left to right through the accelerator region. Because the electron is negatively charged, it will tend to move from the region of low potential (the plate on the left) to the region of high potential (the plate on the right) in Figure 18.8B.

The potential difference ΔV between the plates is sometimes a more useful quantity to know than the electric field. Electrons enter the accelerator with a very small velocity, so for simplicity we'll assume their initial velocity is $v_i = 0$. We can use ΔV to find their final velocity v_f as they leave the accelerator by using the principle of conservation of energy. Conservation of energy tells us that

$$KE_i + PE_{\text{elec}, i} = KE_f + PE_{\text{elec}, f}$$

The initial kinetic energy is zero (since $v_i = 0$), so we have

$$KE_f = \tfrac{1}{2}mv_f^2 = PE_{\text{elec}, i} - PE_{\text{elec}, f} = -\Delta PE_{\text{elec}} \tag{18.14}$$

The change in potential energy is related to the change in the electric potential. Using Equation 18.10 gives

$$\Delta PE_{\text{elec}} = q \, \Delta V = -e \, \Delta V$$

Inserting this result into Equation 18.14 leads to

$$\tfrac{1}{2}mv_f^2 = -\Delta PE_{\text{elec}} = e \, \Delta V \tag{18.15}$$

$$v_f = \sqrt{\frac{2e \, \Delta V}{m}}$$

The potential difference in a TV tube is typically 15,000 V; inserting this value along with the values of the charge and mass of an electron, we find

$$v_f = \sqrt{\frac{2e \, \Delta V}{m}} = \sqrt{\frac{2(1.60 \times 10^{-19} \text{ C})(1.5 \times 10^4 \text{ V})}{9.11 \times 10^{-31} \text{ kg}}} = 7.3 \times 10^7 \text{ m/s}$$

So, these electrons are moving quite rapidly.[1] This high velocity and high kinetic energy enables the electrons to produce light when they strike the specially coated screen at the opposite face of the TV tube.

Relating the Electric Potential to the Electric Field

In Figure 18.8, we assumed the electric field is constant in the accelerator region, which is a good assumption in that case. In many situations, though, the electric field varies with position, so let's now ask how the electric field and the electric potential are related in these more general cases. We consider a very small region of space so that the electric field is approximately constant and apply the approach sketched in Figure 18.9, where a test charge q moves a distance Δx parallel to $\vec{E}$. The force on this charge is $F = qE$, and the work done by the field on the charge is $W = F \Delta x = qE \Delta x$. The change in the electric potential energy during this process is then (from Eq. 18.3)

$$\Delta PE_{elec} = -W = -qE \Delta x \qquad (18.16)$$

Using Equation 18.10 leads to an electric potential difference

$$\Delta V = \frac{\Delta PE_{elec}}{q}$$

which gives

$$\Delta V = -E \Delta x \qquad (18.17)$$

Equation 18.17 is a general relation between the electric field E and the variation in the electric potential V with position. If we know E, we can use Equation 18.17 to calculate how the potential changes as we move from place to place. Strictly speaking, this relation only holds for small steps Δx during which the field is constant. We can, however, combine the potential changes ΔV from many such small steps to find the change in the voltage over a large distance. We can also use Equation 18.17 to deduce the electric field from knowledge of how the potential changes with position. Rearranging that equation gives

$$E = -\frac{\Delta V}{\Delta x} \qquad (18.18)$$

In words, this says that the magnitude of the electric field is largest in regions where V is changing rapidly (and ΔV is large). Conversely, the electric field is zero in regions where V is constant. Notice that because of the negative sign in Equation 18.18 the electric field is directed from regions of high potential to regions of low potential.

The relation between the electric field and changes of the potential with position in Equation 18.18 involves the component of the electric field in the direction parallel to the displacement Δx. The electric field is a vector, so if we want to find E in a particular direction, we must consider how the potential V changes as we (or a test charge) move along that direction.

We now have a relation between the electric field and the variation of the potential V with position. This relation holds in simple cases when the electric field is constant as in the TV tube in Figure 18.8 and in more complicated cases in which the electric field varies with position.

EXAMPLE 18.4 | Electric Field in a TV Tube

Consider the accelerator region of the TV tube in Figure 18.8. If the potential between the two plates is 15,000 V and the plate separation is $L = 2.0$ cm, what is the electric field in this region? Assume the electric field is constant in the accelerator.

---

[1]Our result for v_f is about 25% of the speed of light. We show in Chapter 27 that Newton's mechanics begins to break down at such high speeds. The corrections can be calculated using the theory of special relativity, and they amount to a few percent for these electrons.

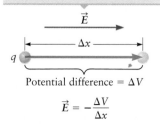

The electric field is proportional to the rate of change of V with position.

Figure 18.9 The magnitude and direction of the electric field are related to how the electric potential V changes with position.

Relation between E and changes in V with position

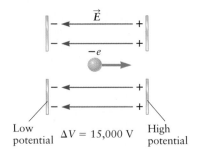

Low potential $\Delta V = 15{,}000\ \mathrm{V}$ High potential

Figure 18.10 Example 18.4.

We can find the electric field using the relation between the electric potential and E in Equation 18.18,

$$E = -\frac{\Delta V}{\Delta x} \tag{1}$$

SKETCH THE PROBLEM

Figure 18.10 shows the problem. Because the electron has a negative charge, it moves from left to right, from low to high electric potential. The corresponding charge on the plates is also shown.

IDENTIFY THE RELATIONSHIPS

The electric field is constant in the accelerator, so we can take ΔV to be the total potential difference, and Δx is the distance between the plates; we thus have $\Delta V = 15{,}000\ \mathrm{V}$ and $\Delta x = L = 2.0\ \mathrm{cm}$. Inserting these values in Equation (1), we find

$$E = -\frac{\Delta V}{\Delta x} = -\frac{1.5 \times 10^4\ \mathrm{V}}{0.020\ \mathrm{m}} = \boxed{-7.5 \times 10^5\ \mathrm{V/m}}$$

What does it mean?

The negative sign indicates that the field is directed from high to low potential as shown in Figure 18.10. The electric field vector always points from regions of high V to low V because $\vec{E}$ is parallel to the direction a positive test charge would move if it were placed at that location.

The Electron-volt as a Unit of Energy

The volt is a unit of electric potential, and the product of charge times electric potential has units of energy (Eq. 18.10). We often deal with the motion of electrons or protons, both of which have a charge of magnitude e. Also, we are often concerned with the energy gained or lost as an electron or ion moves through a potential difference measured in volts as with the example in Figure 18.10. For these reasons, it is convenient to define a unit of energy called the **electron-volt**, the energy gained or lost when an electron travels through a potential difference of 1 V. The electron-volt is abbreviated as "eV"; it is not part of the SI system of units, but we can relate the electron-volt to the joule. Using Equation 18.10, we have

$$V = \frac{PE_{\mathrm{elec}}}{q}$$

which can be rearranged as

$$PE_{\mathrm{elec}} = qV$$

In words, this equation reads

$$\mathrm{energy} = (\mathrm{charge})(\mathrm{potential})$$

Applying this definition to that of the electron volt as a unit of energy, we find

$$1\ \mathrm{eV} = e\,\Delta V = (1.60 \times 10^{-19}\ \mathrm{C})(1\ \mathrm{V})$$

$$1\ \mathrm{eV} = 1.60 \times 10^{-19}\ \mathrm{J} \tag{18.19}$$

In Example 18.4 with the TV tube, the electrons gained a kinetic energy $e\,\Delta V = 15{,}000\ \mathrm{eV}$ as they traveled through the accelerator.

CONCEPT CHECK 18.2 | Electron-volts and Potential Changes

In Figure 18.10, an electron moved from left to right through a region in which the electric potential changed by $\Delta V = 15{,}000$ V. The corresponding change in potential energy was $\Delta PE_{elec} = -e\,\Delta V = -15{,}000$ eV. Suppose the electron is replaced by an oxygen ion O^{2-}. What is the change in potential energy of the oxygen ion as measured in electron-volts, (a) -7500 eV, (b) $-15{,}000$ eV, or (c) $-30{,}000$ eV?

Electric Potential Due to a Point Charge

We have described the electric properties of a point charge thoroughly because this case is useful in many situations. In Equation 18.6, we found the electric potential energy of two point charges. We can combine that with the definition of electric potential (Eq. 18.10) to obtain the electric potential a distance r away from a single point charge q:

$$V = \frac{kq}{r} \tag{18.20}$$

Because the constants k and ε_0 are related by $k = 1/(4\pi\varepsilon_0)$, we can also write the potential from a point charge as

$$V = \frac{q}{4\pi\varepsilon_0 r} \tag{18.21}$$

We show this result as the solid curve in Figure 18.11 for a positive charge ($q > 0$) and as the dashed curve for a negative charge ($q < 0$).

These results reinforce what we learned about potential energy and the behavior of like and unlike charges. Suppose a charge $q > 0$ is placed at the origin; we'll call it a "source charge" because it produces a potential given by Equations 18.20 and 18.21. We then place a positive test charge nearby. Because both the source charge and the test charge are positive, they repel, and the test charge experiences a force that carries it "downhill" along the solid potential curve in Figure 18.11, away from the origin. On the other hand, if the source charge is negative, the potential is described by the dashed curve in Figure 18.11. A positive test charge would be attracted to this source charge (unlike charges) and would again be carried "downhill," this time along the dashed potential curve.

Only Changes in Electric Potential Matter

Only *changes* in potential energy are important, so we must always be clear about where the "reference point" for potential energy is chosen. Because electric potential is proportional to electric potential energy, we must also pay attention to the reference point for electric potential. For example, when dealing with a point charge and Equation 18.20, the standard convention is to choose $V = 0$ at an infinite distance from the source charge, that is, when $r = \infty$. In many other problems, the usual choice is to take the Earth as $V = 0$ because (from Chapter 17) the Earth is a good conductor of charge. That is the origin of the term ***electrical ground*** or, simply, ***ground*** and the convention that the ground is a point where $V = 0$.

Electric Potential and Field Near a Metal

Consider a solid metal sphere as in Figure 18.12A that carries an excess positive charge q, with the charge at rest. To understand the behavior of the electric potential and field, recall from Chapter 17 that the excess electric charge on a metal resides at the surface and the field inside the metal is zero. What about the electric field outside the sphere? Since the metal is spherical, the field must have spherical symmetry; hence, E outside the sphere must depend only on the distance r from the center of the sphere. In Chapter 17, we showed in Example 17.6 that the field outside any spherical ball of charge is given by

$$E = \frac{kq}{r^2} = \frac{q}{4\pi\varepsilon_0 r^2} \tag{18.22}$$

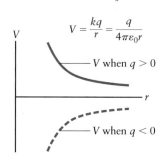

Figure 18.11 Electric potential near a point charge. The solid curve is for a positive point charge, and the dashed curve is for a negative point charge.

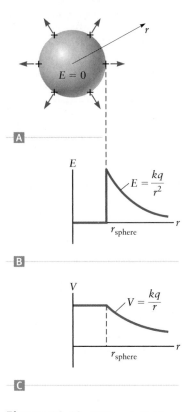

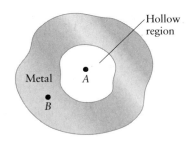

Figure 18.12 A Electric field lines near a charged metal sphere. B Magnitude of the electric field as a function of the distance r from the center of the sphere. C Electric potential as a function of r.

Figure 18.13 Concept Check 18.3.

where q is the total charge and r is the distance from the center of the ball. This result is precisely the same as that for the electric field due to a point charge. Equation 18.22 holds only outside the metal sphere; the field inside is zero. The magnitude of the field, both inside and outside, is plotted in Figure 18.12B as a function of the distance r from the center of the sphere.

We can use this result for the electric field to derive the electric potential. Because the field outside the metal sphere is identical to the field of a point charge, the potential outside the sphere is the same as the potential due to a point charge. We thus have (from Eq. 18.20)

$$V = \frac{kq}{r} \quad \text{for } r > r_{\text{sphere}} \tag{18.23}$$

The field inside the sphere is zero, so a test charge placed inside the metal would experience no force. The work done in moving a test charge around inside the metal is thus also zero. From the relation between work and electric potential energy (Eq. 18.3), we know that both the electric potential energy and the electric potential are *constant inside a metal*. Neither the potential energy nor the electric potential can jump discontinuously (see Insight 18.2), so the potential at all locations in the metal sphere must equal the potential at the outer edge of the sphere, which we can obtain from Equation 18.23. Hence,

$$V = \frac{kq}{r_{\text{sphere}}} \quad \text{everywhere inside a metal sphere} \tag{18.24}$$

Figure 18.12C shows the electric potential as a function of distance r from the sphere's center.

This example reviews and illustrates several important points. (1) The excess charge on a metal always resides at the surface, as we established in Chapter 17. (2) Because the electric field inside a metal is zero, the potential is constant throughout a piece of metal. (3) The electric field lines approach perpendicular to the surface of the metal sphere, Figure 18.12A, as required by the spherical symmetry of this electric field. Moreover, electric field lines *always* approach perpendicular to a metal surface of any shape, spherical or not. Why? Suppose an electric field at the surface of a metal somehow had a component parallel to the surface. This parallel field component would produce a force on any charge at the surface, causing it to move (because excess charge moves very easily in a conductor). This statement, however, contradicts our assumption that all the excess charge is in static equilibrium. Thus, for a piece of metal with any shape, electric field lines near a metal in equilibrium must approach perpendicular to the metal's surface.

CONCEPT CHECK 18.3 | Electric Field and Potential Inside a Metal

Consider a metal shell with a hollow region inside (Fig. 18.13). If the shell has a total charge Q, which of the following statements is true? (More than one statement may be correct.)

(a) The electric field is zero inside the hollow part of the shell (e.g., at point A).
(b) The electric potential is zero inside the hollow part of the shell (e.g., at point A).
(c) The electric field is zero in the metal part of the shell (e.g., at point B).
(d) The electric potential is zero in the metal part of the shell (e.g., at point B).

Electric Field Near a Lightning Rod

We can use our results for the electric field near the surface of a metal to understand the operation of a lightning rod. The stories about Benjamin Franklin's experiments with kites during thunderstorms are famous (even though we don't know for sure if these stories are true!). Franklin understood the connection between lightning and electricity and also showed the value of attaching a lightning rod (which he also invented) to the top of a building as in Figure 18.14A. When lightning strikes the rod, a large amount of charge (electrons) moves onto the rod; the rod is connected to the ground by a metal wire, allowing this charge to move safely to a region of low potential (the ground) through the metal wire rather than travel destructively through other portions of the building. Let's now consider why a lightning strike is "attracted" to a lightning rod.

Figure 18.14B gives a sketch of the electric field lines near a positively charged lightning rod. The electric field lines are perpendicular to the surface of the metal rod as we have just shown above in connection with Figure 18.12A. The field is largest near the sharp tip of the rod and smaller near the flat sides. That is another general property of the electric field near a metal: E is always largest near sharp metal corners and smaller near flat, metal surfaces. We can explain this by applying Equation 18.23 for the electric potential near a metal sphere to the simplified model of a lightning rod shown in Figure 18.14C. Here the rod is modeled as two metal spheres connected by a metal wire. The smaller sphere (radius r_1) represents the tip of the lightning rod, and the larger sphere (radius r_2) represents the flatter body of the rod. We also assume a net positive charge on the system. A portion Q_1 of this charge sits on the surface of the smaller sphere, and the rest of the charge Q_2 sits on the large sphere. The potential due to each sphere can be obtained from Equation 18.23. Strictly speaking, the result in Equation 18.23 applies only to a single sphere far from any other charges. However, if the two spherical parts of our model lightning rod are not too close together we can use Equation 18.23 to calculate the potential of each sphere. Because they are connected by a wire, the two spheres are part of a single piece of metal and must therefore have the same electric potential. Applying Equation 18.23 gives

$$V(\text{sphere 1}) = \frac{kQ_1}{r_1} = V(\text{sphere 2}) = \frac{kQ_2}{r_2}$$

Canceling common factors, we find that the charges on the two spheres are related by

$$\frac{Q_1}{r_1} = \frac{Q_2}{r_2} \qquad\qquad (18.25)$$

We can combine Equations 18.2 and 18.3 to get a relation between a force and the corresponding potential energy. When moving a particle between two points A and B that are a distance Δx apart, we have

$$F\,\Delta x = W = -\Delta PE$$

Solving for F, we get

$$F = -\frac{\Delta PE}{\Delta x}$$

If the two points are close together, Δx is very small. The change in potential energy ΔPE must therefore also be very small; otherwise, the force would be extremely large. This argument applies as the displacement becomes infinitesimal (in the limit $\Delta x \to 0$) and means that the electric potential energy cannot jump discontinuously when a particle moves between two points A and B an infinitesimal distance apart. Since electric potential is proportional to electric potential energy (Eq. 18.10), the electric potential also cannot jump discontinuously.

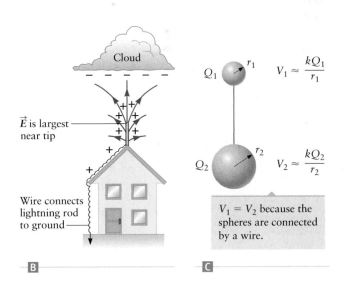

E is largest near tip

Cloud

Wire connects lightning rod to ground

Q_1 r_1 $\qquad V_1 \approx \dfrac{kQ_1}{r_1}$

Q_2 r_2 $\qquad V_2 \approx \dfrac{kQ_2}{r_2}$

$V_1 = V_2$ because the spheres are connected by a wire.

Figure 18.14 A Lightning rods on the roof of a building. B For the lightning rod to function properly, a wire must connect the rod to the ground. C A simplified model of a lightning rod. The small metal sphere represents the sharp tip of the rod.

We can use this equation to find the relative magnitudes of the electric fields produced by the two spheres. The electric field outside each sphere is given by $E = kQ/r^2$ (Eq. 18.22). For the field just outside sphere 1,

$$E_1 = \frac{kQ_1}{r_1^2}$$

and just outside sphere 2, we get

$$E_2 = \frac{kQ_2}{r_2^2}$$

The ratio of these fields is thus

$$\frac{E_1}{E_2} = \frac{kQ_1/r_1^2}{kQ_2/r_2^2} = \frac{Q_1 r_2^2}{Q_2 r_1^2}$$

Inserting the ratio of the charges from Equation 18.25, we find

$$\frac{E_1}{E_2} = \frac{Q_1}{Q_2} \frac{r_2^2}{r_1^2} = \frac{r_1}{r_2} \frac{r_2^2}{r_1^2}$$

$$\frac{E(\text{small sphere})}{E(\text{large sphere})} = \frac{E_1}{E_2} = \frac{r_2}{r_1} \qquad (18.26)$$

Hence, the field is *larger* near the surface of the *smaller* sphere (because $r_2 > r_1$), and the field is largest near the sharp edges of a charged piece of metal.

This result explains how a lightning rod works. The large electric field near the tip causes nearby air molecules to be ionized in an effect known as **dielectric breakdown** (discussed in more detail in Section 18.5). The resulting mobile electrons and ions in the air near the tip are then able to carry charge from the atmosphere (the lightning) to the rod. The key point is that dielectric breakdown occurs preferentially near the sharp tip of the lighting rod, so the large current from a lightning bolt is carried only to the rod and not to the rest of the building.

Shielding Out the Electric Field

Our results for the electric field and potential near and inside the metal sphere in Figure 18.12 can be used to design a region in which the electric field is zero. This process is called "shielding out" the electric field. We already know how to make the electric field zero: we simply go inside a metal. This approach is sketched in Figure 18.15, where we show a piece of metal that contains a hollow cavity. A charge Q placed near the metal produces an electric field in its vicinity, but the metal prevents the field from penetrating into the cavity. The electric field in the cavity is zero no matter what charges or electric fields are present outside the metal.

A physical picture of this shielding is given in Figure 18.15. The field from a positive charge Q pulls electrons to the surface of the metal nearest Q, leaving a net positive charge on the surface farthest from Q. The fields of these induced charges resting on the outer surface of the metal precisely cancel the field from Q inside the metal and inside the cavity. This method of shielding has many applications, one of which you may have already experienced. During a thunderstorm, large and potentially deadly electric fields are associated with every "bolt" of lightning, but when you are in your car during a thunderstorm, you are protected from these large fields. The shell of your car acts as the piece of metal in Figure 18.15 while you sit in the cavity, and the shielding effect of the cavity protects you from the large electric fields of a lightning strike. The same principle is used to construct shielded cables used in high-end stereo systems. The shielding keeps stray electric fields produced by other appliances from penetrating inside the cable and causing "noise" in the signals carried to the stereo speakers.

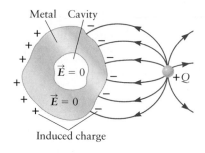

Figure 18.15 Shielding of the electric field in a cavity inside a metal.

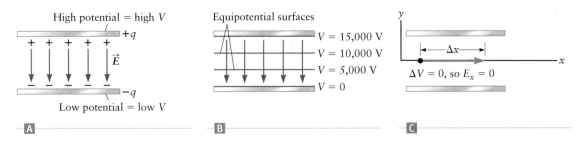

Figure 18.16 **A** Electric field between two parallel, charged metal plates. **B** In this case, the contours of equal V, called equipotential surfaces, are planes parallel to the metal plates. **C** The electric potential is constant along the x direction, so $E_x = 0$.

18.3 | EQUIPOTENTIAL LINES AND SURFACES

We have used drawings of electric field lines extensively to show how the electric field varies in space around various charge distributions. Another very useful way to visualize the electric field is through plots of *equipotential surfaces*. These are contours on which the electric potential is constant. Consider again the electric field in the region between two large, flat plates that carry charges $+q$ and $-q$ as sketched in Figure 18.16A. To be specific, assume they are the accelerator plates in the TV tube in Figure 18.8 and the potential difference between the two plates is 15,000 V. The electric field between the plates is constant and is directed from the plate with positive charge to the plate with negative charge. According to our relation between $\vec{E}$ and the electric potential (Eq. 18.17), the electric potential varies as we move from one plate to the other. We can now draw an imaginary surface between the plates such that $V = 10,000$ V at all points on this surface. Likewise, we can imagine a surface on which the potential is 5000 V (or any other value of interest). Several such surfaces are indicated in Figure 18.16B, labeled by their values of the potential. In this example, these surfaces are planes parallel to the charged plates.

Equipotential surfaces (also called equipotential *lines* in two-dimensional plots) are a very useful way to visualize the electric field and how it varies in space. Equipotential surfaces and lines are always perpendicular to the direction of the electric field. This property follows from the relationship between $\vec{E}$ and V. In Section 18.2, we showed that for a displacement along a particular direction, the change in V is related to the component of the electric field in that direction. If we consider a displacement along the x direction as in Figure 18.16C, the component of the field along x is

$$E_x = -\frac{\Delta V}{\Delta x} \qquad (18.27)$$

By definition, V is constant and $\Delta V = 0$ for motion parallel to an equipotential surface. From Equation 18.27, the electric field component parallel to the surface is thus zero. Hence, $\vec{E}$ is always perpendicular to an equipotential surface.

CONCEPT CHECK 18.4 | Equipotential Surfaces and the Magnitude of E

Figure 18.17 shows equipotential surfaces near a hypothetical charge distribution. Is the magnitude of the electric field largest at (a) point A, (b) point B, or (c) point C? *Hint*: Because $E = -\Delta V/\Delta x$, the field is largest in regions and directions where the potential is changing most rapidly.

Some equipotential surfaces in the neighborhood of a point charge are shown in Figure 18.18. In this case, the electric field lines emanate radially outward from the charge. The equipotential surfaces are perpendicular to $\vec{E}$ and are thus a series of concentric spheres, with different spheres corresponding to different values of V.

Insight 18.3
EQUIPOTENTIAL SURFACES AND GRAVITY
The meaning of an equipotential surface can also be understood by analogy with the gravitational potential energy at different places on a mountain. As you climb up a mountain, your gravitational potential energy PE_{grav} increases, whereas PE_{grav} decreases when you go downhill. However, PE_{grav} stays constant if you follow a path that stays at a constant altitude. Such a path lies on a surface of constant gravitational potential energy and is thus an equipotential surface for gravitation.

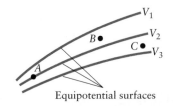

Figure 18.17 Concept Check 18.4.

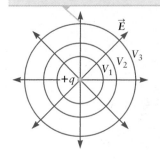

Figure 18.18 The equipotential surfaces around a point charge are spheres.

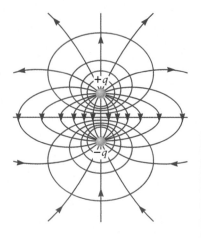

Figure 18.19 Equipotential surfaces near a dipole.

Another example of equipotential surfaces is shown in Figure 18.19. For this dipole consisting of charges $+q$ and $-q$, we plot both the electric field lines (from Example 17.4) and the equipotential surfaces. The result is a two-dimensional plot, so these surfaces appear as equipotential lines. Equipotential surfaces and lines are *always* perpendicular to the electric field.

EXAMPLE 18.5 Equipotential Surfaces Around a Line of Charge

Consider a uniform line of positive charge as shown in Figure 18.20A. Make a plot showing some of the equipotential surfaces.

RECOGNIZE THE PRINCIPLE

Equipotential surfaces and lines are always perpendicular to the electric field. So, we first determine the direction of $\vec{E}$ and then find surfaces perpendicular to the field.

SKETCH THE PROBLEM

In Chapter 17, we showed that the electric field produced by a line of charge emanates radially outward from the line as sketched in Figure 18.20B.

IDENTIFY THE RELATIONSHIPS AND SOLVE

Equipotential surfaces must always be perpendicular to the electric field lines, so the equipotential surfaces are cylinders centered on the line of charge as shown in parts C and D of Figure 18.20.

What does it mean?
Because equipotential surfaces are, by their definition, always perpendicular to $\vec{E}$, they are often the same or very similar to the Gaussian surfaces we would choose in applying Gauss's law.

Figure 18.20 Example 18.5.
Ⓐ A long line of positive charge.
Ⓑ End view of electric field lines near the line of positive charge.
Ⓒ The equipotential surfaces are cylinders centered on the line of charge.
Ⓓ End view of the electric field lines and equipotential surfaces.

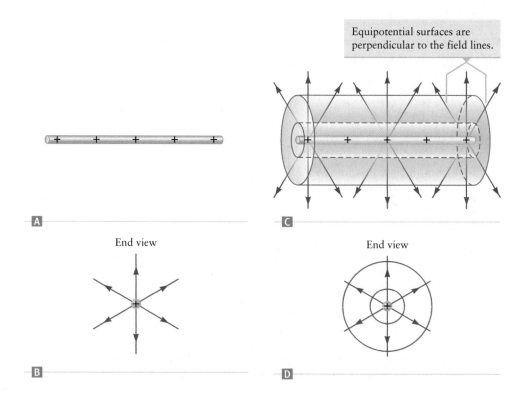

Equipotential surfaces are perpendicular to the field lines.

End view

End view

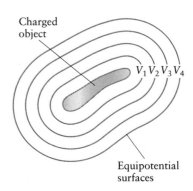
CONCEPT CHECK 18.5 | Field Lines and Equipotential Surfaces for Negative Charges

How would the sketches of the electric field and the equipotential surfaces in Figure 18.20 change if the charge on the line is negative instead of positive?

CONCEPT CHECK 18.6 | Relating the Electric Field to the Equipotential Surfaces

Figure 18.21 shows a family of equipotential surfaces. Sketch qualitatively the corresponding electric field lines. If $V_1 > V_2 > V_3 > V_4$, is the object in Figure 18.21 positively charged or negatively charged?

18.4 | CAPACITORS

You have seen the example of two parallel, charged plates several times in our discussions of electric fields and potential. We now use this arrangement yet again to introduce the notion of *capacitance*. Figure 18.22 shows two flat, metal plates separated by a distance d, and for simplicity the region between the plates is a perfect vacuum. Recall from Chapter 17 that this is called a *parallel-plate capacitor*. It can be used to store electric charge and energy, as we will now demonstrate.

Let's connect the two metal plates of our capacitor to metal wires that can be used to carry charge onto or off the plates. Now suppose we place a charge $+Q$ onto the top plate and a charge $-Q$ on the bottom one. These charges attract each other and therefore sit on the inner surfaces of the two plates as indicated Figure 18.22.

In Chapter 17, we showed that a single plane with area A and charge Q produces an electric field of magnitude $E(\text{single plane}) = Q/(2\varepsilon_0 A)$. If the charge is positive, the field points away from the plane, whereas if the charge is negative, the field points toward the plane as sketched in Figure 18.22A. In the region between the plates (Fig. 18.22B), these fields add, giving a total field

$$E = \frac{Q}{\varepsilon_0 A} \quad (18.28)$$

which is the electric field inside a parallel-plate capacitor (Eq. 17.18). From our relation between the electric field and potential (Eq. 18.17), there must be a potential difference "across" the two plates. For a parallel-plate capacitor, the positively charged plate has a higher potential than the negatively charged plate, and the magnitude of the potential difference between the two plates is

$$\Delta V = Ed$$

where d is the distance between the plates. Inserting E from Equation 18.28, we find

$$\Delta V = Ed = \frac{Qd}{\varepsilon_0 A} \quad (18.29)$$

Hence, the potential difference across the capacitor plates is proportional to the charge Q. We now define a quantity C called the capacitance with

$$\Delta V = \frac{Q}{C} \quad (18.30)$$

Comparing Equations 18.29 and 18.30, C in this case is given by

$$C = \frac{\varepsilon_0 A}{d} \quad \text{(parallel-plate capacitor)} \quad (18.31)$$

where A is the area of a single plate and d is the plate separation.

The parallel-plate capacitor design in Figure 18.22 is a convenient one to analyze, but there are many other designs with different shapes and different spacing

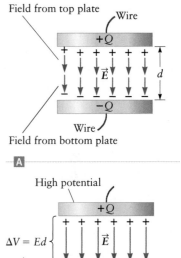
Field from top plate   Wire

$+Q$
$+ \ + \ + \ + \ + \ +$
$\vec{E}$ d
$-Q$

Wire
Field from bottom plate

A

High potential

$+Q$
$+ \ + \ + \ + \ + \ +$
$\Delta V = Ed$ $\vec{E}$
$-Q$

Plate area = A Low potential

B

Figure 18.22 **A** A parallel-plate capacitor consists of two parallel metal plates. Wires connected to each plate allow charge to be placed on or taken off the plates. **B** The electric field between the plates is related to the potential difference between the plates.

Definition of capacitance

Capacitance of a parallel-plate capacitor

geometries. All capacitors, though, employ two metal plates of some sort, and in *all* cases the charge on the capacitor plates is proportional to the potential difference across the plates as in Equation 18.30. The actual value of the capacitance depends on the design; Equation 18.31 applies to the parallel-plate design, whereas other plate geometries lead to other formulas for C (see Problem 46).

According to Equation 18.30, capacitance is the ratio of electric charge to the difference in electric potential across the plates, so the SI unit of capacitance is coulombs per volt, or C/V. This combination of units is called the farad (F) in honor of Michael Faraday, 1791–1867.

Storing Energy in a Capacitor

Capacitors find many applications in electronic circuits, including radios, computers, and MP3 players. All these uses depend on the ability of a capacitor to store electric charge and on the relation between charge and voltage (= potential difference), Equation 18.30.

When there is a nonzero potential difference between the two plates of a capacitor, energy is stored in the device. It is similar to the potential energy present in a system of two point charges as in Figure 18.2. To calculate the potential energy associated with the charges $+Q$ and $-Q$ on the two plates of a capacitor, we imagine that the capacitor is initially uncharged ($Q = 0$). We then transfer small amounts of charge ΔQ (perhaps by moving one electron at a time) from one plate to the other in Figure 18.23A. The capacitor plates are initially uncharged, so the potential difference is initially $\Delta V = 0$ and the electric field between the plates is zero. It therefore takes no energy to transfer the first packet of charge. As we continue to transfer charge between the plates, however, the total charge on the plates grows and, because Q and ΔV are proportional (Eq. 18.30), the potential difference across the plates grows too. This dependence is sketched in Figure 18.23B.

To move an amount of charge ΔQ through a potential difference ΔV requires an energy $(\Delta Q)(\Delta V)$, which corresponds to the area shown by the thin rectangle in Figure 18.23B. The total energy stored in the capacitor is equal to the energy required to move all the packets of charge ΔQ from one plate to the other, and it corresponds to the total area under the ΔV–Q line in Figure 18.23B. The area of this triangle is $\frac{1}{2}Q\,\Delta V$, where ΔV is now the final potential difference and Q is the final charge. Hence, the energy stored in a parallel-plate capacitor is

Energy stored in a capacitor

$$PE_{cap} = \tfrac{1}{2}Q\,\Delta V \tag{18.32}$$

Using the definition of capacitance $\Delta V = Q/C$ (Eq. 18.30), we can write Equation 18.32 as

$$PE_{cap} = \frac{1}{2}C(\Delta V)^2 = \frac{1}{2}\frac{Q^2}{C} \tag{18.33}$$

Figure 18.23 **A** When charge is transferred from one capacitor plate to the other, the value of Q increases. **B** The voltage across the capacitor plates increases with Q. The area under the graph of ΔV versus Q is equal to the potential energy stored in the capacitor PE_{elec}.

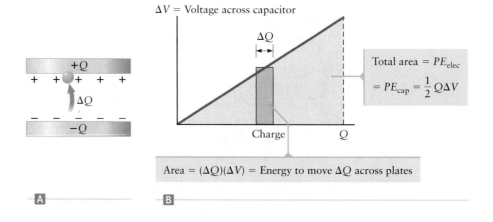

The relations for the stored energy in Equations 18.32 and 18.33 apply to all types of capacitors, not just parallel-plate capacitors. The energy stored in a capacitor is *potential energy*. It can be extracted from the capacitor and transformed into other forms of energy, or it can be used to do mechanical work.

EXAMPLE 18.6 | Capacitance and Energy Storage in a Typical Capacitor

Consider a parallel-plate capacitor that is about the size of your fingernail (Fig. 18.24). The plates are squares with edges of length $L = 1.0$ cm, separated by $d = 10$ μm $= 1.0 \times 10^{-5}$ m, which is about the diameter of a human hair. (a) Find the capacitance. (b) If the potential across the capacitor is $\Delta V = 12$ V, what is the energy stored?

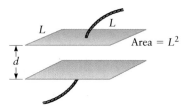

Figure 18.24 Example 18.6. A parallel-plate capacitor with two square plates.

RECOGNIZE THE PRINCIPLE

These calculations involve a direct application of our results for a parallel-plate capacitor and the energy stored in a capacitor. The values we find will be interesting.

SKETCH THE PROBLEM

The capacitor and its dimensions are shown in Figure 18.24.

IDENTIFY THE RELATIONSHIPS AND SOLVE

(a) The area of the capacitor plates is $A = L^2$. Inserting this information along with the given values of L and d into Equation 18.31 gives

$$C = \frac{\varepsilon_0 A}{d} = \frac{\varepsilon_0 L^2}{d} = \frac{[8.85 \times 10^{-12} \text{ C}^2/(\text{N} \cdot \text{m}^2)](0.010 \text{ m})^2}{(1.0 \times 10^{-5} \text{ m})}$$

$$C = \boxed{8.9 \times 10^{-11} \text{ F}}$$

(b) The energy stored in a capacitor is (Eq. 18.33)

$$PE_{cap} = \tfrac{1}{2}C(\Delta V)^2$$

Inserting our result for C along with the given value of the potential difference ($\Delta V = 12$ V), we get

$$PE_{cap} = \tfrac{1}{2}C(\Delta V)^2 = \tfrac{1}{2}(8.9 \times 10^{-11} \text{ F})(12 \text{ V})^2 = \boxed{6.4 \times 10^{-9} \text{ J}}$$

What does it mean?
The value of C for this capacitor is only 89 picofarads (1 picofarad $= 10^{-12}$ F). Typical capacitors used in electronic circuits are much less than 1 F, but the value of C for our fingernail-size capacitor is still quite small compared with most capacitors. We'll discuss ways to increase the value of C in the next section. The energy stored in this capacitor is also very small. As a comparison with familiar mechanical energies, PE_{cap} is much less than the change in the gravitational potential energy when a mosquito ($m \approx 10^{-6}$ kg) falls about 1 cm! One practical use of a capacitor is to store energy. To store useful amounts of energy, we must design capacitors with much larger values of C.

EXAMPLE 18.7 | Using a Capacitor as a Computer Memory Element

Modern computer memories use parallel-plate capacitors to store information, and these capacitors are the basic elements of a random-access memory (RAM) chip. Assume one of these capacitors has plates with an area of $L \times L$, where $L = 1.0$ μm ($= 1.0 \times 10^{-6}$ m), and a plate separation of $d = 10$ nm ($= 1.0 \times 10^{-8}$ m). (a) Find the capacitance of such a capacitor. (b) Calculate the amount of charge that must be placed onto the plates to obtain a potential difference of 5.0 V across the plates. (c) How many electrons does this charge correspond to?

RECOGNIZE THE PRINCIPLE

Our RAM capacitor is just a parallel-plate capacitor, and we can find its capacitance using $C = \varepsilon_0 A/d$ (Eq. 18.31). Using the given value of ΔV with this value of C, we can then calculate the charge on the capacitor plates.

SKETCH THE PROBLEM

Our simplified model of a computer memory capacitor is described by Figure 18.24.

IDENTIFY THE RELATIONSHIPS AND SOLVE

(a) The capacitance of a parallel-plate capacitor is (Eq. 18.31)

$$C = \frac{\varepsilon_0 A}{d} = \frac{[8.85 \times 10^{-12}\ \text{C}^2/(\text{N} \cdot \text{m}^2)](1.0 \times 10^{-6}\ \text{m})^2}{1.0 \times 10^{-8}\ \text{m}} = \boxed{8.9 \times 10^{-16}\ \text{F}}$$

(b) Potential, charge, and capacitance are related by $Q = C\,\Delta V$ (Eq. 18.30), so obtaining a voltage of 5.0 V requires a charge

$$Q = C\,\Delta V = (8.9 \times 10^{-16}\ \text{F})(5.0\ \text{V}) = \boxed{4.5 \times 10^{-15}\ \text{C}}$$

(c) The charge on a single electron has a magnitude of e, so this value of Q corresponds to

$$N = \frac{Q}{e} = \frac{4.5 \times 10^{-15}\ \text{C}}{1.60 \times 10^{-19}\ \text{C/electron}} = \boxed{2.8 \times 10^4\ \text{electrons}}$$

What does it mean?

The number of electrons found in part (c) is not very large. In fact, the latest computer memory capacitors are somewhat smaller than we have assumed here, so they employ an even smaller number of electrons.

Capacitors in Series

Figure 18.25 shows a capacitor-type combination of three metal plates. This is just our usual parallel-plate capacitor (Fig. 18.22) with an added a metal plate in the middle, and for simplicity we assume this middle plate is separated from the top and bottom plates by a vacuum. Let's find the capacitance of this structure; our calculation will lead to some useful results for our work with electric circuits in Chapter 19.

A calculation of the capacitance relies on the definition of C as the charge on the capacitor plates divided by voltage (Eq. 18.30). We imagine that charges $+Q$ and $-Q$ are placed on the outermost plates (Fig. 18.25A), and we want to find the total potential difference ΔV_{total} between these two plates. The ratio $Q/\Delta V_{\text{total}}$ is then the

Figure 18.25 A Three parallel metal plates act as two capacitors sharing a common central plate. B When a charge $+Q$ is added to the top plate and $-Q$ to the bottom plate, these charges move to the "inner" surfaces as shown and induce charges $-Q$ and $+Q$ on the two surfaces of the inner plate.

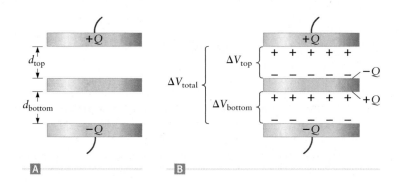

capacitance. To calculate ΔV_{total}, we must know the potential difference between each pair of plates. This potential difference will depend on the charges on the outer plates, which are just $+Q$ and $-Q$, and on the charges on the inner plate, which we now need to find.

The inner plate is neutral, with a net (excess) charge of zero. The charge $+Q$ on the top plate, however, will attract the mobile negative charges (electrons) in the inner plate. This attraction will give a net negative charge on the upper surface of the inner plate, leaving a net positive charge on the lower surface of the inner plate; the different surfaces of the three plates will then be charged as shown in Figure 18.25B. Comparing this figure with our sketch of a single capacitor (two plates) in Figure 18.22, we see that the three plates in Figure 18.25 act as two "consecutive" capacitors, each with charges $\pm Q$. To find the total capacitance of this structure, we need to know the potential difference ΔV_{total} between the top and bottom plates. This total potential difference equals the potential difference across the top two plates ΔV_{top} plus the potential difference across the bottom two plates ΔV_{bottom}. The top two plates form a capacitor, so we can write $\Delta V_{top} = Q/C_{top}$, where C_{top} is the capacitance of the upper capacitor, with a similar result for ΔV_{bottom}. We have

$$\Delta V_{total} = \Delta V_{top} + \Delta V_{bottom} = \frac{Q}{C_{top}} + \frac{Q}{C_{bottom}} \tag{18.34}$$

Capacitance is defined as the charge divided by the total potential difference, so

$$C_{total} = \frac{Q}{\Delta V_{total}} = \frac{Q}{\dfrac{Q}{C_{top}} + \dfrac{Q}{C_{bottom}}}$$

$$C_{total} = \frac{1}{\dfrac{1}{C_{top}} + \dfrac{1}{C_{bottom}}} \tag{18.35}$$

This result can be rewritten in an equivalent (and easier to remember) form:

$$\frac{1}{C_{total}} = \frac{1}{C_{top}} + \frac{1}{C_{bottom}} \tag{18.36}$$

An arrangement of two "consecutive" capacitors as in Figure 18.25 is referred to as *capacitors in series*. The result in Equation 18.36 tells us that when any two capacitors (which we can denote by C_1 and C_2) are placed in series, they are completely *equivalent* to a *single* capacitor whose capacitance C_{equiv} is given by

$$\frac{1}{C_{equiv}} = \frac{1}{C_1} + \frac{1}{C_2} \quad \text{(capacitors in series)} \tag{18.37}$$

We have derived this result for the case of two parallel-plate capacitors, but it applies to *any* type of capacitor. This result for capacitors in series will be very useful in our work with electric circuits in Chapter 19.

The result for capacitors in series from Equation 18.37 is shown schematically in Figure 18.26. Here we introduce the standard notation for drawing a capacitor in an electric circuit. The two parallel lines represent the two plates of a parallel-plate capacitor; this notation is used to represent all types of capacitors.

Capacitors in Parallel

Figure 18.26 shows one particular way of connecting two capacitors. Another way is illustrated in Figure 18.27, which shows two capacitors C_1 and C_2 connected in *parallel*. The capacitors C_1 and C_2 in Figure 18.27A are equivalent to a single capacitor C_{equiv}. To find the value of this equivalent capacitance, we imagine that charges Q_1 and Q_2 are placed on the two capacitors as shown in that figure. The

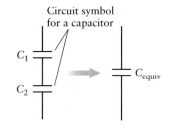

Figure 18.26 In a circuit diagram, capacitors are shown as two parallel lines, representing the plates of the capacitor. The top and middle plates in Figure 18.25 behave as a single capacitor, and the middle and bottom plates behave as a second capacitor. These two capacitors are equivalent to a single capacitance C_{equiv}.

Equivalent capacitance of capacitors in series

Figure 18.27 Two capacitors connected in parallel behave as a single equivalent capacitor C_{equiv}.

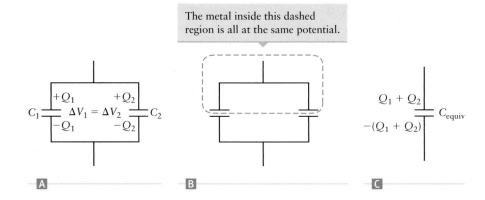

The metal inside this dashed region is all at the same potential.

A B C

voltage across each capacitor is related to its charge and capacitance, and, from Equation 18.30, we have

$$C_1 = \frac{Q_1}{\Delta V_1} \quad \text{and} \quad C_2 = \frac{Q_2}{\Delta V_2} \qquad (18.38)$$

The top plates of these two capacitors are connected by a metal wire. A piece of metal in equilibrium is an equipotential object, so all the metal inside the upper dashed contour in Figure 18.27B is at the same potential. Because they are connected by a metal wire, the two upper plates of C_1 and C_2 are thus at the same potential. Likewise, the lower plate of C_1 is at the same potential as the lower plate of C_2. Hence, because of the way they are connected, the potential difference is the same across the two capacitors, and $\Delta V_1 = \Delta V_2$. For our equivalent capacitor (Fig. 18.27C), the total capacitance is equal to the ratio of the total charge and the potential difference across the plates with (from Eq. 18.30)

$$C_{equiv} = \frac{Q}{\Delta V}$$

The total charge on the equivalent capacitor is $Q = Q_1 + Q_2$, so we have[2]

$$C_{equiv} = \frac{Q_1 + Q_2}{\Delta V} = \frac{Q_1}{\Delta V} + \frac{Q_2}{\Delta V} \qquad (18.39)$$

The voltage across the equivalent capacitor is also equal to the voltage across the individual capacitors $\Delta V = \Delta V_1 = \Delta V_2$. Using this information along with Equation 18.38 leads to

$$C_{equiv} = \frac{Q_1}{\Delta V_1} + \frac{Q_2}{\Delta V_2}$$

Equivalent capacitance of capacitors in parallel

$$C_{equiv} = C_1 + C_2 \quad \text{(capacitors in parallel)} \qquad (18.40)$$

Equation 18.40 is the general rule for how capacitors combine in parallel to act as a single equivalent capacitor.

Combinations of Three or More Capacitors

Equations 18.37 and 18.40 describe how two capacitors combine in series or in parallel. What about combinations of three (or more) capacitors? To see how to deal with such cases, consider three capacitors in parallel as shown in Figure 18.28A. We can first combine capacitors C_1 and C_2 to get an equivalent capacitor C_{12} (Fig. 18.28B); using Equation 18.40, $C_{12} = C_1 + C_2$, which reduces the problem to two capacitors (C_{12} and C_3) in parallel. We can then apply Equation 18.40 again to obtain the effective capacitance of C_{12} and C_3 to get the equivalent capacitance of

---

[2]Notice that the charges Q_1 and Q_2 on the two capacitors need *not* be equal. As we saw with the lightning rod in Figure 18.14, two metal objects at the same potential need not have the same charge.

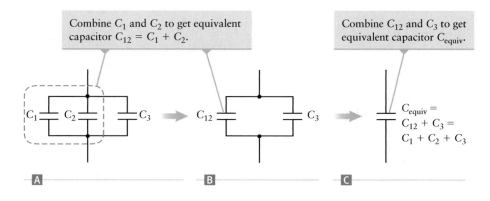

Combine C_1 and C_2 to get equivalent capacitor $C_{12} = C_1 + C_2$.

Combine C_{12} and C_3 to get equivalent capacitor C_{equiv}.

C_1 C_2 C_3 → C_{12} C_3 → $C_{equiv} = C_{12} + C_3 = C_1 + C_2 + C_3$

A **B** **C**

Figure 18.28 Three (or more) capacitors connected in parallel behave as a single equivalent capacitor.

all three capacitors $C_{equiv} = C_{12} + C_3 = C_1 + C_2 + C_3$. The general combination rule for many capacitors in parallel is

$$C_{equiv} = C_1 + C_2 + C_3 + \cdots \quad \text{(many capacitors in parallel)} \qquad (18.41)$$

For multiple capacitors in series, a similar approach gives

$$\frac{1}{C_{equiv}} = \frac{1}{C_1} + \frac{1}{C_2} + \frac{1}{C_3} \cdots \quad \text{(many capacitors in series)} \qquad (18.42)$$

EXAMPLE 18.8 | Combining Capacitors

Four capacitors are connected as shown in Figure 18.29A. If all have the same capacitance ($C_1 = C_2 = C_3 = C_4 = C$), what is the equivalent capacitance of this combination?

RECOGNIZE THE PRINCIPLE

We can apply the rules for combining capacitors in series and in parallel. We first combine C_1 and C_2 in parallel (using Eq. 18.42) and so forth for C_3 and C_4. We are left with two capacitors in series (Fig. 18.29B), which can be analyzed using the rule for a series combination.

SKETCH THE PROBLEM

Figure 18.29 shows the problem. We first combine C_1 and C_2 to get C_{12}, and C_3 and C_4 to get C_{34}. We then combine these results to get C_{equiv}.

IDENTIFY THE RELATIONSHIPS AND SOLVE

Figure 18.29B shows an equivalent arrangement. On the left, we have combined C_1 and C_2 in parallel to get an equivalent capacitance C_{12}. According to Equation 18.41, we have

$$C_{12} = C_1 + C_2 = 2C$$

Likewise, the parallel combination of C_3 and C_4 is

$$C_{34} = C_3 + C_4 = 2C$$

The final step is to combine C_{12} and C_{34} in series (Fig. 18.29C) to get the total capacitance. Applying the rule for combining capacitors in series (Eq. 18.42) gives

$$\frac{1}{C_{equiv}} = \frac{1}{C_{12}} + \frac{1}{C_{34}} = \frac{1}{2C} + \frac{1}{2C}$$

We thus find

$$C_{equiv} = \boxed{C}$$

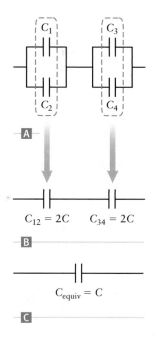

C_1 C_3

C_2 C_4

A

$C_{12} = 2C$ $C_{34} = 2C$

B

$C_{equiv} = C$

C

Figure 18.29 Example 18.8.

What have we learned?
We can deal with most combinations of capacitors by applying the parallel and series combination rules to obtain a single equivalent capacitance.

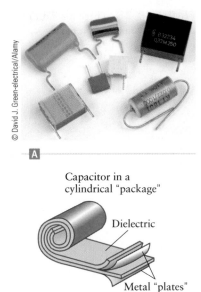

A

Capacitor in a
cylindrical "package"

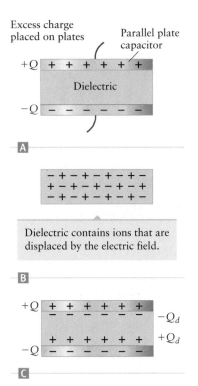

Dielectric

Metal "plates"

B

Figure 18.30 **A** Capacitors can
be constructed in many different
ways and come in different shapes
and sizes. **B** Many capacitors consist of two metal foils ("plates")
wrapped up in a cylinder, with a
dielectric between the foils.

Excess charge
placed on plates

Parallel plate
capacitor

$+Q$ + + + + + +

Dielectric

$-Q$ − − − − − −

A

− + − + − + − + −
+ − + − + − + − +
− + − + − + − + −

Dielectric contains ions that are
displaced by the electric field.

B

$+Q$ + + + + + +
− − − − − − $-Q_d$

+ + + + + + $+Q_d$
$-Q$ − − − − − −

C

Figure 18.31 **A** When a dielectric is placed between the plates of
a capacitor, the value of the capacitance increases. **B** Most dielectrics
are ionic. **C** The electric field inside
the capacitor induces charge $\mp Q_d$
on the surfaces of the dielectric.

18.5 | DIELECTRICS

Most real capacitors are not as simple as the parallel-plate capacitor analyzed in
Examples 18.6 and 18.7. Figure 18.30A shows a number of real capacitors. They all
contain two metal "plates" (obscured by their outer casings in Fig. 18.30A) separated by a thin insulating region, but notice that many of the casings are cylindrical.
These cylinders contain two thin, metal foils (the "plates") wrapped like a jellyroll.
In the parallel-plate capacitors we have considered so far, the region between the
plates is a vacuum. In most real capacitors, however, the space between the plates
is filled with a solid material called a *dielectric* as illustrated in Figures 18.30B and
18.31A.

The Term *Dielectric* Is Used to Describe Any Insulating Material.

Inserting a dielectric material between the plates of a capacitor changes the value of
the capacitance in proportion to a factor called the *dielectric constant* of the material, denoted by κ. If C_{vac} is the value of the capacitance with a vacuum in the gap
between the plates, the capacitance with the dielectric present is

$$C_d = \kappa C_{vac} \tag{18.43}$$

Note that κ is a dimensionless factor, so it has no units. Except for some very
unusual materials, κ is greater than or equal to 1, so the presence of the dielectric
increases the capacitance.

Why Does a Dielectric Change the Capacitance?

Charge on the plates of a capacitor produces an electric field between the plates.
This electric field extends into a dielectric placed between the plates, causing electric forces on charges in the dielectric. "Good" dielectrics—that is, materials with
large values of κ—are usually highly ionic, meaning that they contain positive and
negative ions bound together as sketched in Figure 18.31B. The electric force on
each ion causes the ion to shift slightly according to the direction of $\vec{E}$. In the dielectric in Figure 18.31, the negative ions move slightly toward the top capacitor plate
(because they are attracted to the charge $+Q$) and the positive ions shift a small
amount toward the bottom plate (because they are attracted to the charge $-Q$).
This slight shift in the ions' positions leads to a small amount of negative charge
on the top surface of the dielectric and a corresponding small amount of positive
charge on the bottom surface of the dielectric.

An equivalent way to view a capacitor with dielectric inside is shown in Figure
18.31C. The metal plates have charges $\pm Q$ on their inner surfaces, while the surfaces of the dielectric carry charge $\mp Q_d$. If we lump these two together, we then
have a total charge $+Q - Q_d$ at the top of the capacitor and $-Q + Q_d$ at the bottom. The dielectric thus reduces the magnitude of the charge on the capacitor.

The field between the plates of a parallel-plate capacitor containing a dielectric
is given by

$$E_d = \frac{Q}{\kappa \varepsilon_0 A} \tag{18.44}$$

where Q is the charge on just the metal plates. Since $\kappa \geq 1$, the electric field inside
the capacitor is smaller with the dielectric present than when the gap is filled with a
vacuum (compare with Eq. 18.28). Hence, the potential difference across the plates
is also smaller, again by a factor of κ. Capacitance is defined as the ratio of the
external charge placed on the capacitor plates divided by the potential difference.
Because the potential difference ΔV is reduced by a factor of κ, the ratio $C = Q/\Delta V$
is larger by the same factor, which takes us to Equation 18.43.

These results apply to *any* type of capacitor. Adding a dielectric (1) increases the
capacitance by a factor equal to the dielectric constant κ and (2) reduces the electric

Table 18.1 Values of the Dielectric Constant and the Dielectric Breakdown Field for Some Common Materials

SUBSTANCE	κ	E(breakdown), V/m
Mica	5	100×10^6
Glass	6	14×10^6
Paper	4	16×10^6
Plexiglas	3.4	40×10^6
Quartz	3.8	40×10^6
Mylar	3	300×10^6
Teflon	2	60×10^6
Strontium titanate	230	8×10^6
Air (at room temperature and pressure)	1.0006	3×10^6
Helium gas (at room temperature and pressure)	1.000065	
Water	80.4	
Glycerine	42.5	
Benzene	2.28	

field inside the capacitor by the same factor. The actual value of the dielectric constant depends on the material. Materials that are very ionic tend to have the largest values of κ. Table 18.1 lists the values of κ for solids, liquids, and gases; all three phases of matter exhibit dielectric behavior.

Effects of Very Large Electric Fields

As more and more charge is added to a capacitor, the electric field between the plates increases in proportion to Q (Eq. 18.44). For a capacitor containing a dielectric, this field can be so large that it literally rips the ions in the dielectric apart, an effect called **dielectric breakdown**. The resulting free ions are able to move through the material much like electrons in a metal. Because there is a large electric field present (the field that caused the dielectric breakdown), these ions move rapidly toward the oppositely charged capacitor plate and destroy the capacitor.

The value of the electric field at which dielectric breakdown occurs depends on the material (see Table 18.1) and sets a practical limit on the maximum voltage that can be placed across a capacitor in applications. Dielectric breakdown can occur whenever the electric field is large (not just in capacitors), so it comes into play in other phenomena such as lightning, as we'll discuss in Section 18.6.

EXAMPLE 18.9 | Dielectric Breakdown of a Capacitor

Figure 18.30A shows capacitors used in practical devices such as a laptop computer or an MP3 player. A typical capacitor of this type has a capacitance of around 1.0 nF ($= 1.0 \times 10^{-9}$ F). Assume it is a parallel-plate capacitor with a plate area of $A = 1.0$ cm^2 that uses mica as the dielectric. Find **(a)** the spacing between the plates and **(b)** the voltage at which dielectric breakdown occurs (i.e., the voltage at which the capacitor will fail).

RECOGNIZE THE PRINCIPLE

For a parallel-plate capacitor containing a dielectric, the capacitance is given by

$$C_d = \kappa C_{\text{vac}} = \frac{\kappa \varepsilon_0 A}{d} \tag{1}$$

(a combination of Eqs. 18.31 and 18.43). Given the value of C along with the plate area A and the dielectric constant, we can solve Equation (1) to find the spacing between the plates d. We can then calculate the electric field between the plates and compare it with the dielectric breakdown field for mica.

SKETCH THE PROBLEM

Figure 18.31A describes our capacitor; the dielectric in this case is mica.

IDENTIFY THE RELATIONSHIPS AND SOLVE

(a) Rearranging Equation (1) to solve for the spacing d between the plates, we find

$$d = \frac{\kappa \varepsilon_0 A}{C_d}$$

From Table 18.1, the dielectric constant of mica is $\kappa = 5$. Inserting this and the given values of A and C_d, we get

$$d = \frac{\kappa \varepsilon_0 A}{C_d} = \frac{(5)[8.85 \times 10^{-12}\ \text{C}^2/(\text{N} \cdot \text{m}^2)](0.010\ \text{m})^2}{1.0 \times 10^{-9}\ \text{F}}$$

$$d = 4 \times 10^{-6}\ \text{m} = \boxed{4\ \mu\text{m}}$$

(b) The dielectric breakdown field of mica is $E = 100 \times 10^6$ V/m (from Table 18.1). Using the relation between the electric field and potential difference between the plates (Eq. 18.17) leads to a breakdown voltage of magnitude

$$\Delta V = |-Ed| = (100 \times 10^6\ \text{V/m})(4 \times 10^{-6}\ \text{m}) = \boxed{400\ \text{V}}$$

What does it mean?

Our result for d in part (a) is much smaller than the thickness of a human hair! Even so, pieces of mica this thin are easily obtained. The voltages used in consumer electronics (such as an MP3 player) are typically less than 10 or 20 V, so this capacitor will not be in danger of failure.

18.6 | ELECTRICITY IN THE ATMOSPHERE

One of nature's most impressive displays involves the motion of electric charge called lightning. During a lightning strike, large amounts of electric charge move between a cloud and the surface of the Earth or between two clouds (Fig. 18.32). This charge motion occurs when a large electric field is established between the cloud and the Earth or between two clouds, causing dielectric breakdown of the air. Molecules in the air are ripped apart to form free electrons, leaving behind positively charged ions. These charges are then accelerated by the electric field, and some travel to the Earth, where they arrive with very high velocities. This rapid motion of charge heats nearby air molecules, leading to the sound (thunder) associated with lightning.

Most of the charge motion in lightning involves electrons because they are much lighter than the positive ions that are also produced in dielectric breakdown of air. Which way do these electrons move? Do they travel from a cloud to the Earth, or vice versa? Experiments show that the electric field of a lightning strike is directed from the Earth to the cloud[3] as shown in Figure 18.33. Recall that $\vec{E}$ points in the direction a positive test charge would move. Thus, once dielectric breakdown occurs in the air, electrons travel from the cloud to the Earth's surface. A typical lightning bolt contains a charge of about $|Q| = 20$ C, corresponding to

© Lionel Brown/Image Bank/Getty Images

Figure 18.32 Example of a lightning strike.

[3]In fair weather, there is a small electric field in the opposite direction (downward, toward the Earth).

590   CHAPTER 18 | ELECTRIC POTENTIAL

$$N = \frac{|Q|}{e} = \frac{20\text{ C}}{1.60 \times 10^{-19}\text{ C/electron}} \approx 1 \times 10^{20}\text{ electrons/bolt} \quad (18.45)$$

You might think that these electrons would take the shortest path and travel along a straight line to the Earth, but they actually follow a jagged path, giving a lightning bolt its familiar appearance (Fig. 18.32). The jaggedness is due to the complicated nature of dielectric breakdown in air.

Perhaps the most basic question concerning lightning is, Where does the electric field in Figure 18.33 come from in the first place? Interestingly, this problem is still not very well understood. So far, no one has a complete theory of how motion of air and clouds during a storm leads to the large electric field necessary to cause lightning, but the correct qualitative explanation is probably as follows. During a thunderstorm, the atmosphere is very unsettled, with large temperature and pressure differences between low and high altitudes. These differences cause air molecules, water droplets, dust, and small ice crystals to move vertically through the atmosphere. During the later phases of a thunderstorm, the water droplets and ice crystals move downward, acquiring charge as they "rub" against molecules and ions in the air (Fig. 18.33B). This process is similar to what we found in Chapter 17 when we saw that rubbing a piece of amber with fur causes the amber to acquire a negative charge. In a thunderstorm, rubbing gives the water droplets and ice crystals a negative charge that they carry to the bottom of the cloud, leaving the top charged positively. As negative charge accumulates at the bottom of the cloud, it repels electrons from the Earth's surface because the ground is a good conductor of charge and electrons can move freely from place to place in the soil. This process causes the Earth's surface to be positively charged (Fig. 18.33A), producing an electric field qualitatively similar to the electric field between two parallel plates of a capacitor. This electric field produces the dielectric breakdown, which is the lightning bolt.

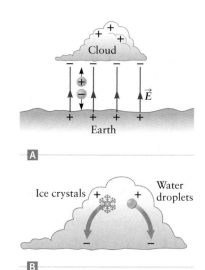

Figure 18.33 ◻ During an electrical storm, the top of a storm cloud is positively charged, while the bottom is negatively charged. The resulting electric field induces a positive electric charge at the Earth's surface. ◻ The cloud's charge is produced by the motion of water droplets and ice crystals as they are pushed along by the wind.

EXAMPLE 18.10 ® **Electric Field and Potential of a Lightning Strike**

Consider a thunderstorm involving the bottom surface of a cloud that is $h = 500$ m above the surface of the Earth. If a lightning bolt travels from this cloud to the Earth, estimate the electric potential of the bottom of the cloud just prior to the lightning bolt, relative to "ground" potential $V = 0$ at the Earth's surface.

RECOGNIZE THE PRINCIPLE

A lightning bolt occurs when there is dielectric breakdown of the air. So, just before the lightning bolt occurs, the electric field between the cloud and the Earth must equal the dielectric breakdown field of air.

SKETCH THE PROBLEM

Figure 18.34 shows the problem.

IDENTIFY THE RELATIONSHIPS

Using the relation between the electric field and the potential (Eq. 18.18) gives

$$E = -\frac{\Delta V}{\Delta x} = -\frac{V_{\text{cloud}} - V_{\text{Earth}}}{h}$$

The Earth is at "ground" potential ($V_{\text{Earth}} = 0$), so

$$V_{\text{cloud}} = -Eh$$

SOLVE

Inserting the dielectric breakdown field of air (Table 18.1) and the height of the cloud, we find

$$V_{\text{cloud}} = -E_{\text{breakdown}}h = -(3 \times 10^6\text{ V/m})(500\text{ m}) \approx \boxed{-2 \times 10^9\text{ V}}$$

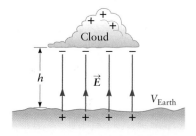

Figure 18.34 Example 18.10.

EKG voltage at different electrodes

Time

B

Figure 18.35 **A** A person having an EKG. **B** In an EKG, the electric potential as a function of time is recorded from electrodes at many different spots on the body. This recording shows when and how strongly different parts of the heart muscle are contracting. Typical EKG potentials are around 1 mV (10^{-3} V), about 10,000 times smaller than the potential produced by the battery of an MP3 player.

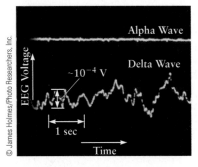

Alpha Wave

Delta Wave

EEG Voltage

$\sim 10^{-4}$ V

1 sec

Time

Figure 18.36 Typical EEG signals. The top signal was taken when the person was awake and alert and is called an alpha wave. The bottom EEG was recorded when the person was sleeping (a delta wave). Typical EEG potentials are about 10^{-4} V, somewhat smaller than in an EKG.

What does it mean?
The negative sign for V_{cloud} means that the cloud is negatively charged. Typical household voltages are between 1 V and 120 V, so the potential associated with lightning is *much* larger than the voltages found around the house!

18.7 | ⊗ BIOLOGICAL EXAMPLES AND APPLICATIONS

In the 1790s, an Italian scientist, Luigi Galvani, discovered that electricity could make muscles twitch. Galvani originally observed that sparks from static electricity can cause muscular contraction in a frog. Galvani's work was the first evidence that the nerves and muscles in frogs use an electrical potential and the movement of charge as part of normal physiological function. The same is true in other animals, including humans.

The heart is perhaps the most important muscle in the body. Like the muscles in a frog's leg, the heart can also be made to twitch using an externally applied potential. For example, a device called a *defibrillator* is used to shock the heart into a normal pattern of beating. A common diagnostic procedure is the *electrocardiogram*, called an ECG or EKG. This procedure uses a collection of electrodes to monitor the potential at various places on a person's chest (Fig. 18.35A). The electric potentials the body sends to various parts of the heart form a rhythmic pattern that causes different chambers of the heart to contract at different times, and the different electrode potentials in an EKG can reveal if a person's heart is (or is not) following a healthy rhythm (Fig. 18.35B). There is no way to use physics to calculate what a healthy rhythm should look like, but your doctor can tell from experience.

A related procedure is the *electroencephalogram* (EEG), in which electrodes on the scalp are used to detect potentials generated in the brain. These potentials oscillate with time, and the frequency of an EEG signal depends on the person's activity. The brain of an awake and alert person generates what are called alpha waves with frequencies of about 10–14 Hz, whereas during deep sleep one finds delta waves with a typical frequency of 2–4 Hz. (Fig. 18.36). Physiologists have also identified beta and gamma waves, which are generated in other brain states and which have other frequencies. The absence of any EEG signal is usually taken as evidence that a person is "brain dead."

18.8 | ELECTRIC POTENTIAL ENERGY REVISITED: WHERE IS THE ENERGY?

A certain amount of potential energy is associated with two point charges as a result of the electric interaction between them. We have described this potential energy as a property of the system of the two charges. Another way to view and think about this electric potential energy, however, is that potential energy is stored *in the electric field* itself. That is, whenever an electric field is present in a particular region of space, potential energy is located in that region. In almost all respects, this viewpoint is equivalent to the one we have taken in previous sections, so it does not affect any of our results to this point. Let's now calculate this field energy in a familiar geometry.

Consider two flat, metal plates separated by a vacuum; this structure is just a parallel-plate capacitor, and we have already worked out many of its properties. If the plates have an area A and are separated by a distance d, the electric field between the plates is (Eq. 18.28)

$$E = \frac{Q}{\varepsilon_0 A} \qquad (18.46)$$

where the charge on the plates is $\pm Q$. The magnitude of the potential difference between the plates is $\Delta V = Ed$. We also found that the electric potential energy stored in the capacitor is (Eq. 18.32)

$$PE_{\text{elec}} = \tfrac{1}{2} Q \, \Delta V$$

Combining these last two results, we get

$$PE_{\text{elec}} = \tfrac{1}{2} Q E d$$

We now eliminate Q using Equation 18.46, leading to

$$PE_{\text{elec}} = \tfrac{1}{2}(\varepsilon_0 E A) E d = \tfrac{1}{2}\varepsilon_0 E^2 (Ad)$$

The factor (Ad) is the volume between the plates, so we can also write

$$PE_{\text{elec}} = \tfrac{1}{2}\varepsilon_0 E^2 (\text{volume}) \qquad (18.47)$$

In words, this says that the electric field energy in the region between the capacitor plates is equal to $\tfrac{1}{2}\varepsilon_0 E^2$ multiplied by the volume of the region. Hence, the energy per unit volume, denoted by u_{elec}, is

$$u_{\text{elec}} = \tfrac{1}{2}\varepsilon_0 E^2 \qquad (18.48)$$

We call u_{elec} the **energy density** in the electric field.

Although our derivation is for a parallel-plate capacitor, this result for u_{elec} applies for any arrangement of charges, including point charges. It tells us that potential energy is present wherever an electric field is present. Philosophically, this result is a major break from our previous view that potential energy is due to the interaction of charges. One might argue that electric fields are caused by charges, so the field E in Equation 18.48 is actually due to a charge. If this argument is correct, aren't we back with our original view that electric potential energy is always a property of a system of charges? In most cases, that is true, but in one extremely important situation it is not: when dealing with an electromagnetic wave. We know from our work on waves in Chapter 12 that waves transport energy. Electromagnetic waves transport energy from place to place, even through regions of space where *no charges are present* or even close by. The energy carried by such a wave is carried by electric and magnetic fields, and the electric part of this energy is given by Equation 18.48. (A portion of the energy is also contained in the magnetic field.) So, to explain the energy carried by an electromagnetic wave, it is essential that energy be carried by the field. Our original view that electric potential energy always involves pairs of charges is thus *not* the entire story. We'll come back to this problem in Chapter 23 when we discuss electromagnetic waves.

SUMMARY | Chapter 18

KEY CONCEPTS AND PRINCIPLES

Electric potential energy

The electric force is a conservative force, so there is an **electric potential energy** PE_{elec}. The potential energy of two point charges separated by a distance r is

$$PE_{\text{elec}} = \frac{kq_1 q_2}{r} \qquad \text{(18.6) (page 566)}$$

(Continued)

Electric potential

The *electric potential* V (also sometimes called just the *potential*) is proportional to the electric potential energy. If the electric potential energy of a charge q at a particular location is PE_{elec}, the electric potential at that point is

$$V = \frac{PE_{elec}}{q} \qquad \textbf{(18.10)} \text{ (page 571)}$$

with V measured in units of *volts* (V).

Relation between the electric field and the electric potential

Suppose the potential changes by an amount ΔV over a distance Δx. The component of the electric field along this direction is then

$$E = -\frac{\Delta V}{\Delta x} \qquad \textbf{(18.18)} \text{ (page 573)}$$

The electric field thus has units of volts per meter (V/m).

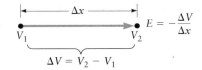

Capacitors

Two parallel metal plates form a *capacitor*. The *capacitance* C of this structure determines how easily charge can be stored on the plates. The charge on a capacitor is related to the magnitude of the *potential difference* between the plates by

$$\Delta V = \frac{Q}{C} \qquad \textbf{(18.30)} \text{ (page 581)}$$

This relation holds for any type of capacitor. For a parallel-plate capacitor, the capacitance is

$$C = \frac{\varepsilon_0 A}{d} \quad \text{(parallel-plate capacitor)} \qquad \textbf{(18.31)} \text{ (page 581)}$$

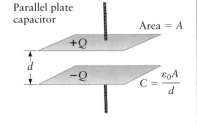

Energy stored in a capacitor and in an electric field

The electric potential energy stored in a capacitor is

$$PE_{cap} = \tfrac{1}{2} Q\, \Delta V = \tfrac{1}{2} C (\Delta V)^2 = \frac{1}{2} \frac{Q^2}{C} \qquad \textbf{(18.32)}, \textbf{(18.33)} \text{ (page 582)}$$

The energy per unit volume stored in an electric field E is

$$u_{elec} = \tfrac{1}{2} \varepsilon_0 E^2 \qquad \textbf{(18.48)} \text{ (page 593)}$$

APPLICATIONS

Equipotential surfaces

An *equipotential surface* is a surface on which the electric potential is constant. The electric field $\vec{E}$ at a particular location is always perpendicular to the equipotential surface at that spot.

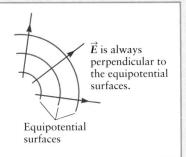

(*Continued*)

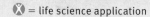
QUESTIONS

SSM = answer in Student Companion & Problem-Solving Guide

✕ = life science application

1. If the electric field is zero in a particular region of space, what does that tell you about the electric potential in that region? Is the potential zero, or constant, or something else? Explain.

2. Will the *electric field* always be zero at any point where the *electric potential* is zero? Why or why not?

3. Will the *electric potential* always be zero at any point where the *electric field* is zero? Why or why not?

4. **SSM** Make sketches of the equipotential surfaces around (a) a point charge, (b) an infinite line of charge, (c) an infinite plane of charge, (d) a finite line of charge, and (e) a charged plate of finite size.

5. An electron is released from rest at the origin and moves along the $+x$ direction. Other experiments show that the electric field is uniform (i.e., constant). Make a sketch showing the direction of the electric field and the equipotential surfaces.

6. A particle of positive charge is released from rest and is found to move as a result of an electric force. Does the particle move to a region of higher or lower potential energy? Does the particle move to a region of higher or lower electric potential?

7. Repeat Question 6 for an electron.

8. A charged particle is released from rest in an electric field. Its motion does not follow the electric field lines. Why? Are there configurations of the electric field that would lead to motion along electric field lines? If so, describe some of them and explain what they have in common.

9. Two particles are at locations where the electric potential is the same. Do these particles necessarily have the same electric potential energy? Explain.

10. ✕ Many species of fish generate electric fields and use these fields to sense their surroundings. If another entity with a dielectric constant different from that of water comes near the fish, the electric field lines generated by the fish are altered, and sensory organs on the fish can detect these slight changes. The electric field of most of these fishes is that of a dipole. Figure Q18.10 shows lines of equal potential around such a fish as seen looking down on the top of the fish. Assume the head of the fish is positively charged. (a) Sketch the electric field associated with these equipotential lines. (You may want to make a photocopy of the figure to proceed.) (b) If a sphere of metal (perhaps a lead fishing weight) were placed near the side of the fish, sketch how the field lines in the vicinity of the metal would change.

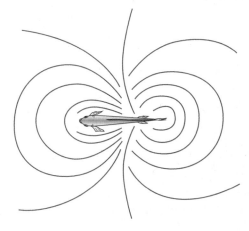

Figure Q18.10

11. How much work is required to move a charge from one spot on an equipotential surface to another location on the same equipotential surface?

12. **SSM** A charged particle obtains a speed v when accelerated through a potential difference of 5 V. How many times faster would it be going if accelerated though a potential difference of 20 V?

13. Consider the energy density in the electric field at a distance d away from a positive point charge. If the positive point charge is then replaced with a negative point charge of equal magnitude, how does the energy density change at that same point?

14. To increase the energy stored in a capacitor, what might you do? Explain your reasoning.
(a) Increase the charge on the capacitor.
(b) Insert a dielectric between the plates while holding the voltage fixed.
(c) Move the plates closer together while keeping the charge fixed.
(d) All the above.
(e) None of the above.

15. Compared with the applied electric field, is the electric field inside a dielectric material (a) smaller by a factor of κ, (b) larger by a factor of κ, or (c) larger by a factor of κ^2? Or, does it (d) depend on the shape of the dielectric? Explain your reasoning.

16. Derive the result for the equivalent capacitance of many capacitors in series in Equation 18.42.

17. When one speaks of a fully charged capacitor, with what is it charged? How is the word *charged* being used here?

18. How does a charged capacitor differ from a capacitor that is not charged? Does the charged capacitor have a different net charge? Explain.

19. A parallel-plate capacitor is connected to a battery such that a constant electric potential difference is produced between the plates. If the plates are moved farther apart, which of the following quantities change? Do they get larger or smaller?
(a) The magnitude of the electric field
(b) The amount of charge on a plate
(c) The capacitance
(d) The energy stored in the capacitor

20. A certain amount of charge $\pm q$ is placed on the plates of a capacitor, and the plates are then disconnected from the outside world. If the capacitor plates are pulled apart, does the amount of electric potential energy stored in the capacitor increase or does it decrease? Explain your answer in terms of the principle of conservation of energy.

PROBLEMS

18.1 ELECTRIC POTENTIAL ENERGY

1. Two point particles of charge $Q_1 = 45\ \mu C$ and $Q_2 = 85\ \mu C$ are found to have a potential energy of 40 J. What is the distance between the charges?

2. Two particles with $Q_1 = 45\ \mu C$ and $Q_2 = 85\ \mu C$ are initially separated by a distance of 2.5 m and then brought closer together so that the final separation is 1.5 m. What is the change in the electric potential energy?

3. The nucleus of a helium atom contains two protons. In a simple model of this nucleus, the protons are viewed as point particles separated by a distance of 1.0 fm (1.0×10^{-15} m). What is the electric potential energy of the two protons?

4. Two point charges $Q_1 = 3.5\ \mu C$ and $Q_2 = 7.5\ \mu C$ are initially very far apart. They are then brought together, with a final separation of 2.5 m. How much work does it take to bring them together?

5. A simple model of a hydrogen atom pictures the electron and proton as point charges separated by a distance of 0.050 nm, and in Example 18.1 we calculated the associated electric potential energy. If the electron is initially at rest, how much work is required to "break apart" these two charges, that is, to separate them by a very large distance?

6. Consider an electric dipole consisting of charges $+Q$ and $-Q$ as sketched in Figure P18.6. How much work is required to place a charge q at the origin? Assume q starts from very far away.

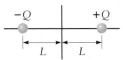

Figure P18.6
Problems 6 and 16.

7. Two point charges are located as shown in Figure P18.7, with charge $q_1 = +2.5$ C at $x = -3.0$ m, $y = 0$,

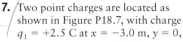

Figure P18.7

and charge $q_2 = +4.0$ C at $x = +1.0$ m, $y = +2.0$ m. An electron is now taken from a point very far away and placed at the origin. How much work must be done on the electron to move it to the origin?

8. ★ The electric field in a particular region of space is found to be uniform, with a magnitude of 400 N/C and a direction parallel to the y axis. (a) What is the change in the electric potential energy of a charge $q = 3.5\ \mu C$ if it is moved from a location $(x, y) = (20$ cm, 45 cm) to (5 cm, 30 cm)? (b) What is the change in the electric potential energy if the charge is moved the same distance along the x axis?

9. ✿ Three point charges $Q_1 = 2.5\ \mu C$, $Q_2 = 4.5\ \mu C$, and $Q_3 = -3.5\ \mu C$ are arranged as shown in Figure P18.9. What is the total electric potential energy of this system?

10. SSM ★ Consider again the three charges in Figure P18.9 with $Q_1 = 2.5\ \mu C$, $Q_2 = 4.5\ \mu C$, and $Q_3 = -3.5\ \mu C$. A fourth charge $q = -5.0\ \mu C$ is brought from very far away and placed at the origin. (a) How much work is required in this process? (b) How much work is required to move q from the origin to a distance that is very far away?

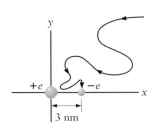

Figure P18.9 Problems 9 and 10.

11. A proton is placed at the origin. An electron that is initially very far away is then taken on the path shown in Figure P18.11 and eventually stops a distance of 3.0 nm ($= 3.0 \times 10^{-9}$ m) from the proton. How much work was done on the electron to move it along this path?

12. ★ An electron and a proton are a distance $r = 7.5 \times 10^{-9}$ m

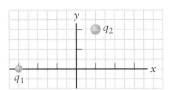

Figure P18.11

apart. How much energy is required to increase their separation by a factor of two?

13. [star] Two point charges $Q_1 = +1.2\ \mu C$ and $Q_2 = +3.2\ \mu C$ are initially separated by 1.0 m and held fixed. The charges are then released, and both move in response to the electric force between them. What is the sum of the kinetic energies of the two particles when they are very far apart?

14. [star] The two electrons in a helium atom are separated by about 0.10 nm and the separation between either electron and the nucleus is about 0.050 nm. What is the total electric potential energy of the atom? Assume the nucleus is a point charge with $q = +2e$ and for simplicity do not include the potential energy associated with the Coulomb repulsion of the two protons in the nucleus. Compare your answer with the ionization energy of a helium atom and with the energy required to remove both electrons from a helium atom.

15. [star] An oxygen ion O^{2-} is a distance $r = 5.0 \times 10^{-10}$ m from a H^+ ion. How much energy is required to separate them completely? Treat both ions as point charges.

18.2 ELECTRIC POTENTIAL: VOLTAGE

16. Find the electric potential V at the center of the electric dipole in Figure P18.6. Express your answer in terms of Q and L.

17. An electron is moved from an initial location where the potential is $V_i = 30$ V to a final location where $V_f = 150$ V. What is the change in the electron's potential energy? Express your answer in joules and in electron-volts.

18. The electric potential difference between two infinite, parallel metal plates is V. If the plates are separated by a distance $L = 3.0$ mm and the electric field between the plates is $E = 250$ V/m, what is V?

19. [star] For the situation described in Problem 17, what is the average electric field along a line 10 cm long that connects the initial and final locations of the electron? Be sure to give both the magnitude and direction of $\vec{E}$.

20. [star] A proton moves from a location where $V = 75$ V to a spot where $V = -20$ V. (a) What is the change in the proton's kinetic energy? (b) If we replace the proton with an electron, what is the change in kinetic energy?

21. Consider a very small charged sphere that contains 35 electrons. At what distance from the center of the sphere is the potential equal to -60 V?

22. [star] How much work (as measured in joules) is required to push an electron through a region where the potential change is $+45$ V? Assume the electron moves from a region of low potential to a region of higher potential.

23. [star] The electrons in a TV picture tube are accelerated through a potential difference of 30 kV. (a) Do the electrons move from a region of high potential to a region of low potential, or vice versa? (b) What is the change in the kinetic energy of one of the electrons? (c) If the initial speed is very small, what is the final speed of an electron?

24. [star] The electric field in a particular region of space is constant with a magnitude of 300 V/m and is along the y direction. (a) Sketch how the electric potential varies along y, starting from the origin and ending at $y = 4$ m. Assume the potential is zero at the origin. (b) Sketch how V varies along the x direction, starting at the origin and ending at $x = 5$ m.

25. [SSM] [star] Four point charges, each with $Q = 4.5\ \mu C$, are arranged at the corners of a square of edge length 1.5 m. What is the electric potential at the center of the square?

26. [star] The electric potential varies with x as sketched in Figure P18.26. Make a qualitative plot of the component of the electric field along the x direction as a function of x.

27. An infinite plane of charge (Fig. P18.27) has a charge per unit area of $\sigma = 2.0\ \mu C/m^2$. What is the change in the electric potential between points that are 2.5 m (initial) and 4.5 m (final) from the plane?

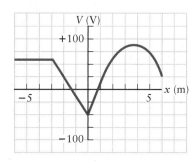

Figure P18.26

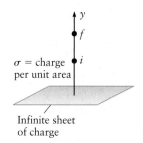

Figure P18.27 Problems 27, 28, and 36.

28. [star] Consider again the plane of charge in Figure P18.27 with $\sigma = 2.0\ \mu C/m^2$. Calculate the change in potential experienced by an electron that starts a distance 2.5 m from the plane and moves a distance 5.5 m in a direction perpendicular to the y axis. Give an intuitive explanation of your answer.

29. Ten electrons are placed on a metal sphere. If the potential of the sphere is -35 V, what is the radius of the sphere?

30. [star] A point charge Q is located at the center of a very thin, spherical metal shell. The net excess charge of the shell is zero. If the radius the shell is r_{shell}, what is the electric potential of the shell?

31. [diamond] A point charge Q is located at the center of a spherical metal shell that has an inner radius r_1 and an outer radius r_2. The net excess charge of the shell is zero. (a) What is the excess charge on the inner surface of the shell (at $r = r_1$)? (b) What is the electric potential at $r = r_1$?

32. [SSM] [star] Two infinite, parallel, uniformly charged plates with charge densities $+\sigma$ and -3σ are separated by a distance L (Fig. P18.32). If the plate on the right is at $V = 0$, what is the potential of the plate on the left? Express your answer in terms of σ and L.

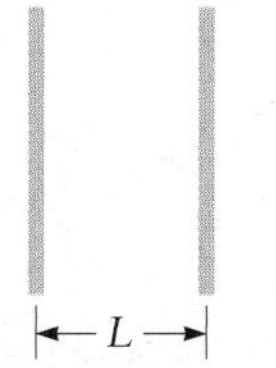

33. [star] A circular ring of charge has a radius R and a charge per unit length λ. What is the electric potential at the center of the ring? Take $V = 0$ at infinity.

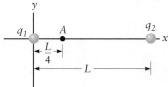

Figure P18.32

34. [star] Consider a solid sphere of radius $R = 0.55$ m that is uniformly charged with $\rho = -2.5\ \mu C/m^3$. What is the electric potential a distance 2.5 m from the center of the sphere?

35. [star] Two point particles with charges q_1 and q_2 are separated by a distance L as shown in Figure P18.35. The electric potential

Figure P18.35

is zero at point A, which is a distance $L/4$ from q_1. What is the ratio q_1/q_2?

18.3 EQUIPOTENTIAL LINES AND SURFACES

36. ☆ Make a sketch of the equipotential surfaces near the plane of charge in Figure P18.27. *Hint*: How is the direction of $\vec{E}$ related to the orientation of an equipotential surface?

37. SSM ☆ The equipotential planes in a particular region of space are shown in Figure P18.37. What is the approximate magnitude and direction of the electric field at the origin? Consider only the components of the field along x and y.

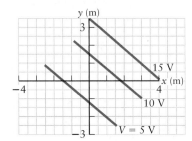

Figure P18.37 Problems 37 and 38.

38. ☆ If an electron placed at the point $(x, y) = (1.0 \text{ m}, 2.0 \text{ m})$ in Figure P18.37 is released from rest, in what direction will it move?

18.4 CAPACITORS

39. ☆ The dimensions of a parallel-plate capacitor are all increased by a factor of three. By what factor does the capacitance change?

40. Make a sketch of the equipotential surfaces between the plates of a parallel-plate capacitor.

41. A parallel-plate capacitor has square plates of edge length 1.0 cm and a plate spacing of 0.010 mm. If the gap between the plates is a vacuum, what is the capacitance?

42. A typical capacitor in an MP3 player has $C = 0.10 \ \mu F$. If a charge $\pm 5.0 \ \mu C$ is placed on the plates, what is the voltage across the capacitor?

43. A voltage of 12 V is placed on a capacitor with $C = 100$ pF (picofarads). (a) What is the charge on the capacitor? (b) How much energy is stored in the capacitor?

44. ☆ ⓡ Design a parallel-plate capacitor with $C = 5.0$ F. That is, find values of the plate area and the plate spacing that will give this capacitance. Assume there is a vacuum between the plates. Be sure to choose a practical value for the separation between the plates. Do you think that you could lift this capacitor?

45. ✪ A charge $\pm Q$ is placed on the plates of a capacitor. The separation between the plates is changed, and it is found that the voltage across the plates decreases. (a) Have the plates been moved closer together or farther apart? (b) If the new voltage is smaller than the original voltage by a factor of nine, what is the ratio of the new plate spacing to the original spacing?

46. ✪ Consider a hollow metal cylinder of radius 1.0 mm that has a thin wire of radius 0.50 mm running down the center as sketched in Figure P18.46. This structure is a capacitor, with

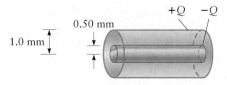

Figure P18.46

the wire acting as one "plate" and the cylinder acting as the other plate. Calculate the capacitance using the following steps. (a) Place charges $\pm Q$ on the wire and the cylinder and calculate the electric field between them. Assume the cylinder has a length L and assume L is very long. (b) Estimate qualitatively the electric potential difference between the wire and the cylinder, which can be done using a plot of the electric field as a function of radius. (c) Use the potential difference to find the capacitance C for $L = 10$ cm.

47. Three capacitors, all with capacitance C, are connected in parallel as shown in Figure P18.47. What is the equivalent capacitance of this combination?

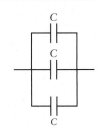

Figure P18.47

48. Four capacitors, all with capacitance C, are connected in series as shown in Figure P18.48. What is the equivalent capacitance of this combination?

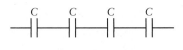

Figure P18.48

49. ☆ Three capacitors are connected as shown in Figure P18.49. What is the equivalent capacitance?

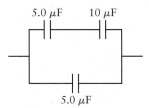

Figure P18.49

50. ✪ Design a combination of identical capacitors, each with capacitance C, for which the equivalent capacitance is $3C/4$.

51. SSM ☆ Four capacitors are connected as shown in Figure P18.51. What is the equivalent capacitance?

52. ✪ Two capacitors with $C_1 = 1.5 \ \mu F$ and $C_2 = 2.5 \ \mu F$ are connected in parallel. If the combined charge on both capacitors is 25 μC, what is the voltage across the capacitors?

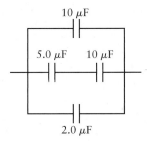

Figure P18.51

53. ✪ An amount of charge $\pm Q$ is placed on the plates of a parallel-plate capacitor so that the potential across the plates is ΔV_{init}. The capacitor is then disconnected, and the separation between its plates is increased by a factor of three. If the charge on the plates is $\pm 35 \ \mu C$ while the final capacitance is 2.0×10^{-9} F, what was ΔV_{init}?

54. ✪ Some charge is placed on a capacitor with $C_1 = 35 \ \mu F$ so that $\Delta V = 12$ V. The capacitor is then attached in parallel to a second capacitor with $C_2 = 55 \ \mu F$. What is the voltage across the two capacitors?

55. ✪ A charge $\pm Q$ is placed on the plates of a parallel-plate capacitor. The plate spacing is then increased from L to $2L$. (a) By what factor does the electric potential energy stored in this capacitor change? (b) Energy must be conserved, so where did this extra energy come from (or go to)?

18.5 DIELECTRICS

56. A parallel-plate capacitor has square plates of edge length 1.0 cm and a plate spacing of 0.10 mm. If the gap between the plates is filled with mica, what is the capacitance?

57. Consider a capacitor with the same dimensions as the capacitor in Example 18.9, but now suppose there is air between the plates. What is the maximum safe operating voltage of this capacitor? That is, what is the voltage at which there will be dielectric breakdown?

58. The space between the plates of a capacitor is filled with paper. By what factor does the paper change the capacitance relative to that found when the plates are filled with air?

59. In Example 18.7, we analyzed a parallel-plate capacitor as might be found in an integrated circuit (RAM chip), but we omitted that such a capacitor would usually contain a dielectric between the plates. This dielectric is composed of SiO_2 and has a dielectric constant close to that of glass. What is the capacitance of the RAM capacitor in Example 18.7 when it is filled with SiO_2?

60. ✪ ✪ ® A cell membrane is composed of lipid molecules and is approximately 10 nm thick. If the dielectric constant of a lipid is $\kappa \approx 5$, what is the approximate capacitance of a spherical cell that has a diameter of 10 μm?

61. Consider a parallel-plate capacitor with an area of 1.0 cm², with a plate spacing of 0.20 mm, and filled with mica. At what voltage will this capacitor exhibit dielectric breakdown?

62. ✪ ✪ ® When you walk across a carpeted floor while wearing socks on a dry day, your socks (and hence you) become charged by rubbing with the carpet. When your finger approaches a metal doorknob, you notice that a spark jumps across the air gap between your finger and the doorknob. What is the approximate magnitude of the electric field between your finger and the doorknob just before the spark jumps? What is the potential difference between your finger and the doorknob?

63. ✪ The dielectric in a capacitor is changed to a material with a dielectric constant that is larger by a factor of five. If the charge on the capacitor is held fixed, by what factor does the energy stored in the capacitor change? Explain why the energy is different in the two cases.

64. SSM ✪ A parallel-plate capacitor initially has Mylar between its plates and carries charge $\pm Q$. A different dielectric is then inserted between the plates without changing the charge. If the energy stored in the capacitor decreases to 30% of its initial value, what is the value of κ for the dielectric?

18.6 ELECTRICITY IN THE ATMOSPHERE

65. SSM ✪ ® Experiments have shown that in good weather the Earth has, on average, a negative charge of approximately 10^{-13} C on every square centimeter of surface area. What is the approximate total excess charge on the entire Earth? How many electrons does that correspond to? For simplicity, assume that charge is spread evenly over the entire surface of the Earth, including the oceans.

66. ✪ ® Use the data from Problem 65 to estimate the electric field at the Earth's surface.

67. ® Using the data from Problems 65 and 66, find the approximate electric potential 1.5 m above the Earth's surface (about eye level). Take the ground to be at $V = 0$.

68. Using the data from Problems 65 and 66, find the approximate separation between two equipotential surfaces with $\Delta V = 100$ V near the Earth's surface. Compare this distance to your height.

18.7 BIOLOGICAL EXAMPLES AND APPLICATIONS

69. ✪ ✪ A defibrillator containing a 20-μF capacitor is used to shock the heart of a patient in serious condition by attaching it to the patient's chest. Just prior to discharging, the capacitor has a potential difference of 10,000 V across its plates. (a) What is the energy released into the patient? (b) If the energy is discharged over 20 ms, what is the power output of the defibrillator?

18.8 ELECTRIC POTENTIAL ENERGY REVISITED: WHERE IS THE ENERGY?

70. Suppose the energy stored in the electric field in a particular region of space is 50 J in a region of volume 10 mm³. What is the average electric field strength in this region?

71. A parallel-plate capacitor with C = 10 μF is charged so as to contain 1.2 J of energy. If the capacitor has a vacuum between plates that are spaced by 0.30 mm, what is the energy density (the energy per unit volume)?

72. SSM ✪ Figure P18.72 shows a point charge. At which point (A, B, or C) is the energy density (the energy per unit volume) largest?

73. ✪ If the charge Q in Figure P18.72 is doubled, by what factor does the energy density change at point A?

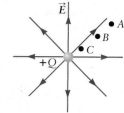

Figure P18.72
Problems 72 and 73.

---

74. ✪ Consider two isolated, charged conducting spheres: a large sphere and a second smaller sphere with a radius four times smaller than that of the large sphere, but with four times as much charge. (a) Compute the ratio of the electric potential at the surface of the large sphere to that of the small sphere. (b) Calculate the ratio (large to small) of the magnitudes of the electric fields at the surface of each sphere.

75. ✪ A proton is directed such that it comes within 1.8×10^{-15} m of a carbon nucleus. (a) How much kinetic energy must the proton have initially to get this close to (i.e., "collide" with) the carbon nucleus? (b) What velocity does that correspond to for the proton? (c) What potential would be needed to give the proton this much kinetic energy? Assume the target carbon nucleus is completely ionized (missing all electrons).

76. ✪ Two metal spheres, each of mass 10 g and initially at rest, are dropped from a height of 5.0 m in an evacuated chamber.

One sphere has a charge of +100 μC, and the other has a charge –100 μC. Find the difference in final speeds of the two spheres (the speeds just before each one hits the ground). Recall that there is an electric field of about 150 V/m near the Earth's surface.

77. **Big cap.** The unit of capacitance is the farad. A capacitor with C = 1 F is a very large one. (a) To get a sense of how big a unit that is, find the area of a hypothetical 1.0-F parallel-plate air capacitor, where the plates are separated by a distance d = 2.0 mm. (b) Find the plate area if we fill the capacitor gap with the excellent dielectric strontium titanate. Give your answers in units of square meters, square kilometers, and square miles.

78. ✪ A homemade capacitor is made from a sandwich of a piece of standard printer paper (8.5 in. by 11 in., 20-lb bond, 0.0038 in. thick) between two equal area sheets of aluminum foil. (a) Calculate the capacitance of this device. (b) What maximum potential can be put across it? (c) How much energy can be stored in it at maximum voltage?

79. ✪ Ⓡ Some of the first capacitors constructed were called Leiden jars. These capacitors were actual jars with a layer of foil on the inside and outside of the glass as shown in Figure P18.79. A conducting sphere topped a metal rod that in turn connected to a dangling chain that made contact with the inner layer of foil, all held in place by a nonconducting lid made of wood or cork. The Leiden jar was developed in 1745 by Pieter van Musschenbroeck of the University of Leiden, the Netherlands, but Benjamin Franklin's experiments with these jars led to an understanding of how energy

© John Jenkins/www.sparkmuseum.com

Figure P18.79

is stored within the jar. The working assumption at the time was that electricity had a fluid nature, so the shape of a jar for storage fit well with this initial hypothesis. One of Franklin's jars measured 4 in. in diameter, with foil on the bottom and up the sides to a height of 9 in. (a) Find the approximate capacitance of his Leiden jar if the wall of the jar is $\frac{1}{8}$ in. thick. (Find the area of foil coverage and assume an equivalent parallel-plate capacitor.) (b) What is the maximum energy that could be stored in such a device?

80. ✪ Benjamin Franklin made use of a "bank" of Leiden jars (Fig. P18.80) for some of his experiments. (a) Are these 35 Leiden jar capacitors arranged in series or in parallel? *Hint:* Close inspection of the photo shows that the central rods of the jars are all connected together. (b) Calculate the maximum capacitance obtained by such an arrangement using the capacitance for one such jar found in Problem 79. (c) What is the maximum amount of energy that can be stored in this bank of jars?

© The American Philosophical Society Library

Figure P18.80 European-made bank of Leiden jars used by Benjamin Franklin in his 1747 experiments.

81. ✪ ⊗ **The electric eel.** The knifefish species *Electrophorus electricus* shown in Figure P18.81 is not really an eel, but it can grow up to 2 m long and generate a potential difference of 600 V between a region just behind its head and its tail. The fish can stun or kill by sending large quantities of charge through its target. The charge comes from Hunter's organ, which in this instance is made up of 70 chains of 4000 stacked cells called electrocytes. Each disk-shaped electrocyte has a diameter of approximately 1.8 mm and a height of about 0.4 mm, and cells are stacked as shown in the detail of Figure P18.81. Assume the oily innards of the fish have a dielectric constant of $\kappa = 105$. (a) Model an electrocyte as a disk-shaped parallel-plate capacitor and find the capacitance of such a cell. (b) If the electrocytes in a chain are stacked so that the capacitance combines in paral-

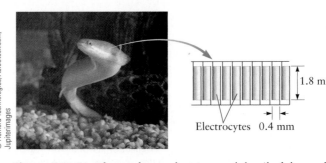

© Hemera Technologies/AbleStock.com/ Jupiterimages

Figure P18.81 *Electrophorus electricus* and detail of the stack of electrocytes in Hunter's organ.

lel and the chains are in turn also combined in parallel, what is the total capacitance of Hunter's organ? (c) Use this model to find how much charge is on the head and tail of the fish when the maximum voltage is generated.

82. ✪ **Big shock.** The world's most advanced capacitor bank (Fig. P18.82), located at the Rossendorf Research Center in Dresden, Germany, can store 50 MJ of energy and release it in less than 5 ms. This high-energy pulse of electricity is used to produce the world's most powerful magnetic fields. (a) If the potential difference at the terminals of the capacitor bank is 22 kV, how much charge is on the collective plates of the capacitor bank? (b) What is the total capacitance of the bank?

Courtesy of Forschungsztrum Dresden-Rossendorf (FZD)

Figure P18.82

83. ✪ (a) A capacitor with C = 15 F is used to store energy. If this capacitor stores an amount of energy equal to the kinetic energy of a baseball ($m = 0.22$ kg) moving at 45 m/s (about 100 mi/h), what is the voltage across the capacitor? (b) If this structure is a parallel-plate capacitor with a plate area of 12 m², what is the electric field between the plates? Assume there is a vacuum between the plates.

84. SSM ✪ ⊗ A useful model of a water molecule is sketched in Figure P18.84, with point charges of 5.2×10^{-20} C at the H sites and a point charge -10.4×10^{-20} C at the O site. The H–O bond length is 9.6×10^{-11} m, and the bond angle is $\theta = 104°$. (a) What is the total electric potential energy of this model of a water molecule? (b) How much work would be required to completely disassociate the molecule?

Figure P18.84

Electric Currents and Circuits

Electric circuits are at the heart of MP3 music players and other devices and are thus an indispensable part of everyday life. (©Andreas Rentz/Getty Images News)

In Chapters 17 and 18, we considered electric forces, fields, and energy in many different situations involving static (stationary) arrangements of electric charge. We now turn our attention to the *motion* of charges and the concept of *electric current*. When charge can flow in a closed path (a "loop"), we describe the path as an electric *circuit*. Electric circuits are at the heart of all modern electronic devices such as televisions, CD players, and computers. In this chapter, we'll also examine some basic properties of circuit elements called resistors and capacitors. Once again, conservation principles will be very useful, and the principles of conservation of energy and charge will lead to general rules for analyzing electric circuits. In Chapter 20, we'll then study how the moving charges that form electric currents give rise to magnetic phenomena.

When charges move from one place to another, we say that there is an **electric current**. The term *current* probably comes from the analogy with the flow of water in a river. The strength of a river's current (Fig. 19.1A) depends on both the speed of the water molecules and the number of water molecules involved in the flow. In a similar way, we can consider the flow of electrons in a wire as sketched in Figure 19.1B. Electrons can move freely through a metal (much like the water in a river), and we will always use the term *wire* to refer to a metal wire.

The strength or magnitude of a current is measured by the amount of water that moves down the river (Fig. 19.1A) or the amount of electric charge that moves along the wire (Fig. 19.1B) in a particular amount of time. Electric current is denoted by the symbol I and defined as the amount of charge Δq that passes a given point on the wire per unit time Δt:

$$I = \frac{\Delta q}{\Delta t} \tag{19.1}$$

For historical reasons, current is defined in terms of net positive charge flow (Fig. 19.1C), even though electrons are responsible for the current in a metal. We'll discuss how to use Equation 19.1 to deal with a current carried by electrons shortly.

Current is measured in units of charge divided by time, or coulombs per second (C/s). This unit has been given the name **ampere** (A), in honor of André-Marie Ampère (1775–1836), who played a major role in discovering of the laws of magnetism. We have

SI unit of electric current: 1 A = 1 C/s $\tag{19.2}$

The ampere is a primary unit in the SI system. Other units involving electricity, such as the coulomb and the volt, are defined in terms of the ampere. For example, the coulomb is defined as the amount of charge carried by a current of 1 A in 1 s.

In Figure 19.1, we have drawn the electric current as confined to a wire (or river). The wire is not a necessary part of the picture, however, because an electric current is present whenever there are moving charges. For the case of a lightning bolt, in which charges move through the atmosphere, we can still discuss and measure the electric current carried by the bolt, even though the moving charges are not confined to a wire.

Which Way Does the Charge Move?

In our definition of current in Equation 19.1, Δq is the *net* electric charge that passes a particular point during a time interval Δt. This amount of charge could consist of a few particles each carrying a large electric charge or of many particles each having a small amount of charge. A current can also be produced by the motion of a combination of positive and negative charges such as a collection of positive and negative ions in solution. By convention, an electric current's direction is defined as shown in Figure 19.2. If the current is carried by positive charges moving with a velocity $\vec{v}$, the direction of the current is *parallel* to $\vec{v}$. On the other hand, if current is carried by negative charges, the direction of I is *opposite* to the charges' velocity. Thus, the direction of the current carried by electrons in a metal wire (an extremely common and important case) is *opposite* to the electron velocity (Fig. 19.2A).

A positive electric current can thus be caused by positive charges moving in one direction *or* by negative charges moving in the opposite direction. A measurement of the current alone cannot distinguish between these two alternatives.[1] Current in a metal is carried by electrons. In a liquid or gas, there will usually be both posi-

Definition of electric current

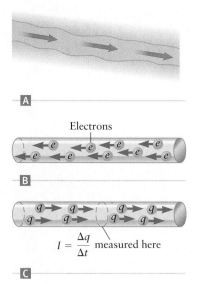

Electrons

$I = \dfrac{\Delta q}{\Delta t}$ measured here

Figure 19.1 **A** When water flows, it forms a current as water molecules move along a river or stream. **B** When electrons move through a wire, they form an electric current as charge moves along the wire. **C** The electric current I is defined as the rate at which net positive charge passes by a point in the wire.

---

[1] A phenomenon called the *Hall effect* does distinguish between these two possibilities. The Hall effect relies on magnetic forces, as we describe in Chapter 20.

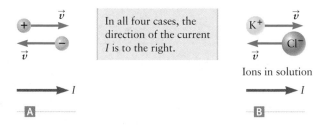

In all four cases, the direction of the current I is to the right.

Ions in solution

Figure 19.2 **A** When a positive charge moves to the right, the current is also directed to the right. When a negative charge moves to the left, the current is to the right. In both cases, the net positive charge on the right increases. **B** In a solution containing positive and negative ions, motion of these ions produces an electric current.

tive and negative mobile ions present; if so, the current is carried by a combination of positive ions moving in one direction and negative ions moving in the opposite direction (Fig. 19.2B).

EXAMPLE 19.1 | How Many Electrons Does It Take to Light a Lightbulb?

To emit a normal amount of light, a typical lightbulb requires a current of about $I = 0.50$ A. If this light is turned on for 1 hour, how many electrons pass through the lightbulb?

RECOGNIZE THE PRINCIPLE

According to the definition of electric current, $I = \Delta q/\Delta t$, so the total amount of charge that passes through the lightbulb in a time Δt is $\Delta q = I \, \Delta t$.

SKETCH THE PROBLEM

Figure 19.1B describes the problem.

IDENTIFY THE RELATIONSHIPS AND SOLVE

Our lightbulb is carrying current (i.e., is turned on) for 1 hour, so $\Delta t = 1$ h $= 3600$ s. We thus find

$$\Delta q = I \, \Delta t = (0.50 \text{ A})(3600 \text{ s}) = 1800 \text{ C}$$

which is the total charge in coulombs. To get the number of electrons, we divide by the magnitude of the electron's charge:

$$N = \frac{\Delta q}{e} = \frac{1800 \text{ C}}{1.6 \times 10^{-19} \text{ C/electron}} = \boxed{1.1 \times 10^{22} \text{ electrons}}$$

What does it mean?

Let's compare this number with the total number of atoms in the filament of a lightbulb (assuming an incandescent bulb). A typical filament is a piece of wire with a radius of about 50 μm and a length of about 1 cm. It is composed of tungsten, which contains approximately 6×10^{28} atoms/m^3. The number of atoms in the filament is thus

$$N_{atoms} = \pi r^2 L (6 \times 10^{28} \text{ atoms/m}^3)$$

$$= \pi (5 \times 10^{-5} \text{ m})^2 (0.01 \text{ m})(6 \times 10^{28} \text{ atoms/m}^3)$$

$$N_{atoms} \approx 5 \times 10^{18} \text{ atoms}$$

So, the number of electrons that pass through the filament during 1 hour is *much larger* that the number of atoms in the filament!

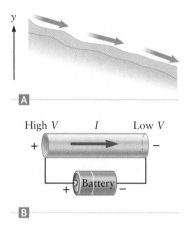

Figure 19.3 Ⓐ Water flows downhill, from a region of high gravitational potential energy to a region of lower potential energy. Ⓑ Electric charges in a wire move from a region of high electric potential energy to a region of lower potential energy. This electric potential difference can be produced by a battery.

19.2 | BATTERIES

Electric Current and Potential Energy

Current is the flow of charge, but how do we produce this charge motion? From our work in mechanics, we know that the current of water in a river (Fig. 19.3A) is connected with gravitational potential energy. The kinetic energy of the moving water is derived from the change in gravitational potential energy as water moves from the high end of the river to the low end.

Electric current is produced in a similar way. For charge to move along a wire, the electric potential energy at one end of the wire must be higher than the electric potential energy at the other end. Suppose the electric current in Figure 19.3B is carried by positive charges moving to the right in the copper cylinder representing the wire. The electric potential energy is related to the potential V by $V = PE_{elec}/q$ (Eq. 18.10). In dealing with current and electric circuits, electrical engineers and physicists often refer to this potential as simply the "voltage." For a positive charge, a region of high potential energy is also a region of high voltage. The current in Figure 19.3B is thus directed from a region of higher voltage to a region of lower voltage, that is, from high to low potential. On the other hand, the current in Figure 19.3B might be carried by electrons (negative charges) moving to the left in the copper wire. Since q is negative, the potential energy PE_{elec} for electrons will be higher at the right end than at the left, and the electrons move from a region of lower voltage to a region of higher voltage in going from right to left in Figure 19.3B. Either way, *the direction of the current I is always from high to low potential*, regardless of whether this current is carried by positive or negative charges.

Figure 19.3B shows the relative voltages at the two ends of the wire using the + and − symbols. The source of this potential difference is often a device called a *battery* (Fig. 19.4A). A typical battery possesses two terminals, labeled "positive" (+) and "negative" (−); these terms refer to the relative electric potential at the terminals. If the battery leads are attached to the ends of a wire as sketched in Figure 19.3B, there will be a potential difference "across" the wire, producing a current.

Construction of a Battery

Alessandro Volta (1745–1827) and his contemporaries discovered and developed the first batteries (see Insight 19.1). Before that time, all studies of electricity used static electricity produced by, for instance, rubbing amber with fur. Although some clever rubbing machines generating quite high electric potentials were constructed, the availability of batteries made possible many new experiments in electricity and magnetism. These experiments and the availability of batteries also led to the first practical applications of electricity such as the telegraph. Batteries illustrate conser-

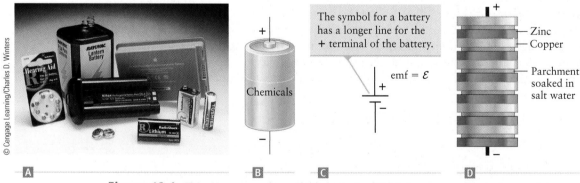

Figure 19.4 Ⓐ Some commonly available batteries. Ⓑ A battery uses chemical potential energy to produce an electric potential difference across its terminals. Ⓒ Circuit diagrams use this symbol to denote a battery. The longer line represents the positive terminal of the battery. Ⓓ Design of the battery constructed by Volta.

vation of energy in an interesting way as they convert chemical energy to electrical energy using *electrochemical reactions*. These reactions are usually not discussed until the end of most introductory chemistry texts, so we will not describe them in detail here. For most of our work, we can take the "black box" view of batteries shown in Figure 19.4B. A typical battery contains two pieces of metal called electrodes, attached by metal wires to the battery terminals. The electrodes are in contact with specially chosen chemicals, and electrochemical reactions involving the electrodes and these chemicals produce an electric potential difference between the two electrodes. Different types of batteries employ different electrochemical reactions and are packaged in many different forms. In drawing an electric circuit, all types of batteries are represented by the symbol shown in Figure 19.4C, with terminals labeled + (positive potential) and − (negative potential).

The potential difference between a battery's terminals is called an *electromotive force*, or *emf* (pronounced "ee-em-eff"). This term was adopted before it was realized that batteries actually produce an *electric potential difference* (and not a force), but unfortunately we are stuck with the term *emf*. When dealing with a potential difference produced by a battery, we follow standard practice and denote this emf by $\mathcal{E}$, and we will often refer to it as simply a voltage. The value of $\mathcal{E}$ produced by a battery depends on the particular chemical reactions it employs and how the electrodes are arranged. Typical values are a few volts, but it is possible to connect batteries in various ways to produce much higher voltages.

Ideal Batteries and Real Batteries

A battery produces an emf, and if we connect the battery terminals to the two ends of a wire (Fig. 19.3B), this emf produces a current. When a battery is attached to a wire as sketched in Figure 19.5, electrons move out of the negative terminal of the battery through the wire and into the positive battery terminal. Hence, charge moves from one battery terminal to the other through this external *circuit*. At the same time, ions inside the battery move between the electrodes as part of the electrochemical reaction we mentioned in connection with Figure 19.4B; this reaction moves charge internally between the electrodes so that no net charge accumulates on the battery terminals while the current is present.

An *ideal battery* has two important properties. First, it *always* maintains a fixed potential difference (a constant emf) between its terminals. Second, this emf is maintained no matter how much current flows from the battery.

This picture of an ideal battery is useful, and we'll employ it in much of our work on electric circuits. *Real* batteries, however, have two practical limitations. One is that the emf decreases when the current is very high. The value of the current at which this decrease becomes significant depends on the battery size and design. The decrease in emf at high current occurs because the electrochemical reactions inside the battery do not happen instantaneously. A certain amount of time is needed for charge to move internally between the electrodes. Another limitation is that a real battery will "run down"; it will not work forever. Under normal operation, there is a certain total amount of charge that can be drawn from the battery, and when this total is reached, the battery ceases to function. In terms of Figure 19.4B, this limit is reached when the chemicals used in the electrochemical reaction are exhausted. At that point, all the available chemical energy has been extracted from the battery.

EXAMPLE 19.2 | Using a Capacitor as a Source of Electric Energy

There are some similarities between a battery and a capacitor. For instance, both can store energy. In a battery, the energy is stored as chemical energy, and in a capacitor, the energy is stored as electric potential energy. In both cases, this energy can be used to produce an electric current. In some applications in which very high currents are

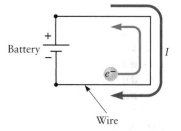

Figure 19.5 Because electrons are negatively charged, their velocity is directed opposite to the current.

needed for a very short time (such as a camera flash), capacitors are preferred over batteries. A typical capacitor used in this way might have a capacitance of 1.0 F and an initial potential difference of 3.0 V between its plates. Compare the energy stored in this capacitor with the energy stored in an AA battery, which is about 7000 J.

RECOGNIZE THE PRINCIPLE

From Chapter 18, we know that the energy stored in a capacitor is given by

$$PE_{elec} = \tfrac{1}{2}C(\Delta V)^2$$

(Eq. 18.33). The values of C and ΔV are given, so we can calculate PE_{elec}.

SKETCH THE PROBLEM

No sketch is necessary.

RECOGNIZE THE PRINCIPLE AND SOLVE

Inserting the given values of capacitance and potential difference leads to

$$PE_{elec} = \tfrac{1}{2}C(\Delta V)^2 = \tfrac{1}{2}(1.0\text{ F})(3.0\text{ V})^2 = \boxed{4.5\text{ J}}$$

What does it mean?
The energy stored in this capacitor is *much* less—by a factor of about 1500—than the amount of energy stored in an AA battery. A very large amount of chemical energy can be stored in a small volume, if one uses the proper chemicals! Although the capacitor stores less energy, its advantage in this application is that its energy can be extracted very quickly, which is essential for a camera flash.

19.3 | CURRENT AND VOLTAGE IN A RESISTOR CIRCUIT

Figures 19.3B and 19.5 show simple electric *circuits*, arrangements in which charge is able to flow around a closed path. In these examples, this path starts at one battery terminal, passes through a wire, continues to the other battery terminal, and then returns to the original terminal as it travels through the interior of the battery. In the previous section, we discussed the part of the path that is within the battery. Now let's focus on the current external to the battery and consider how this current is related to the electric potential difference provided by the battery.

Figure 19.6A shows an atomic-scale view of the charge motion within the wire. Assuming we are dealing with a metal wire, these charges are electrons and their motion is similar to the motion of gas molecules studied in our work on kinetic theory in Chapter 15. The electrons frequently collide with one another and with the stationary ions within the metal, resulting in the zigzag trajectories sketched in Figure 19.6A. Similarly, the molecules in a gas move in a zigzag manner as they collide with other gas molecules (Fig. 19.7A). If no electric field is present, the net result of these zigs and zags is that the average electron displacement after many collisions is zero; there is no net movement of charge and hence no current.

When a battery is connected to the ends of the wire, an electric potential difference is generated between those ends. To this point, we have denoted a potential difference as ΔV, using Δ (the Greek letter delta) to emphasize that we are dealing with a "difference" measured between two locations. When working with circuits, however, it is common practice to simply use V (instead of ΔV). This should cause no confusion as long as we choose a reference point in the circuit at which we take the potential to be zero (our reference potential). For the wire in Figure 19.6B, we can choose this reference point to be at one end of the wire. If the electric potential across the wire has a magnitude V, the electric field within the wire is

MOTION OF ELECTRONS IN A WIRE

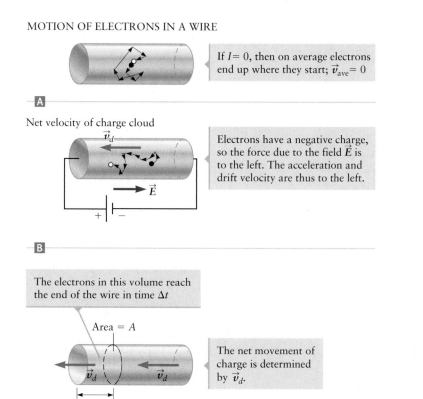

ANALOGY TO GAS MOLECULES

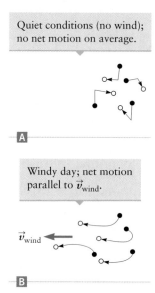

If $I = 0$, then on average electrons end up where they start; $\vec{v}_{ave} = 0$

Net velocity of charge cloud

$\vec{v}_d$

Electrons have a negative charge, so the force due to the field $\vec{E}$ is to the left. The acceleration and drift velocity are thus to the left.

$\vec{E}$

The electrons in this volume reach the end of the wire in time Δt

Area = A

$\vec{v}_d$ $\vec{v}_d$

$\vec{v}_d \Delta t$

The net movement of charge is determined by $\vec{v}_d$.

Quiet conditions (no wind); no net motion on average.

Windy day; net motion parallel to $\vec{v}_{wind}$.

$\vec{v}_{wind}$

Figure 19.7 🅐 Gas molecules in a room follow zigzag trajectories as they collide with other gas molecules, but the net (average) displacement is zero. 🅑 On a windy day, the average velocity of a gas molecule is equal to the wind velocity, even though the instantaneous velocity during the zigs and zags in part A is much greater.

Figure 19.6 The total current in a wire is due to the motion of electrons. 🅐 When $I = 0$, the *average* electron velocity is zero, but the *instantaneous* velocity of any particular electron is not zero. 🅑 An electric field accelerates *all* the electrons in the same direction. 🅒 This acceleration leads to a drift velocity that is the same for all electrons in the wire.

$$E = \frac{V}{L} \tag{19.3}$$

where L is the length of the wire.

Equation 19.3 might seem to contradict Chapter 17, where we argued that the electric field inside a metal is zero. However, the electric field inside a metal is zero only when the charges in the metal are in static equilibrium, which is *not* the case here. For a metal to carry a current, there must be a nonzero electric field inside, and this electric field provides the force that maintains the current.

Drift Velocity and Current

The electric field in the wire in Figure 19.6B produces a force that pushes all the electrons to the left. Hence, the zigzag "swarm" of electrons now moves to the left, and the velocity of this swarm is called the **drift velocity**, $\vec{v}_d$. All electrons in the wire experience this extra "drifting" motion, giving a net movement of negative charge to the left in Figure 19.6B and thus an electric current to the right. The drift motion of the electrons is similar to the motion of gas molecules in a gentle wind. In that case, the molecular motion is a combination of the zigzag trajectories in Figure 19.7A with the wind's velocity, producing a "cloud" of molecules moving as shown in Figure 19.7B. The wind velocity of the molecules is thus similar to the drift velocity of the electrons.

To calculate the electric current in a wire, it is useful to focus on just the drift motion as shown in Figure 19.6C. The current is equal to the amount of charge that passes out the end of the wire per unit time. In a time Δt, all electrons move (on average) a distance $v_d \Delta t$, so all electrons that are within this distance from the left end will leave the wire in a time Δt. If the density of electrons per unit volume is n, the number of electrons that leave is equal to the number of electrons in the gray volume in Figure 19.6C. For a wire with cross-sectional area A, the number of electrons that exit the wire is thus

$$N = nAv_d\,\Delta t \tag{19.4}$$

Each electron carries a charge $-e$, so the total charge that leaves the wire is

$$\Delta q = N(-e) = -nAev_d\,\Delta t \tag{19.5}$$

The current is given by $I = \Delta q/\Delta t$, so we can divide both sides of Equation 19.5 by a factor of Δt to get

$$I = -neAv_d \tag{19.6}$$

Ohm's Law

Equation 19.6 tells how the current is related to the drift velocity, but what we really want to know is how I is related to the potential difference across the ends, that is, to the battery's emf. To answer this question, we must consider the drift velocity v_d in a little more detail. Recall our discussion of the drag forces on a bacterium in a fluid (Chapter 3). An electron moving through a metal (Fig. 19.6A) is similar to a bacterium moving through a fluid. The bacterium undergoes many collisions with fluid molecules, just as electrons experience many collisions in the course of their zigzag trajectories. In both cases, there is a drag force[2] that is directed opposite to the velocity

$$\vec{F}_{\text{drag}} = -b\vec{v}$$

In Chapter 3, we showed that this drag force leads to a terminal velocity proportional to the force that pushes the *E. coli* through a fluid (Eq. 3.25). In the same way, the drag force on electrons leads to a drift velocity proportional to the force pushing the electrons through the metal. This electric force is proportional to E, so the drift velocity is proportional to the magnitude of the electric field

$$v_d \propto E \tag{19.7}$$

According to Equation 19.3, E is itself proportional to the voltage across the wire, so the drift velocity is also proportional to V:

$$v_d \propto V \tag{19.8}$$

Inserting Equation 19.8 into our expression for the current (Eq. 19.6) gives

$$I \propto V \tag{19.9}$$

The constant of proportionality between I and V in Equation 19.9 involves a quantity called the electrical **resistance** of the wire R, which is defined by

Ohm's law

$$I = \frac{V}{R} \tag{19.10}$$

This relation between resistance, current, and voltage is called **Ohm's law**, and it applies to a very wide range of materials. According to Equation 19.10, R can be measured in units of voltage/current = volts/amperes = V/A. This combination of units has the name "ohm" and is denoted by the symbol Ω.

The value of a wire's resistance depends on its composition, size, and shape. It is useful to define a quantity ρ called the **resistivity**, where the value of ρ depends on the material used to make the wire, but not on its size or shape. The resistance of a cylindrical wire of length L and cross-sectional area A (Fig. 19.8) is related to the resistivity by

$$R = \rho\frac{L}{A} \tag{19.11}$$

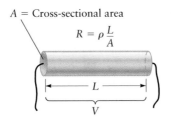

A = Cross-sectional area

$$R = \rho\frac{L}{A}$$

Figure 19.8 The resistance R of a cylindrical wire is related to the resistivity ρ by $R = \rho L/A$.

[2]See Chapter 3 and Equation 3.23.

Table 19.1 Resistivities of Some Common Materials at Room Temperature

MATERIAL	$\rho \ (\Omega \cdot m)$
Metals:	
Copper	1.7×10^{-8}
Aluminum	2.7×10^{-8}
Gold	2.2×10^{-8}
Silver	1.6×10^{-8}
Lead	22×10^{-8}
Insulators:[a]	
Glass	1 to 1000×10^{9}
Rubber	1 to 100×10^{12}
Semiconductors[a]	
Silicon	0.1 to 100
Germanium	0.001 to 1

[a]The resistivities of insulators and semiconductors depend strongly on their purity.

The value of ρ is different for different metals; copper has a different resistivity than aluminum or gold, for example. Values of ρ can also depend on the metallurgical history (how the particular metal is heat treated, etc.). Typical resistivities of some common materials are listed in Table 19.1.

Ohm's law (Eq. 19.10) predicts a linear relationship between current and voltage. However, even though it is called a "law," it is not a fundamental law of physics in the sense of Newton's laws of mechanics. While the relation between electric current and voltage for many materials (including metals) is accurately described by Ohm's law, there are some important exceptions. For example, devices called diodes and transistors do not obey Ohm's law. These devices are essential for making amplifiers, electronic switches, and computer memory elements, and are thus extremely important in many applications. Even though Ohm's law does not apply to every type of circuit element, the lessons we'll learn from studying resistors and Ohm's law in this chapter are the foundation of all circuit theory.

EXAMPLE 19.3 | The Resistance of a Piece of Copper Wire

The wires commonly used in household wiring are made of copper and have a typical radius $r = 1.0$ mm. (a) Find the resistance of a piece of this wire that has a length $L = 5.0$ m. (b) Assume a battery with an emf of 12 V is attached to the ends of this wire. What is the current?

RECOGNIZE THE PRINCIPLE

We can use the relation between resistance and resistivity, $R = \rho L/A$, to find the resistance. Given the voltage across the wire, Ohm's law then gives the current.

SKETCH THE PROBLEM

Figure 19.8 describes the problem.

IDENTIFY THE RELATIONSHIPS AND SOLVE

(a) We use $R = \rho L/A$ (Eq. 19.11) to find the resistance of the wire. The wire's cross-sectional area is $A = \pi r^2 = 3.1 \times 10^{-6}$ m^2. Using this area and the length of the wire along with the resistivity of copper from Table 19.1, we find

$$R = \rho \frac{L}{A} = (1.7 \times 10^{-8} \; \Omega \cdot \text{m}) \frac{5.0 \text{ m}}{3.1 \times 10^{-6} \text{ m}^2} = \boxed{0.027 \; \Omega}$$

(b) The corresponding current through this wire can be found from Ohm's law (Eq. 19.10):

$$I = \frac{V}{R} = \frac{12 \text{ V}}{0.027 \; \Omega} = \boxed{440 \text{ A}}$$

What have we learned?
We'll compare these results for resistance and current to other common cases shortly. However, you should suspect that the result in part **(b)** is a very large current.

Wires and Resistors

Figure 19.9 shows a photo of some common types of electrical resistors, also referred to as simply *resistors*. The metal wires we considered above are types of resistors, but resistors that follow Ohm's law can be made in other shapes and from other materials. Each of these objects has a particular resistance R that determines how the current through and voltage "across" the resistor are related. If we connect the two ends of a resistor to the terminals of a battery, the voltage V across the resistor is equal to the battery's emf, hence, $V = \mathcal{E}$. Inserting into Ohm's law (Eq. 19.10), we find a current $I = \mathcal{E}/R$ through the resistor. This electric circuit is represented schematically in the circuit diagram in Figure 19.10A. The battery is represented using the symbol ⊣⊢ as shown in Figure 19.4C, and the resistor is represented in a circuit diagram by the symbol —W—. By convention, these symbols represent batteries and resistors in circuit diagrams. Table 19.2 lists the symbols for these and several other common circuit elements.

The straight lines in Figure 19.10A represent metal wires that attach the resistor to the battery's terminals. (The wires may not be straight in the actual circuit as shown in Fig. 19.10B, but bending a wire in this way does not significantly affect its resistance.) Metal wires are commonly referred to as simply "wires" or "leads." We saw in Example 19.3 that a metal wire has a nonzero resistance. Hence, strictly speaking, we should also refer to a wire as a resistor. However, the value of R for a typical metal wire is *much smaller* than for the resistors used in electronic devices, so when analyzing circuits it is a very good approximation to assume all wires are ideal having $R = 0$. We can account for the small but nonzero resistance of a real wire when necessary.

Dots in a circuit diagram represent connections between different circuit elements. In Figure 19.10A, they represent the points at which the wires attach to the battery, and where the wires attach to the resistor. It is sometimes useful to show the dots when one wants to be clear about where the different parts of a circuit begin and end. When that is not important or is obvious, it is usually simpler to omit the dots.

Speed of an Electric Current

If we attach a battery to a metal wire at time $t = 0$, when will electrons start to "flow" out of the end of the wire? From Figure 19.6, you might expect that this time is determined by the drift velocity, but that is *not* correct. To see why, let's

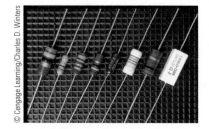

Figure 19.9 Resistors come in a wide variety of shapes and sizes. Larger ones are generally able to carry higher currents without overheating.

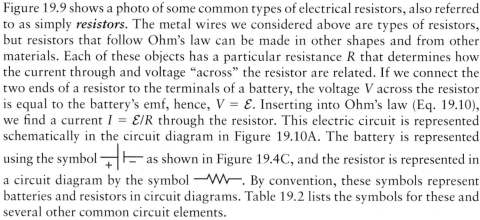

A

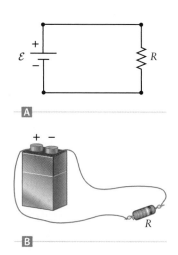

B

Figure 19.10 **A** Circuit diagram for a battery connected to a resistor. **B** Sketch of the actual physical circuit represented by the diagram in **A**.

Table 19.2 Symbols for Some Commonly Used Circuit Elements

CIRCUIT ELEMENT	SYMBOL	ACTUAL APPEARANCE	CIRCUIT ELEMENT	SYMBOL	ACTUAL APPEARANCE
Battery	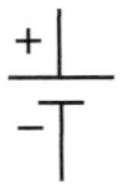	 © Cengage Learning/ Charles D. Winters	Lightbulb		 © David Graeme-Baker/Alamy
Resistor		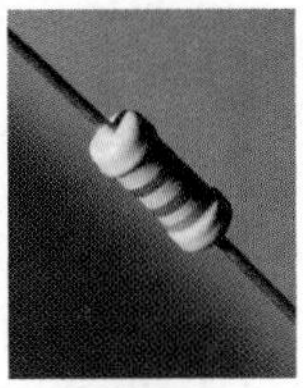 ©INSADCO Photography/Alamy	Wire		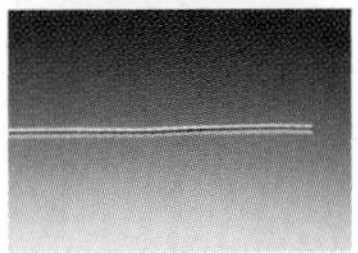 © Cengage Learning/ Charles D. Winters
Capacitor		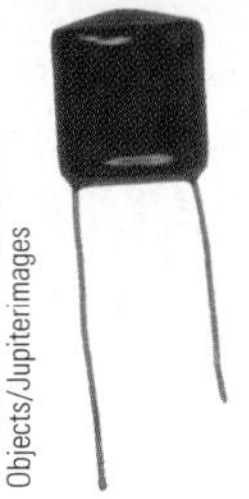 © Hemera Technologies/Photo Objects./Jupiterimages	Connection between wires		 © Cengage Learning/ Charles D. Winters
Switch	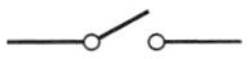	 © Hemera Technologies/Photo Objects./Jupiterimages	Electrical ground	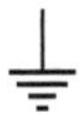	

calculate v_d in a typical case. The relation between the current and the drift velocity is (Eq. 19.6)

$$I = \frac{\Delta q}{\Delta t} = -neAv_d$$

Solving for v_d leads to

$$v_d = -\frac{I}{neA} \tag{19.12}$$

For a household wire such as the copper wires that connect to a lightbulb, the current is typically $I = 1$ A. The cross-sectional area is about $A = 3 \times 10^{-6}$ m^2 (a cylindrical wire with a radius of 1 mm), and the density of electrons in copper is about[3] 2×10^{26} electrons/m^3. Inserting these values into Equation 19.12 gives

$$v_d = -\frac{I}{neA} = -\frac{1 \text{ A}}{(2 \times 10^{26} \text{ electrons/m}^3)(1.60 \times 10^{-19} \text{ C})(3 \times 10^{-6} \text{ m}^2)}$$

$$v_d = -0.01 \text{ m/s}$$

where the negative sign means that the drift velocity is directed opposite to the current since the electron has a negative charge. The magnitude of this drift velocity is quite small; you can easily walk at a speed greater than v_d. A typical wire can

---

[3]This value assumes there is one current-carrying electron for each copper atom in the wire, which is a good approximation—accurate to within a factor of two or so—for copper and most metals.

be many meters long, so electrons moving at this drift velocity take many seconds to complete their path through a circuit. You might therefore expect to notice a considerable delay from the moment you "throw a switch" to the time a lightbulb begins to emit light.

You have no doubt done this experiment, so you know that there is in fact *no* perceptible time delay from the time you push a switch and a light turns on. How do we reconcile this observation with the very small value of the drift velocity? Consider an analogy with the flow of water through a garden hose when the hose is initially full of water. When water from a faucet enters one end of the hose, water leaves the other end almost immediately. Similarly, when the first electrons enter one end of a wire, other electrons leave the other end almost immediately. The first electrons travel through the entire length of the wire during a time determined by the drift velocity, but charges begin to leave the opposite end of the wire almost instantaneously once a potential difference is established across the ends of the wire. In fact, the speed of an electric current is equal to the speed of electromagnetic radiation in the wire, which is nearly the speed of light in a vacuum. (We'll return to this subject in Chapter 23.)

19.4 | DC CIRCUITS: BATTERIES, RESISTORS, AND KIRCHHOFF'S RULES

An electric circuit is a combination of connected elements such as batteries and resistors forming a complete path through which charge is able to move. Calculating the current in a circuit is a job electrical engineers call *circuit analysis*. In this chapter, we consider only the simplest "DC" circuits. The letters DC stand for "direct current," in contrast to circuits with an "AC" or "alternating current." Roughly speaking, an AC circuit is one in which the current oscillates at a particular frequency, while all other cases fall into the category of DC circuits. The current in a DC circuit is constant (independent of time) or approaches a constant if we wait long enough. Our main goal for the rest of this chapter is to learn how to analyze DC circuits.

What Is a "Circuit"?

In the circuit in Figure 19.10, two wires connect the battery to the resistor. Why do we need two wires? Why can't we carry charge from the battery to the resistor with a single wire? The reason two wires are required can be understood from analogy with the "water pipe circuit" in Figure 19.11, in which a water pump is used to push water through a pipe. This pipe carries water from the outlet (high-pressure) side of the pump and returns it to the inlet (low-pressure side), and the water current has the same magnitude throughout the circuit. This water circuit could be broken by inserting a wall into the pipe, which prevents water from flowing around the circuit and makes the current zero. Notice that the current is zero on *both* sides of the wall. The pressure produced by the water pump is now felt as a pressure difference between the two sides of the wall.

The same ideas apply to the electric circuit in Figure 19.11C. The electric current can be viewed as the motion of positive charge[4] out the high-potential (positive) terminal of the battery; this charge travels around the circuit passing through the resistor and returns at the low-potential (negative) terminal of the battery. As with the water circuit, the electric current is the same at all points in the circuit: I has the *same* value when this charge is "heading into" and "returning from" the resistor. If we now cut the wire (Fig. 19.11D), charge cannot return to the battery and we have an **open circuit**. The current is now zero *throughout the circuit*, and the potential

---

[4]For a metal wire, the charge would be carried by electrons (negative charges) going in the opposite direction, but that does not change our argument or conclusions.

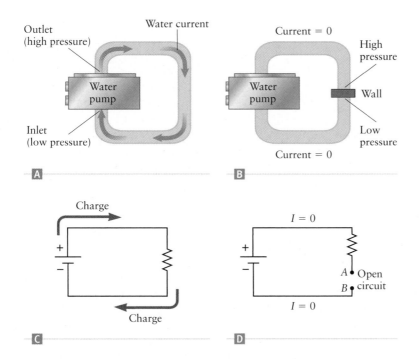

Figure 19.11 **A** The pressure difference produced by a water pump can cause a current of water around a closed loop of pipe. **B** If a wall is inserted into the pipe, the current of water *everywhere* stops. A pressure difference then exists across the wall. **C** A battery uses an emf to produce a current of charge around a complete circuit. **D** If a wire in the circuit is cut, the current stops *everywhere* in the circuit.

difference generated by the battery appears across the open circuit (points *A* and *B* in Fig. 19.11D), just as the pressure difference from the water pump appears across the wall in Figure 19.11B.

The important message from Figure 19.11 is that *there must be a complete circuit for charge flow.* Simply connecting one wire from a battery to a resistor cannot produce a current. *There must always be a return path* for the current to return to the voltage source. This fact is illustrated in Figure 19.12A, which shows a bird sitting on an electric power line. A power line acts as a voltage source for your house and is at high potential. When a bird comes in contact with a power line, the bird is also at a high potential. If the bird does not simultaneously touch the ground, however, there is no return path to ground through the bird, and the current through the bird is negligible. We thus have essentially an open circuit with no current (Fig. 19.12B). As we'll discuss in Section 19.8, the health hazards of electricity are determined by the current; as long as very little current flows through the bird, its health is not endangered.

One-Loop Circuits: Kirchhoff's Loop Rule

The circuit in Figure 19.13 shows a single ideal battery with an emf $\mathcal{E}$ connected to a single resistor with resistance R. From Ohm's law (Eq. 19.10), the current in this circuit is

$$I = \frac{\mathcal{E}}{R} \tag{19.13}$$

In words, we say that the battery produces an electric potential difference $\mathcal{E}$ "across" the resistor. We also refer to $\mathcal{E}$ as the "voltage across the resistor."

The electric current in Figure 19.13A consists of charges that travel along a complete loop, passing through the battery, through the attached wires and resistor, and back to the battery. Let's consider the electric potential energy of a test charge q as it moves through the circuit. Electric potential energy is related to the potential by $PE_{\text{elec}} = qV$, so we are really asking how the electric potential varies around the circuit. Suppose the test charge begins at the negative battery terminal. We can use this spot as the reference point for the potential and thus assign $V = 0$ at this point. We also recall the convention that $V = 0$ at an electrical "ground," so we call that our "ground point." In a circuit diagram, ground is indicated by the triangular symbol indicated in Figure 19.13B. Our charge q thus begins at the lower left in the figure

Figure 19.12 If a bird on a power line does not touch the ground, the circuit is not completed and the current is zero. The bird is spared.

Current through a resistor

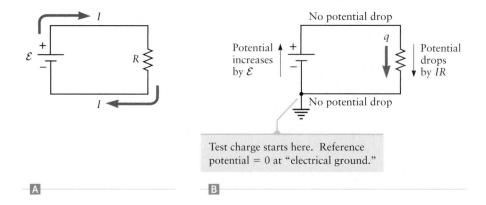

Figure 19.13 A This battery produces a clockwise current through the resistor. B The electric potential changes as we move from one part of the circuit to another. The sum of all these voltage changes must be zero when we move completely around the circuit.

with a potential $V = 0$ and then travels through the battery, where the electrochemical reaction increases its potential by an amount $\mathcal{E}$, the battery's emf. When the charge reaches the wire on the top edge of the circuit, it thus has a potential $\mathcal{E}$. We assume the wires in this circuit are ideal, so their resistance is zero and the potential is the same everywhere within this top wire. The potential at the "top" end of the resistor is thus still $\mathcal{E}$. The charge q then passes through the resistor, moving from a point of high potential (the top end of the resistor) to a point of lower potential (the bottom of the resistor). According to Equation 19.10, the potential *decreases* by IR, so the voltage "drops" an amount IR as the charge moves between these two ends of the resistor. The final step takes the charge through the bottom wire back to the battery, and because it is another ideal resistanceless wire, there is no potential change along this wire.

At the completion of these four steps, the charge has returned to its starting point. The net change of the potential energy and hence also the net change in the electric potential must be *zero* when the charge returns to where it began, at the bottom of the battery. We thus have

$$\underset{\substack{\text{change in the} \\ \text{potential around} \\ \text{the circuit}}}{} = \underset{\substack{\text{potential increase} \\ \text{from the battery}}}{+\mathcal{E}} - \underset{\substack{\text{potential drop} \\ \text{across the resistor}}}{IR} = 0 \qquad (19.14)$$

This leads directly to

$$I = \frac{\mathcal{E}}{R}$$

which is just Equation 19.13.

This simple example shows that conservation of energy is at the heart of circuit analysis. The steps in Figure 19.13B amount to the statement that *the change in potential energy of a charge as it travels around a complete circuit loop must be zero*. This statement is known as **Kirchhoff's loop rule.** Because the electric potential energy of a charge is proportional to the potential ($PE_{\text{elec}} = qV$), Kirchhoff's loop rule also means that the change in the electric potential around a closed-circuit path is zero. Kirchhoff's loop rule is one of our main tools for analyzing all types of circuits.

Kirchhoff's loop rule

Dissipation of Energy in a Resistor

In Figure 19.13, the test charge gained energy (an increase in potential) when it passed through the battery and lost energy (a decrease in potential) as it passed through the resistor. If the potential energy of a charge decreases when it moves through a resistor, where does this energy go? The answer is that the energy of the charge is converted to heat energy inside the resistor. We say that this energy is *dissipated* as heat. In Figure 19.13B, the potential of the charge q decreases (i.e., drops) an amount $V = IR$ as it travels through the resistor. The energy of this charge thus decreases by an amount

$$\text{energy decrease} = qV$$

From our definition of current, $q = I\,\Delta t$, so

$$\text{energy decrease} = (I\,\Delta t)V$$

This energy shows up as heat energy in the resistor; hence,

$$\text{dissipated energy} = (I\,\Delta t)V$$

The amount of energy dissipated per unit time is the dissipated *power* P, so we get

$$P = \frac{\text{dissipated energy}}{\Delta t} = VI \qquad (19.15)$$

Power dissipated in a circuit element

Using our relation between voltage, current, and resistance ($I = V/R$, Eq. 19.10), Equation 19.15 can be written as

$$P = I^2 R = \frac{V^2}{R} \qquad (19.16)$$

Power dissipated in a resistor

When there is current in a resistor, a certain amount of power given by Equations 19.15 and 19.16 is dissipated as heat. This energy comes from the chemical energy of the battery. So, the circuit in Figure 19.13 converts chemical energy (in the battery) to heat energy (in the resistor). We should also note that Equation 19.15 applies to *all* circuit elements, not just resistors. The power dissipated or transferred from a current I to a circuit element is always equal to the product VI, where V is the voltage across the circuit element.

> **Insight 19.2**
> **ENERGY DISSIPATED IN A RESISTOR**
> The energy dissipated in a resistor is converted to heat, showing up as a temperature increase of the resistor and its surroundings. It is similar to the conversion of mechanical energy into heat via friction.

Resistors in Series

Figure 19.14A shows a slightly more complicated circuit with a battery connected to two resistors. This arrangement, in which the current passes first through one resistor (R_1) and then another (R_2), is called *resistors in series*. All the charge that flows from the battery must pass through both resistors consecutively. Let's apply Kirchhoff's loop rule to this circuit, starting from a reference point (ground) at the negative terminal of the battery, where $V = 0$. If a charge q begins there and travels clockwise around the circuit, its potential first increases by $\mathcal{E}$ as it passes through the battery. The potential then decreases by an amount IR_1 as it passes through the first resistor and decreases by IR_2 when it passes through the second resistor. According to Kirchhoff's loop rule, the sum of these changes around the closed circuit loop must be zero:

$$
\begin{array}{ccccc}
& \overbrace{\text{potential increase}}^{\text{in battery}} & \overbrace{\text{potential drop}}^{\text{across } R_1} & \overbrace{\text{potential drop}}^{\text{across } R_2} & \\
\begin{array}{c}\text{change in the}\\ \text{potential around} =\\ \text{the circuit}\end{array} & +\mathcal{E} & - \quad IR_1 & - \quad IR_2 & = 0
\end{array}
$$

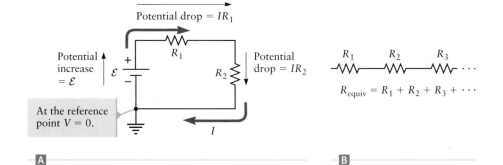

Figure 19.14 A To apply Kirchhoff's loop rule, we add up all the changes in electric potential as we move around the circuit. B When resistors are connected in series, they are equivalent to a resistor of value R_{equiv} as given in Equation 19.18.

Solving for the current, we find

$$I(R_1 + R_2) = \mathcal{E}$$

$$I = \frac{\mathcal{E}}{R_1 + R_2} \tag{19.17}$$

For the circuit with a single resistor (Fig. 19.13), we found $I = \mathcal{E}/R$ (Eq. 19.14). The combination of two resistors in Figure 19.14 is thus equivalent to a single resistor with a resistance of $R_{equiv} = R_1 + R_2$. In general, we can connect any number of resistors in series as in Figure 19.14B and they will be equivalent to a single resistor with

$$R_{equiv} = R_1 + R_2 + R_3 + \cdots \quad \text{(resistors in series)} \tag{19.18}$$

This result will be very useful when we analyze more complex circuits.

CONCEPT CHECK 19.1 | Potential in a One-Loop Circuit

Consider the one-loop circuit in Figure 19.15. What is the potential at point A?
 (a) $V_A = \mathcal{E}$ (b) $V_A = -\mathcal{E}$ (c) $V_A = 0$

CONCEPT CHECK 19.2 | Potential in a One-Loop Circuit (Revisited)

One of the wires in Figure 19.15 is cut, resulting in the open circuit shown in Figure 19.16. What is the potential difference between points B and C?
 (a) $V_B - V_C = \mathcal{E}$ (b) $V_B - V_C = -\mathcal{E}$ (c) $V_B - V_C = 0$

Combining resistors in series

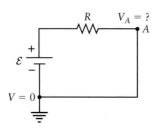

Figure 19.15 Concept Check 19.1.

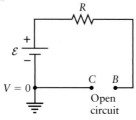

Figure 19.16 Concept Check 19.2.

Insight 19.3

WHAT IS AN "EQUIVALENT" CIRCUIT?

When we say that a resistor R_{equiv} is "equivalent" to a certain arrangement of several resistors, we mean that if that arrangement of resistors is replaced by R_{equiv}, the current through the rest of the circuit is unchanged. This notion of equivalence applies to many types of circuit elements, including capacitors.

| **EXAMPLE 19.4** | **Batteries in Series** |

Consider the circuit in Figure 19.17A, in which two batteries with emfs $\mathcal{E}_1$ and $\mathcal{E}_2$ are connected in series to a single resistor. Use Kirchhoff's loop rule to find the current in this circuit.

RECOGNIZE THE PRINCIPLE

To apply Kirchhoff's loop rule, we start at a particular point in the circuit (the reference point at ground is a convenient choice) and move around the circuit, adding up the changes in the electric potential. According to Kirchhoff's loop rule, the total change in the potential around one complete loop is zero.

SKETCH THE PROBLEM

Figure 19.17B shows how the potential changes as we move around the circuit.

IDENTIFY THE RELATIONSHIPS

We start at the reference ground point, at the negative terminal of battery 1, and move clockwise in Figure 19.17B. There is an increase $\mathcal{E}_1$ in the potential when we pass through battery 1 and another increase $\mathcal{E}_2$ as we pass through battery 2, followed by a potential drop of magnitude IR as we move through the resistor.

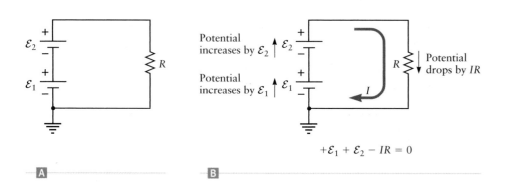

$$+\mathcal{E}_1 + \mathcal{E}_2 - IR = 0$$

Figure 19.17 Example 19.4. Combining batteries in series.

SOLVE

According to Kirchhoff's loop rule, the sum of these changes in potential must be zero:

$$\text{change in potential around the circuit} = \mathcal{E}_1 + \mathcal{E}_2 - IR = 0$$

Solving for the current, we find

$$I = \frac{\mathcal{E}_1 + \mathcal{E}_2}{R}$$

What does it mean?

This combination of two *batteries in series* is equivalent to a single battery with emf $\mathcal{E}_{equiv} = \mathcal{E}_1 + \mathcal{E}_2$.

Multibranch Circuits

Figure 19.18 shows a circuit in which the current from the positive battery terminal can return to the negative terminal along either of two paths: one path goes through resistor R_1, and the other path is through R_2. These different current paths are called *branches*, and this is an example of a multibranch circuit. The particular arrangement in Figure 19.18 is also referred to as *resistors in parallel*. The currents through these different branches need not be equal (unless $R_1 = R_2$), so now we have to solve for two unknown currents. In Figure 19.18A, we label the current through resistor 1 as I_1 and the current through resistor 2 as I_2. The current through the other parts of the circuit (the wires that connect to the battery terminals) is I_3. These three currents are not independent. Figure 19.18A shows expanded views of the points where these three currents are connected. Such points are called circuit "junctions" or "nodes." Electric charge cannot be created or destroyed at a junction (or anywhere else), so *the amount of current entering a junction must be equal to the amount of current leaving it.* This fact is called *Kirchhoff's junction rule*, and, for this particular circuit, it means that

Kirchhoff's junction rule

$$I_3 = I_1 + I_2 \qquad (19.19)$$

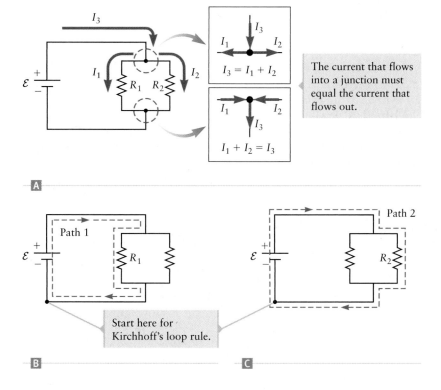

Figure 19.18 ⓐ A circuit with two resistors connected in parallel. According to Kirchhoff's junction rule, the total current entering a junction must equal the total current leaving the junction. ⓑ and ⓒ Kirchhoff's loop rule still holds, so the sum of all changes in potential along any complete path (either path 1 or path 2) around a circuit must be zero.

The current that flows into a junction must equal the current that flows out.

Start here for Kirchhoff's loop rule.

Insight 19.4

APPLYING KIRCHHOFF'S LOOP RULE

In all our applications of Kirchhoff's loop rule so far, we have taken the starting point at the ground point of the circuit, but that is not absolutely necessary. Because we add up changes in the potential as we make a complete loop around the circuit, we could take any point as a starting point without affecting the final result.

To analyze this circuit and solve for all the currents, we must also apply Kirchhoff's loop rule. We start, as usual, from the negative terminal of the battery and sum the changes in electric potential as we move around a closed loop. Beginning from the negative battery terminal, two closed loops take us around this circuit and back to the battery as shown in parts B and C of Figure 19.18. Following the dashed loop in part B (path 1) in which we move through resistor 1, the changes in potential are

$$\text{change in potential around path 1} = \overbrace{+\mathcal{E}}^{\text{battery emf}} - \overbrace{I_1R_1}^{\text{potential drop across } R_1} = 0 \tag{19.20}$$

The dashed loop in Figure 19.18C (path 2) takes us through resistor 2, and the changes in potential in this case are

$$\text{change in potential around path 2} = \overbrace{+\mathcal{E}}^{\text{battery emf}} - \overbrace{I_2R_2}^{\text{potential drop across } R_2} = 0 \tag{19.21}$$

Solving for the two branch currents I_1 and I_2, we get

$$I_1 = \frac{\mathcal{E}}{R_1} \qquad I_2 = \frac{\mathcal{E}}{R_2} \tag{19.22}$$

The total current through the battery is then (according to Eq. 19.19)

$$I_3 = I_1 + I_2 = \frac{\mathcal{E}}{R_1} + \frac{\mathcal{E}}{R_2} \tag{19.23}$$

Kirchhoff's Rules

Kirchhoff's two rules are at the heart of *all* problems in circuit analysis. These two rules can be summarized as follows.

Kirchhoff's rules for circuit analysis

1. *Kirchhoff's loop rule:* **The total change in the electric potential around any closed circuit path must be zero.**
2. *Kirchhoff's junction rule:* **The current entering a circuit junction must equal the current leaving the junction.**

Kirchhoff's rules are not new laws of physics, but are actually applications of more fundamental ideas. The loop rule follows from the principle of conservation of energy, while the junction rule is true because charge is conserved. These two general and fundamental conservation principles lead directly to Kirchhoff's very practical rules for circuit analysis.

So far, we have discussed Kirchhoff's rules using DC circuits containing batteries and resistors, but these rules apply to *all* types of circuits involving *all* types of circuit elements. The general approach to circuit analysis can be summarized as follows.

PROBLEM SOLVING General Rules for Circuit Analysis

1. **RECOGNIZE THE PRINCIPLE.** All circuit analysis is based on Kirchhoff's loop and junction rules.
2. **SKETCH THE PROBLEM.** Start with a circuit diagram and identify all the different circuit branches. The currents in each of these branches are usually the unknown variables in the problem.
3. **IDENTIFY** the different possible closed loops and junctions in the circuit.
   - Apply Kirchhoff's loop rule to each closed loop to get a set of equations involving the branch currents (such as I_1, I_2, and I_3 in Fig. 19.18) and quantities such as the battery emfs and the resistances in each loop.

- Apply Kirchhoff's junction rule to each junction to get equations relating the different branch currents.

4. **SOLVE** the resulting equations for the current in each branch.

5. Always *consider what your answer means* and check that it makes sense.

When we analyzed the circuit in Figure 19.18, we had three unknowns, the current in the three branches I_1, I_2, and I_3. To solve for these three unknowns, we needed three equations, which we obtained by applying the junction rule once and the loop rule twice. This circuit contains two junctions at the top and bottom of resistors R_1 and R_2, though. Could we apply the junction rule a second time to get another equation? Indeed, we could apply the junction rule to both of these junctions, obtaining the equations $I_3 = I_1 + I_2$ (Eq. 19.19) for the upper junction and $I_1 + I_2 = I_3$ for the lower junction. These two equations are identical, so the second application of the junction rule gave no new mathematical information. In general, if a circuit contains N junctions, the junction rule can be applied $N - 1$ times to obtain $N - 1$ independent equations.

Let's also consider how many times we can apply the loop rule to the circuit in Figure 19.18. Figure 19.19 shows three possible loops around the circuit, and next to each path is the resulting loop equation. We used the loop equations from paths 1 and 2 in our previous analysis of this circuit (Eqs. 19.20 and 19.21). The third loop equation—obtained using path 3 in Figure 19.19—is just the difference of the loop equations for paths 1 and 2. Hence, this third loop equation is not an independent equation. In general, a loop equation will be independent of other loop equations if it contains at least one circuit element that is not involved in the other loops.

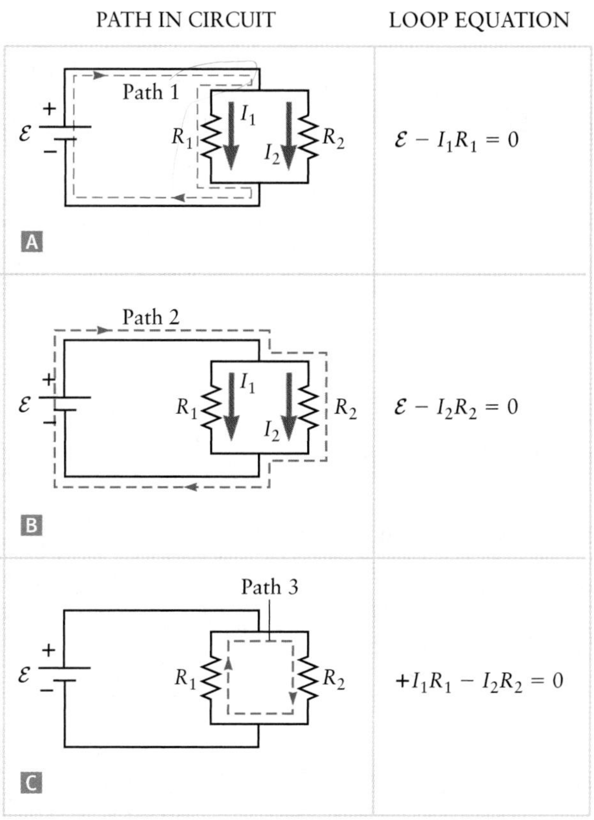

PATH IN CIRCUIT	LOOP EQUATION
A	$\mathcal{E} - I_1 R_1 = 0$
B	$\mathcal{E} - I_2 R_2 = 0$
C	$+I_1 R_1 - I_2 R_2 = 0$

Figure 19.19 Kirchhoff's loop rule can be applied for all three of these loops around the circuit. The loop equation in **C** is the difference of the loop equations in **A** and **B**, however, so there are only two independent loop equations.

Using Kirchhoff's Rules

Circuit elements can be connected in many different ways, so many different conceivable loops and branches are possible. While all such circuits can be attacked and solved using the above "General Rules for Circuit Analysis," it is useful to keep the following points in mind.

When applying Kirchhoff's loop rule, you must pay attention to the sign of the voltage drop across a circuit element. For example, when navigating clockwise around the loop in Figure 19.20A, the potential *drops* as we move through the resistor. The potential change across the resistor is thus $-IR$ because we are moving in a direction parallel to the direction of the current I. On the other hand, when applying Kirchhoff's loop rule, we are also free to move counterclockwise around the loop (Fig. 19.20B). The potential change in moving through the resistor is then $+IR$ because we are now moving from the low-potential end to the high-potential end. When writing Kirchhoff's loop equation, we can go around a loop in either direction; changing this direction changes the sign of every term in Kirchhoff's loop equation, so we are left mathematically with the same equation.

The signs of the potential changes in Figure 19.20 depend on the assumed direction of the current. Here and in our other examples with Kirchhoff's loop rule, we assume a direction for the current and stick with it as we write the loop rule equations. After deriving these equations, we can solve for the value of I. If I is found to be positive, our assumed direction for the current was correct, whereas if I has a negative value, the current is opposite to our assumed direction.

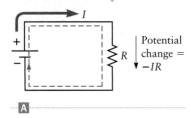

Potential drops by *IR* when moving with the current.

A

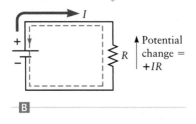

Potential increases by *IR* when moving against the current.

B

Figure 19.20 **A** The potential *drops* when passing through a resistor in the direction parallel to the current. **B** The potential *increases* when moving through a resistor in the direction opposite ("against") the current.

Resistors in parallel

Figure 19.21 Many resistors connected in parallel are equivalent to a resistance of value R_{equiv} as given by Equation 19.25.

When applying Kirchhoff's rules to find the currents through all the branches in a circuit, the number of equations must (according to the rules of algebra) equal the number of unknown currents. The repeated use of Kirchhoff's rules can sometimes lead to "repeated" equations as discussed earlier in connection with Figure 19.19. When solving for branch currents, we must use Kirchhoff's rules to get a set of independent equations, with the number of equations equal to the number of unknowns.

Resistors in Parallel

When we analyzed the circuit in Figure 19.14A, we found that two resistors R_1 and R_2 in series are equivalent to a single resistor of value $R_{equiv} = R_1 + R_2$. The resistors R_1 and R_2 can thus be replaced by this equivalent resistor without affecting the rest of the circuit; in particular, the current in the circuit will be the same. The resistors in Figure 19.21 are connected in parallel, and it is also possible to replace the parallel resistors R_1 and R_2 by a single equivalent resistor R_{equiv}. To find the value of this equivalent resistor, we recall our result for the total current through the battery in the parallel circuit (Eq. 19.23):

$$I_3 = \frac{\mathcal{E}}{R_1} + \frac{\mathcal{E}}{R_2} = \mathcal{E}\left(\frac{1}{R_1} + \frac{1}{R_2}\right)$$

For the equivalent circuit in Figure 19.21B, we have

$$I = \frac{\mathcal{E}}{R_{equiv}}$$

where I is again the current through the battery. For the circuits in parts A and B of Figure 19.21 to be truly equivalent, the current through the battery must be the same, so we have

$$\mathcal{E}\left(\frac{1}{R_1} + \frac{1}{R_2}\right) = \frac{\mathcal{E}}{R_{equiv}}$$

which gives

$$\frac{1}{R_{equiv}} = \frac{1}{R_1} + \frac{1}{R_2} \tag{19.24}$$

This result can be extended to any number of resistors in parallel (Fig. 19.21C). In the general case, the value of the equivalent resistance is

$$\frac{1}{R_{equiv}} = \frac{1}{R_1} + \frac{1}{R_2} + \frac{1}{R_3} + \cdots \quad \text{(resistors in parallel)} \tag{19.25}$$

We'll demonstrate this handy result in the next example.

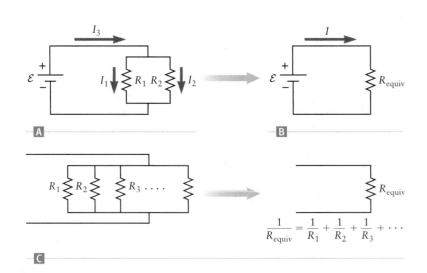

EXAMPLE 19.5 | Resistors in Series and in Parallel

Consider the circuit in Figure 19.22A. Find the current through resistor R_1.

RECOGNIZE THE PRINCIPLE

We could attack this problem using Kirchhoff's rules, applying the loop and junction rules to derive equations involving the currents through the different branches. In this case, however, we can instead use the rules for combining resistors in series and in parallel (Eqs. 19.18 and 19.25). These equations were derived from Kirchhoff's rules, so those rules are still the foundation of our calculation.

SKETCH THE PROBLEM

The three parts of Figure 19.22 show how the parallel and series addition rules can be applied to obtain a single equivalent resistor R_{equiv}.

IDENTIFY THE RELATIONSHIPS AND SOLVE

Resistors R_2 and R_3 in Figure 19.22A are connected in parallel, so we can replace them with a single resistor $R_{parallel}$ as shown in Figure 19.22B. That parallel combination of R_2 and R_3 is equivalent to a single resistor with (Eq. 19.25)

$$\frac{1}{R_{parallel}} = \frac{1}{R_2} + \frac{1}{R_3}$$

which can be rearranged to get

$$R_{parallel} = \frac{R_2 R_3}{R_2 + R_3}$$

We now have an equivalent circuit (Fig. 19.22B) containing two resistors R_1 and $R_{parallel}$ in series. We can replace these two resistors by a single new resistor R_{equiv} as shown in Figure 19.22C. To find the value of R_{equiv}, we use the rule for combining resistors in series (Eq. 19.18) to get

$$R_{equiv} = R_1 + R_{parallel}$$

Using our result for $R_{parallel}$ gives

$$R_{equiv} = R_1 + R_{parallel} = R_1 + \frac{R_2 R_3}{R_2 + R_3}$$

We have now arrived at a much simpler equivalent circuit that contains a battery with a single resistor R_{equiv}. The current in this circuit is given by our familiar result

$$I = \frac{\mathcal{E}}{R_{equiv}}$$

What have we learned?
Complicated combinations of resistors can often be analyzed by applying the rules for series and parallel resistors as we have done here. For some circuits, though, one has to start directly with Kirchhoff's rules instead as we'll see in Example 19.6.

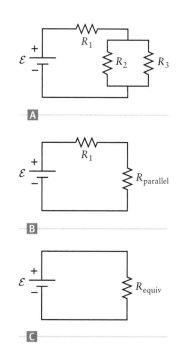

Figure 19.22 Example 19.5.
A Using the rule for combining resistors in parallel, we can replace R_2 and R_3 by an equivalent resistor $R_{parallel}$ in **B**. We can then combine R_1 and $R_{parallel}$ in series to get a resistance value R_{equiv} shown in **C** that is equivalent to the three original resistors.

CONCEPT CHECK 19.3 | Combining Resistors

Three resistors are connected as shown in Figure 19.23. What is the equivalent resistance of this combination?
  (a) R (b) $3R$ (c) $R/3$

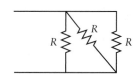

Figure 19.23 Concept Check 19.3.

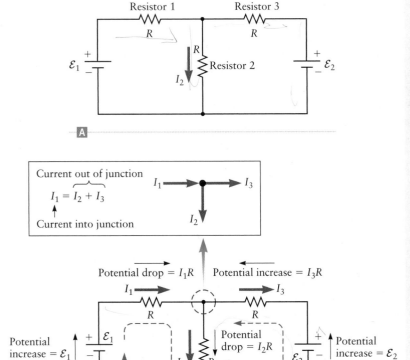

EXAMPLE 19.6 | A Multiloop Circuit

Consider the circuit in Figure 19.24A. Apply Kirchhoff's rules to find the current I_2 though the central branch. For simplicity, assume all three resistors have the same resistance R.

RECOGNIZE THE PRINCIPLE

This circuit cannot be simplified using the rules for series and parallel combinations of resistors because two of the branches contain both a battery and a resistor. The rules for resistors in series or in parallel in Equations 19.18 and 19.25 were derived assuming the branches contain only resistors. We must therefore start with Kirchhoff's two rules, following Step 1 of the approach outlined in the "General Rules for Circuit Analysis" given earlier.

SKETCH THE PROBLEM

Step 2: We begin by labeling the currents through all branches of the circuit in Figure 19.24B and assigning positive directions for each current as indicated by arrows in the figure. We do not know the values of the emfs, so we cannot know at this point if the currents will be in the directions indicated by these arrows, but we can proceed with these assumed directions and solve for I_1, I_2, and I_3. A negative value for one of the currents means that current is directed opposite to the corresponding arrow in Figure 19.24B.

Figure 19.24 Example 19.6.
A A multiloop circuit with three identical resistors. **B** We label the currents through the three branches I_1, I_2, and I_3 and apply Kirchhoff's rules to solve for these currents.

IDENTIFY THE RELATIONSHIPS

Step 3 of the strategy: We express Kirchhoff's loop rule for the two loops indicated by the dashed paths in the figure. Starting at the negative terminal of battery 1, we travel around path 1 in a clockwise direction. There is a potential increase of $\mathcal{E}_1$ at the battery, a drop of $I_1 R$ as we pass through resistor 1, and another potential drop $I_2 R$ when passing through resistor 2. These potential changes must add up to zero, so

$$\mathcal{E}_1 - I_1 R - I_2 R = 0 \tag{1}$$

We next travel around loop 2, starting at the negative terminal of battery 2 and moving around this loop in the counterclockwise direction, which takes us first through the battery with a potential increase $\mathcal{E}_2$. We then pass through a resistor "against" the assigned direction of I_3. We are moving from the low-potential end of the resistor to the high-potential end, so now we have a potential *increase* of $I_3 R$. Continuing around loop 2, we have a potential drop $I_2 R$ on passing through the resistor in the central branch that carries current I_2. Summing all the potential changes along path 2, Kirchhoff's loop rule gives

$$\mathcal{E}_2 + I_3 R - I_2 R = 0 \tag{2}$$

We now have two equations, (1) and (2), and three unknowns (the currents through the three branches). We need one more equation (because the number of equations must equal the number of unknowns), and we can get it using Kirchhoff's junction rule. Applying this rule at the junction indicated at the top in Figure 19.24B, we find

$$I_1 = I_2 + I_3 \tag{3}$$

which gives us three equations. We can now solve for all three unknown currents.

SOLVE

Step 4: Adding Equations (1) and (2) gives

$$\mathcal{E}_1 - I_1R - I_2R + \mathcal{E}_2 + I_3R - I_2R = 0 \qquad (4)$$

Collecting the terms involving R leads to

$$\mathcal{E}_1 + \mathcal{E}_2 + R(-I_1 - I_2 + I_3 - I_2) = 0$$

$$\mathcal{E}_1 + \mathcal{E}_2 = (I_1 + 2I_2 - I_3)R$$

We can use Equation (3) to eliminate I_3 and find

$$\mathcal{E}_1 + \mathcal{E}_2 = (I_1 + 2I_2 - I_3)R = (I_1 + 2I_2 - I_1 + I_2)R$$

$$\mathcal{E}_1 + \mathcal{E}_2 = 3I_2R$$

Solving for the current I_2 gives our final result:

$$\boxed{I_2 = \frac{\mathcal{E}_1 + \mathcal{E}_2}{3R}}$$

What have we learned?

If we want to find the currents in the other branches, we could substitute this result for I_2 into Equation (1) to obtain I_1 and so forth, which would give us the complete solution for this circuit.

EXAMPLE 19.7 ℞ **Parallel and Series Combinations of Resistors (Revisited)**

You are given an unlimited number of identical resistors with $R = 10$ kΩ. **(a)** How could you use several of these resistors to construct an equivalent resistance with $R_{equiv} = 60$ kΩ? **(b)** How could you use several of these resistors to construct an equivalent resistance with $R_{equiv} = 2.5$ kΩ?

RECOGNIZE THE PRINCIPLE

We begin with some observations about the rules for combining resistors in series and in parallel. The general series combination rule is (Eq. 19.18)

$$R_{equiv} = R_1 + R_2 + R_3 + \cdots \qquad (1)$$

In words, this says that for a purely series combination of resistors, the equivalent resistance is always *larger* than *any* of the individual resistors. The parallel-resistor combination rule is (Eq. 19.25)

$$\frac{1}{R_{equiv}} = \frac{1}{R_1} + \frac{1}{R_2} + \frac{1}{R_3} + \cdots \qquad (2)$$

In words, this says that the equivalent resistance is always *smaller* than *any* of the individual resistances.

SKETCH THE PROBLEM

No figure is necessary.

IDENTIFY THE RELATIONSHIPS AND SOLVE

(a) The problem asked for a total resistance of 60 kΩ, but the components have $R = 10$ kΩ; according to Equation (1), we must connect six of the component resistors in series.

(b) If we connect n of these resistors in parallel, Equation (2) becomes

$$\frac{1}{R_{equiv}} = n\left(\frac{1}{R}\right)$$

Solving for n and inserting the desired value of the equivalent resistance, we get

$$n = \frac{R}{R_{equiv}} = \frac{10 \text{ k}\Omega}{2.5 \text{ k}\Omega} = 4$$

We should thus connect four resistors in parallel to get an equivalent resistance of 2.5 kΩ.

What does it mean?
Connecting resistors in series always gives a total (equivalent) resistance that is larger than the resistance of any of the component resistors. A parallel connection always gives an equivalent resistance smaller than any of the component resistances.

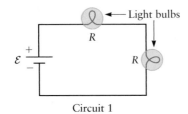
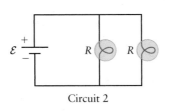

← Light bulbs

Circuit 1

Circuit 2

Figure 19.25 Concept Check 19.4.

CONCEPT CHECK 19.4 | Comparing Series and Parallel Circuits
Incandescent lightbulbs, like the ones used in flashlights, act as resistors, and their light intensity depends on the current: a high current produces a high light intensity, while a small current gives a small light intensity. Figure 19.25 shows two circuits with identical batteries and identical lightbulbs (each with resistance R). In which circuit do the lightbulbs shine brightest?

Real Batteries

In Section 19.2, we mentioned that an ideal battery always maintains a constant voltage across its terminals; the value of this voltage is just the emf of the battery. An ideal battery maintains this constant voltage no matter how much current it provides to a circuit. Real batteries fall short of this ideal. The voltage provided by a real battery decreases at high currents, in part because of the time associated with the electrochemical reactions and the movement of charge inside the battery (Fig. 19.4B). This behavior can also be understood using the equivalent circuit picture in Figure 19.26A.

A real battery is equivalent to an ideal battery in series with a resistor $R_{battery}$, called the **internal resistance** of the battery. The current through the battery passes through this internal resistance, and the effect can be seen by analyzing the circuit in Figure 19.26B. The internal resistor $R_{battery}$ and the external resistor R are connected in series; applying our relation for resistors in series (Eq. 19.18), the current through this circuit is

$$I = \frac{\mathcal{E}}{R + R_{battery}}$$

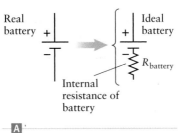

A

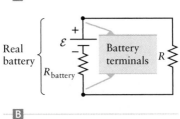

B

Figure 19.26 **A** All real batteries have a nonzero internal resistance $R_{battery}$. **B** We can think of a real battery as an ideal battery in series with a resistance $R_{battery}$ and apply Kirchhoff's rules to this circuit model.

Usually, we are interested in the voltage (i.e., the potential difference) that appears across the battery terminals. The voltage across the real battery's terminals is also equal to the potential drop across the external resistor, which is IR. We thus have

$$\text{potential of real battery} = IR = \mathcal{E}\left(\frac{R}{R + R_{battery}}\right) \tag{19.26}$$

If the internal resistance of the battery is small compared to the external resistor ($R_{battery} \ll R$), the factor in parentheses in Equation 19.26 is approximately 1 and the voltage across the terminals is nearly equal to $\mathcal{E}$, the value for an ideal battery. However, if the internal resistance is not small compared with R, the voltage across the battery terminals can be much less than $\mathcal{E}$. Hence, in practical applications, the internal resistance of the battery should be small compared with the resistance of the circuit to which the battery is attached.

19.5 | DC CIRCUITS: ADDING CAPACITORS

In Chapter 18, we introduced capacitors and showed that when charges $\pm q$ are placed onto the plates of a capacitor, the potential difference V across the capacitor plates has a magnitude (Eq. 18.30)

$$V = \frac{q}{C} \qquad (19.27)$$

Here, as in all our work on circuits in this chapter, V is the potential *difference* across the circuit element (in this case, the capacitor). In Chapter 18, we denoted this potential difference by ΔV, but now we follow the convention widely used in circuit analysis and drop the "Δ."

Applying Kirchhoff's Rules to a Circuit with a Capacitor

Kirchhoff's rules for circuit analysis apply to all kinds of circuits, including those with capacitors. Figure 19.27 shows an *RC* circuit containing a resistor and a capacitor in series with a battery. A capacitor is represented in a circuit diagram as two parallel plates (Fig. 18.26); this symbol is designed to remind you of a parallel-plate capacitor, even though most real capacitors do not have this simple geometry. To apply Kirchhoff's loop rule, we start at the negative terminal of the battery and proceed clockwise around the circuit. The potential increases by $\mathcal{E}$ as we pass through the battery and then decreases by an amount IR (a potential drop) as we move across the resistor. The potential across a capacitor is given by Equation 19.27; because the top capacitor plate in Figure 19.27 is assumed to be positively charged ($+q$), the potential drops as we move from the top to the bottom capacitor plate (as we travel clockwise around the circuit). Collecting these terms and writing Kirchhoff's loop rule for this circuit, we get

$$
\begin{array}{c}
\text{change in} \\
\text{potential around} \\
\text{the circuit}
\end{array}
=
\overbrace{+\mathcal{E}}^{\substack{\text{potential increase} \\ \text{in battery}}}
-
\overbrace{IR}^{\substack{\text{potential drop} \\ \text{across } R}}
-
\overbrace{\frac{q}{C}}^{\substack{\text{potential drop} \\ \text{across } C}}
= 0
$$

As in our previous applications of Kirchhoff's rules, we want to solve this equation to find the current in the circuit. Now, though, we have a new complication: the value of q will change with time because current carries charge onto or off the capacitor plates. As a result, both I and q will be time dependent.

RC Circuits: A Qualitative Analysis

We can get a complete solution for I and q in an *RC* circuit using the relation between current and charge $I = \Delta q / \Delta t$ (Eq. 19.1) along with Kirchhoff's loop rule, but the mathematics is more complicated than we wish to tackle in this book. We will therefore resort to a qualitative analysis based on the general relationship between current and the charge on a capacitor.

Figure 19.28 shows a circuit that contains a single capacitor along with a resistor, a battery, and a *switch* denoted by S. A simple switch can be open, in which case it does not allow any current to pass, or closed, allowing current to pass with no resistance, like an ideal wire. Suppose the capacitor is initially uncharged and the switch is open (Fig. 19.28A). In this case, the current is zero and there is no voltage across the capacitor because $q = 0$. We now close the switch at $t = 0$ and consider what happens. Immediately after the switch is closed, there is a clockwise current as indicated in Figure 19.28B, carrying a positive charge to the top plate of the capacitor. The charge accumulating on the top plate repels charge from the bottom plate, causing charge to flow into the negative battery terminal and leaving the

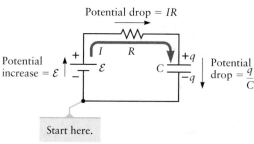

Figure 19.27 The circuit symbol for a capacitor consists of two parallel lines representing the two plates of a parallel-plate capacitor. When charges $\pm q$ are placed on the capacitor plates, an electric potential difference $V = q/C$ is established between the two plates.

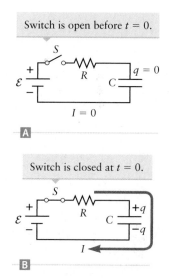

Figure 19.28 ⒜ An *RC* circuit containing a battery, a resistor, a capacitor, and a switch. When the switch is open, no current can pass through it or through the rest of the circuit. ⒝ When the switch is closed, a current is established, taking charge to (or from) the plates of the capacitor.

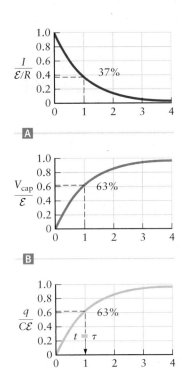

$t = \tau$

Figure 19.29 Variation of **A** the current, **B** the voltage across the capacitor, and **C** the charge on the capacitor as functions of time for the circuit in Figure 19.28. The switch was closed at $t = 0$.

bottom plate of the capacitor with a negative charge. At any particular instant, the capacitor plates are charged $\pm q$, and according to Equation 19.27 there is a nonzero voltage across the capacitor.

As time passes, the current carries more and more charge onto the capacitor, and we say that the capacitor is "charged" by the current. As q increases, so does the voltage across the capacitor. Figure 19.29 shows how the current through the circuit and the voltage across the capacitor vary with time. The voltage across the capacitor increases and asymptotically approaches $\mathcal{E}$, the battery emf. At the same time, the current through the circuit decreases to zero. Mathematically, the current in the circuit is described by

$$I = \frac{\mathcal{E}}{R} e^{-t/\tau} \qquad (19.28)$$

and the voltage across the capacitor is

$$V_{cap} = \mathcal{E}(1 - e^{-t/\tau}) \qquad (19.29)$$

From our relation between the voltage across and charge on a capacitor (Eq. 19.27), we get

$$q = CV_{cap} = C\mathcal{E}(1 - e^{-t/\tau}) \qquad (19.30)$$

which is also plotted in Figure 19.29. The quantity τ is called the **time constant** and is given by

$$\tau = RC \qquad (19.31)$$

The results in Equations 19.28 through 19.31 along with Figure 19.29 show how the charge on and voltage across a capacitor vary with time as the capacitor is "charged up." The amount of time required for this process depends on the time constant τ and hence depends on the product of R and C. After the switch is closed for an amount of time equal to τ—that is, for a period equal to "one time constant"—the current decreases to $e^{-t/\tau} = e^{-1} = 0.37 = 37\%$ of its value at $t = 0$. Likewise, V_{cap} and q increase to $1 - e^{-1} = 0.63 = 63\%$ of their asymptotic values after this time.

Although we will not derive the results for the current, voltage, and charge in Equations 19.28 through 19.31, we can still use Kirchhoff's loop rule to understand the circuit behavior just after the switch is closed (Fig. 19.30A). The initial charge on the capacitor is zero, and just after the switch is closed the charge will still be very small. If q is small, the voltage across the capacitor is also small, that is, $V_{cap} \approx 0$ just after closing the switch. In terms of a circuit analysis, the capacitor at this moment is nearly equivalent to a wire because there is almost no potential drop as we pass from one side of the capacitor to the other. The equivalent circuit thus looks as shown Figure 19.30B, containing a battery and a resistor, and the current is $I = \mathcal{E}/R$. This current is the initial ($t = 0$) value in Figure 19.29A (and Eq. 19.28).

We can analyze the circuit behavior when t is large—that is, a long time after the switch is closed—in a similar way. As t increases, the charge on the capacitor in Figure 19.30 becomes larger and larger. Eventually, it reaches a value for which the voltage across the capacitor is $V_{cap} \approx \mathcal{E}$, the battery emf. At this time, the equivalent circuit appears as shown in Figure 19.30D; the battery with emf $\mathcal{E}$ is now connected to the capacitor across which there is a voltage also equal to $\mathcal{E}$. However, the *polarity* of the capacitor voltage is such that it *opposes* the battery emf. We essentially have two voltage sources connected in opposition, and the net result is no current in the circuit. That is why the current approaches zero when t is large in an RC circuit (Fig. 19.29A).

Discharging a Capacitor

We have just analyzed how the current and charge in an RC circuit vary when a capacitor is charged. Let's now consider the behavior when a capacitor is discharged. An appropriate circuit is shown in Figure 19.31; it is identical to our original circuit

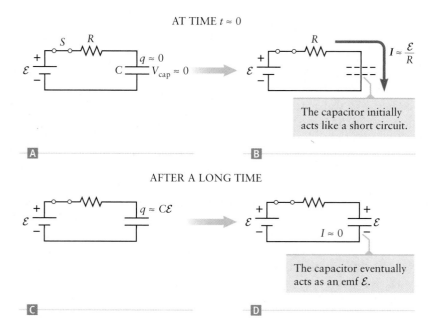

AT TIME $t \approx 0$

The capacitor initially acts like a short circuit.

A

B

AFTER A LONG TIME

The capacitor eventually acts as an emf $\mathcal{E}$.

C

D

Figure 19.30 **A** When the charge on the capacitor $q = 0$, the voltage across the capacitor is zero. The circuit is then equivalent to **B**, in which the only potential changes are across the resistor and the battery. **C** When $q > 0$, there is a voltage drop across the capacitor, which must be included when applying Kirchhoff's loop rule. **D** After a long time, the potential drop across the capacitor has a magnitude $\mathcal{E}$ and "opposes" the battery potential, so the current is zero.

(Fig. 19.28) except for an added second switch S_2. The capacitor can be charged by keeping switch S_1 closed for a very long time with switch S_2 open (Fig. 19.31A), and we have seen that the charge on the capacitor is then $q = C\mathcal{E}$ (Eq. 19.30). Now let's open switch S_1 and close S_2. This produces a counterclockwise current through the loop on the right in Figure 19.31B, carrying charge from the (initially) positive capacitor plate at the top to the (initially) negative plate at the bottom. After a long time, the charge on both plates is zero and the capacitor is "discharged." If switch S_2 is closed at $t = 0$ (i.e., we start our clock when the capacitor begins to discharge), the discharge process is described by

$$I = -\frac{\mathcal{E}}{R} e^{-t/\tau} \tag{19.32}$$

$$V_{cap} = \mathcal{E} e^{-t/\tau} \tag{19.33}$$

$$q = C\mathcal{E} e^{-t/\tau} \tag{19.34}$$

These results are plotted in Figure 19.32, where we see the same exponential behavior found during the charging process. Notice that the current is *negative* as the capacitor discharges. The current's direction is *opposite* to the direction found when the capacitor was charged (Fig. 19.28). The time constant for discharging is again given by $\tau = RC$ (Eq. 19.31). The same time constant describes both charging and discharging.

First, S_1 is closed and the capacitor is "charged."

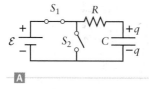

A

Switch S_1 is opened and S_2 is closed. The capacitor is then discharged.

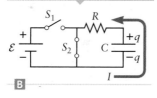

B

Figure 19.31 **A** Charging and **B** discharging a capacitor.

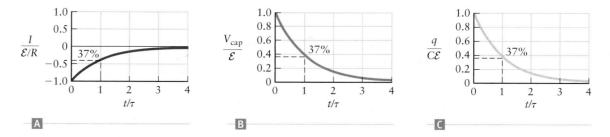

A

B

C

Figure 19.32 For the circuit in Figure 19.31, **A** the current through the capacitor, **B** the voltage across the capacitor, and **C** the charge on the capacitor as the capacitor discharges. Switch S_2 was closed at $t = 0$.

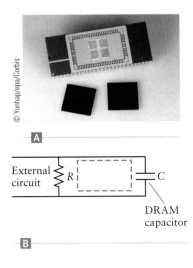

A

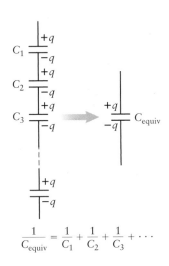

External circuit — R — C

DRAM capacitor

B

Figure 19.33 **A** A real capacitor such as the ones used in a DRAM circuit can be modeled **B** as an ideal capacitor in parallel with an ideal resistor. Typical DRAM integrated circuits are a few cm on a side.

C_1 $+q$ $-q$
C_2 $+q$ $-q$
C_3 $+q$ $-q$

$\rightarrow$ $+q$ $-q$ C_{equiv}

$+q$ $-q$

$$\frac{1}{C_{equiv}} = \frac{1}{C_1} + \frac{1}{C_2} + \frac{1}{C_3} + \cdots$$

Figure 19.34 When many capacitors are connected in series, they behave as a single equivalent capacitance given by Equation 19.35.

Equivalent capacitance of capacitors in series

EXAMPLE 19.8 | Discharge Time for a Capacitor

A key quantity when charging and discharging a capacitor is the time constant τ. The value of τ can vary widely because values of R in practical circuits can range from less than 1 Ω to more than 10^{12} Ω, and the capacitance can vary from as much as 1 F to 10^{-12} F or even less. Capacitors are central elements of a computer's dynamic random-access memory (DRAM) integrated circuit chip (Fig. 19.33A). In normal operation, a DRAM capacitor is connected to a resistor (Fig. 19.33B), and this RC combination is attached to other circuitry on the same computer chip that produces a current to charge or discharge the capacitor. The charge on the capacitor is used to store information in binary form; the presence of a certain amount of charge denotes a "1" in the memory element, and the absence of charge denotes a "0." The RC circuit loop in Figure 19.33B is identical to the loop on the right in Figure 19.31B, so the resistance in parallel with the capacitor provides a discharge path. This discharging is actually a bad thing because immediately after charge is placed on the DRAM capacitor it starts to discharge, and the information stored in the memory element is eventually lost. Consider a DRAM chip for which $C = 10^{-15}$ F and $R = 10^{13}$ Ω. Calculate the discharge time for this RC circuit.

RECOGNIZE THE PRINCIPLE

The solution requires only that we evaluate the time constant τ.

SKETCH THE PROBLEM

The problem is described in Figure 19.33B.

IDENTIFY THE RELATIONSHIPS AND SOLVE

The time constant $\tau = RC$ (Eq. 19.31), so we have

$$\tau = RC = (10^{13}\ \Omega)(10^{-15}\ \text{F}) = 10^{-2}\ \text{s} = \boxed{0.01\ \text{s}}$$

What does it mean?

The voltage across the capacitor in an RC circuit decays in a time approximately equal to τ, so a DRAM capacitor will "remember" its contents for only about 0.01 s. For this reason, a DRAM chip contains circuitry that recharges the capacitor (the technical term is "refresh") at regular intervals. Typical refresh rates are once every 0.001 s, which is much shorter than τ, and the capacitor is recharged before the memory information is lost. The refresh circuit requires a power supply to recharge the capacitors, so when your computer is turned off the DRAM information is lost. It is possible to make memory chips containing much larger resistances R so that the capacitors do not need to be refreshed. Called *static* RAM chips, these are more expensive to make than other chips, but are useful in devices that must be able to hold their memory even when no power is available. This type of memory chip is used in many MP3 players.

Capacitor Combinations

Two or more capacitors may be connected in either series or in parallel arrangements. We considered such problems in Chapter 18, and it is interesting to compare those results with the rules for combining resistors in Section 19.4.

Several capacitors connected in series are shown in Figure 19.34. Let's assume an external circuit has placed charge $+q$ on the upper plate of C_1 and charge $-q$ on the bottom plate of the lowest capacitor. In Chapter 18, we saw that charges $\mp q$ are then induced on the corresponding inner capacitor plates. We also showed (Eq. 18.42) that this arrangement is equivalent to a single capacitor with

$$\frac{1}{C_{equiv}} = \frac{1}{C_1} + \frac{1}{C_2} + \frac{1}{C_3} + \cdots \quad \text{(capacitors in series)} \qquad (19.35)$$

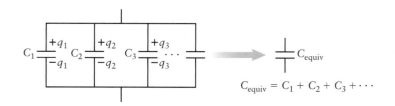

$$C_{equiv} = C_1 + C_2 + C_3 + \cdots$$

Figure 19.35 When many capacitors are connected in parallel, they behave as a single equivalent capacitance given by Equation 19.36.

which has the same form as the rule for combining resistors in *parallel* (Eq. 19.25). This form also tells us that the equivalent capacitance C_{equiv} is *smaller* than *any* of the individual capacitances $C_1, C_2, \ldots$.

A combination of two or more capacitors connected in parallel is also equivalent to a single capacitor (Fig. 19.35). In Chapter 18, we derived the rule for combining capacitors in parallel (Eq. 18.41):

$$C_{equiv} = C_1 + C_2 + C_3 + \cdots \quad \text{(capacitors in parallel)} \qquad (19.36)$$

Equivalent capacitance of capacitors in parallel

Hence, the equivalent capacitance is just the sum of the individual component capacitances and is therefore *greater* than *any* of the individual values. Notice that the rule for combining capacitors in parallel has the same form as the rule for combining resistors in *series* (Eq. 19.18).

CONCEPT CHECK 19.5 | What Is the *RC* Time Constant?

Suppose the *RC* time constant for the combination in Figure 19.36A is 2.4 ms. A second capacitor is then placed in parallel with the original capacitor as in Figure 19.36B. If these capacitors are identical, what is the new value of the time constant τ?

    (a) 2.4 ms   (b) 4.8 ms   (c) 1.2 ms

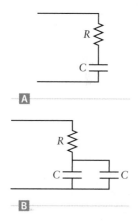

Figure 19.36 Concept Check 19.5.

19.6 | MAKING ELECTRICAL MEASUREMENTS: AMMETERS AND VOLTMETERS

We can use the circuit analysis rules described in Sections 19.4 and 19.5 to calculate the current in a particular circuit branch or the voltage across a particular circuit element, but how does one actually measure these currents and voltages? A device that measures current is called an ***ammeter***, while voltage is measured with a ***voltmeter***. Ammeters and voltmeters are both circuit elements; let's now consider how to incorporate them into a circuit analysis.

The purpose of an ammeter is to measure the current through a circuit branch, which can be accomplished by arranging for all the current to pass through the ammeter. Hence, an ammeter must be connected *in series* with the desired circuit branch as shown in Figure 19.37A. An ideal ammeter should have absolutely no effect on the circuit. That is, an ideal ammeter should measure the current without changing its value. From the properties of resistors in series, this means that an ammeter must have a very low resistance, ideally zero. Although an ammeter with a resistance of exactly zero is difficult to construct, in practice it is only necessary that the ammeter's resistance be much smaller than the resistance of the rest of the circuit branch. For the circuit in Figure 19.37A, the resistance of the ammeter should be much less than the resistance R_3.

The purpose of a voltmeter is to measure the voltage *across* a particular circuit element (or elements). The voltmeter must therefore be connected in a *parallel* arrangement as shown in Figure 19.37B. Here a voltmeter is connected in parallel with resistor R_3 so that it can measure the voltage across this resistor. The voltmeter should measure this voltage without affecting its value; a voltmeter should therefore

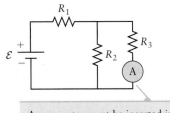

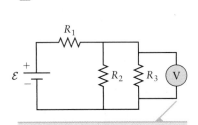

An ammeter must be inserted in series with the branch whose current is to be measured.

A

A voltmeter is inserted in parallel with the circuit element whose voltage drop (or increase) is to be measured.

B

Figure 19.37 **A** An ammeter must be placed in series to measure the current in a particular branch of a circuit. This ammeter is placed in series with R_3 so that it can measure the current through R_3. **B** A voltmeter must be placed in parallel with the circuit element of interest. This voltmeter is in parallel with R_3 so that it can measure the voltage across that resistor.

have a resistance much larger than the circuit element to which it is attached so that very little current passes through it. An ideal voltmeter has an infinite resistance so that the current through the voltmeter is exactly zero. Real voltmeters never have an infinite resistance, but in practice it is possible to make the internal resistance of a voltmeter very large; values of 10^{12} Ω or even more can be achieved.

19.7 | *RC* CIRCUITS AS FILTERS

It is often desirable to reduce or "filter out" time-dependent fluctuations in a voltage signal. Circuits that do so are called *filters* and can be constructed with a simple *RC* combination. Suppose the input voltage in Figure 19.38A is initially constant. From our work with *RC* circuits, we know that the capacitor will initially have a charge q such that the voltage across the capacitor is equal to the input voltage V_{in}. Now suppose at a certain time V_{in} increases very quickly as shown in Figure 19.38B. This increase causes the charge on the capacitor to change to a new value, and this charging is described by the time constant τ of the *RC* circuit. If the time constant is long, the voltage across the capacitor increases much more slowly than the abrupt change of the input voltage. In this way, the voltage across the capacitor, which serves as the output voltage for the circuit, removes rapid changes or fluctuations in the voltage and replaces them with more gradual variations.

A filter circuit is useful in many applications. For example, the noise in a radio signal ("static") can be removed using an *RC* filter, leaving the music signal largely unaffected. The amount of filtering depends on the value of $\tau = RC$, which can be varied over a wide range by the appropriate choice of R and C.

Electric Currents in Nerves ⊗

An interesting example of an *RC* filter is found in the nervous system of the human body and in many other animals. Many nerves are long and thin, much like a wire, carrying electric currents from one part of the body to another. Nerve fibers consist of a thin layer of lipid molecules enclosing a conducting solution containing a mix of ions (K^+, Na^+, Cl^-, and others) as sketched in Figure 19.39A. This figure describes the simplest nerve fibers as found in squids and other invertebrates. The conducting solution inside the fiber acts as a resistor and the lipid layer acts as a capacitor, so this structure and biochemical composition causes a nerve fiber to behave as an *RC* filter circuit, just as in Figure 19.38.

Sensory cells provide the input signal to some nerve fibers. For example, when a squid is touched by a predator, the squid's sensory cells produce an abrupt increase in the input voltage to a nerve fiber, similar to that sketched for V_{in} in Figure 19.38B. The output signal from the *RC* nerve filter increases more slowly than the input. It is in the squid's best interest for V_{out} to increase as rapidly as possible so that the sensory signal will reach the brain quickly and the squid can react rapidly. The *RC* time

Figure 19.38 **A** An *RC* filter circuit takes an input voltage signal and produces an output voltage. **B** Comparison of the input (solid) and output (dashed) voltages. Shown is a low-pass filter that "filters out" rapid changes in the input voltage and only passes slow changes to the output.

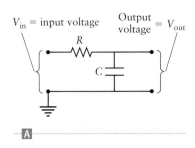

A

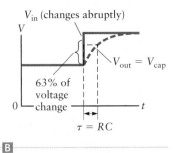

B

constant should therefore be as small as possible. In squids and in other animals, nerves that carry signals for which a rapid response is needed have values of R and C that make the time constant very small, guaranteeing a rapid response. The value of τ is estimated for a realistic nerve fiber in Example 19.9.

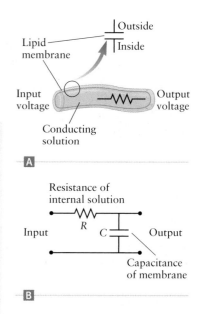

A

B

Figure 19.39 **A** A nerve fiber is electrically equivalent to the RC filter circuit in **B**. The membrane acts like a parallel-plate capacitor, and the resistance is provided by the ionic solution inside the nerve fiber.

| EXAMPLE 19.9 | ✖ Filtering by a Squid Nerve Fiber |

Consider a squid nerve fiber with a radius of $r = 1$ mm and a length of 5 cm. What is the time constant τ for this RC filter? The resistivity of the internal solution is about $\rho = 2\ \Omega \cdot$ m, the thickness of the lipid layer is $d = 10$ nm, and the dielectric constant of the lipid is $\kappa = 10$.

RECOGNIZE THE PRINCIPLE

The time constant is $\tau = RC$ (Eq. 19.31), so to find τ we must first calculate the values of R and C.

SKETCH THE PROBLEM

Figure 19.39 shows how the conducting fluid inside a nerve fiber acts as a resistor and how the lipid layer can be modeled as a capacitor, thus giving an RC circuit.

IDENTIFY THE RELATIONSHIPS

The resistance of the nerve fiber can be found using $R = \rho L / A$ (Eq. 19.11). Inserting the resistivity along with the given values of the length and cross-sectional area ($A = \pi r^2$) of the fiber into Equation 19.11, we obtain

$$R = \rho \frac{L}{A} = (2\ \Omega \cdot \text{m}) \frac{0.05\ \text{m}}{\pi (0.001\ \text{m})^2} = 3 \times 10^4\ \Omega$$

To estimate the capacitance, we approximate the lipid layer as a parallel-plate capacitor. This capacitor's "plate" area is the cylindrical surface area of the fiber, $A_{\text{cap}} = 2\pi r L$. (This cylindrical surface area is *not* the same as the cross-sectional area A of the fiber, which we used to find the resistance.) Using the expression for the capacitance of a parallel-plate capacitor filled with dielectric (Eqs. 18.31 and 18.43), we find

$$C = \frac{\kappa \varepsilon_0 A}{d} = \frac{(10)[8.85 \times 10^{-12}\ \text{C}^2/(\text{N} \cdot \text{m}^2)][2\pi (0.001\ \text{m})(0.05\ \text{m})]}{1.0 \times 10^{-8}\ \text{m}} \quad (1)$$

$$C = 3 \times 10^{-6}\ \text{F}$$

SOLVE

The time constant is then

$$\tau = RC = (3 \times 10^4\ \Omega)(3 \times 10^{-6}\ \text{F}) = \boxed{0.09\ \text{s}}$$

What does it mean?
According to our results for an RC filter in Figure 19.38, this time constant is the approximate response time of a squid nerve fiber.

Designing a Better Nerve Fiber ✖

The time constant calculated in Example 19.9 is a rough measure of how fast the nerve transmits its information to the squid's brain. To keep this reaction time short, τ should be small. The squid's nerve fiber attains a fairly small time constant

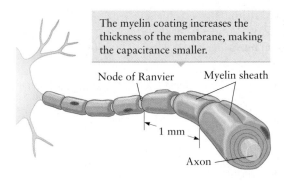

The myelin coating increases the thickness of the membrane, making the capacitance smaller.

Node of Ranvier   Myelin sheath

1 mm

Axon

Figure 19.40 Some vertebrate nerve fibers have very thick membranes due to a coating of myelin that increases the plate spacing of the membrane capacitor (Fig. 19.39) and hence reduces the capacitance. They are called myelinated fibers.

by making the resistance R small, and it has a small resistance because its radius is so large. Another way to keep the time constant short is to make the capacitance C small; that approach is used in many of the sensory and muscle nerve fibers in vertebrates, such as humans. Vertebrate nerve fibers are often coated with a relatively thick layer of a protein called myelin (Fig. 19.40). This coating increases the distance between the capacitor plates—the value of d in Equation (1) of Example 19.9—making C smaller. As a result, the value of the time constant is also reduced, leading to a fast response time for these nerve fibers. A faster response time is always better for the nervous system (and for survival).

19.8 | ✖ ELECTRIC CURRENTS IN THE HUMAN BODY

Electric Current and Your Health

Your body is a moderately good conductor of electricity. Taken as a whole, your body's resistance measured for conduction from the fingers of one hand to the other is about 1500 Ω when the body is dry and typically 500 Ω when wet. The electric potential at a wall socket is 120 V, so if you inadvertently touch a high-potential house wire with one hand and ground with the other, the current through your body is approximately

$$I = \frac{V}{R} \approx \frac{120 \text{ V}}{1500 \text{ }\Omega} = 0.080 \text{ A} = 80 \text{ mA} \quad (\text{dry})$$

$$\approx \frac{120 \text{ V}}{500 \text{ }\Omega} = 240 \text{ mA} \quad (\text{wet}) \qquad (19.37)$$

This current is carried through different parts of the body; some flows in the skin, and some can flow through your heart or other organs.

The effect of current in your body is described in Table 19.3. A current of 1 mA is felt as a "tingling" sensation, but has no harmful effect.[5] Your body uses electric currents to control muscle contraction, and external currents also affect muscles. External currents of 10 mA to 20 mA can cause your muscles to contract involuntarily, which is what happens when a person "can't let go" of a high-potential wire. Your heart is a muscle, and currents above about 100 mA can cause your heart to beat irregularly or even stop. Comparing these values with the currents calculated in Equation 19.37, we see that the current produced by contact with a wall socket can be a definite health hazard.

Large currents can be beneficial in certain situations. For example, your heart can sometimes produce a series of rapid and irregular beats, a condition called fibrillation. A large and brief external current applied by a medical device called a defibrillator can "shock" the heart back into a normal rhythm.

Table 19.3 Effect of Electric Current on the Human Body

Current (mA)	Effect on Body
1	Threshold for sensing (can "feel it tingle")
5	Harmless
10–20	Involuntary muscle contraction
50	Pain
100–200	Disrupts the heart
300	Burns

19.9 | HOUSEHOLD CIRCUITS

Electric circuits are everywhere around us; for example, your house is filled with devices (computers, lights, televisions, etc.) connected to the electric wiring in the house, which in turn is connected to the electric power lines provided by your utility company. In this chapter, we have focused on DC electric circuits involving batteries. However, the voltage provided by your utility company is not constant; rather, it is an AC voltage oscillating at a frequency of 60 Hz. This is the frequency

---

[5] Your tongue is especially sensitive to small currents. Some people are known to test for a dead battery using their tongue (not recommended!).

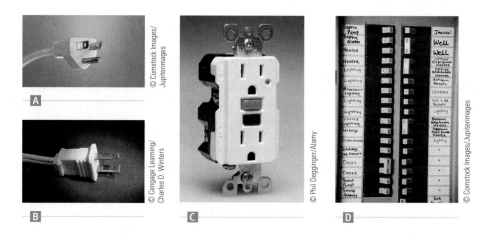

in the United States; the frequency may be different in other countries. We'll discuss AC circuits in Chapter 22. Here we will mention some electrical safety issues that apply to both AC and DC circuits.

Safety in Home Circuits

Figure 19.41 shows some components used in household circuits, including different types of electrical connectors ("plugs") that can be used to connect to an outlet. Most modern outlets and plugs (Fig. 19.41A) contain three connections: two are flat strips, and the third is round. The round one is connected to ground potential, whereas the other two connections carry current to and from devices in your house. The connector in Figure 19.41B is called a polarized plug; it has two flat connectors of different size so that it can be inserted into the wall outlet in only one way. The larger of the two flat strips connects to a low potential that is very close to ground potential.

In normal operation, an appliance such as a hair dryer is plugged into a wall outlet (Fig. 19.41C); current passes from the outlet, through the hair dryer, and back to the outlet, completing the electric circuit. The magnitude of this current can be large enough to pose a health hazard if it were to flow through your body. For example, a wire in the hair dryer might accidentally touch the handle while you are holding it, allowing a large amount of current to pass to the hair dryer handle and through your body. To prevent damage to the household circuitry and to a user, safety devices are inserted into the circuit connecting the wall outlet to the utility company's power line. Older houses employ *fuses*, in which current passes through a thin metal strip. This strip acts as a resistor with a small resistance, so under normal conditions only a small amount of power is dissipated in the fuse. If a failure causes the current to become very large, heat dissipated in the fuse is sufficient to melt and break the metal strip, stopping the current. Newer houses employ *circuit breakers* (Fig. 19.41D), which also stop the current when it exceeds a predetermined limit. A circuit breaker—unlike a fuse, which must be replaced after its metal strip melts—contains a switch that can be "reset" for continued use after the offending circuit has been repaired.

Electrical devices also contain safety features. The ground connection from the wall outlet is usually connected via the round piece of the plug to the outer case of an appliance, so even if there is an internal failure, a large and dangerous voltage does not appear on any surfaces that can be touched by a user. Older appliances sometimes do not use a three-pronged plug, so they cannot make use of this ground connection. In these appliances, the low-potential side of the polarized plug is often connected to the case, providing almost as much protection as a connection to ground. What do we mean by "almost"? When a device without a ground connection fails (perhaps a hair dryer), there may be a large amount of current I through the low-potential plug in the wall outlet (Fig. 19.42). The resistance R of the wires to your hair dryer is small but not zero, so this current causes the voltage at the low-potential side of the hair dryer to increase to a value IR relative to ground. If

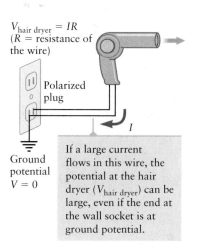

$V_{\text{hair dryer}} = IR$
(R = resistance of the wire)

Polarized plug

Ground potential $V = 0$

I

If a large current flows in this wire, the potential at the hair dryer ($V_{\text{hair dryer}}$) can be large, even if the end at the wall socket is at ground potential.

Figure 19.42 A hair dryer with a polarized plug is not as safe as one with a three-pronged plug. If the hair dryer with a polarized plug fails, it is possible for a large voltage to be present on the normally grounded part of the circuit.

I is very large, this voltage may be large enough to pose a health hazard, especially if it is connected to the hair dryer's handle so that a large current could then flow through a person holding the hair dryer. This arrangement is thus not as safe as using a three-pronged plug, which connects the handle of the hair dryer to ground potential through a separate wire. This separate "ground wire" in the hair dryer is designed so that it should never carry a large current, even if the dryer fails, and the handle should never be at a dangerous potential.

19.10 | TEMPERATURE DEPENDENCE OF RESISTANCE AND SUPERCONDUCTIVITY

In Section 19.3, we explained the atomic-scale origin of resistivity: when electrons move through a metal, they collide, especially with ions in the metal (Fig. 19.6). As temperature increases, these ions vibrate with larger and larger amplitudes, causing more frequent collisions and an increase in the resistance. For many metals near room temperature, this temperature dependence of the resistivity can be described by

$$\rho = \rho_0[1 + \alpha(T - T_0)] \tag{19.38}$$

where α is called the **temperature coefficient of the resistivity**. In Equation 19.38, ρ_0 is the resistivity at a reference temperature T_0. Usually, one picks T_0 to be room temperature, so ρ_0 is just the resistivity listed in Table 19.1. According to Equation 19.38, the resistivity varies linearly with temperature; because the resistance of a wire is proportional to ρ (Eq. 19.11), the resistance of a metal also varies linearly with temperature. This linear variation usually breaks down at very low temperatures, and the resistivity of metals such as copper approaches a nonzero value at very low temperatures (Fig. 19.43A).

For some metals, the resistivity does not quite follow the low-temperature behavior found for copper. In some cases, such as mercury and lead, the resistivity drops abruptly and is zero below a certain temperature as sketched in Figure 19.43B. The value of this "critical" temperature depends on the metal. Metals for which the resistivity is zero below a critical temperature are called **superconductors**. Such behavior is most commonly found in metals cooled to within a few degrees of absolute zero, although there are a few examples of superconductivity at temperatures above 100 K. These materials have many important applications; for example, they can be used to construct very powerful magnets as we'll discuss in Chapter 20.

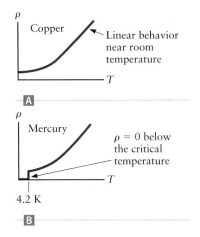

Figure 19.43 Qualitative behavior of the resistivity of **A** copper and **B** mercury as a function of temperature.

SUMMARY | Chapter 19

Electric current
Electric current I is the amount of charge that passes a particular spot per unit time.

$$I = \frac{\Delta q}{\Delta t} \qquad \textbf{(19.1)} \text{ (page 602)}$$

where I is measured in units of **amperes** (A), with 1 A = 1 C/s.

(Continued)

Electrical resistance
The **resistance** R of a material of length L and cross-sectional area A is

$$R = \rho \frac{L}{A} \qquad \textbf{(19.11)} \text{ (page 608)}$$

where ρ is the **resistivity** of the material (Table 19.1). For a circuit element called a **resistor**, the voltage across the circuit element and the current through the element are related by **Ohm's law**:

$$V = IR \qquad \textbf{(19.10)} \text{ (page 608)}$$

Kirchhoff's rules
Kirchhoff's rules tell how to determine the current throughout an electrical circuit. These two rules are based on the principles of conservation of energy and charge:

1. **Kirchhoff's loop rule**: The total change in the electric potential around any closed circuit path must be zero.

2. **Kirchhoff's junction rule**: The current entering a circuit junction must equal the current leaving the junction.

KIRCHHOFF'S LOOP RULE

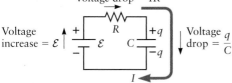

KIRCHHOFF'S JUNCTION RULE

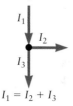

$I_1 = I_2 + I_3$

Voltage drop across a capacitor
If a capacitor carries a charge $\pm q$, the voltage across the capacitor is

$$V = \frac{q}{C} \qquad \textbf{(19.27)} \text{ (page 625)}$$

where C is the capacitance.

APPLICATIONS

Batteries
Batteries act as sources of energy in an electric circuit. The electric potential difference between the terminals of a battery is called the **electromotive force** or **emf** $\mathcal{E}$ produced by the battery.

Charge carriers
The current in a metal wire is carried by mobile electrons that move at the **drift velocity**.

(Continued)

Circuit analysis
Kirchhoff's rules tell how resistors and capacitors behave when connected in *series* and in *parallel* arrangements.

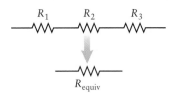

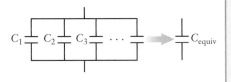

Series resistors: $R_{equiv} = R_1 + R_2 + \cdots$ Parallel resistors: $\dfrac{1}{R_{equiv}} = \dfrac{1}{R_1} + \dfrac{1}{R_2} + \cdots$

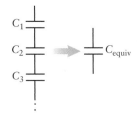

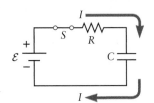

Series capacitors: $\dfrac{1}{C_{equiv}} = \dfrac{1}{C_1} + \dfrac{1}{C_2} + \cdots$ Parallel capacitors: $C_{equiv} = C_1 + C_2 + \cdots$

Charging and discharging a capacitor
When the switch is closed in an *RC* circuit, the current through the circuit and the voltage across the capacitor vary with time according to

$$I = \frac{\mathcal{E}}{R} e^{-t/\tau}$$ **(19.28)** (page 626)

$$V_{cap} = \mathcal{E}(1 - e^{-t/\tau})$$ **(19.29)** (page 626)

where $\tau = RC$ is the *time constant* for the circuit.

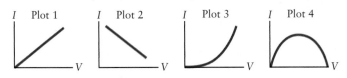

QUESTIONS

SSM = answer in Student Companion & Problem-Solving Guide   X = life science application

1. Discuss how Kirchhoff's rules for circuit analysis are related to conservation principles.

2. Explain why a complete circuit is required to have a nonzero current in a circuit.

3. In a metal, is an electric current carried by (a) protons, (b) electrons, or (c) both protons and electrons?

4. In a solution of salt water (NaCl dissolved in water), is an electric current carried mainly by (a) electrons, (b) protons, (c) Na^+, (d) Cl^-, (e) Na^+ and Cl^-, or (f) electrons and protons?

5. Is the ratio of the voltage across a conductor to the current through the conductor called the (a) power, (b) capacitance, or (c) resistance?

6. Is the product of the voltage across a conductor and the current through the conductor called the (a) power, (b) capacitance, or (c) resistance?

7. In Chapter 17, we argued that the electric field inside a metal in equilibrium is always zero. In this chapter, we found that there is a nonzero electric field inside a current-carrying wire. How can both of these statements be correct?

8. Which of the plots in Figure Q19.8 shows how the current depends on voltage for an ordinary resistor?

I Plot 1   *I* Plot 2   *I* Plot 3   *I* Plot 4

Figure Q19.8

9. SSM An incandescent lightbulb contains a filament that has a certain electrical resistance *R*. The brightness of the bulb depends on the current and increases as the current through the filament is increased. Consider the following situations in Figure Q19.9 on the next page.
   (a) Two identical lightbulbs are connected as shown to a single ideal battery as in circuit 2. Will the brightness of one of these bulbs be greater than, less than, or the same as the brightness of the bulb in circuit 1?
   (b) Repeat part (a) for the bulbs in circuits 3, 4, and 5, comparing their brightness to that of the bulb in circuit 1. Explain your answers.

10. If two circuit elements are connected in series, which of the following statements is true?

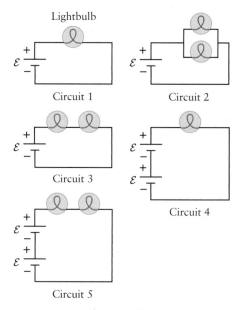

Lightbulb

Circuit 1

Circuit 2

Circuit 3

Circuit 4

Circuit 5

Figure Q19.9

(a) The voltage across the circuit elements is the same.
(b) The current through the circuit elements is the same.
(c) The current is largest through the first circuit element.

11. If three circuit elements are connected in parallel, which of the following statements is always true?
   (a) The voltage across the circuit elements is the same.
   (b) The current through the circuit elements is the same.
   (c) The power is the same in all three circuit elements.

12. You are given four resistors, each with a resistance R. Devise two ways to connect these resistors to get a total equivalent resistance that is greater than R.

13. You are given four resistors, each with a resistance R. Devise two ways to connect these resistors to get a total equivalent resistance less than R.

14. You are given four resistors, all with resistance R. Devise a way to connect these resistors to get an equivalent resistance of $2.5R$.

15. Show that the RC time constant in Equation 19.31 does indeed have units of time.

16. You are given four capacitors, each with a capacitance C. Devise two ways to connect these capacitors to get a total equivalent capacitance greater than C.

17. You are given four capacitors, each with capacitance C. Devise two ways to connect these capacitors to get a total capacitance less than C.

18. SSM In Example 19.5, we used the rules for combining resistors in series and in parallel to analyze the circuit in Figure 19.22. Take a different approach and use Kirchhoff's rules to derive equations for the currents through the different circuit branches. Show that these equations lead to the same result found in Example 19.5.

19. Figure Q19.19 shows two circuits which contain the same circuit elements, but the positions of the resistors are interchanged. Will that affect the current? Justify your answer using Kirchhoff's rules.

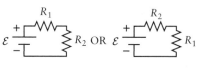

Figure Q19.19

20. In Example 19.2, we discussed how a capacitor can be used to store electrical energy. Investigate how capacitors are used in real camera flash circuits and discuss why they are better for this application than batteries.

21. In Example 19.1, we stated an estimate for the number of tungsten atoms in the filament of a lightbulb. Carry out a calculation of this number and compare your result with our estimate.

22. If one terminal of a battery is connected to an object, does any charge flow from the battery to the object? If there is charge flow, explain why it is very brief.

23. There will be a nonzero current in the circuit in Figure Q19.23. This current will carry electrons from one end of the resistor through the wire at the top to the positive terminal of the battery. We know that electrons always move from a region of low electric potential to a region of higher potential. We also know that an ideal wire has zero resistance, so there can be no potential drop across an ideal wire. Explain how to reconcile these two facts.

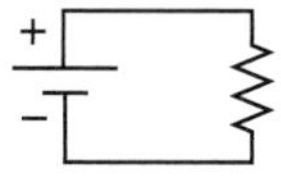

Figure Q19.23

24. When the author built his current house, he ran wires in the walls to be used for stereo speakers. He ran wires from several different rooms to a central location, but he forgot to mark the wires so he could tell which wires emerge in which room. Explain how he was able to solve this problem and identify which wire is which.

PROBLEMS

19.1 ELECTRIC CURRENT: THE FLOW OF CHARGE

1. A current of 3.5 A flows through a wire. How many electrons pass a particular point on the wire in 12 s?

2. ⓧ As a treatment for chronic back pain, a medical patient may be fitted with a device that passes a small electrical current (10 mA) through the muscles in the lower back as shown in Figure P19.2. If the current is supplied in short pulses 0.50 s in length, how many electrons pass through the muscles during one pulse?

3. A current of 0.75 A flows through a lightbulb for 1 h. How many electrons pass through the lightbulb in this time?

Figure P19.2

© Michael Krasowitz/Photographers Choice/Getty Images

4. Electrons move between points A and B in Figure P19.4 at a rate of 15 electrons per second. What is the current? Give the magnitude and direction of I.

5. SSM During a thunderstorm, a lightning "bolt" carries current between a cloud and the ground below. If a particular bolt carries a total charge of 20 C in 1.0 ms, what is the magnitude of the current? How many electrons are involved in this process?

Figure P19.4

19.2 BATTERIES

6. ★ Car batteries are typically rated in ampere-hours, or A · h. (a) Show that ampere-hours are actually a unit of charge and determine the conversion from ampere-hours to coulombs. (b) If a 100 A · h battery can deliver a steady current of 5.0 A until it is completely depleted, what is the total time the battery can deliver current without being recharged? (c) How much total charge will the battery deliver in this time?

7. SSM ★ ✪ Lithium-iodine batteries are particularly useful in situations in which small amounts of current are required over a long period of time. One such application is in a cardiac pacemaker (Fig. P19.7), where changing a battery requires a patient to undergo an operation. Typical pacemakers require 0.50 μA of continuous current. If a lithium-ion battery is capable of supplying 0.50 A · h of charge, what is the lifetime of the battery? Does this lifetime seem sufficiently long?

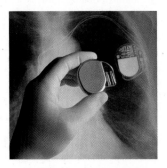

© Photodisc/Alamy

Figure P19.7 View of a cardiac pacemaker, both inside and outside a patient.

8. ★ A 14.4-V battery for a cordless drill can supply 2.0 A of current. (a) If the drill is continuously operated for 5 min, what is the total charge passing through the battery? (b) What total amount of work does the battery do on the charge during these 5 min?

19.3 CURRENT AND VOLTAGE IN A RESISTOR CIRCUIT

9. Calculate the resistance of a piece of copper wire that is 1.0 m long and has a diameter of 1.0 mm.

10. ★ If the length of a wire is increased by a factor of 4, by what factor does the resistance change?

11. ★ If the diameter of a piece of wire is reduced by a factor of 2.5, by what factor does the resistance change?

12. A wire has a diameter of 1.0 mm and a length of 30 m, and is found to have a resistance of 1.2 Ω. What is the resistivity of the wire?

13. ★ ✪ The electric eel (*Electrophorus electricus*) can generate a potential difference of 600 V between a region just behind its head and its tail. Suppose an unsuspecting fish swims into this area, resulting in a lethal current of 1.0 A passing through the length of its body. If the fish is approximately 25 cm in length and 5.0 cm in diameter, what is its average resistivity?

14. An aluminum wire has a diameter of 0.50 mm and a length of 7.5 m. What is its resistance?

15. It is possible to get copper wires with diameters as small as about 1 μm. If such a wire has a resistance of 5 Ω, how long is it?

16. If the current through a resistor is increased by a factor of 3, by what factor does the voltage change?

17. ★ If the current through a resistor is increased by a factor of 3, by what factor does the power change?

18. A glass capillary tube with a diameter of 0.50 mm and length 10 cm is filled with a salt solution with a resistivity of 0.10 Ω · m. What is the resistance?

19. SSM ★ A copper wire is made with the same diameter and length as the capillary tube in Problem 18. What is the ratio of the resistance of the capillary tube to the resistance of the copper wire?

20. There is a potential difference of $V = 20$ V across a resistor with $R = 50$ Ω. What is the current through the resistor?

21. If a current of 0.10 A flows through a resistor and the potential drop across the resistor is 1.5 V, what is the resistance?

19.4 DC CIRCUITS: BATTERIES, RESISTORS, AND KIRCHHOFF'S RULES

22. ★ A resistor with $R = 50$ Ω is connected to a battery with emf $\mathcal{E} = 3.0$ V. (a) What is the current through the battery? (b) What is the current through the resistor? (c) What is the power dissipated in the resistor? (d) How much energy is dissipated in the resistor in 3 minutes?

23. ★ The three resistors in Figure P19.23 all have the same resistance. When the switch in Figure P19.23 is closed, (a) by what factor does the current in resistor 1 change? (b) By what factor does the potential difference across resistor 2 change? (c) By what factor does the current in resistor 2 change?

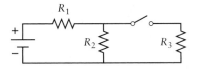

Figure P19.23

24. ✪ Two resistors with $R_1 = 1500$ Ω and $R_2 = 3500$ Ω are connected in parallel as shown in Figure P19.24. If the battery emf is $\mathcal{E} = 12$ V, what is the current through each of the resistors?

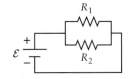

25. What is the equivalent resistance of the two resistors in Problem 24?

Figure P19.24 Problems 24, 25, and 43.

26. Five resistors, all with resistance R, are connected in parallel. What is the equivalent resistance of this combination?

27. ★ Two resistors with $R_1 = 1500$ Ω and $R_2 = 3500$ Ω are connected in series as shown in Figure P19.27. (a) If the battery emf is $\mathcal{E} = 12$ V, what is the current through each of the resistors? (b) What is the equivalent resistance of the two resistors?

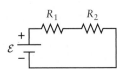

Figure P19.27 Problems 27, 29, 32, and 45.

28. Seven resistors, all with resistance R, are connected in series. What is the equivalent resistance of this combination?

29. ★ Consider the circuit in Figure P19.27 with $R_1 = 1500$ Ω and $R_2 = 3500$ Ω. If $\mathcal{E} = 12$ V, what is the power dissipated in R_2?

30. ✪ A typical lightbulb in your house is rated 100 W, which means that the bulb dissipates 100 W when connected to a DC voltage of 110 V. (a) What is the current? (b) If the voltage is reduced to 55 V, what is the dissipated power?

31. ✪ An electric heater consumes electrical energy at a rate of $P = 250$ W when connected to a battery with an emf of 120 V. (a) What is the current in the heater? (b) If you wish to increase the power consumption to 500 W, by what factor should the heater resistance be changed?

32. ☆ If the current in the circuit in Figure P19.27 is 0.15 A and the resistances are $R_1 = 1500\ \Omega$ and $R_2 = 2500\ \Omega$, what is the emf of the battery?

33. ✪ Two appliances are connected in parallel to a 120-V battery and draw currents of $I_1 = 2.0$ A and $I_2 = 3.5$ A. If these appliances are instead connected in series to the same battery, what is the total current in the circuit?

34. ✪ The circuit in Figure P19.34 shows three identical lightbulbs attached to an ideal battery with $\mathcal{E} = 18$ V. If the resistance of each bulb is $200\ \Omega$, what are (a) the current through bulb 1 and (b) the current through bulb 2? (c) If bulb 2 burns out, what happens to the brightness of bulbs 1 and 3?

 (i) The brightness of both bulbs stays the same.
 (ii) Bulbs 1 and 3 get brighter.
 (iii) Bulbs 1 and 3 both get dimmer.
 (iv) Bulb 1 gets dimmer while bulb 3 gets brighter.
 (v) Bulb 1 gets brighter while bulb 3 gets dimmer.

 Hint: The brightness of a bulb increases as its current increases.

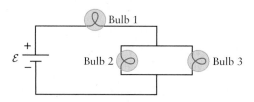

Figure P19.34

35. ✪ You are given four resistors, each with resistance R. Devise a way to connect these resistors so that the total equivalent resistance is R. No cheating; you must use all the resistors so that there is current through each one if a battery is connected.

36. ✪ Consider the circuit in Figure P19.36. The resistors are identical with $R_1 = R_2 = R_3 = 2000\ \Omega$, and the battery voltage is $\mathcal{E}_1 = 9.0$ V. What is the current through resistor R_1?

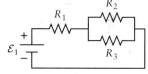

Figure P19.36 Problems 36, 37, 38 and 39.

37. ☆ Consider the circuit in Figure P19.36 and assume all the resistors are identical. Only one of the following statements is true. Which one?
 (a) The power dissipated is the same in all three resistors.
 (b) The power dissipated in R_1 is greater than the power dissipated in R_2.
 (c) The power dissipated in R_1 is less than the power dissipated in R_2.
 (d) The power dissipated in R_1 is equal to the power dissipated in R_2.
 (e) The power dissipated in R_1 is equal to sum of the powers dissipated in R_2 and R_3.

In Problems 38 through 44, assume the resistance values are $R_1 = 2400\ \Omega$, $R_2 = 1400\ \Omega$, $R_3 = 4500\ \Omega$, and $R_4 = 6000\ \Omega$ and the battery emfs are $\mathcal{E}_1 = 1.5$ V and $\mathcal{E}_2 = 3.0$ V.

38. SSM ☆ Analyze the circuit in Figure P19.36. (a) Express the resistors in parallel as a single equivalent resistor. What is the value of this resistor? (b) Combine this equivalent resistor with R_1 to get the total equivalent resistance of the circuit. What is the value of this resistor? (c) The circuit is now "reduced" to a

battery in series with a single equivalent resistance. Use your result from part (b) to find the current through the battery and resistor R_1.

39. ☆ Use your results from Problem 38 to find the current through R_2 in Figure P19.36. *Hint*: First find the voltage across the parallel combination of resistors.

40. ✪ Use the approach from Problem 38 to analyze the circuit in Figure P19.40. What is the current through the battery and resistor R_1? *Hint*: First redraw the circuit so that the parallel and series resistor combinations are more obvious.

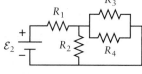

Figure P19.40

41. ✪ Use Kirchhoff's rules to analyze the circuit in Figure P19.41. (a) Let I_1 be the branch current through R_1 and I_2 be the branch current through R_2. Write Kirchhoff's loop rule relation for a loop that travels through battery 1, resistor 1, and battery 2. (b) Write Kirchhoff's loop rule relation for a loop that travels through battery 2 and resistor 2. (c) You should now have two equations and two unknowns (I_1 and I_2). Solve for the two branch currents.

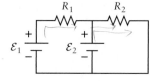

Figure P19.41
Problems 41 and 42.

42. ☆ Consider again the circuit in Figure P19.41. Write the Kirchhoff's loop rule relation for a loop that passes through battery 1, resistor 1, and resistor 2. Show that the resulting equation can be derived by combining the two loop equations from Problem 41.

43. ✪ Consider the resistors in parallel in Figure P19.24. If $R_1 = 2500\ \Omega$ and $R_2 = 6500\ \Omega$, what is the *ratio* of the powers dissipated in the two resistors?

44. ✪ Use Kirchhoff's rules to analyze the circuit in Figure P19.44. (a) Let I_1 be the branch current though R_1, I_2 be the branch current through R_2, and I_3 be the branch current through R_3. Write Kirchhoff's loop rule relation for a loop that travels through battery 1, resistor 1, and resistor 3. (b) Write Kirchhoff's loop rule relation for a loop that travels through battery 2, resistor 2, and resistor 3. (c) Apply Kirchhoff's junction rule to the junction at A to get a relation between the three branch currents. (d) You should now have three equations and three unknowns (I_1, I_2, and I_3). Solve for the three branch currents.

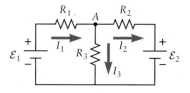

Figure P19.44

45. ☆ Consider the resistors in series in Figure P19.27. If $R_1 = 2500\ \Omega$ and $R_2 = 3500\ \Omega$, what is the *ratio* of the powers dissipated in the two resistors?

46. ✪ Consider the cube of resistors in Figure P19.46. If all the resistors have the same value, what is the equivalent resistance between points A and B? *Hint*: Start by considering a Kirchhoff's rule analysis and sketch the various branch currents. Can you find some relation(s) between these currents?

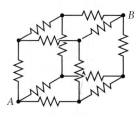

Figure P19.46

47. ✪ The lightbulbs in Figure P19.47 are all identical. Which bulb will be brightest? Which one will be dimmest?

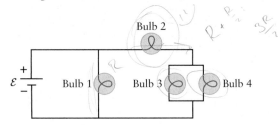

Figure P19.47

19.5 DC CIRCUITS: ADDING CAPACITORS

48. SSM ✪ Consider the *RC* circuit in Figure P19.48, with $R_1 = 1000\ \Omega$, $R_2 = 3000\ \Omega$, and $C = 1.0\ \mu F$. When the switch is closed, the current will vary with time as sketched in Figure 19.29. What is the time constant for the circuit in Figure P19.48? *Hint*: Use equivalent circuit ideas to "reduce" this circuit to a form that looks like Figure 19.28.

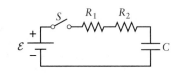

Figure P19.48 Problems 48 and 49.

49. ✪ For the circuit in the Problem 48, what is the current through the circuit (a) just after the switch is closed and (b) after the switch has been closed for a very long time?

50. What is the equivalent capacitance of the two capacitors in Figure P19.50? Take $C_1 = 7.5\ \mu F$ and $C_2 = 3.5\ \mu F$.

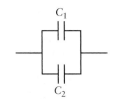

Figure P19.50

51. Five capacitors, all with the same capacitance C, are connected in parallel. What is the value of the equivalent capacitor?

52. ✪ What is the equivalent capacitance of the three capacitors in Figure P19.52? Take $C_1 = 2.5\ \mu F$, $C_2 = 3.5\ \mu F$, and $C_3 = 1.5\ \mu F$.

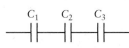

Figure P19.52

53. Seven capacitors, all with the same capacitance C, are connected in series. What is the value of the equivalent capacitor?

54. ✪ What is the equivalent capacitance of the four capacitors in Figure P19.54? Take $C_1 = 2.5\ \mu F$, $C_2 = 4.5\ \mu F$, $C_3 = 3.3\ \mu F$, and $C_4 = 1.5\ \mu F$.

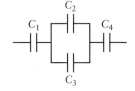

Figure P19.54

In the circuits in Figures P19.55, P19.58, and P19.60, assume the resistance values are $R_1 = 1500\ \Omega$ and $R_2 = 2400\ \Omega$, with capacitances $C_1 = 45\ \mu F$ and $C_2 = 25\ \mu F$, and with $\mathcal{E} = 5.5\ V$. The switches are labeled *S*.

55. ✪ What is the current in the circuit in Figure P19.55 the moment after the switch is closed? Assume the capacitor is initially uncharged.

56. ✪ If the switch in Figure P19.55 is closed for a very long time, what is the charge on the capacitor?

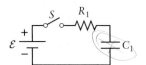

Figure P19.55 Problems 55, 56, and 57.

57. ✪ What is the time constant for the circuit in Figure P19.55? Sketch how the current through the circuit and the voltage across the capacitor vary with time after the switch is closed.

58. ✪ The capacitor in Figure P19.58 is initially uncharged, and both switches are open. What is the current through the battery the instant after switch S_1 is closed?

59. ✪ Consider again the circuit in Figure P19.58. After switch S_1 is closed for a very long time, switch S_1 is opened and switch S_2 is simultaneously closed. (a) What is the current through switch S_2 the instant after it is closed? (b) Make a sketch of how this current varies with time. (c) What is the current through S_2 after it is closed for a very long time?

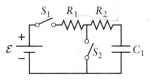

Figure P19.58 Problems 58 and 59.

60. ✪ The capacitors in Figure P19.60 are initially uncharged when the switch is closed. Make a sketch of how the current through the circuit varies with time after the switch is closed. What is the time constant for this circuit?

61. ✪ Consider the circuit in Figure P19.60 and assume the switch has been closed for a very long time. What is the current through the circuit? What are the charges on the two capacitors?

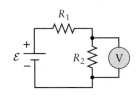

Figure P19.60 Problems 60 and 61.

19.6 MAKING ELECTRICAL MEASUREMENTS: AMMETERS AND VOLTMETERS

62. ✪ Although an ideal voltmeter has an infinite internal resistance, this theoretical ideal is usually not met in practice. The voltmeter in Figure P19.62 has an internal resistance of $10^9\ \Omega$ and is used to measure the voltage across a resistor with $R_2 = 150\ k\Omega$ as shown. How much does this nonideal voltmeter affect the circuit? That is, attaching the voltmeter changes the voltage across R_2. What is the fractional change in the voltage across R_2 when the voltmeter is attached?

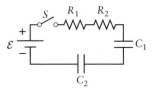

Figure P19.62

63. ✪ Although an ideal ammeter has an internal resistance of zero, this theoretical ideal is usually not met in practice. The ammeter in Figure P19.63 has an internal resistance of $0.010\ \Omega$ and is used to measure the current in a circuit containing a resistor with $R = 150\ \Omega$ as shown. How much does this nonideal ammeter affect the circuit? That is, what is the fractional change in the current when the ammeter is attached?

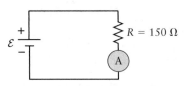

Figure P19.63

64. Suppose an ammeter has an internal resistance of $1.0\ m\Omega$. Find the current in the ammeter when it is properly connected to a 2.0-Ω resistor and a 12-V source.

65. SSM Suppose a voltmeter has an internal resistance of $50\ k\Omega$. Determine the current through the meter when it is properly connected across a 50-Ω resistor that is connected to a 10-V source.

66. ✪ For the circuit in Figure P19.66, draw the proper placement of an ammeter or voltmeter

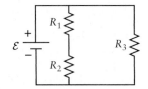

Figure P19.66

to measure (a) the current through and voltage across R_1, (b) the current through and voltage across $(R_1 + R_2)$, (c) the current through and voltage across R_3, and (d) the total current through the battery.

67. ✪ Consider the circuit in Figure P19.67. Find the potential difference between points A and B. Express your answer in terms of $\mathcal{E}$ and R.

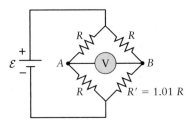

Figure P19.67

68. ✪ Suppose an ammeter and voltmeter are used to determine a resistor's resistance. If the ammeter is connected in series with the resistor and the voltmeter is connected across the resistor only, show that the correct resistance of the resistor is given by $R = V/[I − (V/R_V)]$, where V is the voltage measured by the voltmeter, I is the current measured by the ammeter, and R_V is the internal resistance of the voltmeter.

69. ✪ It is possible to devise clever circuit arrangements that behave as ideal meters. The circuit in Figure P19.69 is able to measure the voltage across resistor R_4 *without* affecting the original circuit. The resistor R_1 is an adjustable resistor whose value can be varied (e.g., by turning a knob). Analyze this circuit and show that for a certain value of the ratio $R_2/(R_1 + R_2)$, this voltmeter circuit has *no* effect on the current in the original circuit. Find the value of this ratio. Express your answer in terms of $\mathcal{E}_1$, $\mathcal{E}_2$, R_3, and R_4.

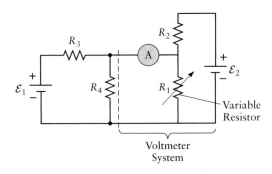

Figure P19.69

ADDITIONAL PROBLEMS

77. Consider a lightbulb that dissipates 75 W when attached to a battery with an emf of 120 V. If this lightbulb is attached to a battery with a emf of 9.0 V, how many electrons pass through the bulb each second? Assume the resistance of the lightbulb is independent of the current through the bulb.

78. ✪ Suppose that electrical energy costs about $0.060/(kW · hour) (that is what you pay your local electric company). In a recent month, the author's electric bill was $150. (a) If the electric potential provided by the electric company was 120 V, what was the average current in all circuits in the author's house? (b) How many electrons did the author "purchase" from the electric company during this month?

79. ✪ The photo in Figure P19.79 shows an electric circuit containing resistors, batteries, and a capacitor. Draw a circuit diagram that corresponds to this circuit.

19.7 RC CIRCUITS AS FILTERS

70. ✪ ✪ Ⓡ Consider the nerve fiber discussed in Example 19.9. Assume the fiber is coated with a layer of myelin that is 100 nm thick. If the dielectric constant of myelin is the same as for a lipid layer, what is the time constant of this fiber? Ignore the gaps in the myelin sheath in Figure 19.40. These gaps are called nodes of Ranvier.

19.8 ELECTRIC CURRENTS IN THE HUMAN BODY

71. ✪ Cell membranes contain openings called ion channels that allow ions to move from the inside of the cell to the outside. Consider a typical ion channel with a diameter of 1.0 nm and a length of 10 nm. If this channel has a resistance of 1.0×10^9 Ω, what is the resistivity of the solution in the channel?

72. SSM ✪ ✪ Consider a single ion channel (Problem 71) with $R = 10^9$ Ω. During an action potential, this channel is open for approximately 1.0 ms for the flow of Na^+ ions with a potential difference across the cell membrane of about 70 mV. How many Na^+ ions travel through the channel?

73. ✪ ✪ A defibrillator is used by emergency medical staff to shock the heart of an accident victim by applying a 10,000-V potential difference across the victim's chest. (a) If the defibrillator has an internal resistance of 10 Ω and the victim's heart has a resistance of 300 Ω, what is the current that passes through the victim's heart the moment the defibrillator is discharged? (b) If the defibrillator delivers this current over a 10-ms time interval, how many electrons pass through the heart during this process?

19.10 TEMPERATURE DEPENDENCE OF RESISTANCE AND SUPERCONDUCTIVITY

74. The temperature coefficient of resistivity for copper is $\alpha = 3.9 \times 10^{-3}$ K^{-1}. If a copper wire has a resistance of 350 Ω at 20 K, what is its resistance at 420 K?

75. SSM ✪ The filament of an incandescent lightbulb is a thin tungsten wire. Suppose a lightbulb has a resistance of 50 Ω at room temperature. During normal operation, the filament reaches a temperature of about 3000 K. (a) What is its resistance at that temperature? The temperature coefficient of resistivity for tungsten is $\alpha = 4.5 \times 10^{-3}$ K^{-1}. (b) If the bulb's brightness depends only on the current, and a larger current gives a greater brightness, will the bulb's brightness be greater or smaller after it has warmed up?

76. ✪ Platinum wires are often used as thermometers due to the change of the resistivity of platinum with temperature. The temperature coefficient of resistivity for platinum is $\alpha = 3.9 \times 10^{-3}$ K^{-1}. Suppose a platinum resistance thermometer has a resistance of 100 Ω at room temperature (293 K). If the temperature is changed so that the thermometer has a resistance of 55 Ω, what is the temperature?

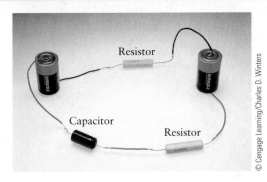

Figure P19.79

80. ✴ Figure P19.80 shows a realistic sketch of a circuit. Draw a circuit diagram that corresponds to this circuit.

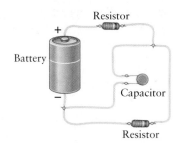

Figure P19.80

81. ✿ ✖ Consider a bird as it flies to and lands on a power line with a potential $V = 880$ V relative to ground potential. Just before the bird lands on the power line, the bird's potential is zero; after it lands, its potential is 880 V. Long after it lands, the current through the bird is zero, but there is a nonzero current just after the bird lands (called a "transient" current). (a) How much charge q must flow onto the bird (from the power line) to make the bird's potential equal 880 V? Estimate q by approximating the bird as one side (one "plate") of a capacitor, with the other plate at ground potential. Assume the value $C = 1.0\ \mu$F. (b) If the bird's resistance is 1000 Ω, what is the time constant τ for the circuit? (c) The current I in this RC circuit flows for a time approximately equal to τ. Find the maximum value of I.

82. ✿ Design a parallel-plate capacitor that is able to store the same amount of energy as in the AA battery mentioned in Example 19.2. Give the area of the plates, the plate separation, and the composition of the dielectric and determine the voltage across the capacitor.

83. ✴ A battery is connected to a single resistor (Fig. P19.83), and the current is measured with an ammeter. A second resistor having the same resistance is placed in parallel with the first resistor. Does the current through the ammeter increase or decrease?

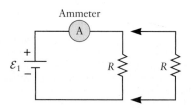

Figure P19.83

84. ✿ A marine battery has an emf of approximately 12.5 V and an internal resistance of 0.075 Ω. The starter motor for the outboard engine requires 40 A while cranking the engine. (a) Determine the potential difference across the starter motor when it is cranking. (b) What is the resistance of the starter motor?

85. SSM ✴ Several ammeters are placed in the circuit in Figure P19.85.
(a) Arrange the values of the potentials $(V_1, V_2, ...)$ from highest to lowest.
(b) Arrange the values of the currents measured by the ammeters $(I_1, I_2, ...)$ from highest to lowest.

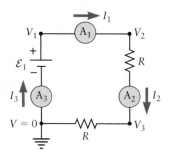

Figure P19.85

86. ✿ Seven capacitors are connected to a 100-V battery as shown in Figure P19.86. (a) What is the total equivalent capacitance of the circuit? (b) Determine the total charge stored in the capacitors when they are fully charged. (c) Determine the total energy stored in the capacitors when they are fully charged. (d) Suppose the battery is then disconnected and replaced by a 100-Ω resistor. Determine the current through the resistor 10 s after the connection to the resistor is made.

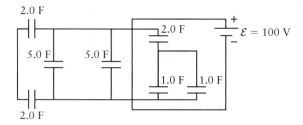

Figure P19.86

87. ✿ Three resistors R_1, R_2, and R_3 are initially connected in parallel to a 12-V source. The total current supplied to the three resistors is 500 μA. Assume R_1 has a resistance of 40,000 Ω and R_2 has a current running through it of 100 μA. If the 12-V supply is kept in place but R_3 is moved from its current location and placed in series with R_1 and R_2, which are still kept in parallel, what will be the total current through the circuit?

88. ✿ A 70-Ω electrical appliance and a 200-Ω electrical appliance are plugged into the outlets of a house. By what factor will the power used change if an additional 400-Ω electrical device is plugged in?

89. ✴ A 100-W lightbulb and a 60-W lightbulb are connected to a 120-V source. Determine the voltage across each lightbulb if they are connected (a) in series and (b) in parallel.

90. ✿ ✖ Ⓡ With the level of childhood obesity reaching epidemic proportions in the United States, one inventor decided to create a device to make children exercise to generate the electrical energy required for them to watch television. Attaching a small electric generator to a stationary exercise bike, the "Cyclevision" (Fig. P19.90) is able to generate enough electrical energy to supply a typical television. Assuming the generator is 80% efficient at converting the mechanical energy of the exercise bike to electrical energy, estimate the number of calories burned while watching 4 h of television (the average daily amount for a child in the United States).

Figure P19.90

91. ✿ Ⓡ **Lightning strike!** As discussed in Chapter 18, lightning is one of nature's most impressive displays. For a typical storm, a lightning bolt is generated when the average electric field is about 10% of the breakdown field for air. Consider a case in which the bottom of the storm cloud is 500 m above the Earth's surface and the cloud has an area of 200 × 200 m². At the moment of a lightning strike, determine (a) the potential difference between the cloud and the Earth's surface and (b) the total charge accumulated on the bottom of the cloud. (c) Estimate the current in the lightning strike. Assume the current flows for 0.1 ms.

92. ✿ Most recreational vehicles and boats have the capability of operating electrical appliances using a 12-V battery, which either requires the use of special appliances designed to run on 12-V DC or the use of an inverter that converts direct current to alternating current as used in our homes. Consider one such 12-V DC circuit in which the following appliances and devices are being operated simultaneously on the same circuit: a 40-W minirefrigerator, a 150-W hair dryer, a 200-W blender, and three 15-W lightbulbs. Determine (a) the current through each of these appliances/devices and (b) the minimum fuse required to handle this load. (Fuses are rated according to the maximum current they will allow.)

93. ✪ Ⓡ **Hydroelectric power.** Niagara Falls has one of the largest hydroelectric power-generating facilities in the world, providing one fourth of New York State's and all Ontario's electrical energy. The Canadian portion of the falls is called Horseshoe Falls based on its shape (Fig. P19.93). Upstream from the falls, water is diverted to the Canadian hydroelectric power-generating plants, which generate 2,000,000 kW of electrical power. The diverted water reenters the river downstream. Assuming the generators are 100% efficient at turning the mechanical energy of the flowing water into electrical energy, estimate the volume flow rate of the water diverted through the generating plants.

Figure P19.93 Horseshoe Falls.

94. ✪ **Resistor puzzle 1.** Consider the configuration of resistors shown in Figure P19.94. Each connection can have a resistor that is disconnected or whose value is 5.0 Ω or 10.0 Ω. To determine the value of each resistor or disconnection point, you measure the following with an ohmmeter: $R_{AB} = 5.0$ Ω, $R_{AC} = 4.0$ Ω, $R_{CD} = 6.0$ Ω, $R_{AD} = 6.0$ Ω, $R_{BD} = 11.0$ Ω, and $R_{BC} = 9.0$ Ω. From these measurements, determine the value of each resistor or if it is disconnected.

95. ✪ **Resistor puzzle 2.** Consider again the configuration of resistors shown in Figure P19.94. As in Problem 94, each connection can have a resistor that is disconnected or whose value is 5.0 Ω or 10.0 Ω. This time, you measure a resistance of 6.25 Ω between points A and B, between A and C, between C and D, and between B and D. The resistance between points A and D is

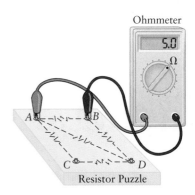

Figure P19.94 Problems 94 and 95.

measured to be 5.0 Ω. From these measurements, determine the value of each resistor or if it is disconnected.

96. ✪ Ⓡ For reasons of safety, the interior light of most modern automobiles is designed to come on when you lift the exterior door handle and then stay on for a short period of time. Assuming the timing circuit for the light is controlled by an RC circuit, estimate the value of the capacitor used in this circuit.

97. ✪ Ⓧ Ⓡ For muscle rehabilitation, patients can be fitted with a small electrical device designed to help strengthen muscle tissue. With this device, a small electrical current is passed through the muscle tissue, stimulating muscle contraction in that region. With involuntary muscle stimulation requiring a current of 10 mA, estimate the minimum required potential difference applied over the length of a bicep muscle needed to stimulate muscle contraction. Assume a bicep muscle has a diameter of 10 cm and a resistivity of 150 Ω · m.

98. ✪ Electrical wire comes in different sizes (different diameters), which are referred to according to their "gauge" number. For example, an 18-gauge wire has a diameter of approximately 1.02 mm and is often used for extension cords. Suppose an 8.0-m extension cord composed of 18-gauge copper wire is carrying 8.0 A of current to a compound miter saw. Determine the voltage drop along the length of the extension cord, remembering that the current must travel down the length of the cord and back.

99. ✪ Ⓡ **Immersion heater.** Tired of having his morning cup of coffee go cold, an enterprising young physics major decides to build an immersion heater out of batteries and a 1.0-m length of 32-gauge (0.20-mm diameter) copper wire. By coiling up the copper wire so that it can fit in a coffee cup and connecting the coil to several 9-V batteries connected in series, the coil of wire will heat up and in turn heat the coffee in the cup. Estimate the number of batteries this student will need to heat up a cup of coffee from room temperature to a reasonable drinking temperature (90°C) in a reasonable amount of time.

Chapter 20

Magnetic Fields and Forces

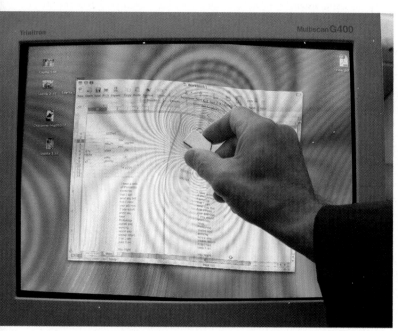

Placing a magnet next to a computer monitor or TV picture tube can distort the image and may do permanent damage to the display. In this chapter, we explain why. (© Cengage Learning/Charles D. Winters)

In Chapters 17 and 18, we studied electric fields and the forces they exert on charged particles. An electric field can be produced by a charge or a collection of charges, and electric fields lead to forces on an electric charge. We now consider the same kinds of questions for magnetic fields and forces. Magnetic fields are produced by *moving* electric charges and these fields exert forces on other *moving* charges. We have two main jobs in this chapter: (1) to understand the sources of magnetic fields and the fields they produce and (2) to see how to calculate the magnetic force on a charged particle. These two problems are intertwined. We begin with a qualitative discussion of how a magnetic field can be produced, and then we describe magnetic forces. We later return to a more quantitative treatment of how to calculate the magnetic field.

Magnetic forces act on charged particles, which suggests that magnetism and electricity are connected in some way, and that is indeed the case. In Chapters 21 and 22, we'll see that electric fields can in fact serve as a source of magnetic fields and vice versa. This close connection gives rise to electromagnetic waves and light, topics we'll explore in Chapters 23–26.

20.1 | SOURCES OF MAGNETIC FIELDS

The first observations of magnetic fields involved **permanent magnets**. The ancient Greeks, Romans, and Chinese all discovered that the mineral magnetite (also called lodestone) has the ability to attract or repel pieces of other magnetic materials, including iron and other pieces of magnetite. Permanent magnets are used in applications ranging from compass needles and "refrigerator magnets" to speakers, motors, and computer hard disks. Figure 20.1 shows a **bar magnet**, a permanent magnet made in the shape of a "bar," as well as small pieces of iron (called iron "filings") sprinkled around the bar magnet. These filings are themselves small, needle-shaped, permanent magnets. The needle directions indicate the direction of the magnetic field $\vec{B}$ near the bar magnet. Figure 20.1 shows some of the magnetic field lines deduced from the pattern of iron filings.

Figure 20.1 raises two questions. First, how can we understand this pattern of field lines? Second, why do the iron filings align along the direction of $\vec{B}$? Figure 20.2 compares the pattern of magnetic field lines near a bar magnet with the pattern of electric field lines near an electric dipole, and we see that the two patterns are quite similar. We have also indicated the **magnetic poles** at the ends of the bar magnet in Figure 20.2A. The magnetic poles are called north (N) and south (S), and are analogous to the positive and negative electric charges in Figure 20.2B. This analogy helps explain why the iron filings line up parallel to the magnetic field lines in Figure 20.1. Each iron filing is a small bar magnet, with its own north and south poles. The north pole of an iron filing is attracted to the south pole of the large bar magnet and is repelled from the north pole. At the same time, the south pole of an iron filing is attracted to the north pole of the large bar magnet and is repelled from the south pole. These magnetic forces cause the iron filing to align parallel to the magnetic field line and hence parallel to $\vec{B}$ as shown in Figure 20.2A. This is just like the situation with electric charges and electric fields studied in Chapter 17 and shown in Figure 20.2B.

The unit of magnetic field is a derived unit of measure in the SI system called the **tesla** (T). We'll show below how it is related to other SI units. The magnetic field close to one of the poles of a strong bar magnet has a magnitude of about 1 T. The Earth also acts approximately as a bar magnet, with a magnetic field of approximately 50 μT (= 5×10^{-5} T) near the Earth's surface. The Earth's magnetic field is responsible for the alignment of a compass needle, as we'll discuss in Section 20.9.

The magnetic field lines outside a permanent magnet "emanate" from the north pole and are directed toward the south pole, and the strength (magnitude) of the field decreases as one moves farther from the poles. We have already noted that the pattern of magnetic field lines of a bar magnet is similar to the pattern of electric field lines near two electric charges. However, the two field patterns in Figure 20.2 *differ* in one very important way: the magnetic field lines *inside* the bar magnet point from the south pole toward the north pole, whereas between the two charges of the electric dipole the electric field lines are in the *opposite* direction. As a result, the magnetic field lines form closed "loops." While electric field lines start on positive charges and end on negative changes, **magnetic field lines always close on themselves**, forming closed loops. This general property of magnetic fields is not limited to the magnetic field produced by a bar magnet.

Permanent magnets come in many shapes and sizes; the "horseshoe" magnet is another common example. A horseshoe magnet can be made by simply bending a

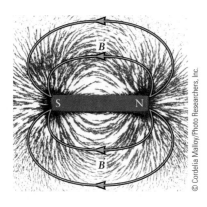

Figure 20.1 Iron filings are small slivers of iron. When placed near a bar magnet, they align along the magnetic field lines produced by the magnet. A few of these field lines are shown here. The letters "N" and "S" label the poles of the bar magnet.

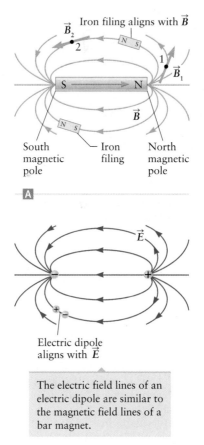

Electric dipole aligns with $\vec{E}$

The electric field lines of an electric dipole are similar to the magnetic field lines of a bar magnet.

Figure 20.2 **A** Magnetic field lines near a bar magnet. **B** Electric field lines near an electric dipole.

Figure 20.3 A horseshoe magnet has a north pole and a south pole, and the lines of $\vec{B}$ form closed loops.

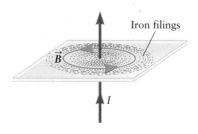

Iron filings

Figure 20.4 The magnetic field lines produced by a straight current-carrying wire form circles around the wire. The field lines can be visualized using iron filings. The magnitude of the field decreases as one moves away from the wire.

Right-hand rule number 1 relates the directions of the current and the magnetic field $\vec{B}$ produced by that current.

Figure 20.5 Ⓐ The direction of $\vec{B}$ is given by right-hand rule number 1. When the thumb of your *right* hand is aligned parallel to the current, your fingers will curl in the direction of $\vec{B}$. Ⓑ If the current is reversed, the lines of $\vec{B}$ circulate in the opposite direction.

bar magnet, and some of the resulting magnetic field lines are shown in Figure 20.3. There are north and south poles at the ends of the horseshoe, and the field lines again form closed loops as they circulate through the horseshoe. Permanent magnets with horseshoe shapes are used in many applications, including motors and generators. Key properties of these magnets are that the magnetic field is largest in the horseshoe gap and that it is directed across the gap from the north pole toward the south pole.

In the introduction to this chapter, we mentioned that magnetic fields are produced by moving electric charges. This statement also applies to the bar magnets and horseshoe magnets in Figures 20.1 through 20.3. The moving charges in these cases are electrons that move ("orbit") around the atoms in the magnetic material. We'll discuss the nature and size of the resulting currents, along with other properties of permanent magnet materials, in Section 20.8.

Magnetic Fields Produced by an Electric Current

We have already mentioned that moving charges produce a magnetic field. Since an electric current consists of moving charges, an electric current will also produce a magnetic field. Figure 20.4 shows long, straight wires carrying an electric current. Each wire passes through a hole in a piece of paper, and iron filings have been sprinkled on the paper. We have seen how iron filings line up parallel to the magnetic field lines produced by a bar magnet; their alignment in Figure 20.4 indicates the pattern of magnetic field lines near each wire. For the case of a straight wire, the magnetic field lines form circles as sketched in the figure and the direction of $\vec{B}$ is always tangent to these field line circles. In addition, the "strength" of the magnetic field (the magnitude of $\vec{B}$) decreases as one moves away from the wire.

So, the magnetic field lines circulate around a current-carrying wire, but how do we determine in which direction they circulate? The direction of the magnetic field produced by a current is given by the **right-hand rule** illustrated in Figure 20.5. To apply the right-hand rule, point the thumb of your *right* hand in the direction of the current, with your thumb parallel to the wire. Curling the fingers of your *right* hand around the wire then gives the direction of $\vec{B}$. *We call this method **right-hand rule number 1**.* (We'll encounter another right-hand-rule procedure below.)

If the direction of the current is reversed, the direction of $\vec{B}$ is also reversed. Figure 20.5B shows the same wire as in Figure 20.5A but with the current in the opposite direction. Applying right-hand rule number 1 shows that the lines of $\vec{B}$ now encircle the wire in the opposite sense.

Right-hand rule number 1 gives the direction of $\vec{B}$ but does not tell us its magnitude. We'll learn how to find the magnitude of $\vec{B}$ in Section 20.7.

Plotting Magnetic Fields and Field Lines

Since the magnetic field lines produced by a current-carrying wire encircle the wire, the relationship between I and $\vec{B}$ is inherently three dimensional, and three-

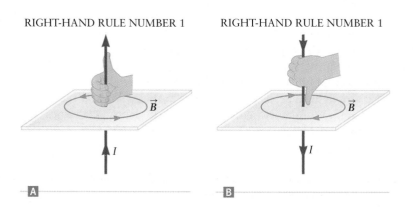

RIGHT-HAND RULE NUMBER 1   RIGHT-HAND RULE NUMBER 1

Ⓐ   Ⓑ

dimensional perspective plots likes those in Figures 20.4 and 20.5 will be extremely useful in our studies of magnetic fields. In many cases, however, we'll consider two-dimensional plots in which $\vec{B}$ points either into or out of the plane of the drawing. For example, Figure 20.6 shows the magnetic field of a long, straight wire. The three-dimensional representation of Figure 20.6A (similar to Fig. 20.5) shows how the field lines encircle the wire, whereas Figure 20.6B shows a two-dimensional plot of the same field. In the region above the wire, the magnetic field $\vec{B}$ points *out of* the x–y plane, whereas the field lines point *into* the x–y plane below the wire. In Figure 20.6, a large "dot" (•) denotes the tip of the $\vec{B}$ vector when it points out of the plane and a "cross" (×) denotes the tail of the $\vec{B}$ vector when it points into the plane.

Figure 20.6C shows a similar two-dimensional plot of the magnetic field produced by a current-carrying wire, but here the current is in the $-x$ direction. The magnetic field lines again encircle the wire, but are now directed out of the plane in the area below the wire and into the plane above the wire.

In Figures 20.5 and 20.6, we have drawn long, straight wires, but right-hand rule number 1 can also be used to find the direction of the magnetic field lines from a short piece of wire as sketched in Figure 20.7A. We again put the thumb of our right hand along the direction of I, and curling our fingers then gives the direction of $\vec{B}$. The magnetic field lines again form circles (as for a long wire), but now the magnitude of $\vec{B}$ falls off as we go farther from the short wire along the x axis in either direction.

The electric current in a short current-carrying wire can be modeled as a collection of positive electric charges moving with a velocity $\vec{v}$ parallel to the current direction (Fig. 20.7B). Each of these individual moving charges produces a magnetic field, and the direction of this magnetic field is given by right-hand rule number 1. Place your thumb of your right hand along the direction of $\vec{v}$; your fingers will then curl in the direction of the magnetic field (Fig. 20.7B). Figure 20.7C shows what happens to the magnetic field if the direction of the current is reversed. Here we have positive charges with $\vec{v}$ directed to the left, and applying right-hand rule number 1 shows that the field lines circulate opposite to those in Figure 20.7B. A positive charge moving to the left produces the same magnetic field as a negative charge moving to the right. Figure 20.7D shows such a moving negative charge; its magnetic field lines are the same as those produced by the positive charge in Figure 20.7C. These examples all illustrate how magnetic fields are produced by moving electric charges.

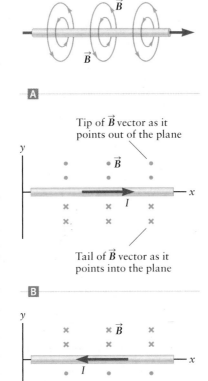

Figure 20.6 A Magnetic field near a long current-carrying wire. B and C In a two-dimensional plot, lines of $\vec{B}$ directed out of the plane are indicated by dots (the tip of the $\vec{B}$ vector), whereas lines of $\vec{B}$ pointing into the plane are denoted by crosses (the tail of the $\vec{B}$ vector).

CONCEPT CHECK 20.1 | Direction of the Magnetic Field

Part 1: A proton travels at a constant velocity, with $\vec{v}$ directed along the $+y$ direction as shown in Figure 20.8A. At a particular moment, the proton is at the origin. Is the direction of the magnetic field at point A (a) along the $+x$ direction, (b) along the $-x$ direction, (c) along the $+z$ direction, or (d) along the $-z$ direction?

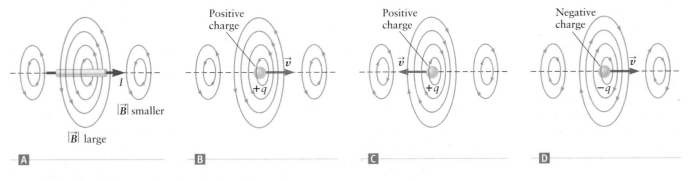

Figure 20.7 A Magnetic field lines produced by a short section of a current-carrying wire. B–D Magnetic fields produced by moving charges.

Figure 20.8 Concept Check 20.1.

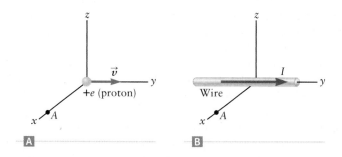

Part 2: The proton is replaced by a long, straight wire carrying a current in the $+y$ direction as shown in Figure 20.8B. Is the direction of the magnetic field at point A (a) along the $+x$ direction, (b) along the $-x$ direction, (c) along the $+z$ direction, or (d) along the $-z$ direction?

Magnetic Fields and Superposition

So far, we have considered the magnetic field lines produced by a bar magnet, by a current in a straight wire, and by a moving electric charge. The qualitative pattern of the magnetic field lines in most other situations can be understood using these results along with the *principle of superposition.*

The principle of superposition

> *Principle of superposition:* **The total magnetic field produced by two or more different sources is equal to the sum of the fields produced by each source individually.**

We can apply the principle of superposition to find the direction of the magnetic field produced by a "current loop." Consider a circular loop of wire carrying a current I as shown in Figure 20.9A. This case might seem very special and idealized, but current loops are actually found in many applications. We treat the loop as many small pieces and apply right-hand rule number 1 to find the field from each piece. The magnetic field lines encircle each segment of the wire, and for the current direction shown here, $\vec{B}$ is directed upward through the center of the loop and downward outside the loop. We then use the principle of superposition and add up the fields from all these pieces, giving the overall pattern of magnetic field lines in Figure 20.9B. The lines of $\vec{B}$ circulate through the current loop, all passing inside the loop and then circling around the outside.

The overall pattern of field lines from a current loop is very similar to the magnetic field produced by a bar magnet (Fig. 20.2A). The similarity between the magnetic fields of a current loop and a bar magnet is no accident. The field of a bar magnet and other permanent magnets is due to atomic-scale current loops, which we'll describe in Section 20.8.

Current out of plane

Current into plane

Figure 20.9 **A** Application of right hand rule number 1 to find direction of the field lines near the two sides of a current loop. **B** Magnetic field lines produced by a full current loop.

EXAMPLE 20.1 | Magnetic Field from Two Parallel Wires

Use the principle of superposition to sketch *qualitatively* the magnetic field lines near two long, straight wires, each parallel to the x axis and lying in the x–y plane. Assume each wire carries a current I, but with the currents in *opposite* directions.

RECOGNIZE THE PRINCIPLE

According to the principle of superposition, the total field is the sum of the fields from each wire. We use this principle along with our previous results for the magnetic field produced by a single wire (Figs. 20.5 and 20.6). Recall that for a single wire the field lines encircle the wire according to right-hand rule number 1 and the magnitude of $\vec{B}$ decreases as one moves away from the wire.

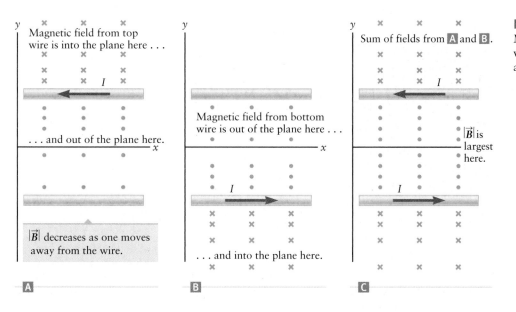

y

Magnetic field from top wire is into the plane here . . .

× × ×
× × ×
× × I ×

← I

• • •
• • •
• • •

. . . and out of the plane here.

x

• • •
• • •

$|\vec{B}|$ decreases as one moves away from the wire.

A

y

× × ×

Magnetic field from bottom wire is out of the plane here . . .

• • •
• • •
• • •

x

• • •
• • •
• • •
I • • •

→ I

× × ×
× × ×
× × ×

. . . and into the plane here.

× × ×

B

y

Sum of fields from **A** and **B**.

× × ×
× × ×
× × I ×

← I

• • •
• • •
• • •
• • •

$|\vec{B}|$ is largest here.

• • •
• I •

→ I

× × ×
× × ×
× × ×
× × ×

C

Figure 20.10 Example 20.1. Magnetic field due to **A** the top wire alone, **B** the bottom wire alone, and **C** both wires.

SKETCH THE PROBLEM

Figure 20.10A shows the magnetic field produced by one of the wires, and Figure 20.10B shows the field produced by the other. The field lines encircle both wires and pass perpendicularly through the plane of the drawing (the x–y plane).

IDENTIFY THE RELATIONSHIPS

The sum of the fields from the two wires is shown qualitatively in Figure 20.10C. In the region between the wires, the fields from both wires pass upward out of the plane of the drawing, so the total field in this region is also directed out of the plane. In the region above the top wire, the field from the top wire is directed into the plane while the field from the bottom wire is out of the plane, so these fields are in opposite directions.

SOLVE

The top wire is closest to points in the upper region of Figure 20.10C, so the field of the top wire is larger (in magnitude) than the field from the bottom wire in that region. The sum of the two fields at the top of Figure 20.10C therefore points into the plane. The same reasoning shows that the field in the region below the bottom wire in Figure 20.10C is also directed into the plane.

What have we learned?

The principle of superposition can be used to find the pattern of magnetic field lines in virtually all situations. Most cases can be analyzed as a collection of bar magnets, straight wires, or current loops; the total magnetic field is then the sum of the fields produced by all the individual parts of the collection.

20.2 | MAGNETIC FORCES INVOLVING BAR MAGNETS

In Section 20.1, we described the magnetic fields produced by various sources. Let's now consider the magnetic *forces* that these fields produce, beginning with permanent magnets. Figure 20.11A shows a bar magnet situated in a constant magnetic field; when we say that $\vec{B}$ is constant, we mean that the magnetic field has a constant direction and a constant magnitude. To describe the total force on the bar magnet, we must include two forces, acting on the two poles of the bar. The force on the north pole is directed parallel to $\vec{B}$, while the force on the south pole is antiparallel

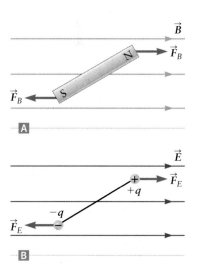

Figure 20.11 **A** Magnetic forces on a bar magnet placed in a uniform magnet field. **B** Electric forces on an electric dipole placed in a uniform electric field.

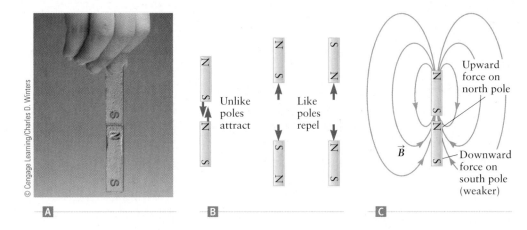

Figure 20.12 **A** and **B** Forces between two bar magnets. Unlike magnetic poles attract, and like poles repel. **C** The force on the north pole of the bottom magnet is upward, parallel to the field $\vec{B}$ of the upper magnet. The force on the south pole of the bottom magnet is downward. Since the field from the top magnet becomes smaller as one moves downward, the total (net) force on the bottom magnet is upward.

(opposite) to $\vec{B}$. For the case of a constant magnetic field, these two forces are equal in magnitude and in opposite directions. Although the total force is thus zero, these two forces produce a nonzero torque on the bar magnet, which acts to align the bar magnet along the field lines. That is why iron filings (Figs. 20.1 and 20.4) align with the magnetic field. (Recall that each iron filing is a small bar magnet.)

It is interesting to compare the magnetic forces and torque on the bar magnet in Figure 20.11A with the electric forces on an electric dipole as sketched in Figure 20.11B. The figure shows an electric dipole formed by positive and negative charges $+q$ and $-q$ attached to the ends of an uncharged rod. This electric dipole is placed in a constant electric field $\vec{E}$, which exerts a force directed to the right on $+q$ and a force directed to the left on $-q$. These forces thus produce a torque on the electric dipole, just like the torque on the bar magnet in Figure 20.11A. This similarity illustrates that a bar magnet possesses a ***magnetic moment***. This comparison also suggests that we can view the north and south poles of the bar magnet as a sort of "magnetic charge." The analogy with electric charge then implies that the north pole of one bar magnet should attract the south pole of another magnet (Fig. 20.12A), just as positive and negative electric charges attract each other. Likewise, we expect that "like" magnetic poles (two north poles or two south poles) will repel each other, just as found with two "like" electric charges (Fig. 20.12B). Such forces between the poles of permanent magnets do indeed occur and are one reason bar magnets make popular toys.

The forces between magnetic poles and between two bar magnets can also be analyzed and understood in terms of how the magnetic field produced by one magnet exerts a force on the other magnet. Figure 20.12C shows two bar magnets as well as the magnetic field produced by the top magnet. The central "axes" of the two magnets are aligned, so the magnetic field at the north pole of the lower magnet is directed upward, producing an upward force on this pole. The magnetic field at the bottom end of the lower magnet is also directed upward, which gives a downward force on the south pole. The field is weaker at the lower end, however, because it is farther from the upper magnet. As a result, the total force on the lower bar magnet is upward, and the two magnets attract.

The similar behavior of electric charges and magnetic poles, and electric and magnetic forces, in Figures 20.11 and 20.12 can help you visualize and understand magnetic fields and forces. However, there are some important differences between the electric and magnetic cases. In particular, north and south magnetic poles *always* occur in pairs. We'll see why it is not possible to obtain an isolated magnetic pole when we study what goes on inside a bar magnet in Section 20.8.

20.3 | MAGNETIC FORCE ON A MOVING CHARGE

Magnetic forces act on electric charges, such as individual electrons, protons, or ions. The magnetic force depends on the velocity of the charge, and it is nonzero only if the charge is in motion. This is very different from gravity and the electric force, both of which are completely independent of velocity. The dependence of magnetic forces on velocity has some deep philosophical implications that we'll discuss in Section 20.11.

Figure 20.13A shows a particle with positive charge q moving with velocity $\vec{v}$. Here $\vec{v}$ is along the x direction, and the magnetic field $\vec{B}$ is parallel to y. The magnetic force on this charge is along the $+z$ direction, with a magnitude

$$F_B = qvB \sin \theta \qquad (20.1)$$

where θ is the angle between $\vec{v}$ and $\vec{B}$. In Figure 20.13, this angle is 90°, so $\sin \theta$ in Equation 20.1 is equal to 1. Notice that Equation 20.1 gives only the *magnitude* of the magnetic force. The *direction* of the force can be found using another right-hand rule illustrated in Figure 20.13B. Point the fingers of your right hand in the direction of $\vec{v}$ and curl your fingers in the direction of $\vec{B}$, taking the smallest angle between $\vec{v}$ and $\vec{B}$. Your thumb then points in the direction of the magnetic force $\vec{F}_B$. This procedure gives the direction of the force when q in Equation 20.1 is positive. For a positive charge ($q > 0$) with $\vec{v}$ and $\vec{B}$ directed as shown in Figure 20.13, the magnetic force is along the $+z$ direction. On the other hand, if q is negative, such as for an electron, the force direction is reversed and in this example is along $-z$. For a negative charge q, the value of F_B in Equation 20.1 is negative, so you can think of this as just reversing the direction of the magnetic force.

We'll call the procedure for finding the direction of the magnetic force on a moving charge in Figure 20.13B **right-hand rule number 2**. This procedure is similar to right-hand rule number 1 (Fig. 20.5), which gives the direction of the magnetic field from a current. You should practice using both rules.

Compared with other fundamental forces such as the gravitational and electric forces, the magnetic force in has several remarkable features. One is the dependence of $\vec{F}_B$ on velocity that we already noted. Also, the direction of $\vec{F}_B$ is always perpendicular to both the magnetic field and the particle's velocity. That is quite different from the gravitational and electric fields for which the force and the field are always parallel. Indeed, we originally introduced these fields as being simply proportional in magnitude and parallel in direction to the gravitational and electric forces. We now find that fields and forces (at least magnetic ones) need *not* be related in such a simple way.

CONCEPT CHECK 20.2 | Direction of the Magnetic Force

Part 1: Figure 20.14 shows a positive charge q moving with a nonzero speed v. If the magnetic field is directed to the right, is the direction of the magnetic force (a) into the plane of the drawing, (b) out of the plane of the drawing, (c) to the left, or (d) is the magnetic force zero?

Part 2: If the particle stops, is the direction of the magnetic force (a) into the plane of the drawing, (b) out of the plane of the drawing, (c) to the left, or (d) is the magnetic force zero?

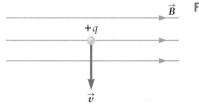

Figure 20.14 Concept Check 20.2.

Magnetic force on a charged particle

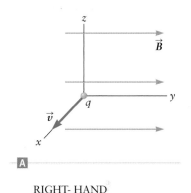

A

RIGHT-HAND RULE NUMBER 2

If $q > 0$, the force will be along $+z$.

Right-hand rule number 2 gives the direction of magnetic force on a moving charged particle.

B

Figure 20.13 **A** A charged particle moving in a magnetic field. **B** The direction of the magnetic force exerted on the particle is given by right-hand rule number 2. Place the fingers of your right hand in the direction of $\vec{v}$ and wrap them in the direction of $\vec{B}$. If $q > 0$, your thumb then points in the direction of the magnetic force.

Magnetic Force on an Electron: How Big Is It?

To get a feeling for the size of magnetic forces, consider an electron moving near a bar magnet. The field very close to a bar magnet can be as large as 1 T or 2 T, so let's take $B = 1$ T for this example. The electron's speed depends on the situation; let's assume a typical terrestrial speed of 50 m/s (about 100 mi/h). The charge of an electron is $q = -e = -1.60 \times 10^{-19}$ C. For simplicity, we assume the angle θ between $\vec{v}$ and $\vec{B}$ is 90° and insert these values for q, v, and B into Equation 20.1. The magnitude of the force is

$$F_B = qvB \sin\theta = (1.60 \times 10^{-19} \text{ C})(50 \text{ m/s})(1 \text{ T})\sin(90°)$$

$$= 8 \times 10^{-18} \text{ C} \cdot \text{m} \cdot \text{T/s} \qquad (20.2)$$

Here, F_B must have units of force, so in the SI system the combination of units in Equation 20.2 must be equal to newtons (N). In terms of just the units, Equation 20.2 is thus

$$1 \text{ N} = 1 \text{ C} \cdot \text{m} \cdot \text{T/s}$$

We can now rearrange to solve for the tesla in terms of the primary SI units:

$$1 \text{ T} = 1 \frac{\text{N} \cdot \text{s}}{\text{C} \cdot \text{m}} = 1 \frac{(\text{kg} \cdot \text{m/s}^2) \cdot \text{s}}{\text{C} \cdot \text{m}} \qquad (20.3)$$

$$1 \text{ T} = 1 \frac{\text{kg}}{\text{C} \cdot \text{s}} \qquad (20.4)$$

Returning to Equation 20.2, we find that the force on an electron moving at 50 m/s near a typical bar magnet is

$$F_B = 8 \times 10^{-18} \text{ N} \qquad (20.5)$$

To judge if this force is large or small, let's compare F_B with the gravitational force on the electron. For an electron near the Earth's surface,

$$F_{\text{grav}} = mg = (9.11 \times 10^{-31} \text{ kg})(9.8 \text{ m/s}^2) = 8.9 \times 10^{-30} \text{ N}$$

Comparing with F_B in Equation 20.5, we see that the gravitational force on an electron is smaller than the magnetic force in this case by a factor of

$$\frac{F_{\text{grav}}}{F_B} = \frac{8.9 \times 10^{-30}}{8 \times 10^{-18}} \approx 1 \times 10^{-12}$$

So, this magnetic force is *much* larger that the gravitational force. Of course, this comparison depends on the electron's velocity and the magnetic field, but one typically finds that magnetic forces are very much larger than gravitational ones.

Motion of a Charged Particle in a Magnetic Field

Let's now consider the direction of $\vec{F}_B$ and how this force affects the motion of a charged particle in a little more detail. For simplicity, we assume the magnetic field is constant, so the magnitude and direction of $\vec{B}$ are the same everywhere. Figure 20.15A shows a charged particle moving parallel to the direction of $\vec{B}$. In this case, the angle between $\vec{v}$ and $\vec{B}$ is $\theta = 0$; the factor $\sin\theta$ in Equation 20.1 is then zero, so the magnetic force in this case is also zero. *If a charged particle has a velocity parallel to $\vec{B}$, the magnetic force on the particle is zero.*

Figure 20.15B shows a charged particle moving perpendicular to $\vec{B}$. Now we have $\theta = 90°$, so $\sin\theta = 1$. The magnitude of the magnetic force is thus $F_B = qvB$, and the force is perpendicular to $\vec{v}$. In Chapter 5, we learned that when a particle experiences a force of constant magnitude perpendicular to its velocity, the result is *circular motion* as sketched in Figure 20.15B. Hence, *if a charged particle is moving perpendicular to a constant magnetic field, the particle will move in a circle.* This circle lies in the plane perpendicular to the field lines. We can calculate the radius of

---

MOTION OF A POSITIVELY
CHARGED PARTICLE

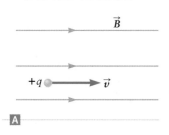

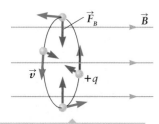

$\theta = 90°$, the charged particle moves in a circle that lies in a plane perpendicular to the plane of this page.

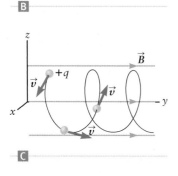

Figure 20.15 **A** When the velocity of a charged particle is parallel to the magnetic field $\vec{B}$, the magnetic force on the particle is zero. **B** When $\vec{v}$ makes a right angle with $\vec{B}$, the particle will move in a circle. **C** When the velocity has nonzero components parallel and perpendicular to $\vec{B}$, the particle will move along a helical path.

the circle by recalling that for a particle to move in a circle of radius r, there must a force of magnitude mv^2/r directed toward the center of the circle (Eq. 5.6). Here the force producing circular motion is the magnetic force, so we have

$$F_B = \frac{mv^2}{r}$$

Inserting F_B from Equation 20.1 and using $\sin \theta = 1$ leads to

$$qvB = \frac{mv^2}{r}$$

Solving for r gives

$$r = \frac{mv^2}{qvB} = \frac{mv}{qB} \tag{20.6}$$

Let's calculate the value of r for the electron we considered in connection with Equation 20.5. In that example, we had $v = 50$ m/s and $B = 1$ T. Inserting those values into Equation 20.6, we find

$$r = \frac{mv}{qB} = \frac{(9.11 \times 10^{-31}\ \text{kg})(50\ \text{m/s})}{(1.60 \times 10^{-19}\ \text{C})(1\ \text{T})} \tag{20.7}$$

$$r = 3 \times 10^{-10}\ \text{m} \tag{20.8}$$

which is *quite* a small radius, approximately the radius of an atom. A faster-moving electron (larger v) will move in a circle with a larger radius.

EXAMPLE 20.2 | ® Magnetic Fields and Television Picture Tubes

Some televisions and computer monitors produce images using an electron beam that strikes a fluorescent screen as shown in Figure 20.16. The electrons are steered to specific spots on the screen by magnetic fields, forming spots of different brightness and color according to the desired image. The magnetic field from an external bar magnet placed near the television can deflect the electron beam in an unwanted direction and thereby distort or ruin the image. (You can also produce "artistic" results; see the photo at the beginning of this chapter.) Assuming the electrons travel at a speed of approximately 5×10^7 m/s, find the approximate magnetic field necessary to disrupt a television image.

RECOGNIZE THE PRINCIPLE

We wish to find the field needed to just deflect the electrons a significant amount, so we do not need to calculate the precise electron trajectory. Instead, we know that an electron traveling perpendicular to a magnetic field is deflected into a circular path with a radius r given by Equation 20.6, so we'll calculate the value of B that gives a radius comparable to the size of a TV picture tube.

SKETCH THE PROBLEM

Figure 20.16 shows the problem. A beam of electrons passes through a region in the "neck" of the tube where there are current loops that produce a magnetic field directed perpendicular to the electron velocity.

IDENTIFY THE RELATIONSHIPS

A typical television picture tube has a size of about 50 cm, so we take that for our value of r. We have already discussed how the radius of a charged particle's circular motion is related to B. From Equation 20.6, we have

$$r = \frac{mv}{qB}$$

Insight 20.1
DEFLECTION COILS IN A TV
The television tube in Figure 20.16 uses magnetic fields in the "neck" of the tube to control the electron deflection. The electron velocity is to the right in this figure, and there are two deflecting fields; one field is vertical as shown in the figure, and another (not shown) is perpendicular to the plane of the drawing. The magnetic force $\vec{F}_B$ is perpendicular to both $\vec{v}$ and $\vec{B}$, so the vertical deflecting field in Figure 20.16 produces a *horizontal* deflection force that is perpendicular to the plane of the drawing. When the horizontal magnetic field is added, there is also a vertical deflection force. Controlling these two magnetic fields controls the overall deflection and determines the location at which the electron beam strikes the TV screen.

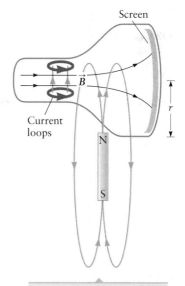

The external magnetic field from a bar magnet can deflect the electrons in a TV picture tube.

Figure 20.16 Example 20.2. Electron motion in a TV picture tube.

Rearranging to solve for B gives

$$B = \frac{mv}{qr}$$

We don't need to worry here about the charge on the electron being negative; it affects the direction in which the electron moves along its circular trajectory, but does not change the radius of the circle.

SOLVE

Inserting the given value of v, our estimate for r, and the known values of the charge and mass of an electron, we find

$$B = \frac{mv}{qr} = \frac{(9.11 \times 10^{-31}\text{ kg})(5 \times 10^{7}\text{ m/s})}{(1.60 \times 10^{-19}\text{ C})(0.50\text{ m})} = \boxed{6 \times 10^{-4}\text{ T}}$$

What does it mean?
This answer is about 10 times larger than the Earth's field. Hence, it does not take a very strong magnet to disrupt the image on a television or a computer monitor; almost any bar magnet will do. We don't recommend that you try this experiment at home because it can do permanent damage to your TV or monitor!

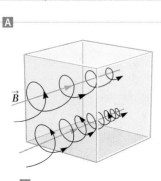

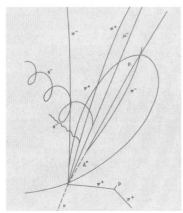

A

B

Figure 20.17 A Photo of the paths of charged particles as they move through a bubble chamber. B Schematic particle paths in three dimensions. In some cases, the particles lose energy as they move, causing the trajectories to spiral inward. As the particle speed decreases, the radius of the spiral trajectory also decreases.

In most cases, the velocity of a charged particle is not perfectly parallel or perpendicular to $\vec{B}$. Figure 20.15C shows the particle trajectory in such cases: the particle moves in a helical (spiral) path, with the axis of the helix parallel to $\vec{B}$. The helical motion can be understood as a combination of the results in parts A and B of Figure 20.15. For the component of $\vec{v}$ parallel to the magnetic field, there is no magnetic force and this component of the particle's velocity is constant. The perpendicular velocity component produces results as in Figure 20.15B: circular motion in a plane perpendicular to the field lines. The combination of circular motion in the x–z plane and a constant velocity along y gives a helical trajectory (Fig. 20.15C). Hence, *a charged particle will spiral around the magnetic field lines.*

A dramatic example of this spiral motion is shown in Figure 20.17A. This photograph is from a bubble chamber, a device used to study the trajectories of electrons, protons, and other particles. When charged particles travel through certain fluids at very high speed, they leave behind a trail of ionized fluid molecules, visible in the photograph as tracks of bubbles. The entire bubble chamber is inside a large magnet (Fig. 20.17B) whose magnetic field causes the tracks to be helical. Measuring the radius of these helical tracks gives information about a particle's charge, mass, and velocity through the relation in Equation 20.6.

CONCEPT CHECK 20.3 | Bubble Chamber Trajectories

Part 1: Suppose an electron and a proton with the same velocity are traveling in the same direction when they enter the bubble chamber in Figure 20.17. Each moves in a helical trajectory inside the chamber. How does the radius of the electron's helix compare to the radius of the proton's helix?
  (a) The helix followed by the electron has a smaller radius.
  (b) The helix followed by the electron has a larger radius.
  (c) The electron and proton helices have the same radius.

Part 2: Sometimes the particle path in a bubble chamber is not a perfect helix, but instead the particle "spirals inward." (A few such paths are visible in Fig. 20.17.) What causes these paths to deviate from a simple helix?
  (a) The particle loses energy and slows down as it travels through the bubble chamber.

(b) The particle gains energy and speeds up as it travels through the bubble chamber.

(c) Other forces (such as gravity) act on the particle.

EXAMPLE 20.3 | Direction of the Magnetic Force on a Moving Charge

Consider a positively charged particle traveling parallel to a current-carrying wire as sketched in Figure 20.18A. What is the *direction* of the magnetic force on this particle?

RECOGNIZE THE PRINCIPLE

Right-hand rule number 2 gives the direction of the magnetic force on a moving charge. To apply this rule, we must first establish the direction of the magnetic field produced by the wire, which we can find using right-hand rule number 1.

SKETCH THE PROBLEM

Figure 20.18 shows the direction of the current and the direction of $\vec{v}$, the particle's velocity.

IDENTIFY THE RELATIONSHIPS AND SOLVE

From our work with current-carrying wires and right-hand rule number 1 (Figs. 20.5 and 20.6), the magnetic field lines encircle the wire and are directed *into the plane* of the drawing at the particle's location (Fig. 20.18B). To find the magnetic force on this particle, we now apply right-hand rule number 2, placing our fingers along $\vec{v}$ and curling them in the direction of $\vec{B}$ as shown in Figure 20.18B. The thumb then points in the direction of the force, *toward* the wire as shown in the figure.

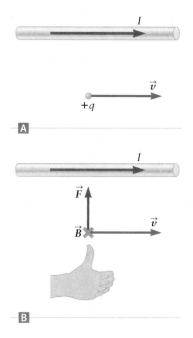

Figure 20.18 Example 20.3. Finding the direction of $\vec{F}_B$.

What have we learned?

To find the magnetic force on a moving charged particle, we must first find the direction of the magnetic field; here we used right-hand rule number 1 to find the direction of $\vec{B}$. We then applied right-hand rule number 2 to find the direction of the magnetic force. To deal with magnetic forces and fields, you *must* be able to apply the right-hand rules. Practice is strongly recommended.

CONCEPT CHECK 20.4 | Finding the Direction of $\vec{B}$

A positively charged particle is moving with a velocity $\vec{v}$ as shown in Figure 20.19. If the magnetic force on this particle is directed out of the plane of the drawing, is the possible direction of $\vec{B}$ (a) along direction 1 in the figure, (b) along direction 2, (c) along direction 3, or (d) directed into the plane of the drawing?

Applying the Right-Hand Rules

We have introduced two right-hand rules and applied them in several example calculations. Let's state them once more for future reference.

Right-hand rule number 1: Finding the direction of $\vec{B}$ from an electric current.

1. Place the thumb of your right hand along the direction of the current.

2. Curl your fingers; they will then give the direction of $\vec{B}$ as the field lines encircle the current.

Right-hand rule number 2: Finding the direction of the magnetic force on a moving charge q.

1. Point the fingers of your right hand along the direction of $\vec{v}$.

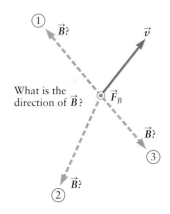

Figure 20.19 Concept Check 20.4.

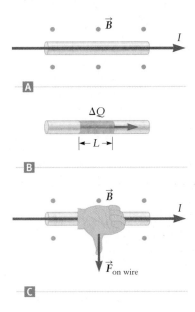

Figure 20.20 **A** A current-carrying wire in an external magnetic field. **B** The current in the wire is due to the motion of charge ΔQ along the wire. **C** The magnetic force on this moving charge produces a force on the wire. The direction of the force is given by right-hand rule number 2. Place the fingers of your right hand in the direction of the current and wrap your fingers in the direction of $\vec{B}$. Your thumb then points in the direction of the magnetic force.

Magnetic force on a current-carrying wire. The direction is given by right-hand rule number 2.

2. Curl your fingers in the direction of $\vec{B}$. Always curl through the smallest angle that connects $\vec{v}$ and $\vec{B}$.

3. If q is positive, the magnetic force on q is parallel to your thumb. If q is negative, the magnetic force is in the opposite direction.

20.4 | MAGNETIC FORCE ON AN ELECTRIC CURRENT

An electric current consists of a collection of moving charges, so our result for the magnetic force on a single moving charge (Eq. 20.1) can be used to find the magnetic force on a current-carrying wire. This force is important in many applications, including electric motors. Consider a current-carrying wire placed in an external magnetic field as sketched in Figure 20.20A, with $\vec{B}$ constant and perpendicular to the wire. This magnetic field is produced by some external source; it is *not* produced by the current in the wire. We can calculate the magnetic force on the wire due to this external field by adding up the magnetic forces on all the moving charges in the wire.

Let's focus on the segment of the wire of length L in Figure 20.20B. The current in this segment is

$$I = \frac{\Delta Q}{\Delta t} \tag{20.9}$$

where ΔQ is the electric charge that passes by one end of the wire segment in a time Δt; this is just our usual relation between charge and current from Chapter 19. Applying Equation 20.1, the magnetic force on this moving charge is

$$F_B = qvB = (\Delta Q)vB \tag{20.10}$$

The velocity of the charge is just

$$v = \frac{L}{\Delta t}$$

Inserting this into Equation 20.10 gives

$$F_B = (\Delta Q)vB = (\Delta Q)\frac{L}{\Delta t}B$$

Using the relation between current and charge in Equation 20.9, we get

$$F_B = \frac{\Delta Q}{\Delta t}LB = ILB$$

This force on the moving charge is really a force on the wire, so we arrive at

$$F_{\text{on wire}} = ILB$$

In Figure 20.20, we assumed the magnetic field is perpendicular to the wire. If $\vec{B}$ makes an angle θ with the wire, one finds

$$F_{\text{on wire}} = ILB \sin \theta \tag{20.11}$$

The direction of $F_{\text{on wire}}$ is given by (you guessed it) a right-hand rule as illustrated in Figure 20.20C. Begin with the fingers of your right hand in the direction of the current and curl them in the direction of the field. Your thumb then points in the direction of $\vec{F}_{\text{on wire}}$, downward in Figure 20.20C. The external magnetic field pulls this wire downward as long as the current is directed to the right. The magnetic force on a current is due to the force on a moving charge, so the right-hand rule used here is another example of right-hand rule number 2, with the direction of $\vec{v}$ for a moving charge replaced by the direction of the current.

The wire in Figure 20.21 carries a current along the $+z$ direction. If there is an external magnetic field in the $+y$ direction, is the direction of the magnetic force on the wire along (a) $+x$, (b) $-x$, (c) $+y$, (d) $-y$, (e) $+z$, or (f) $-z$?

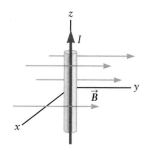

Figure 20.21 Concept Check 20.5.

EXAMPLE 20.4 | Magnetic Force between Two Wires

Consider two parallel wires, each carrying a current I as sketched in Figure 20.22A. Find the direction of the magnetic force that the top wire exerts on the bottom one. Then find the direction of the force that the bottom wire exerts on the top one.

RECOGNIZE THE PRINCIPLE

To find the force exerted on the bottom wire, we must first determine the direction of the magnetic field produced by the top wire, which we can do by using right-hand rule number 1. We can then apply right-hand rule number 2 to find the force produced by this field on the bottom wire.

SKETCH THE PROBLEM

Figure 20.22B shows the direction of the magnetic field produced by the top wire at the location of the bottom wire as well as the application of right-hand rule number 2 to find the force on the bottom wire. Figure 20.22C shows the application of right-hand rule number 2 to find the force on the top wire.

IDENTIFY THE RELATIONSHIPS AND SOLVE

From right-hand rule number 1, the magnetic field produced by the top wire is directed into the plane of the drawing near the bottom wire as shown in Figure 20.22B. (See also Fig. 20.6.) Applying right-hand rule number 2 then gives a magnetic force on the bottom wire directed upward, and the bottom wire is *attracted* to the top wire.

Now consider the force on the top wire. The magnetic field produced by the bottom wire in the vicinity of the top wire is directed out of the plane of the drawing (Fig. 20.22C), and applying right-hand rule number 2 shows that the force exerted on the top wire is directed downward, toward the bottom wire. Hence, the force on the top wire is downward and the force on the bottom wire is upward. Two wires carrying parallel currents are attracted to each other.

What does it mean?

The magnetic forces exerted by each wire on the other are an action–reaction pair of forces. Hence, in accord with Newton's third law (the action–reaction principle), these forces must be in opposite directions, which is precisely what we have found. You should expect that these forces must have equal magnitudes; we'll show this in Section 20.7.

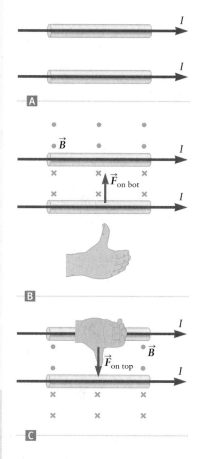

Figure 20.22 Example 20.4. ◼A Two parallel current-carrying wires. ◼B The magnetic force on the bottom wire. ◼C The magnetic force on the top wire.

Consider again the two wires in Figure 20.22 where the currents are parallel and both directed to the right. Suppose the currents are instead in opposite directions, with I_{top} to the right and I_{bottom} to the left. Which of the following statements is correct?

  (a) The force on the bottom wire is still upward in Figure 20.22, but the force on the top wire is now also upward.
  (b) The force on the bottom wire is now downward and the force on the top wire is upward, so the overall effect is a repulsive force.
  (c) There is no change.

Figure 20.23 Torque on a current loop. **A** The magnetic forces on the sides of this loop tend to rotate the loop clockwise as viewed from the front, along the axis. **B** When the plane of the loop is perpendicular to $\vec{B}$, the torque is zero. **C** The torque on a current loop depends on the angle θ the field makes with the direction perpendicular to the loop.

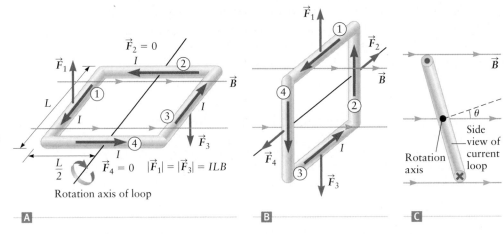

20.5 | TORQUE ON A CURRENT LOOP AND MAGNETIC MOMENTS

In Section 20.2, we saw that when a bar magnet is placed in a uniform magnetic field, there is a torque on the magnet (Fig. 20.11). A magnetic field can also produce a torque on a current loop. Figure 20.23 shows a square current loop with sides of length L carrying a current I in a constant magnetic field. In Figure 20.23A, the field is parallel to the plane of the loop and the forces on all four sides of the loop are shown. These forces can all be found using right-hand rule 2. In sides 2 and 4, the current is either parallel or antiparallel to $\vec{B}$, so the forces are zero. The forces on sides 1 and 3, denoted $\vec{F}_1$ and $\vec{F}_3$, respectively, are in opposite directions in Figure 20.23A and produce a torque around the loop's axis. The lever arms for these two forces are both $L/2$, and magnitude of this torque is

$$\tau = F_1\left(\frac{L}{2}\right) + F_3\left(\frac{L}{2}\right)$$

The force on each wire segment is $F_1 = F_3 = ILB$ (from Eq. 20.11), so

$$\tau = F_1\left(\frac{L}{2}\right) + F_3\left(\frac{L}{2}\right) = 2ILB\left(\frac{L}{2}\right)$$

$$\tau = IL^2B \tag{20.12}$$

Figure 20.23B shows the forces on the loop when its plane is perpendicular to the magnetic field. Now there is a nonzero force on all four sides of the loop; the four forces are of equal magnitude $F_1 = F_2 = F_3 = F_4 = ILB$ but are in different directions, and the total force on the loop is zero. The total torque is now also zero. Comparing parts A and B of Figure 20.23, we find that the torque on a current tends to align the plane of the loop perpendicular to $\vec{B}$.

So far, we have considered the torque when the plane of the loop is perpendicular or parallel to the field. When the angle between these two directions is θ (Fig. 20.23C), the torque is

$$\tau = IL^2B \sin\theta$$

The factor of L^2 is just the area of the square loop. For loops with different shapes, including circular, the torque on a loop of area A is

$$\tau = IAB \sin\theta \tag{20.13}$$

The torque on a current loop is very similar to the torque on a bar magnet; both the loop and the magnet tend to "line up" with the magnetic field. Both act as a ***magnetic moment***, with the direction of the magnetic moment either along the axis of the bar magnet or perpendicular to the current loop as shown in Figure 20.24.

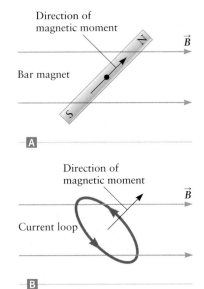

Figure 20.24 **A** The magnetic moment of a bar magnet is along the axis of the magnet, whereas **B** the magnetic moment of a current loop is perpendicular to the loop. In both cases, a magnetic field produces a torque that tends to align the magnetic moment with the field.

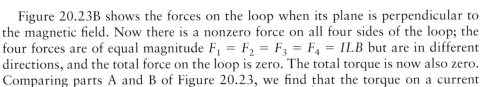

The magnetic torque tends to align the magnetic moment parallel to $\vec{B}$, and the strength of the torque depends on the magnitude of the magnetic moment. For a current loop, the magnitude of the magnetic moment equals IA, the product of the current and loop area.

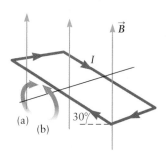

CONCEPT CHECK 20.7 | Torque on a Current Loop

Figure 20.25 shows a current loop in a magnetic field. Is there a torque on the loop and, if so, in what direction?
  (a) There is a torque that tends to rotate the loop in direction (a), clockwise.
  (b) There is a torque that tends to rotate the loop in direction (b), counter-clockwise.
  (c) The torque is zero.

Figure 20.25 Concept Check 20.7.

20.6 | MOTION OF CHARGED PARTICLES IN THE PRESENCE OF ELECTRIC AND MAGNETIC FIELDS

We considered the magnetic force on a charged particle in a magnetic field in Section 20.3 and found that various types of motion, including circular and helical trajectories, are possible. Also of interest is the motion of an electric charge in the presence of a *combination* of magnetic and electric fields. We'll analyze a few such cases and describe some applications in this section.

The Mass Spectrometer

Scientists often want to separate ions according to their mass or charge. For example, archaeologists use a technique called carbon dating to determine an object's age. This technique is based on the observations that there are two isotopes of carbon atoms—denoted ^{12}C and ^{14}C (called "carbon 12" and "carbon 14," respectively)—and that the relative amounts of ^{12}C and ^{14}C in certain types of materials such as an animal bone can be used to determine the bone's age. (See Section 30.5.) One way to distinguish ^{12}C from ^{14}C is through the difference in their masses, which can be done using a mass spectrometer.

Figure 20.26 shows the essential parts of a mass spectrometer. The charges are typically ions such as $^{12}C^+$ or $^{14}C^+$ that enter from the left, and for simplicity we'll assume they all have the same speed v. The ions then pass into a region in which the magnetic field is perpendicular to the velocity. In Figure 20.26, the magnetic field is out of the plane and perpendicular to the plane of the drawing, producing a magnetic force that causes the ions to move in a circle (as in Fig. 20.15B). The radius of this circle is (Eq. 20.6)

$$r = \frac{mv}{qB} \tag{20.14}$$

where v is the velocity and q is the charge of the ion.

Suppose the incoming ions are of two types, $^{12}C^+$ and $^{14}C^+$, with different masses m_{12} and m_{14}. According to Equation 20.14, the different values of m lead to different values of r, so the two types of ions follow different circular arcs through the mass spectrometer. A mass spectrometer has an ion detector positioned at a certain radius r as shown in Figure 20.26. The value of either v or B is then adjusted until the detector gives a maximum ion current, indicating that the ion's circular trajectory has the radius r of the detector. For our hypothetical example with $^{12}C^+$ and $^{14}C^+$, the mass spectrometer could be used to measure the ion currents for $^{12}C^+$ and $^{14}C^+$ separately and thereby determine the relative amounts of the two isotopes in an object, which would then be used to find the object's age.

Another application of mass spectrometers is to "fingerprint" unknown molecules. Molecules of unknown composition are first ionized—by bombarding them

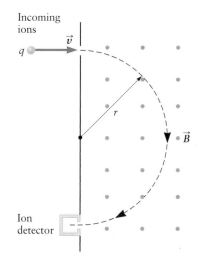

Figure 20.26 Design of a mass spectrometer.

with electrons, for instance—producing several different types of ions that are then fed to a mass spectrometer. The spectrometer then measures the values of v, B, and r corresponding to each ion. Equation 20.14 can be rearranged to give

$$\frac{q}{m} = \frac{v}{rB}$$

Because v, B, and r are all measured, the mass spectrometer gives q/m, the ratio of charge to mass for the ions. Chemists use this information to deduce the composition of each ion and hence the composition of the original molecule.

EXAMPLE 20.5 | ⊗ **Using a Mass Spectrometer**

A mass spectrometer (Fig. 20.26) is used in a carbon dating experiment. The incoming ions are a mixture of $^{12}C^+$ and $^{14}C^+$ having a velocity $v = 1.0 \times 10^5$ m/s, and the magnetic field is $B = 0.10$ T. The ion detector is first positioned to find the value of r for $^{12}C^+$ and then moved to find the value of r for $^{14}C^+$. How far must the detector move?

RECOGNIZE THE PRINCIPLE

The location of the detector depends on r, the radius of the ion's circular path. The velocity v and field B are given, and the ions $^{12}C^+$ and $^{14}C^+$ each have a charge $+e$. Equation 20.14 gives the value of r in terms of v, B, and m, the mass of the ion. If we can also find the mass m of each ion, we can therefore calculate r.

SKETCH THE PROBLEM

Figure 20.26 shows the problem. The ions follow a circular arc from where they enter the mass spectrometer to where they enter the detector.

IDENTIFY THE RELATIONSHIPS

The value of r in the mass spectrometer depends on the mass of the ion, so we need to find the masses of $^{12}C^+$ and $^{14}C^+$. The ion $^{12}C^+$ contains six protons, six neutrons, and five electrons. Consulting the table of fundamental constants on the inside front cover, we find a total mass of

$$m_{12} = 6m_p + 6m_n + 5m_e$$

$$m_{12} = 6(1.67 \times 10^{-27} \text{ kg}) + 6(1.67 \times 10^{-27} \text{ kg}) + 5(9.11 \times 10^{-31} \text{ kg})$$

$$m_{12} = 2.0 \times 10^{-26} \text{ kg}$$

A similar calculation gives the mass of $^{14}C^+$:

$$m_{14} = 2.3 \times 10^{-26} \text{ kg}$$

SOLVE

From Equation 20.14, the radius in the mass spectrometer for $^{12}C^+$ is

$$r_{12} = \frac{m_{12}v}{qB}$$

and the charge is $q = +1.60 \times 10^{-19}$ C. Inserting the given values, we find

$$r_{12} = \frac{m_{12}v}{qB} = \frac{(2.0 \times 10^{-26} \text{ kg})(1.0 \times 10^5 \text{ m/s})}{(1.60 \times 10^{-19} \text{ C})(0.10 \text{ T})} = 0.13 \text{ m}$$

The same approach gives

$$r_{14} = 0.14 \text{ m}$$

The ion detector must move a distance equal to the difference in the diameters of the circular trajectories, so it must move a distance

$$\Delta r = 2(r_{14} - r_{12}) \approx \boxed{0.02 \text{ m}}$$

or about 2 cm.

What does it mean?
The difference between r_{12} and r_{14} is not large, so to get a more accurate value we would need to employ more significant figures in our calculation. Even so, we can see that the trajectories of the two ions would differ by about 2 cm, which can easily be measured in a modern mass spectrometer.

Hall Effect: How Do We Know If the Charge Carriers Are Positive or Negative?

An electric current is produced by moving electric charges. In Chapter 19, we saw that the value of the current is proportional to the product of the charge q and velocity v of the charge carriers (Eq. 19.6), or

$$I \propto qv \tag{20.15}$$

Suppose we reverse the direction of the velocity, changing from $+v$ to $-v$, and at the same time change the charge from $+q$ to $-q$. The new current is

$$I_{new} \propto (-q)(-v) = qv$$

which equals the old value. A particular value of the current can thus be produced by a positive charge moving to the right (Fig. 20.27A) or by a negative charge moving to the left (Fig. 20.27B). Because the current in a metal is carried by electrons (negative charges), the picture in Figure 20.27B should apply, but how do we really know? Is there some measurement we can perform that distinguishes between these two possibilities?

The answer to this question was given more than a century ago by Edwin Hall.[1] He suggested placing a current-carrying wire in a magnetic field directed perpendicular to the current. Suppose the current is produced by positive charges moving to the right (Fig. 20.28A). If $\vec{B}$ is directed into the plane of the drawing, the magnetic force on the moving charges is directed upward, causing the positively charged carriers to be deflected toward the wire's upper edge. The result is an excess of positive charge on the top edge of the wire and a deficit of positive charge on the bottom, giving an electric potential difference between the top and bottom edges. In Figure 20.28A, the upper edge has a positive electric potential relative to the bottom, which can be measured by attaching a voltmeter to the wire's top and bottom edges.

Now consider what happens if the current is carried instead by negative charges (electrons) moving to the left (Fig. 20.28B). Applying right-hand rule number 2 for negative charges, we find that the magnetic force is again directed toward the top edge of the wire. The magnetic force is thus in the *same* direction as with the positive charge carriers in Figure 20.28A, and the electrons in Figure 20.28B are

Insight 20.2 ⊗
APPLICATIONS OF MASS SPECTROMETERS TO GENOMICS
Mass spectrometers are now being used in work on genomics and proteomics. For example, a protein of unknown sequence is first "cut" into fragments (e.g., peptides) using biochemical techniques. The fragments are then analyzed with a mass spectrometer that gives the ratios of q/m for the different ions (using Eq. 20.14). This information, together with the pattern of mass spectrometer intensities, is compared with mass spectrometer data on known peptides. With knowledge of the peptide "identities," the sequence of the original protein can then be determined.

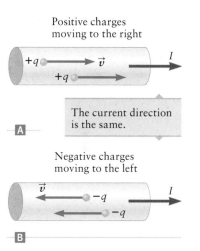

Figure 20.27 The Hall effect gives a way to measure if a current is caused by the motion of **A** positive charges traveling to the right or **B** negative charges traveling to the left. In both cases, the current is directed to the right.

Figure 20.28 Hall's experiment. Motion of electric charges in a wire if the current is carried by **A** positive charges traveling to the right, or **B** negative charges traveling to the left.

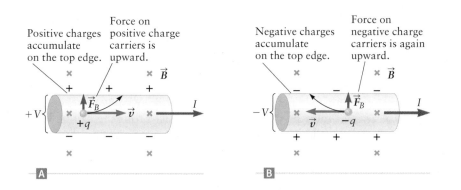

[1] Hall had the idea for this experiment when he was a student! This question is still a good homework problem.

deflected toward the upper edge of the wire, leading to a *negative* potential on that edge relative to the bottom.

Hall's experiment thus gives a way to determine the sign of the charge carriers: we simply measure the potential difference between the two edges of the wire. Hence, we can indeed distinguish between a positive charge moving in one direction and a negative charge moving in the opposite direction. That is how we know that the current in metals such as copper and gold is carried by electrons.

20.7 | CALCULATING THE MAGNETIC FIELD: AMPÈRE'S LAW

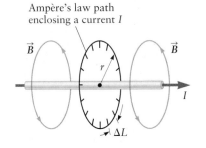

Ampère's law path enclosing a current I

Figure 20.29 Ampère's law relates the magnetic field along a closed path to the current enclosed by that path. The path shown here is a circle of radius r.

Ampère's law

In Chapter 17, we discussed two approaches for calculating the electric field, Coulomb's law and Gauss's law. There are also several ways to calculate the magnetic field produced by a current. One approach, which treats each small piece of a wire as a separate source of $\vec{B}$, is similar in spirit to Coulomb's law but is mathematically too complicated for us to deal with here. Another approach, which uses *Ampère's law*, is very useful when the magnetic field lines have a simple symmetry. This approach is similar to Gauss's law for electric fields, which is most useful when the electric field is highly symmetric.

Consider a long, straight current-carrying wire encircled by magnetic field lines as shown in Figure 20.29, and imagine traveling around a closed path that also encircles the wire. Ampère's law relates the magnetic field *along* this path to the electric current *enclosed by* the path. Let's travel along the path in Figure 20.29 taking steps of length ΔL, and let $B_\parallel$ be the component of the magnetic field parallel to these steps. Ampère's law states that

$$\sum_{\substack{\text{closed} \\ \text{path}}} B_\parallel \Delta L = \mu_0 I_{\text{enclosed}} \tag{20.16}$$

where I_{enclosed} is the total current passing through the surface surrounded by the chosen closed path. The quantity μ_0 is a constant of nature called the *permeability of free space* whose value is

$$\mu_0 = 4\pi \times 10^{-7} \, \text{T} \cdot \text{m/A} \tag{20.17}$$

Ampère's law thus relates the magnetic field *on the perimeter* of a region to the current that *passes through* the region. It is similar to Gauss's law for the electric field (Chapter 17), which relates the electric field *on a surface* to the electric charge *enclosed by* the surface.

Applying Ampère's Law: The Magnetic Field Produced by a Straight Wire

Although Ampère's law has a fairly simple mathematical form, it can be hard to apply in some cases because the sum $\sum B_\parallel \Delta L$ on the left-hand side of Equation 20.16 involves the field at all points along the Ampère's law path. If the value of $B_\parallel$ varies as one moves along the path, Ampère's law can be impossible to apply in practice. On the other hand, in some cases $B_\parallel$ is constant all along the chosen path; Ampère's law is then very handy.

Consider the magnetic field near a long, straight wire. We have already discussed how the magnetic field lines in this case form circles. From the symmetry in Figure 20.29, $B_\parallel$ must have the same value at all points along a circular path that is centered on the wire. We can thus write Ampère's law as

$$\sum_{\substack{\text{closed} \\ \text{path}}} B_\parallel \Delta L = B_\parallel \sum_{\substack{\text{closed} \\ \text{path}}} \Delta L = \mu_0 I_{\text{enclosed}}$$

The key here is that $B_\parallel$ is the same all along the path, so we can move it in front of the summation. If the circular path has a radius r, the total length of the path is $\sum_{\text{closed path}} \Delta L = 2\pi r$ and Ampère's law gives

$$\sum_{\substack{\text{closed} \\ \text{path}}} B_{\parallel} \Delta L = B_{\parallel} \sum_{\substack{\text{closed} \\ \text{path}}} \Delta L = B_{\parallel}(2\pi r) = \mu_0 I_{\text{enclosed}} \qquad (20.18)$$

The right-hand side of Ampère's law involves the total current passing through the chosen surface, which is just the current I in the wire. Inserting that into Equation 20.18 gives

$$B_{\parallel}(2\pi r) = \mu_0 I$$

Although $B_{\parallel}$ is the component of the field along the circular path, in this case it is also the total field, so we can denote it simply as B and find

$$B = \frac{\mu_0 I}{2\pi r} \qquad (20.19)$$

Magnetic field near a long, straight wire

With this result, we now have the full solution for the magnitude and direction of the magnetic field produced by a long, straight wire.

Let's use the result in Equation 20.19 to find the magnitude of the field produced by a current in a typical case. Suppose the current is $I = 1$ A (a common value in household appliances) and we are a distance $r = 1$ cm $= 0.01$ m from the wire. Inserting these values in Equation 20.19, we have

$$B = \frac{\mu_0 I}{2\pi r} = \frac{(4\pi \times 10^{-7} \text{ T} \cdot \text{m/A})(1 \text{ A})}{2\pi(0.01 \text{ m})} = 2 \times 10^{-5} \text{ T}$$

which is about half as large as the Earth's magnetic field. A substantial current of 1 A thus produces only a modest magnetic field, even at this small distance from the wire. This result for B also suggests that the direction of a compass needle can be noticeably affected by the magnetic field from the current in a nearby wire.

EXAMPLE 20.6 | Force between Two Current-Carrying Wires

Consider two long, straight, parallel wires each carrying a current $I = 2.0$ A, with the currents in the same direction. (a) Find the magnetic field at one wire produced by the other wire if the wires are separated by a distance $r = 1.0$ mm. (b) What is the magnitude of the magnetic force exerted by one of these wires on a 1.0-m-long section of the other wire?

RECOGNIZE THE PRINCIPLE

To find the force, we must first find the magnetic field produced by one wire at the location of the other. We can find the magnitude of B using Ampère's law applied to a wire with a current I (Fig. 20.29). Once we have B, we can use our previous result for the magnetic force on a current.

SKETCH THE PROBLEM

Figure 20.22B describes the problem; it shows the current directions and the magnetic field produced by the top wire.

IDENTIFY THE RELATIONSHIPS AND SOLVE

(a) From our Ampère's law result (Eq. 20.19), the magnetic field due to the top wire is

$$B = \frac{\mu_0 I_{\text{top}}}{2\pi r} \qquad (1)$$

where I_{top} is the current in the top wire. Inserting the given values of I_{top} and $r = 1.0$ mm, we get

$$B = \frac{\mu_0 I_{\text{top}}}{2\pi r} = \frac{(4\pi \times 10^{-7} \text{ T} \cdot \text{m/A})(2.0 \text{ A})}{2\pi(1.0 \times 10^{-3} \text{ m})} = \boxed{4.0 \times 10^{-4} \text{ T}}$$

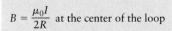

$B = \dfrac{\mu_0 I}{2R}$ at the center of the loop

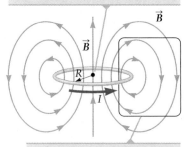

$B_\parallel$ is not constant on this path (or on any other closed path around a current loop).

Figure 20.30 In some cases, such as this current loop, Ampère's law does not give a (mathematically) simple way to calculate $\vec{B}$. For a circular current loop, the magnitude of the magnetic field at the center of the loop is $B = \mu_0 I / 2R$.

Magnetic field at the center of a current loop

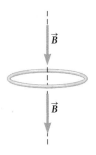

Figure 20.31 Concept Check 20.8.

(b) The field from the top wire is into the plane of the drawing at the location of the bottom wire (Fig. 20.22B), so the angle between $\vec{B}$ and the current in the bottom wire is $\theta = 90°$. The force on the bottom wire is (Eq. 20.11)

$$F_{\text{on bot}} = I_{\text{bot}} L B \sin \theta$$

where I_{bot} is the current in the bottom wire and $L = 1.0$ m is the section of the bottom wire on which we are asked to find the force. Substituting for B from part (a) and inserting $\sin \theta = 1$ (since $\theta = 90°$) gives

$$F_{\text{on bot}} = I_{\text{bot}} L B \sin \theta = (2.0\text{ A})(1.0\text{ m})(4.0 \times 10^{-4}\text{ T})(1) = \boxed{8.0 \times 10^{-4}\text{ N}}$$

What does it mean?

Although these two wires carry substantial currents and are quite close together, the magnetic force between them is still very modest. The wires in a typical household appliance or in the walls of your house carry similar currents and are sometimes spaced this closely. Our result for the force on the bottom wire shows that the magnetic force will not do any damage in such a case.

Field from a Current Loop

The mathematics of Ampère's law is manageable when one can find a path along which the magnetic field is constant. Unfortunately, that is not possible in many cases such as the current loop sketched in Figure 20.30, which shows the pattern of field lines described in Section 20.1. There is no closed path on which $B_\parallel$ (the component of the magnetic field along the path) is constant, so there is no simple way to apply Ampère's law to find $B_\parallel$ in this situation. Other methods for calculating $\vec{B}$ can be applied, and it is found that the field at the center of a circular current loop of radius R is

$$B = \frac{\mu_0 I}{2R} \qquad (20.20)$$

The field at the center of the loop is perpendicular to the plane of the loop (Fig. 20.30) as given by right-hand rule number 1. The field lines encircle the loop as we already saw in Figure 20.9.

CONCEPT CHECK 20.8 | Direction of the Field Near a Current Loop

The current in the circular loop in Figure 20.31 produces a magnetic field on the axis that points downward as shown. As viewed from above, is the direction of the current in the loop (a) clockwise or (b) counterclockwise?

Field Inside a Solenoid

The pattern of magnetic field lines produced by a circular current loop is very similar to the field produced by a bar magnet. One handy feature of the current loop is that its field is adjustable simply by varying the current I in the loop. One disadvantage is that for reasonable values of I the field of a single current loop is much smaller than the field of a bar magnet, but this disadvantage can be overcome by stacking many current loops together. Figure 20.32A shows such a stack of current loops and the pattern of field lines they produce. This field can be understood in terms of the field of a single current loop (Fig. 20.30) along with the principle of superposition. By stacking the loops very close along the vertical axis, the field along this axis is much larger than the field of a single loop.

Usually, it is not practical to stack many separate current loops. Instead, a long piece of wire is wrapped in a very tight helix as shown in Figure 20.32B. Each loop of the helix forms a nearly circular loop, so each turn of the helix produces a field

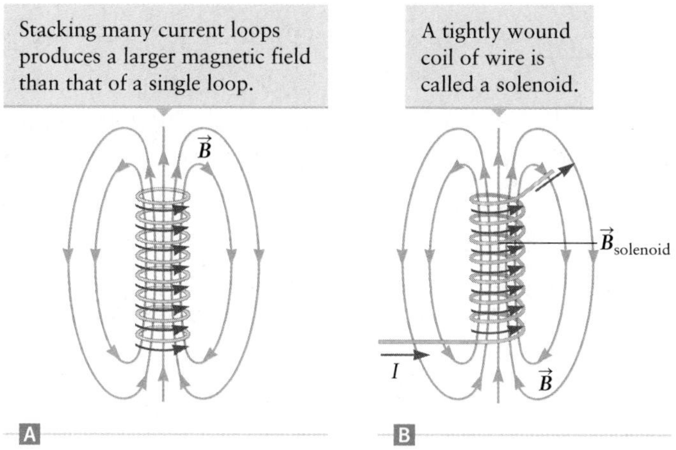

Stacking many current loops produces a larger magnetic field than that of a single loop.

$\vec{B}$

A tightly wound coil of wire is called a solenoid.

$\vec{B}_{solenoid}$

I

$\vec{B}$

A

B

Figure 20.32 A Stacking many current loops produces a magnetic field similar to that of a bar magnet. B Winding a wire in a "tight" helix produces a field similar to that in A.

Insight 20.3
AMPÈRE'S LAW IS ALWAYS TRUE (BUT MAY BE DIFFICULT TO APPLY).
In Figure 20.30, there is no path around the current loop for which the magnetic field component $B_\parallel$ is constant along the entire path. For this reason, Ampère's law does not lead to a simple way of finding the magnetic field in this case, but it is *still true* in such cases. Even though $B_\parallel$ may vary in a complicated fashion as one moves along a closed path, the left-hand side of Ampère's law (Eq. 20.16) is always equal to μ_0 multiplied by the total current through the surface enclosed by the path.

nearly identical to the circular loops in Figure 20.32A. Such a helical winding of wire is called a *solenoid*. The field lines circulate around the outside of the solenoid, leaving from one end and reentering at the other (Fig. 20.32B). Outside the solenoid, the field lines "expand" to fill a very large volume, so $\vec{B}$ is very much smaller outside than inside the solenoid. For a very long solenoid, it is a good approximation to assume the field is constant inside and zero outside. With this approximation, let's apply Ampère's law along the path shown in Figure 20.33. This path is a rectangle with one side (part 1) inside the solenoid and parts 2, 3, and 4 outside. We therefore split the sum on the left-hand side of Ampère's law into four parts:

$$\sum_{\substack{\text{closed} \\ \text{path}}} B_\parallel \Delta L = \sum_{\text{part 1}} B_\parallel(1) \Delta L + \sum_{\text{part 2}} B_\parallel(2) \Delta L + \sum_{\text{part 3}} B_\parallel(3) \Delta L + \sum_{\text{part 4}} B_\parallel(4) \Delta L$$

(20.21)

where $B_\parallel(1)$ is the field along part 1, and so on. Because the field outside the solenoid is approximately zero, $B_\parallel(2) \approx B_\parallel(3) \approx B_\parallel(4) \approx 0$ and we get

$$\sum_{\substack{\text{closed} \\ \text{path}}} B_\parallel \Delta L = \sum_{\text{part 1}} B_\parallel(1) \Delta L$$

(20.22)

Here, $B_\parallel(1)$ is just the field inside the solenoid, which is a constant along part 1 of the path; we can denote it as B_{solenoid}. If the solenoid has a length L, Equation 20.22 leads to

$$\sum_{\substack{\text{closed} \\ \text{path}}} B_\parallel \Delta L = \sum_{\text{part 1}} B_{\text{inside}} \Delta L = B_{\text{solenoid}} \sum_{\text{part 1}} \Delta L = B_{\text{solenoid}} L$$

Inserting this result into Ampère's law (Eq. 20.16) gives

$$B_{\text{solenoid}} L = \mu_0 I_{\text{enclosed}}$$

(20.23)

where I_{enclosed} is the total current passing through the surface enclosed by our Ampère's law path. A helical coil with N windings cuts through this surface N times, so if I is the current through the coil, then $I_{\text{enclosed}} = NI$. Inserting into Equation 20.23 leads to

$$B_{\text{solenoid}} L = \mu_0 NI$$

Solving for B_{solenoid}, we get

$$B_{\text{solenoid}} = \frac{\mu_0 NI}{L}$$

(20.24)

This result for B_{solenoid} is a good approximation for a solenoid whose length L is much greater than its diameter.

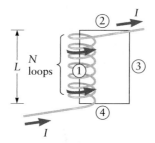

Figure 20.33 Application of Ampère's law to find the magnetic field inside a solenoid; we use the black rectangle as our Ampère's law path. The field outside a solenoid is very small (approximately zero), so only the inside field contributes to the sum of the field components in Ampère's law (Eq. 20.21).

Magnetic field inside a long solenoid

EXAMPLE 20.7 | Designing a Solenoid

You are given the task of designing a solenoid that must produce a field in its interior that is 20 times larger than the Earth's field ($B_{Earth} \approx 5 \times 10^{-5}$ T). This solenoid is to be 10 cm long and 1.0 cm in diameter, with $N = 500$ turns. What current must the wire be able to carry without failing?

RECOGNIZE THE PRINCIPLE

We have already used Ampère's law to find the field inside a solenoid (Eq. 20.24). We can use that result to find the current required to give the desired field in our solenoid.

SKETCH THE PROBLEM

Figure 20.33 describes our solenoid.

IDENTIFY THE RELATIONSHIPS

Equation 20.24 gives $B_{solenoid}$ in terms of the number of turns N, the length L, and the current I. Setting $B_{solenoid}$ equal to $20 \times B_{Earth}$, we find

$$B_{solenoid} = \frac{\mu_0 N I}{L} = 20 B_{Earth}$$

SOLVE

Solving for I and inserting the given values of the various quantities gives

$$I = \frac{L(20 B_{Earth})}{\mu_0 N} = \frac{(0.10 \text{ m})(20)(5 \times 10^{-5} \text{ T})}{(4\pi \times 10^{-7} \text{ T} \cdot \text{m/A})(500)} = \boxed{0.16 \text{ A}}$$

What does it mean?

This current can be safely carried by a wire with a diameter no smaller than about 0.2 mm. A tightly wrapped helix with 500 turns each 0.2 mm thick would be about 10 cm long, so this solenoid could indeed be built. If, however, we wanted a solenoid of this size to produce a field as large as that found near a common bar magnet (for which $B = 1$ T), a *much* larger current would be required. In fact, the current required in that case would probably melt the wire of the solenoid! For this reason, it is hard to use a solenoid like this one to obtain a field as large as can be attained near a typical bar magnet. A redesign of the solenoid is necessary, a problem we will consider further in the next section.

20.8 | MAGNETIC MATERIALS: WHAT GOES ON INSIDE?

We have mentioned several times that magnetic poles *always come in pairs. It is not possible to separate the poles of a magnet.* To understand why magnetic poles always come in pairs, we must consider the atomic origin of permanent magnetism. We have already noted that the pattern of magnetic field lines produced by a bar magnet is similar to the field produced by a current loop. The next example compares the magnitude of the field of a typical permanent magnet with the field near a current loop.

EXAMPLE 20.8 | Comparing the Magnetic Fields of a Permanent Magnet and a Current Loop

Let's calculate how much current is required to make the field at the center of a current loop comparable to the magnetic field near one of the poles of a permanent magnet. The field near a typical bar magnet is about $B_{\text{bar mag}} = 1$ T. Consider a single circular wire loop of radius $R = 0.50$ cm carrying a current I. What value of I is required to produce a field equal to $B_{\text{bar mag}}$ at the center of the loop?

RECOGNIZE THE PRINCIPLE

The magnetic field at the center of a current loop depends on the current and the radius of the loop and is given by Equation 20.20. We can apply this relation to a current loop with the given radius and then find the value of the current needed to give a field equal to $B_{\text{bar mag}}$.

SKETCH THE PROBLEM

Figures 20.2 and 20.9 show the fields of a bar magnet and a current loop.

IDENTIFY THE RELATIONSHIPS

The magnetic field at the center of the loop is (Eq. 20.20)

$$B = \frac{\mu_0 I}{2R}$$

Setting this result equal to $B_{\text{bar mag}}$ and rearranging to solve for I, we get

$$I = \frac{2R B_{\text{bar mag}}}{\mu_0}$$

SOLVE

Inserting the given values for R and $B_{\text{bar mag}}$ gives

$$I = \frac{2(0.0050 \text{ m})(1 \text{ T})}{4\pi \times 10^{-7} \text{ T} \cdot \text{m/A}} = \boxed{8000 \text{ A}} \tag{1}$$

What does it mean?

This current is *very* large. A current loop made from a copper wire of any reasonable size would not be able to safely carry this much current; the wire would melt instead. We have already suggested that a permanent magnet contains atomic-scale current loops. It is amazing that the atomic-scale currents can be this large.

Permanent Magnets: A Microscopic View

How can a permanent magnet generate such a strong magnetic field, especially when no obvious current is present? Atoms consist of a nucleus surrounded by one or more electrons, and in a classical picture the electrons are in constant motion (as in the Bohr model of the atom; Chapter 29). The motion of an electron around a nucleus can be pictured as a tiny current loop, with a radius of approximately the radius of the atom (Fig. 20.34A). The direction of the resulting magnetic field for a single atom is determined by the orientation of the current loop along with right-hand rule number 1 (Fig. 20.9).

An electron also produces a magnetic field due to an effect called ***electron spin.*** It is useful to imagine the electron as a spinning ball of negative charge. A spinning charge acts as a circulating electric current, producing a magnetic field as shown in Figure 20.34B. In analogy with the properties of a current loop, we also say that the electron has a ***spin magnetic moment*** just as the current loop in Figure 20.24 has a magnetic moment. And, like a bar magnet, when an electron is placed in a magnetic field it will tend to align its spin magnetic moment with the magnetic field.

This picture of an electron is very simplified; a correct explanation of electron spin requires quantum theory, which confirms that an atom can produce a magnetic field in the two ways sketched in Figure 20.34: (1) through the electron's orbital

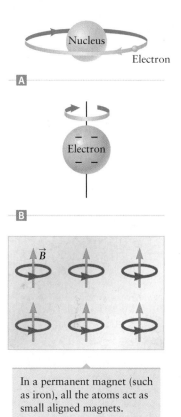

In a permanent magnet (such as iron), all the atoms act as small aligned magnets.

Figure 20.34 🄰 An electron in an atom acts as a small current loop, producing a magnetic field. 🄱 A spinning charged particle also acts as a current loop, with the different regions of charge circulating around the rotation axis. 🄲 The total magnetic field of a permanent magnet is the sum of the fields from parts A and B and from the alignment of the fields produced by many atoms.

The net effect is the same as that of one large current loop around the perimeter.

Figure 20.35 A collection of aligned atomic current loops in a "slice" of magnetic material is equivalent to one large current loop around the outside and hence one large magnet.

North and south magnetic poles cannot be separated.

current loop and (2) from the electron's spin. The total magnetic field produced by a single atom is the sum of these two fields.

Figure 20.34C shows a collection of atomic current loops, with each current loop representing a single atom. This collection of "small" current loops then acts as one "large" current loop (Fig. 20.35), producing the magnetic field of a magnetic material. In Example 20.8, we found that such a large loop must carry a very large current if it is to give the necessary field. Although the current in each atomic loop is very small, there are *many* atoms in a typical bar magnet, and the sum of all these atomic current loops does indeed give a net current similar to the value of I found in Example 20.8.

Lining Up the Atomic Magnets to Make a Permanent Magnet

Figure 20.34 shows how the electrons in an atom are capable of producing a magnetic field. Not all atoms will actually be magnetic, however, because the current loops of different electrons may point in different directions so that their magnetic fields cancel. The total magnetic field produced by a solid depends on how all the atomic magnetic fields are aligned. In some solids, the atomic magnets are not aligned; their individual fields point in different directions and the total field is zero. In materials that form permanent magnets, the atomic magnetic fields align to give a net magnetic field. A permanent magnet thus consists of many atomic-scale "bar magnets" all pointing in the same direction, giving rise to a large magnetic field.

This picture of a permanent magnet as a collection of aligned atomic-scale current loops explains why it is not possible to isolate the north and south magnetic poles of a magnet. One might think that the north and south poles of a permanent magnet could be separated by simply cutting the magnet in half. Figure 20.36 shows why this approach fails. Although one can certainly cut a bar magnet in half, each of the resulting pieces still produces the magnetic field of a "complete" bar magnet and hence still possesses a north pole and a south pole. It is thus not possible to produce an isolated magnetic pole.

Induced Magnetism and Magnetic Domains

Although permanent magnets can be formed from certain materials such as iron and magnetite, not all pieces of iron or magnetite behave as bar magnets. How can

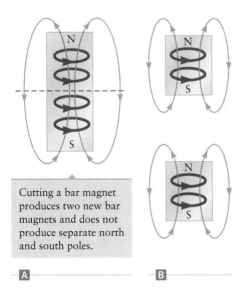

Cutting a bar magnet produces two new bar magnets and does not produce separate north and south poles.

Figure 20.36 Ⓐ Cutting a bar magnet in half Ⓑ produces two separate bar magnets.

Atomic magnets

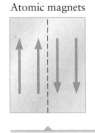
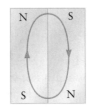

The atomic magnets in different parts of a permanent magnet can point in different directions. Here the result is two side-by-side bar magnets.

Ⓐ   Ⓑ

Figure 20.37 Ⓐ If the atomic magnets within a permanent magnet are aligned in opposite directions, the magnetic field outside can be very small. Ⓑ It is equivalent to having two bar magnets side by side, but with opposite orientations. To an observer on the outside, this material would appear to be nonmagnetic.

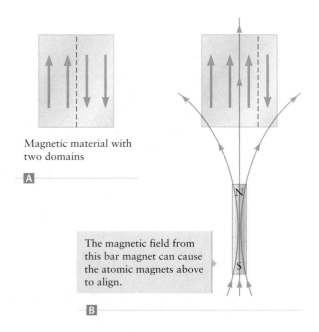

Magnetic material with
two domains

A

The magnetic field from
this bar magnet can cause
the atomic magnets above
to align.

B

Figure 20.38 **A** Oppositely aligned magnetic domains. **B** A weak magnetic field can cause the atomic magnets in a magnetic material to reorient. Some domains then grow at the expense of the others. The material at the top now behaves like a permanent magnet.

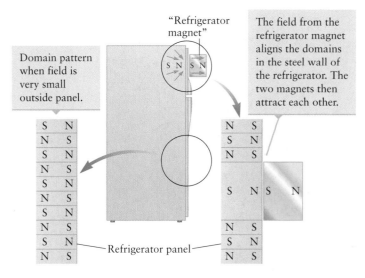

"Refrigerator magnet"

Domain pattern
when field is
very small
outside panel.

The field from the refrigerator magnet aligns the domains in the steel wall of the refrigerator. The two magnets then attract each other.

Refrigerator panel

Figure 20.39 A "refrigerator magnet" is a permanent magnet that "sticks" to a refrigerator panel through a magnetic force. The refrigerator panel contains many domains oriented in opposite directions. The field of the refrigerator magnet aligns the domains so that these two magnets then attract each other.

that be? Our atomic-scale view of a permanent magnet in Figure 20.34C shows the atomic bar magnets all aligned in the same direction, but it is also possible for the atomic magnets in different regions within a magnetic material to point in different directions. One possibility is shown in Figure 20.37A, which shows a material in which the atomic magnets on one side all point up and those on the other side all point down. This arrangement is equivalent to two bar magnets, side by side as in Figure 20.37B. The magnetic field lines are also shown and are confined almost totally within the material. An observer outside the material thus sees a very small field, so this material would not appear to be a permanent magnet.

The two oppositely aligned regions in Figure 20.37 are called *magnetic domains*, and they have some interesting properties. Figure 20.38A shows a material in which the two domain regions are about equal in size. Application of a magnetic field from a bar magnet can cause one of these regions to grow at the expense of the other. In Figure 20.38B, the domain aligned with the magnetic field grows at the expense of the other domain, and the material itself now acts like a bar magnet. This behavior occurs whenever you stick a "refrigerator magnet" to your refrigerator. A refrigerator panel is made of steel and usually has magnetic domains that alternate in direction as shown in Figure 20.39. It is similar to the two oppositely directed domains in Figure 20.37A, so your refrigerator appears to be nonmagnetic. When a magnet is brought near, however, the domains in the refrigerator panel align (Fig. 20.39); the panel now acts like a bar magnet and attracts the nearby magnet. That is why a refrigerator magnet sticks to a refrigerator due to a magnetic force, even though the refrigerator by itself does not appear to be magnetic.

20.9 | THE EARTH'S MAGNETIC FIELD

The largest magnet on the Earth is the Earth itself. About 400 years ago, William Gilbert proposed that a compass needle orients along a north–south direction in

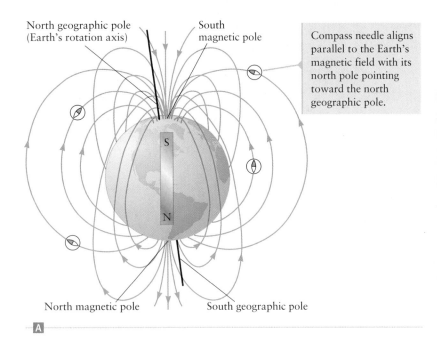

North geographic pole
(Earth's rotation axis)

South
magnetic pole

Compass needle aligns parallel to the Earth's magnetic field with its north pole pointing toward the north geographic pole.

North magnetic pole

South geographic pole

A

© Cengage Learning/Charles D. Winters

B

Figure 20.40 △ Magnetic field of the Earth. ⊟ A compass needle is a small bar magnet and thus aligns parallel to the magnetic field. If there are no other sources of magnetic field nearby, the north end of a compass needle points toward the Earth's north geographic pole.

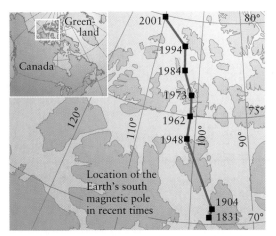

Figure 20.41 The Earth's magnetic poles move from year to year. The south magnetic pole (which is located near the north geographic pole) was near latitude 70° in northern Canada only 100 years ago.

response to the Earth's magnetic field (Fig. 20.40A). A compass needle (Fig. 20.40B) is a small bar magnet mounted so that it can swivel freely about its center. The Earth's magnetic field provides a torque on this magnet. From experiments with compass needles and some insightful reasoning, Gilbert suggested that the Earth acts as a very large magnet oriented as shown in Figure 20.40A.

A compass needle aligns with its north *magnetic* pole pointing approximately toward the Earth's north *geographic* pole, which is why the poles of a bar magnet were given the names "north" and "south." Because a north magnetic pole of one magnet (the compass needle) is attracted to the south magnetic pole of a second magnet, the Earth's *north geographic pole* is actually a *south magnetic pole*!

Gilbert was not able to explain what causes the Earth's magnetic field. That problem is still not completely solved, but several clues point to a qualitative explanation. First, we know that the location of the Earth's south magnetic pole does not quite coincide with the north geographic pole. In fact, the Earth's south magnetic pole actually moves slowly from day to day. Currently, it moves about 40 km/year, and within the last 100 years it has even been located in northern Canada (Fig. 20.41). Second, geological studies show that the Earth's field has completely reversed direction many times during the planet's history. The last such geomagnetic reversal occurred about 780,000 years ago. So, in the years immediately prior to that time, a compass needle would have pointed toward the Earth's *south geographic* pole.

This behavior of the Earth's magnetic field is probably caused by electric currents in the core. Recall (Chapter 12) that the Earth's core contains a liquid. This liquid conducts electricity, and the spin of the Earth about its axis causes the liquid to circulate much like the current in a conducting loop (Fig. 20.42). Circulation within the Earth's core is thought to have a complicated flow pattern that varies with time, and these variations are believed to cause the wandering of the Earth's magnetic poles. Scientists still do not have a complete understanding of these phenomena, however.

When we use a compass, we usually hold it "horizontally" so that the needle lies in a horizontal plane. However, the Earth's magnetic field has both horizontal and vertical components (Fig. 20.40). Your friend is an arctic explorer and pitches his tent directly over the Earth's south magnetic pole. Will his compass needle point (a) toward the Earth's north geographic pole, (b) downward, or (c) upward?

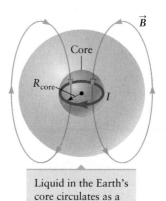

Liquid in the Earth's core circulates as a large current loop.

Figure 20.42 The Earth's field is believed to be caused by a current in the liquid core.

EXAMPLE 20.9 | ® Estimating the Electric Current in the Earth's Core

The strength of the Earth's magnetic field has an average value on the surface of about 5×10^{-5} T. Model this magnetic field by taking the Earth's core to be a current loop, with a radius equal to the radius of the core. What electric current must this current loop carry to generate the Earth's observed magnetic field? The Earth's core has a radius of approximately $R_{core} = 3 \times 10^6$ m (Fig. 20.42). *Hint*: Approximate the current in the core as a single current loop.

RECOGNIZE THE PRINCIPLE

To get an approximate value of the current, we treat the core as a simple current loop of radius of R_{core}; Equation 20.20 then gives the resulting magnetic field at the center of the loop. The field at the core's center will be larger than the field at the Earth's surface, but it will give an approximate value for the field seen by a compass at the surface.

SKETCH THE PROBLEM

Figure 20.42 shows the problem.

IDENTIFY THE RELATIONSHIPS

The field produced by our model current loop is

$$B = \frac{\mu_0 I}{2R_{core}}$$

(Eq. 20.20). We can rearrange this equation to solve for the current:

$$I = \frac{2R_{core}B}{\mu_0}$$

SOLVE

Inserting the radius of the core and setting B equal to the Earth's field (both given above), we find

$$I = \frac{2R_{core}B}{\mu_0} = \frac{2(3 \times 10^6 \text{ m})(5 \times 10^{-5} \text{ T})}{4\pi \times 10^{-7} \text{ T} \cdot \text{m/A}} = \boxed{2 \times 10^8 \text{ A}}$$

What does it mean?
This current is enormous. A scientific explanation of the Earth's field must explain how such a large current is maintained. Physicists still do not have a full solution for this problem.

Cosmic Rays and the Earth's Magnetic Field

We usually think of "space"—the regions far above the Earth's atmosphere between the Earth, the Moon, and the Sun—as empty, but these regions actually contain many charged particles, including electrons and protons, that are emitted by the

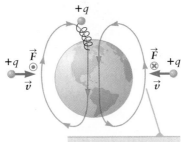

Cosmic ray particles at the poles spiral in along the field lines.

+q

+q $\vec{F}$ $\vec{F}$ +q
 $\vec{v}$ $\vec{v}$

Cosmic rays at the equator are deflected away from the Earth.

Figure 20.43 Ⓐ The motion of a cosmic-ray particle is affected by the Earth's magnetic field. Ⓑ The northern lights are caused by cosmic rays that ionize atoms and molecules as the cosmic rays travel through the atmosphere.

Sun or that reach us from outside our solar system. Such particles are called *cosmic rays*. Because they are charged, their motion is affected by the Earth's magnetic field. Figure 20.43A shows two positively charged cosmic-ray particles that approach the Earth's equator. Let's consider the magnetic force on the particle that approaches from the right. In this region, the magnetic field is directed parallel to the Earth's surface in a geographic south-to-north direction; applying right-hand rule number 2 shows that the magnetic force on this particle is directed perpendicular to the plane of this drawing and parallel to the Earth's surface in a west-to-east direction as shown in Figure 20.43A (into the plane of the drawing). Hence, the Earth's field deflects positively charged cosmic rays near the equator so that they tend to move in an easterly direction, away from the Earth's surface.

Cosmic-ray particles also bombard the Earth at the poles. In polar regions, the magnetic field is perpendicular to the surface, so these particles follow a helical path and spiral around the field lines (as in Fig. 20.15C). The Earth's magnetic field thus "steers" cosmic rays so that they tend to strike at the poles as opposed to locations at lower latitudes. As they travel through the atmosphere, these particles collide with and ionize air molecules, causing the molecules to emit light, much like the light in a fluorescent lightbulb. This phenomenon is observed as the *aurora borealis*, also called the "northern lights" (Fig. 20.43B). A similar phenomenon at the south geographic pole is called the *aurora australis* ("southern lights").

20.10 | APPLICATIONS OF MAGNETISM

So far, we have mentioned only a few applications of magnetism, including mass spectrometers and refrigerator magnets. In this section, we describe how magnetism is used by doctors, engineers, archaeologists, and even bacteria.

Blood-Flow Meters ⊗

An arrangement similar to the Hall effect experiment is used by surgeons to measure the blood velocity in arteries. Blood contains many ions that move at some speed v along the artery (Fig. 20.44A), just like the moving charges in the Hall effect conductor in Figure 20.28. When a magnetic field is directed perpendicular to the artery, the ions are deflected toward the sides of the artery, with positive ions deflected to one side and negative ions deflected to the opposite side (Fig. 20.44B). The result is an electric potential difference across the artery that can be measured with a voltmeter. Flowmeters based on this principle are used to measure the blood velocity. One disadvantage with these devices is that the voltmeter leads must be placed very close to the walls of the artery, so they are typically used in surgery when the artery is exposed. We'll consider the operation of a flowmeter and estimate the magnitude of the induced emf in Question 18 and Problems 92 and 93.

Figure 20.44 Ⓐ Blood flowing through an artery contains many ions. Ⓑ The magnetic force deflects the positive and negative ions to opposite sides of the artery. A blood-flow meter uses the electric potential difference across the artery to measure the speed of the blood.

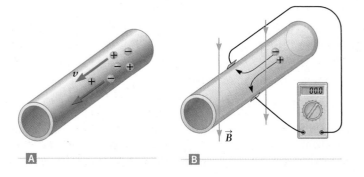

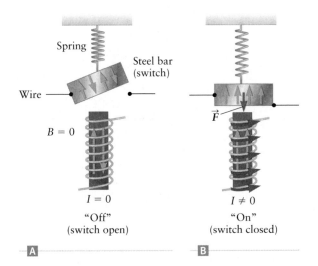

Spring

Steel bar
(switch)

Wire

$B = 0$

$I = 0$

"Off"
(switch open)

$\vec{F}$

$I \neq 0$

"On"
(switch closed)

A

B

Figure 20.45 A relay uses the magnetic field from a solenoid to close a mechanical switch. This solenoid has been filled with a magnetic material whose domains are aligned by the field from the solenoid's current, producing a much larger field than is possible from the current alone.

Relays and Speakers

We have already seen how a solenoid (Fig. 20.32) produces a magnetic field when an electric current passes through it. In Example 20.7, we designed such an "electromagnet" and found that even with a substantial current the field is rather modest. The field produced by a solenoid can be made much larger if the solenoid is filled with a magnetic material. The small field produced by the current can align the atomic magnets in the material (the magnetic domains described in Section 20.8), giving a large magnetic field due mainly to the magnetic material. In this way, the large field just outside the ends of a solenoid can be controlled with a relatively modest current.

The notion that a small magnetic field or current can control a large magnetic field is used to make an electromechanical switch called a *relay*, sketched in Figure 20.45. A relay is a type of "remote control" switch that uses a solenoid to exert a magnetic force on the movable part of a switch (a steel bar). A small current through the solenoid can thus control a much larger current through the switch. Although a switch is perhaps not a very "exciting" application, relays are widely used in a safety device called a ground fault interrupter (GFI), which is found in virtually all new homes; GFIs are discussed in more detail in Chapter 21.

A stereo speaker also employs a solenoidal coil wrapped around a magnetic material (Fig. 20.46). The current in the speaker's solenoid is the music "signal" played by the speaker. This current varies between positive and negative values according to the amplitude and frequency of the music. As the current varies, so does the magnetic alignment of the material inside the solenoid coil. Next to the solenoid is a stationary permanent magnet, so as the solenoid's magnetic poles vary back and forth (switching between north and south), the force on the solenoid varies and causes the solenoid to vibrate. The speaker "cone" is attached to the solenoid, and the vibrations of the cone produce sound.

Making an Electric Motor

In Figure 20.23, we described how a magnetic field can produce a torque on a current loop. If the loop is attached to a rotatable shaft, we have the "beginnings" of an electric motor. However, in a practical motor, the simple current loop is replaced by a solenoid containing a magnetic material (Fig. 20.47). This solenoid is similar to that used in a relay or a speaker (Figs. 20.45 and 20.46). When current is passed through the solenoid, the material inside forms a bar magnet, with the orientation of the poles determined by the direction of the current. The north and south poles of this magnet are attracted to the poles of a second stationary magnet (shaped like a horseshoe magnet), producing a torque on the solenoidal magnet and causing

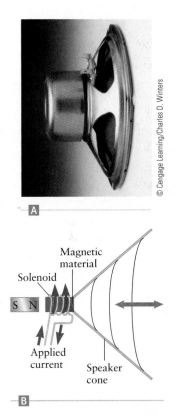

A

Magnetic
material

Solenoid

S N

Applied
current

Speaker
cone

B

Figure 20.46 A speaker contains a solenoid. The current through the solenoid varies with time according to the sound, producing a time-varying force on the solenoid. The speaker cone is attached to the solenoid, so both vibrate. The cone produces the sound that you hear.

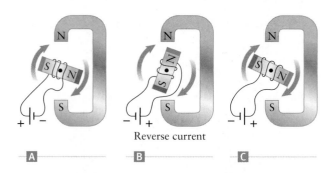

Figure 20.47 A DC electric motor uses the torque on a solenoid magnet to cause a shaft to rotate.

Reverse current

A   B   C

the shaft to rotate (Fig. 20.47A). The torque on the solenoidal magnet is thus very similar to the torque on a current loop. The advantage of using the solenoid is that it gives a much larger torque, for the same reason that a bar magnet produces a much larger magnetic field than a current loop.

To make a working motor, we need to add one more "twist" to the design. When the solenoidal magnet rotates just past the horseshoe magnet, the direction of the current is reversed, switching the poles on the rotating magnet (Fig. 20.47B). These poles are now repelled from the poles of the horseshoe magnet, so the shaft continues to rotate. The current is reversed again every time the solenoid passes the poles of the horseshoe magnet, thus providing a continuing torque on the solenoidal magnet, and the motor's shaft continues to rotate.

Electric motors are a very important and widespread application of magnetism. A closely related application is the ***electric generator***. A generator produces an electric current by rotating a coil between the poles of a magnet; hence, a generator is a sort of "motor in reverse" as we'll discuss in Chapter 21. A combination of electric motor plus generator is at the heart of today's fuel-efficient "hybrid" automobiles such as the Toyota Prius and the Honda Insight.

Magnetic Bacteria ⊗

In addition to physicists and engineers, even bacteria seem to have an application for magnetism. ***Magnetotactic*** bacteria possess small grains of iron called magnetosomes, which are usually arranged as a "necklace" as shown in Figure 20.48. Magnetically, each grain acts like a bar magnet, and the north and south poles of the grains align just as found for bar magnets.

Bacteria use their magnetosomes to orient themselves with respect to the Earth's field. In the northern hemisphere, the Earth's field has a downward component as shown in Figure 20.40A; magnetic bacteria "prefer" to swim downward—toward the bottom of a lake, for instance—where the concentrations of oxygen and other nutrients are most favorable. It is believed that these bacteria use their magnetosomes to determine up from down.

Magnetic Dating

In Section 20.9, we mentioned that the direction of the Earth's field varies slightly from year to year. Occasionally, the direction even reverses completely. The most recent reversal occurred 780,000 years ago, and the dates of many previous reversals are also known (Fig. 20.49A). Archaeologists and geologists use these field reversals to "date" past events. When a volcano erupts, it produces molten rock, some of which solidifies to form magnetite and other magnetic materials. When molten, these materials are not magnetic because their atomic magnets point in random directions. Magnetite and other materials only become permanent magnets when they cool to normal surface temperatures. The orientation of these permanent magnets is set by the direction of the Earth's field at the time of cooling; hence, these materials record the direction of the Earth's field when they last cooled.

To understand how magnetic reversals are used for dating, consider the hypothetical deposits of material in Figure 20.49B. The material nearest the surface

Magnetosomes

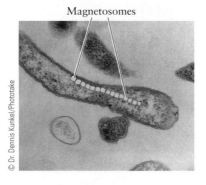

© Dr. Dennis Kunkel/Phototake

Figure 20.48 Some bacteria contain small grains of iron (shown here in yellow). It is believed that the bacteria use these iron magnets as a sort of compass to help determine the direction in which they swim.

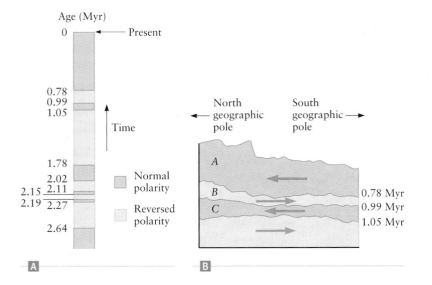

Age (Myr)

North geographic pole ← | → South geographic pole

0.78 Myr
0.99 Myr
1.05 Myr

Normal polarity

Reversed polarity

A

B

(layer *A*) is magnetized parallel to the current magnetic field lines, with $\vec{B}$ pointing toward the north geographic pole. As one digs deeper, one finds a layer (*B*) magnetized in the opposite direction and then a third layer (*C*) magnetized parallel to layer *A*. This layering can be understood in terms of the geomagnetic reversals in Figure 20.49A. The transition from layer *A* to layer *B* dates from the most recent reversal 780,000 years ago, whereas the transition from layer *B* to layer *C* occurred 990,000 years ago. Hence, archaeological artifacts found in layer *B* must have an age between 780,000 and 990,000 years.

The dates of more than 25 geomagnetic reversals going back more than five million years are known, so this approach can be used to date many artifacts of interest to archaeologists, anthropologists, and geologists.

20.11 | THE PUZZLE OF A VELOCITY-DEPENDENT FORCE

A remarkable feature of the magnetic force exerted on a moving charged particle is that it depends on the particle's velocity. Neither gravity (Newton's law of universal gravitation, Eq. 5.16) nor the electric force (Coulomb's law, Eq. 17.3 or 17.5) depend on velocity. The velocity dependence of the magnetic force has a very interesting consequence. Consider a particle of charge $+q$ moving with a velocity $\vec{v}$ perpendicular to a magnetic field as in Figure 20.50A. A stationary observer watching the particle's motion would see a force on the particle as indicated in the figure. (The observer would notice this force from the particle's acceleration.) This observer (observer 1) would explain the force using $F_B = qvB \sin \theta$ and right-hand rule number 2.

Now suppose this scene is viewed by a second observer (observer 2) who is moving at velocity $\vec{v}$. Observer 2 is moving along with the charged particle, so *relative to observer 2* the particle is not moving (Fig. 20.50B). This observer would also notice a force on the particle because she would also see that the particle is accelerated. But, according to this observer, the velocity of the particle is zero, so she would predict that the magnetic force on the particle is zero!

Our two observers would both agree that the particle is accelerated, but only observer 1 would say that there is a magnetic force acting on the particle. How does observer 2 interpret this scene? Such puzzles will always arise when the force depends on velocity. Einstein's theory of relativity provides the way out of this dilemma. In Chapter 27, we show that according to observer 2 there is an *electric field* present, so she will say that the force on the particle is an *electric force*. That is, observer 1 will claim that the force in Figure 20.50 is magnetic, whereas observer

+q $\vec{v}$

$\vec{F}_B$

Observer 1

A

$\vec{v}$

$\vec{F}_E$

$\vec{v}$

Observer 2

B

Figure 20.50 When a charged particle moves through a magnetic field, a force is exerted on the particle. The interpretation of this force is different for different observers, however. ◭ According to observer 1, the charge $+q$ is has a velocity $\vec{v}$ and therefore experiences a nonzero magnetic force given by Equation 20.1. ◯ Observer 2 moves at the same velocity as charge $+q$, so relative to this observer this charge is at rest ($v = 0$). According to observer 2, the magnetic force on q is zero.

2 will measure a force of the same magnitude and direction but will claim that it is electric. There is thus a deep connection between electric and magnetic forces. This connection is a crucial part of the theory of *electromagnetism*, which we'll explore in the next chapter.

SUMMARY | Chapter 20

Sources of magnetic field

There are two ways to produce a *magnetic field*: a permanent magnet and an electric current. A bar magnet and a current loop produce similar patterns of magnetic field lines.

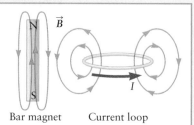

Bar magnet   Current loop

Permanent magnets and magnetic poles

The atoms in a permanent magnet material act as small bar magnets; their magnetic fields are produced by the circulating current of the atomic electrons and the spin magnetism of the electrons. A bar magnet contains a *north magnetic pole* and a *south magnetic pole*.

Ampère's law

Ampère's law can be used to calculate the magnetic field produced by a current. This law relates the field directed along a closed path on the edge of a surface with the current that passes through the surface:

$$\sum_{\substack{\text{closed} \\ \text{path}}} B_{\parallel}\,\Delta L = \mu_0 I_{\text{enclosed}} \qquad \textbf{(20.16)} \text{ (page 662)}$$

where B is measured in *teslas* (T). The constant μ_0 is the *permeability of free space* and has the value

$$\mu_0 = 4\pi \times 10^{-7} \text{ T} \cdot \text{m/A} \qquad \textbf{(20.17)} \text{ (page 662)}$$

When applied to a long, straight wire carrying a current I, Ampère's law gives the field a distance r from the wire as

$$B = \frac{\mu_0 I}{2\pi r} \quad \text{magnetic field near a long wire} \quad \textbf{(20.19)} \text{ (page 663)}$$

Right-hand rule number 1

To find the direction of the field produced by a current,

1. Place the thumb of your right hand along the direction of the current.
2. Curl your fingers; they will then give the direction of $\vec{B}$ as the field lines encircle the current.

RIGHT-HAND RULE NUMBER 1

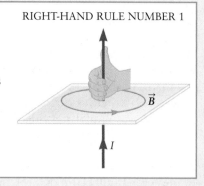

(Continued)

Magnetic force on a moving charge

The magnitude of the magnetic force exerted on a charge moving with velocity $\vec{v}$ is given by

$$F_B = qvB \sin\theta \qquad \textbf{(20.1)} \text{ (page 651)}$$

where θ is the angle between $\vec{v}$ and $\vec{B}$.

Right-hand rule number 2

To find the direction of the magnetic force on a moving charge,

1. Point the fingers of your right hand along the direction of $\vec{v}$.
2. Curl your fingers in the direction of $\vec{B}$. Always curl through the smallest angle that connects $\vec{v}$ and $\vec{B}$.
3. If q is positive, the magnetic force on q is parallel to your thumb. If q is negative, the magnetic force is in the opposite direction.

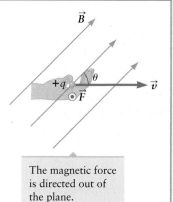

The magnetic force is directed out of the plane.

APPLICATIONS

Magnetic fields due to a current

The magnetic field at the center of a current loop is

$$B = \frac{\mu_0 I}{2R} \quad \text{magnetic field of a current loop} \quad \textbf{(20.20)} \text{ (page 664)}$$

A *solenoid* consists of N tightly wound turns of wire of finished length L. The field inside a solenoid is equal to

$$B_{\text{solenoid}} = \frac{\mu_0 NI}{L} \quad \text{magnetic field inside a solenoid} \quad \textbf{(20.24)} \text{ (page 665)}$$

A permanent magnet material is sometimes inserted inside a solenoid; in this case, the total field will be the field due to the current (Eq. 20.24) plus the field from the permanent magnet.

Hall effect

The **Hall effect** involves a current-carrying wire that is placed in a magnetic field perpendicular to the current. The magnetic force on the current leads to an electric potential difference between the edges of the wire, which can be used to determine the sign (positive or negative) of the charges that carry the current.

Earth's magnetic field

The Earth is a large magnet, with a south magnetic pole located near the north geographic pole (and a north magnetic pole located near the south geographic pole). The Earth's field is produced by large, circulating electric currents in the core. The Earth's magnetic field aligns compass needles. The location of the Earth's magnetic poles varies with time, and throughout geological history (see Section 20.10) the Earth's field has reversed its direction many times.

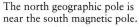

The north geographic pole is near the south magnetic pole.

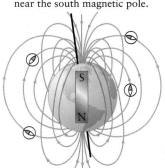

SSM = answer in Student Companion & Problem-Solving Guide   ⊗ = life science application

1. The photo in Figure Q20.1 shows a beam of electrons moving from right to left in a glass tube in which a vacuum has been established. This beam is deflected downward by a magnetic field. What is the direction of this magnetic field at the center of the tube?

Figure Q20.1

2. A long, straight, current-carrying wire is placed in a region where there is a magnetic field $\vec{B}$ that has a constant direction. If there is no magnetic force on the wire, what can you say about the direction of $\vec{B}$ relative to the wire?

3. Consider the moving cosmic-ray particle on the right in Figure 20.43A and suppose it has a negative charge. What is the direction of the magnetic force on this particle?

4. A positively charged particle moves with a velocity $\vec{v}$ in a region where there is a magnetic field $\vec{B}$ and a nonzero magnetic force $\vec{F}$ on the particle. Consider how these three vectors are oriented with respect to one another. (a) Which pairs of these vectors are always perpendicular? (b) Which pairs can have any angle between them?

5. Pairs of wires that allow current to flow into and out of a circuit are often twisted together. Figure Q20.5 shows a common type of cable with such twisted pairs. What is the advantage of pairing the wires that carry input currents with wires carrying output currents?

Figure Q20.5 An Ethernet cable.

6. SSM Make a qualitative sketch of the magnetic field near two bar magnets that are placed side by side as shown in Figure Q20.6.

S	N
N	S

Figure Q20.6

7. Figure Q20.7 shows the direction of $\vec{B}$ in a particular region of space. The density (i.e., spacing) of the crosses and dots indicate qualitatively the magnitude of $\vec{B}$. What is the possible source of this field?

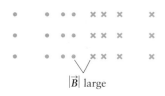

$|\vec{B}|$ large

Figure Q20.7

8. In our discussion of the solenoid in Figure 20.32B, we claimed that the field outside the solenoid is much smaller than the field inside and that it is a good approximation to assume the field outside a very long solenoid is zero. Explain how this approximation can be consistent with the property that magnetic field lines always form closed loops.

9. Two parallel wires are oriented perpendicular to the page as shown in Figure Q20.9. The wires carry equal currents, and the direction of the current is into the page for each. The vertices of an equilateral triangle are marked by the two wires and a point P as shown. (a) What is the direction of the magnetic field produced by the two wires at point P? (b) A third wire, parallel to the other two and carrying the same current in the same direction, is placed at point P. Determine the direction of the force on the wire at this point.

Figure Q20.9
Questions 9 and 10.

10. Repeat Question 9, but this time assume the current I_2 is directed out of the page and opposite to current I_1.

11. ⊗ The medical technique called *magnetic resonance imaging* (MRI) uses the magnetic field produced by a large solenoid to obtain images from inside living tissue. These solenoids are large enough that a person can fit inside. Assuming the inside field is 1 T, design an MRI solenoid. Estimate the necessary length and radius of the solenoid assuming the wire can carry currents as large as 100 A. How many turns must this solenoid have to produce the desired field?

12. The rail gun. The simple configuration in Figure Q20.12 provides a means for launching projectiles, as long as the projectile can conduct electricity. Two copper rails are connected to a battery such that the current flows as indicated in the figure when a conducting rod is laid across the rails. What will happen to the conducting rod? If it begins to move, in what direction does it go? Consider (a) the direction of the magnetic field produced in the vicinity of the rod by the currents in the rails and rod and (b) the direction of the current in the rod. How could one make the rod move in the opposite direction?

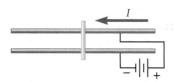

Figure Q20.12

13. A bubble chamber is a device that makes the path of a charged particle visible. When a charged particle with sufficient kinetic energy passes through a substance such a liquid, it will knock electrons off the atoms in its path. If the liquid is at a temperature above its boiling point, these ionized atoms become the nucleation point for a bubble to form, thereby marking the path of the particle. For the images in Figure Q20.13, the magnetic field was directed out of the page. (a) Is the curve of the track in image 1 (on the left) consistent with that of a high-speed electron or of a high-speed proton? (b) The particle trajectory in image 1 spirals inward. Explain why. *Hint*: Each ion-creating encounter is a collision.

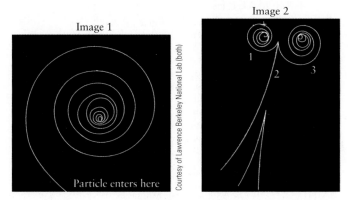

Figure Q20.13 Questions 13 and 14, and Problem 98.

14. [SSM] The bubble chamber image on the right in Figure Q20.13 (image 2) shows three particles emanating from a point near the top of the image. The bubble chamber is in a magnetic field oriented so that the field lines point out of the page. (a) Identify the sign of the charge on each particle. (b) Rank the initial speeds of the particles from least to greatest. How do you know? Assume the particles all have the same mass.

15. The Earth's magnetic field is produced by the flow of electrons within its molten interior. If the flow is in the form of a ring of current in the plane of the equator, in which direction would the electrons have to flow to produce the observed polarity of the Earth's field?

16. Two 20-penny nails are wrapped with insulated wire in the shape of a solenoid. They are then laid down end to end with the heads almost touching as shown in Figure Q20.16. If a battery is connected to the solenoid of each nail as shown, will the nails be attracted or will they repel each other? Explain.

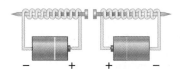

Figure Q20.16

17. Current is made to flow through a spring-shaped (helical) solenoid coil of thin wire. Will the loops of the coil tend to contract or tend to spread apart? Why?

18. ⊗ Blood plasma has a high content of Na^+ and Cl^- ions, and as the blood follows through a vessel, the ions travel along at the same speed. A magnetic field is applied such that the field is perpendicular to the vessel as shown in Figure Q20.18. (a) What effect will the magnetic field have on the distribution of positive ions? (b) What effect with the magnetic field have on how the negative ions are distributed? (c) Consider the overall distribution of charge in the presence of the magnetic field. Will an electric field be present? If so, in what direction is it oriented?

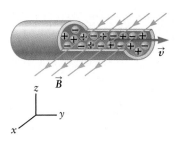

Figure Q20.18

19. Figure Q20.19 shows in part A an electric charge q moving past a stationary bar magnet and in part B a bar magnet moving past a stationary electric charge. In which case will there be a magnetic force on the charge? Will there be a force on the charge in the other case? Explain.

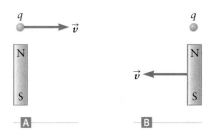

Figure Q20.19

20. A square current loop with current I is placed in a uniform magnetic field. Explain why the total force on the loop is zero, no matter how the loop is oriented relative to the field direction.

21. Figure Q20.21 shows a straight wire of length L carrying a current I in a magnetic field of magnitude B. In which case is the magnetic force on the wire largest? In which case is it smallest? Order the situations from largest to smallest according to the magnitude of the force.

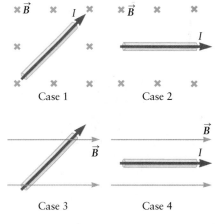

Figure Q20.21

22. Does a current-carrying wire placed in a magnetic field always experience a magnetic force? Explain.

23. A wire carries a current in the $-z$ direction. If there is a magnetic force in the $+x$ direction, what might be the direction of the magnetic field, (a) $+x$, (b) $-x$, (c) $+y$, (d) $-y$, (e) $+z$, or (f) $-z$?

20.1 SOURCES OF MAGNETIC FIELDS

1. The two long wires in Figure P20.1 carry parallel currents of equal magnitude I. Give the direction of the total magnetic field at points A, B, C, and D.

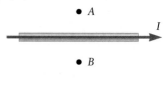
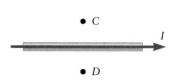

Figure P20.1 Problems 1 and 2.

2. Suppose the currents in Figure P20.1 are in opposite directions: the current in the top wire is from left to right and the current in the bottom wire is from right to left. Find the direction of the magnetic field at points A, B, C, and D.

3. Four long, straight wires each carrying a current I are oriented perpendicular to the plane of Figure P20.3, forming a square as shown. What is the direction of $\vec{B}$ at the center of the square (point A)?

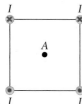

4. An electron is traveling near a current-carrying wire as shown in Figure P20.4. If the magnetic force $\vec{F}_B$ is directed as shown, what is the direction of the current in the wire?

Figure P20.3

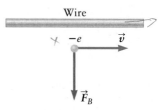

Figure P20.4

5. SSM ★ The magnetic field produced by the current loop in Figure P20.5 is along the $-z$ direction. What is the direction of the current in the loop? That is, is it clockwise or is it counterclockwise as viewed from above?

6. ✪ Use right-hand rule number 1 to find the direction of the magnetic field at point A near a wire that makes a right-angle bend as sketched in Figure P20.6.

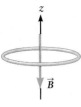

Figure P20.5

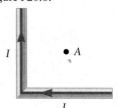

Figure P20.6

20.2 MAGNETIC FORCES INVOLVING BAR MAGNETS

7. SSM ★ A bar magnet is placed in a uniform magnetic field (Fig. P20.7). (a) What is the direction of the total force on the bar magnet? (b) What is the direction (clockwise or counterclockwise) of the torque on the bar magnet about an axis perpendicular to the plane and passing through point A?

8. ★ In Figure P20.8, what is the direction of the force on bar magnet 3?

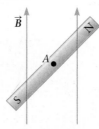

Figure P20.7

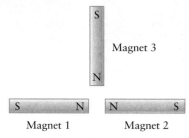

Figure P20.8

9. In Figure P20.9, bar magnet 1 is fixed in place and bar magnet 2 is free to move. (a) What is the direction of the total force on bar magnet 2? (b) What is the direction of the torque (clockwise or counterclockwise) on bar magnet 2 about an axis perpendicular to the plane of the drawing and through the bar magnet's center?

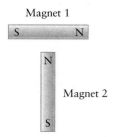

Figure P20.9

20.3 MAGNETIC FORCE ON A MOVING CHARGE

10. ★ The charged particle in Figure P20.10 has a velocity along $+y$ and experiences a force along the $+x$ direction. Does this particle have a positive charge or a negative charge?

11. A proton moves at a speed of 1000 m/s in a direction perpendicular to a magnetic field with a magnitude of 0.75 T. What is the magnitude of the magnetic force on the proton?

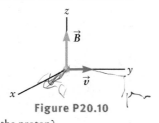

Figure P20.10

12. Repeat Problem 11, but now assume the particle is an electron.

13. A proton with $v = 300$ m/s is moving through a region in which the magnetic field is $B = 2.5$ T. If the magnitude of the force on the proton is 6.4×10^{-17} N, what angle does the proton's velocity make with $\vec{B}$?

14. ★ Ⓡ Consider a proton that approaches the Earth from outer space. If the proton is moving perpendicular to the Earth's surface near the equator, what is the direction of the magnetic force on the proton? The direction of the Earth's magnetic field is shown in Figure 20.40A.

15. ☆ An electron is moving in the x–y plane with its velocity along the $-y$ direction. If the magnetic force on the electron is in the $+z$ direction when $\vec{B}$ lies along one of the coordinate axes, what is the direction of the magnetic field?

16. ☆ A proton is at the origin and is moving in the $+z$ direction. What is the direction of the magnetic field due to the proton at a location on the x axis at $x = +1$ m?

17. SSM ☆ An electron is traveling near a current-carrying wire as shown in Figure P20.17. The electron's velocity is parallel to the $+x$ direction and the wire is parallel to the y axis. What is the direction of the force on the electron from the wire's magnetic field?

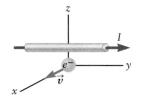

Figure P20.17

18. ☆ A particle of unknown charge has a velocity along the $+y$ direction. There is a constant magnetic field along the $-z$ axis, and it is found that the magnetic force on the particle is along $+x$. Does the particle have positive charge or a negative charge?

19. The particle in Figure P20.19 has a positive charge, and its initial velocity is to the right when it enters a region where $\vec{B}$ is directed into the plane of the drawing. Will the particle be deflected toward the top of the figure or toward the bottom?

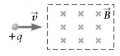

Figure P20.19

20. ☆ The particle in Figure P20.20 has a negative charge, and its velocity vector lies in the x–y plane and makes an angle of 75° with the y axis. If the magnetic field is along the $+x$ direction, what is the direction of the magnetic force on the particle?

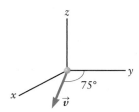

Figure P20.20

21. A particle of charge $q = 7.2\ \mu$C is moving at an angle of 25° with respect to the x axis with a speed of 250 m/s. There is a constant magnetic field of magnitude 0.55 T parallel to x. Find the magnitude of the magnetic force on the particle.

22. SSM ☆ A particle of charge $q = -8.3\ \mu$C has a velocity of 600 m/s that lies in the x–y plane and makes an angle of 65° with respect to the x axis as shown in Figure P20.22. If there is a constant magnetic field of magnitude 1.5 T parallel to y, what are the magnitude and direction of the magnetic force on the particle?

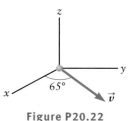

Figure P20.22

23. ☆ An alpha particle is a helium nucleus that consists of two protons bound with two neutrons. An alpha particle moving through a perpendicular magnetic field of 1.5 T experiences an acceleration of 2000 m/s². What is the alpha particle's speed?

24. ☆ Figure P20.24 shows an alpha particle (a helium atom, but without the two electrons) moving with a velocity in the $-x$ direction. If this particle experiences a force along the $+z$ direction, what can you conclude about the direction of $\vec{B}$?

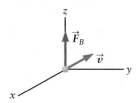

Figure P20.24

25. ☆ An oxygen ion O^{-2} is moving with a speed of 300 m/s in a direction perpendicular to a magnetic field of magnitude B. If the acceleration of the O^{-2} ion is 1.5×10^9 m/s², what is B?

26. ✿ A particle of charge 23 μC and mass 1.3×10^{-16} kg moves at a speed of 2000 m/s, and its velocity vector lies in the x–y plane. There is a magnetic field of magnitude 0.88 T along the $+z$ direc-tion. The magnetic force on the particle causes the particle to move in a circle that lies in the x–y plane. (a) Find the radius of this circle. (b) Find the period of the motion. (c) If the particle is moving parallel to $+y$ at $t = 0$, is the circular motion clockwise or is it counterclockwise as viewed from along the $+z$ axis?

27. ✿ Repeat part (c) of Problem 26, but now assume the particle's velocity is along the $-x$ direction at $t = 0$.

28. ☆ A proton with velocity $\vec{v}_A$ (which lies in the x–z plane) enters a region in which the magnetic field is along z as shown in Figure P20.28. Describe qualitatively the trajectory of the proton.

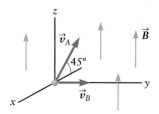

Figure P20.28 Problems 28–31.

29. ☆ Repeat Problem 28, but now assume the particle is an electron.

30. A proton has an initial velocity $\vec{v}_B$ when it enters a region in which $\vec{B}$ is parallel to the z direction (Fig. P20.28). (a) Show that the proton will move in a circle that lies in the x–y plane. (b) If $v_B = 500$ m/s and $B = 0.80$ T, what is the radius of this circle?

31. If the particle in Problem 30 is an electron instead of a proton, what is the radius of the trajectory?

32. ☆ Ⓡ A cosmic ray of charge $+5e$ approaches the Earth's equator with a speed of 2.2×10^7 m/s. If the acceleration of this particle is 8.0×10^{10} m/s², what is the approximate mass of the cosmic-ray particle?

33. Ⓡ A nitrogen ion N^+ is traveling through the atmosphere at a speed of 500 m/s. What is the approximate force on this ion from the Earth's magnetic field if the ion is moving perpendicu-lar to the Earth's field?

34. ☆ A positively charged particle undergoes circular motion as a result of the magnetic force produced by a field that is perpendicular to the plane in Figure P20.34. Does this particle move clockwise or counterclockwise?

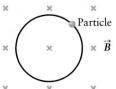

Figure P20.34

35. ✿ Ⓡ Cosmic rays undergo a helical (i.e., spiral) trajectory as they travel to the Earth. Consider a cosmic ray that is an iron ion Fe^+ trav-eling at 1×10^7 m/s. What is the approximate minimum radius of the helix when the ion is near the Earth's surface?

36. ☆ A bubble chamber is a device used to study the trajectories of elementary particles such as electrons and protons. Figure P20.36 shows the trajectories of some typical particles, one of which is indicated by a white arrow. Suppose this particle is moving in the direction indicated by the green arrow; the

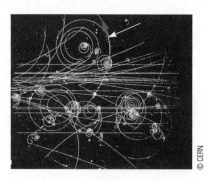

Figure P20.36 Problems 36 and 37.

particle follows a curved path due to the presence of a magnetic field perpendicular to the plane of the photo. If $\vec{B}$ is directed into the plane, does this particle have a positive charge or a negative charge?

37. ☆ How would the trajectory in Figure P20.36 change if the charge on the particle were increased by a factor of 3? How would it change if the mass were decreased by a factor of 10?

38. ✪ A charged particle moves in the x–y plane, and a magnetic field B is directed into the plane as shown in Figure P20.38. The magnetic force on the particle causes the particle to move clockwise along a circle of radius R. (a) Does the particle have a positive charge or a negative charge? (b) The charge of the particle is increased by a factor of three, and the particle is found to move in a circle of radius R′. What is the ratio R′/R ? (c) The mass of the particle is decreased by a factor of four, and the particle is found to move in a circle of radius R′. What is the ratio R′/R ?

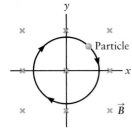

Figure P20.38

39. ✪ An electron and a proton are both moving in a circle in the x–y plane due to the force produced by a magnetic field perpendicular to the plane (Fig. P20.39). (a) If the electron moves clockwise and the proton moves counterclockwise, is the magnetic field directed into the plane or out of the plane of the drawing? (b) If the circular paths followed by the electron and proton have the same radii, what is the ratio of their speeds?

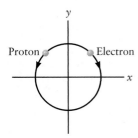

Figure P20.39

40. A positively charged particle moves as shown in Figure P20.40. Which statement best describes the direction of $\vec{B}$?
(a) It is into the plane of the figure.
(b) It is out of the plane of the figure.
(c) It is along the +x direction.
(d) It is along the +y direction.
(e) The direction of $\vec{B}$ varies as the particle moves.

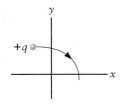

Figure P20.40

41. ☆ Figure P20.41 shows three identical particles, each with a positive charge and all moving at the same speed. In which case is $\vec{B}$ largest? Smallest? Assume $\vec{B}$ is perpendicular to the plane of the drawing.

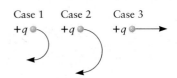

Figure P20.41

42. ☆ A charged particle moves as shown in Figure P20.42. Does the particle have a positive charge or a negative charge?

Figure P20.42

43. ☆ Figure P20.43 shows two cases in which a positively charged particle is moving near a bar magnet. What is the direction of the magnetic force in each case?

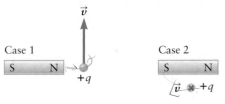

Figure P20.43

44. ☆ Figure P20.44 shows an electron moving near a current loop. If the electron's velocity is parallel to the y axis, what is the direction of the magnetic force on the electron?

20.4 MAGNETIC FORCE ON AN ELECTRIC CURRENT

45. A long, straight wire of length 1.4 m carries a current of I = 3.5 A. If a magnetic field of magnitude B = 1.5 T is directed perpendicular to the wire, what is the magnitude of the force on the wire?

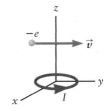

Figure P20.44

46. A long, straight wire carries a current of I = 100 A in a region where the magnetic field has a magnitude B = 10 T, but it is found that the force on the wire is zero. Explain how that can be.

47. ☆ A long, straight wire of length 0.75 m carries a current of I = 1.5 A in a region where B = 2.3 T. If the force on the wire is 1.4 N, what is the angle between the field and the wire?

48. ☆ Repeat Problem 47, but assume the force is 5.6 N. *Hint:* Is this possible?

49. ☆ Ⓡ The electric current carried in a typical household circuit is about 1 A, and the Earth's magnetic field is about 5×10^{-5} T. What is the approximate magnitude of the magnetic force on such a wire if its length is 1 m? Could you detect this force if you were holding the wire? Compare this force to the weight of some terrestrial object (such as a mosquito).

50. ☆ Ⓡ Consider a straight piece of copper wire of length 2 m and diameter 1 mm that carries a current I = 3.5 A. There is a magnetic field of magnitude B directed perpendicular to the wire, and the magnetic force on the wire is just strong enough to "levitate" the wire (i.e., the magnetic force on the wire is equal to its weight). Find B. *Hint:* The density of copper is 9000 kg/m³.

51. ☆ A right-triangular current loop carries a current I, and there is a constant magnetic field B directed perpendicular to one edge of the loop (Fig. P20.51). If θ = 37°, L = 30 cm, and B = 2.3 T, what is the total force on the loop?

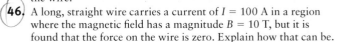

Figure P20.51 Problems 51 and 52.

52. ✪ Consider again the force on the triangular current loop in Figure P20.51, but now work out the force as a function of L, θ, and B. Explain your answer.

53. A wire carries a current along the +z direction and experiences a force along −y. If the magnetic field is perpendicular to the wire, what is the direction of $\vec{B}$?

54. ☆ A wire carries a current along the +z direction and experiences a force that lies in the x–y plane as shown in Figure P20.54. What might be the direction of $\vec{B}$?

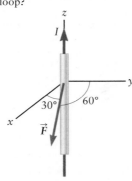

Figure P20.54

55. SSM ★ A long current-carrying wire is parallel to the y axis with the current $I = 4.5$ A along the $+y$ direction. A magnetic field of magnitude $B = 1.2$ T lies in the y–z plane, directed 30° away from the y axis as shown in Figure P20.55. (a) What is the direction of the magnetic force on the wire? (b) What is the magnitude of the magnetic force on a 1.0-m-long section of the wire?

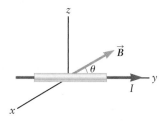

Figure P20.55

56. A square current loop of edge length $L = 0.25$ m with $I = 4.5$ A is pulled from a region where the magnetic field is zero into a region where the magnetic field is perpendicular to the loop with a magnitude $B = 2.5$ T (Fig. P20.56). (a) What is the direction of the magnetic force on the loop? (b) What is the magnitude of the magnetic force on the loop? (c) If the loop's velocity is reversed, what is the direction of the magnetic force?

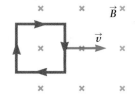

Figure P20.56

20.5 TORQUE ON A CURRENT LOOP AND MAGNETIC MOMENTS

57. ★ Figure P20.57 shows several different current loops in a magnetic field. In Case 1 the current is clockwise when viewed from the upper left. In Case 2 it is clockwise when viewed from the upper right. In Case 3 it is clockwise when viewed from above. What is the direction of the torque on each current loop? Express your answer as either clockwise or counterclockwise as viewed along an axis perpendicular to the plane of the drawing.

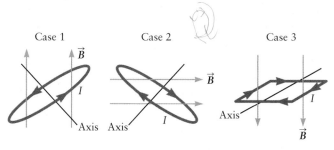

Figure P20.57 Problems 57 and 58.

58. SSM ★ Consider the square current loop, case 3 in Figure P20.57. It has an edge length $L = 0.33$ m and carries a current $I = 7.5$ A. The magnetic field $B = 0.22$ T is perpendicular to the plane of the loop. What is the magnitude of the torque on the loop?

59. ✪ Consider a square current loop ($L = 25$ cm) that carries a current $I = 5.6$ A (Fig. P20.59). A constant magnetic field $B = 0.25$ T makes an angle $\theta = 30°$ with the direction normal to the plane of the loop and perpendicular to the line connecting points A and B. (a) Find the total magnetic force on the loop. (b) Assume the loop is attached to a rotatable rod at points A and B. Find the torque on the loop.

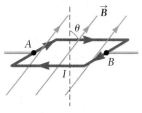

Figure P20.59

60. The plane of a circular current loop is oriented at an angle θ relative to a magnetic field parallel to z (Fig. P20.60). If the current loop is free to rotate about an axis perpendicular to z, which of the following is true?
(a) The plane of the loop will rotate so as to be perpendicular to $\vec{B}$.
(b) The plane of the loop will rotate so that the field produced by the current loop is parallel to $\vec{B}$.
(c) Either statement (a) or statement (b) is possible, depending on the direction of the current.

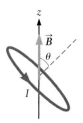

Figure P20.60

61. ★ A ball of positive charge rotates about a vertical axis (Fig. P20.61). (a) If the rotation is clockwise as viewed from above, what is the direction of the magnetic moment of the ball? (b) If the charge on the ball is negative, what is the direction of the magnetic moment?

Figure P20.61

20.6 MOTION OF CHARGED PARTICLES IN THE PRESENCE OF ELECTRIC AND MAGNETIC FIELDS

62. ✪ **Velocity selector.** Consider a charged particle moving through a region in which the electric field is perpendicular to the magnetic field, with both fields perpendicular to the initial velocity of the particle (Fig. P20.62). Such a device is called a **velocity selector** because for one particular value of the particle speed v_{select} the particle will travel straight through the device without being deflected, whereas particles with speeds that are either less than or greater than v_{select} will be deflected. Consider a case with $E = 500$ V/m and $B = 0.25$ T. (a) When protons are used, what speed is selected? (b) What is the speed selected for Ca^{2+} ions? (c) If you were to use your selector with F^- ions, what change would be necessary?

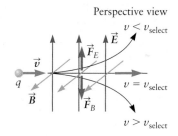

Figure P20.62 Problems 62 and 63.

63. ✪ You wish to design a velocity selector (see Problem 62 and Fig. P20.62) that will allow protons to pass through only if they have a speed of 500 m/s. (a) If the magnetic field is $B = 0.050$ T, what electric field do you need? (b) You decide to produce this electric field using a parallel-plate capacitor with a plate spacing of 1.0 mm. What is the electric potential across the plates? (c) Will your velocity selector also work for Li^+ ions? For Mg^{2+} ions? Explain why or why not.

64. ★ A mass spectrometer can separate ions according to their charges and masses. One simple design for such a device is shown in Figure 20.26. Ions of mass m, charge q, and speed v enter a region in which the magnetic field B is constant and perpendicular to the plane. The ions then travel in a circular arc and leave the spectrometer a distance $L = 2r$ from their entry point. Consider a hypothetical mass spectrometer used to study the isotopes of hydrogen. Find r for H^+, D^+ (deuterium), and T^+ (tritium).

65. SSM ★ Consider a mass spectrometer used to separate the two isotopes of uranium, $^{238}U^{3+}$ and $^{235}U^{3+}$. Assume the ions enter the magnetic field region with a speed of 6.0×10^5 m/s. What value of B is required to give a separation of 1.0 mm when the ions leave the spectrometer?

66. ✪ Some mass spectrometers use an electric field to give the incoming ions a certain known kinetic energy as they enter the magnetic field region. (a) If the ions have a charge q and a kinetic energy KE, what is the value of r in Figure 20.26 in terms of m, q, B, and KE? (b) How does the final kinetic energy of an ion (just before it reaches the ion detector) compare with the initial kinetic energy (just as the ion enters the region where the magnetic field is nonzero)?

67. ✪ Consider the Hall effect experiment in Figure P20.67 and suppose the current is carried by electrons. In Section 20.6, we saw that the magnetic field exerts a force on the moving electrons directed from the bottom edge of

Figure P20.67

the wire toward the top. This force causes a net negative charge to accumulate along the top edge, leaving a net positive charge along the bottom. The result is an electric field (called the Hall field) directed from bottom to top. The Hall field opposes the magnetic force on the electrons. The edge charge will grow until the electric force from the Hall field balances the magnetic force. Use this fact to calculate the magnitude of the Hall field. Express your answer in terms of the drift speed v of the electrons and the magnitude B of the magnetic field.

20.7 CALCULATING THE MAGNETIC FIELD: AMPÈRE'S LAW

68. A long, straight wire carries a current of 5.4 A. What is the magnitude of the magnetic field at a distance of 10 cm from the wire?

69. The Earth's magnetic field is approximately 5×10^{-5} T. If the field produced by the current in a long straight wire 1.0 mm from the wire is equal to the Earth's field, what is the current in the wire?

70. Two long, straight, parallel wires of length 4.0 m carry parallel currents of 3.5 A and 1.2 A. (a) If the wires are separated by a distance of 3.5 cm, what is the magnitude of the force between the two wires? (b) Is this force attractive or repulsive? (c) If the currents are in opposite directions (antiparallel), how do the answers to parts (a) and (b) change?

71. ✪ Consider a long wire that makes a right angle (90°) bend as shown in Figure P20.71. If the wire carries a current I, what is the magnetic field at point A? *Hint*: Use the result for the field from an infinitely long wire (Eq. 20.19) to find the field produced by a "semi"-infinite wire (i.e., half of an infinite wire).

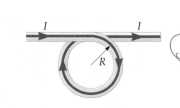

Figure P20.71   **Figure P20.72**

72. SSM ✪ A wire is formed so as to make a circular loop connecting two long, straight sections as shown in Figure P20.72. If the radius of the loop is $R = 20$ cm and the wire carries a current $I = 2.4$ A, what are the magnitude and direction of the magnetic field at the center of the loop?

73. Consider a single circular loop of wire that carries a current $I = 1.5$ A. If the field due to the current in the loop at the loop's center is equal to the Earth's field, what is the radius of the loop?

74. ✪ Ⓡ A magnetic field $B = 1$ T is considered to be a strong field. If you want to produce this field using a single loop of wire wound around a pencil, what must the current be?

75. The magnetic field near a particular permanent magnet is 2.0 T. Consider a long wire carrying a current I. If the field produced by the current at a location 1.0 cm from the wire is equal to the field of this permanent magnet, what is I?

76. ✪ Design a solenoid that will produce a magnetic field of 0.025 T using a current $I = 30$ A. If the radius of the solenoid is 10 cm and the solenoid is 1.0 m long, how many turns of wire are required?

77. ✪ A long, straight wire (Fig. P20.77) carries a current $I_1 = 3.5$ A and is a distance $h = 15$ cm from a rectangular current loop that carries a current $I_2 = 1.5$ A. (a) If the loop has a length $L = 45$ cm and a width $w = 20$ cm, what are the magnitude and direction of the force exerted by the wire on the loop? (b) What is the force exerted by the current loop on the wire?

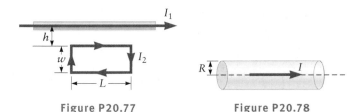

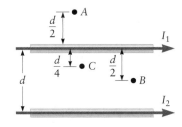

Figure P20.77   **Figure P20.78**

78. ✪ According to Ampère's law, the magnetic field produced by a long, straight wire depends only on the current enclosed by the Ampère's law path (Eq. 20.16). Use this fact to compute the magnetic field *inside* a wire (Fig. P20.78). Assume the wire has a radius $R = 1.0$ mm and carries a current $I = 2.5$ A. (a) Find the field a distance $R/2$ from the axis of the wire. (b) What is the ratio of the field in part (a) to the field at the surface of the wire?

79. ✪ Consider two long, straight, parallel wires (Fig. P20.79) carrying currents $I_1 = I_2$. At which point A, B, or C in the figure will the total magnetic field be largest?

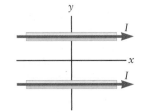

Figure P20.79 Problems 79 and 80.

80. ✪ Consider again the two wires in Figure P20.79. For what value of the ratio I_1/I_2 is the magnetic field zero at (a) point A, (b) point B, and (c) point C. Do your answers to this problem depend on the value of d?

81. ✪ Two long, straight wires lie in the x–y plane and are parallel to the x direction as shown in Figure P20.81. The wires are both a distance of 2.0 m from the origin, and each wire carries a current of 25 A in the $+x$ direction. Find the magnitude of the magnetic field at the point $x = 0$, $y = 1.5$ m, $z = 0$.

Figure P20.81

82. ✪ Four very long wires are directed perpendicular to the plane of Figure P20.82. The points where these wires cross the plane

of the drawing form a square of edge length 0.45 m, and each wire carries a current of magnitude 2.5 A. Find the magnetic field at the center of the square (i.e., at the origin).

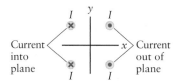

Figure P20.82

83. ✪ Three very long wires are directed perpendicular to the plane of Figure P20.83. Each passes through the x–y plane a distance L from the origin, and each carries a current I that is directed out of the plane of the drawing. An electron is at the origin and has a velocity directed along $+x$. What is the direction of the magnetic force on the electron?

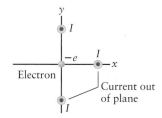

Figure P20.83

84. ⚡ ℝ A typical lightning bolt carries a peak current of 100 kA. (a) Estimate the peak magnetic field a distance of 500 m from where the lightning bolt strikes the ground. (b) If you have a device that can measure magnetic fields as small as $B = 100$ pT $(10^{-10}$ T), how far can you be from the lightning bolt and still detect its magnetic field?

85. A high-voltage power line carries a current of 5000 A. If the power line is a height 20 m above the ground, what is the magnetic field strength at ground level? How does it compare with the Earth's magnetic field?

86. A solenoid has a length of 15 mm and a radius of 0.25 mm, and consists of 5000 circular turns. If a current of 0.35 A is passed through the solenoid, what is the magnetic field at the center (inside the solenoid)?

87. Consider the solenoid discussed in Example 20.7. What value of current is required to give a field of magnitude $B = 1.0$ T? Would this value be practical? Why or why not?

20.8 MAGNETIC MATERIALS: WHAT GOES ON INSIDE?

88. ✪ ℝ The U.S. Navy can detect a submarine by measuring the magnetic field produced by its metal hull (usually made of a magnetic metal). For the purpose of calculating its magnetic field, model a submarine as one large current loop located at the middle of the submarine carrying 8000 A, as calculated in Example 20.8 and with a diameter equal to the diameter of submarine. Assume that far from the submarine the magnitude of the magnetic field varies as $B_0(r/r_0)^{-3}$, where B_0 is the field at the center of the current loop, r_0 is the radius of the loop, and r is the distance from the center of the loop (and also the center of the submarine). What is the approximate value of the magnetic field a distance of 1000 m from the submarine?

89. ✪ ℝ Based on the results of Problem 88, the U.S. Navy has developed ways to make the magnetic field from its own submarines as small as possible. One way is by wrapping a solenoid around the hull and passing a current through the solenoid so that the field produced by the solenoid approximately cancels the field from the hull. Take the diameter of the submarine to be 20 m, and its length to be 50 m. Assume the field of the submarine alone to be 0.0010 T. Assuming a solenoid with 1000 turns, what is the approximate current required in the solenoid?

20.9 THE EARTH'S MAGNETIC FIELD

90. ℝ The Earth's magnetic field has a magnitude of approximately 5.0×10^{-5} T, and in New York City this field is directed toward the north geographic pole. If a proton is traveling in an easterly direction with a speed of 600 m/s, what are the magnitude and direction of the magnetic force on the proton?

91. SSM ⚡ ℝ Suppose the Earth's magnetic field is produced by a single loop of wire that has a radius of R_E, where R_E is the radius of the Earth, and assume this loop lies in the equatorial plane. Estimate the current that must flow in this loop to produce a field equal to the Earth's field at the center of the loop.

20.10 APPLICATIONS OF MAGNETISM

92. ✪ ✖ **Blood-flow meter.** Blood plasma has a high content of Na^+ and Cl^- ions, and as the blood follows through a blood vessel, the ions travel along at the same speed as the fluid (Fig. P20.92, left). One method of measuring the blood's velocity is to place the vessel in a magnetic field and then measure the potential difference across the width of the vessel perpendicular to the magnetic field direction. Unlike the Hall effect, both positive and negative charges here are moving in the same direction, so the magnetic force pushes the positive charges to the top and the negative charges to the bottom as seen in Figure P20.92, right. (a) Consider the equilibrium condition where the magnetic force on the charges balances the electric force on the charges. Find an expression for the velocity of the blood v in terms of the width of the blood vessel, d, the magnitude of the supplied magnetic field, B, and the resulting measured electric potential, V. (b) In designing an actual measuring device for a typical blood-flow velocity of 30 cm/s, an engineer uses an available permanent magnet that supplies a uniform 0.20-T field. We then need to find how sensitive a voltmeter we would need to measure the flow in a small blood vessel. What potential difference would develop across a 1.2-cm-diameter blood vessel with this configuration?

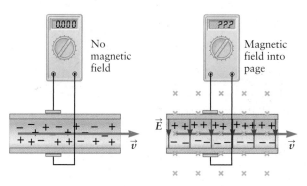

Figure P20.92 Problems 92 and 93.

93. SSM ✪ ✖ Consider a blood-flow sensor placed around the artery in Figure P20.92. Assume the artery has a radius of 3.0 mm and the magnetic field is $B = 0.050$ T. (a) If the blood speed is 25 cm/s, what is the magnitude of the induced emf? (b) Which side of the blood vessel in Figure P20.92 is at a positive potential (relative to the other side)? (c) How does the induced emf change if the artery is smaller, with $r = 0.50$ mm? (d) Explain why this type of blood-flow meter is most useful for large arteries.

20.11 THE PUZZLE OF A VELOCITY-DEPENDENT FORCE

94. ✪ This problem leads you through a study of electric and magnetic fields as viewed by two different observers, that is, in two different reference frames.

Suppose a charge q_1 is moving with a constant velocity $\vec{v}_1$ (Fig. P20.94) and that its trajectory takes it to within a distance

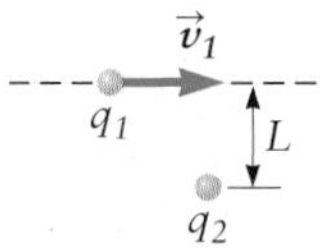

Figure P20.94

L of stationary charge q_2. (a) What are the magnitude and the direction of the resulting electric current? (b) Equation 20.19 gives the magnetic field produced by a current-carrying wire. Assume this relation can also be applied here (it can!) and use it to find the magnetic field produced by the current in part (a). What is the resulting magnetic field a distance L from the path of q_1? (c) Now assume charge q_2 has a velocity $\vec{v}_2$ parallel to $\vec{v}_1$. Use the result from part (b) to show that the magnitude of the magnetic force on q_2 when the two charges are at their closest is $F = (\mu_0 q_1 q_2 v_1 v_2/2\pi L)$. (d) What is the direction of the force in part (c) if q_1 and q_2 are both positive charges? If one is positive and one is negative?

An observer watching q_1 and q_2 travel by would say that the force found in part (c) is a magnetic force. An observer in a moving reference frame would see things differently. If an observer is moving along with q_1 at a velocity $\vec{v}_1$, he or she would say that q_1 is not moving, so the current and hence the magnetic field and the magnetic force are zero. However, this observer would still find that there is a force between the two charges as given in part (c) and would attribute it to an extra electric field E_{new}. (e) Find the magnitude and direction of E_{new}. *Hint*: The Coulomb's law force produced by E_{new} must be equal to the force found in part (c). This electric field is a relativistic effect, and we'll discuss the ideas behind it in Chapter 27.

ADDITIONAL PROBLEMS

95. ✪ Electric power is transmitted from the generation plant to a municipality via high-voltage power lines such as those connected to the tower in Figure P20.95. Consider a case with towers typically placed at 0.50-km intervals, with three separate lines, spaced 20 m apart and operating at a high voltage (760 kV) with typical maximum currents of 500 A. (a) Find the force on a tower-to-tower length of one of the outside wires from the other two. (b) If the wire has a cross-sectional area of 200 mm^2 and an average density of 4000 kg/m^3, compare the weight of a section of the wire to the force found in part (a).

Figure P20.95

© Comstock/Jupiterimages

96. ✪ Two aluminum rods of length 75 cm and mass 50 g are suspended by nonconducting cables of length 10 cm as shown in Figure P20.96. The rods are part of a circuit (not shown) that supplies a current I through each rod as shown. If the angle between the two end cables is 4.0°, what is the magnitude of the current I? Assume the rods are very long.

Figure P20.96

97. ✪ A conducting rod of mass 20 g and length 45 cm has holes drilled near each end so that it is free to slide in a vertical direction along two conducting rails as shown in Figure P20.97. A current of 200 A is made to flow through an identical rod at the bottom of the assembly and return through the sliding rod (notice the placement of an insulator). Determine the height of the sliding rod above the base rod when in equilibrium. Assume the horizontal rods are very long.

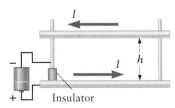

Figure P20.97

98. ✪ An electron enters a bubble chamber and produces a spiral track like that shown on the left in Figure Q20.13. The magnetic field is directed into the page, and the electron had an initial velocity of 2.7×10^8 m/s. (a) If the initial radius of the electron's trajectory is 5.0 cm, what is its velocity when it has spiraled into a radius of 2.0 cm? (b) What is the correspond-

ing change in kinetic energy? Give your answer in joules and electron-volts. (c) Calculate the magnitude of the magnetic field present in the bubble chamber. (d) What is the change in force on the electron in going from the initial radius to the final radius?

99. ✪ The *cyclotron* is a device that provides a beam of high-velocity charged particles. Consider the electron-accelerating cyclotron depicted in Figure P20.99. Two semicircular magnets provide a uniform magnetic field of 5.0 mT between the poles; in addition, an electric potential difference of 220 V is placed on each D-shaped section so that an accelerating electric field is produced in the gap between the semicircles. An electron will accelerate as it crosses the gap and enter the magnetic field, producing a circular orbit so that the electron is directed back into the gap. By the time the electron returns to the gap, circuitry will have reversed the direction of the electric field so that the electron is again accelerated. Ignore relativistic effects in the following. (a) How much does the energy of an electron increase each time it crosses the gap? If the radius of each D-shaped magnet is 5.0 cm and the electrons start in the center of the magnets at one edge of the gap, what are the (b) maximum speed and maximum kinetic energy that an electron can attain? (c) At what rate must the accelerating electric potential be reversed to make the device work?

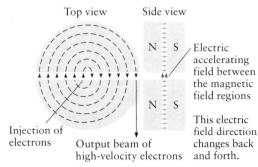

Figure P20.99

100. SSM ✪ You are standing directly under a high-voltage power line that is 4.5 m above, and a detailed map you have indicates that the power line runs exactly north to south. Your compass needle does not align with the power line, however, but instead makes an angle of 32° toward the east with respect to the power line's direction. (a) Knowing that the magnitude of the Earth's magnetic field is approximately 5.0×10^{-5} T, determine the magnitude and direction of the magnetic field produced by the power line at ground level. (b) How much current is being carried in the power line? (c) Is the current flowing toward the north or toward the south?

Magnetic Induction

An electric guitar uses magnetic induction to detect the motion of the guitar strings. This chapter explains how that works. (© Shepard Sherbell/Corbis)

Are electricity and magnetism separate phenomena, or are they related in some way? Perhaps the most obvious hint that they are connected is that electric and magnetic forces both act only on particles carrying an electric charge. Another clue is that moving electric charges (an electric current) create a magnetic field. But how do we know there is *really* an important connection between $\vec{E}$ and $\vec{B}$? After all, charged particles have mass, so they experience gravitational forces. Does that mean gravity is also connected to electricity and magnetism? (The answer is no, gravity does not play a role in electromagnetism.) In this chapter, we'll see that a changing magnetic field creates an electric field. This effect, called **magnetic induction,** links $\vec{E}$ and $\vec{B}$ in a truly fundamental way. Magnetic induction is also the key to many practical applications of magnetism, including motors, generators, loudspeakers, and cell phones.

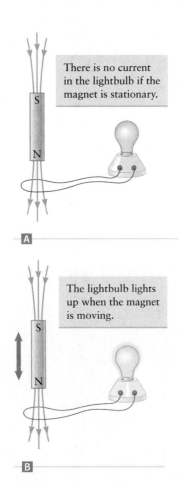

There is no current in the lightbulb if the magnet is stationary.

The lightbulb lights up when the magnet is moving.

Figure 21.1 **A** When a bar magnet is placed near a loop of wire, a magnetic field penetrates the loop. **B** When the bar magnet is in motion, an electric current is induced in the loop, lighting the lightbulb.

21.1 | WHY IS IT CALLED *ELECTROMAGNETISM*?

The discovery of electric charges and forces by the Greeks and other early civilizations involved static electricity, whereas the first observations of magnetic effects employed permanent magnets composed of the mineral magnetite as mentioned in Chapter 20. There was no obvious connection between the two effects until a chance observation by Danish physicist Hans Christian Ørsted in 1820. While giving a class lecture, he discovered that an electric current in a wire can exert a force on a compass needle. (Such magnetic forces were discussed in Chapter 20.) An electric current is produced by an electric field and a compass needle is a small bar magnet, so Ørsted's lecture experiment also showed that an electric field can lead to a force on a magnet. He therefore concluded that *an electric field can produce a magnetic field.*

When this connection between $\vec{E}$ and $\vec{B}$ was announced in 1820 it was a complete surprise and stimulated many further experiments to probe the connection between electricity and magnetism. One scientist engaged in this work was Michael Faraday, who reasoned that if an electric field can produce a magnetic field, perhaps a magnetic field can produce an electric field. Faraday attempted to observe such an "induced" electric field using an experiment like the one sketched in Figure 21.1A in which a bar magnet is positioned close to a loop of wire. This loop is connected to a lightbulb, so if there is current in the loop, the bulb will light up and thus act as a detector of the electric current. (Faraday didn't use a lightbulb—they weren't invented yet—instead, he used an ammeter to detect the current, with the same results.)

Faraday positioned a bar magnet near the loop of wire in Figure 21.1A so that the magnetic field would penetrate through and around the loop. If this magnetic field produces an electric field, the electric field would cause an electric current in the loop, lighting the bulb. Faraday was probably very disappointed that this experiment was not successful. No current was detected in the wire loop when the bar magnet was at rest near the loop. He did, however, discover a variation on this experiment in which the bulb does light up as shown in Figure 21.1B. If the bar magnet is *in motion* either toward or away from the loop, the bulb emits light, indicating an electric field within the wire. This electric field is produced *only when the magnet is moving.* If the magnet is stationary (as in Fig. 21.1A), the current and hence also the electric field are always zero.

Faraday devised another approach to this experiment in which a solenoid is positioned near a loop of wire containing a lightbulb (Fig. 21.2). Recall from Chapter 20 that when an electric current passes through a solenoid, a magnetic field is produced with field lines similar to those produced by a bar magnet. Faraday wanted to know if this magnetic field could also produce a current in his wire loop. He passed an electric current I through the solenoid by connecting it to a battery through a switch. Faraday found that when the current through the solenoid is constant (Fig. 21.2A), there is no current (the bulb did not light), but the bulb does light up the moment the switch is opened or closed (Fig. 21.2B). An electric current is produced

Figure 21.2 **A** When a current-carrying coil is placed near a loop of wire, the magnetic field produced by the coil penetrates the loop. When this magnetic field is constant, no current is induced in the loop. **B** When the magnetic field is changing with time (i.e., just as the switch is closed or opened), an electric current is induced in the loop.

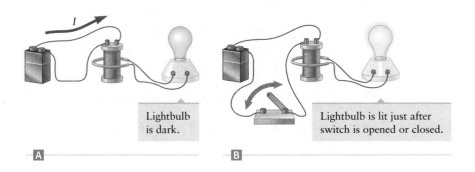

Lightbulb is dark.

Lightbulb is lit just after switch is opened or closed.

during those instants in which the current through the solenoid is *changing* from zero to some nonzero value (when the switch was closed) or from a nonzero value to zero (as the switch was opened).

Faraday's experiments with the bar magnet (Fig. 21.1) and the solenoid (Fig. 21.2) show that an electric current is produced in the wire loop only when the magnetic field at the loop is changing. Hence, *a changing magnetic field produces an electric field.* An electric field produced in this way is called an *induced* electric field, and this phenomenon is called **magnetic induction.**

21.2 | MAGNETIC FLUX AND FARADAY'S LAW

Faraday's experiments in Figures 21.1 and 21.2 demonstrated qualitatively the effect of magnetic induction. Faraday also developed a quantitative theory of the effect, now called **Faraday's law,** that tells how to calculate the induced electric field in different situations. To use this law, we must first introduce a quantity called **magnetic flux.**

Suppose there is a magnetic field in some region of space and let A be the area of a particular surface as sketched in Figure 21.3A. If lines of $\vec{B}$ cross this surface, there is a magnetic flux through the surface. Magnetic flux is very similar to the concept of electric flux that we encountered in Chapter 17 in connection with Gauss's law.

Let's now calculate the magnetic flux in a few specific cases. Figure 21.3A shows a case in which the magnetic field has a constant magnitude B (the same everywhere in this region of space). The surface area A is flat, and $\vec{B}$ is perpendicular to the surface. The magnetic flux, denoted by Φ_B, is given by

$$\Phi_B = BA \quad \text{(when } \vec{B} \text{ is perpendicular to a surface)} \tag{21.1}$$

Qualitatively, you can think of the magnetic flux as a "count" of the number of field lines that pass through the surface.

Two other common situations are shown in parts B and C of Figure 21.3. In both cases, $\vec{B}$ has a constant magnitude and a constant direction. In Figure 21.3B, the magnetic field lines cut through the indicated surface, making an angle θ with the direction perpendicular (normal) to the surface. Here the magnetic flux is

$$\Phi_B = BA \cos \theta \quad \text{(when } \vec{B} \text{ makes an angle } \theta \text{ relative to the normal to a surface)} \tag{21.2}$$

Since θ is not zero, the flux in Figure 21.3B is smaller than when the field lines cut perpendicularly through the surface as in Figure 21.3A (for the same values of B and A). In Figure 21.3C, the magnetic field is parallel to the plane of the surface; no lines of $\vec{B}$ cut through the surface, so the magnetic flux is zero:

$$\Phi_B = 0 \quad \text{(when } \vec{B} \text{ is parallel to a surface)} \tag{21.3}$$

This case—with $\vec{B}$ parallel to the surface—is actually a special case of Equation 21.2, with $\theta = 90°$. Likewise, when $\vec{B}$ is perpendicular to the surface (Fig. 21.3A), $\theta = 0$ and the expression for the flux in Equation 21.2 reduces to Equation 21.1.

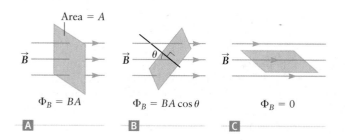

Figure 21.3 **A** When a magnetic field passes through a surface with an area A, there is a magnetic flux through the surface. If $\vec{B}$ is perpendicular to the surface and the field's magnitude is the same at all points on the surface, the magnetic flux is $\Phi_B = BA$. **B** If $\vec{B}$ makes an angle θ with a line drawn perpendicular to a surface of area A, the magnetic flux through this surface is $\Phi_B = BA \cos \theta$. **C** If the magnetic field is parallel to the surface so that $\theta = 90°$, the magnetic flux through the surface is zero.

For simplicity, in the three examples in Figure 21.3 we have assumed the magnetic field is the same at all points on the surface and we are dealing with a flat surface. However, Φ_B can be defined in a similar way for any surface. Any complicated surface can be broken into small regions that are each approximately flat. The magnetic flux through each small region is equal to the magnitude of $\vec{B}$ multiplied by the surface area and the cosine of θ, the angle between $\vec{B}$ and the direction perpendicular to that surface region. Hence, Equation 21.2 can be applied to find the flux through any small region. The total flux is then the sum of the fluxes through all the individual pieces of the surface being considered.

When calculating electric flux in our applications of Gauss's law in Chapter 17, we were interested in closed surfaces, that is, surfaces that completely enclose a particular region of space. In magnetic induction, however, the surfaces of interest are *open* surfaces, like the ones in Figure 21.3. Such open surfaces always possess an edge or perimeter, which (as we will soon see) plays a big role in magnetic induction.

The units of magnetic flux can be deduced from Equation 21.2. The right-hand side of this equation has units of magnetic field multiplied by area, so magnetic flux has units of $T \cdot m^2$. This unit is named the ***weber*** (Wb):

$$\text{unit of magnetic flux: } 1\ \text{Wb} = 1\ T \cdot m^2 \tag{21.4}$$

The weber is the official SI unit of magnetic flux and is used in some texts, but many physicists just use the combination of units $T \cdot m^2$.

Faraday's Law

Faraday's induction experiments in Figures 21.1 and 21.2 demonstrate that a changing magnetic field produces a current in a wire loop. ***Faraday's law*** tells us how to calculate the potential difference that produces this induced current. This law is written in terms of the magnetic flux and the electromotive force (emf) induced in the wire loop. Recall from Chapter 19 that an emf (denoted by $\mathcal{E}$) is not a force; rather, it is an electric potential difference, like the one that appears across the terminals of a battery. Faraday's law states that a changing magnetic flux produces an electric potential difference $\mathcal{E}$ given by

Faraday's law

$$\mathcal{E} = -\frac{\Delta\Phi_B}{\Delta t} \tag{21.5}$$

This simple formula contains two important ideas. The first is that the *magnitude* of the induced emf equals the rate of change of the magnetic flux:

$$|\mathcal{E}| = \left|\frac{\Delta\Phi_B}{\Delta t}\right| \tag{21.6}$$

The second important "idea" is the negative sign in Equation 21.5. In fact, this negative sign is so important that it has its own name: ***Lenz's law***. We will devote Section 21.3 to a discussion of this negative sign and explain why it deserves to be called a law of physics. Before we do that, we'll consider some properties and situations that depend on just the magnitude of the induced emf (Eq. 21.6). After we understand that part of Faraday's law, we will be better prepared to understand the negative sign in Equation 21.5 and the importance of Lenz's law.

Let's consider the meaning of the two sides of Equation 21.6 in the context of the experiment in Figure 21.4. Here the magnetic field produced by a bar magnet passes through a loop of wire that encloses an area A. This arrangement is similar to the one in Figure 21.1 except for two small changes. First, suppose the wire loop is very small, so the magnetic field is constant across the area of the loop and the field is perpendicular to the plane of the loop. The magnetic flux is then simply $\Phi_B = BA$ (Eq. 21.1). However, the value of B changes as the bar magnet is moved closer or farther away from the loop, so the flux changes with time as the magnet moves. Second, we have replaced the lightbulb in Figure 21.1 with a voltmeter, and for simplicity we assume the voltmeter's input terminals are very close together and close to the wire loop so that the perimeter of the loop still defines our area A.

Let's now consider the terms in Faraday's law (Eq. 21.6) and what they mean.

1. The quantity $\mathcal{E}$ on the left-hand side is the induced emf that appears in the wire loop. The value of $\mathcal{E}$ would be indicated by the voltmeter in the wire loop circuit and has units of volts (V). This emf is an induced electric potential difference and is related to the electric field directed along and inside the wire loop. This induced potential difference produces the current in the lightbulb in Figures 21.1 and 21.2.

2. The right-hand side of Faraday's law contains the change in magnetic flux $\Delta\Phi_B$ during a short time interval Δt, divided by Δt. The ratio $\Delta\Phi_B/\Delta t$ is just the rate of change of the magnetic flux with time. In words, Faraday's law says that an emf (i.e., a voltage or electric potential difference) is produced in a circuit by *changes* in the magnetic flux through the circuit. A constant flux, no matter how large, does not produce an induced voltage, in accord with Faraday's unsuccessful experiments (Figs. 21.1A and 21.2A). Magnetic induction requires that the flux through the wire loop change with time. In Faraday's successful experiments (Figs. 21.1B and 21.2B), this requirement was accomplished by arranging for the magnitude of $\vec{B}$ to change with time. According to Equation 21.2, however, the magnetic flux also depends on the area A and the angle θ. Changes in A and θ can thus also produce a changing flux and hence an induced emf.

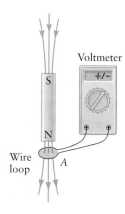

Voltmeter

Figure 21.4 If a voltmeter is attached to the wire loop in Figure 21.1 in place of the lightbulb, it can be used to measure the emf (i.e., the voltage) induced in the loop.

What about the *direction* of the induced current in a wire loop? The lightbulbs in Figures 21.1 and 21.2 were either lit or not, indicating current or no current, but that does not tell the current direction. This direction can be read from the *sign* of the voltmeter reading in Figure 21.4, which is at the heart of Lenz's law. The voltmeter reading (positive or negative) specifies the direction of the induced emf and hence the direction of the induced current and electric field, all of which are important in practical applications of Faraday's law.

EXAMPLE 21.1 | Magnitude of $\mathcal{E}$ Produced by Moving a Bar Magnet

Faraday's law tells us that a changing magnetic flux produces an induced emf, but how large is this emf in typical cases? Consider a wire loop of area $A = 1 \text{ cm}^2$ initially far from any sources of magnetic field. A strong bar magnet with $B = 1$ T near its poles is suddenly inserted into the loop, filling the loop completely (Fig. 21.5). If the time required to insert the magnet is $\Delta t = 0.1$ s, what is the approximate value of the emf induced in the loop?

RECOGNIZE THE PRINCIPLE

The magnitude of an induced emf $\mathcal{E}$ is given by Faraday's law, with $\mathcal{E}$ equal to the change of the magnetic flux divided by the time over which this change of flux occurs.

SKETCH THE PROBLEM

Figure 21.5 shows the initial and final positions of the bar magnet, which correspond to the initial and final flux through the loop.

IDENTIFY THE RELATIONSHIPS

The change in magnetic flux is $\Delta\Phi_B = \Phi_f - \Phi_i$ (the final flux minus the initial flux). Because the wire loop is initially far from any magnetic fields, the initial flux through the loop is approximately zero and $\Phi_i = 0$. After inserting the bar magnet, the field through the loop is $B \approx 1$ T, so the final flux (according to Eq. 21.1) is $\Phi_f = BA$ and the change in flux is

$$\Delta\Phi_B = \Phi_f - \Phi_i = BA \tag{1}$$

This flux change occurs over a time interval of $\Delta t = 0.1$ s (as given above).

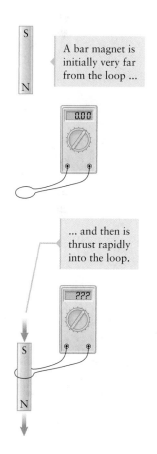

A bar magnet is initially very far from the loop ...

0.00

... and then is thrust rapidly into the loop.

???

Figure 21.5 Example 21.1. If a bar magnet is inserted into this wire loop, the voltmeter will measure a nonzero induced electric potential.

SOLVE

Inserting the change in flux from Equation (1) into Faraday's law, along with the value of Δt, we find the magnitude of the emf to be

$$|\mathcal{E}| = \left| \frac{\Delta \Phi_B}{\Delta t} \right| = \frac{BA}{\Delta t} = \frac{(1 \text{ T})(1 \times 10^{-4} \text{ m}^2)}{(0.1 \text{ s})} = \boxed{0.001 \text{ V}}$$

What does it mean?

This voltage is rather modest. The emf of a typical flashlight battery is 1.5 V, for instance, and the voltage in a typical household circuit is about 120 V, so this induced emf would not be a practical replacement for the battery in a radio or MP3 player. An induced voltage of this size can be measured quite easily, however, and could be used in other applications. (See Example 21.9.)

CONCEPT CHECK 21.1 | Faraday's Law

Suppose the area A of the wire loop in Example 21.1 (Fig. 21.5) is increased and the bar magnet is also enlarged so that it still fills the loop. If the rest of the experiment (the values of B and Δt) stays the same, how is the magnitude of the induced emf affected?

  (a) The induced emf is larger.
  (b) The induced emf is smaller.
  (c) The induced emf does not change because the magnetic field is the same.

Key Features of Faraday's Law

According to Faraday's law, the emf induced along a closed path such as a wire loop depends on changes in the magnetic flux through the area enclosed by that path. This fact leads to a number of important consequences.

1. Only *changes* in Φ_B matter. The magnetic flux might be very large, but if it changes only a small amount, the induced emf will still be small.
2. Rapid changes in magnetic flux produce larger values of $\mathcal{E}$ than do slow changes. This dependence on the *frequency* at which the flux changes means that the induced emf plays a key role in alternating-current (AC) circuits (Chapter 22).
3. The magnitude of $\mathcal{E}$ is proportional to the rate of change of the flux $\Delta\Phi_B/\Delta t$. If this rate of change is constant, then $\mathcal{E}$ is constant. However, achieving a constant rate of change $\Delta\Phi_B/\Delta t$ requires that Φ_B must eventually become very large, which is usually not possible in practice. So, in most applications, the rate of change of Φ_B is not constant and the induced emf varies with time, leading again to AC voltages (Chapter 22).
4. The induced emf is present even if there is no current in the path enclosing an area of changing magnetic flux. Hence, the voltmeter in Figure 21.4 will register an induced voltage whether or not a current is present in the loop.

Magnetic Flux through a Changing Area

Changes in the magnetic flux can be produced by changes in the magnitude of $\vec{B}$, by changes in area A of the surface, and by changes in the angle θ that the surface makes with respect to $\vec{B}$. An example in which this area is changing is sketched in Figure 21.6A, which shows a metal bar of length L that can slide on two frictionless, horizontal metal rails. The metal rails are connected through a resistor on the left, so the rails plus the bar form a closed electric circuit. A magnetic field is directed perpendicular to the plane of the rails and bar, with a constant magnitude B and direction throughout. Suppose the bar is being pushed to the right by a force from an external agent (perhaps a person is pulling on it), making the bar move with a constant speed v. Let's use Faraday's law to find the induced emf in this circuit.

We apply Faraday's law using the closed path indicated by the dashed loop in Figure 21.6A. The magnetic field is constant and perpendicular to the enclosed area, so the magnetic flux is (Eq. 21.1)

$$\Phi_B = BA = BwL \qquad (21.7)$$

where w is the distance from the left side of the circuit to the bar. Because the bar is moving to the right, this distance and hence the magnitude of the flux increase with time. The speed v of the bar is constant and is $v = \Delta w/\Delta t$, so the flux in Equation 21.7 changes with time according to

$$\frac{\Delta \Phi_B}{\Delta t} = BL\frac{\Delta w}{\Delta t} = BLv \qquad (21.8)$$

Inserting Equation 21.8 into Faraday's law (Eq. 21.6), we find the magnitude of the induced emf,

$$|\mathcal{E}| = \frac{\Delta \Phi_B}{\Delta t} = BLv \qquad (21.9)$$

This induced emf acts much like a battery in the "equivalent" circuit shown in Figure 21.6B, with an emf given by Equation 21.9. The magnitude of the current I in this circuit is $I = \mathcal{E}/R$ (Eq. 19.10). Inserting our result for $\mathcal{E}$, we have

$$I = \frac{BLv}{R} \qquad (21.10)$$

Recall from Chapter 19 that in circuits like the one in Figure 21.6B, power is dissipated in the resistor. The dissipated power is given by $P = I^2R$ (Eq. 19.16); hence,

$$P = I^2R = \left(\frac{BLv}{R}\right)^2 R = \frac{B^2L^2v^2}{R} \qquad (21.11)$$

Conservation of Energy and the Induced emf

Let's now summarize in words what the result in Equation 21.11 means.

1. Since the metal bar in Figure 21.6A is moving, the area of the loop indicated by the dashed red line increases with time, producing a change in the magnetic flux through this loop.
2. According to Faraday's law of induction, an emf is induced in the circuit composed of the metal bar plus rails.
3. This induced emf acts just like a battery with an emf (or voltage) $\mathcal{E}$, producing an electric current in the circuit given by Equation 21.10.
4. This current leads to a power dissipation P in the circuit.

Where does this power come from? In mechanics, power equals the product of force and velocity, so to answer this question we must consider the forces acting on the metal bar. Actually, two forces act on the metal bar in Figure 21.6. One is a magnetic force due to the current in the bar. We have already calculated the magnitude of this current in Equation 21.10. The direction of this current is specified by Lenz's law; in Section 21.3, we'll show in detail how Lenz's law gives the current direction, from the lower end of the rod toward the upper end, sketched in Figure 21.6D. The magnetic field is perpendicular to the bar, so from Equation 20.11, the magnitude of the magnetic force on the bar is

$$F_B = ILB \qquad (21.12)$$

We now apply right-hand rule number 2 from Chapter 20 to find the direction of this force. An upward current in Figure 21.6D corresponds to the motion of positive charges moving upward in the figure. If you place the fingers of your right hand upward along the bar and then curl them toward the direction of the magnetic field, your thumb must point to the left in the figure as indicated by the vector $\vec{F}_B$ in parts

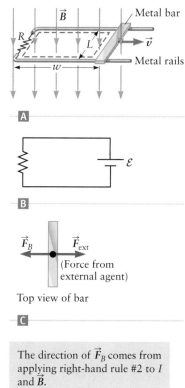

Figure 21.6 **A** The magnetic flux through a loop changes with time if the area of the loop changes. As the movable bar on the right slides along the rails, the enclosed area changes. **B** The equivalent circuit is a battery connected to a resistor. The battery voltage represents the induced emf in the sliding bar. **C** Forces acting on the sliding bar. **D** The emf induced in the bar produces a current through the circuit. The resulting magnetic force $\vec{F}_B$ on the sliding bar opposes its motion.

C and D of Figure 21.6. Hence, the magnetic force on the bar *opposes* the motion of the bar.

Now we consider the second force acting on the bar. This force is from the external agent mentioned at the start of the problem, and we have assumed this agent acts so as to make the bar move to the right with a constant velocity. According to Newton's laws of mechanics, if the metal bar is moving with a constant velocity, the total force on the bar must be zero. Hence, the force of the external agent $\vec{F}_{ext}$ must be equal in magnitude and opposite in direction to the magnetic force in Equation 21.12. We thus have

$$F_{ext} = ILB \tag{21.13}$$

and this external force must be directed to the right (Fig. 21.6C). This force must be provided by the external agent to keep the bar moving at a constant velocity. The power exerted by the external agent is then (Chapter 6) $P_{ext} = F_{ext}v$. Inserting F_{ext} from Equation 21.13 leads to

$$P_{ext} = F_{ext}v = ILBv$$

We know the current I from Equation 21.10. Inserting that gives

$$P_{ext} = \left(\frac{BLv}{R}\right)LBv = \frac{B^2L^2v^2}{R} \tag{21.14}$$

which is identical to the electrical power dissipated in the resistor, Equation 21.11. Hence, the mechanical power "put into" the bar by the external agent is converted into electrical power that is "delivered to" the resistor. From a work–energy perspective, we should expect that these two must be equal, and our results in Equations 21.11 and 21.14 show that to be the case. While energy is converted from one form (mechanical) to another (electrical), we have shown that the total energy is conserved.

In Chapter 6, we saw how the principle of conservation of energy can be deduced for mechanical systems from Newton's laws of motion. Now we have found that conservation of energy is also obeyed by electromagnetic phenomena. Indeed, it is believed that the principle of energy conservation is obeyed by *all* physical processes and by all laws of physics. Whenever new laws of physics are proposed, one of the first tests is to check if the proposed law conserves energy. In this way, the principle of conservation of energy is at the very heart of physics.

EXAMPLE 21.2 | **Using Faraday's Law to Design a Generator**

A device that converts mechanical energy into electrical energy is called a ***generator***. The system in Figure 21.6A is a very simple type of generator. A person pulling on the rod exerts a force on the sliding bar (doing mechanical work), converting mechanical energy into electrical energy as shown by the current in the circuit. Suppose we make a generator as in Figure 21.6A, with a bar of length $L = 1$ m and a magnetic field $B = 1$ T. To be useful for producing power for your house, the generator's output voltage (the induced emf) should be approximately 100 V. (The voltage in household wiring is typically 120 V.) How fast must the bar travel to produce this value for the induced emf? Is this way of making a generator practical?

RECOGNIZE THE PRINCIPLE

The combination of the bar and rails in Figure 21.6 acts as a generator since an emf is induced across the ends of the moving bar. According to Faraday's law, the magnitude of the emf induced across the bar equals the rate of change of the magnetic flux $\Delta\Phi_B/\Delta t$, which we calculated above in our discussion of Figure 21.6. We found that $\Delta\Phi_B/\Delta t$ is proportional to the speed of the bar v. So, given the desired value of $\mathcal{E}$, we can find the required value of v.

SKETCH THE PROBLEM

Figure 21.6 describes the problem.

IDENTIFY THE RELATIONSHIPS AND SOLVE

Faraday's law led to the magnitude of the induced emf in Equation 21.9:

$$|\mathcal{E}| = \frac{\Delta\Phi_B}{\Delta t} = BLv$$

Rearrange to solve for v gives

$$v = \frac{|\mathcal{E}|}{BL}$$

Inserting the given values of $\mathcal{E}$, B, and L, we have

$$v = \frac{100 \text{ V}}{(1 \text{ T})(1 \text{ m})} = \boxed{100 \text{ m/s}}$$

What does it mean?

Is this approach for generating electric power practical? This velocity is about 200 mi/h. If we turn the generator on for 10 s, the horizontal rails would have to be at least $vt = (100 \text{ m/s})(10 \text{ s}) = 1000 \text{ m}$ long! That would be an extremely impractical generator.

Designing a Practical Electrical Generator

The electrical generator in Example 21.2 could certainly be built, producing a constant emf as long as the bar slides at a constant speed, yet we concluded in Example 21.2 that such a DC generator is not a practical device. How can we make the rate of change of the flux $\Delta\Phi_B/\Delta t$ large enough to give a useful emf? The key is to use rotational motion instead of the linear motion of the bar in Figure 21.6. Generators based on this idea are found in all electric power plants, including nuclear power plants, hydroelectric plants, and those that burn coal or natural gas.

A simple generator based on rotational motion is sketched in Figure 21.7. A permanent magnet produces a constant magnetic field in the region between its two poles (north and south), and a wire loop is located in this region. This loop has a fixed area but is mounted on a rotating shaft, so the angle θ between the field and the plane of the loop changes as the loop rotates. According to the definition of magnetic flux, Φ_B is largest when the loop is perpendicular to the field (Fig. 21.7A) and zero when the plane of the loop is parallel to the field (Fig. 21.7B).

In practice, this generator would be driven mechanically by turning the shaft, which could be accomplished with a steam turbine (as in most power plants) or with a waterfall or windmill. If the shaft rotates with a constant angular velocity, the magnetic flux through the loop varies sinusoidally with time, leading to an induced emf that also varies sinusoidally with time (Fig. 21.7C).

The generator sketched in Figure 21.7 sounds reasonable, but is it really practical? Let's estimate the magnitude of the emf that could be produced by such an AC (alternating-current) generator. If the shaft rotates with a period T, the loop will turn from its perpendicular orientation in Figure 21.7A to the parallel orientation in Figure 21.7B in a time $T/4$ because this rotation is one-fourth of a complete turn of the rotating shaft. During this time, the magnetic flux changes from $\Phi_i = BA$ (Fig. 21.7A) to $\Phi_f = 0$ (Fig. 21.7B). The change in magnetic flux is

$$\Delta\Phi_B = \Phi_f - \Phi_i = -BA \tag{21.15}$$

during a time $\Delta t = T/4$. Inserting Equation 21.15 into Faraday's law (Eq. 21.6) gives

$$|\mathcal{E}| = \left|\frac{\Delta\Phi_B}{\Delta t}\right| = \frac{BA}{T/4} = \frac{4BA}{T} \tag{21.16}$$

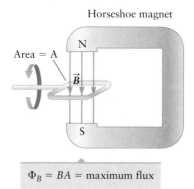

Horseshoe magnet

Area = A

$\Phi_B = BA$ = maximum flux

A

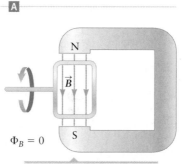

$\Phi_B = 0$

The flux is zero when the loop is in this orientation.

B

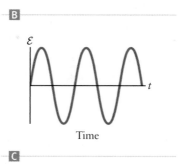

$\mathcal{E}$

t

Time

C

Figure 21.7 Design for an electrical generator based on Faraday's law of induction. **A** A wire loop is inserted between the poles of a magnet, creating a magnetic flux through the loop. If the loop is perpendicular to $\vec{B}$, the flux has a magnitude BA. **B** When the loop is parallel to $\vec{B}$, the flux is zero. If the loop rotates about its axis, the magnetic flux will vary with time, producing a time-varying induced voltage in the loop. **C** The emf induced in the loop as a function of time.

Figure 21.8 Different ways to produce an induced emf.

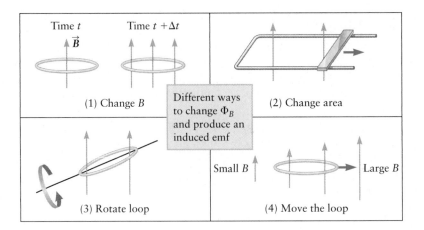

Time t Time $t + \Delta t$

$\vec{B}$

(1) Change B

Different ways to change Φ_B and produce an induced emf

(2) Change area

(3) Rotate loop

Small B Large B

(4) Move the loop

For a realistic generator, we might have $B = 1$ T (typical for a permanent magnet), and we take the area of the loop to be $A = 1$ m^2 as a fair comparison with the DC generator in Example 21.2. The period T depends on the angular speed of the shaft; a rotation rate of 1000 rpm is common (a car engine runs typically in the 2000–4000 rpm range). The corresponding period is

$$T = \frac{1 \text{ min}}{1000 \text{ rev}} \times \frac{60 \text{ s}}{1 \text{ min}} = 0.060 \text{ s}$$

Inserting all these results into Equation 21.16 gives

$$|\mathcal{E}| = \frac{4BA}{T} = \frac{4(1 \text{ T})(1 \text{ m}^2)}{0.060 \text{ s}} = 70 \text{ V}$$

This voltage is quite useful, so this design would indeed be practical. The voltage could easily be increased by increasing the rotation rate of the shaft or by making the area larger.

The basic design in Figure 21.7 is used in most real electrical generators, but it is still important to recall that because $\Phi_B = BA \cos \theta$, we can produce a change in magnetic flux and hence an induced emf in four different ways (Fig. 21.8): (1) if the magnitude of the magnetic field B changes with time, (2) if the area A changes with time, (3) if the loop rotates so that the angle θ between $\vec{B}$ and the loop changes with time (as in the generator in Fig. 21.7), or (4) if the loop moves from one region to another and the magnetic field is different in the two regions.

CONCEPT CHECK 21.2 | Designing a Better Generator

The generator in Figure 21.7 employs a loop that rotates with a period T. Suppose the rotation rate is increased so that the loop is made to rotate faster and the period is reduced by a factor of two. How will that affect the magnitude of the induced emf?

(a) The magnitude of the induced emf will not change.
(b) The magnitude of the induced emf will increase by a factor of two.
(c) The magnitude of the induced emf will decrease by a factor of two.

21.3 | LENZ'S LAW AND WORK–ENERGY PRINCIPLES

In Section 21.2, we applied Faraday's law to calculate the magnitude of the induced emf in several different situations. Let's now consider the *sign* of this emf. Faraday's law reads (Eq. 21.5)

$$\mathcal{E} = -\frac{\Delta \Phi_B}{\Delta t} \tag{21.17}$$

Algebraically, the sign of $\mathcal{E}$ is determined by the sign of $\Delta\Phi_B$ together with the negative sign on the right-hand side of Equation 21.17, so it would appear that determining the sign of the induced emf $\mathcal{E}$ is straightforward. However, there is a mathematical complication: the magnetic flux is produced by a field that is perpendicular to the surface of interest, whereas the induced current is along the perimeter of this surface (Fig. 21.9). The emf on the left-hand side of Faraday's law (Eq. 21.17) thus leads to a current *perpendicular* to the flux direction on the right-hand side. Hence, the two sides of Faraday's law involve quantities with different and perpendicular directions, which complicates the meaning of the negative sign. There are ways to deal with this complication, but it is simpler to make use of a principle called *Lenz's law*. This principle gives an easy way to determine the sign of the induced emf $\mathcal{E}$, and we'll see that it is closely connected with the principle of conservation of energy.

In many situations, the induced emf of Faraday's law produces a current leading to an induced magnetic field. Lenz's law is a statement about the direction of this induced field:

Lenz's law

The magnetic field produced by an induced current always opposes any changes in the magnetic flux.

Applying Lenz's Law

Let's apply Lenz's law to an arrangement in which the magnetic field passes through a metal loop with the field initially directed upward through the loop (Fig. 21.10A). We also suppose the magnitude of the field $|\vec{B}|$ increases with time (Fig. 21.10B) and ask two questions: (1) Is there a current induced in the loop? (2) What is the direction of this current?

Since we are concerned with the current in the loop, we let it be the perimeter of our Faraday's law surface. Here, $\vec{B}$ is perpendicular to the plane of the loop, so there is a magnetic flux directed upward through this surface. The magnitude $|\vec{B}|$ is increasing with time, so Φ_B is also increasing. This changing flux produces an induced emf, so there will indeed be an induced current in the loop. According to Lenz's law, the magnetic field produced by this induced current must *oppose the change* in flux, so the induced magnetic field must be *downward*. From our work on current loops (and right-hand rule number 1 in Chapter 20), the induced current must be *clockwise* in Figure 21.10C when looking at the loop from above to give a field in this direction.

The crucial feature of Lenz's law is that the induced field *opposes changes* in the flux as illustrated in Figure 21.11, which again shows a circular wire loop in a perpendicular magnetic field as in Figure 21.10. Now, however, we assume the magnitude of

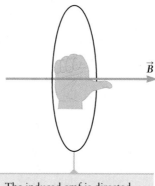

The induced emf is directed along the perimeter of the flux surface. The induced current is thus perpendicular to $\vec{B}$.

Figure 21.9 The magnetic field that produces the magnetic flux in Faraday's law is perpendicular to the plane in which the induced emf is sensed. Lenz's law gives the direction of the induced emf, that is, either clockwise or counterclockwise around the perimeter of the surface of interest.

Lenz's law

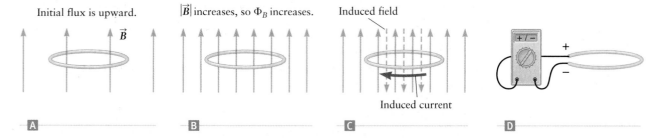

Initial flux is upward. $|\vec{B}|$ increases, so Φ_B increases.   Induced field   Induced current

Figure 21.10 ◻A◻ This magnetic field is directed through the plane of the loop, giving a flux "upward" through the loop. ◻B◻ If the magnitude of B increases with time, the flux increases. ◻C◻ According to Lenz's law, the induced emf in the loop leads to a magnetic field that opposes this change in flux. A downward magnetic field is created by a clockwise induced current. ◻D◻ Even if the loop is broken, there is still an induced emf $\mathcal{E}$, which could be measured by attaching a voltmeter to the two open ends of the loop. The voltmeter reading indicates the polarity (+ or −) of $\mathcal{E}$.

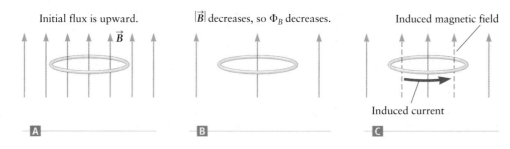

Initial flux is upward.

$|\vec{B}|$ decreases, so Φ_B decreases.

Induced magnetic field

Induced current

A   B   C

Figure 21.11 ◻A $\vec{B}$ is directed upward through the loop, so the magnetic flux is also in the "upward" direction. ◻B If the magnitude of the field decreases with time, the flux through the loop decreases. ◻C The induced current opposes this change, so $I_{induced}$ produces a field that is upward. The required direction for the induced current is counterclockwise as viewed from above (according to right-hand rule number 1).

the field is decreasing with time, so the "upward" flux is getting smaller. This change in flux again produces an induced emf, causing an induced current around the loop. The induced magnetic field produced by this current opposes the *change* in flux, so now the induced field is directed upward and the induced current is counterclockwise when viewed from above the loop.

We have stated Lenz's law in terms of the magnetic field produced by the induced current, but what if there is no induced current? For example, Figure 21.10D shows a loop that contains a small break, so current cannot flow all the way around the loop. Since there can be no current through this loop, it cannot produce a magnetic field to oppose the changing flux, yet there is still an induced emf. This emf could be measured by attaching a voltmeter across the open ends of the loop. In such cases, Lenz would say that the induced emf always attempts to oppose the change in flux.

PROBLEM SOLVING | Applying Lenz's Law: Finding the Direction of the Induced emf

1. **RECOGNIZE THE PRINCIPLE.** The induced emf always opposes changes in flux through the Lenz's law loop or path.

2. **SKETCH THE PROBLEM,** showing a closed path that runs along the perimeter of a surface crossed by magnetic field lines.

3. **IDENTIFY** if the magnetic flux through the surface is increasing or decreasing with time.

4. **SOLVE.** Treat the perimeter of the surface as a wire loop; suppose there is a current in this loop and determine the direction of the resulting magnetic field. Find the current direction for which this induced magnetic field opposes the change in magnetic flux. This current direction gives the sign (i.e., the "direction") of the induced emf.

5. Always *consider what your answer means* and check that it makes sense.

EXAMPLE 21.3 | Using Lenz's Law to Find the Direction of an Induced Current

Consider the sliding metal bar in Figure 21.12. The magnetic field is constant and is directed into the plane of the drawing. If the bar is sliding to the right, use Lenz's law to find the direction of the induced current.

RECOGNIZE THE PRINCIPLE

According to Lenz's law, the induced current produces a field that opposes the change in flux through the circuit loop.

SKETCH THE PROBLEM

Following our "Applying Lenz's Law" problem-solving strategy (step 2), the sketch in Figure 21.12 shows a dashed, rectangular path. We are interested in the current induced in this circuit, so we must consider the flux through this rectangle.

IDENTIFY THE RELATIONSHIPS

Step 3: The magnetic field is directed into the plane of this drawing, so the flux through the rectangular surface is directed into the plane. The area of this chosen sur-

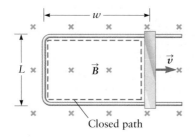

Closed path

Figure 21.12 Example 21.3.

face is wL, and the magnetic flux through the surface is $\Phi_B = BwL$. Because the bar is sliding to the right, Φ_B is increasing and is downward.

SOLVE

The induced emf produces an induced current that opposes the downward increase in the flux, so the induced magnetic field must be directed *upward*. Applying right-hand rule number 1 (Chapter 20), this field direction is produced by a *counterclockwise* induced current.

> *What does it mean?*
> The flux through a given area may be "upward" or "downward," and its magnitude may be increasing or decreasing with time. The induced emf always *opposes any changes* in the flux.

CONCEPT CHECK 21.3 | Applying Lenz's Law

Suppose the metal bar in Figure 21.12 is moving to the left instead of to the right as assumed in Example 21.3. How would this change affect the direction of the induced current?

    (a) There would be no change. The current through the loop will still be counterclockwise, directed from the bottom of the moving bar toward the top.
    (b) The direction of the current will be reversed, directed from top to bottom through the bar.
    (c) The current will be zero.

EXAMPLE 21.4 | A Child's Toy Based on Magnetic Induction

Consider the simple toy shown in Figure 21.13A, consisting of two wire loops. The lower loop is connected through a switch to a battery. Both loops are mounted on a vertical wooden rod, and the upper loop is able to slide freely along the rod. The current through the lower loop is initially zero. (The switch is open.) The switch is then closed, producing a counterclockwise current through the lower loop when viewed from above. **(a)** What is the direction of the current induced in the upper loop? **(b)** Will the upper loop be attracted to or repelled from the lower loop?

RECOGNIZE THE PRINCIPLE

When the current is switched on in the lower loop, there will be a change in flux through the upper loop. According to Lenz's law, the current induced in the upper loop will oppose this change in flux.

SKETCH THE PROBLEM

Following the "Applying Lenz's Law" problem-solving strategy (step 2), the sketch in Figure 21.13B shows the magnetic field produced by the counterclockwise current in the lower loop. This field is upward through the upper loop.

IDENTIFY THE RELATIONSHIPS

Step 3: The current in the lower loop produces a magnetic field that gives an upward flux through the upper loop. Hence, the flux increases upward when the switch is closed.

SOLVE

(a) Continuing with the "Applying Lenz's Law" problem-solving strategy, (step 4), the field induced in the upper loop must be downward to oppose the change in flux. This requires a *clockwise* induced current in the upper loop as viewed from above as sketched in Figure 21.13B.

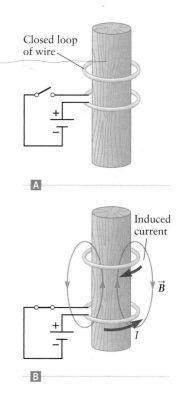

Figure 21.13 Example 21.4.

(b) We now have two current-carrying loops. We can determine the direction of the force on the upper current loop in several different ways. For example, the currents through the two loops are in opposite directions, so the loops act as two oppositely directed bar magnets and thus *repel* each other. In addition, in Concept Check 20.6 we found that the force between two parallel current-carrying wires is repulsive if the currents are in opposite directions. The two current loops in Figure 21.13 are like parallel wires, so they repel for the same reason.

What does it mean?

The force on the upper loop will give it an upward acceleration and hence a nonzero velocity and a nonzero kinetic energy. Electrical energy from the lower loop is transferred to the mechanical energy of the upper loop, but how does this energy actually get to the upper loop? We'll discuss the amazing answer in Section 21.8.

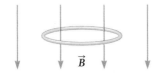

Figure 21.14 Concept Check 21.4.

CONCEPT CHECK 21.4 | Lenz's Law and a Decreasing Magnetic Field

Consider a loop of wire in a horizontal plane with a magnetic field B directed downward through the loop (Fig. 21.14). If the magnitude of the field is decreasing, is the direction of the current induced in the loop (a) counterclockwise as viewed from above or (b) clockwise as viewed from above?

Lenz's Law and Conservation of Energy

Lenz's law involves the negative sign on the right-hand side of Faraday's law (Eqs. 21.5 and 21.17). This negative sign is *very* important. To see why, we return to the example of a metal bar sliding on two horizontal rails in Figure 21.15. A constant magnetic field is directed upward, and for simplicity there is no friction between the bar and the rails. The bar is initially at rest and is then given a push so that it starts to move with a very small velocity $\vec{v}$ to the right.

Let's use Lenz's law to analyze the resulting motion. The surface of interest is enclosed by the rails and the metal bar, and has an area A. The magnetic flux through this area is $\Phi_B = BA$. The bar is moving to the right, so this area and the flux through it are increasing. According to Lenz's law, the induced current opposes this change in flux; hence, the induced current is *clockwise* as viewed from above, producing a downward induced magnetic field. The resulting current through the bar will be as shown in Figure 21.15A. Because this current-carrying bar is in a magnetic field, there is a magnetic force $F_B = ILB$ exerted on the bar, in the direc-

Figure 21.15 When a conductor such as this metal bar moves through a magnetic field, the induced emf produces a current through the loop formed by the bar and rails. There is then a magnetic force on the sliding bar. **A** The negative sign in Faraday's law guarantees that the force on the bar opposes its motion; that is, $\vec{F}_B$ is directed opposite to the bar's velocity $\vec{v}$. **B** If Faraday's law did not contain a negative sign, the force would be parallel to $\vec{v}$, causing the bar to accelerate. This situation does *not* occur!

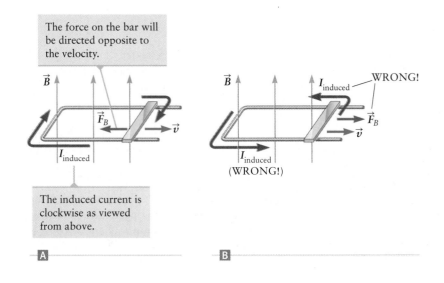

The force on the bar will be directed opposite to the velocity.

The induced current is clockwise as viewed from above.

WRONG!

A   **B**

tion given by right-hand rule number 2 (Chapter 20). Figure 21.15A shows that this force is directed to the left, so it *opposes the motion* of the bar. The force associated with magnetic induction will thus tend to bring the bar to a stop.

Now imagine what the behavior would be if the "polarity" of Lenz's law were reversed, that is, if we changed the negative sign in Faraday's law to a positive sign. This seemingly innocent change would have a very profound effect. The induced magnetic field would then add to the change in the magnetic flux, and the direction of the induced current would be as shown in Figure 21.15B. The induced current drawn here is counterclockwise (as viewed from above), and $\vec{F}_B$ would be in the same direction as the velocity of the bar. Hence, giving the bar a slight push and a small velocity would lead to a force $\vec{F}_B$ that *increases* this velocity. A larger $\vec{v}$ would then lead to a larger force, then to a larger velocity, and so forth, and the bar would accelerate to the right in Figure 21.15B, with no limit to the velocity! This acceleration would violate the principle of conservation of energy. We thus conclude that the correct version of Lenz's law, which states that the induced magnetic field always *opposes* changes in the flux, is actually *a consequence of the principle of conservation of energy*!

Mathematically, Lenz's law is just the negative sign in Faraday's law, so the principle of conservation of energy is contained in Faraday's law. This result is amazing because nowhere in Faraday's law or in the other laws of electricity and magnetism is there any explicit mention of energy or its conservation. When these laws are analyzed, however, we find that energy is indeed conserved in all electric and magnetic phenomena, suggesting that the principle of conservation of energy is in some sense a "deeper" principle of physics than Faraday's law or even Newton's laws of mechanics. In fact, as previously mentioned, physicists believe that *all* laws of physics must satisfy the principle of conservation of energy.

21.4 | INDUCTANCE

So far, when calculating the magnetic flux we have assumed that we can ignore the contribution of the induced field to the flux. This assumption is often correct, but in some situations we can't ignore the induced flux. Consider a coil of wire (a solenoid) connected to a battery through a switch as in Figure 21.16. Most real circuits and coils of wire have some nonzero electrical resistance R, shown as a separate resistor in the circuit. If there are no magnets or other sources of magnetic field nearby, the magnetic flux through the coil from all external sources is zero.

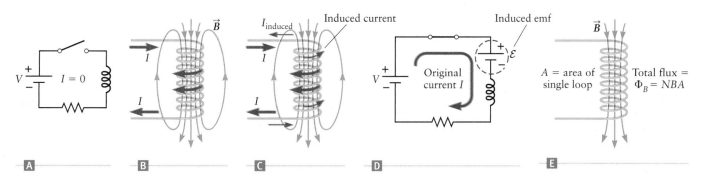

Figure 21.16 ◼A A simple circuit containing a battery, switch, solenoid coil, and resistor. When current enters the coil just after the switch is closed, the rapidly increasing magnetic field lines thread through the coil as shown in ◼B. This results in an increasing magnetic flux, inducing an opposing emf and current in the coil as shown in ◼C ◼D The induced emf can be viewed as a battery ($\mathcal{E}$) with a polarity that opposes changes in current through the coil. A circuit element that acts in this way is called an inductor. ◼E If each loop has an area A, the flux through each loop of a solenoid is BA. With N loops stacked up together, the total flux through all the loops is $\Phi_B = NBA$. The total flux determines the induced emf in Faraday's law.

Suppose the switch is initially open, so the current in the circuit is initially zero. If we close the switch, what happens?

When the switch is closed, there is a sudden change in the current in the coil. Figure 21.16B shows just the coil from Figure 21.16A and the current I produced by the battery. This current produces a magnetic field as sketched in the figure with a direction given by right-hand rule number 1 (Chapter 20). The magnetic field was zero before the switch was closed and is nonzero afterwards, so the magnetic flux through the coil increases with time. According to Faraday's law, this change in magnetic flux induces an emf in the coil. Lenz's law dictates that the induced emf must oppose the increase in flux, so the induced magnetic field is directed opposite to the magnetic field in Figure 21.16B. The induced current thus opposes the original applied current I produced by the battery (Fig. 21.16C). Figure 21.16D shows an equivalent way to understand what is happening. The induced emf can be viewed as a battery with a potential $\mathcal{E}$ and a polarity *opposite* to the polarity of the real battery in the circuit. This opposing emf $\mathcal{E}$ produces the opposing current I_{induced} in Figure 21.16C.

The coil in Figure 21.16 is a type of circuit element called an ***inductor***. Many inductors are constructed as small solenoids, but almost any coil or loop of wire will act as an inductor. Whenever the current through an inductor changes such as when the switch in Figure 21.16 is closed, a voltage is induced in the inductor that *opposes this change*. This phenomenon is called ***self-inductance*** because the changing current through a coil produces an induced current in the same coil. The induced current opposes the original applied current, as dictated by Lenz's law.

Inductance of a Solenoid

Let's now use Faraday's law to give a more quantitative description of inductance, based again on the solenoid in Figure 21.16. The induced emf across the solenoid is

$$\mathcal{E} = -\frac{\Delta \Phi_B}{\Delta t} \tag{21.18}$$

We assume the solenoid contains N loops (also called "turns"), and has a length ℓ and a cross-sectional area A. The magnetic field through the solenoid is (Eq. 20.24)

$$B = \frac{\mu_0 N I}{\ell} \tag{21.19}$$

This fields runs through the entire solenoid, whose loops are stacked as shown in Figure 21.16E. The area of a single loop is A, so the magnetic flux through one loop is $\Phi_1 = BA$. The N loops are stacked on top of one another, and there is flux Φ_1 through each loop. The total flux through the solenoid is thus

$$\Phi_B = NBA = \frac{\mu_0 N^2 I A}{\ell} \tag{21.20}$$

Here, N and A are constants, so changes in flux are produced by changes in the current I. Inserting this result for the flux into Faraday's law (Eq. 21.18), we get

$$\mathcal{E} = -\left(\frac{\mu_0 N^2 A}{\ell}\right)\frac{\Delta I}{\Delta t}$$

The factor in parentheses is called the ***inductance*** L of the solenoid:

$$L = \frac{\mu_0 N^2 A}{\ell} \quad \text{(inductance of a solenoid)} \tag{21.21}$$

The voltage across the solenoid is thus

Definition of inductance L

$$\mathcal{E} = -L\frac{\Delta I}{\Delta t} \tag{21.22}$$

Although the result for inductance in Equation 21.21 applies only for the particular case of a solenoid, the basic notion of inductance is contained in Equation 21.22. By its definition, the inductance L relates the induced emf to changes in the current. This result applies to all coils or loops of wire. Whenever a current produces a magnetic flux through a circuit element, the element will possess an inductance L. The value of L depends on the physical size and shape of the circuit element and on factors such as the number of loops and the area of each loop, just as we found for a solenoid. When the current in an inductor changes with time, the emf across the inductor is proportional to the inductance as given by Equation 21.22. The negative sign in Equation 21.22 indicates the polarity of the voltage. It is conventional practice to speak in terms of the potential "drop" across the inductor, which is given by

$$V_L = L \frac{\Delta I}{\Delta t} \qquad (21.23)$$

Potential drop across an inductor

Because V_L is defined as the voltage *drop*, there is no negative sign in Equation 21.23. This result for V_L will be needed when we construct a circuit theory that deals with inductors.

The unit of inductance is the *henry* (H). Equation 21.23 relates the unit of inductance to the units of voltage, current, and time. Rearranging Equation 21.23 as $L = V_L(\Delta t/\Delta I)$, the units are related by

$$1 \text{ H} = 1 \text{ V} \cdot \text{s/A} \qquad (21.24)$$

EXAMPLE 21.5 │ Inductance of a Typical Coil

In a practical electric circuit, inductors are often coils of wire in the form of a solenoid. Consider a solenoid of length $\ell = 2.0$ cm consisting of 1000 turns of wire. Each turn is a circular loop with radius $r = 0.50$ cm. Calculate the value of L for this inductor.

RECOGNIZE THE PRINCIPLE

The inductance of a solenoid depends on its length, its cross-sectional area, and the number of turns. We can find the value of the inductance L by applying Equation 21.21. The point of this calculation is to get a feeling for the value of L in typical cases, which will be useful when we analyze circuits with inductors in Section 21.5.

SKETCH THE PROBLEM

Figure 21.16E describes the problem.

IDENTIFY THE RELATIONSHIPS AND SOLVE

Inserting the given values of the length and number of turns, with a loop area $A = \pi r^2$, we find

$$L = \frac{\mu_0 N^2 A}{\ell} = \frac{(4\pi \times 10^{-7} \text{ T} \cdot \text{m/A})(1000)^2 \pi (5.0 \times 10^{-3} \text{ m})^2}{0.020 \text{ m}}$$

$$L = 4.9 \times 10^{-3} \text{ H} = \boxed{4.9 \text{ mH}}$$

What does it mean?
This value of L is typical for common circuit inductors. An inductance with $L = 1$ H is very large, whereas a value of 1 microhenry (10^{-6} H) is rather small. An inductance of 1 mH (millihenry) is typical.

Insight 21.1
THE CONCEPT OF
***SELF*-INDUCTANCE**
Self-inductance poses an interesting conceptual question: How can a current "act on" or affect itself? After all, the electric field of an electron does not act on the electron that produced the field in the first place. With an inductor, changes in current are acting "back" on the current itself through the induced emf.

An electric current is produced by the motion of many charges, and the induced emf is actually due to the force exerted by one charge (from the current in one part of the inductor) on another charge (the current in a different part of the inductor). When viewed in this way, self-inductance is just the effect of one electric charge on another.

Mutual Inductance

In an inductor (i.e., a "coil"), the magnetic field produced by the coil produces an induced current in that coil. It is also possible for the magnetic field produced by

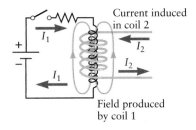

Figure 21.17 Two coils wound around each other exhibit a mutual inductance. Changes in the current in one coil (coil 1) induce an emf in the other (coil 2).

one coil to produce an induced current in a second coil. Figure 21.17 shows two solenoidal coils, one wrapped around the other. The inner coil (1, shown in black) is connected to a battery and switch. When the switch is closed and the current I_1 increases, this coil produces a magnetic field that gives an increasing flux in the outer coil (2, shown in gray). According to Faraday's law, the changing flux in coil 2 produces an induced emf in that coil. Hence, the current I_1 in coil 1 produces an emf in coil 2, even though the two coils are not "directly" connected by any wires. They are, however, "indirectly" connected through the magnetic flux. This effect is called ***mutual inductance*** because it involves the "mutual" behavior of two separate coils. Mutual inductance is used in devices called transformers, which we'll discuss in Chapter 22.

21.5 | *RL* CIRCUITS

We can apply Kirchhoff's circuit rules from Chapter 19 to circuits that contain inductors by including the voltage drop across an inductor given in Equation 21.23. The basic behavior of such circuits can be described using the following rules.

General behavior of *RL* circuits

Qualitative behavior of DC circuits with inductors

DC circuits may contain resistors, inductors, and capacitors. The voltage source in a DC circuit is a battery or some other source that provides a constant voltage across its output terminals.

1. Immediately after any switch is closed or opened, the induced emfs keep the current through all inductors equal to the values they had the instant before the switch was thrown.
2. After a switch has been closed or opened for a very long time, the induced emfs are zero.

The first rule above can be understood from the process shown in Figure 21.16. *Whenever there is an attempt to change* the current through an inductor (i.e., by closing or opening a switch), the induced emf across the inductor always opposes this change. Momentarily, this opposition prevents the current through the inductor from changing, and the current just after a switch is thrown is equal to the current just before. This result applies only for the *instant* after a switch is thrown, and is simply a statement of Lenz's law as applied to inductors.

After some time, the current through the inductor will change, eventually approaching a constant value. If I is constant, $\Delta I / \Delta t$ in Equation 21.23 is zero and all currents and magnetic fields are then constant. Hence, the voltage across the inductor is zero, leaving us with DC circuit rule 2.

These two rules apply only to DC circuits; in these circuits, the only time-dependent changes are initiated by the closing or opening of switches. Another important class of circuits, called AC circuits, contain power sources whose electric potentials vary sinusoidally with time. We'll discuss those circuits in Chapter 22, when we generalize our circuit rules to apply to both AC and DC circuits.

Let's use our two circuit analysis rules to study the behavior of the DC circuit sketched in Figure 21.18, which contains a battery, a switch, two resistors, and an inductor. The presence of the resistors and inductor make it an "*RL* circuit." The general circuit symbol for an inductor is a solenoid-like coil. In Figure 21.18A, the switch has been open for a very long time, so the current is zero through both branches of the circuit. At $t = 0$, the switch is closed, inducing a potential across the inductor that is equivalent to a battery with emf $\mathcal{E} = V_L$ (Fig. 21.18B). To find the values of I_1 (the current through resistor 1) and I_2 (the current through resistor 2) just after the switch is closed, we apply rule 1 from the boldfaced list above. At the instant after $t = 0$, the induced emf keeps the current through the inductor at the value it had before the switch was closed. Hence, $I_2 = 0$ just after $t = 0$; at that moment, the entire circuit is equivalent to the lower circuit in Figure 21.18B, and

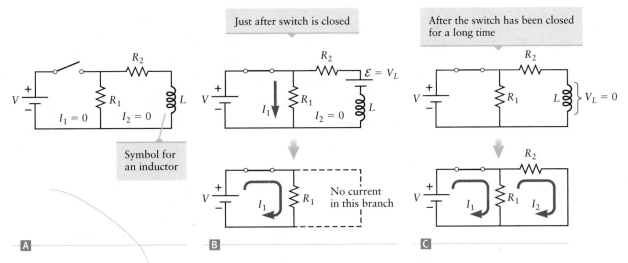

Figure 21.18 **A** An *RL* circuit. **B** Just after the switch is closed, the emf induced in the inductor opposes any increase in current through it. At this instant, the current I_2 is zero. The equivalent circuit *just after* the switch is closed is shown below. **C** After the switch has been closed for a long time, the magnetic flux through the inductor is constant (not changing with time), so the induced emf is zero. The inductor then acts as a wire (a short circuit).

the circuit behaves as if L and R_2 have been removed. The result is a circuit with a battery connected to a single resistor carrying current $I_1 = V/R_1$.

Now let's find the currents after the switch in Figure 21.18 has been closed for a very long time. We apply rule 2 highlighted above and find that the voltage across the inductor is now zero, which is equivalent to replacing the inductor by a simple wire that has zero resistance (lower circuit in Fig. 21.18C). The voltage across each resistor is now the battery emf V, so the currents in the two circuit branches are

$$I_1 = \frac{V}{R_1} \quad \text{and} \quad I_2 = \frac{V}{R_2} \tag{21.25}$$

EXAMPLE 21.6 | **Current in an *RL* Circuit**

Consider again the circuit in Figure 21.18. Now assume the switch has been closed for a very long time, so the currents through R_1 and R_2 are as given in Equation 21.25 with the directions shown in Figure 21.18C. Find the directions of the currents through R_1 and R_2 the instant after the switch is reopened.

RECOGNIZE THE PRINCIPLE

Whenever a switch is opened or closed, the induced emf across an inductor always acts to keep the current the same as it was just before the switch was thrown. This principle is circuit rule 1 of "Qualitative behavior of DC circuits with inductors," page 704.

SKETCH THE PROBLEM

The current I_2 through the inductor remains at the value it had just prior to the change of the switch (Eq. 21.25), and it will also keep the same direction. Hence, at the instant the switch is opened the current through the branch containing R_2 and the inductor is directed clockwise as sketched in Figure 21.19B.

IDENTIFY THE RELATIONSHIPS AND SOLVE

Applying the equivalent circuit in Figure 21.19B, this current must flow "upward" through R_1, opposite to the direction found when the switch was closed in Figure 21.18B.

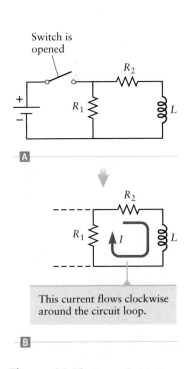

Figure 21.19 Example 21.6.

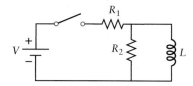

Figure 21.20 Concept Check 21.5.

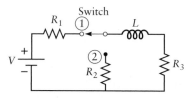

Figure 21.21 Concept Check 21.6.

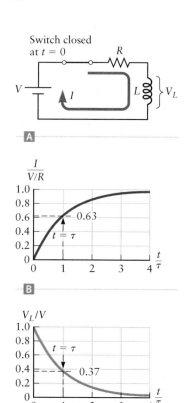

Figure 21.22 **A** An *RL* circuit with the switch closed. **B** Current through the circuit as a function of time. The switch was closed at $t = 0$. **C** Voltage across the inductor as a function of time. When this voltage is positive, it *opposes* the flow of current through the inductor.

CONCEPT CHECK 21.5 | Analyzing an *RL* Circuit (1)

Consider the circuit in Figure 21.20. The current is initially zero in both branches. The switch is then closed. Is the current through R_1 just after closing the switch (a) $I = V/R_1$, (b) $I = V/R_2$, (c) $I = V/(R_1 + R_2)$, or (d) $I = 0$?

CONCEPT CHECK 21.6 | Analyzing an *RL* Circuit (2)

Consider the circuit in Figure 21.21. The switch is in position 1 for a very long time and then changed to position 2. What is the direction of the current through the inductor just after the switch is changed?
   (a) It is to the right.
   (b) It is to the left.
   (c) The current is zero just after the switch is thrown.
   (d) The direction depends on the values of R_1, R_2, and R_3.

Quantitative Behavior of an *RL* Circuit

Our qualitative rules for dealing with inductors in DC circuits tell how to find the current the moment just after a switch is closed or opened and how to find the current after a switch has been closed or open for a very long time. The current between these two times can also be calculated using Kirchhoff's rules for circuit analysis. Figure 21.22A shows the simplest *RL* circuit, with a battery connected through a switch to a resistor R and an inductor L. The switch is open for a long time and then is closed at $t = 0$. The quantitative behavior of the current through and voltage across the inductor is shown in parts B and C of Figure 21.22. The current starts at $I = 0$ because the induced emf keeps the current through the inductor momentarily at the value it had before the switch was thrown (in accord with our circuit rule 1). After a very long time, the voltage across the inductor falls to zero (in accord with rule 2) and the current approaches $I = V/R$. Mathematically, the current is given by

$$I = \frac{V}{R}(1 - e^{-t/\tau}) \qquad (21.26)$$

where e^x is the exponential function and τ is called the **time constant** of the circuit. For an *RL* circuit in which a single resistor is in series with a single inductor as in Figure 21.22, the time constant is

$$\tau = \frac{L}{R} \qquad (21.27)$$

The corresponding behavior of the voltage across the inductor is

$$V_L = Ve^{-t/\tau} \qquad (21.28)$$

The time constant τ is the approximate time scale over which the current and voltage change in an *RL* circuit. From the properties of the exponential function (see Appendix B), the current in Equation 21.26 reaches approximately 63% of its final value when $t = \tau$ and the voltage across the inductor (Eq. 21.28) falls to approximately 37% of its initial value after this time. After three time constants

$(t = 3\tau)$, the current reaches 95% of its final value. The time constant τ in Equation 21.27 applies only to an RL circuit. It is *different* from the time constant we encountered for RC circuits in Chapter 19.

EXAMPLE 21.7 | Value of τ for a Typical RL Circuit

Consider an RL circuit with $R = 4000\ \Omega$ and $L = 3.0$ mH; these values are typical of those used in many electronic circuits such as radios and computers. Find the time constant for a circuit that uses these components.

RECOGNIZE THE PRINCIPLE

The time constant for an RL circuit is just the ratio L/R. The purpose of this example is to determine a "typical" value of the time constant τ. Is it hours, seconds, or microseconds? The value of τ is important in applications (see below).

SKETCH THE PROBLEM

No figure is necessary.

IDENTIFY THE RELATIONSHIPS AND SOLVE

We evaluate the time constant τ from Equation 21.27 using the given values of R and L:

$$\tau = \frac{L}{R} = \frac{3.0 \times 10^{-3}\ \text{H}}{4000\ \Omega} = 7.5 \times 10^{-7}\ \text{s} = \boxed{0.75\ \mu\text{s}}$$

What does it mean?

For this circuit, the time constant is approximately 1 microsecond (1 μs). In many applications of RL circuits (Chapter 22), this time constant is associated with the period of an oscillation, so the period of this circuit would correspond to a frequency of about

$$f = \frac{1}{\tau} \approx \frac{1}{1 \times 10^{-6}\ \text{s}} = 1 \times 10^{6}\ \text{Hz}$$

(1 megahertz). This frequency lies in the middle of the "AM band" on your radio, so such an RL circuit could be useful in an AM radio.

21.6 | ENERGY STORED IN A MAGNETIC FIELD

The RL circuit in Figure 21.23A is the same circuit we analyzed in Figures 21.18 and 21.19. We know that if the switch has been closed for a long time, currents I_1 and I_2 are established through the resistors as given in Equation 21.25. If the switch is then opened at $t = 0$, there is a current in the circuit loop formed by R_1, R_2, and L (Fig. 21.23B); the time dependence of this current is shown qualitatively in Figure 21.23C. After the switch is opened, the battery is completely disconnected from this circuit. Current thus exists even with the battery removed, and electrical

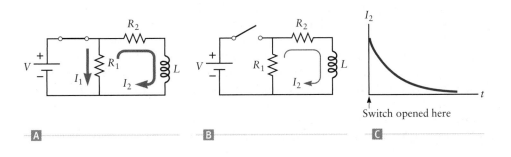

Figure 21.23 ◰ If the switch in this RL circuit is closed for a long time, there will be current in the inductor. ◱ When the switch is opened, the current through the inductor decays with time as plotted in ◲.

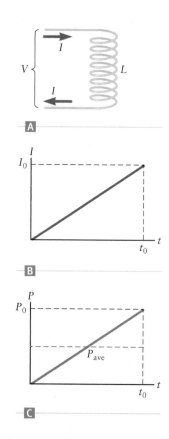

Figure 21.24 **A** Voltage across and current through an inductor. **B** If the voltage across the inductor is constant, the current increases linearly with time (Eq. 21.30). **C** The power delivered to the inductor also increases linearly with time.

Energy stored in an inductor

energy is dissipated as heat in the resistors during this time. Where does this energy come from if the battery has been disconnected? The answer is that this energy was originally (at $t = 0$) stored in the magnetic field of the inductor.

One way to analyze the energy stored in this magnetic field is to consider the energy required to establish the current in an inductor in the first place. Suppose the voltage across the inductor in Figure 21.24A has a constant value V. To keep V constant, the current must increase with time. We can see that from Equation 21.23; the voltage drop across the inductor is $V = L\,\Delta I/\Delta t$. Solving for the change in current, we have

$$\Delta I = \frac{V}{L}\Delta t \tag{21.29}$$

Hence, if V is constant and the current starts from zero at $t = 0$, Equation 21.29 means that the current through the inductor I must increase linearly with t (Fig. 21.24B), and

$$I = \frac{V}{L}t \tag{21.30}$$

The power P dissipated in a circuit element is equal to the product of the voltage across that element and the current (Eq. 19.15). Hence, the instantaneous power in the inductor—the rate at which energy is delivered to the inductor at time t—is

$$P = VI = V\left(\frac{V}{L}t\right) = \left(\frac{V^2}{L}\right)t \tag{21.31}$$

This linear relation between P and t is plotted in Figure 21.24C. We can use the result of Equation 21.31 to find the total energy that has been delivered to the inductor when the current reaches a particular value I_0 at time t_0. We denote this energy as PE_{ind} since we will soon see that it is a kind of potential energy. This total potential energy PE_{ind} is equal to the average of the power in Figure 21.24C multiplied by the total time t_0. The instantaneous power P varies linearly with time, so the average power is half the maximum power. Using Equation 21.31, we get

$$P_{ave} = \frac{1}{2}\frac{V^2 t_0}{L}$$

$$PE_{ind} = P_{ave}t_0 = \frac{V^2 t_0^2}{2L} \tag{21.32}$$

From Equation 21.30, we find $I_0 = Vt_0/L$, which we rearrange to give $t_0 = I_0 L/V$. Inserting into our expression for PE_{ind} leads to

$$PE_{ind} = \frac{V^2}{2L}\left(\frac{I_0 L}{V}\right)^2 = \frac{1}{2}LI_0^2$$

where I_0 is the final current through the inductor; we can just call it I and then write our final result as

$$PE_{ind} = \tfrac{1}{2}LI^2 \tag{21.33}$$

Equation 21.33 is the general result for the magnetic energy PE_{ind} stored in an inductor. It has a form that is very similar to the electric energy stored in a capacitor, $PE_{cap} = \tfrac{1}{2}CV^2$ (Eq. 18.33). In the case of a capacitor, we saw that this energy is stored in the electric field that exists between the capacitor plates. In a similar way, the energy PE_{ind} is stored *in the magnetic field* of the inductor, and we can rewrite Equation 21.33 to make this more apparent. The inductance of a solenoid is $L = \mu_0 N^2 A/\ell$ (Eq. 21.21). Inserting into our expression for PE_{ind} (Eq. 21.33) gives

$$PE_{ind} = \frac{1}{2}LI^2 = \frac{1}{2}\left(\frac{\mu_0 N^2 A}{\ell}\right)I^2$$

which can be rearranged as

$$PE_{ind} = \frac{1}{2}\left(\frac{\mu_0 N^2 A}{\ell}\right)I^2 = \frac{1}{2\mu_0}\left(\frac{\mu_0 NI}{\ell}\right)^2(A\ell)$$

We have rewritten the potential energy in this way because the factor involving I in parentheses is the solenoid's magnetic field ($B = \mu_0 NI/\ell$, Eq. 21.19), whereas the last factor $A\ell$ is the volume of the solenoid (the length times the cross-sectional area). So we get our final result,

$$PE_{mag} = \frac{1}{2\mu_0}B^2(\text{volume}) \qquad (21.34)$$

Energy stored in a magnetic field

In Equation 21.34, we have now denoted the energy as PE_{mag}. It is the energy contained in the magnetic field of the inductor and applies not only to an inductor but to any region of space that contains a magnetic field. This energy is stored *in the magnetic field*. Because a magnetic field can exist in a vacuum, this potential energy can exist even in "empty" regions containing no matter! This fact, which also applies to electric fields and electric potential energy, will be important in understanding electromagnetic waves (Chapter 23).

Another way to express the potential energy in Equation 21.34 is to say that there is a certain *energy density* (the energy per unit volume) in the magnetic field. From Equation 21.34, we find

$$\text{energy density} = u_{mag} = \frac{PE_{mag}}{\text{volume}} = \frac{1}{2\mu_0}B^2 \qquad (21.35)$$

This expression for the magnetic energy density is very similar in form to the energy density contained in an electric field. In Chapter 18 we found that $u_{elec} = \frac{1}{2}\varepsilon_0 E^2$ (Equation 18.48), so there is an interesting parallel between our results for magnetic and electric energies. In both cases, the energy density is proportional to the square of the field.

EXAMPLE 21.8 | ✕ ℝ Magnetic Field Energy: How Large Is It?

To get a feeling for the magnitude of the energy stored in a magnetic field, consider the energy stored in a magnet used for magnetic resonance imaging (MRI) (Fig. 21.25). A typical MRI magnet is a solenoid providing a field of about $B = 1.5$ T. **(a)** What is the approximate energy stored in this magnetic field? **(b)** If the current through the magnet during normal operation is $I = 100$ A, what is its inductance?

RECOGNIZE THE PRINCIPLE

The motivation for this example is to understand how the potential energy stored in a magnetic field compares with cases familiar from mechanics such as the gravitational potential energy or the kinetic energy of a moving object. To find the energy stored in an MRI magnet, we need to know its magnetic field and its inside volume. The volume is not given, but we can estimate its value because a person must be able to fit inside (Fig. 21.25).

SKETCH THE PROBLEM

Figure 21.25 shows a typical MRI magnet. We can estimate the volume from this photo.

IDENTIFY THE RELATIONSHIPS

We can find the stored magnetic energy using our result for the potential energy PE_{mag} (Eq. 21.34).

SOLVE

(a) From personal experience (and the photo in Fig. 21.25), we estimate that a typical MRI magnet is about 1.5 m long with an inside diameter of about 0.5 m so that

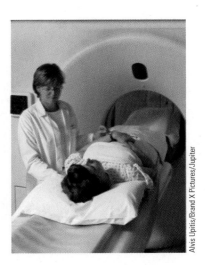

Figure 21.25 Example 21.8. The magnets used for magnetic resonance imaging (MRI) are very large, with a large amount of stored magnetic energy.

a person can fit inside. We have volume $= A\ell = (\pi r^2)\ell$, and inserting our estimates for r and ℓ gives

$$\text{volume} = (\pi r^2)\ell = \pi(0.25 \text{ m})^2(1.5 \text{ m}) = 0.3 \text{ m}^3$$

Using this volume in Equation 21.34 along with the given value of B, we find

$$PE_{mag} = \frac{1}{2\mu_0} B^2(\text{volume}) = \frac{1}{2(4\pi \times 10^{-7} \text{ T} \cdot \text{m/A})}(1.5 \text{ T})^2(0.3 \text{ m}^3)$$

$$PE_{mag} = \boxed{3 \times 10^5 \text{ J}}$$

(b) The magnetic energy found in part (a) is stored in an inductor because the MRI magnet is just a solenoid. Therefore, $PE_{mag} = PE_{ind}$ is related to the inductance through Equation 21.33. Using our calculated value of PE_{mag} gives

$$\tfrac{1}{2}LI^2 = PE_{mag}$$

$$L = \frac{2PE_{mag}}{I^2} = \frac{2(3 \times 10^5 \text{ J})}{(100 \text{ A})^2} = \boxed{60 \text{ H}}$$

This inductance is large, much larger than we found for the small coil in Example 21.5.

What does it mean?

The amount of energy found in part (a) is quite large; it is approximately equal to the kinetic energy of a car of mass 1000 kg with a velocity of 50 mi/h (23 m/s)! An MRI magnet is carefully designed to be sure this energy is dealt with safely. In addition, patients undergoing an MRI exam as well as MRI technicians must take precautions such as removing all magnetic objects from their pockets.

21.7 | APPLICATIONS

In this section, we consider a few important applications of Faraday's law.

Bicycle Odometers

Most serious bicyclists use an odometer to monitor their speed and distance traveled. Figure 21.26A shows a photo of the odometer control unit, and a sketch of the working elements is shown in Figure 21.26B. A small, permanent magnet is

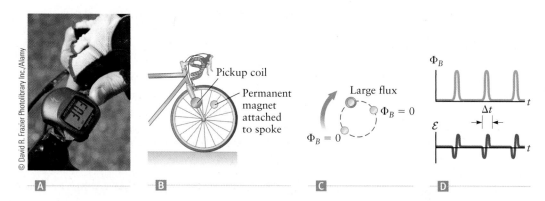

Figure 21.26 A This bicycle odometer uses Faraday's law to count the rotations of a bicycle wheel. B A small, permanent magnet is attached to a spoke, and a pickup coil is mounted on the bicycle frame. C When the magnet passes near the pickup coil, the time-varying magnetic flux induces an emf in the coil. D Qualitative sketch of the flux through the pickup coil as a function of time. This time-dependent flux produces an induced emf that varies with time as a series of pulses. The odometer counts these pulses.

attached to one of the spokes of a wheel, and a "pickup" coil is mounted on the axle support. On the author's bike, this pickup unit is contained in a piece of opaque plastic with a small coil of wire inside. As the wheel turns, the permanent magnet on the spoke travels in a circular path that takes it past the pickup coil once during each revolution of the wheel. When this magnet is close to the pickup coil (Fig. 21.26C), its magnetic field penetrates the coil, giving a large flux, whereas the flux through the coil is zero when the magnet is far from the coil. A qualitative sketch of the flux through the coil as a function of time is shown in the upper graph of Figure 21.26D. Because a pulse occurs each time the magnet passes over the pickup coil, there is one "pulse" in this plot for each revolution of the wheel.

From Faraday's law, we know that the time-varying magnetic flux through the pickup coil induces an emf in the coil. A corresponding plot of this induced emf is shown in the lower graph of Figure 21.26D. Because $\mathcal{E} = -\Delta\Phi_B/\Delta t$, the induced emf is negative while the flux is increasing (Lenz's law) and positive as Φ_B falls back to zero. The pickup coil is connected to an electronic circuit and a computer that keeps track of the number of pulses in the emf signal, and also measures the time between pulses. This information is then displayed as the bicycle's speed and distance traveled.

EXAMPLE 21.9 | ® Design of a Bicycle Odometer

Let's calculate the emf (voltage) induced in the odometer pickup coil in Figure 21.26 to show that this odometer design is actually practical. For a typical permanent magnet, the field very near one of the poles is about 1 T. In practice, the permanent magnet attached to the spoke must not be too close to the frame of the bicycle (so they don't hit each other), which reduces the field at the pickup coil. To be conservative, we assume a field $B = 0.01$ T at the coil. The odometer pickup on the author's bicycle has an area of about $A = 0.1$ cm^2 (about 3 mm on a side). We also assume the coil of wire inside consists of $N = 100$ loops.

RECOGNIZE THE PRINCIPLE

The odometer signal is due to Faraday's law: the voltage induced in the pickup coil is equal to the rate of change of the magnetic flux as the odometer magnet moves past. To find $\Delta\Phi_B/\Delta t$, we need to know B and the area of the coil, which are both given. We also need to know Δt, which is the approximate time the magnet "overlaps" with the coil. We can estimate Δt from the angular speed of the bicycle wheel.

SKETCH THE PROBLEM

Figure 21.26 describes the problem. Following our "Applying Lenz's Law" problem-solving strategy, Figure 21.26C shows the pickup coil. We will use this area when calculating the induced emf.

IDENTIFY THE RELATIONSHIPS

To find the change in flux, we need to calculate Φ_B when the odometer magnet is directly over the pickup coil and when it is far away. When the magnet far away, the flux is zero; when it is over the coil, the magnetic flux through one loop of the coil is BA. So, for N loops, the total flux is $\Phi_B = NBA$. Inserting our approximate values for N, B, and A, we get

$$\Phi_B = NBA = (100)(0.01 \text{ T})(1 \times 10^{-5} \text{ m}^2) = 1 \times 10^{-5} \text{ T} \cdot \text{m}^2 \qquad (1)$$

which is the flux at the peaks in Figure 21.26D and also the change in flux $\Delta\Phi_B$ when the magnet passes by the pickup coil.

To apply Faraday's law, we must now estimate the time Δt over which this flux change takes place. This time interval is indicated in Figure 21.26D; it depends on the speed of wheel, and it decreases as the bicycle travels faster. A slowly moving bicycle might travel at 5 mph, which is about 2 m/s. The wheels of the author's bike have a radius of about $R = 50$ cm, so the angular velocity at this speed is (from Eq. 8.43)

$$\omega = \frac{v}{R} = \frac{2 \text{ m/s}}{0.5 \text{ m}} = 4 \text{ rad/s}$$

The magnet on the spoke is approximately halfway from the edge of the wheel at $R_{\text{magnet}} = 0.25$ m, so its speed is

$$v_{\text{magnet}} = \omega R_{\text{magnet}} = (4 \text{ rad/s})(0.25 \text{ m}) = 1 \text{ m/s}$$

A reasonable diameter for the permanent magnet is about $d = 1$ cm (a value taken from the author's bike), so the time it takes for the magnet to move past the sensor is approximately

$$\Delta t = \frac{d}{v_{\text{magnet}}} = \frac{0.01 \text{ m}}{1 \text{ m/s}} = 0.01 \text{ s} \tag{2}$$

SOLVE

Inserting these results for $\Delta \Phi_B$ from Equation (1) and Δt into Faraday's law gives the approximate magnitude of the induced emf,

$$|\mathcal{E}| = \frac{\Delta \Phi_B}{\Delta t} = \frac{1 \times 10^{-5} \text{ T} \cdot \text{m}^2}{0.01 \text{ s}} = \boxed{0.001 \text{ V}}$$

What does it mean?
A voltage of this size can be easily measured (as your electrical engineering friends can verify), so this odometer design is indeed practical.

CONCEPT CHECK 21.7 | Changing Speeds with the Bicycle Odometer

If the speed of the bicycle in Example 21.9 is increased, how does the amplitude of the induced emf in the pickup coil change?
  (a) The induced emf gets smaller.
  (b) The induced emf gets larger.
  (c) The induced emf does not change.

Ground Fault Interrupters

A ground fault interrupter (GFI) is a safety device used in many household circuits (Fig. 21.27A). A GFI is often built into an electric power outlet in bathrooms, kitchens, and other places where there is an increased risk of accidental contact with a substantial electric potential. A GFI uses Faraday's law together with an electromechanical relay (Fig. 20.45). Suppose a hair dryer is plugged into a wall socket that contains a GFI as in Figure 21.27B. This connection contains two current-carrying wires, with current directed from the wall socket to the hair dryer through one wire and back to the wall socket through the other wire. It is just like the circuits we have drawn using batteries; all have two wires connecting the battery to the rest of the circuit. During normal (safe) operation of the hair dryer, the two connecting wires carry equal currents, but in the event of an accident, some current might pass through your body and then to ground. In that case, the current in the return wire in Figure 21.27B is much smaller than the current in the outgoing wire. A GFI senses this difference and immediately turns off power to the entire circuit so that no permanent harm is done.

A GFI accomplishes its task by using a relay to control the current through the circuit. In Chapter 20, we saw that relays use the current through a coil to exert a force on a magnetic metal bar. This bar is part of a switch, so the current through the coil controls whether the switch is open or closed. The relay in a GFI contains

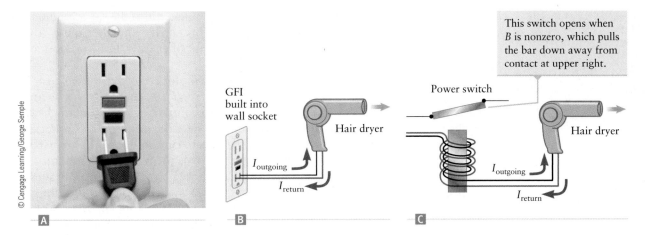

This switch opens when B is nonzero, which pulls the bar down away from contact at upper right.

GFI built into wall socket

Hair dryer

$I_{outgoing}$

I_{return}

Power switch

Hair dryer

$I_{outgoing}$

I_{return}

A **B** **C**

Figure 21.27 **A** Ground fault interrupters (GFIs) use a specially designed inductor to detect circuit failure. **B** Most GFI units are built into a wall socket, so you plug an appliance, such as a hair dryer, directly into the GFI. **C** Two coils of wire, one carrying the outgoing current and another carrying the return current, are used to power a relay in the GFI. If an appliance fails, the current through the return coil drops to zero and the relay is activated, turning off current to the appliance.

two separate coils; the current through one coil is from the outgoing wire to the hair dryer, and the current through the other coil is from the return circuit. The two coils are wound in opposition so that their magnetic fields are in opposite directions. During normal (safe) operation, the currents in the two coils are equal and their fields cancel, giving zero magnetic field in the relay. The relay is designed so that its switch is closed in this case, allowing current in the hair dryer circuit. If there is an accident, however, the current in the return coil is smaller than the outgoing current, leading to a nonzero magnetic field from the coil. The relay switch then opens (Fig. 21.27C), turning off the current to the hair dryer.

The inductor in a GFI thus employs two separate coils of wire wound in opposite directions. This type of winding is used in other applications in which one wants to make the inductance of a device as small as possible.

Electric Guitars

An electric guitar (Fig. 21.28A) uses Faraday's law to sense the motion of the strings. A guitar string is made from steel, a magnetic material. The string passes near a "pickup" coil wound around a permanent magnet. The field from this magnet causes nearby regions of the string to be magnetized as shown in Figure 21.28B. This induced magnetism in the steel of the guitar string is just like the induced magnetism in the steel panel of a refrigerator or of any other magnetic material containing magnetic domains (see Fig. 20.39). The magnetized region of the string then produces its own magnetic field, which gives a magnetic flux through the pickup coil. As the string vibrates, this flux changes with time, leading to an induced emf (from Faraday's law) in the coil. This emf is sent to an amplifier, and the resulting signal can then be played through speakers.

Generators, Motors, and Hybrid Cars

In Chapter 20, we showed how a magnetic field can exert a torque on a current loop that can be used to make a motor and thus produce mechanical energy (Fig. 20.47). The "opposite" process takes place in a generator, which uses an externally applied mechanical torque on a current loop to produce an electric current and thus produce electrical energy. A motor and generator therefore provide examples of the conservation of energy and the conversion of energy from one type to another, mechanical to electrical and vice versa. So, it is no surprise that an ideal generator

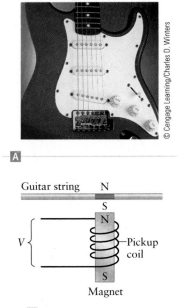

A

Guitar string N

S

N

V

Pickup coil

S

Magnet

B

Figure 21.28 **A** The strings of an electric guitar are made of steel, a magnetic material. **B** When the string vibrates, it produces a time-varying magnetic flux through the pickup coil (called a guitar "pickup"), producing a voltage in the coil that is converted into sound by the guitar's amplifier. In part A the pickups are the small circular regions, with three pickups under each string.

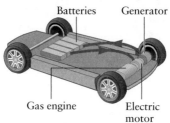

Figure 21.29 A hybrid car such as this Toyota Prius contains a gasoline powered motor along with an electric motor, a generator, and batteries.

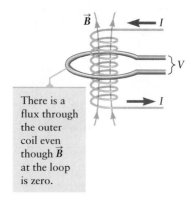

There is a flux through the outer coil even though $\vec{B}$ at the loop is zero.

Figure 21.30 This long, thin solenoid produces a magnetic flux through the larger-diameter loop that encircles it. If this flux varies with time, an emf is induced in the outer loop, even though there is essentially no magnetic field there.

can take a certain amount of mechanical energy and convert it to the same amount of electrical energy (and vice versa for the motor). This conversion back and forth is used in so-called hybrid cars to minimize the amount of energy "lost" to friction and hence improve fuel economy.

A hybrid car contains two motors (Fig. 21.29); one motor is a conventional gasoline powered motor (also called an "engine"; see Chapter 16), and the other is an electric motor powered by batteries. A hybrid car also contains a generator attached to the wheels, usually the front wheels, which are the ones connected to the motors. Starting from a standstill, either the gas motor or electric motor (or both) are used to accelerate the car. At low speeds, the electric motor is the more efficient motor, so the gas motor usually shuts off automatically for those periods. When the brakes are applied or the car travels downhill, the generator is switched on and the emf it produces is used to recharge the batteries. In an ordinary car, applying the brakes causes the car's kinetic energy to be converted to heat energy through friction. A hybrid car "recaptures" this energy, storing it in the batteries so that it can be used later. A hybrid car is thus a good practical example of the conversion between mechanical energy and electrical energy.

Of course, this conversion can never be perfect; no generator is perfect, so all the mechanical energy of braking can never be converted to electrical energy. Even so, the hybrid approach can cut a car's gasoline consumption by 30% or even more, which is a significant savings.

21.8 | THE PUZZLE OF INDUCTION FROM A DISTANCE

A solenoid carrying a current produces a large magnetic field in its interior and a very small field outside. By making the solenoid very long, the field outside can be made extremely small. Suppose a very long solenoid is inserted at the center of a single loop of wire as sketched in Figure 21.30. This example is of mutual inductance (Fig. 21.17) because the field of the inner solenoid produces a flux through the outer loop and changes in the current through the inner solenoid thus induce an emf in the loop. In Chapter 20, however, we saw that for a long solenoid, the field outside is very small. Hence, with the very long solenoid in Figure 21.30, the field from the solenoid *at the outer loop* is essentially zero. Even so, the field inside the solenoid at the center of the loop still produces a magnetic flux through the inner portion of the loop.

The solenoid field is proportional to the current I (Eq. 21.19), and let's assume I is changing and so this field is changing with time. The flux through the outer loop then changes with time, inducing an emf in the outer loop. This emf can be used to power an electric circuit (we might attach the loop to a lightbulb), so there is electrical energy associated with this induced emf. You should suspect that this energy was somehow transferred from the solenoid to the outside loop, but how is this energy transferred across the empty space between the two conductors? What's more, there is no magnetic field from the solenoid at the outside loop, so how does the solenoid have any influence at all on that loop?

The answer to this puzzle is that energy is carried from the solenoid to the outer loop by an ***electromagnetic wave.*** Just as with mechanical waves (Chapters 11 and 12), electromagnetic waves transport energy from one place to another. We'll explore electromagnetic waves and how they propagate in Chapter 23.

SUMMARY | Chapter 21

KEY CONCEPTS AND PRINCIPLES

Magnetic flux and Faraday's law

A changing magnetic flux produces an electric field. This phenomenon is called **induction**. The induced electric potential is given by **Faraday's law**,

$$\mathcal{E} = -\frac{\Delta \Phi_B}{\Delta t} \qquad \textbf{(21.5)} \text{ (page 690)}$$

where $\mathcal{E}$ is the **induced emf** along a closed path and Φ_B is the **magnetic flux** through the area enclosed by that path.

The magnetic flux due to a field of magnitude B through a surface of area A is given by

$$\Phi_B = BA \cos \theta \qquad \textbf{(21.2)} \text{ (page 689)}$$

where θ is the angle that $\vec{B}$ makes with the direction normal to the surface.

A changing magnetic flux can be produced if B is changing, the size of the area is changing, or the angle between $\vec{B}$ and the area is changing. In all three cases, the change in flux produces an induced emf according to Faraday's law.

Magnetic flux
$\Phi_B = BA \cos \theta$

Lenz's law

Lenz's law says that the emf induced by a changing magnetic flux always tends to produce a magnetic field that opposes the change in flux. The key to Lenz's law is that the induced emf always opposes externally imposed changes.

Energy stored in a magnetic field

The energy stored in a magnetic field B in a particular volume of space is

$$PE_{\text{mag}} = \frac{1}{2\mu_0} B^2 (\text{volume}) \qquad \textbf{(21.34)} \text{ (page 709)}$$

APPLICATIONS

Inductors

An **inductor** is a circuit element such as a coil or solenoid that generally consists of one or more loops of wire. When applied to an inductor with inductance L, Faraday's law gives the voltage drop across the inductor as

$$V_L = L \frac{\Delta I}{\Delta t} \qquad \textbf{(21.23)} \text{ (page 703)}$$

There is a nonzero voltage across an inductor only when the current is changing with time. The voltage induced across the inductor opposes changes in the applied current.

DC circuits with inductors

The qualitative behavior of a DC circuit containing inductors can be described using two general rules:

1. Immediately after any switch is closed or opened, the induced emfs keep the current through all inductors equal to the values they had the instant before the switch was thrown.
2. After a switch has been closed or opened for a very long time, the induced emfs are zero.

(Continued)

Time constant for an RL circuit

The time dependence of an RL circuit is characterized by the *time constant* τ.

$$\tau = \frac{L}{R}$$ (21.27) (page 706)

Quantitatively, when the switch in an LR circuit is closed at $t = 0$, the current varies as

$$I = \frac{V}{R}(1 - e^{-t/\tau})$$ (21.26) (page 706)

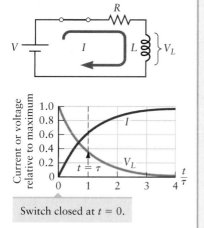

Switch closed at $t = 0$.

Energy stored in an inductor

The energy stored in an inductor (in its magnetic field) is

$$PE_{ind} = \tfrac{1}{2}LI^2$$ (21.33) (page 708)

QUESTIONS

SSM = answer in Student Companion & Problem-Solving Guide   ⊗ = life science application

1. Describe an example in which the magnetic flux through a surface is zero and the magnetic field is not zero.

2. An aluminum disk is held between the poles of a powerful magnet and then pulled out. The person pulling on the disk feels an opposing force due to currents induced through Faraday's law. Explain why the magnitude of the force becomes larger as the disk is pulled faster.

3. SSM Modern coin vending machines use a configuration of magnets to help sort appropriate currency from fake or foreign coins. When a coin is dropped through the slot, it passes within a few millimeters of the magnets as it falls. Sensors (light sensitive) detect the passage of the coin as it breaks a beam at the top and bottom of the magnet assembly. How might the magnets help reject fake or foreign coins? How would the time of passage of a zinc coin compare with that of a coin made with copper in the presence of the magnetic field? (Modern pennies are copper-plated zinc, and modern dimes and quarters have a layer of copper sandwiched between the face and reverse.)

4. The inductance L of a solenoid is proportional to the square of the number of turns, whereas the energy stored in the solenoid PE_{ind} is proportional to L. (a) If the length of the solenoid is doubled, by what factors do L and PE_{ind} change? (b) Explain physically why the factors in part (a) are larger than 2.

5. The switch in Figure Q21.5 is initially closed for a very long time. Explain why Faraday's law can cause sparks when the

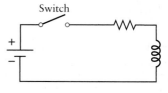

Figure Q21.5

switch is then opened. *Hint:* Assume the switch contacts involve two pieces of metal that are pulled apart.

6. Many trains use "magnetic brakes." Explain how Faraday's law could be used to give a "braking force." *Hint:* Consider the sliding bar in Figure 21.6 and imagine that it is part of the train.

7. A moving charged particle produces an electric current and hence an electric field. Suppose a positively charged particle passes over a circular current loop as sketched in Figure Q21.7. Make a qualitative sketch of the current through the loop as a function of time. Be sure to indicate the time at which the particle is directly over the center of the current loop.

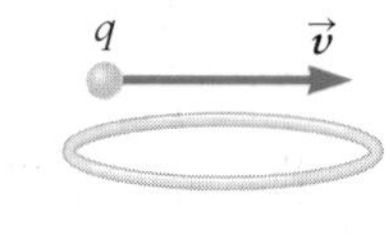

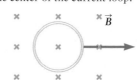

Figure Q21.7

8. A conducting loop is pulled at constant velocity through a region in which the magnetic field is a constant, both inside and outside the loop (Fig. Q21.8). Explain why the induced emf is zero.

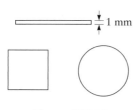

Figure Q21.8

9. Faraday and many other scientists expected that there should be a certain symmetry between electric and magnetic phenomena. Is there a magnetic analog of electric charge? That is, is there such a thing as "magnetic charge"?

10. You are given a length of wire L and told to form it into a current loop so as to produce the maximum possible emf when the loop is placed in a time-varying magnetic field. You investigate the three loop geometries in Figure Q21.10: a long, thin rectangle; a square; and a circle. Which one will give the largest emf?

Figure Q21.10

11. Consider a circular path as sketched in Figure Q21.11. This path forms the boundary for a disk and for a hemisphere as well as many other possible surfaces. If the magnetic flux through the disk is $\Phi_{B,\text{disk}}$ and the flux through the hemisphere is $\Phi_{B,\text{hemi}}$, how are these two fluxes related? Give a physical reason to explain your answer.

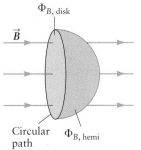

Figure Q21.11

12. A loop of copper passes through regions of magnetic field with a constant velocity as indicated in Figure Q21.12. (a) Determine the direction of the current (if any) induced in the copper loop at each point, 1 through 5, along its path. Assume the field is uniform and of constant magnitude inside the dashed boundary and zero outside. (b) Rank the currents according to their magnitude from greatest to least for points 1 through 5. Indicate ties, if any.

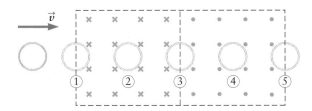

Figure Q21.12 Questions 12 and 13.

13. Consider again the copper loop in Figure Q21.12, but this time assume the loop has a constant acceleration. Again determine the direction of the current (if any) induced in the copper loop at each point 1 through 5 along its path. Again assume the field is uniform inside the dashed boundary and zero outside.

14. **SSM** A popular classroom demonstration of Lenz's law involves a vertical copper tube and a permanent magnet that fits very loosely inside the tube. The instructor first shows that a permanent magnet is not attracted to the copper of the tube. The magnet is then dropped down the tube and takes a surprisingly long time to come out the bottom end of the tube. When repeating the demonstration, a student looks down the tube and sees that the magnet actually moves at a constant and rather slow velocity and does not come in contact with the tube. Explain how this demonstration illustrates Lenz's law. Why doesn't the magnet come to a complete stop?

15. The demonstration described in Question 21.14 often has some additional components, sometimes including tubes of aluminum and lead with the exact dimensions of the copper tube. The magnet falls at a different rate through the different tubes. Why? Rank the tubes from least to greatest according to the time interval for the magnet to fall through the length of each. Table 19.1 may be useful.

16. You are hired by an amusement park and asked to design a ride similar to that sketched in Figure Q21.16. A horizontal metal rod slides along two sloped, frictionless rails with a constant

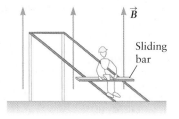

Figure Q21.16

magnetic field directed as shown. A rider sits on the bar and slides down along the rails. Magnetic induction leads to a force on the bar. Use Lenz's law to show that this force is directed upward along the rails and hence keeps the rider from reaching excessive speeds.

17. When a large, time-varying current flows through a coil, a piece of steel placed within the coil will quickly become red hot (Fig. Q21.17). Why? If a piece of copper were placed in the coil, would the effect be the same? Explain.

Figure Q21.17

18. **Induction range.** An alternative to gas or electric stove burners, which have an energy transfer efficiency of less than 50%, are magnetic induction cooktops (Fig. Q21.18), which can be 90% efficient. The "burner" surface does not get hot at all; only the pan gets hot. The main component of an induction range is a coil just under the surface. Describe how this device might work, taking into account that steel and iron cookware must be used. Although the surface never gets hot, is this method of heating safer than gas or conventional electric heating elements? (Consider the effect on a fork or spoon left misplaced while someone is cooking near the coil.)

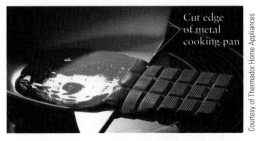

Figure Q21.18 Chocolate in the metal cooking pan is melted while the chocolate in direct contact with the induction burner remains cool and solid.

19. In Chapter 19, we explained how an RC circuit could be used as a filter. The RL circuit in Figure Q21.19 can also be used as a filter. Give a qualitative argument to explain why this circuit will filter out rapid fluctuations in the input voltage.

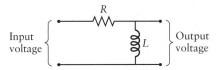

Figure Q21.19

20. In an electric circuit, the largest induced voltage usually occurs just after a switch is opened or closed. Explain why.

21. The induced emf is an electric potential difference, so it should have units of volts (V). Show that the right-hand side of Faraday's law (Eq. 21.5) does indeed have the correct units.

22. Use the result for the current in an RL circuit in Equation 21.26 to derive the voltage across the inductor as given in Equation 21.28.

23. The roads near many traffic lights contain buried sensors that detect when a car is present (waiting at a stop light). These sensors are large loops of wire just beneath the surface of the road. (a) Explain how they can use Faraday's induced emf to detect the presence of a car. *Hint:* Cars contain a lot of steel, which is magnetic. (b) Explain why these sensors sometimes fail to detect a motorcycle and always fail to detect bicycles.

SSM	= solution in Student Companion & Problem-Solving Guide	⊗ = life science application
★	= intermediate ✿ = challenging	ℝ = reasoning and relationships problem

21.2 MAGNETIC FLUX AND FARADAY'S LAW

1. A rectangular loop of wire that is 35 cm wide and 15 cm long is placed in a region where the magnetic field is $B = 1.2$ T and directed perpendicular to the plane of the loop. What is the magnitude of the magnetic flux through the loop?

2. Suppose the wire loop in Problem 1 lies in the x–y plane and that the magnetic field now makes an angle of 60° with the z axis. Find the magnitude of the magnetic flux.

3. ★ The magnetic flux through a current loop is $\Phi_B = 5.5$ T · m². If the magnetic field has a magnitude $B = 0.85$ T, what is the smallest possible value for the area of the loop?

4. ★ The magnetic field in a particular region is independent of position, but varies in time as sketched in Figure P21.4. If a loop of area 0.24 m² is oriented perpendicular to this magnetic field, what is the approximate magnitude of the flux through the loop at $t = 0.25$ s?

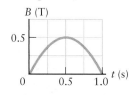

Figure P21.4
Problems 4, 5, and 10.

5. ✿ Consider again the current loop described in Problem 4. What is the approximate magnitude of the induced emf at (a) $t = 0$ and (b) $t = 0.7$ s? (c) How are the signs of these two emfs related?

6. ★ A bar magnet is thrust into a current loop as sketched in Figure P21.6. Before the magnet reaches the center of the loop, what is the direction of the induced current as seen by the observer on the right, clockwise or counterclockwise?

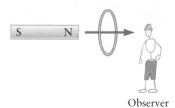

Observer

Figure P21.6 Problems 6, 7, and 8.

7. ★ Consider again the bar magnet and loop in Figure P21.6. Make a qualitative sketch of the current through the loop as a function of time. Assume the bar magnet starts on the far left and continues to the far right and take a clockwise current as viewed from the right as positive.

8. ★ Repeat Problem 7, but assume the leading end of the bar magnet is now the south magnetic pole.

9. SSM ★ ⊗ ℝ Magnetic resonance imaging (MRI) devices use a very large magnet that produces a large magnetic field. Consider an MRI magnet that can accommodate a person. Typically, the person lies down and the field is parallel to her spine (Fig. 21.25). If the magnetic field has a magnitude $B = 1.2$ T, what is the approximate flux through the person?

10. ✿ ℝ A current loop with a resistance $R = 350\ \Omega$ and an area $A = 0.20$ m² is oriented perpendicular to a magnetic field that varies in time as sketched in Figure P21.4. What is the approximate current induced in the loop at (a) $t = 0.30$ s and (b) $t = 0.50$ s?

11. ✿ ℝ Consider a magnetic field that varies sinusoidally with time with an amplitude (peak value) of 0.30 T and a frequency of 400 Hz. If a current loop of area $A = 2.5$ m² is oriented perpendicular to this field, what is the approximate value of the maximum induced emf in the loop?

12. A single loop of wire is placed in a time-varying magnetic field, and an induced emf of 3.5 V is observed. Suppose 25 identical loops with the same orientation are now connected in series. What is the value of the total induced emf?

13. ✿ A square current loop (edge length 45 cm) is found to have an induced emf of 1.2 V at $t = 0$. (a) Assuming the magnetic field is perpendicular to the loop and that the magnitude of the field varies linearly with time, what is the change in the magnetic field from $t = 0$ to $t = 0.10$ s? (b) Explain why it is only possible to compute the change in B and not its absolute magnitude.

14. SSM ★ A metal bar of length $L = 1.5$ m slides along two horizontal metal rails as shown in Figure P21.14. There is a magnetic field of magnitude $B = 2.3$ T directed vertically. (a) If the bar is moving at a speed $v = 0.54$ m/s, what is the emf induced across the ends of the bar? (b) Which end of the bar is at the higher potential?

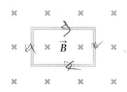

Figure P21.14 Problems 14 and 22.

15. ✿ The Earth's magnetic field is typically 5×10^{-5} T, but fluctuates with time. Suppose the fluctuations in the Earth's field are sinusoidal with a frequency of 10^4 Hz and an amplitude of 5% of the magnitude of the Earth's field. What is the approximate magnitude of the induced emf in a current loop of area 1 cm²?

21.3 LENZ'S LAW AND WORK– ENERGY PRINCIPLES

16. A metal loop is placed in a perpendicular magnetic field as sketched in Figure P21.16. (a) We wish to find the induced current through the loop. What closed path and which corresponding area should we choose for the application of Faraday's law? (b) What is the direction of the magnetic flux through the area chosen in part (a)? That is, is the flux upward (out of the plane of the figure) or downward (into the plane)? (c) If the magnitude of B is increasing with time, is this flux increasing or decreasing with time? (d) What is the direction of the magnetic field produced by the induced current? (e) Will the induced current in the loop be clockwise or counterclockwise?

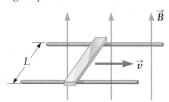

Figure P21.16
Problems 16 and 17.

17. SSM ★ The loop in Figure P21.16 is rotated out of the plane while B is kept constant. (a) Does the flux through the loop increase or decrease with time? (Assume the loop has not had time to rotate by more than 90°.) (b) Will the induced current be clockwise or counterclockwise?

18. ★ A circular loop of wire lies in the x–y plane so that the axis of the loop lies along z (Fig. P21.18). A magnetic field of magnitude B is parallel to the z axis. This field is positive (i.e., along the $+z$ direction) and is

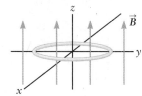

Figure P21.18

the same everywhere within the area of the loop. If B is decreasing with time, what is the direction of the induced current in the loop as viewed from above?

19. A long, straight, current-carrying wire passes near a wire loop (Fig. P21.19). The current in the straight wire is directed to the right, and its magnitude is decreasing. (a) What is the direction of the magnetic flux through the loop? (b) Is the magnitude of the flux through the loop increasing or decreasing with time? (c) What is the direction of the magnetic field produced by the induced current in the loop? (d) What is the direction of the current induced in the loop? (e) This induced current leads to a magnetic force on the loop. What is the direction of this force? Is it directed toward the wire or away from it?

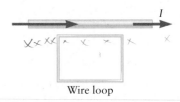

Wire loop

Figure P21.19 Problems 19, 20, and 21.

20. ⬛ Consider again the situation in Figure P21.19, but suppose the current I is constant and the loop is pulled away from the wire. (a) What is the direction of the current induced in the loop? (b) This induced current leads to a magnetic force on the loop. What is the direction of this magnetic force? Is it directed towards the wire or away from it?

21. ✪ In Problem 19, there is a current I_{ind} induced in the loop. If there is a resistance R in the loop, there will be a certain amount of power $P = I_{ind}^2 R$ dissipated in the resistor. Does this energy come from (a) the force exerted on the loop by the magnetic field of the wire, (b) the voltage source that produces the current in the long straight wire, or (c) the gravitational potential energy of the loop?

22. Suppose the metal rails in Figure P21.14 are now connected on the left so that a complete conducting path (through the bar and the rails) is formed. (a) If we wish to find the induced current through the bar, what closed path should we choose for the application of Faraday's law? (b) What is the direction of the flux through the path chosen in part (a)? That is, is the flux upward or downward relative to the plane of the rails? (c) Is the magnitude of the flux increasing or decreasing with time? (d) What is the direction of the magnetic field produced by the induced current? (e) If the resistance of the loop is 1500 Ω, what are the magnitude and the direction of the current through the bar?

23. A metal bar is pulled through a region in which the magnetic field is perpendicular to the plane of motion (Figure P21.23). There is an induced emf across the bar, with the top of the bar positive relative to the bottom. Is $\vec{B}$ directed into or out of the plane of the drawing?

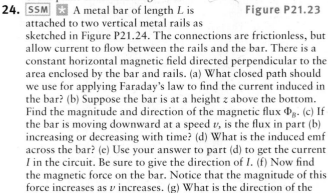

Figure P21.23

24. SSM ⬛ A metal bar of length L is attached to two vertical metal rails as sketched in Figure P21.24. The connections are frictionless, but allow current to flow between the rails and the bar. There is a constant horizontal magnetic field directed perpendicular to the area enclosed by the bar and rails. (a) What closed path should we use for applying Faraday's law to find the current induced in the bar? (b) Suppose the bar is at a height z above the bottom. Find the magnitude and direction of the magnetic flux Φ_B. (c) If the bar is moving downward at a speed v, is the flux in part (b) increasing or decreasing with time? (d) What is the induced emf across the bar? (e) Use your answer to part (d) to get the current I in the circuit. Be sure to give the direction of I. (f) Now find the magnetic force on the bar. Notice that the magnitude of this force increases as v increases. (g) What is the direction of the force on the bar? (h) If the bar has a mass m, there will also be a gravitational force on the bar. Eventually, the bar will reach a

speed v at which the gravitational and magnetic forces cancel so that the total force on the bar is zero. Find v.

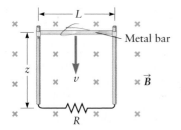

Figure P21.24

25. ✪ A metal bar of length $L = 0.60$ m slides freely on two frictionless, horizontal metal rails as shown in Figure P21.25. The entire circuit has a resistance $R = 200\ \Omega$. There is a magnetic field $B = 1.5$ T directed upward, perpendicular to the plane of the drawing and the rectangular area enclosed by the bars. An external force is applied to the bar such that it moves to the right with a speed $v = 25$ m/s. What are the magnitude and direction of the current induced in the bar?

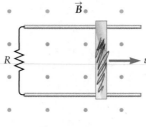

Figure P21.25

26. ✪ A circular metal hoop is dropped over a bar magnet as sketched in Figure P21.26. (a) Make a qualitative sketch of how the current induced in the hoop varies as a function of its height z. Start with the hoop well above the north pole of the magnet and let it fall all the way to large negative values of z ($z \ll L/2$). (b) Make a qualitative plot of the magnetic force on the hoop as a function of z. Interpret your result in terms of Lenz's law.

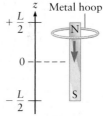

Figure P21.26
Problems 26 and 27.

27. ✪ Consider again the metal hoop in Figure P21.26. If a person grabs the hoop and pulls it upward, what is the direction of the magnetic force on the hoop? Give answers when the hoop is at (a) $z = +L/2$ and (b) $z = 0$.

28. ✪ Two circular current loops lie parallel to the x–y plane and are separated by a short distance along z (Fig. P21.28). The currents through these loops are positive if they are counterclockwise as viewed from the $+z$ direction and are negative if they are clockwise. (a) If the current I_1 through the bottom loop is increasing with time, what is the direction of the induced current in the top loop? (b) If the current I_2 through the top loop is decreasing with time, what is the direction of the induced current in the bottom loop?

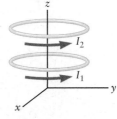

Figure P21.28

29. ✪ A bar magnet is dropped through a current loop as shown in Figure P21.29. (a) When the bar magnet is above the plane of the loop, the current induced in the current loop is counterclockwise as viewed from above. Is the

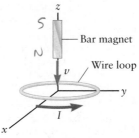

Figure P21.29

orientation of the bar magnet north pole up or south pole up? (b) What is the direction of the resulting magnetic force on the bar magnet? (c) Suppose the orientation of the bar magnet is inverted. How does your answer to part (b) change?

21.4 INDUCTANCE

30. Calculate the inductance of a solenoid that is 1.5 cm long and 3.0 mm in diameter, with 300 turns of wire.

31. An inductor carries a current of 2.5 A, and the magnetic energy stored in the inductor is 0.0035 J. What is the inductance?

32. An inductor with $L = 1.5$ mH is connected to a circuit that produces a current increasing steadily from 1.5 A to 5.6 A over a time of 2.3 s. What is the voltage across the inductor?

33. [SSM] ✴ An inductor has the form of a coil with 2000 turns and a diameter of 1.5 mm. The inductor is placed in a magnetic field perpendicular to the plane of the coil and increasing at a rate of 0.35 T/s. The current in the inductor is zero at $t = 0$, and then increases to 6.5 mA at $t = 1.0$ s. What is the inductance?

34. ✴ If the number of turns in a solenoid is increased by a factor of three, (a) by what factor does the magnetic field inside the solenoid change? (b) By what factor does the inductance change?

35. ✴ An MRI magnet has an inductance of $L = 5.0$ H. What is the total flux through the magnet's coils when the current is $I = 100$ A?

36. The flux through an inductor with $L = 5.0$ mH changes at a rate of $\Delta\Phi_B/\Delta t = 0.0045$ T·m²/s. What is the rate of change $\Delta I/\Delta t$ of the current through the inductor?

37. ✪ Figure P21.37 shows a top view of a mutual inductor (compare with Fig. 21.17). The two coils are shown in different colors, and the top and bottom wire leads are also indicated. (a) If the current in coil 1 is directed as shown and I_1 is increasing, what is the sign of the induced emf V_2 in coil 2? ($V_2 = V_A - V_B$) (b) Both coils have a length $\ell = 12$ cm and diameter 1.5 mm. Coil 1 has 300 turns, and coil 2 has 3000 turns. If $\Delta I/\Delta t = 3.0$ A/s, what is the magnitude of V_2?

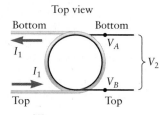

Top view

Figure P21.37

21.5 *RL* CIRCUITS

38. ✪ Consider an *RL* circuit (Fig. P21.38) with $V = 9.0$ V, $L = 5.5$ mH, and $R = 1500$ Ω. After the switch has been left open for a long time, the switch is closed at $t = 0$. (a) What is the current the instant after the switch closed? (b) What is the current a very long time after the switch is closed? (c) How long does it take the current in the circuit to reach 50% of its final (long time) value?

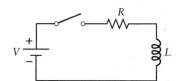

Figure P21.38 Problems 38–41.

39. ✴ Consider again the circuit in Figure P21.38. The switch is first closed for a very long time so that the current through the inductor is constant. The switch is then opened suddenly. What is the voltage across the inductor the instant after the switch is opened?

40. For the circuit in Figure P21.38, if $R = 4000$ Ω, what must the value of L be to have a time constant of 1.5 s?

41. [SSM] ✴ The switch in Figure P21.38 is open for a very long time and then closed at $t = 0$. What is the voltage across the resistor just after the switch is closed?

42. ✪ The switch in Figure P21.42 is open for a very long time and then closed at $t = 0$. (a) What is the current through R_1 and through R_2 immediately after the switch is closed? (b) What is the voltage across the inductor immediately after the switch is closed? (c) After the switch is closed for a very long time, what is the current through R_1, the current through R_2, and the voltage across L? (d) The switch is now opened at $t = t_{\text{open}}$. What are the magnitude and direction of the current through R_1 immediately after t_{open}? (e) What is the voltage across L just after t_{open}?

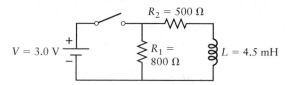

Figure P21.42 Problems 42 and 48.

43. Find the RL time constant for decay of the current in the circuit in Figure 21.23B. Express your answer in terms of the values of the resistances R_1 and R_2, and the inductance L.

44. ✪ Consider three inductors L_1, L_2, and L_3 connected in series as shown in Figure P21.44. These three inductors act as one equivalent inductance L_{equiv}. To find L_{equiv}, we first note that because the current flows through the inductors sequentially, the factor $\Delta I/\Delta t$ is the same for each. (a) What is the voltage across each inductor? Express your answers in terms of L_1, L_2, L_3, and $\Delta I/\Delta t$. (b) Use the result from part (a) to find the total voltage across all three inductors. (c) The voltage in part (b) is also equal to the voltage across the equivalent inductance L_{equiv}. Use this fact to find L_{equiv} in terms of L_1, L_2, and L_3. The general result for many inductors in series is $L_{\text{equiv}} = L_1 + L_2 + L_3 + \cdots$.

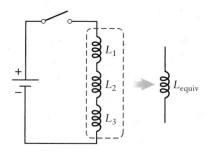

Figure P21.44

45. ✪ Consider two inductors L_1 and L_2 connected in parallel as shown in Figure P21.45. These two inductors act as one equivalent inductance L_{equiv}. To find L_{equiv}, we first notice that because they are connected in parallel, the voltage across L_1 and L_2 must be the same, but the rate at which the current changes with time $\Delta I/\Delta t$ is different for the two inductors. Use these facts to write the total voltage across the two inductors in the form of Equation 21.23, where L is now the equivalent inductance, and find L_{equiv}. *Note:* The general result for many inductors in parallel is $1/L_{\text{equiv}} = 1/L_1 + 1/L_2 + 1/L_3 + \cdots$.

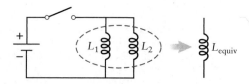

Figure P21.45

21.6 ENERGY STORED IN A MAGNETIC FIELD

46. What is the energy stored in the inductor in Figure P21.46 after the switch has been closed for a very long time?

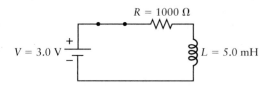

Figure P21.46

47. ✪ ⓧ Consider an MRI (magnetic resonance imaging) magnet that produces a magnetic field $B = 1.5$ T at a current $I = 140$ A. Assume the magnet is a solenoid with a radius of 0.30 m and a length of 2.0 m. (a) What is the number of turns of the solenoid? (b) What is its inductance? (c) How much energy is stored in this magnet? (d) If all the energy in part (b) were converted to the kinetic energy of a car ($m = 1000$ kg), what would the speed of the car be?

48. SSM ✪ Consider again the circuit in Figure P21.42. (a) Suppose the switch is closed for a very long time. How much energy is stored in the inductor? (b) If the switch is then opened, where does this energy go? Explain.

49. ✪ Consider a region of space in which the energy density stored in the magnetic field (Eq. 21.35) is equal to the energy density in the electric field (Eq. 18.48). What is the ratio of B/E? Express your result in terms of the fundamental constants ε_0 and μ_0.

21.7 APPLICATIONS

50. ✪ ⓡ A single circular loop of wire of radius $r = 2.0$ cm rotates at a frequency of 60 Hz in a constant magnetic field of magnitude $B = 1.5$ T. What is the approximate maximum emf induced in this generator?

51. ✪ ⓡ Use the result from Problem 50 to design a generator that produces an induced emf of 500 V. *Hint:* You must choose values for the number of loops and the area of each loop.

52. SSM ✪ ⓡ The pickup coil in an electric guitar contains 10,000 turns and has a diameter of 2.0 mm. When the pickup coil is very close to a guitar string, there is a field of magnitude $B = 0.10$ T in the coil. If this steel string is vibrating at a frequency of 200 Hz, what is the approximate maximum emf induced in the coil? For simplicity, assume the induced flux varies back and forth between zero and the value found when B in the coil is at its maximum value (found when the string is very close to touching the coil).

53. ✪ A guitar string can vibrate at different frequencies, depending on how it is held (Chapter 12). Consider a string that is being played at either 400 Hz or 800 Hz. If everything else is kept the same, how do the induced emfs at the two frequencies compare? That is, which one is larger, and by what factor? *Note to musicians:* Electric guitar amplifiers are designed to compensate for this frequency variation.

54. ✪ ⓡ A metal detector uses a coil to find pieces of metal that are buried or otherwise obscured. Many metals are magnetic, and when a magnet is moved near a metal detector's coil, an emf is induced in the coil. Assume a coil of diameter 10 cm with 500 turns. This coil is scanned over a large piece of iron that is just below soil's surface. Estimate the magnitude of the emf induced in the coil. Assume the field near the iron is $B = 0.30$ T.

21.8 THE PUZZLE OF INDUCTION FROM A DISTANCE

55. In Section 21.8, we mentioned that an emf is induced in a distant coil or current loop via an electromagnetic wave. For the case in Figure 21.30, this electromagnetic wave travels from the inner solenoid to the outer loop. We will see in Chapter 23 that in a vacuum this wave travels at a speed $c = 3.00 \times 10^8$ m/s. Calculate the time Δt it takes this wave to travel a distance of 1.0 mm as might be typical in an electric circuit containing several inductors. Your result for Δt should show that in this case the induced emf appears nearly instantaneously.

56. ✪ A long, straight wire carrying a current I is next to a wire loop, with the straight wire lying in the plane of the loop (Fig. P21.56). The current I is increasing with time. What is the direction of the induced current in the loop, clockwise or counterclockwise as viewed in the figure?

57. ✪ A current loop (loop 1) lies in a horizontal plane. A second loop (loop 2) of the same size is positioned above loop 1 and is also oriented horizontally (Fig. P21.57). The current in loop 1 is constant and counterclockwise as viewed from above. If loop 2 falls toward loop 1, what is the direction of the induced current in loop 2?

Figure P21.56

Figure P21.57

58. SSM ✪ A current loop is placed in a magnetic field that varies with time according to the graph in Figure P21.58. Assume the loop is oriented perpendicular to the magnetic field. Sketch the graph for the induced emf in the loop for the same time intervals.

Figure P21.58

59. ✪ Figure P21.59 shows four configurations of a wire loop placed near a long, straight wire with a time-varying current. Determine the direction of (a) the induced magnetic field in the wire loop (into or out of the page) and (b) the induced current (clockwise or counterclockwise) for each configuration (1) through (4).

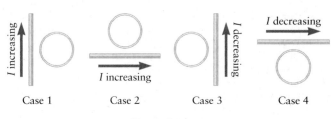

Figure P21.59

60. ✪ A 3.0-V flashlight bulb is connected to a square loop of wire that measures 60 cm on a side as shown in Figure P21.60. Assume the bulb will only light if the potential across its terminals is at least 3.0 V. The loop is oriented perpendicular to a uniform magnetic field of 2.5 T. The magnetic field is then made to decrease to zero over a period of time Δt such that the bulb is just barely lit. (a) Find how long the bulb will light at full output. (b) In which direction will the current flow around the loop?

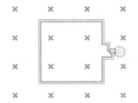

Figure P21.60

61. ✪ The 3.0-V flashlight bulb assembly of Problem 60 is now attached to a wood dowel with a crank handle as shown in Figure P21.61, and the loop is rotated in a uniform magnetic field of 2.5 T. (a) Assume the crank arm rotates with a period T. What value of T is just small enough to make the bulb light? For simplicity, assume that if the loop makes 1/4 turn in a time $\Delta t_{1/4}$, the induced emf is equal to $\Delta\Phi/\Delta t_{1/4}$. (b) If the magnitude of the field is reduced to 0.25T, what is the smallest value of T that will work?

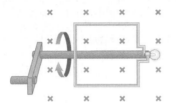

Figure P21.61

62. ✪ The switch in the circuit shown in Figure P21.62 has been open for a long time. The resistors in the circuit each have a value of 50 Ω, the inductors each have a value of 40 mH, and $V = 12$ V. (a) What is the total equivalent inductance of the circuit between points A and B? (b) What is the current through R_1 the instant after the switch is closed? (c) What is the current through R_1 long after the switch has been closed? *Hint:* See Problems 44 and 45.

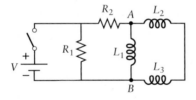

Figure P21.62

63. ✪ Consider an RL circuit with resistance 45 Ω and inductance 250 mH placed in series. (a) If a 10-V battery is connected to the circuit, how long would it take the current to reach 10 mA? (b) What is the voltage drop across the inductor at this time?

64. ✪ Consider an RL circuit with resistance 100 Ω and inductance 200 mH. (a) If the battery is removed from the circuit and replaced by a connecting wire to complete the circuit, how long would it take for the potential difference across the resistor to drop to 37% of its initial value? (b) How long would it take to reach just 50%? (c) How long to get down to 10%?

65. ✪ The switch in Figure P21.65 is closed, and 5.5 μs later the current is measured to be 0.10 A. (a) What is the inductance L? (b) What is the maximum current in this circuit? (c) When is the current half of this maximum?

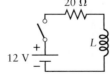

Figure P21.65

66. ✪ The circuit in Figure P21.66 has two identical resistors and three identical inductors. If $L = 30$ mH, what value must R have to give a time constant of 50 μs? *Hint:* See Problems 44 and 45.

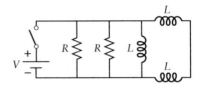

Figure P21.66

67. ✪ An inductor ($L = 80$ mH) is subjected to a time-varying current according to the graph in Figure P21.67. Plot the corresponding emf as a function of time. What is the highest potential difference produced? What is the lowest? Assume the potential of the supply voltage is positive so that a decrease in current generates a positive drop across the inductor.

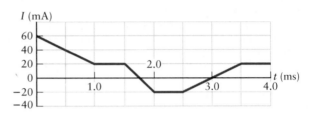

Figure P21.67

68. ✪ An inductor ($L = 50$ μH) is subjected to a time-varying current. The emf then produced by the inductor is graphed in Figure P21.68. The initial current is 40 mA. Plot the corresponding current as a function of time. Assume the potential of the supply voltage is positive so that a decrease in current generates a positive drop across the inductor.

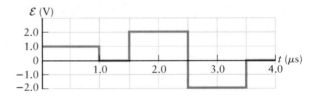

Figure P21.68

69. ✪ Ⓡ The Earth's magnetic field is approximately 5 × 10⁻⁵ T. Suppose a square coil of edge length 0.10 m is rotated with a frequency $f = 400$ Hz about an axis perpendicular to the Earth's field (Fig. P21.69). The induced emf in this coil will vary with time according to $V = V_0 \sin(2\pi ft)$. What is the approximate value of V_0?

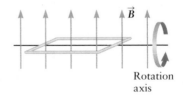

Figure P21.69

Alternating-Current Circuits and Machines

We discussed circuits containing resistors and capacitors in Chapter 19 and those with inductors in Chapter 21. The source of electrical energy in those circuits was a battery, which provides a constant voltage at its output terminals (Fig. 22.1A). If a battery-powered circuit contains only resistors, the current will be independent of time. In circuits containing capacitors (*RC* circuits) or inductors (*RL* circuits), the current can vary with time but always approaches a constant value a long time after closing or opening a switch. Such circuits are called **DC circuits,** where "DC" stands for **direct current,** which is another way of referring to a constant current. In an **AC circuit,** the battery is replaced by a device that produces an electric potential (i.e., a voltage) that varies with time (Fig. 22.1B). Such a device is called an AC voltage source, where "AC" stands for **alternating current.**

Electrical energy is delivered to your house using AC voltages. The flow of this energy is controlled by transformers (shown here) and by other devices described in this chapter. (© Lester Lefkowitz/ Photographer's Choice/Getty)

Figure 22.1 🅐 A battery produces a constant electric potential (the emf) at its terminals. 🅑 The electric potential (voltage) at the terminals of an AC voltage source varies sinusoidally with time.

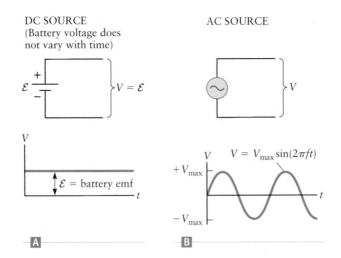

An AC voltage source is represented in a circuit diagram by the symbol ―⊗―. The voltage at the output terminals of this device varies sinusoidally with time, with a frequency f and an amplitude V_{max}, also called the peak value, as shown in Figure 22.1B.

The electrical energy used in your house is provided by an AC source. If you were to measure the voltage at the terminals of an electrical socket in a house in the United States, you would find a time dependence like that shown in Figure 22.1B with a frequency of 60 Hz. The famous inventor Thomas Edison originally proposed the use of DC electrical systems for houses and cities, but around 1880 other engineers, including one named George Westinghouse, pushed for an AC system. In the end, Westinghouse's proposal won out, and a 60-Hz system was adopted in the United States in the 1890s.

In this chapter, we'll explore the advantages of AC power. We begin with a study of AC circuits containing the basic circuit elements: resistors, capacitors, and inductors. After describing some of the properties of AC circuits, we'll then be able to understand why we use alternating current instead of direct current in our homes.

22.1 | GENERATION OF AC VOLTAGES

One reason an AC system was adopted for the large-scale distribution of electrical energy is that generating an AC voltage is simpler than constructing and maintaining a large DC source. Most sources of AC electrical energy employ a generator based on magnetic induction. Figure 21.7 showed the general design of an AC generator. At a hydroelectric facility, water is used to turn a mechanical shaft; a steam turbine serves a similar function at a nuclear or coal-burning power plant (Fig. 22.2). This shaft is part of the generator and holds a coil with many loops of wire. The coil is positioned between the poles of a permanent magnet, so the magnetic flux through the coil varies with time as the shaft turns. According to Faraday's law of induction (Chapter 21), this changing flux induces a voltage in the coil, which is the generator's output. Faraday discovered his law of induction in 1831, and the first AC generators based on this principle were constructed in 1832! It did not take long to make practical use of Faraday's discovery.

The approach in Figure 21.7 produces an AC voltage with a frequency equal to the rotation frequency of the shaft, and electrical energy is produced as long as the shaft is rotating. In contrast, for a DC source such as a battery, one would have to replace the battery often (or replenish the chemicals in the battery), and for many practical applications this battery would need to be quite large. That is one reason an AC power system is simpler to implement than a DC one.

Devices that are commonly referred to as "generators" of electrical energy actually just convert the mechanical energy of the rotating shaft into electrical energy (an electric potential difference that can produce a current in a circuit). The principle of conservation of energy still applies, so for an ideal generator the amount of electrical energy dissipated in an attached circuit is equal to the mechanical energy used to rotate the generator shaft. In the same way, when we refer to the circuit element —⊗— as a "source" of electrical energy, you should realize that this device simply enables the transfer of electrical energy from a generator to an attached circuit.

Figure 22.2 Large AC generators are used to produce AC voltages at hydroelectric dams and at other types of electric power plants.

| EXAMPLE 22.1 | Generating an AC Voltage

Let's analyze the large AC generators that are used in facilities such as the Hoover Dam power station near Las Vegas. Estimate the amplitude of the voltage (V_{max} in Figure 22.1B) produced by rotating a coil containing one loop of wire in the generator. Assume the generator's magnetic field has a magnitude $B = 1.0$ T (typical of a large magnet) and the coil has an area $A = 5.0$ m^2. (Generators like the one in Fig. 22.2 are large.) Because this voltage is destined for household use, assume a frequency $f = 60$ Hz.

RECOGNIZE THE PRINCIPLE

Generators are based on Faraday's law. Because the coil rotates, the magnetic flux through it changes with time. This time-varying flux induces a voltage in the coil, which is the AC voltage produced by the generator.

SKETCH THE PROBLEM

Figure 22.3 shows how the magnetic flux changes as the loop in our coil rotates.

IDENTIFY THE RELATIONSHIPS AND SOLVE

The magnetic flux through a generator coil is largest when the face of the coil is perpendicular to the magnetic field (Fig. 22.3A). This maximum flux is $\Phi_B = BA$, where B is the magnitude of the magnetic field and A is the area of the coil (see Eq. 21.1). As the coil rotates one half turn, the flux alternates in sign from $+BA$ to $-BA$ (Fig. 22.3B) because the angle θ between the field and the coil's surface varies between 0° and 180°. The total flux change is

$$\Delta\Phi_B = -2BA$$

If the rotation period of the shaft is T, this flux change occurs in a time $\Delta t = T/2$. The rotation period is related to the frequency by $T = 1/f = 1/60$ s.

From Faraday's law (Eq. 21.6), the magnitude of the emf induced in the coil is

$$|\mathcal{E}| = \left|\frac{\Delta\Phi_B}{\Delta t}\right|$$

Strictly speaking, this equation applies only for small flux changes $\Delta\Phi_B$ over small time intervals Δt, but we can still use it to get an approximate value for the generator emf. Inserting our estimates for $\Delta\Phi_B$ and Δt, we get

$$|\mathcal{E}| = \left|\frac{\Delta\Phi_B}{\Delta t}\right| = \frac{2BA}{T/2} = \frac{2(1.0 \text{ T})(5.0 \text{ m}^2)}{(1/120) \text{ s}}$$

$$|\mathcal{E}| = \boxed{1200 \text{ V}}$$

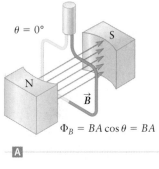

$\theta = 0°$

$\vec{B}$

$\Phi_B = BA\cos\theta = BA$

A

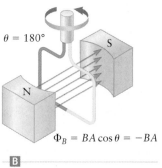

$\theta = 180°$

$\Phi_B = BA\cos\theta = -BA$

B

Figure 22.3 Example 22.1. As the wire loop makes a half turn around the rotation axis, the flux changes from $\Phi_B = +BA$ to $\Phi_B = -BA$.

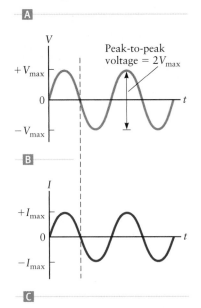

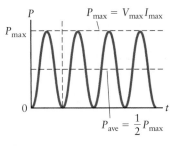

Figure 22.4 🄰 Circuit with an AC voltage source connected to a resistor. 🄱 Voltage at the output terminals of the source as a function of time. 🄲 Current in the circuit as a function of time. 🄳 Instantaneous power dissipated in the resistor as a function of time. The average power (i.e., averaged over many cycles of the oscillation) is one-half the maximum power.

CONCEPT CHECK 22.1 | Generating a Larger Voltage

Consider the Hoover Dam generator discussed in Example 22.1. Suppose you wish to produce an AC voltage with an amplitude of 50,000 V. Which of the following approaches would work? Explain your answer.
  (a) Connect many generators in series.
  (b) Connect many generators in parallel.
  (c) Add more turns to the coil.

AC Circuits and Simple Harmonic Motion

An old-fashioned but very descriptive unit of frequency is the "cycle per second." In terms of the time dependence of an AC voltage in Figure 22.1B, a frequency of 1 Hz means that the voltage completes one full cycle each second. The voltage variation in Figure 22.1B should remind you of the simple harmonic oscillators in Chapter 11. Because of this similarity, the voltage and current in an AC circuit are said to "oscillate." We'll also find (Sections 22.5 and 22.6) a close connection between circuits with capacitors and inductors (called *LC* circuits) and simple harmonic oscillators.

22.2 | ANALYSIS OF AC RESISTOR CIRCUITS

Now that we've seen how AC voltages can be generated, let's consider how to calculate the current in an AC circuit. We begin with the simplest possible circuit consisting of an AC generator and a resistor R as sketched in Figure 22.4A.

The voltage across the output terminals of an AC source varies with time. If f is the frequency and V_{max} is the amplitude of the AC voltage, the output voltage has the form

$$V = V_{max} \sin(2\pi ft) \qquad (22.1)$$

In Equation 22.1, V is the *instantaneous* potential difference between the terminals of the AC source and is plotted as a function of time in Figure 22.4B. In this circuit, the voltage across the output terminals is also equal to the voltage across the resistor. To find the resulting current I through the resistor, we can use **Ohm's law** (Eq. 19.10):

$$I = \frac{V}{R} \qquad (22.2)$$

In Chapter 19, we applied Ohm's law for resistors in DC circuits, but it also holds for resistors in an AC circuit, where V and I are the instantaneous voltage and current. You can think of an AC voltage source as equivalent to a battery whose output voltage varies in time according to Equation 22.1. At each instant in time, the current through this AC circuit is given by $I = V/R$, just as for a DC circuit containing a battery and a resistor. As time passes, the output voltage V of the AC source varies sinusoidally, so the current through the resistor is similarly time dependent. Inserting V from Equation 22.1 into Ohm's law (Eq. 22.2), we find

$$I = \frac{V_{max}}{R} \sin(2\pi ft) \qquad (22.3)$$

which is shown in Figure 22.4C. The current thus oscillates with time, with the same frequency as the voltage. The voltage amplitude V_{max} is constant, so we can denote the current amplitude as I_{max} and write

$$I = I_{max} \sin(2\pi ft) \quad \text{with } I_{max} = \frac{V_{max}}{R} \tag{22.4}$$

Root-Mean-Square Values

The result in Equation 22.4 might be the current in your television or some other appliance. The appliance manufacturer usually specifies how much current and voltage the device is designed for. However, the voltage and current vary with time, so what numbers should the manufacturer provide? One possibility is to provide some kind of average value, but from Figure 22.4B and C we see that the average values of V and I taken over many oscillation cycles are both zero, so these values are not useful as specifications. That is one reason the notion of "root-mean-square" was adopted. Figure 22.5 shows a plot of the AC voltage $V = V_{max} \sin(2\pi ft)$ as well as how the square of the voltage varies with time. Since V^2 is always positive, its average is not zero. The average of V^2 is related to the voltage amplitude V_{max} by (see Fig. 22.5B)

$$(V^2)_{ave} = \tfrac{1}{2} V_{max}^2 \tag{22.5}$$

We now define a quantity called the root-mean-square (rms) voltage, denoted as V_{rms}, through the relation $V_{rms} = \sqrt{(V^2)_{ave}}$. Here the word *mean* indicates an average, and "root" refers to a square root. We thus have

$$\underbrace{V_{rms}}_{\text{square } root} = \underbrace{\sqrt{(V^2)_{ave}}}_{\substack{\text{square}}} \qquad \boxed{\textit{mean (average)}} \tag{22.6}$$

In words, V_{rms} is the square *root* of the *mean* (average) of the voltage *squared*. Inserting Equation 22.5 into the definition of the root-mean-square gives

$$V_{rms} = \sqrt{\frac{1}{2} V_{max}^2} = \frac{V_{max}}{\sqrt{2}} \tag{22.7}$$

Definition of rms voltage

So, the rms value of the voltage is equal to the voltage amplitude V_{max} divided by $\sqrt{2}$; hence, $V_{rms} \approx 0.71 \times V_{max}$. (See Fig. 22.6.)

The root-mean-square value can be defined for any quantity that varies with time, including the current in an AC circuit. So, for the oscillating current in Equation 22.4 we have

$$I_{rms} = \sqrt{\frac{1}{2} I_{max}^2} = \frac{I_{max}}{\sqrt{2}} \approx 0.71 \times I_{max} \tag{22.8}$$

Definition of rms current

The root-mean-square values of the voltage and current are widely used to specify the properties of an AC circuit. For example, in the United States the voltage in a household circuit has the value "120 V AC." This voltage is actually the rms value of the voltage across the terminals of a wall socket.

Power in an AC Circuit

The instantaneous power dissipated in a resistor is equal to the product of the instantaneous voltage across the resistor and the instantaneous current through the resistor, which is just an extension of Equation 19.15 to AC circuits. Hence, the instantaneous power dissipated in the resistor in Figure 22.4A is

$$P = VI \tag{22.9}$$

where V and I are the results in Equations 22.1 and 22.4. Because V and I in this circuit both vary with time, the power P is also a function of time. Inserting our results for V and I into Equation 22.9 gives

$$P = [V_{max} \sin(2\pi ft)][I_{max} \sin(2\pi ft)] = V_{max} I_{max} \sin^2(2\pi ft) \tag{22.10}$$

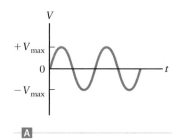

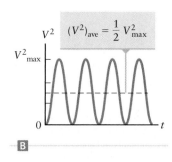

Figure 22.5 Variation of an AC voltage V and V^2 as functions of time.

which is plotted in Figure 22.4D. The instantaneous power oscillates between a maximum value of $V_{max}I_{max}$ and a minimum value of zero. These minima occur at moments when the instantaneous voltage and current are zero.

The instantaneous power in Figure 22.4D might be the power dissipated in your refrigerator or desktop computer, or anything else that is plugged into a wall socket. These and other household devices come with a power rating provided by the man-ufacturer—this rating is a single number that tells you about the power dissipated in the device. The power rating is useful because it gives you an idea of how much it will cost to run the device. How does the power rating relate to the instantaneous power curve in Figure 22.4D? Since the AC voltage and hence the instantaneous power oscillate rapidly (in the United States, the voltage has a frequency of 60 Hz), it makes sense to average the power over many cycles of its oscillation. The value of the **average power** P_{ave} can be seen from Figure 22.4D, which shows that the average power equals half the maximum power. According to Equation 22.10, the maximum power is $P_{max} = V_{max}I_{max}$ so the average power is

Average power dissipated in a resistor in terms of the voltage and current amplitudes

$$P_{ave} = \frac{V_{max}I_{max}}{2} \tag{22.11}$$

Using our results for the rms voltage V_{rms} and current I_{rms} (Eqs. 22.7 and 22.8), we can write the average power in Equation 22.11 as

$$P_{ave} = V_{rms}I_{rms} \tag{22.12}$$

One advantage of writing the average power in this way is that Equation 22.12 has precisely the same mathematical form as the power in a DC circuit (Eq. 19.15).

When dealing with the power dissipated in a resistor, the relation for the power (Eq. 22.12) can be simplified using Ohm's law, $I = V/R$. Since the Ohm's law relation between current, voltage, and resistance (Eq. 22.2) holds for instantaneous values, it also holds for rms values. We can thus write

$$I_{rms} = \frac{V_{rms}}{R}$$

Combining this relation with Equation 22.12 leads to

Average power dissipated in a resistor in terms of the rms voltage or current

$$P_{ave} = \frac{V_{rms}^2}{R} = I_{rms}^2 R \tag{22.13}$$

EXAMPLE 22.2 | Power and Voltage in a Household Circuit

The standard AC voltage found in homes in the United States has a frequency of 60 Hz and is often referred to as "120-V power," which is the value of the rms volt-age; that is, $V_{rms} = 120$ V. What is the corresponding value of V_{max} for the voltage at an electrical outlet?

RECOGNIZE THE PRINCIPLE

The amplitude V_{max} of an AC voltage is also the "peak" value of the voltage (Fig. 22.6). V_{max} and V_{rms} are just two ways of describing the same AC voltage. They are related through the basic time dependence of an AC voltage $V = V_{max} \sin(2\pi f t)$ and Equation 22.7. The point of this example is to get an intuitive understanding of the relation between rms and peak values.

SKETCH THE PROBLEM

Figure 22.6 shows an AC voltage and indicates V_{max}.

IDENTIFY THE RELATIONSHIPS AND SOLVE

Applying Equation 22.7, we have

$$V_{rms} = \frac{V_{max}}{\sqrt{2}}$$

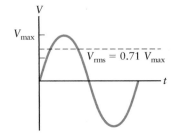

Figure 22.6 Example 22.2.

Solving for V_{max} and inserting the given value of V_{rms} gives

$$V_{max} = \sqrt{2} V_{rms} = \sqrt{2}(120 \text{ V}) = \boxed{170 \text{ V}}$$

What does it mean?

Figure 22.6 could represent a plot of an AC voltage with $V_{max} = 170$ V (the value found in this example). A DC voltage equal to the rms value of 120 V would give the same average dissipated power in the resistor.

EXAMPLE 22.3 | Voltage, Current, and Power in a Lightbulb

As a circuit element, we can assume that an incandescent lightbulb functions just like a resistor (Fig. 22.7), so the relations between the voltage, current, and power in a resistor apply also to a lightbulb. Household lightbulbs are specified by their power rating. For example, one might be rated as a "60-watt" bulb. This power rating refers to the average dissipated power P_{ave}, so $P_{ave} = 60$ W when a 60-W bulb is used in a normal household "120-V" circuit. Find the resistance of a 60-W lightbulb.

RECOGNIZE THE PRINCIPLE

The AC voltage in a "120-V" household circuit has the rms value $V_{rms} = 120$ V. When connected to a lightbulb, it gives the same average power as if we had instead connected a battery with a constant emf of $V_{DC} = 120$ V. For a given value of the voltage, the average power dissipated in a resistor depends on the value of the resistance R. For a DC voltage, the relation is $P_{ave} = V_{DC}^2/R$, whereas for an AC voltage, it is $P_{ave} = V_{rms}^2/R$. The point of this example is to get a feeling for the value of the resistance of a common household "appliance."

SKETCH THE PROBLEM

Figure 22.7 describes the problem.

IDENTIFY THE RELATIONSHIPS

For an AC circuit, the average power dissipated in a resistor is (Eq. 22.13) $P_{ave} = V_{rms}^2/R$.

SOLVE

Solving for R, we get

$$R = \frac{V_{rms}^2}{P_{ave}}$$

For the AC voltage in a household circuit, $V_{rms} = 120$ V, which we can use along with the given value of P_{ave} for the lightbulb to get

$$R = \frac{V_{rms}^2}{P_{ave}} = \frac{(120 \text{ V})^2}{60 \text{ W}} = \boxed{240 \text{ } \Omega}$$

What does it mean?

The average power is important because it determines the amount of energy used by the lightbulb over a long time, and that is how the power company computes your bill. The instantaneous intensity of a lightbulb depends on the instantaneous power dissipated in the filament. According to Figure 22.4D, the power oscillates, so the intensity must also oscillate. Hence, an ordinary lightbulb must "flicker." Why don't we notice this flickering? (*Hint*: See Concept Check 22.2.)

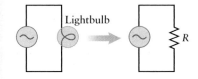

Figure 22.7 Example 22.3. In circuit analysis, an incandescent lightbulb can be modeled as a resistor.

Table 22.1 | Useful Quantities and Notation for Describing AC Circuits

VOLTAGE	CURRENT	POWER
V = instantaneous voltage $= V_{max} \sin(2\pi ft)$	I = instantaneous current	P = instantaneous power = VI
V_{max} = voltage amplitude	I_{max} = current amplitude	P_{max} = power amplitude
V_{rms} = rms voltage = $\dfrac{V_{max}}{\sqrt{2}}$	I_{rms} = rms current = $\dfrac{I_{max}}{\sqrt{2}}$	—
$V_{ave} = 0$	$I_{ave} = 0$	P_{ave} = average power = $V_{rms}I_{rms}$

When dealing with AC circuits, it is important to distinguish between instantaneous and average values of the voltage, current, and power. The amplitudes of the AC variations and the rms values are also important. Table 22.1 collects these terms and summarizes the relations between them.

CONCEPT CHECK 22.2 | Flickering Lightbulbs

In Example 22.3, we noted that the intensity of an ordinary incandescent lightbulb flickers with time in the same way that the instantaneous power in a resistor varies with time (Fig. 22.4D). In the United States, the AC voltage across the lightbulb has a frequency of f = 60 Hz. Is the frequency of the flickering (a) 30 Hz, (b) 60 Hz, (c) 120 Hz, or (d) 240 Hz? How does the frequency of the flickering compare with the response time of your eyes?

CONCEPT CHECK 22.3 | Traveling in Europe

In most European countries, the AC voltage available at a wall outlet has an rms value of about 220 V and a frequency of 50 Hz. (Both of these values are different from those used in the United States.) Is the peak value (the amplitude) of this AC voltage (a) 220 V, (b) 170 V, or (c) 310 V?

Describing AC Voltages and Currents: Phasors

The sinusoidal variation of the voltage in an AC circuit,

$$V = V_{max} \sin(2\pi ft) \tag{22.14}$$

is very similar to what we encountered for circular motion (Chapter 5) and simple harmonic motion (Chapter 11). This similarity leads to a useful graphical way to analyze AC circuits. Imagine an arrow of length V_{max} as sketched in Figure 22.8. The arrow's tail is tied to the origin, while the tip moves along a circle of radius V_{max}. If this arrow makes an angle θ with the horizontal axis, the vertical component of the "voltage arrow" is

$$V = V_{max} \sin \theta \tag{22.15}$$

If the tip of the arrow undergoes uniform circular motion, the angle θ varies with time according to

$$\theta = 2\pi ft \tag{22.16}$$

where f is the rotation frequency. Inserting this result into Equation 22.15, we get $V = V_{max} \sin(2\pi ft)$, which is just the voltage in an AC circuit, Equation 22.14. The rotating arrow representing the voltage in Figure 22.8 is called a **phasor**. A phasor is *not* a vector; the voltage in an AC circuit is *not* a vector in the way that quantities such as position and velocity are vectors. Phasor diagrams simply provide a convenient way to illustrate and think about the time dependence in an AC circuit.

The current in an AC circuit can also be represented by a phasor. Figure 22.9A shows a resistor connected to an AC voltage source. A phasor diagram for this cir-

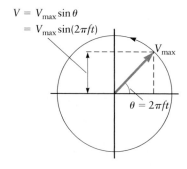

Figure 22.8 Phasor diagram of the voltage in an AC circuit. The voltage is represented by a "phasor," an arrow of length V_{max} that rotates in the coordinate plane. The instantaneous voltage in the circuit is equal to the vertical component of this arrow.

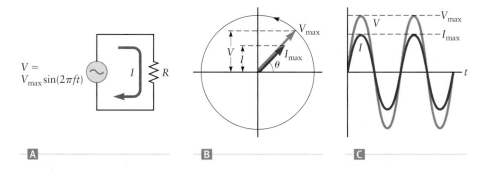

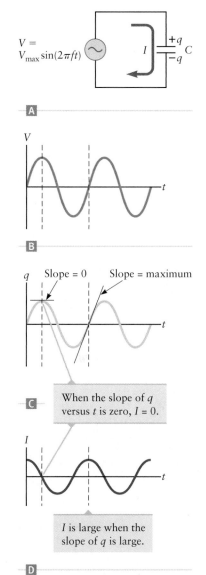

cuit is given in Figure 22.9B, which shows the voltage arrow as it rotates around the origin as well as the current as a rotating arrow of length I_{max}. In Equation 22.4, we found that $I = I_{max} \sin(2\pi ft) = I_{max} \sin\theta$, so the current arrow rotates in synchrony with the voltage arrow. These two arrows—that is, these two phasors—always make the same angle θ with the horizontal axis as time passes. Because the voltage and current phasors rotate together, we say that they are *in phase*. This phase relationship can also be seen from plots of V and I as functions of time in Figure 22.9C. When the instantaneous voltage V is at one of its maxima, the instantaneous current I is also a maximum. Likewise, the minima of V and I occur at the same times, and the voltage and current also pass through zero at the same time.

For the resistor circuit in Figure 22.9A, the phase relation is very simple: the voltage and the current always oscillate "in synchrony" with each other. The current and voltage relationships in circuits with capacitors and inductors are usually not this simple, and phasor diagrams are extremely useful for understanding those circuits.

22.3 | AC CIRCUITS WITH CAPACITORS

Figure 22.10A shows an AC circuit containing a single capacitor with capacitance C. We want to calculate the current when the capacitor is attached to an AC generator with $V = V_{max} \sin(2\pi ft)$. The plates of the capacitor carry charges $+q$ and $-q$, related to the potential difference V across the capacitor by $q = CV$ (which is just Eq. 18.30 with the potential difference or "voltage" across the capacitor denoted simply by V instead of ΔV). This relation between q and V holds at all times, so q is the *instantaneous* charge on the capacitor, just as V is the instantaneous voltage across the capacitor. Since V is also equal to the voltage produced by the AC source, we have

$$q = CV = CV_{max} \sin(2\pi ft) \tag{22.17}$$

Parts B and C of Figure 22.10 show the voltage and charge as functions of time. The capacitor's voltage and charge are *in phase* with each other; that is, their maxima occur at the same times.

To understand the time dependence of the current, we recall that the electric current is defined as (Chapter 19)

$$I = \frac{\Delta q}{\Delta t} \tag{22.18}$$

The instantaneous current I equals the rate at which charge flows onto the capacitor plates in a short time interval Δt. On a graph of q versus t, the current I is the *slope* of the q–t plot. The qualitative behavior of this slope is illustrated in Figure 22.10C, which shows one value of t when the slope $\Delta q/\Delta t$ is zero along with another value of t when the slope is a maximum. By estimating the slope in a graphical way, we can construct the qualitative plot of the current I as a function of time shown in Figure 22.10D. The current has its largest value when the charge on the plates is zero. On the other hand, the current is zero when the charge has its largest values

Figure 22.10 ◢ An AC circuit with a capacitor. ◳ Voltage, ◳ charge on the capacitor, and ◳ current as functions of time for the circuit in part A.

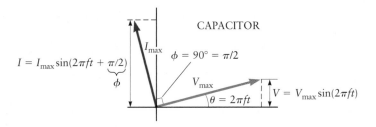

Figure 22.11 Phasor diagram showing the voltage and current through a capacitor. The current is represented by a rotating arrow of length I_{max}. The instantaneous current is the component of this arrow along the vertical axis. The current arrow always makes an angle of 90° ($\pi/2$) with the voltage, so the instantaneous current and voltage are always out of phase by 90°.

(in magnitude). Comparing the general plots of q and I, we see that although q is a sine function (Eq. 22.17), the current is proportional to a cosine function:

$$I = I_{max} \cos(2\pi ft) \tag{22.19}$$

The sine and cosine functions have the property[1] that for any angle α, $\cos(\alpha) = \sin(\alpha + \pi/2)$. We can therefore write the current in the equivalent forms

$$I = I_{max} \sin(2\pi ft + \pi/2)$$

$$I = I_{max} \sin(2\pi ft + \phi) \quad \text{with } \phi = \pi/2 \tag{22.20}$$

A phasor diagram for voltage and current in the capacitor circuit in Figure 22.10A is given in Figure 22.11. We show the voltage as a rotating arrow (a phasor) making an angle $\theta = 2\pi ft$ with the x axis. The tip of this voltage phasor moves in a circle as in the resistor circuit (compare with Eq. 22.15 and Fig. 22.9B). The current in the capacitor circuit is represented by an additional rotating arrow of length I_{max} (which we have not yet calculated), and according to Equation 22.20 this phasor makes an angle $2\pi ft + \pi/2 = \theta + \pi/2$ with the horizontal axis. The current is therefore *out of phase with the voltage* (parts B and D of Fig. 22.10). Unlike the behavior we found in the resistor circuit, the maximum values of the voltage and current *do not occur at the same time*. The angle $\pi/2$ is called the *phase angle ϕ* between V and I. Since the angle $\pi/2$ corresponds to 90°, for this circuit the voltage and current are $\phi = 90°$ out of phase.

The graphical analysis in Figure 22.10 shows that the current and voltage are out of phase for a capacitor. This is quite different from a resistor, for which I and V are always in phase (Fig. 22.9). Why is a capacitor different from a resistor? They are different because the *charge* on a capacitor is proportional to the voltage (Eq. 22.17), whereas for a resistor the *current* is proportional to V (Eq. 22.2). This difference, together with the relation between current and charge $I = \Delta q/\Delta t$ (Eq. 22.18), makes the behavior of the AC capacitor circuit different from that of an AC resistor circuit.

Equation 22.20 and the phasor diagram in Figure 22.11 give the phase of the current through a capacitor. To find the peak value (the amplitude) I_{max}, we must calculate the slope of the q–t plot in Figure 22.10. This calculation requires more mathematics than we can include here. The result is

Amplitude of the current in a circuit with a capacitor

$$I_{max} = \frac{V_{max}}{X_C} \tag{22.21}$$

The factor X_C is called the *reactance* of the capacitor and is given by

Reactance of a capacitor

$$X_C = \frac{1}{2\pi fC} \tag{22.22}$$

The relation between the peak current and voltage for a capacitor in Equation 22.21 has the same mathematical form as the corresponding relation for a resistor $I = V/R$. Here, X_C (for a capacitor) and R (for a resistor) are both measures of how much the circuit "resists" a current. Both resistance and reactance are measured in ohms (Ω), but X_C differs from R in one very crucial way: the reactance of a capaci-

---

[1]See Appendix B.4 for a review of the properties of sine and cosine functions.

tor X_C depends *on frequency*. As a result, the peak value of the current through a capacitor, I_{max} in Equation 22.21, depends on frequency. The reason for this frequency dependence can be traced back to the relation between current and charge $I = \Delta q / \Delta t$. Current I is the rate of change of the charge q on the capacitor. If the frequency is increased, the charge oscillates more rapidly and Δt is smaller, giving a larger current. That is why the peak current in a capacitor is larger at high frequencies and why the reactance X_C becomes smaller at high frequencies.

| EXAMPLE 22.4 | Properties of a Typical Capacitor

Figure 22.12 shows a photo of some typical capacitors, like ones commonly found in a radio or MP3 player, and some typical resistors, each with $R = 100 \ \Omega$. Consider a capacitor with a capacitance $C = 1000$ pF $(1.0 \times 10^{-9}$ F$)$. Find the frequency at which the reactance of the capacitor is equal to the resistance of one of these resistors.

RECOGNIZE THE PRINCIPLE

The reactance X_C of a capacitor depends on its capacitance C and the frequency. By adjusting the frequency, we can attain a reactance of 100 Ω. We'll see in Section 22.7 how this frequency dependence is used in applications of *RC* circuits; for example, it is used to tune a radio to different stations.

SKETCH THE PROBLEM

No figure is needed.

IDENTIFY THE RELATIONSHIPS

The reactance of a capacitor is (Eq. 22.22) $X_C = 1/(2\pi f C)$. We want to find the value of f at which this reactance is equal to R, so we set

$$X_C = \frac{1}{2\pi f C} = R$$

SOLVE

Solving for the frequency and substituting $R = 100 \ \Omega$, we get

$$f = \frac{1}{2\pi R C} = \frac{1}{2\pi (100 \ \Omega)(1.0 \times 10^{-9} \ \text{F})} = 1.6 \times 10^6 \ \text{Hz} = \boxed{1.6 \ \text{MHz}}$$

What does it mean?
At this frequency, the capacitor's reactance X_C and the resistor's resistance R are the same, but that is not true at any other frequency. The reactance of a capacitor, and hence the current in an AC circuit containing a capacitor, depends on frequency. For this combination of R and C, the frequency at which $X_C = R$ lies at the upper end of the AM radio band.

Figure 22.12 Example 22.4. Photo of typical capacitors (larger cylindrical objects) and resistors (smaller objects at lower right) from a small radio.

© Cengage Learning/Charles D. Winters

| CONCEPT CHECK 22.4 | The Effect of Frequency on a Capacitor's Reactance

Consider an AC circuit containing a single capacitor C. If the frequency is increased by a factor of 4, how does the amplitude of the current change?
  (a) The current increases by a factor of 4.
  (b) The current increases by a factor of 16.
  (c) The current decreases by a factor of 4.

Power in a Circuit with Capacitors

The instantaneous power dissipated in a resistor is $P = VI$ (Eq. 22.9). The derivation of this relation in Chapter 19 (Eq. 19.15) calculated the power from the change in the potential energy of a charge as it moved through an electric potential difference

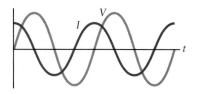

Energy flows into the capacitor.

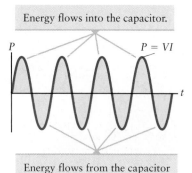

$P = VI$

Energy flows from the capacitor back to the generator.

Figure 22.13 Instantaneous power P is the product of V (a sine wave) and I (a cosine); the result is another sine wave. At some portions of the cycle P is positive, whereas at other times it is negative. The average power over one complete cycle is zero.

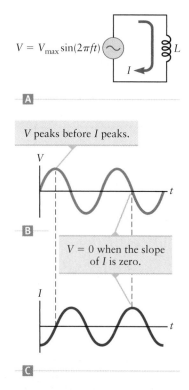

$V = V_{max} \sin(2\pi ft)$

A

V peaks before I peaks.

$V = 0$ when the slope of I is zero.

B

C

Figure 22.14 **A** An AC circuit with an inductor. **B** Voltage across the inductor as a function of time. **C** Current through the inductor as a function of time.

V across a circuit element. In Chapter 19, that circuit element was a resistor, but the general argument applies to *any* circuit element. Let's now use that result for P to find the instantaneous power in a circuit in which an AC voltage source is connected to a capacitor. Inserting the results for V (Eq. 22.14) and I (Eq. 22.19) into our expression for P, we find

$$P = VI = V_{max}I_{max} \sin(2\pi ft)\cos(2\pi ft) \qquad (22.23)$$

This result is plotted in Figure 22.13, which shows how P oscillates between positive and negative values. These oscillations and the negative values of P can be traced to the oscillating behaviors of the sine and cosine functions in Equation 22.23. The *average* value of the power over many oscillation cycles is $P_{ave} = 0$. The voltage and current across the capacitor both oscillate with time, so there are portions of each cycle during which current is flowing into the generator and "against" the generator voltage. That is, at times during each cycle the voltage across the capacitor is positive while the current is negative. When the current and voltage have opposite signs, the power $P = VI$ is negative. At these moments, energy is being transferred from the capacitor back to the generator. In fact, during half of each cycle, energy flows from the generator to the capacitor (P is positive in Fig. 22.13), whereas during the other half of the cycle energy flows from the capacitor back into the generator (P is negative). The net result—the average power—is zero.

We say that electrical energy is *dissipated* in a resistor because the electrical energy supplied by the voltage source is converted to heat energy and hence is lost from the circuit. In a purely capacitive circuit, however, the energy is not converted to heat. Instead, energy is *stored* in the capacitor as electric potential energy, and this stored energy is later returned to the AC source. Hence, in a capacitor circuit we should really speak of the power P that is *delivered to* the capacitor. When P is positive, energy is being added to the capacitor, whereas when P is negative, energy is being removed from the capacitor. That is possible because a capacitor can store energy, but a resistor cannot.

22.4 | AC CIRCUITS WITH INDUCTORS

We next consider a circuit consisting of an AC generator and a single inductor as shown in Figure 22.14A. Our job again is to calculate the current in this circuit. In Chapter 21, we learned that the voltage across an inductor is proportional to the rate of change in the current. If L is the inductance and ΔI is the change in the current over a short time interval Δt, the voltage drop across the inductor is

$$V = L\frac{\Delta I}{\Delta t} \qquad (22.24)$$

This voltage is also equal to output of the AC voltage source attached to the inductor, so $V = V_{max} \sin(2\pi ft)$. In words, Equation 22.24 states that V is proportional to the slope of the I–t relation. Because we are given V (Fig. 22.14B), we can infer how I varies with t; to do so, we must construct an I–t plot with a slope that satisfies Equation 22.24. We have sketched the answer qualitatively in Figure 22.14C. Here, $V = 0$ at times when the slope of I is zero, and V is largest and positive when I has its largest positive slope.

From Figure 22.14, we can see that V oscillates according to a sine function, whereas I oscillates with time according to a cosine function

$$I = -I_{max} \cos(2\pi ft) \qquad (22.25)$$

We can use the trigonometric identity $\cos \alpha = -\sin(\alpha - \pi/2)$ to write Equation 22.25 as

$$I = I_{max} \sin(2\pi ft - \pi/2)$$

$$I = I_{max} \sin(2\pi ft + \phi) \qquad \text{with } \phi = -\pi/2 \qquad (22.26)$$

Figure 22.15 shows the current and voltage relationship for this inductor circuit in a phasor diagram. The current phasor makes an angle of $\phi = \pi/2 = -90°$ with the voltage, with the negative sign arising from Equation 22.26 and from our standard convention of measuring positive angles in the counterclockwise direction in the coordinate plane. In words, we say that the current is 90° out of phase from the voltage. This phasor diagram is very similar to what we found with the capacitor. For both capacitors and inductors, the current is 90° out of phase with the voltage. For a capacitor the phase angle is $\phi = +90°$, whereas for an inductor it is $\phi = -90°$.

The current and voltage are out of phase in an inductor because the voltage across it is proportional to the rate of change of the current, $V = L(\Delta I/\Delta t)$ (Eq. 22.24). This is in contrast to the case with a resistor, for which the current is proportional to the voltage (Eq. 22.2) so that I and V are in phase with each other.

Equation 22.26 and the phasor diagram in Figure 22.15 give the phase of the current through an inductor. An exact calculation of the peak current I_{max} shows that

$$I_{max} = \frac{V_{max}}{X_L} \tag{22.27}$$

where X_L is called the **reactance** of the inductor and is given by

$$X_L = 2\pi f L \tag{22.28}$$

The relation between the peak current and voltage for an inductor is very similar to the corresponding relations for a capacitor (Eq. 22.21) and a resistor (Eq. 22.2). Again, X_L is a measure of how strongly an inductor "resists" a current. Just as for a capacitor, the reactance of an inductor depends on frequency. Unlike in a capacitor, however, the reactance of an inductor X_L becomes larger as the frequency is increased. (Recall from Eq. 22.22 that X_C decreases at high frequencies.) This behavior of X_L can be traced back to the basic properties of an inductor. According to Lenz's law (Chapter 21), an inductor opposes changes in the flux through its coils. Increasing the frequency increases the rate of change of the flux and hence increases this opposing emf. For this reason, the current through an inductor decreases at high frequencies.

We have collected the results for the peak current in AC circuits with a resistor, a capacitor, or an inductor in Table 22.2.

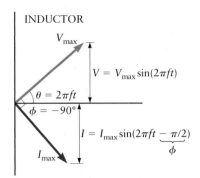

INDUCTOR

Figure 22.15 Phasor diagram of the voltage and current in an AC circuit with an inductor. The current through an inductor is 90° out of phase with the voltage.

CONCEPT CHECK 22.5 | Reactance of an Inductor

An inductor has a reactance $X_L = 2000 \ \Omega$ when measured at a frequency of 1500 Hz. The frequency is then changed, and the reactance is found to be 6000 Ω. Is the new frequency (a) 500 Hz, (b) 1000 Hz, (c) 3000 Hz, or (d) 4500 Hz?

Table 22.2 Properties of AC Circuits

CIRCUIT ELEMENT	RESISTANCE OR REACTANCE TO CURRENT FLOW	CURRENT THROUGH CIRCUIT ELEMENT	PHASE RELATION BETWEEN CURRENT AND VOLTAGE	AVERAGE POWER
Resistor —⌁⌁—	R	$I = V/R$	I and V are in phase	$P_{ave} = V_{rms}I_{rms} = \dfrac{V_{max}I_{max}}{2}$
Capacitor —⊣⊢	$X_C = \dfrac{1}{2\pi f C}$	$I = V/X_C$	I and V are out of phase by 90°	$P_{ave} = 0$
Inductor —◟◟◟◟—	$X_L = 2\pi f L$	$I = V/X_L$	I and V are out of phase by 90°	$P_{ave} = 0$

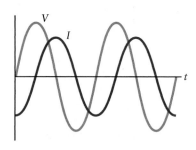

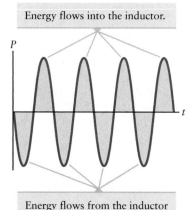

Energy flows into the inductor.

Energy flows from the inductor back to the generator.

Figure 22.16 The instantaneous power P delivered to the inductor is the product of V (a sine wave) and I (a cosine); the result is another sine wave, and the power delivered to an inductor in an AC circuit varies sinusoidally with time. At some portions of the cycle P is positive, whereas at other times it is negative. The average power over one complete cycle is zero.

Power in a Circuit Containing an Inductor

We can calculate the instantaneous power in a circuit containing an inductor using the result for the current in Equation 22.25 together with the general result for power in a circuit element $P = VI$. We find

$$P = VI = -V_{max}I_{max} \sin(2\pi ft)\cos(2\pi ft) \tag{22.29}$$

This result is shown in Figure 22.16; it is very similar to our result for the power in a circuit with a capacitor (Fig. 22.13). The power delivered to the inductor oscillates between positive and negative values, and the average power over a complete cycle is zero. The generator delivers energy to the inductor during part of the AC cycle, and the inductor delivers an equal amount of energy back to the generator during other parts of the cycle. Energy is stored in the inductor as magnetic potential energy by virtue of the inductor's magnetic field.

22.5 | *LC* CIRCUITS

We have analyzed the simplest AC circuits containing only a single resistor, capacitor, or inductor. Most useful circuits contain multiple circuit elements, so now let's see what happens when resistors, capacitors, and inductors are connected together in different ways. In this section, we consider one particularly important example, the *LC* circuit.

Figure 22.17 shows the simplest possible *LC* circuit, consisting of a single inductor connected to a single capacitor. There is no AC generator in this circuit, but some excess charge is placed on the capacitor at $t = 0$ while the current is zero at that moment. That is, we start with a charge $+q$ on one plate and $-q$ on the other (Fig. 22.17A); a capacitor with a nonzero charge stores an energy $PE_{cap} = q^2/(2C)$, so we have "placed" this amount of energy in the circuit. The charge on the capacitor produces a voltage $V_C = q/C$ across the capacitor, and since the capacitor's leads are connected directly to the inductor, V_C is also the voltage across the inductor. In qualitative terms, the capacitor tries to "push" a current through the inductor as the charge on one capacitor plate attempts to flow through the inductor to the other plate. An inductor always resists changes in current, however; it does so through an induced emf and Lenz's law. Hence, at $t = 0$ there is an induced emf across the inductor opposing any current, and $I = 0$ at that instant.

This state of affairs with $I = 0$ is only maintained for an instant at $t = 0$; the current is nonzero after $t = 0$, just as for *RL* circuits (Chapter 21). So, at times $t > 0$ charge moves from one capacitor plate to the other and current passes through the inductor. Eventually, the charge on each capacitor plate falls to zero (Fig. 22.17B) and the corresponding voltage across the capacitor is $V_C = q/C = 0$. At this point you might expect that the current through the inductor would fall to zero, but the inductor again *opposes change* in the current, so the induced emf now acts to *maintain* the current at a nonzero value! This current continues to transport charge from one capacitor plate to the other even after reaching the point when $q = 0$, causing the capacitor's charge and voltage to reverse sign (Fig. 22.17C). The voltage across the capacitor then peaks at a negative value, leading to a current in the opposite direction through the inductor (Fig. 22.17D) as charge again moves from one plate to the other. Eventually, the charge on the capacitor plates returns to the original values found at $t = 0$ (Fig. 22.17E). The circuit has now returned to its original state (compare parts A and E of Fig. 22.17), and the process starts over again. The voltage and current in this circuit thus *oscillate* between positive and negative values. In fact, this circuit behaves as a *simple harmonic oscillator* with the current and charge given by

$$q = q_{max} \cos(2\pi ft) \quad \text{and} \quad I = I_{max} \sin(2\pi ft) \tag{22.30}$$

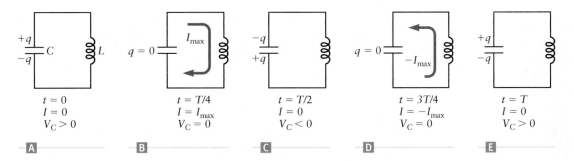

Figure 22.17 In an *LC* circuit with an initial charge, the current in the circuit and the voltage across the capacitor oscillate with a period *T*. This circuit behaves as a simple harmonic oscillator.

These results are identical to the time dependence of the position and velocity of a simple harmonic oscillator such as a mass on a spring as we studied in Chapter 11 (Eqs. 11.9 and 11.10).

Energy Conservation and the Oscillations in an *LC* Circuit

Capacitors and inductors can both store energy; the energy in a capacitor is stored in its electric field and depends on the charge, $PE_{cap} = \frac{1}{2}(q^2/C)$, while the energy in an inductor is stored in its magnetic field and depends on the current, $PE_{ind} = \frac{1}{2}LI^2$. Hence, as the charge and current oscillate in Figure 22.17, the energies stored in the capacitor and inductor also oscillate. We can compute these energies using the results for q and I from Equation 22.30 and find

$$PE_{cap} = \frac{1}{2}\frac{q^2}{C} = \frac{1}{2}\frac{q_{max}^2}{C}\cos^2(2\pi ft) \qquad (22.31)$$

and

$$PE_{ind} = \frac{1}{2}LI^2 = \frac{1}{2}LI_{max}^2\sin^2(2\pi ft) \qquad (22.32)$$

These results are shown in Figure 22.18. The energy oscillates back and forth between the capacitor and its electric field (PE_{cap}) and the inductor and its magnetic field (PE_{ind}). This behavior is very similar to that of a mechanical simple harmonic oscillator such as a pendulum or a mass on a spring. In a mechanical oscillator, the energy oscillates back and forth between kinetic and potential energy.

From the principle of conservation of energy, we expect that the *total* energy in the circuit $PE_{cap} + PE_{ind}$ must be constant. Hence, the maximum value of PE_{cap} must equal the maximum value of PE_{ind} as shown in Figure 22.18. Then, from Equations 22.31 and 22.32,

$$(PE_{cap})_{max} = (PE_{ind})_{max}$$

$$\frac{1}{2}\frac{q_{max}^2}{C} = \frac{1}{2}LI_{max}^2$$

$$I_{max} = \frac{1}{\sqrt{LC}}q_{max} \qquad (22.33)$$

This equation tells us how the amplitudes of the current and charge oscillations in an *LC* circuit are related.

Frequency of Oscillations in an *LC* Circuit

Equation 22.33 relates the amplitudes of the current and charge oscillations in an *LC* circuit. We also want to know the *frequency* of these oscillations. In the *LC*

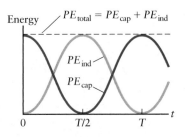

Figure 22.18 Plots of the energy in an *LC* circuit as a function of time, where PE_{cap} is the electric potential energy stored in the capacitor and PE_{ind} is the energy stored in the magnetic field of the inductor. The total energy $PE_{cap} + PE_{ind}$ is constant.

Figure 22.19 Metal detectors use changes in the resonant frequency of an LC circuit to sense when a piece of metal is nearby.

Frequency of an LC oscillator

circuit in Figure 22.17, the instantaneous voltages across the capacitor and inductor are always equal. Hence, the magnitude of the voltage across the capacitor must equal the magnitude of the voltage across the inductor. In terms of the reactances, these voltages are $V_C = IX_C$ (Eq. 22.21) and $V_L = IX_L$ (Eq. 22.27). Equating these magnitudes leads to

$$|V_C| = |IX_C| = |V_L| = |IX_L|$$

The reactances are always positive, so we can cancel the factors of I to get

$$X_C = X_L \qquad (22.34)$$

This condition was derived by assuming the current in the LC circuit is oscillating and hence applies *only* at the oscillation frequency. In words, Equation 22.34 says that at this oscillation frequency the reactances of the capacitor and inductor are equal.

Inserting the results for X_C and X_L from Equations 22.22 and 22.28 leads to

$$X_C = \frac{1}{2\pi f C} = X_L = 2\pi f L$$

The frequency here is the **resonant frequency** of the circuit. It is the oscillation frequency of the charge and current in Figure 22.17. Solving for this frequency, which we now denote as f_{res}, we get

$$f_{res} = \frac{1}{2\pi\sqrt{LC}} \qquad (22.35)$$

The oscillation frequency of an LC circuit thus depends on the values of L and C.

This frequency plays an important role in many applications of AC circuits. For example, a metal detector uses changes in the inductance of an LC circuit to sense the presence of a metal object. A metal detector used by beachcombers (Fig. 22.19) has a large coil of wire for its inductor. When the coil is placed near a metal object, the inductance of the coil changes, changing the resonant frequency of the circuit. The electronic unit of the metal detector is designed to register an audible signal when that happens. Metal detectors used for screening people at airports work in the same way.

22.6 | RESONANCE

Our analysis of an LC circuit in Section 22.5 explains how this circuit can act as a simple harmonic oscillator. Let's now consider a more realistic circuit containing an inductor, capacitor, and resistor along with an AC voltage source (Fig. 22.20). Our goal is to understand how the current in this LCR circuit depends on the amplitude of the AC voltage and on frequency. We will also show how phasors can be used to analyze the behavior of an AC circuit.

To analyze the LCR circuit in Figure 22.20, we appeal to Kirchhoff's loop rule (Chapter 19) and move around the circuit, adding up the voltage changes across the AC voltage source (V_{AC}), the resistor (V_R), the capacitor (V_C), and the inductor (V_L). According to the loop rule, the total change in the potential must be zero, so we have

$$V_{AC} = V_L + V_C + V_R \qquad (22.36)$$

The simple form of Equation 22.36 hides one crucial point—these voltages are not all in phase—so we must take account of both their magnitudes and the phase angles when we add them.

Figure 22.21 shows a series of phasor diagrams for the current in our LCR circuit and the voltage across each circuit element. The inductor, capacitor, and resistor are in series, so the current is the same through each one, and all the current phasors in Figure 22.21 have the same orientation. Figure 22.21A is for the

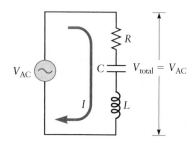

Figure 22.20 An LCR circuit.

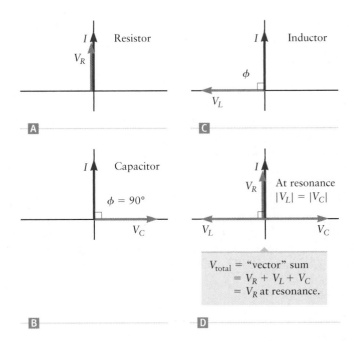

resistor; in this case, the current and voltage are in phase with each other and the phasor arrows for I and V_R are parallel. This diagram is similar to Figure 22.9B except that we have placed these arrows along the vertical axis. Parts B and C of Figure 22.21 show the corresponding phasor diagrams for the voltages across the capacitor and inductor. As we saw in Sections 22.3 and 22.4, these voltages are 90° out of phase with the current. Figure 22.21D shows all three voltage phasors; the total voltage across all three circuit elements is the sum of these three phasors. The key point is that the phasors for V_L and V_C are in *opposite directions*, so they tend to cancel. In fact, we found in deriving Equation 22.34 that the two phasors V_C and V_L have the same magnitude (i.e., the same "length") when the frequency equals the resonant frequency f_{res}. Hence, at this frequency the voltages across the capacitor and inductor *cancel*. Only the resistor is left to "resist" the flow of current at this frequency. This perfect cancellation only occurs at f_{res}, so the current is highest when the frequency equals f_{res}.

Figure 22.22 shows how the current through an *LCR* circuit varies with frequency. Here, I is largest at f_{res} and decreases at both higher and lower frequencies. This is an example of **resonance**, and it is very similar to the behavior of mechanical harmonic oscillators that we discussed in Chapter 11 (Fig. 11.28). By itself, an *LC* circuit acts as a simple harmonic oscillator with a resonant frequency given by Equation 22.35. Adding a resistor to the circuit (Fig. 22.20) adds damping and gives a damped harmonic oscillator, but the system still oscillates at the frequency f_{res}. When this oscillator is "driven" by attaching it to an AC voltage source, the resulting current depends on frequency. The current is largest when the frequency of the AC source matches the resonant frequency of the circuit, hence the peak in I in Figure 22.22 when $f = f_{res}$.

Applications of Resonance in Electronic Circuits

The resonant behavior of an *LCR* circuit is used in radios, televisions, and other similar applications. The input to a radio comes from an antenna that "picks up" signals from many different stations at many different frequencies. This composite signal is the AC voltage source for an *LCR* circuit. This circuit responds most strongly—that is, it gives the largest current—at frequencies that are near its resonant frequency. When you tune your radio, you are changing the value of the capacitance in an *LCR* circuit so that the resonant frequency matches the frequency of the station you want to listen to. The resulting current is then dominated by the

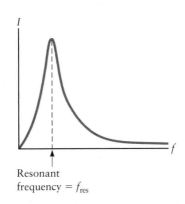

Figure 22.22 Current I in an *LCR* circuit as a function of frequency.

frequency of the desired station; this current is amplified and ultimately sent to your speakers or earphones. In this way, *LCR* circuits are used to construct devices that are frequency selective.

EXAMPLE 22.5 Resonant Frequency of an *LC* Circuit

AM radio stations broadcast at frequencies in a range centered near 1.0 MHz (1.0×10^6 Hz). If the inductance in an *LC* circuit is $L = 50$ μH (5.0×10^{-5} H), what is the value of the capacitance that gives a resonant frequency at the middle of the AM range?

RECOGNIZE THE PRINCIPLE

We want to choose the value of the capacitance so that the resonant frequency equals the AM frequency $f = 1.0 \times 10^6$ Hz. The purpose of this problem is to see what particular values of L and C are needed to get the desired resonant frequency. The answer will tell us if an *LC* circuit is really practical for this application.

SKETCH THE PROBLEM

No figure is needed.

IDENTIFY THE RELATIONSHIPS

The resonant frequency of an *LC* circuit is given by (Eq. 22.35)

$$f_{res} = \frac{1}{2\pi\sqrt{LC}}$$

Rearranging to solve for C, we get

$$\sqrt{LC} = \frac{1}{2\pi f_{res}}$$

$$LC = \frac{1}{4\pi^2 f_{res}^2}$$

$$C = \frac{1}{4\pi^2 f_{res}^2 L}$$

SOLVE

Inserting the given values of f_{res} and L yields

$$C = \frac{1}{4\pi^2(1.0 \times 10^6 \text{ Hz})^2(5.0 \times 10^{-5} \text{ H})} = \boxed{5.1 \times 10^{-10} \text{ F}}$$

What does it mean?
This value for the capacitance is easily achieved with typical capacitors (see Example 22.4), so an *LC* circuit with a resonant frequency in the AM band is quite practical. Similar circuits are used in televisions, MP3 players, cell phones, and other applications in which one needs to distinguish between electrical signals according to their frequency.

22.7 | AC CIRCUITS AND IMPEDANCE

We introduced phasors as a convenient way to represent the phase relationship of current and voltage in a capacitor or inductor. We also used phasors in our discussion of resonance in *LCR* circuits. Now let's use them to find the current in some other AC circuits. Figure 22.23A shows an AC voltage source connected to an inductor. A typical inductor is constructed using a coil of wire, and real wires have

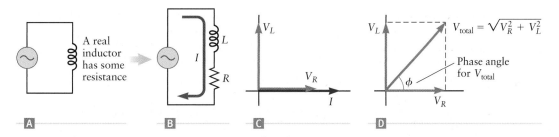

Figure 22.23 Real inductors have at least a small amount of resistance, so 🅐 an AC circuit with an inductor can be modeled as 🅑 a circuit with an ideal inductor in series with a resistor. 🅒 Phasor diagram for the current in the circuit and the voltages across the resistor and the inductor. 🅓 The total voltage across the inductor and the resistor is the phasor sum of the voltages in part C.

a nonzero resistance. A real inductor can therefore be modeled as an ideal inductor L in series with a resistor R as sketched in Figure 22.23B. We can calculate the current I in this circuit using phasors. The voltage across the inductor is $V_L = IX_L$ (Eq. 22.27), and the voltage across the resistor is $V_R = IR$ (Eq. 22.2). These two circuit elements are connected in series, so their currents are equal and the sum of their voltages equals the AC source voltage. It is tempting to just add our expressions for V_L and V_R, but that would *not* give the correct result. The voltages V_L and V_R must be added as phasors, so we must consider their phases when we take this sum.

Figure 22.23C is a phasor diagram showing the current I along with the voltages across L and R. For simplicity, we have placed the current phasor along the horizontal axis; this phasor describes the current through both the inductor and resistor because they are in series. The voltage across the resistor is then a phasor that also lies along the horizontal direction. (Compare with Fig. 22.9B.) The voltage across the inductor makes a phase angle $\phi = 90°$ with respect to the current in the "counterclockwise" sense (as in Figure 22.15), so the phasor representing V_L is along the upward vertical direction in this diagram. The total voltage V_{total} across the inductor and resistor combined is the sum of these two voltage phasors. This sum is sketched in Figure 22.23D, which shows that the total voltage has an amplitude

$$V_{\text{total}} = \sqrt{V_R^2 + V_L^2} \tag{22.37}$$

Inserting our results for V_R and V_L gives

$$V_{\text{total}} = \sqrt{V_R^2 + V_L^2} = \sqrt{(IR)^2 + (IX_L)^2} = I\sqrt{R^2 + X_L^2}$$

$$V_{\text{total}} = I\sqrt{R^2 + (2\pi fL)^2} \tag{22.38}$$

We now define a quantity called the **impedance** Z of the circuit through the relation

$$V_{\text{total}} = IZ \tag{22.39}$$

Impedance of an AC circuit

From Equation 22.38, we find

$$Z = \sqrt{R^2 + (2\pi fL)^2} \quad \text{(impedance for an } RL \text{ circuit)} \tag{22.40}$$

The impedance is a measure of how strongly a circuit "impedes" current. For a single resistor, the impedance is simply equal to the resistance. Likewise, for a single capacitor or inductor, the impedance is equal to the reactance of the capacitor or inductor. For combinations of these circuit elements, the impedance is given by a more complicated combination of these quantities. The result in Equation 22.40 applies only for a circuit consisting of a single resistor in series with a single inductor. The impedance of other circuits can be calculated by adding phasors, as we'll see in other examples below.

Figure 22.23D contains another important lesson. In all the AC circuits we have analyzed so far, the phase angle between the current and voltage was either zero or

$\pm\pi/2$. For an LR circuit, the phase angle has a value between zero and $\pi/2$, depending on the relative values of V_R and V_L, which themselves depend on R, L, and f. For most AC circuits, the phase angle depends on frequency and also on the specific values of the resistances, inductances, and capacitances in the circuit. We'll explore this topic further in Problems 52 and 53.

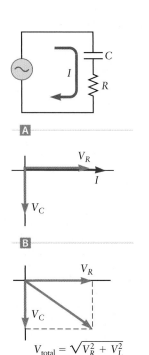

Figure 22.24 Example 22.6.

| EXAMPLE 22.6 | Phasor Analysis of an *RC* Circuit

Consider the circuit in Figure 22.24, with a capacitor in series with a resistor. Given that the frequency of the AC source is f, find an expression for the impedance of this circuit. Express your answer in terms of C, R, and f.

RECOGNIZE THE PRINCIPLE

The impedance is proportional to the total voltage through Equation 22.39. To find the total voltage across the capacitor plus the resistor, we must add their voltages as phasors. The phasor diagrams in this case are similar to the case with an inductor (Fig. 22.23) except that the phasor representing V_C is downward along the vertical axis (Fig. 22.24B); compare this figure with the phasor diagram for a capacitor alone in Figure 22.11.

SKETCH THE PROBLEM

Figure 22.24 shows the circuit along with the corresponding phasor diagrams for the current and voltage across the capacitor and resistor.

IDENTIFY THE RELATIONSHIPS

We add the phasors representing V_R and V_C as shown in Figure 22.24C. The phasor arrows form a right triangle, so we have

$$V_{total} = \sqrt{V_R^2 + V_C^2}$$

Inserting $V_R = IR$ (Eq. 22.2) and $V_C = IX_C = I/(2\pi f C)$ (Eqs. 22.21 and 22.22) gives

$$V_{total} = \sqrt{V_R^2 + V_C^2} = \sqrt{(IR)^2 + \left(\frac{I}{2\pi f C}\right)^2} = I\sqrt{R^2 + \left(\frac{1}{2\pi f C}\right)^2}$$

SOLVE

Comparing this expression with the definition of impedance $V = IZ$ (Eq. 22.39), we find that for this circuit the impedance is

$$Z = \sqrt{R^2 + \left(\frac{1}{2\pi f C}\right)^2}$$

What does it mean?
The impedance of this circuit decreases at high frequencies, so for a given value of the AC voltage, the current is larger at high frequencies than at low frequencies. This is because the reactance X_C of the capacitor becomes smaller at high frequencies, reducing the impedance of the entire circuit.

Impedance of an *LCR* Circuit

A phasor analysis for an *LCR* circuit is shown in Figure 22.21D at the resonant frequency only. We can find the impedance for the general case using a method like that in Figures 22.23 and 22.24. Beginning anew in Figure 22.25A, we again place the current phasor along the horizontal axis, so the phasor for the resistor's voltage is also in the horizontal direction. The phasor representing the voltage across the inductor is then upward along the vertical axis, and the phasor for the capacitor's voltage is downward. Collecting the horizontal and vertical components in Figure 22.25B, we add them to get the total voltage across the circuit:

$$V_{\text{total}} = \sqrt{(IR)^2 + I^2(X_L - X_C)^2} = \sqrt{(IR)^2 + I^2\left(2\pi fL - \frac{1}{2\pi fC}\right)^2}$$

$$V_{\text{total}} = I\sqrt{R^2 + \left(2\pi fL - \frac{1}{2\pi fC}\right)^2}$$

The impedance is defined through $V_{\text{total}} = IZ$, so for this circuit we get

$$Z = \sqrt{R^2 + \left(2\pi fL - \frac{1}{2\pi fC}\right)^2} \qquad (22.41)$$

If the voltage provided by the AC generator is $V_{\text{total}} = V_{\max} \sin(2\pi ft)$, the amplitude of the current is $I_{\max} = V_{\max}/Z$.

Impedance and Resonance in an *LCR* Circuit

Let's examine more carefully our result for the impedance of a general *LCR* circuit (Eq. 22.41) and see how it and the current in the circuit behave near resonance.

From the general definition of impedance, the amplitude of the current is $I_{\max} = V_{\max}/Z$. Since the impedance varies with frequency, the current amplitude is also frequency dependent and the maximum value of the current occurs when the impedance Z is a minimum. This minimum in Z occurs when the factors involving L and C cancel. We have

$$Z = \sqrt{R^2 + \left(2\pi fL - \frac{1}{2\pi fC}\right)^2}$$

The minimum impedance thus occurs when

$$2\pi fL - \frac{1}{2\pi fC} = 0$$

Solving for the frequency gives

$$f = \frac{1}{2\pi\sqrt{LC}}$$

which is precisely the resonant frequency we found for an *LC* circuit in Section 22.5 (Eq. 22.35). The maximum current thus occurs at the resonant frequency, the frequency at which the *LCR* circuit responds most strongly to an applied AC voltage.

CONCEPT CHECK 22.6 | Phase Angle in an *LCR* Circuit at Resonance

Is the phase angle between the current and voltage in an *LCR* circuit at resonance (a) 0, (b) 90°, or (c) 180°? (*Hint*: Consider the phasor diagrams in Figs. 22.21 and 22.25.)

22.8 | FREQUENCY-DEPENDENT BEHAVIOR OF AC CIRCUITS: A CONCEPTUAL RECAP

We can make better sense of the circuit theory in this chapter, with its many new concepts including reactance and impedance, if we have a clear qualitative picture of why the frequency is so important in AC circuits.

Circuit Behavior at Low and High Frequencies

Resistors in an AC circuit behave very much like resistors in a DC circuit. For both types of circuits, the current in a resistor is related to the voltage by $I = V/R$ independent of the frequency, and the current is always *in phase* with the voltage (Fig. 22.9). In contrast, the current through a capacitor or an inductor is *frequency dependent*, a key aspect of AC circuits that makes them useful in applications such

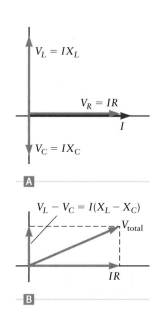

Figure 22.25 **A** Phasor diagram of the current in a series *LCR* circuit and the voltages across all three circuit elements for the general case in which $X_L \neq X_C$. Notice how the voltages across the capacitor and inductor are out of phase and hence in opposite directions. (Compare with Fig. 22.21.) **B** Phasor sum of the voltages.

Table 22.3 Qualitative Behavior of Capacitors and Inductors at Low and High Frequencies

CIRCUIT ELEMENT	RESISTANCE OR REACTANCE	BEHAVIOR AT VERY LOW FREQUENCIES	BEHAVIOR AT VERY HIGH FREQUENCIES
Resistor —⩗—	R	Independent of frequency	Independent of frequency
Capacitor ⊣⊢	$X_C = \dfrac{1}{2\pi f C}$	$X_C \approx \infty$; very little current passes	$X_C \approx 0$; current passes freely
Inductor —⬯⬯⬯—	$X_L = 2\pi f L$	$X_L \approx 0$; current passes freely	$X_L \approx \infty$; very little current passes

as the radio tuner discussed in Section 22.6 and the filter circuits described later in this section.

The frequency-dependent behavior of AC circuits is due to the frequency dependence of the reactances X_C and X_L. For a capacitor,

$$I_{max} = \frac{V_{max}}{X_C} \quad \text{(for a capacitor)} \tag{22.42}$$

and X_C is a measure of how strongly the capacitor "resists" current. When the frequency is low, the oscillation period is long and more charge is "pushed" on and off a capacitor during each AC cycle. This larger charge produces a larger electric field in opposition; hence, the reactance X_C is largest at low frequencies. From Equation 22.42, the current through a capacitor is *smallest* at *low frequencies*. At very low frequencies, a capacitor thus approaches an "open circuit" and passes very little current, whereas at very high frequencies, a capacitor resembles a short circuit, passing current like a simple ideal wire with no resistance.

For an inductor, we have

$$I_{max} = \frac{V_{max}}{X_L} \quad \text{(for an inductor)} \tag{22.43}$$

and X_L is a measure of how strongly an inductor "resists" current. Inductance is a result of the laws of Faraday and Lenz because an inductor opposes changes in current. The rate of change of I is largest at high frequencies, making the reactance X_L largest at high frequencies. From Equation 22.43, this fact makes the current through an inductor *smallest* at *high frequencies*. At very low frequencies, an inductor thus passes current freely, like a simple wire with no resistance, whereas at very high frequencies, an inductor approaches an open circuit and passes little or no current.

These qualitative observations are summarized in Table 22.3. We can use these results to get a general sense of the behavior of a circuit that contains multiple circuit elements. Consider the circuit in Figure 22.26A, an RL circuit for which we indicate the "input" and "output" voltage signals. Although we show an AC generator providing the input voltage V_{in}, this voltage might instead be provided by an antenna (as in a radio or TV) or another AC circuit. The voltage across the inductor is the "output" voltage V_{out} and is typically connected to another portion of the circuit that is not shown here. Let's use our qualitative rules from Table 22.3 to estimate the output voltage at very low and at very high frequencies.

When the input frequency is very low, the reactance of the inductor is very small. The inductor thus acts as a simple wire, so we have redrawn the circuit with the inductor replaced by a wire in Figure 22.26B. From our work with resistors and wires in Chapter 19, we know that there is zero voltage drop across the ends of the wire, so V_{out} in Figure 22.26B is zero. Hence, even though the input voltage is generally not zero, the output voltage of this circuit is very small at low frequencies.

In the limit of very high frequencies, an inductor acts as an open circuit (Table 22.3) as sketched in Figure 22.26C. Because an open circuit passes no current, the current through the resistor and the voltage drop across it are approximately zero.

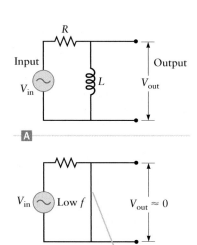

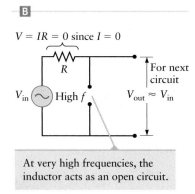

Figure 22.26 Qualitative analysis of A an LR circuit. B At very low frequencies, the reactance of the inductor X_L is very small and the inductor acts as a short circuit (i.e., a simple wire). C At very high frequencies, X_L is very large, so the inductor acts as an open circuit. The original circuit thus removes low frequencies from the output and passes high frequencies.

The output voltage is then equal to the input voltage, $V_{out} = V_{in}$. We thus find that the output voltage of the circuit in Figure 22.26A is zero at low frequencies but equal to the input voltage at high frequencies. This circuit thus acts as a *high-pass filter*, allowing high-frequency signals to pass through to the output while low frequencies are blocked. In the end-of-chapter problems, we'll explore how other combinations of inductors, resistors, and capacitors can be used to make other types of filters (e.g., low-pass filters).

Insight 22.1

APPLICATIONS OF A LOW-PASS FILTER

A low-pass filter can be constructed by replacing the inductor in Figure 22.26A with a capacitor (see Example 22.7). A low-pass filter is used in radios and MP3 players. A music (audio) signal often contains "static," which comes from unwanted high-frequency components added to the desired music. These high frequencies can be filtered out using a low-pass filter.

| EXAMPLE 22.7 | ℝ An *RC* Circuit at Low and High Frequencies |

Figure 22.27A shows an *RC* circuit. Use the qualitative frequency-dependent behavior of a capacitor as given in Table 22.3 to estimate the output voltage at (a) very low and (b) very high frequencies.

RECOGNIZE THE PRINCIPLE

The qualitative behavior at low and high frequencies can be found using the frequency dependence of the reactance of a capacitor (Table 22.3), which we can use to draw "equivalent circuits" for low and high frequencies. At low frequencies, the capacitor is replaced by an open circuit (Fig. 22.27B), whereas at high frequencies, the capacitor is replaced by a short circuit (Fig. 22.27C).

SKETCH THE PROBLEM

Figure 22.27 shows the circuit, along with equivalent circuits at low and high frequencies.

IDENTIFY THE RELATIONSHIPS AND SOLVE

(a) At very low frequencies, the capacitor acts approximately as an open circuit (Fig. 22.27B) and the current is very small. The output voltage then equals the input voltage, and we have

$$V_{out} = \boxed{V_{in}} \quad \text{at low frequencies}$$

(b) At very high frequencies, the capacitor acts as a short circuit, so we replace it by a wire in Figure 22.27C. The output voltage is equal to the voltage across this short circuit, which is zero. Hence,

$$V_{out} = \boxed{0} \quad \text{at high frequencies}$$

What does it mean?
This circuit prevents high frequencies from reaching the output (since $V_{out} = 0$ at high frequencies), so this circuit acts as a *low-pass filter*.

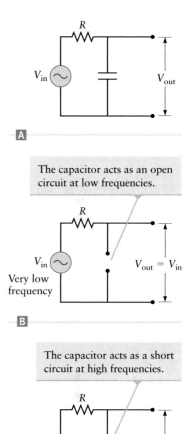

Figure 22.27 Example 22.7. Qualitative analysis of the frequency dependence of an *RC* circuit.

When Are We at Low or High Frequency?

The results from Table 22.3 are extremely useful in gaining a qualitative understanding of an AC circuit's behavior at low and high frequencies, but how do we know when a particular frequency is low enough or high enough to apply the limiting values of X_C and X_L from Table 22.3? For an *RL* circuit (Fig. 22.26), the input frequency must be compared with the corresponding *RL* time constant. The time constant of an *RL* circuit is given by $\tau_{RL} = L/R$ (Eq. 21.27). Qualitatively, you can think of this time constant as the period of an oscillation and define a corresponding frequency by

$$f_{RL} = \frac{1}{\tau_{RL}} = \frac{R}{L} \tag{22.44}$$

The high-frequency limit in Table 22.3 applies when the input or source frequency in an *RL* circuit is much greater than f_{RL}, whereas the low-frequency limit corresponds to frequencies much lower than f_{RL}. For qualitative estimates, any frequency

FILTERS AND STEREO SPEAKERS

Many stereo speakers actually contain two separate speakers. One speaker (the tweeter) is designed to perform well at high frequencies, whereas the other (the woofer) is designed for low frequencies. The input signal to the speaker system passes through an AC circuit called a crossover network, which is a combination of low-pass and high-pass filters. The output of the high-pass filter is sent to the tweeter, while the low-pass filter is connected to the woofer. In this way, the signal sent to each speaker contains only those frequencies for which the speaker is most efficient.

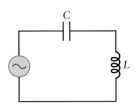

Figure 22.28 Concept Check 22.7.

greater than about $10 \times f_{RL}$ falls into the high-frequency limit, whereas the low-frequency limit is below about $f_{RL}/10$.

The approach is similar for an AC circuit containing a capacitor and a resistor. In this case, we have an RC time constant given by $\tau_{RC} = RC$ (Eq. 19.31), which leads to a corresponding frequency

$$f_{RC} = \frac{1}{\tau_{RC}} = \frac{1}{RC} \qquad (22.45)$$

The high- and low-frequency limits for a capacitor circuit then apply when the input or source frequency is much higher or much lower[2] than f_{RC}.

As a final case, we consider an AC circuit containing a capacitor and an inductor. In this case, the resonant LC frequency $f_{res} = 1/(2\pi\sqrt{LC})$ (Eq. 22.35) determines the boundary between the high- and low-frequency limits in Table 22.3.

CONCEPT CHECK 22.7 | Frequency Dependence of an LC Circuit

Consider the LC circuit in Figure 22.28. The voltage from the AC source is $V = V_{max} \sin(2\pi ft)$, with V_{max} held fixed while the frequency f is adjustable. Which of the following statements is correct?

(a) At very high frequencies, the voltage across the inductor is greater than the voltage across the capacitor.

(b) At very high frequencies, the voltage across the inductor is less than the voltage across the capacitor.

(c) The voltage across the inductor is equal to the voltage across the capacitor at all frequencies.

22.9 | TRANSFORMERS

Transformers are devices that can either increase or decrease the amplitude of an applied AC voltage. The ability to "transform" an AC voltage from small to large, or from large to small, is used by electric companies to transmit electrical energy efficiently over long distances. Transformers are also found in many household devices, including televisions and radios, and are used in the "power bricks" that connect your laptop computer to a wall socket (Fig. 22.29).

A simple transformer consists of two solenoidal coils with the loops arranged so that all or most of the magnetic field lines and flux generated by one coil pass through the other coil. This can be accomplished in several ways, two of which are illustrated in Figure 22.30. In Figure 22.30A, two coils are placed end to end, whereas Figure 22.30B shows one coil wrapped tightly around the other. The wires are covered with a nonconducting layer ("insulation") so that current cannot flow directly from one coil to the other. Even so, an AC current in one coil will induce an AC voltage across the other one. According to Faraday's law, the magnitude of the voltage drop across a coil is related to the magnetic flux through the coil by

$$V = \frac{\Delta \Phi}{\Delta t} \qquad (22.46)$$

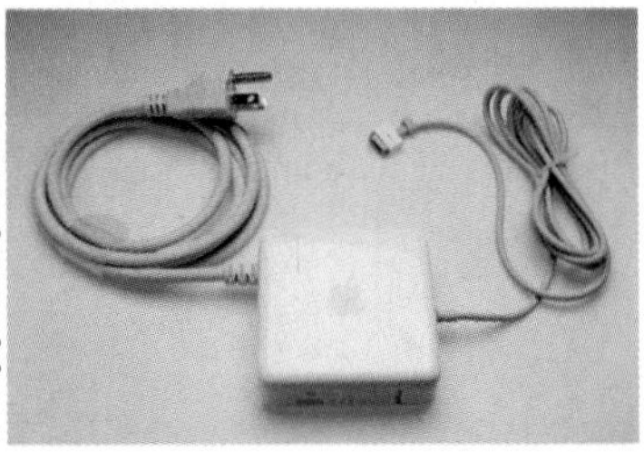

In a transformer, an AC voltage source is typically attached to one of the coils—denoted as the "input" coil—and the other is called the "output" coil. Equation 22.46 applies for both coils, so we can write

$$V_{in} = \frac{\Delta \Phi_{in}}{\Delta t} \qquad V_{out} = \frac{\Delta \Phi_{out}}{\Delta t} \qquad (22.47)$$

The coils in Figure 22.30 are designed so all the field lines that pass through one coil also pass through the other. If the coils are identical (having the same diameters,

Figure 22.29 A "power brick" used with a laptop computer employs a transformer.

---

[2]As with an inductor, the high- and low-frequency limits for a capacitor are approximately $10 \times f_{RC}$ and $f_{RC}/10$, respectively.

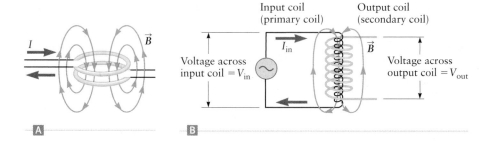

Figure 22.30 Two simple transformers.

Input coil
(primary coil)

Output coil
(secondary coil)

Voltage across
input coil = V_{in}

$\vec{B}$

Voltage across
output coil = V_{out}

A

B

lengths, and numbers of turns), the flux is the same through each coil and $\Phi_{in} = \Phi_{out}$. The fluxes in Equation 22.47 are thus the same, so the voltages are also equal:

$$V_{in} = \frac{\Delta\Phi}{\Delta t} = V_{out} \qquad (22.48)$$

We have thus managed to "transfer" a voltage from the input coil to the output coil without any direct connection between the two. However, there is an indirect connection through the magnetic flux.

With the transformers in Figure 22.30, the output voltage is equal to the input voltage, but that could have been accomplished more simply by just connecting the input and output wires. How can we *change* the voltage? Consider again the two coils in either part of Figure 22.30, but this time let the output coil have twice as many turns as the input coil. If the flux through the input coil is Φ_{in}, the flux through the output coil will be $\Phi_{out} = 2\Phi_{in}$ because there are twice as many loops in the output coil. We can use this relation between the fluxes in Equation 22.47 to get a relation between the input and output voltages:

$$V_{out} = \frac{\Delta\Phi_{out}}{\Delta t} = 2\frac{\Delta\Phi_{in}}{\Delta t} = 2V_{in}$$

Hence, the output voltage is twice as large as the input voltage.

We can apply the same analysis to the general case of an input coil with N_{in} turns and an output coil with N_{out} turns. For a coil with N turns and a total flux Φ, the flux through a single turn is $\Phi_1 = \Phi/N$; this relation applies to both the input coil and the output coil. For the transformers in Figure 22.30 (and for most practical transformers), the flux through one turn of the output coil equals the flux through one turn of the input coil. The input and output fluxes are then related by

$$\Phi_{out} = \frac{N_{out}}{N_{in}}\Phi_{in}$$

Using this result with Equation 22.47, the output voltage is

$$V_{out} = \frac{N_{out}}{N_{in}}V_{in} \qquad (22.49)$$

Since the ratio N_{out}/N_{in} can be larger or smaller than unity, we can thus "transform" the input voltage to a different value, either larger or smaller than the input voltage. That is why the device is called a transformer. Transformers are based on Faraday's law, so, to apply Equation 22.47, the flux must be changing with time. Hence, transformers cannot be used to "transform" a DC voltage.

The transformer designs in Figure 22.30 are simplest to understand, but the different arrangement shown in Figure 22.31 is used in most practical applications. The central regions of both coils are filled with a magnetic material such as iron. The field from the input coil aligns the atomic magnets in the magnetic material, producing a much larger flux than would be produced by the coil's field alone. This larger flux then yields larger voltages at both the input coil and the output coil. The ratio V_{out}/V_{in} is not affected by the presence of magnetic material in the transformer, however, and this ratio is still given by Equation 22.49.

Output voltage of a transformer

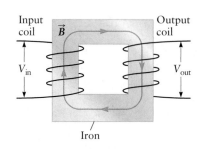

Figure 22.31 Many transformers use a magnetic material such as iron to "carry" the flux from the input coil to the output coil.

A transformer has an output coil with three times as many turns as the input coil. According to Equation 22.49, the output voltage is thus three times larger than the input voltage. If the input frequency is f, is the output frequency (a) f, (b) $3f$, or (c) $f/3$?

Applications of Transformers

One of the most important uses of transformers involves the "transmission" of electric power over large distances such as from a power plant to a city many miles away. Transformers are also used in many household appliances to convert the AC voltage at a wall socket (120-V rms) to the smaller voltages needed in radios, computers, and other devices that commonly require DC voltages around 9 V or 12 V. These DC voltages are produced in a two-step process by "power bricks" like the one in Figure 22.29 or by power supplies built into an appliance. The first step uses a transformer to convert the 120 V from a wall socket to a value near 9 V or 12 V. The second step is to convert this smaller AC voltage to direct current, using a circuit called a "rectifier." We don't have space here to describe how a rectifier works, but its output is a DC voltage, similar to the output of a battery. That is why a power brick can substitute for batteries for your laptop computer.

Transforming Voltage and Power

According to Equation 22.49, the output voltage of a transformer can be much larger than the input voltage by just arranging for the number of turns in the output coil to be much larger than the number of turns in the input coil. Electrical power delivered to a circuit is the product of the voltage and current (Eq. 22.9). So, if V_{out} is much larger than V_{in}, what happens to the output power? Is the output power also much larger than the input power? We can apply the principle of conservation of energy to answer these questions.

A transformer is like an inductor because it stores energy in the magnetic field produced by the two coils. This energy is *stored*; it is not "lost" or dissipated as heat as in a resistor. According to energy conservation principles, the energy delivered through the input coil must either be stored in the transformer's magnetic field or transferred to the output circuit. Over many AC cycles, the stored energy is constant, so the power delivered to the input coil must *equal* the output power. The input power is $P_{in} = V_{in}I_{in}$, whereas the output power is $P_{out} = V_{out}I_{out}$. Hence, if V_{out} is transformed to a value larger than V_{in}, the output current is simultaneously transformed to a smaller value than the input current.

This argument applies only to ideal transformers. The coils of a real transformer always have a small electrical resistance, causing some power dissipation. Hence, for a real transformer, the output power is always less than the input power, but usually by only a small amount.

22.10 | MOTORS

In Chapter 20, we described how the magnetic torque on a current loop can be used to make a DC motor (Section 20.10). Another way to make a motor is shown in Figure 22.32; this approach also uses the torque on a current loop, but now we use an AC voltage source to "power" the motor. The AC source is connected to a coil (called the input coil) wound around a horseshoe magnet. The input coil induces a magnetic field that circulates through the horseshoe magnet in the direction parallel to the field from the coil. As the current in the input coil oscillates, the direction of

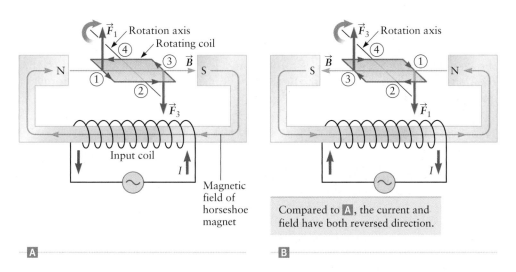

Figure 22.32 Design of an AC motor.

Compared to **A**, the current and field have both reversed direction.

the magnetic field from the horseshoe magnet alternates back and forth as indicated in Figure 22.32.

A second coil of wire is mounted between the poles of the horseshoe magnet and attached to a rotating shaft. The rotating coil is connected to a source of DC voltage (i.e., a battery), so this coil carries a DC current. Figure 22.32A shows the rotating coil in a particular position and the north pole of the horseshoe magnet on the left. From Chapter 20, when a current-carrying wire—that is, sections of the rotating coil—is placed in a magnetic field, it experiences a force. Applying right-hand rule number 2 to calculate the direction of this force, we find the results shown in Figure 22.32A. The force $\vec{F}_1$ on section 1 of the coil is directed upward, whereas the force $\vec{F}_3$ on section 3 is down. There is no force on sections 2 and 4 because the current in those sections is parallel or antiparallel to the magnetic field. The forces $\vec{F}_1$ and $\vec{F}_3$ produce a torque on the coil, causing it and the shaft to rotate. Hence, the motor turns.

After a short time, the coil rotates to the position in Figure 22.32B and at the same time the AC current in the input coil changes direction so that the north pole of the horseshoe magnet is now on the right. Applying right-hand rule number 2 to calculate the forces on the current loop, we find that $\vec{F}_1$ is now downward and $\vec{F}_3$ is upward. Comparing the two parts of Figure 22.32, we see that the torques on the coil are in the same direction; both torques are clockwise as viewed from the front of the drawing. Hence, the motor shaft continues to turn in a clockwise manner.

As the coil continues to rotate, the field from the horseshoe magnet continues to change direction so that the torque exerted on the coil is always clockwise. In this way, the oscillations of an AC current and field make the shaft rotate. Practical AC motor designs are more complicated than shown here, with typically three separate magnets (mounted at different angles) and a slightly more complicated rotating coil, but the principle is the same as shown in Figure 22.32.

EXAMPLE 22.8 | Torque Produced by an AC Motor

Consider an AC motor like the one in Figure 22.32 and let the rotating coil be square with sides of length $L = 10$ cm, carrying a current $I = 1.0$ A. If the field of the horse-shoe magnet is $B = 1.5$ T, what is the torque on the coil when it is oriented as shown in Figure 22.32A?

RECOGNIZE THE PRINCIPLE

The torque exerted on the coil is produced by the forces $\vec{F}_1$ and $\vec{F}_3$ as shown in more detail in Figure 22.33. In each case, we have a current-carrying wire in a

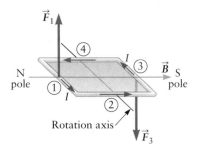

Figure 22.33 Example 22.8. Torque on the rotating coil of the AC motor in Figure 22.32A. The coil is composed of four straight sections.

perpendicular magnetic field, so from Chapter 20 each force has a magnitude $F = ILB$. The directions of these forces are given by right-hand rule number 2 and are shown in Figure 22.33; each gives a torque with a lever arm $L/2$ (the distance from the rotation axis to the point where the forces act). From Chapter 8 (Eq. 8.9), the torque associated with $\vec{F}_1$ is thus

$$\tau_1 = F_1 \times (\text{lever arm}) = \frac{F_1 L}{2}$$

In a similar way, the torque from $\vec{F}_3$ is

$$\tau_3 = F_3 \times (\text{lever arm}) = \frac{F_3 L}{2}$$

These torques combine to give the total torque on the shaft.

SKETCH THE PROBLEM

Figure 22.33 shows the problem.

IDENTIFY THE RELATIONSHIPS

The torques τ_1 and τ_3 both tend to make the shaft rotate clockwise, so the magnitude of the total torque is the sum

$$\tau_{\text{total}} = \tau_1 + \tau_3 = \frac{F_1 L}{2} + \frac{F_3 L}{2} = \frac{L}{2}(F_1 + F_3) \tag{1}$$

To compute the total torque from Equation (1), we must find the forces F_1 and F_3. The force on a wire in a perpendicular magnetic field B is $F = ILB$ (from Chapter 20). Both sections of the wire have length L and carry the same current I, so we find

$$F_1 = ILB = F_3$$

Inserting this expression into Equation (1), we get

$$\tau_{\text{total}} = \frac{L}{2}(F_1 + F_3) = \frac{L}{2}(2ILB) = IL^2 B$$

SOLVE

Inserting the given values of the various quantities gives

$$\tau_{\text{total}} = IL^2 B = (1.0\ \text{A})(0.10\ \text{m})^2(1.5\ \text{T}) = \boxed{0.015\ \text{N} \cdot \text{m}}$$

What does it mean?

This torque is not very large, so this design would make a very weak motor. In more practical motors, the coil would consist of many turns (hundreds or thousands or more), giving a much larger torque.

22.11 | WHAT CAN AC CIRCUITS DO THAT DC CIRCUITS CANNOT?

At the beginning of this chapter, we mentioned that Thomas Edison proposed to wire cities and houses with DC circuits, whereas George Westinghouse and others pushed for an AC system. Westinghouse won the argument, but why? What advantages does alternating current have over direct current for such applications?

One advantage discussed in Chapter 21 is that large AC generators are easily built using a rotating shaft and permanent magnets. The biggest advantage of AC circuits over DC circuits, however, is in the systems that distribute electrical power across long distances. Most electrical power is generated at large facilities such as hydroelectric dams or coal-burning power plants. The electrical power produced at these facilities must then be distributed to distant cities as shown schematically in Figure 22.34A. The power plant acts as an AC generator, sending current through power lines (large, overhead wires) to the destination. These wires have a small but nonzero electrical resistance R_{line}, so if the power line carries an rms current I_{rms}, there is some power $P_{ave} = I_{rms}^2 R_{line}$ dissipated in the power line resistance. Because electrical energy costs money, the power company prefers to minimize the energy dissipated in the power line. The resistance of the power line usually can't be changed easily; hence, the only way to reduce the dissipated power is to make I_{rms} as small as possible. The power delivered to the destination city is $P_{city} = V_{city} I_{city}$, so if the power company decreases the current in the power line, it must increase the voltage so as to transmit the same amount of power to the city. This voltage increase is accomplished using a transformer.

The normal household voltage is "120-V" alternating current, an rms value. In Example 22.2, we showed that this value actually corresponds to a voltage amplitude of $V_{max} = 170$ V. A much larger AC voltage is used for power transmission. Typical power line voltages have $V_{max} = 500,000$ V or even higher. When the power line reaches your house, a transformer reduces the AC voltage amplitude to the value $V_{max} = 170$ V required by your household appliances (Fig. 22.34B). Even with a power line voltage of 500,000 V, typically 5% to 10% of the energy that leaves a power plant is dissipated in the resistance of the power line. The fraction lost would be much greater if the power line were operated at lower AC voltage amplitudes.

Transformers can readily change AC voltage amplitudes, but there is no simple way to accomplish the same changes with DC voltages. There is no efficient way to increase and decrease DC voltage levels without dissipating significant amounts of energy. That is the main reason alternating current was adopted for the electrical power system.

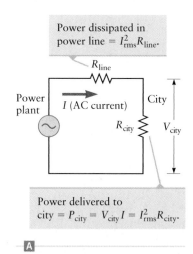

B

Figure 22.34 **A** Circuit diagram showing a power line carrying current from an electric power plant to a city. **B** Photo of a transformer used to transform the high voltage at a power line to a lower voltage suitable for use in a house.

| EXAMPLE 22.9 | Power Dissipated in an Electrical Transmission Line |

An AC power line operates with a voltage $V_{rms} = 500,000$ V and carries an AC current with $I_{rms} = 1000$ A. **(a)** What is the average (rms) power carried by the power line? **(b)** If 10% of this power is dissipated in the power line itself, what is the resistance of the power line? **(c)** This power line is now operated with a voltage $V_{rms} = 250,000$ V and current $I_{rms} = 2000$ A. The product $V_{rms}I_{rms}$ is the same as in (a), so the power carried by the line is the same. What percentage of this power is now dissipated in the power line?

RECOGNIZE THE PRINCIPLE

The total power in any circuit element is $P = VI$, so the power carried by the power line is $P_{ave} = V_{rms}I_{rms}$ (Eq. 22.12). Some of this power is dissipated in the line, and the rest is delivered to the city. The power line can be modeled as a resistor R_{line}, and the power dissipated in this resistance is $P_{line} = I_{rms}^2 R_{line}$. We can use these relations for P_{ave} and P_{line} to (a) find the average power carried by the line, (b) find the line resistance, and (c) analyze the effect of changing the line voltage.

SKETCH THE PROBLEM

Figure 22.34A describes the problem.

IDENTIFY THE RELATIONSHIPS AND SOLVE

(a) The total power carried by the power line is

$$P_{ave} = V_{rms}I_{rms} = (500{,}000 \text{ V})(1000 \text{ A})$$

$$P_{ave} = 5.0 \times 10^8 \text{ W} = \boxed{500 \text{ MW}}$$

(b) Most of this power is delivered (and sold) to the city, whereas 10% is dissipated in the power line. As a resistor, the line dissipates an amount of power $P_{line} = I_{rms}^2 R_{line}$, which is equal to 10% of $P_{ave} = 0.10 \times P_{ave}$. We thus have

$$P_{line} = I_{rms}^2 R_{line} = (0.10)P_{ave}$$

Solving for R_{line},

$$R_{line} = \frac{(0.10)P_{ave}}{I_{rms}^2} = \frac{(0.10)(5.0 \times 10^8 \text{ W})}{(1000 \text{ A})^2} = \boxed{50 \text{ } \Omega}$$

(c) If the power line voltage is reduced by a factor of two (to 250,000 V) while the current is increased by the same factor (to 2000 A), the power delivered is unchanged and we still have $P_{ave} = 5.0 \times 10^8$ W; see part (a). The power dissipated in the line is now

$$P_{line} = I_{rms}^2 R_{line} = (2000 \text{ A})^2(50 \text{ } \Omega) = 2.0 \times 10^8 \text{ W}$$

The percentage of the total power now dissipated in the line is

$$\text{percent dissipated} = \frac{P_{line}}{P_{total}} \times 100 = \frac{2.0 \times 10^8 \text{ W}}{5.0 \times 10^8 \text{ W}} \times 100$$

$$\text{percent dissipated} = \boxed{40\%}$$

What does it mean?
A much larger fraction of the energy is "lost" in the power line in part (c), resulting in much smaller profits for the power company. The fraction of the energy dissipated in the line decreases when the system operates at higher voltages and hence lower currents. That is why power lines are operated with very high AC voltages.

SUMMARY | Chapter 22

KEY CONCEPTS AND PRINCIPLES

Root-mean-square (rms) voltage
The *instantaneous voltage* V in an AC circuit varies in time as

$$V = V_{max} \sin(2\pi ft) \qquad \text{(22.1) (page 726)}$$

where V_{max} is the voltage *amplitude* and f is the frequency. The *root-mean-square* (rms) voltage is

$$V_{rms} = \frac{V_{max}}{\sqrt{2}} \qquad \text{(22.7) (page 727)}$$

(Continued)

AC voltage and current in a resistor: phasors

In an AC circuit containing a resistor, the instantaneous current is

$$I = \frac{V}{R} = \frac{V_{max}}{R} \sin(2\pi ft) \qquad \textbf{(22.2)} \text{ and } \textbf{(22.3)} \text{ (page 726)}$$

This current is **in phase** with the voltage. The time dependences of V and I are represented in a **phasor diagram**.

The **average power** dissipated in the resistor is

$$P_{ave} = V_{rms}I_{rms} \qquad \textbf{(22.12)} \text{ (page 728)}$$

where I_{rms} is the rms current.

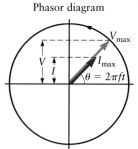

Phasor diagram

$$V = V_{max}\sin(2\pi ft)$$
$$I = I_{max}\sin(2\pi ft)$$

Reactance of a capacitor in an AC circuit

In an AC circuit containing a capacitor, the current and voltage are out of phase as shown in a phasor diagram. The amplitude of the current is

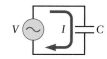

$$I_{max} = \frac{V_{max}}{X_C} \qquad \textbf{(22.21)} \text{ (page 732)}$$

where X_C is the **reactance** of the capacitor given by

$$X_C = \frac{1}{2\pi fC} \qquad \textbf{(22.22)} \text{ (page 732)}$$

The average power delivered to the capacitor is $P_{ave} = 0$.

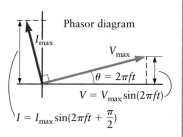

Phasor diagram

$$V = V_{max}\sin(2\pi ft)$$
$$I = I_{max}\sin(2\pi ft + \frac{\pi}{2})$$

Reactance of an inductor in an AC circuit

In an AC circuit containing an inductor, the current and voltage are again out of phase as shown in a phasor diagram The amplitude of the current is

$$I_{max} = \frac{V_{max}}{X_L} \qquad \textbf{(22.27)} \text{ (page 735)}$$

where X_L is the **reactance** of the inductor and is given by

$$X_L = 2\pi fL \qquad \textbf{(22.28)} \text{ (page 735)}$$

The average power delivered to the inductor is $P_{ave} = 0$.

Hence, for both capacitors and inductors, the circuit behavior *depends on the frequency.*

Phasor diagram

$$V = V_{max}\sin(2\pi ft)$$
$$I = I_{max}\sin(2\pi ft - \pi/2)$$

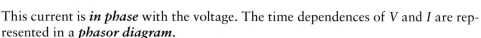

(Continued)

LC circuits

An *LC* circuit acts as a simple harmonic oscillator; both the charge on the capacitor and the current through the circuit oscillate with time:

$$q = q_{max} \cos(2\pi ft) \quad \text{and} \quad I = I_{max} \sin(2\pi ft) \qquad \textbf{(22.30)} \text{ (page 736)}$$

The frequency of the oscillations is

$$f_{res} = \frac{1}{2\pi\sqrt{LC}} \qquad \textbf{(22.35)} \text{ (page 738)}$$

These oscillations are just like the oscillations of the position and velocity of a mechanical simple harmonic oscillator.

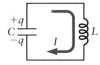

Impedance

The amplitude of the current in a general AC circuit is related to the voltage amplitude by

$$V = IZ \qquad \textbf{(22.39)} \text{ (page 741)}$$

where *Z* is the **impedance** of the circuit. The impedance *Z* can be found from an AC circuit analysis using phasors.

APPLICATIONS

Frequency-dependent behavior of AC circuits

Reactances of capacitors and inductors are frequency dependent. At very low frequencies a capacitor acts as an open circuit and an inductor approximates a simple wire (a short circuit). In the opposite limit, at very high frequencies a capacitor acts as a short circuit and an inductor acts as an open circuit. These frequency dependences are due to the dependences of the reactances (Eqs. 22.22 and 22.28) on frequency and lead to applications such as high- and low-pass filters.

Transformers

A **transformer** changes the amplitude of an AC voltage. It is an essential element of power transmission circuits.

Generators and motors

Alternating-current machines can convert rotational kinetic energy into electrical energy (a **generator**) and electrical energy into rotational motion (a **motor**).

QUESTIONS

SSM = answer in Student Companion & Problem-Solving Guide   ⊗ = life science application

1. Explain in words why (a) the reactance of a capacitor decreases as the frequency increases and (b) the reactance of an inductor increases as the frequency increases.

2. An AC circuit contains a capacitor in series with a resistor. If the frequency of the voltage source is increased while its amplitude is held fixed, does the current amplitude increase, decrease, or stay the same?

3. A resistor is connected in series with an inductor (Fig. Q22.3). If the frequency of the AC voltage source is increased while

its amplitude V_{max} is kept fixed, does the current amplitude increase, decrease, or stay the same?

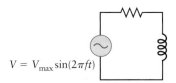

Figure Q22.3

4. An AC source with a frequency of 60 Hz is connected to a lightbulb. The instantaneous brightness of the lightbulb depends on the magnitude of the instantaneous current. Explain why the brightness of the bulb flickers (i.e., oscillates with time) and find the frequency of the flicker. Why don't we notice the flicker?

5. The AC generator described in Section 22.1 uses a rotating coil to produce an AC voltage through Faraday's law of induction. (a) Explain how this approach can produce a voltage whose frequency is equal to the rotation frequency of the shaft. (b) Explain how you could modify the rotating coil so that the frequency of the AC voltage is twice the shaft rotation frequency. (c) Explain how you could modify the magnets in Figure 22.3 so that the frequency of the AC voltage is four times the rotation frequency of the shaft.

6. **Distributed circuits.** When applying Kirchhoff's loop rule to an AC circuit, one assumes the current in the loop is the same through all circuit elements at any particular instant. Changes in current travel at the speed of light (see Chapter 23), so a certain amount of time is actually required for the AC changes to propagate from one part of a circuit to another. (a) Estimate how much time Δt is needed for these changes to travel across a circuit 5 cm in size. Assume the propagation speed in a metal wire is approximately 3×10^8 m/s. (b) Use your value of Δt to find the lowest frequency at which the current can be a maximum in one part of the circuit and a minimum at another part. Cases like this one at high frequencies are called **distributed circuits** and must be analyzed using a generalized version of Kirchhoff's laws, which accounts for the propagation time.

7. Discuss why it is relatively easy to convert the amplitude of an AC voltage to a higher or lower value, but why it is difficult to change its frequency.

8. Most AC voltage sources produce a voltage that varies sinusoidally with time, but some produce a "square wave" as sketched in Figure Q22.8. Use the general definition of rms voltage to find the relation between the amplitude of the square wave and the rms voltage. That is, if the square wave amplitude is V_{sq}, find V_{rms}.

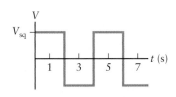

Figure Q22.8 Question 8 and Problem 6.

9. Consider two AC voltage sources connected in series with voltages $V_1 = V_{1,\,max} \sin(2\pi ft)$ and $V_2 = V_{2,\,max} \sin(2\pi ft + \pi/4)$, and $V_{1,\,max} = V_{2,\,max} = 30$ V. Show that these two sources in series are equivalent to a single source with $V_{series} = V_{equiv} \sin(2\pi ft + \phi)$ and find the values of V_{equiv} and ϕ.

10. If the voltage across a capacitor is $V = V_{max} \sin(2\pi ft)$, the charge on the capacitor is also proportional to $\sin(2\pi ft)$ (see Eq. 22.17). The current through the capacitor in this case is given by $I = I_{max} \sin(2\pi ft + \pi/2)$; hence, the phase angle is $\phi = \pi/2$ radians relative to the voltage. Suppose we begin instead by assuming the capacitor voltage is proportional to a cosine function $V = V_{max} \cos(2\pi ft)$. Use a graphical argument to show that the current is now given by a function of the form $I = I_{max} \cos(2\pi ft + \pi/2)$ so that the phase of the current relative to the voltage is again $\phi = \pi/2$ rad.

11. Show that the reactance of a capacitor X_C has units of ohms. Explain why it makes sense to compare the reactance of a capacitor with the resistance of a resistor.

12. The reactance of an inductor is given by $X_L = 2\pi fL$. Show that this reactance has units of ohms. Explain why it makes sense to compare the reactance of an inductor with the resistance of a resistor.

13. **SSM** Consider the circuit in Figure Q22.13. (a) Draw equivalent circuits that apply at high and low frequencies. (b) Estimate the ratio V_{out}/V_{in} at high and low frequencies. (c) Explain how this circuit acts as a filter that blocks high frequencies but lets low frequencies pass. It is one type of "low-pass" filter.

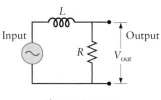

Figure Q22.13

14. A capacitor is connected in parallel with an inductor, and this circuit is connected to an AC voltage source of frequency f. It is found that the impedance of this circuit increases if the frequency is increased by a small amount. Is f above or below the resonant frequency of the circuit? Explain.

15. An LC circuit is called an oscillator. What, physically, is oscillating in such a circuit? What variables are periodic?

16. The equations describing an LC oscillator are very similar in form to those describing the mechanical harmonic oscillator of a mass on a spring. Identify the term in the equation for the mechanical oscillator that plays the role of C. What term plays the role of L? How are these corresponding pairs of variables similar?

17. An LCR circuit with a resonant frequency of 40 Hz is connected to a standard U.S. wall outlet of 120 V AC. Does the current in the circuit "lag" or "lead" the emf? That is, does the current have its maximum before or after the maximum in the voltage?

18. How is a phasor different from a vector? Is the emf a vector quantity?

19. In actual transformers (Fig. Q22.19), the iron cores are actually comprised of thin sheets of metal laminated together with insulating layers in between. Why would the cores be made this way?

Figure Q22.19 Transformer with a laminated core.

20. Although transformers are AC devices, is there some way to test a transformer to see if it works with a battery? Explain.

21. **SSM** **Boosting and bucking transformers.** Sometimes an electrical device may need an input voltage that is a few percent higher or lower than the supply voltage. A *boosting transformer* connects the output coil with the input voltage (here, a 120-V AC source) as shown on the left in Figure Q22.21. (a) Draw the sine waveforms of the input and output coils of the transformer and show how they add to produce the 150-V output. (b) If the leads from the output coil are connected as shown on the right in Figure Q22.21, a *bucking transformer* configuration results. (It has this name because two voltages are combined in "opposition" to make the output.) What output voltage would you expect from the circuit on the right in Figure Q22.21? Again draw the sine waveforms of the input and output coils, and the final output waveform. (Note that when the voltages from the input and output coils are combined, you must consider the phases of these voltages relative to each other.)

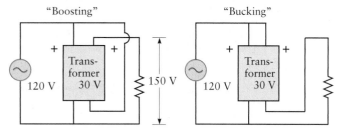

Figure Q22.21

22. Some resistors are made from a coil of wire. What is the best way to model the circuit behavior of such a resistor? Choose one answer and explain.
(a) As an ideal resistor in series with an ideal inductor?
(b) As an ideal resistor in parallel with an ideal inductor?
(c) As an ideal resistor in parallel with an ideal capacitor?

23. For an ideal generator, the electrical energy generated equals the work W done in making the shaft rotate. For a real generator, is the electrical energy greater than, equal to, or less than W? Explain.

22.1 GENERATION OF AC VOLTAGES

1. Ⓡ Figure P22.1 shows the AC voltage produced by a particular source. Estimate (a) the frequency of this voltage, (b) the amplitude of this voltage, and (c) the rms value of this voltage. (d) If this AC voltage were attached to a lightbulb, would you notice the flicker of the bulb?

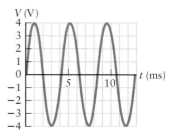

Figure P22.1

2. Some clocks that are powered by an AC voltage use the AC frequency to keep time. The power company makes this possible by arranging to have precisely the correct number of cycles occur every day, which keeps such clocks from gaining or losing time over long periods. (a) How many cycles of AC voltage are there each day in the United States? (b) Suppose a clock designed for the United States (where the AC frequency is 60 Hz) is taken to a different country where the frequency is 55 Hz. After one full day at its new location, this clock will be in error by an amount Δt. Find Δt.

3. SSM ★ Make a graph of an AC voltage that has a frequency of 300 Hz and an rms value of 45 V. Be sure to label the axes of your graph.

4. ★ Suppose the magnetic field B in the generator in Example 22.1 is reduced in magnitude by a factor of three. By what factor will the AC voltage change?
(a) The voltage increases by a factor of three.
(b) The voltage decreases by a factor of three.
(c) The voltage increases by a factor of nine.
(d) The voltage decreases by a factor of nine.
(e) There is no change in the voltage.

5. ★ Suppose the loop in the generator in Example 22.1 is square, and that the length of each side of the loop is increased by a factor of two. By what factor will the AC voltage change?

6. Consider the square wave voltage in Figure Q22.8. What is its frequency?

7. Consider an AC voltage source that produces a voltage given by $V = V_{max} \sin(2\pi f t)$, with $V_{max} = 18$ V and $f = 6.2$ Hz. What is the instantaneous value of the voltage at $t = 1.4$ s?

22.2 ANALYSIS OF AC RESISTOR CIRCUITS

8. In an AC circuit, does the voltage across a resistor differ in phase from the current by a phase angle of (a) zero, (b) 30°, (c) 45°, (d) 90°, or (e) 180°?

9. SSM ★ A resistor with $R = 2.0$ kΩ is connected to an AC generator that operates at 60 Hz with an amplitude of 35 V. Find (a) the amplitude of the current through the resistor, (b) the maximum instantaneous power dissipated in the resistor, (c) the minimum instantaneous power dissipated in the resistor, and (d) the average power dissipated in the resistor. (e) How long will it take this resistor to dissipate 250 J?

10. ★ The value of R for the resistor in Problem 9 is increased by a factor of 2.5 while the current is kept fixed. By what factor does the average power change?

11. The average power dissipated in a 45-kΩ resistor in an AC circuit is 75 W. What is the amplitude of the AC voltage across the resistor?

12. What is the resistance of a lightbulb that has a rating of 300 W? Assume the lightbulb is designed for use in an AC circuit with $V_{rms} = 120$ V. *Hint:* In a circuit analysis, a lightbulb can be treated as a resistor.

13. ★ A heater runs on an AC voltage of 120 V and draws a power of 25 W. If this heater were run instead on a DC voltage of 75 V, how much current would flow through the heater? *Hint:* The heater can be modeled as a resistor.

14. Consider the AC circuit containing two resistors in Figure P22.14. If the amplitude of the AC voltage source is $V_{max} = 25$ V, $R_1 = 300$ Ω and $R_2 = 500$ Ω what is the amplitude of the current through R_1 and R_2?

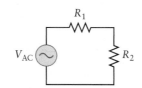

Figure P22.14
Problems 14 and 15.

15. ★ Make a phasor diagram for the circuit considered in Problem 14, showing the voltage across each resistor and the current through the circuit. Do these three phasors all make the same angle with the horizontal axis? Explain why or why not.

16. ★ Make a phasor diagram for the circuit in Figure P22.16 showing the voltage across each resistor and the current through each resistor. Do these phasors make the same angle with the horizontal axis? Express the length of the phasors in terms of the amplitude V_{max} of the AC voltage source, and R_1 and R_2.

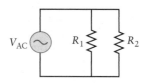

Figure P22.16 Problems 16 and 17.

17. ⭐ Consider the circuit in Figure P22.16 and suppose $R_1 = 20\ \Omega$ and $R_2 = 40\ \Omega$. If the total average power dissipated in the two resistors is 55 W, what is the amplitude V_{max} of the AC voltage?

22.3 AC CIRCUITS WITH CAPACITORS

18. In an AC circuit, does the voltage across a capacitor differ in phase from the current by a phase angle of (a) zero, (b) 30°, (c) 45°, (d) 90°, or (e) 180°?

19. You are given two capacitors with capacitances of 20 pF and 20 μF. If the frequency is 5000 Hz, which capacitor has the larger reactance?

20. ⭐ An AC voltage source with a frequency of 120 Hz and an amplitude $V_{max} = 25$ V is connected across a capacitor with $C = 1.5\ \mu$F (Fig. P22.20). Draw a phasor diagram showing the voltage across and the current through the capacitor. What is the length of the phasor that represents the current across the capacitor? Be sure that your answer has the correct units.

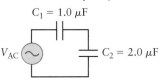

Figure P22.20
Problems 20, 26, and 27.

21. What is the reactance of a 1200-pF capacitor at 400 Hz?

22. ⭐ If the frequency is increased from 35 Hz to 140 Hz, does the reactance of a capacitor increase or decrease? By what factor does it change?

23. A capacitor has a reactance of 450 Ω at 1200 Hz. What is its capacitance?

24. The AC voltage across a capacitor with $C = 1.5\ \mu$F has an amplitude of $V_{max} = 22$ V, and the current through the capacitor has an amplitude of 2.5 mA. What is the frequency?

25. ⭐ Two capacitors are connected to an AC source as shown in Figure P22.25; assume $V_{AC} = V_{max}\sin(2\pi ft)$, with $V_{max} = 18$ V and a frequency $f = 2000$ Hz. (a) What is the amplitude (peak value) of the current through the capacitors? (b) What is the rms current?

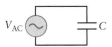

$C_1 = 1.0\ \mu$F

$C_2 = 2.0\ \mu$F

Figure P22.25

26. ⭐ For the circuit in Figure P22.20, find (a) the average power, (b) the maximum instantaneous power, and (c) the minimum instantaneous power delivered to the capacitor. Use $f = 400$ Hz, $V_{max} = 25$ V, and $C = 1.5\ \mu$F.

27. [SSM] ⭐ Consider the AC circuit in Figure P22.20 and assume $V_{AC} = V_{max}\sin(2\pi ft)$ with $V_{max} = 12$ V, $f = 300$ Hz, and $C = 1.5\ \mu$F. The charge on the capacitor will also vary sinusoidally. Find (a) the amplitude of the charge oscillation and (b) the maximum energy stored in the capacitor. (c) At what time t is the energy stored a maximum? (Find just one value of t, but there will be an infinite number of values.)

28. Consider a circuit containing a resistor with $R = 1200\ \Omega$ and a capacitor with $C = 6.5\ \mu$F. At what frequency does the reactance of the capacitor equal the resistance of the resistor?

29. ⭐ Two capacitors with $C_1 = 450$ pF and $C_2 = 820$ pF are connected in parallel. What is the reactance of the equivalent capacitance at 30 kHz?

30. ⭐ Two capacitors with $C_1 = 4.5\ \mu$F and $C_2 = 1.9\ \mu$F are connected in series. What is the reactance of the equivalent capacitance at 350 Hz?

31. ⭐ The phase angle between the voltage and current in an AC circuit is 90°. Is the average power delivered to the circuit from the AC source positive, negative, or zero?

22.4 AC CIRCUITS WITH INDUCTORS

32. In an AC circuit, does the voltage across an inductor differ in phase from the current by a phase angle of (a) zero, (b) 30°, (c) 45°, (d) 90°, or (e) 180°?

33. You are given two inductors with inductances of 40 μH and 80 mH. If the frequency is 5000 Hz, which inductor has the larger reactance?

34. An inductor has a reactance of 5500 Ω at a frequency of 2500 Hz. What is the value of L?

35. ⭐ Find the reactance of an inductor with $L = 15$ mH at 22 kHz. If the frequency is increased to 66 kHz, by what factor does the reactance change?

36. Consider the circuit in Figure P22.36. (a) If the current through the inductor ($L = 25$ mH) at 400 Hz has an amplitude of 55 mA, what is the amplitude of the AC voltage? (b) What is the rms value of the AC voltage across the inductor?

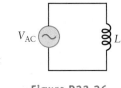

Figure P22.36
Problems 36 and 37.

37. [SSM] ⭐ Draw a phasor diagram that shows the voltage across and current through the inductor in Problem 36. See Figure P22.36.

38. The voltage source in Figure P22.38 has an amplitude of 5.0 V and a frequency of 75 Hz. What is the amplitude of the current through the inductors?

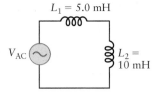

$L_1 = 5.0$ mH

$L_2 = 10$ mH

Figure P22.38

39. ✪ Figure P22.39 shows the voltage across an inductor with $L = 20$ mH. (a) At what time is the energy stored in the inductor a maximum? What is the energy at that moment? (b) At what time is the energy stored in the inductor a minimum? What is the energy at that moment?

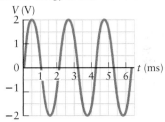

Figure P22.39

40. Consider an inductor with $L = 50$ mH and a resistor with $R = 20$ kΩ. At what frequency is the reactance of the inductor equal to the resistance of this resistor?

22.5 LC CIRCUITS

22.6 RESONANCE

41. ⭐ The resonant frequency of an LC circuit is f_0. (a) If the inductance is increased by a factor of eight, what is the new resonant frequency? (b) How should the capacitance be changed to change the resonant frequency back to f_0?

42. Consider an LC circuit with $L = 15$ mH and $C = 2500$ pF (Fig P22.42). (a) At what frequency is the reactance of the inductor equal to the reactance of the capacitor? (b) What is the resonant frequency of this circuit?

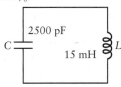

Figure P22.42 Problems 42, 43, and 46.

43. ✪ Consider the LC circuit in Problem 22.42 and Figure P22.42. At $t = 0$, the current is $I = 45$ mA and the charge q on the capacitor is zero. (a) What is the energy stored in the inductor at $t = 0$? (b) At some instant t_1, the current will be zero. What is the charge on the capacitor at t_1? (c) Find t_1. (There are many possible values of t_1; find the one closest to $t = 0$.) (d) Using the value of t_1 from part (c), find the magnitudes of I and q at $t = 3t_1$.

44. An LC circuit has a resonant frequency $f = 50$ kHz, and $L = 0.44$ mH. What is the capacitance of the capacitor?

45. ✪ An LC circuit has $L = 0.50$ mH and $C = 1.2\ \mu$F. If the total energy stored in this circuit is 2.5 μJ (2.5×10^{-6} J), what are

(a) the maximum charge q_{max} on the capacitor and (b) the maximum current? (c) When the charge on the capacitor is $q_{max}/2$, what is the magnitude of the current?

46. ☆ Consider again the *LC* circuit in Problem 42 and Figure P22.42. If the charge on the capacitor at a particular instant is 5.0×10^{-8} C, and the energy stored in the capacitor is equal to the energy stored in the inductor, what is the current at that moment?

47. [SSM] ☆ An *LC* circuit with $L = 55$ mH is used in a radio to determine the frequency of a selected radio station. The station's frequency is equal to the resonant frequency of the circuit. (a) The radio is initially set for a frequency $f = 1.05$ MHz. What is the value of *C*? (b) This capacitance is then increased by 10%. What is the new resonant frequency? (c) Can you calculate the answer to part (b) without first finding the value of *C* in part (a)?

48. ✪ The *LCR* circuit in Figure P22.48 is driven by a voltage source with $V_{AC} = V_{max} \sin(2\pi ft)$, where $V_{max} = 12$ V, $R = 150\ \Omega$, and $C = 25$ nF. (a) If the resonant frequency is $f_{res} = 95$ MHz, what is *L*? (b) What is the amplitude (the peak value) of the current when $f = f_{res}$?

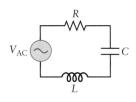

Figure P22.48

49. ☆ You are given a collection of inductors and capacitors and are asked to construct *LC* circuits for several different applications. The inductances range from $L = 10\ \mu$H to 85 mH, and the capacitances are in the range of $C = 200$ pF to 1.5 mF. Is it possible to use these inductors and capacitors (one of each) to make *LC* circuits with the following frequencies? If so, give one set of possible values for *L* and *C*.
(a) $f_{res} = 1.5$ MHz for use in an AM radio
(b) $f_{res} = 95$ MHz for use in an FM radio
(c) $f_{res} = 60$ MHz for use in a television
(d) $f_{res} = 2.0$ GHz for use in a cell phone

22.7 AC CIRCUITS AND IMPEDANCE

50. [SSM] ☆ An inductor with $L = 35$ mH is connected in parallel with a capacitor with $C = 75\ \mu$F. (a) If they are connected to an AC voltage source with $f = 2.5$ kHz, where is the current larger, in the inductor or in the capacitor? (b) At what frequency is the current amplitude the same?

51. When an *LC* circuit is at resonance, is the impedance (a) zero, (b) $2\pi fL$, (c) $1/(2\pi fC)$, or (d) $1/\sqrt{LC}$?

52. ✪ Consider the *RL* circuit in Figure P22.52 with $R = 1500\ \Omega$, $L = 2.5$ mH, and an AC frequency of 70 kHz. What is the phase angle ϕ between the phasors that represent V_{AC} and the current?

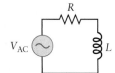

Figure P22.52
Problems 52 and 54.

53. ✪ An *LCR* circuit contains a resistor with $R = 4500\ \Omega$, a capacitor with $C = 5500\ \mu$F, and an inductor with $L = 2.2$ mH, all connected in series. The AC source has a frequency of 75 kHz. (a) What is the impedance of the circuit? (b) What phase angle does the phasor describing the voltage across the voltage source make with the current phasor?

22.8 FREQUENCY-DEPENDENT BEHAVIOR OF AC CIRCUITS: A CONCEPTUAL RECAP

54. ☆ Consider the circuit in Figure P22.52. The voltage of the AC source is $V = V_{max} \sin(2\pi ft)$. If the frequency is extremely low (i.e., if *f* is very small), what is the amplitude of the current in the circuit? Express your answer in terms of V_{max}, *R*, and *L*.

55. ☆ Estimate the impedance of the *LC* circuit in Figure P22.55 at the very high frequency of 30 MHz. Will this impedance be dominated by the capacitor or the inductor? Will this impedance increase or decrease if the frequency is increased?

56. ☆ Estimate the impedance of the *LC* circuit in Figure P22.55 at the very low frequency of 30 Hz. Will this reactance be dominated by the capacitor or by the inductor? Will this reactance increase or decrease if the frequency is increased?

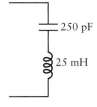

Figure P22.55
Problems 55 and 56.

57. An *LCR* circuit contains a resistor with $R = 450\ \Omega$ and a capacitor with $C = 5500\ \mu$F. If the resonant frequency is 33 kHz, what is the value of the inductance?

58. [SSM] ☆ An *RC* circuit containing a resistor with $R = 2500\ \Omega$ and a capacitor with $C = 1500\ \mu$F is attached to an AC generator with $V_{max} = 3.5$ V and $f = 25$ kHz. (a) What is the impedance of this circuit? (b) What is the amplitude of the current?

59. ☆ An *RL* circuit contains a resistor with $R = 6500\ \Omega$ and an inductor with $L = 1500\ \mu$H. If the impedance of this circuit is 150,000 Ω, what is the frequency?

60. ✪ The circuit in Figure P22.60 acts as a filter. (a) What is the ratio of the output voltage (amplitude) to the input voltage at very high frequencies? (b) What is the ratio of the output voltage to the input voltage at very low frequencies? (c) Is it a low-pass filter or a high-pass filter?

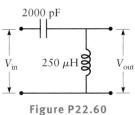

Figure P22.60

61. ✪ The circuit in Figure P22.61 acts as a filter. (a) What is the ratio of the output voltage (amplitude) to the input voltage at very high frequencies? (b) What is the ratio of the output voltage to the input voltage at very low frequencies? (c) Is it a low-pass filter or a high-pass filter?

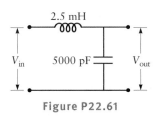

Figure P22.61

62. ✪ ✪ A nerve fiber can be modeled electrically as an *RC* circuit as illustrated in Figure P22.62. The interior of the fiber acts as a resistor, and the resistance is determined by the solution inside the fiber, which has a resistivity of about $\rho = 0.1\ \Omega \cdot$ m. The fiber's membrane acts as a capacitor; its capacitance per unit area of membrane is about 10^{-2} F/m^2. (a) Estimate the *R* and *C* of the equivalent circuit in Figure P22.62. (b) Is it a low-pass filter or a high-pass filter? (c) What frequency separates the high and low frequency limits for this circuit?

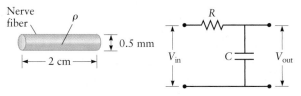

Figure P22.62

22.9 TRANSFORMERS

63. A transformer has an input coil with $N_{in} = 300$ turns and $N_{out} = 1000$ turns. If the voltage across the input coil has an amplitude of 2.0 V, what is the voltage across the output circuit?

64. Consider again the transformer in Problem 63. Suppose the two coils are reversed so that 2.0 V is put across the secondary coil. What is the voltage across the primary coil?

65. **European trip.** In France, the wall sockets provide an AC voltage with $V_{rms} = 230$ V. You want to use an appliance designed to operate in the United States ($V_{rms} = 120$ V) and decide to build a transformer to convert the power line voltage in France to the value required by your appliance. (a) Should you use a "step-down" transformer (to make V_{rms} smaller) or a "step-up" transformer (which makes V_{rms} larger)? (b) If the input coil of your transformer has 2000 turns, how many turns should the output coil have?

66. The transformer in a power brick (as used in a laptop computer) converts the AC voltage from a wall socket ($V_{rms} = 120$ V) to a value near 12 V. (a) If the output coil of the transformer has 4000 turns, how many turns does the input coil have? (b) If this transformer were accidentally connected backward (with the input and output coils reversed), what would the output voltage be?

67. ✪ The power brick for the author's laptop computer has a power rating of 45 W. It uses a step-down transformer to convert the AC voltage from a wall socket ($V_{rms} = 120$ V) to a value near 12 V. (a) What is the rms current in the input coil of the transformer? (b) What is the rms current in the output coil of the transformer?

22.10 MOTORS

68. ✱ Consider the AC motor described in Example 22.8. You want to make this motor more powerful; that is, you want it to produce a larger torque. Evaluate the following proposed design changes. (a) If the number of turns in the coil were increased from one to five, how would the torque change? (b) If the size of the rotating coil were increased to $L = 30$ cm, how would the torque change? (c) If the magnetic field strength were increased

to $B = 2.5$ T, how would the torque change? (d) If all three of these changes were made, by what factor would the total torque increase?

69. SSM ✱ A small electric motor such as the one in a home furnace is rated at "1/4 horsepower." Recall that "horsepower" (abbreviated hp) is a unit of power, with 1 hp ≈ 750 W. (a) How much work does this motor do in 1 s? In 1 h? (b) Assume it is an ideal motor and all the input electrical energy is converted to mechanical energy. If the motor operates with an AC voltage with $V_{rms} = 120$ V, what is the rms current in the motor?

22.11 WHAT CAN AC CIRCUITS DO THAT DC CIRCUITS CANNOT?

70. ✪ You are given the job of evaluating potential power line systems. One system is designed to operate at $V = 3000$ V, and another would operate at $V = 12,000$ V. Suppose the power line at the higher voltage would dissipate 5% of the power that it carries. What percentage of the power would be dissipated in the low-voltage power line?

71. SSM ✱ Consider the power line system in Example 22.9. Suppose the operating voltage is increased to 750,000 V rms and a power of 500 MW is transmitted by the power line. What percentage of the power would now be dissipated in the power line?

72. ✪ When an inductor with $L = 50$ mH is attached to an AC voltage source, the inductive reactance is 8000 Ω. What is the capacitive reactance of a capacitor with $C = 450$ pF when attached to the same voltage source?

73. ✪ Figure P22.73 shows a model for a leaky capacitor, with a resistor in parallel with an ideal capacitor. Find the phase angle between the current through and voltage across the leaky capacitor in Figure P22.73. Express your answer in terms of R, C, and the frequency f.

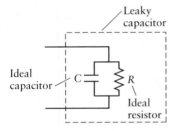

Figure P22.73

74. ✪ (a) At what frequency will a 10-μF capacitor have the same reactance as a 10-mH inductor? (b) What is the reactance at this frequency? (c) If these components were assembled into a series LC circuit, what would the resonance frequency be?

75. ✪ (a) A generator supplies a current to a circuit containing an inductor. It is found that the current through the inductor reaches its maximum value 22 ms after the voltage across the inductor reaches its maximum value. What is the minimum possible operating frequency of the generator? (b) The generator is now connected to a lone capacitor. If the maximum current of the generator occurs at $t = 0$ s, how long after that will the voltage across the capacitor reach its maximum?

76. ✪ **The variac.** An indispensable device for any electronic workbench is a variable transformer or variac. Most variacs use a rotating contact to select a certain number of windings on the secondary coil, usually a solenoid in the shape of a toroid like that in Figure P22.76. Both the primary and secondary coils have 800 windings, but the rotor can connect to any of the individual loops of the secondary coil so that the output voltage is "tapped off" in as few as 100 turns on the secondary coil. (a) If the AC voltage across the primary is 120 V, what are the maximum and minimum voltage outputs from this device? (b) What is the smallest increment of change in voltage possible? (c) The dial on the variac needs calibration. What angle of rotation of the contact corresponds to 25 V? Assume the rotation angle varies from 0° (at the minimum output voltage) to 360° (at the maximum output voltage).

Figure P22.76 A variac with control dial and windings exposed.

77. ✪ Transformers are not limited to two sets of windings. The relationship for winding ratios holds true for multiple sets of coils as well. The AC voltage on the primary coil in Figure P22.77 is $V_P = 120$ V, and the two outputs supply $V_{S1} = 220$ V and $V_{S2} = 12$ V, respectively. If the 220-V secondary coil has 880 turns, what are the number of turns in the 12-V secondary coil and the primary coil?

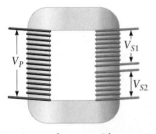

Figure P22.77 A transformer with two secondary coils.

78. ✪ ✖ **Induction loop assistive listening system.** Magnetic induction is used to provide real-time amplification to hearing-impaired audience members. The system taps into the amplification system already in place in the room and routes the amplified electric waveforms from the speaker's microphone into two wire loops around the perimeter of the audience. (Figure P22.78 shows only one of these loops.) The bottom loop is at floor level, and the top loop is near the ceiling. The changing current in the loop induces changes in magnetic flux, which induces currents in smaller loops worn by audience members sitting within the main loop (photo), where the signal is then amplified and provided to headphones or earpieces. (a) What is the inductance of two loops, separated by 2.5 m, that fits the perimeter of a rectangular room 7.0 m wide by 15 m long? (Approximate it as a solenoid.) (b) The frequency of sound from speaking ranges from approximately 500 Hz to 4 kHz. Since the system should be sensitive to this range, what capacitance would give this system a resonance frequency of 2000 Hz? *Hint:* The expression for the inductance of a solenoid was derived in Chapter 20 (Eq. 20.21).

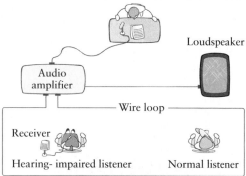

Figure P22.78 Diagram of an induction loop assistive listening system for a lecture hall and an audience member using an induction loop receiver.

79. ✪ **Crystal radio set.** The phenomenon of resonance makes it possible to listen to AM broadcasts with a very simple circuit and never need a battery. When a system is driven at its resonant frequency, even if by minute oscillations of current in the antenna, the resulting amplitude can be many times greater than the driving amplitude. The crystal radio, when *tuned* to a broadcast frequency, can produce a current large enough to operate an earphone speaker. A crystal radio is a simple *LCR* circuit with an antenna wire and a diode (the "crystal," involved in separating the signal from the carrier wave for the earphone). Figure P22.79 shows an illustration and circuit diagram of such a radio. The *LCR* circuit is made to match the broadcast frequency by varying the inductance, in this case by moving

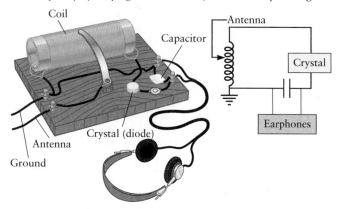

Figure P22.79 A crystal radio set and corresponding circuit diagram.

a wire to select how many turns of the solenoid are included in the circuit. (a) If the capacitance of the circuit is 950 pF, what inductance is needed to receive a broadcast at the highest AM frequency (1710 kHz)? At the lowest (520 kHz)? (b) The solenoid inductor's windings are made on a cylinder 6.0 cm in diameter and 20 cm long. How many turns are needed to provide the maximum and minimum inductances to cover the full AM band range? (c) Calculate the number of windings per length of solenoid (turns per millimeter). (d) AM radio specifications mandate that stations are spaced 10 kHz apart. Will this solenoid have sufficient resolution to separate two stations? (What change in the resonant frequency is produced by including one more winding?)

80. ✪ A series *LCR* circuit is comprised of a resistor, an inductor, and a capacitor of values 50 Ω, 87 mH, and 44 μF, respectively. (a) What is the generator frequency if the total impedance of the circuit measures 57 Ω? (b) What is the impedance of the circuit if it is plugged into a European outlet (50 Hz)? (c) At what frequency is the impedance of the circuit a minimum?

81. SSM ✪ **Simple inductive furnace.** An iron nut of mass 50 g is heated from room temperature to its melting point in 52 s through magnetic induction as shown in Figure P22.81. The circuit has a capacitance of 560 nF so that it will operate at its resonant frequency of 173 kHz with an input rms voltage of 1000 V. (a) If this inductive heater is 90% efficient, what is the power supplied to the inductor of this circuit to heat the iron? *Hint:* The melting temperature of iron is 1810 K, and its specific heat is 440 J/(kg · K). (b) What is the inductance of the coil of thick copper wire? (c) What is the rms current through the coil? Is the thick copper wire needed? (d) The needed capacitance is provided by placing 12 identical capacitors in parallel. What is the capacitance of one of these capacitors?

Figure P22.81

82. ✪ **Fast Internet.** Copper telephone wires were originally designed to carry "commercial speech" using a band of frequencies from 300 Hz to 3400 Hz through a system called the PSTN (Public Switched Telephone Network). The ADSL (Asymmetric Digital Subscriber Line) uses frequencies very much higher than this speech band to carry fast data traffic using frequencies between 25 kHz and 1.1 MHz. To ensure that the higher-frequency transmissions do not interfere with normal phone electronics, filters like that shown in Figure P22.82 must be installed on all phone lines. (a) Calculate the frequencies that would be filtered out with the circuit configuration in the figure. Do they fall in the ADSL transmission range? (b) If the three inductors in series on each line were instead replaced by one equivalent inductor, would it alter the effectiveness of the filter? Why or why not?

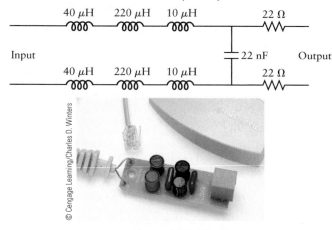

Figure P22.82

Electromagnetic Waves

These radio telescopes detect electromagnetic waves that originate outside our galaxy. (Courtesy NRAO/AUI)

Theoretical understanding of electricity and magnetism developed rapidly during the early 1800s and seemed to be complete by about 1850. The laws of Coulomb and Gauss describing electric forces and fields and of Ampère and Faraday describing magnetic forces and fields were verified in many experiments and provided the basis of new technologies as key parts of the industrial revolution. Despite this success, some deep questions at the heart of electromagnetic theory remained unanswered. One concerned the nature of electric and magnetic field lines and the notion of action at a distance. (A similar question also applies to the gravitational field.) Consider two electric charges q_1 and q_2 located in distant space, far from all other matter. The electric field associated with q_1 is felt by (i.e., produces a force on) q_2. Now suppose q_1 moves at some instant in time to a new location. We expect the electric field lines to move along with q_1, but how fast will these field lines move? Is the motion of q_1 felt *instantaneously* at q_2, or is there some delay, and if so, what is the delay time? These questions are different aspects of the larger issue, "What are electric field lines?"

Physicist James Clerk Maxwell pondered some of these questions in the mid-1800s. His work led to what is probably the greatest physics discovery of the 19th century, the discovery of electromagnetic waves. Electromagnetic waves are used in many applications, including radios, televisions, radar, microwave ovens, cell phones, and the wireless Internet. Life today would certainly be very different without these applications of electromagnetic waves.

The general features of wave motion were introduced in Chapters 12 and 13. We'll rely heavily on that material in our work on electromagnetic waves in this chapter.

23.1 | THE DISCOVERY OF ELECTROMAGNETIC WAVES

In Chapters 20 and 21, we learned how Ampère's law and Faraday's law can be used to calculate magnetic fields and forces in different situations. An intriguing aspect of Faraday's law is that a time-varying magnetic field gives rise to an electric field. That is, a magnetic field *by itself* can produce an electric field. Many physicists then wondered, can an electric field by itself produce a magnetic field? Maxwell hypothesized that it should indeed be possible, which led him to propose a modification of Ampère's law. According to this modification, a time-varying electric field produces a magnetic field. Maxwell's new version of Ampère's law does not affect any of our work with Ampère's law in Chapter 20, and it also fixes a subtle problem with that law that had not previously been noticed. The most important result of Maxwell's hypothesis, however, is that it gives a new way to create a magnetic field; it also gives the equations of electromagnetism an interesting symmetry.

Symmetry of E and B

- The correct form of Ampère's law (due to Maxwell) says that a changing electric flux produces a magnetic field. Since a changing electric flux can be caused by a changing E, we can also say that *a changing electric field produces a magnetic field*.

- Faraday's law says that a changing magnetic flux produces an induced emf, and an emf is always associated with an electric field. Since a changing magnetic flux can be caused by a changing B, we can also say that a *changing magnetic field produces an electric field*.

There is thus a sort of "circularity" as illustrated in Figure 23.1: an electric field can (if it is changing) produce a magnetic field, and a magnetic field can (if it is changing) produce an electric field. Can this "loop" or oscillation involving E and B continue, with the new electric field producing a new magnetic field, producing a new electric field, and so on? The answer is yes; self-sustaining oscillations involving electric and magnetic fields are indeed possible. In fact, such oscillations are an *electromagnetic wave*. Electromagnetic waves are also referred to as *electromagnetic radiation*.

A key feature of the circularity in Figure 23.1 is that both the electric and magnetic fields must be changing with time. Faraday's law requires a time-varying magnetic field to produce an electric field, and Ampère's law (as modified by Maxwell) needs a time-varying electric field to produce a magnetic field. These requirements also affect how an electromagnetic wave can be created. Typically, a time-varying electric field is set up by an AC current

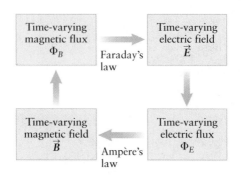

Figure 23.1 In schematic terms, a changing (time-varying) electric field produces a magnetic field, whereas a changing (time-varying) magnetic field produces an electric field.

in a wire or by moving electric charges, leading to the electric and magnetic fields in Figure 23.1. We'll come back to this idea in Sections 23.4 and 23.5.

Maxwell was the first to recognize that the equations of electromagnetism lead to the existence of electromagnetic waves, and he worked out the properties of these waves in great mathematical detail. Maxwell's prediction of the existence of electromagnetic waves was ahead of the experiments of the time. In fact, there was no experimental proof of the existence of electromagnetic waves until Heinrich Hertz demonstrated the generation and detection of radio waves in 1887, eight years after Maxwell's death. Hertz also showed that these waves travel at the speed of light (as predicted by Maxwell). Many other workers contributed to the work on radio, including Guglielmo Marconi, who developed the first practical radio system. (Marconi achieved trans-Atlantic radio communication in 1901.)

A full mathematical discussion of electromagnetic waves is more than we can include in this book, but we can use our understanding of Faraday's law to understand one important property of these waves. According to Faraday's law, a changing magnetic flux through a given area produces an induced emf directed along the perimeter of that area (Fig. 23.2A). This induced emf is associated with an induced electric field through our usual connection between electric potential and field (Chapter 18). Hence, the direction of the induced $\vec{E}$ is *perpendicular* to the magnetic field $\vec{B}$ that produced it. In a similar way, the magnetic field $\vec{B}$ produced by a changing electric field is perpendicular to the electric field that produced it (Fig. 23.2B). As a result, the electric and magnetic fields in an electromagnetic wave are perpendicular to each other. In addition, the direction of propagation of an electromagnetic wave is perpendicular to both $\vec{E}$ and $\vec{B}$. The directions of $\vec{E}$ and $\vec{B}$ and the propagation direction all depend on how the wave is generated, but these three directions are always perpendicular to one another.

Let's now summarize three fundamental properties of all electromagnetic waves.

Properties of electromagnetic waves

- An electromagnetic wave involves *both* an electric field and a magnetic field.
- These fields are *perpendicular* to each other.
- The propagation direction of the wave is *perpendicular to both $\vec{E}$ and $\vec{B}$.*

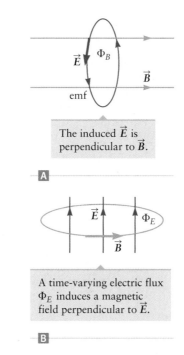

The induced $\vec{E}$ is perpendicular to $\vec{B}$.

A

A time-varying electric flux Φ_E induces a magnetic field perpendicular to $\vec{E}$.

B

Figure 23.2 **A** According to Faraday's law, a changing magnetic field produces an electric field. The induced electric field is perpendicular to the magnetic field. **B** A changing electric field produces a magnetic field, and these two fields are again perpendicular.

23.2 | PROPERTIES OF ELECTROMAGNETIC WAVES

In the previous section, we gave some qualitative arguments concerning the existence and properties of electromagnetic waves. The laws of electricity and magnetism can be used to derive *all* the properties of electromagnetic waves. Here are some of the key results.

Electromagnetic Waves Are Transverse Waves

We have already mentioned that the fields associated with an electromagnetic wave are perpendicular to each other and to the propagation direction. These relationships are illustrated in Figure 23.3A, which shows a snapshot of $\vec{E}$ and $\vec{B}$ for an electromagnetic wave at one instant in time. In this snapshot, the electric field is directed along x, the magnetic field is along y, and the wave propagates in the $+z$ direction.

To interpret Figure 23.3A, first focus on the electric field at the origin. Here, $\vec{E}$ is parallel to the x axis while the wave travels along z, and the tips of the electric field vectors form a pattern similar to a wave on a string. If we took another snapshot a short time later, we would find that the entire electric field pattern has moved a short distance along the z direction while maintaining the same wave profile. The behavior of the magnetic field in Figure 23.3A is similar, but $\vec{B}$ is along the

Figure 23.3 ◭ Electric and magnetic fields of an electromagnetic wave. This plot shows a snapshot of $\vec{E}$ and $\vec{B}$ at a particular moment in time. ◼ The fields $\vec{E}$ and $\vec{B}$ are perpendicular to each other. The propagation direction $\vec{v}$ of the wave is along the direction defined by right-hand rule number 2.

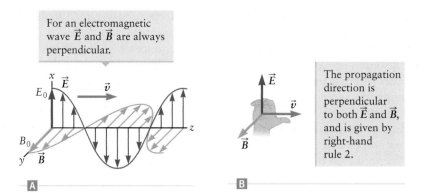

> For an electromagnetic wave $\vec{E}$ and $\vec{B}$ are always perpendicular.

> The propagation direction is perpendicular to both $\vec{E}$ and $\vec{B}$, and is given by right-hand rule 2.

A

B

y direction, so it is perpendicular to both $\vec{E}$ and the propagation direction. Because the fields are both perpendicular to the propagation direction, this wave is a ***transverse wave***.

Electromagnetic waves are transverse waves.

The propagation direction can be determined with right-hand rule number 2 using the procedure shown in Figure 23.3B. Place the fingers of your right hand along the direction of $\vec{E}$ and curl your fingers in the direction of $\vec{B}$. Your thumb then points along the propagation direction.

> **CONCEPT CHECK 23.1** | Finding the Propagation Direction
>
> At a particular instant, the electric field associated with an electromagnetic wave is directed along $+y$ and the magnetic field is along $+z$. Is the propagation direction (a) $+x$, (b) $-x$, (c) $+y$, or (d) $+z$?

Light Is an Electromagnetic Wave

Maxwell found that the speed of an electromagnetic wave can be expressed in terms of two universal constants that appear in the theories of electricity and magnetism: the permittivity of free space ε_0 and the magnetic permeability of free space μ_0. The speed of an electromagnetic wave in a vacuum (free space) is commonly denoted by the symbol c:

Speed of electromagnetic waves in a vacuum

$$\text{speed of an electromagnetic wave} = c = \frac{1}{\sqrt{\varepsilon_0 \mu_0}} \qquad (23.1)$$

When Maxwell inserted the known values of ε_0 and μ_0 into Equation 23.1, he found

$$c = \frac{1}{\sqrt{\varepsilon_0 \mu_0}} = \frac{1}{\sqrt{[8.85 \times 10^{-12} \ \text{C}^2/(\text{N} \cdot \text{m}^2)](4\pi \times 10^{-7} \ \text{T} \cdot \text{m/A})}}$$

$$= 3.00 \times 10^8 \ \text{m/s} \qquad (23.2)$$

This value for the speed c is precisely equal to the speed of light. When Maxwell derived this result in 1864, the nature of light was the subject of much debate. Most physicists suggested it was a wave while a few believed it was a particle, but no one could say what kind of wave or particle it might be. Maxwell had the answer: ***light is an electromagnetic wave***. He thus showed that the equations of electricity and magnetism also provide the theory of light.

It is difficult to overestimate the importance of Maxwell's discovery. Besides explaining the nature of light, his work led to the discovery of other types of electromagnetic waves, including radio waves. This work also led eventually to the work of Einstein and his theory of relativity (Chapter 27).

Electromagnetic Waves Can Travel through a Vacuum, All at the Same Speed

Although Maxwell worked out many of the basic properties of electromagnetic waves, including the value for their speed c, some results must have surprised even him. Recall from Chapter 12 that all mechanical waves travel in a medium of some kind. For waves on a string, the medium is the string, and sound waves travel in a material such as air. These waves "disturb" their medium, and the properties of the medium determine the wave speed. Many physicists searched for the corresponding medium for an electromagnetic wave. These waves can travel through material substances such as air or glass, but they can also travel through a vacuum, which is certainly *not* a material substance. Hence, electromagnetic waves are not like the mechanical waves we studied in Chapter 12. Roughly speaking, you can think of electric and magnetic fields as the "media" through which electromagnetic waves travel. These fields can exist in a vacuum, and electromagnetic waves can also travel through a vacuum.

Maxwell's theory does not specify the frequency or wavelength of an electromagnetic wave. Rather, frequency and wavelength are determined by the way the wave is produced. In Section 23.4, we'll describe examples of electromagnetic waves, including radio waves and X-rays. They *all* travel with speed c through a vacuum.

Electromagnetic Waves Travel Slower Than c in Material Substances

When an electromagnetic wave travels through a material substance such as glass, water, or the Earth's atmosphere, the speed of the wave depends on the properties of the substance. In such cases, the wave speed is always less than c, the speed in a vacuum (Eq. 23.2). We'll see in Chapter 24 that this difference in wave speed explains how a lens can focus light. What's more, the speed of an electromagnetic wave inside a substance depends on the wave's frequency. So, for example, blue light travels through glass at a higher speed than red light. This effect is called *dispersion* and gives rise to rainbows and other interesting effects, as discussed in Chapter 24.

CONCEPT CHECK 23.2 | Time Delay in a Cell Phone Conversation

A cell phone signal that travels from coast to coast in the United States must travel approximately 3000 miles (about 5×10^6 m). This signal is carried by different types of electromagnetic radiation (typically radio and microwaves) at a speed close to c. Is the approximate time required for your cell phone signal to propagate from coast to coast (a) 0.5 s, (b) 3 s, (c) 0.02 s, or (d) 0.0005 s? Can you notice this time delay?

23.3 | ELECTROMAGNETIC WAVES CARRY ENERGY AND MOMENTUM

When we first discussed waves in Chapter 12, we stated that waves transport energy without transporting matter. This general property of all waves is true for electromagnetic waves, too. For mechanical waves such as waves on a string or sound, this energy is contained in the medium through which the wave travels. For example, with a wave on a string, the energy of the wave is associated with the elastic energy and the tension in the string. With an electromagnetic wave, the energy is carried by the electric and magnetic fields associated with the wave. We can understand how these fields carry energy by considering what happens when an electromagnetic wave encounters a charged particle. Figure 23.4 shows the fields associated with an electromagnetic wave along with an electric charge $+q$ that is in the path of the wave. The charge will experience an electric force $\vec{F}_E = q\vec{E}$; as the wave passes by and the electric field oscillates along the x axis, the force on the charge will also oscillate. If this charge is able to move (which is true for electrons and ion cores in a solid), the electric force will do work on the charge, increasing its kinetic energy. In this way, energy is transferred from the wave to the charge; hence, the wave does indeed carry energy. You might also wonder about the magnetic force on the charge in Figure 23.4; in most cases, the

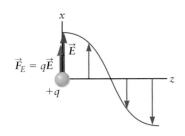

Figure 23.4 When an electromagnetic wave "strikes" an electric charge, the electric field associated with the wave produces a force on the charge. This force oscillates in direction as the wave passes by.

magnetic force is much smaller than the electric force, but we will see below that the magnetic field is crucial for understanding how an electromagnetic wave can carry momentum.

In Chapters 18 and 21, we discussed the potential energy contained in the electric and magnetic fields. For an electric field, the energy is proportional to E^2 (Eq. 18.48), and the potential energy per unit volume (the energy density) is

$$u_{\text{elec}} = \frac{1}{2}\,\varepsilon_0 E^2 \tag{23.3}$$

For a magnetic field, the potential energy is proportional to B^2 (Eq. 21.35), and the magnetic potential energy per unit volume is

$$u_{\text{mag}} = \frac{1}{2\mu_0}\,B^2 \tag{23.4}$$

These results apply also to electromagnetic waves, and the total energy per unit volume carried by an electromagnetic wave is the sum of its electric and magnetic energies, $u_{\text{elec}} + u_{\text{mag}}$. As the wave propagates, the electric and magnetic fields oscillate (Fig. 23.3), so u_{elec} and u_{mag} also oscillate. It also turns out that the electric and magnetic energies are equal, so $u_{\text{elec}} = u_{\text{mag}}$. Applying this relation when the fields have their peak values E_0 and B_0 (Fig. 23.3A) and using Equations 23.3 and 23.4, we get

$$\frac{1}{2}\,\varepsilon_0 E_0^2 = \frac{1}{2\mu_0}\,B_0^2 \tag{23.5}$$

These terms are the peak energies carried by the electric and magnetic fields of the wave. In words, Equation 23.5 also says that the peak values of the fields E_0 and B_0 are proportional. Rearranging, we get

$$E_0 = \frac{1}{\sqrt{\varepsilon_0\mu_0}}\,B_0$$

The factor $1/\sqrt{\varepsilon_0\mu_0}$ is just the speed of light in a vacuum c (Eq. 23.1); hence,

Relation between *E* and *B* for an electromagnetic wave

$$E_0 = cB_0 \tag{23.6}$$

EXAMPLE 23.1 | Electric and Magnetic Fields of a TV Signal

The signal from a radio station is an electromagnetic wave and is detected through its electric field. In a typical case, the electric field needed to give a "good" radio signal is about 1×10^{-2} V/m, the amplitude (peak value) E_0 associated with the wave. What is the corresponding magnetic field amplitude associated with this electromagnetic wave?

RECOGNIZE THE PRINCIPLE

For an electromagnetic wave, the amplitude of the electric and magnetic fields are proportional. Because the electric field amplitude E_0 is given, we can find the magnetic field amplitude B_0. The aim of this problem is to understand the magnitudes of these fields for a typical application.

SKETCH THE PROBLEM

No sketch is needed.

IDENTIFY THE RELATIONSHIPS

The relation between the peak electric and magnetic fields in an electromagnetic wave is $E_0 = cB_0$ (Eq. 23.6).

SOLVE

Inserting the given value of E_0, we find

$$B_0 = \frac{E_0}{c} = \frac{1 \times 10^{-2}\ \text{V/m}}{3.00 \times 10^8\ \text{m/s}} = \boxed{3 \times 10^{-11}\ \text{T}}$$

What does it mean?

This value for B_0 certainly "seems" much smaller than the electric field E_0 (because the numerical values are very different in magnitude), but a careful comparison requires us to consider the relative magnetic and electric forces in a typical case. We'll do so after we discuss how an antenna works in Section 23.5; that analysis will show that the electric force is (in that case) much greater than the magnetic force. Thus, an electromagnetic wave is usually detected through its electric field and not by the effect of its magnetic field.

Intensity of an Electromagnetic Wave

The strength of an electromagnetic wave is usually measured in terms of its intensity, with units of watts per square meter. The intensity is the amount of energy the wave transports per unit time across a surface of unit area, and it equals the energy density (the amount of energy carried by the wave) multiplied by the wave speed (because a faster wave delivers energy at a faster rate). The wave speed is just c, the speed of light. Denoting the intensity by I, we have

$$I = \text{total energy density} \times c \qquad (23.7)$$

The peak electric and magnetic energy densities are given in Equation 23.5. The total energy density carried by the wave is proportional to the sum of these two terms:

$$\text{total energy density} \propto \frac{1}{2}\varepsilon_0 E_0^2 + \frac{1}{2\mu_0}B_0^2 \qquad (23.8)$$

We have already mentioned that the electric and magnetic contributions to the energy are equal, so we can rewrite Equation 23.8 as

$$\text{total energy density} \propto \frac{1}{2}\varepsilon_0 E_0^2 + \frac{1}{2\mu_0}B_0^2 = \varepsilon_0 E_0^2$$

Inserting into Equation 23.7, we get $I \propto \varepsilon_0 E_0^2 c$. In our derivation, we have dealt with E_0 and B_0, which are the peak values of the fields. To find the average intensity, we must average over a complete cycle of oscillation of the fields; this average intensity is half the peak intensity, so our final result for the intensity is

$$I = \frac{1}{2}\varepsilon_0 c E_0^2 \qquad (23.9)$$

This result is the *average* intensity of an electromagnetic wave in terms of the electric field amplitude (the maximum electric field) E_0. The intensity of an electromagnetic wave is thus proportional to the square of the electric field amplitude. Since $E_0 = cB_0$, the intensity is also proportional to the square of the magnetic field amplitude.

Intensity of an electromagnetic wave

| EXAMPLE 23.2 | The Electric Field of Sunlight |

On a typical summer day, the intensity of sunlight at the Earth's surface is approximately $I = 1000$ W/m^2. What is the amplitude of the electric field associated with this electromagnetic wave?

RECOGNIZE THE PRINCIPLE

The intensity of an electromagnetic wave is proportional to E_0^2, where E_0 is the amplitude of the electric field. Given the intensity, we can thus find E_0. This calculation will give an understanding of the size of the electric field associated with an important source of electromagnetic radiation.

SKETCH THE PROBLEM

No sketch is necessary.

Insight 23.1
APPLICATIONS OF SOLAR ENERGY: SOLAR CELLS
Example 23.2 mentions that the intensity of sunlight on a typical sunny day is about 1000 W/m^2. Several approaches can be used to "capture" this energy. A *solar cell* converts the energy from sunlight into electrical energy that can be used to run the circuits in your house or to charge a battery. A typical house uses energy at an average rate of a few thousand watts. Because sunlight has an intensity of about 1000 W/m^2, you might think that a solar collector that captures sunlight over an area of only a few square meters would be sufficient. Current photovoltaic cells, however, capture only about 15% of the energy that strikes them, and one must also allow for nights and cloudy days during which the Sun's intensity is much lower than its peak value. The problem of making better and more practical solar cells is an important engineering challenge.

The electric field amplitude and the intensity are related by $I = \frac{1}{2}\varepsilon_0 c E_0^2$ (Eq. 23.9). Rearranging to solve for E_0 gives

$$E_0 = \sqrt{\frac{2I}{\varepsilon_0 c}}$$

SOLVE

Inserting the value of the intensity of sunlight, we find

$$E_0 = \sqrt{\frac{2(1000 \text{ W/m}^2)}{[8.85 \times 10^{-12} \text{ C}^2/(\text{N} \cdot \text{m}^2)](3.00 \times 10^8 \text{ m/s})}} = \boxed{870 \text{ V/m}}$$

What does it mean?
This electric field is large, especially compared with the field found in a typical electric circuit. This large value for E_0 explains how sunlight can produce large effects, such as sunburn!

Electromagnetic Waves Carry Momentum

In mechanics, a particle's momentum is equal to the mass of the particle times its velocity (Eq. 7.1). An electromagnetic wave does not have a mass, however, so how can it have momentum? The momentum of a particle is related to the force that the particle exerts in a collision with another object. Let's therefore consider the force that an electromagnetic wave exerts when it is absorbed by (i.e., collides with) an object as sketched in Figure 23.5. If the object is at rest before the collision, all the initial momentum is carried by the electromagnetic wave (Fig. 23.5A). Because the wave is absorbed by the object, all the final momentum is carried by the object. The absorption of the wave occurs through the electric and magnetic forces on charges in the object. Figure 23.5B and C shows the effects on a positive charge q (which might be an ion core) at the surface of the object. When the wave is absorbed, momentum is transferred to the charge q. Since total momentum is conserved in a collision, we can write

$$p_i(\text{wave}) = p_f(\text{charge } q) \tag{23.10}$$

To analyze the effect of the electromagnetic wave on q, let's first consider the electric force. The electric field of the wave $\vec{E}$ is upward along $+x$ in Figure 23.5B, so the electric force $\vec{F}_E = q\vec{E}$ is also upward (assuming q is positive). This force accelerates the charge, giving it a velocity that is also upward. Now consider the magnetic force on this moving charge in Figure 23.5C. The magnetic field of the wave gives a magnetic force of magnitude $F_B = qvB$, and, according to right-hand rule number 2, this force is to the right along $+z$. As the electric field $\vec{E}$ oscillates up and down, the velocity in Figure 23.5B also oscillates up and down, but the magnetic force F_B in Figure 23.5C is always to the right because when $\vec{E}$ reverses direction, so does $\vec{B}$. (See Question 17 at the end of the chapter.)

We have thus shown that when a charge q absorbs an electromagnetic wave, there is a force on the charge in the direction of propagation of the original wave. According to the impulse theorem (Eq. 7.5), the force F_B on the charge q is related to the charge's change in momentum Δp along the z direction during a time Δt by

$$F_B = \frac{\Delta p}{\Delta t} \tag{23.11}$$

Hence, the charge's final momentum along the z direction in Figure 23.5 is given through Equation 23.11. According to the principle of conservation of momentum, the final momentum of the charge must equal the initial momentum of the electromagnetic wave (Eq. 23.10), so the wave does indeed have momentum!

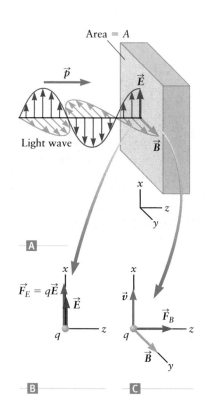

Figure 23.5 **A** An electromagnetic wave carries momentum. **B** and **C** This momentum can be understood in terms of the force exerted on charges in a material from the electric and magnetic fields of the wave. When electromagnetic radiation is absorbed by an object, the momentum of the object increases such that the total momentum of the system (radiation plus object) is conserved.

A full calculation of the momentum p of an electromagnetic wave shows that p is proportional to the total energy E_{total} carried by the wave, with the result

$$p = \frac{E_{total}}{c} \qquad (23.12)$$

Momentum carried by an electromagnetic wave

Radiation Pressure

The analysis in Figure 23.5 shows that when an electromagnetic wave is absorbed by an object, it exerts a force on the object. The total force on the object is proportional to its exposed area. Recalling that pressure is equal to the force per unit area, we can define a quantity called the **radiation pressure** $P_{radiation}$, which is equal to this electromagnetic force divided by the area. If I is the intensity of the wave, an analysis based on the ideas in Figure 23.5 gives

$$P_{radiation} = \frac{F}{A} = \frac{I}{c} \qquad (23.13)$$

Radiation pressure

In words, when an electromagnetic wave is absorbed by an object, the object experiences a radiation pressure $P_{radiation}$. As an example, the radiation pressure from sunlight could be used to propel a spacecraft, as we explore in the next example.

EXAMPLE 23.3 | The Momentum Carried by Sunlight

A spacecraft designer wants to use radiation pressure to propel a spacecraft as sketched in Figure 23.6. Near the Earth, the intensity of sunlight is about 1000 W/m² and the exposed area of the spacecraft is 500 m². If the sunlight is completely absorbed, what is the force on the spacecraft?

RECOGNIZE THE PRINCIPLE

The radiation pressure produced by an electromagnetic wave is proportional to its intensity. From the definition of pressure, the resulting force is then equal to the radiation pressure multiplied by the area of the spacecraft.

SKETCH THE PROBLEM

Figure 23.6 shows the problem.

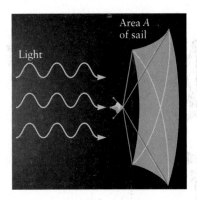

Figure 23.6 Example 23.3. A solar sail absorbs momentum from sunlight and uses the resulting force to accelerate a spacecraft.

IDENTIFY THE RELATIONSHIPS

The radiation pressure is related to the intensity by $P_{radiation} = I/c$, and the force on the spacecraft is $F = P_{radiation}A$. We thus have

$$F = P_{radiation}A = \frac{IA}{c}$$

SOLVE

Inserting the given values of the intensity and the area leads to

$$F = \frac{(1000 \text{ W/m}^2)(500 \text{ m}^2)}{3.00 \times 10^8 \text{ m/s}} = \boxed{1.7 \times 10^{-3} \text{ N}}$$

What does it mean?

This force is rather small; it is roughly equal to the weight of a penny at the Earth's surface. Achieving a larger force would require a larger exposed area (a larger "solar sail"). This approach does have one advantage, however: the fuel is free. Spacecraft that rely on solar sails for propulsion have not yet been built, but the effects of this force on the motions of several spacecraft (including *Mariner 10*) have been observed.

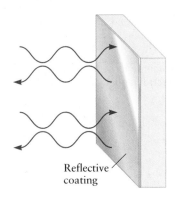

Reflective coating

Figure 23.7 Concept Check 23.3.

When light is absorbed by an object, the radiation pressure is $P_{radiation} = I/c$ (Eq. 23.13). Suppose the exposed surface of an object has instead a reflective coating so that it reflects electromagnetic waves (Fig. 23.7). Which of the following statements is true? *Hint*: Compare absorption (Fig. 23.5) to a completely inelastic collision and reflection (Fig. 23.7) to an elastic collision.

(a) The force on the object due to radiation pressure is the same whether the light is absorbed or reflected.

(b) The force on the object due to radiation pressure is greater by a factor of two when the light is reflected.

(c) There is no force on the object due to radiation pressure when the light is reflected.

23.4 | TYPES OF ELECTROMAGNETIC RADIATION: THE ELECTROMAGNETIC SPECTRUM

All electromagnetic waves travel through a vacuum at the same speed c. The value of c has been measured with *great* precision and has the value

$$c = 2.99792458 \times 10^8 \text{ m/s} \qquad (23.14)$$

Of all the fundamental physical constants, the speed of electromagnetic waves in a vacuum is known the most accurately. In fact, c is actually *defined* to have the value in Equation 23.14, and the value of the SI standard meter is then derived through the value of c and the unit of time, the standard second. In most calculations, we will not need all the significant figures in Equation 23.14; the value $c = 3.00 \times 10^8$ m/s is accurate enough for the work in this book.

Electromagnetic waves are classified according to their frequency f and wavelength λ. The frequency, wavelength, and speed of an electromagnetic wave are related by

$$f\lambda = c \qquad (23.15)$$

which is similar to the general relation we found for mechanical waves in Chapter 12 (Eq. 12.2); as the frequency is increased, the wavelength decreases, and vice versa. Figure 23.8 lists a number of different types of electromagnetic waves, arranged according to their frequencies and wavelengths. In this listing, the frequency increases from bottom to top; frequency and wavelength are inversely related (Eq. 23.15), so the wavelength increases from top to bottom.

This range of all possible electromagnetic waves is called the *electromagnetic spectrum*. There is no strict lower or upper limit for electromagnetic wave frequencies, which vary over an extremely large range, from below a few hertz (Hz) for ultralow-frequency radio waves to above 10^{22} Hz for gamma rays. The range of frequencies assigned to the different types of waves is somewhat arbitrary. For example, the ultraviolet and X-ray regions in Figure 23.8 overlap as do the regions assigned to microwaves and radio waves. The names of these different regions were chosen based on how the radiation in each frequency range interacts with matter and on how it is generated. We now describe some properties of these waves, starting at the low frequency end of the electromagnetic spectrum.

Radio Waves

Radio waves have frequencies from as low as a few hertz to about 10^9 Hz. The corresponding wavelengths vary from 10^8 m (around 10^5 miles!) to a few centimeters. Radio waves are usually produced by an AC circuit attached to an antenna. A simple wire can function as an antenna (as discussed in Section 23.5), but antennas containing multiple conducting elements (Fig. 23.9) are usually more efficient and are much more common than those consisting of a single wire. The frequency of a radio wave generated in this way is equal to the frequency of the current in the

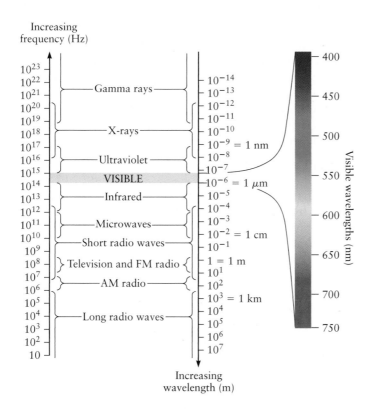

Increasing frequency (Hz)

Frequency (Hz)		Wavelength (m)
10^{23}		10^{-14}
10^{22}	Gamma rays	10^{-13}
10^{21}		10^{-12}
10^{20}		10^{-11}
10^{19}		10^{-10}
10^{18}	X-rays	$10^{-9} = 1$ nm
10^{17}		10^{-8}
10^{16}	Ultraviolet	10^{-7}
10^{15}		$10^{-6} = 1\ \mu$m
10^{14}	VISIBLE	
10^{13}	Infrared	10^{-5}
10^{12}		10^{-4}
10^{11}	Microwaves	10^{-3}
10^{10}		$10^{-2} = 1$ cm
10^{9}	Short radio waves	10^{-1}
10^{8}	Television and FM radio	$1 = 1$ m
10^{7}		10^{1}
10^{6}	AM radio	10^{2}
10^{5}		$10^{3} = 1$ km
10^{4}	Long radio waves	10^{4}
10^{3}		10^{5}
10^{2}		10^{6}
10		10^{7}

Increasing wavelength (m)

Visible wavelengths (nm): 400, 450, 500, 550, 600, 650, 700, 750

Figure 23.8 The electromagnetic spectrum. Frequency increases from bottom to top, whereas wavelength increases from top to bottom. Near the middle of the spectrum is a narrow section corresponding to visible light; this section is expanded at the right, showing also how the color of visible light varies with frequency and wavelength.

attached AC circuit, which would usually contain inductors and capacitors as discussed in Chapter 22. Radio waves can be detected by an antenna similar to the one used for generation. These waves are used to carry conventional radio and television signals as well as signals for some cell phones and pagers.

In our discussion of the induced electric and magnetic fields in Figure 23.1, we emphasized that an electromagnetic wave is produced by *changing* electric and magnetic fields. For radio waves, the AC current in the antenna is produced by time-varying (oscillating) electric fields in the antenna. This time-varying electric field then produces a changing magnetic field and thus the electromagnetic wave. When the AC current in an antenna oscillates with time, the moving charges in the antenna are accelerated as they oscillate. Many other sources of electromagnetic radiation involve similar time-varying electric and magnetic fields produced by *accelerated* charges. In general, *when an electric charge is accelerated, it produces electromagnetic radiation.*

Microwaves

Microwaves have frequencies between about 10^9 Hz and 10^{12} Hz. The corresponding wavelengths vary from a few centimeters to a few tenths of a millimeter. This radiation interacts strongly with molecules such as H_2O and CO, making microwaves useful in the kitchen. Microwave ovens (Fig. 23.10) generate radiation with a frequency near 2.5×10^9 Hz. Waves of this frequency are absorbed strongly by H_2O molecules, causing the atoms in those molecules to vibrate just as masses connected by springs can vibrate. The term *absorption* means that some of or all the microwave energy is transferred to the water molecules, heating your food. Vibrating molecules can also emit electromagnetic radiation because they contain accelerated charges. For example, a vibrating H_2O molecule contains accelerated charges and thus emits electromagnetic radiation.

Infrared Radiation

Infrared radiation falls into the frequency range from around 10^{12} Hz to about 4×10^{14} Hz (wavelengths from a few tenths of a millimeter to a few microns). This

Figure 23.9 A set of parallel conductors (i.e., wires) arranged with the proper spacing acts as an antenna that can efficiently emit or absorb electromagnetic radiation. This antenna is designed for radio waves.

Figure 23.10 A microwave oven uses radiation in the microwave range to heat food.

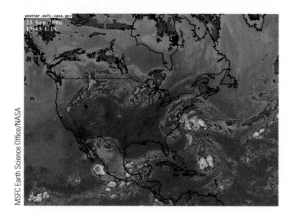

Figure 23.11 Blackbody radiation (typically in the infrared range) indicates the temperature of an object. In this satellite photo, the temperature of the Earth's surface and atmosphere is measured by the wavelength at which the blackbody radiation has the highest intensity. (See Fig. 14.32.) In this satellite image, different temperatures are displayed as different colors.

radiation is what we sense as "heat," and much of the blackbody radiation from objects at temperatures near room temperature falls in the infrared range. Satellite photographs sensitive to infrared radiation are thus useful for monitoring the Earth's weather (Fig. 23.11). Infrared radiation interacts strongly with molecules and is absorbed by most substances. Infrared radiation can be produced by vibrating molecules or when electrons undergo transitions within an atom. We'll discuss these properties of electrons in atoms in Chapter 29.

Visible Light

The radiation we are able to detect with our eyes falls into the rather narrow frequency range from about 4×10^{14} Hz to near 8×10^{14} Hz, corresponding to wavelengths from about 750 nm to 400 nm. This radiation is called *visible light*. The color of visible light varies systematically with its frequency. Light at the low end of this frequency range (long wavelengths) is sensed as the color red; increasing frequencies then correspond to yellow, green, and blue, while the high-frequency (short-wavelength) end of the visible range is violet. A mixture of all these colors such as the light from a flashlight produces "white" light.

The speed of an electromagnetic wave inside a material substance has a value c' that is less than the speed c in a vacuum. For example, visible light travels through air at a speed about 0.03% smaller than c, whereas in water the speed of light is about 33% less than c. This variation of the speed of light from one material to another leads to many effects in optics, including refraction and the focusing of light by a lens (Chapter 24). The speed of light inside a material also depends on the frequency of the radiation. This effect is called *dispersion* because it causes a prism to "disperse" the different colors in white light (Fig. 23.12). Dispersion is also responsible for rainbows.

As with infrared radiation, visible light can be produced when electrons undergo atomic transitions (Chapter 29). Many sources of visible light such as the flashlight in Figure 23.12 produce radiation with a range of frequencies, but it is also possible to produce light consisting essentially of a single frequency. Such radiation is called *monochromatic*, meaning "single color." This term is usually used when describing visible light, but can also be applied to radiation in other parts of the spectrum. Lasers (Fig. 23.13) produce monochromatic radiation. We'll explain how lasers work in Chapter 29.

Ultraviolet Light

Proceeding to frequencies above the range of visible light takes us to *ultraviolet (UV) light*, with frequencies from about 8×10^{14} Hz to 10^{17} Hz; the associated wavelengths are from about 3 nm to 400 nm. Although we cannot detect this radiation with our eyes, it has some important health consequences. The UV portion of the spectrum is commonly subdivided into several regions: UV-A has wavelengths

Figure 23.12 White light from a flashlight is a combination of many different frequencies (i.e., colors). After passing through a prism, the different colors propagate in different directions. Newton used a prism to show that white light contains different colors.

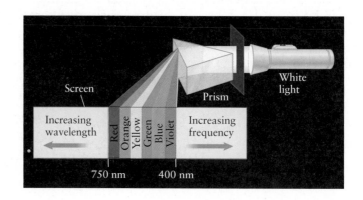

from 315 nm to 400 nm, UV-B falls in the wavelength range of 280 nm to 315 nm, and UV-C lies in the range of 200 nm to 280 nm. Even shorter wavelengths are referred to as the vacuum ultraviolet range because these wavelengths are so strongly absorbed by air and other substances that they can only propagate a significant distance in vacuum. As the wavelength becomes shorter, the damage done to exposed human skin increases. Thus, UV-C radiation has the greatest potential for damaging tissue. Fortunately, the atmosphere absorbs UV radiation most strongly at the shorter wavelengths, so most of the UV-C radiation from the Sun does not reach the Earth's surface.

Ultraviolet radiation has one major health benefit: it stimulates the production of vitamin D in the body. Nearly all the other effects of UV radiation are detrimental. Excessive exposure to UV light (Fig. 23.14) can cause sunburn, skin cancer, and cataracts. Many skin cancers (the type known as carcinoma) are treatable and often not fatal, but malignant melanoma is much more deadly. Persons who receive only two serious sunburns as children are found to have about twice the normal probability of developing melanoma as adults as those who did not receive such sunburns.

X-Rays

X-rays have frequencies from about 10^{17} Hz to around 10^{20} Hz. Wilhelm Röntgen discovered X-rays in 1895 and found that they can be used to form images from inside living tissue. Only one month after he announced this discovery, X-rays were being applied in medicine and other fields, and Röntgen became a worldwide celebrity. (He received the first Nobel Prize in Physics soon after, in 1901.) An early X-ray image of a human hand is reproduced in Figure 23.15A. The soft tissue, bones, and even a ring on one of the fingers are clearly visible. This image is possible because X-rays are weakly absorbed by skin and other soft tissue and strongly absorbed by dense materials such as bone, teeth, and metal. This difference in absorption provides the contrast seen in Figure 23.15B and makes it possible for X-rays to detect broken bones.

The X-ray images in Figure 23.15 are essentially "shadows" of the tissue that the X-rays pass through. All X-ray images were produced in this way until the 1970s, when a technique called computed axial tomography (abbreviated CT or CAT) was developed. Figure 23.16A shows how photographs taken at two different angles can be combined to reveal information about the thickness of an object. For example, if two bones overlap in an X-ray image, the bone behind is obscured. This effect is illustrated in Figure 23.16A, where a basketball is positioned in front of the person;

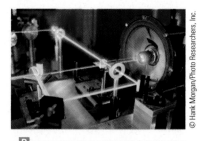

Figure 23.13 A A "key chain" laser. B A research laser. Both lasers produce monochromatic radiation (light of a single color).

Figure 23.14 Sunbathers use ultraviolet radiation from the Sun to get tan. Too much exposure to UV radiation can be hazardous to your health.

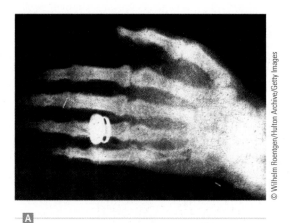

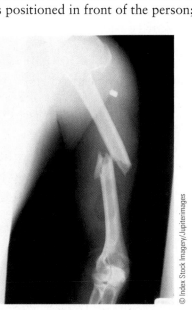

Figure 23.15 X-rays can pass through skin, but are absorbed strongly by dense material such as bone and metal.

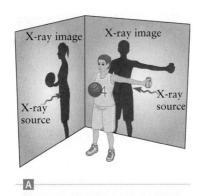

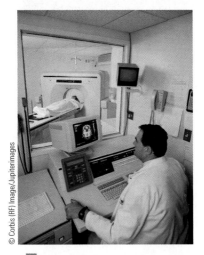

Figure 23.16 A Shadow images taken at different angles can show the thickness of an object and can also reveal when one part of an object is in the shadow of another part. B Example of a CT scan apparatus as found in most hospitals.

here the basketball is obscured in the image on the screen behind. For any single image, there will always be parts of the person's body that are obscured in some way. This problem can be overcome by comparing two images taken from different angles; in Figure 23.16A, the basketball is visible when viewed with the X-ray source at the right. A CT scan (Fig. 23.16B) takes many X-ray images at many different angles and then employs a sophisticated computer analysis to combine these images into a three-dimensional representation of the object. The results can be used to measure accurately the shape and size of bones, tumors, and so forth within the body.

Although X-ray imaging has enormous health benefits, excessive exposure to X-ray radiation can have harmful effects, including burns. At high exposures, X-rays can cause cancer and other serious diseases.

Gamma Rays

Gamma rays lie at the high-frequency end of the electromagnetic spectrum, with frequencies above about 10^{20} Hz and wavelengths less than about 10^{-12} m. (As with other regions of the electromagnetic spectrum, there is no precise boundary between the X-ray and gamma ray parts of the spectrum.) Gamma rays are produced by processes occurring inside the nucleus of an atom, as we'll discuss in Chapter 30. Gamma rays are produced in nuclear power plants and the Sun. Gamma rays also reach the Earth from sources outside our solar system.

EXAMPLE 23.4 ⊗ **Wavelength of X-rays**

The X-rays used in your dentist's office or local hospital have a frequency of approximately $f = 2.0 \times 10^{18}$ Hz. What is the wavelength of these X-rays, and how does that wavelength compare to the size of an atom?

RECOGNIZE THE PRINCIPLE

The frequency and wavelength of an electromagnetic wave are related through $f\lambda = c$. Since the wave speed c is known, we can use the given value of f to find the wavelength. These X-rays are created by processes within the atoms of metals such as copper, molybdenum, and tungsten.

SKETCH THE PROBLEM

No sketch is necessary.

IDENTIFY THE RELATIONSHIPS

Using the relation between speed, frequency, and wavelength (Eq. 23.15), we get $\lambda = c/f$.

SOLVE

Inserting the given value of the frequency and the known value of c yields

$$\lambda = \frac{c}{f} = \frac{3.00 \times 10^8 \text{ m/s}}{2.0 \times 10^{18} \text{ Hz}} = \boxed{1.5 \times 10^{-10} \text{ m}}$$

What does it mean?
The radius of an atom is about 5×10^{-11} m, so the wavelength of these X-rays is approximately the diameter of an atom. As mentioned above, these X-rays are produced by atomic processes (so-called atomic transitions; see Chapter 29). Hence, a process involving electrons in atoms produces X-rays whose wavelengths are comparable to the size of an atom.

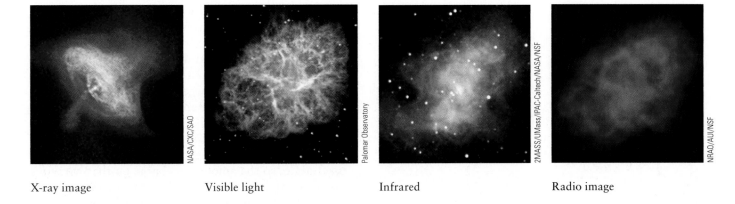

X-ray image   Visible light   Infrared   Radio image

Figure 23.17 Photographs of electromagnetic radiation emitted by the Crab Nebula, taken using different types of radiation. The colors in these images indicate the intensity of the radiation and not its true color.

CONCEPT CHECK 23.4 | Types of Electromagnetic Radiation

Place the following types of electromagnetic radiation in order according to their wavelength. Start with the radiation with the longest wavelength.

    (a) Infrared
    (b) X-rays
    (c) Blue light
    (d) Radio
    (e) Microwaves
    (f) Red light

Astronomy and Electromagnetic Radiation

In the next few sections, we'll describe some of the applications of electromagnetic radiation. These applications use different types of radiation, depending on the needs of the application. For example, most cell phones use microwaves, the optical fibers used to carry signals between computers use infrared and visible light, and tanning salons use UV radiation. One application—astronomy—uses virtually all types of electromagnetic radiation. Many objects in the universe such as stars and supernovas emit radiation over a wide range of frequencies, from radio waves to visible light to X-rays and gamma rays. For example, Figure 23.17 shows images of an object called the Crab Nebula. It is the remnants of a supernova, a star that exploded in the year 1054. The images shown here were constructed using different types of radiation. Comparisons of the different images give astronomers important information about how mass is distributed in the nebula and how it generates radiation.

23.5 | GENERATION AND PROPAGATION OF ELECTROMAGNETIC WAVES

In Section 23.4, we described many different types of electromagnetic waves. Let's now consider in more detail how one particular type of electromagnetic waves—radio waves—can be generated. Figure 23.18 shows an AC voltage source connected to two wires. These wires act as an *antenna*; they are not connected to anything else, so this setup is different from the AC circuits discussed in Chapter 22. As the voltage of the AC source oscillates, the electric potentials of the two wires also oscillate, and we indicate these potentials with the positive and negative signs in Figure 23.18. These signs also indicate electric charges present on the two wires; these charges flow from the voltage source onto and off the wires as the AC voltage alternates from positive to negative, back to positive, back to negative, and so on.

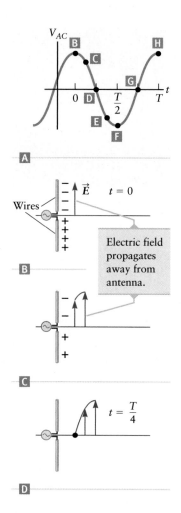

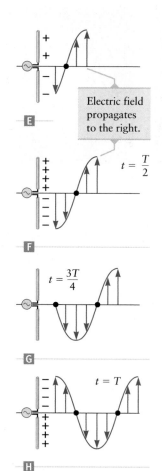

Figure 23.18 Electric field near a dipole antenna. **A** Variation of the AC voltage across the antenna as a function of time. **B** through **H** Snapshots of the electric field produced by the antenna as a function of time. As the electric field at the antenna oscillates, an electromagnetic wave propagates away from the antenna.

Figure 23.18A shows how the potential varies with time (just our usual AC voltage source), and the other parts of the figure show "snapshots" taken at different times as the potential difference between the two antenna wires oscillates. In part B of Figure 23.18, the potential difference is large; there are many negative charges on the upper wire and many positive charges on the bottom wire. These charges produce a large electric field directed upward as shown. As time passes (Fig. 23.18C), the potential difference becomes smaller in magnitude, so the amount of charge on the two wires decreases and the electric field is smaller. At the same time, the large electric field from part B has begun to propagate away from the wires. In Figure 23.18D, the potential difference is zero and the electric field at the wires is zero, and the fields generated at previous times (in parts B and C) continue to propagate away. Parts E through H of Figure 23.18 show the charge on the wires and the corresponding electric fields as time passes and the AC potential oscillates through one complete cycle.

The behavior of the electric field in Figure 23.18 is similar to the behavior of a wave on a string. When one end of a string is vibrated up and down, a wave is generated that propagates away along the string. Here the electric field acts in some ways like the string, and the AC voltage source generates a vibrating electric field at the antenna, similar to shaking the end of a string.

The pictures in Figure 23.18 also explain some other important features of electromagnetic waves. First, the charges move onto and off of the antenna wires in an oscillatory way. From our work on the motion of a simple harmonic oscillator (Chapter 11), we know that these charges are accelerated, confirming the claim we made in Section 23.2 that electromagnetic waves are produced whenever charges have a nonzero acceleration. Second, the charges in Figure 23.18 undergo harmonic motion at a frequency equal to the frequency of the AC voltage source, which is also the frequency of the electromagnetic wave that is generated. Whenever an electromagnetic wave with frequency f is observed, we can say that it was generated by an electromagnetic oscillator (such as an LC circuit) that oscillates at the same frequency f.

CONCEPT CHECK 23.5 | What Is the Direction of $\vec{B}$?

In Figure 23.18, we considered how charge on the wires produces electric fields that then generate an electromagnetic wave. At the same time, the motion of charges on and off the wires also gives rise to the magnetic field associated with the wave. The electromagnetic wave in Figure 23.18 propagates to the right. What is the direction of the magnetic field $\vec{B}$ near the antenna in the snapshot in Figure 23.18B? (a) Directed to the right, (b) directed to the left, (c) directed out of the plane (toward the reader), or (d) directed into the plane (away from the reader)?

Antennas

The basic approach in Figure 23.18, in which an AC voltage source is attached to a set of wires, is used to produce the waves that carry radio, television, and cell phone signals. The wires in Figure 23.18 are one type of antenna, but there are many other designs, some containing many wires (Fig. 23.9); the simple case with two wires is called a **dipole antenna**. The reason for the name "dipole" can be seen from Figure 23.19, which shows a dipole antenna with the wires parallel to the vertical (z)

direction. At any particular moment, the two wires are oppositely charged, forming a sort of electric dipole. For the antenna in Figure 23.19, the electric field is along z and the electromagnetic waves propagate away in the x and y directions (and in other directions in the x–y plane).

In our discussion of antennas and the electric fields they produce (Figs. 23.18 and 23.19), we have emphasized the electric fields near the antennas. There is also an electromagnetic wave propagating inside the antenna wires, with an electric field that oscillates between positive and negative values as one moves away from the AC source. For a very long antenna, these oscillations tend to cancel, so a very long antenna does not necessarily produce a "stronger" electromagnetic wave than a short antenna. This cancellation becomes significant when one half of the antenna is longer than about $\lambda/4$ (one quarter of the wavelength). For this reason, most dipole antennas have a total length of $\lambda/2$.

Although other antennas may be more complicated than the dipole design, the cancellation effect mentioned above still occurs, so the size or length of antenna is usually comparable to the wavelength of the radiation. That is why the antenna for your cell phone is so small (Fig. 23.20A), whereas the antennas for an AM radio station are quite large (Fig. 23.20B).

Figures 23.18 and 23.19 describe how an antenna generates an electromagnetic wave, but the same antennas can be used to detect these waves. Figure 23.21 shows an electromagnetic wave interacting with a dipole antenna that might be in your cell phone. The electric field associated with the wave exerts a force on the electrons in the antenna, producing a current and an induced voltage across the antenna wires. This voltage is then the input to an AC circuit inside your cell phone. That AC circuit amplifies the voltage signal and ultimately sends it to the phone's speakers.

Propagation and Intensity

The sketch of charge motion in Figure 23.18 shows how accelerated charges give rise to an electromagnetic wave. This picture is useful for understanding many radiation sources, but the charge motion is sometimes more complicated. For example, an ordinary lightbulb produces light due to the vibrations of electrons and ion cores in the hot filament of the bulb. Unlike the case in Figure 23.18, these vibrations in the filament are not confined to one particular direction; rather, charges vibrate along all different directions. As a result, the bulb produces light that propagates outward in all directions, unlike the case in Figure 23.18, in which the waves propagate only in the horizontal direction.

A dipole antenna emits waves that propagate perpendicular to the antenna.

Figure 23.19 Radiation from a dipole antenna propagates outward in directions perpendicular to the antenna's axis. There is essentially no radiation generated "off the ends" of the antenna. This antenna radiates strongly in all directions in the x–y plane, but does not emit radiation in the $+z$ or $-z$ directions.

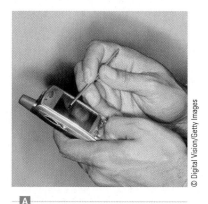

Figure 23.20 In most applications, an antenna's length is comparable to $\lambda/2$. **A** For a cell phone, the wavelength is short, so the antenna (visible at the upper left) is also short. **B** The wavelength of the radiation used by an AM radio station is much longer, so its antenna is very large.

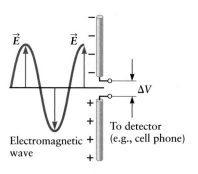

Figure 23.21 A dipole antenna can be used to detect an electromagnetic wave. The electric field of the wave induces a voltage across the wires of the antenna. This induced voltage is then amplified and processed by your cell phone or radio.

In many other cases, a source produces waves that propagate outward in all directions. The "ideal" case of a very small source producing spherical wave fronts is called a "point source." These wave fronts describe how energy propagates outward from the source. In Chapter 12, we saw that the intensity of a spherical wave falls with distance as (Eq. 12.11)

$$I \propto \frac{1}{r^2} \tag{23.16}$$

The intensity falls as one moves away from the source because there is a constant amount of energy and power associated with each wave front, whereas the wave front area ($A = 4\pi r^2$) increases with distance. (See Fig. 12.12, page 389, or Fig. 17.12, page 539.) Intensity is the power per unit area, so the intensity decreases as this constant amount of energy is spread over an ever-increasing area. The decrease of intensity with distance is thus a direct consequence of the principle of conservation of energy.

The relation between intensity and distance in Equation 23.16 applies to many cases, including the falloff of intensity for a lightbulb, the falloff of intensity of sunlight with distance through the universe, and the "strength" (i.e., intensity) of a radio signal from a distant radio station or space satellite.

EXAMPLE 23.5 | Intensity of Sunlight

The intensity of sunlight near the Earth is about 1000 W/m². Since the Earth is very far from the Sun, the Sun acts approximately as a point source of radiation. What is the intensity of sunlight at the surface of Mars?

RECOGNIZE THE PRINCIPLE

Because the Sun acts as a point source of spherical waves, its intensity falls with distance as $1/r^2$. We can use the known distances between the Sun and the Earth and the Sun and Mars to get both the relative intensity and the absolute intensity at Mars.

SKETCH THE PROBLEM

Figure 23.22 describes the problem.

IDENTIFY THE RELATIONSHIPS

We denote the Sun–Earth distance as r_E and the Sun–Mars distance as r_M. Using Equation 23.16, we can write the intensity at Earth as $I_E = K/r_E^2$ and the intensity at Mars as $I_M = K/r_M^2$. The constant of proportionality K is the same in these two cases, so taking the ratio, we have

$$\frac{I_M}{I_E} = \frac{K/r_M^2}{K/r_E^2} = \frac{r_E^2}{r_M^2} \tag{1}$$

From Table A.3 of solar system data in Appendix A, we find $r_E = 1.5 \times 10^8$ km and $r_M = 2.3 \times 10^8$ km.

SOLVE

Rearranging Equation (1) to solve for the intensity at Mars gives

$$I_M = I_E \left(\frac{r_E}{r_M} \right)^2$$

Inserting our values of the intensity at the Earth and the distances r_E and r_M we get

$$I_M = (1000 \text{ W/m}^2) \left(\frac{1.5 \times 10^8 \text{ km}}{2.3 \times 10^8 \text{ km}} \right)^2 = \boxed{430 \text{ W/m}^2}$$

What does it mean?
Since the intensity at the Earth is about 1000 W/m², the intensity at Mars is only 43% of that value. As a result, the surface temperature of Mars is much less than that of the Earth.

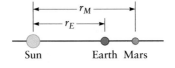

Figure 23.22 Example 23.5.

23.6 | POLARIZATION

When discussing an electromagnetic wave, it is important to know its propagation direction since that tells the direction in which energy is traveling. We know that the electric field must be perpendicular to the propagation direction, but there are still many different possible directions for $\vec{E}$ that are perpendicular to the wave direction. The direction of $\vec{E}$ is important because it determines how the wave interacts with matter (including antennas).

Figure 23.19 shows how a dipole antenna can be used to produce an electromagnetic wave with an electric field $\vec{E}$ directed along $\pm z$. To indicate that the electric field is always parallel to z, we say that this wave is *linearly polarized* along the z direction. Electromagnetic radiation that is linearly polarized is in many ways the simplest case to describe. Most light is not polarized in this way, however. For example, sunlight and the light from a flashlight are both *unpolarized*. Light from these sources is a combination of many individual waves, all with different field amplitudes and polarization directions.

Polarizers

Polarized light can be created using a device called a *polarizer*. One type of polarizer is shown in Figure 23.23. It consists of a thin, plastic film that allows an electromagnetic wave to pass through only if the electric field of the wave is parallel to a particular direction called the axis of the polarizer. The polarizer absorbs radiation with electric fields that are not along the axis. Hence, when unpolarized light (perhaps from a flashlight) strikes a polarizer, the light that comes out is linearly polarized.

To evaluate the intensity of the light transmitted by a polarizer, let's first consider what happens when linearly polarized light strikes a polarizer. In Figure 23.24A, the incident light is polarized parallel to the axis of the polarizer and the outgoing electric field is equal in amplitude to the incoming field, so $E_{out} = E_{in}$; all the incident energy is thus transmitted through. In Figure 23.24B, the incident light is polarized perpendicular to the polarizer axis and no light is transmitted, so $E_{out} = 0$. In Figure 23.24C, the incident light is polarized at an angle θ relative to the axis of the polarizer. The outgoing light is now polarized along the direction defined by the polarizer with an electric field amplitude

$$E_{out} = E_{in} \cos \theta$$

In words, only the component of the electric field along the polarizer axis is transmitted. Since intensity is proportional to the square of the electric field amplitude (Eq. 23.9), we can also write

$$I_{out} = I_{in} \cos^2 \theta \qquad (23.17)$$

which is called *Malus's law*. When $\theta \neq 0$, the outgoing intensity is smaller than the incident intensity and the outgoing light is linearly polarized along the direction of the polarizer axis.

EXAMPLE 23.6 | Electric Field and Intensity Transmitted through a Polarizer

Linearly polarized light is incident on a polarizer, with an angle θ between the incoming electric field and the axis of the polarizer (Fig. 23.24C). Find the value of θ for which the transmitted intensity is half the incident intensity.

RECOGNIZE THE PRINCIPLE

When the incident light is linearly polarized, only the component of the electric field along the polarizer axis is transmitted, and the transmitted intensity is given by Malus's law (Eq. 23.17). We can use this relation to find the value of θ for which I_{out} is half of I_{in}.

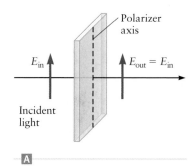

Figure 23.23 When unpolarized light is incident on a polarizer, only the component with the electric field parallel to the axis of the polarizer is transmitted.

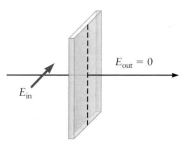

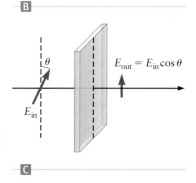

Figure 23.24 When linearly polarized light is incident on a polarizer, the transmitted intensity depends on the angle θ between the incident electric field E_{in} and the polarizer axis.

SKETCH THE PROBLEM

Figure 23.24C describes the problem. It shows the incident and transmitted electric fields.

IDENTIFY THE RELATIONSHIPS

From Malus's law, we have $I_{out} = I_{in} \cos^2 \theta$, which we can rearrange to read

$$\frac{I_{out}}{I_{in}} = \cos^2 \theta$$

SOLVE

We want the transmitted intensity to be half the incident intensity; hence, $I_{out}/I_{in} = \frac{1}{2}$, which leads to

$$\cos^2 \theta = \frac{I_{out}}{I_{in}} = \frac{1}{2}$$

$$\cos \theta = \frac{1}{\sqrt{2}}$$

$$\theta = \boxed{45°}$$

What does it mean?

When the angle θ between the polarization direction of the incident light and the axis of the polarizer is not zero, the transmitted intensity is smaller than the incident intensity. The transmitted light is still linearly polarized, but with a different polarization direction.

Malus's law applies when the incident light is linearly polarized, but it can also be used to understand what happens when unpolarized light strikes a polarizer. Unpolarized light can be thought of as a collection of many separate light waves, each linearly polarized but with different and random polarization directions. Each separate wave is transmitted through the polarizer according to Malus's law. The average outgoing intensity is then the average of I_{out} for all the incident waves. We can thus write

$$I_{out} = (I_{in} \cos^2 \theta)_{ave}$$

The average here is an average over all possible polarization angles for the incoming light waves. The factor $\cos^2 \theta$ varies between 0 and 1, and its average is $\frac{1}{2}$, so for unpolarized light the intensity transmitted through a polarizer is

$$I_{out} = \tfrac{1}{2}I_{in} \quad \text{(unpolarized light)} \tag{23.18}$$

An interesting application of Malus's law is sketched in Figure 23.25A, which shows light that is initially unpolarized before it passes through two polarizers in a row. After passing through polarizer 1, the light intensity is reduced by half (Eq. 23.18) and is now linearly polarized along z, the axis of polarizer 1. The axis of polarizer 2 is perpendicular to z, so the final transmitted intensity is zero.

Figure 23.25B shows a similar arrangement, but now we have inserted a third polarizer in the middle. The axis of polarizer 3 makes a 45° angle with the z axis, so some of the light from polarizer 1 is transmitted through polarizer 3. Moreover, after passing through polarizer 3 this light is polarized at a 45° angle with respect to polarizer 2. Hence, the final transmitted intensity after passing through all three polarizers is nonzero. Comparing this conclusion to the experiment in Figure 23.25A leads to the following "puzzle." In Figure 23.25A, the final transmitted intensity I_{final} is zero. In Figure 23.25B, we have added an extra polarizer, and now I_{final} is no longer zero! Although that may seem surprising, it can be understood by using two facts about polarizers. First, when analyzing light as it passes through

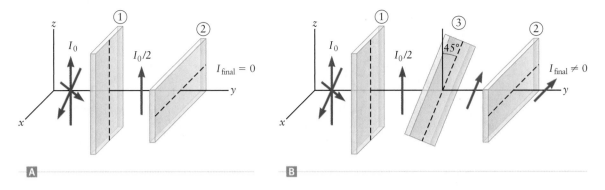

Figure 23.25 **A** Unpolarized light is incident on two polarizers whose axes are perpendicular. The final transmitted intensity is zero. **B** A third polarizer is placed between the polarizers in part A. Now the transmitted intensity is not zero!

several polarizers in succession, you must always analyze the effect of one polarizer at a time. Second, the light transmitted by a polarizer is always linearly polarized, with a polarization direction determined solely by the polarizer axis. The transmitted light wave has no "memory" of its original polarization.

CONCEPT CHECK 23.6 | Polarizers in "Series"

If the intensity of the initial unpolarized light beam in Figure 23.25B is I_0, is the final transmitted intensity (a) $I_0/2$, (b) $I_0/4$, (c) $I_0/8$, or (d) $I_0/16$?

How Does a Polarizer Work?

Polarizers have some interesting applications (including sunglasses and in the displays in watches and other devices) and can be made in several different ways. Most applications use a sandwich structure with certain types of long molecules (usually polyvinyl alcohol dyed with iodine) placed between two thin sheets of plastic.[1] When the molecules are aligned parallel to one another, the sheet acts as a polarizer, with the axis of the polarizer *perpendicular* to the direction of the molecules (Fig. 23.26).

Electrons in the polarizer molecules respond to electric fields as shown in Figure 23.26. When the electric field of an electromagnetic wave is parallel to the molecular chains, the electrons move back and forth along the molecule as the direction of $\vec{E}$ oscillates, producing an AC electric current along the molecular chains. This current leads to the absorption of energy, just like a resistor dissipates (absorbs) energy in an AC electric circuit. Light with a polarization direction parallel to the molecular direction is therefore absorbed. On the other hand, when $\vec{E}$ is perpendicular to the molecular direction, each electron is bound to a particular molecule and moves very little because it cannot "jump" from one molecule to another. The electrons then absorb very little energy, letting light with this polarization pass through. That is why the polarization axis is perpendicular to the molecular direction.

Polarization by Scattering and Reflection of Light

Most applications of polarized light use polarizers like those described in Figure 23.26, but polarized light is also found in nature. As shown in Figure 23.27A, light from the Sun is unpolarized, but sunlight can become polarized after it is scattered by molecules in the atmosphere through a series of events. The molecules act as antennas; the positive ion cores and electrons respond to sunlight by oscillating

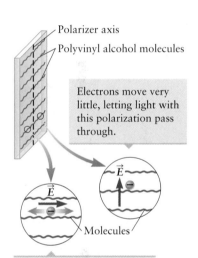

Figure 23.26 A film containing long molecules of polyvinyl alcohol aligned parallel to one another acts as a polarizer. The axis of the polarizer is perpendicular to the molecular direction. When the electric field is parallel to the molecules, electrons move easily along the molecules, producing an electric current that causes energy to be absorbed just like a resistor dissipates energy in an electric circuit. When the electric field is perpendicular to the molecular direction, the electrons move very little and most of the radiation passes through.

---

[1]This type of polarizer was invented by Edwin Land when he was an undergraduate student. He also invented "instant" film cameras (called "Polaroid" cameras).

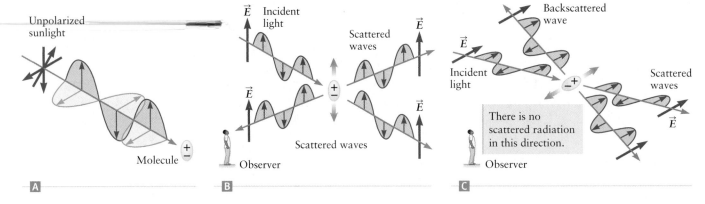

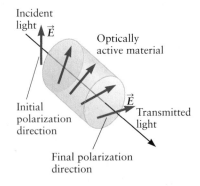

Figure 23.27 Polarization by scattering of light. **A** Sunlight is unpolarized. When light strikes molecules in the atmosphere, the electrons and ion cores oscillate in response to the electric force due to the light's electric field. **B** The vibrating charges act as a dipole antenna and produce scattered electromagnetic waves. The electric field of the scattered waves is parallel to the vibration direction. Here there is a scattered (and polarized) wave in the direction of the observer. **C** A dipole antenna does not emit radiation "off its ends," so radiation with this polarization does not propagate toward the observer in the position shown. Since the only radiation that reaches the observer has the electric field direction shown in part B, the scattered radiation is polarized even if the incident light is not.

back and forth in the direction of the electric field, like the charges in the antenna wires in Figure 23.18. Parts B and C of Figure 23.27 show this motion for light polarized in two different directions. In Figure 23.27B, the polarization direction and the direction of $\vec{E}$ are vertical; the ion cores and electrons thus vibrate up and down and produce new outgoing waves that are also polarized vertically. These outgoing waves are called *scattered waves*.

Figure 23.27C shows how the ion cores and electrons move when the incoming light is polarized horizontally. The ion cores and electrons now oscillate in a horizontal direction and again produce scattered light waves. Because dipole antennas do not radiate "off their ends" (compare with Fig. 23.19), no scattered light reaches the observer looking along the vibration direction.

Now let's combine the results in parts B and C of Figure 23.27 to understand how unpolarized sunlight is scattered. The component of this light polarized vertically (Fig. 23.27B) is scattered to the observer, but none of the horizontally polarized scattered light propagates in the observer's direction. Hence, the light that reaches the observer is said to be *polarized by reflection* (or by scattering). You can observe this polarization by looking at the sky through a polarizer and comparing to the case without a polarizer.

Similar effects can occur whenever unpolarized light is scattered. For example, sunlight scattered by dust particles on a windshield is polarized. This scattered light, called "glare," is usually a nuisance for a driver. It can be removed using the polarizers that are built into most sunglasses. We will return to the polarization of light in Chapter 24.

Optical Activity ⊗

When linearly polarized light passes through certain materials, the polarization direction is rotated as shown in Figure 23.28. This effect is called *optical activity*. Materials that are optically active usually contain molecules with a screwlike or helical structure. DNA is one of many examples; in fact, simple sugars such as fructose and dextrose are optically active molecules. In words, the electric field of linearly polarized light is "dragged along" the screw direction of the molecules. Dextrose and fructose molecules are both helical, but one forms a right-handed helix and the other is left-handed, so they rotate the polarization in different directions.

Figure 23.28 When linearly polarized light passes through an optically active material, the direction of polarization is rotated. The outgoing light is still linearly polarized, but with a different polarization direction. Optically active materials contain molecules with screwlike (helical) shapes. The screw direction determines the rotation direction.

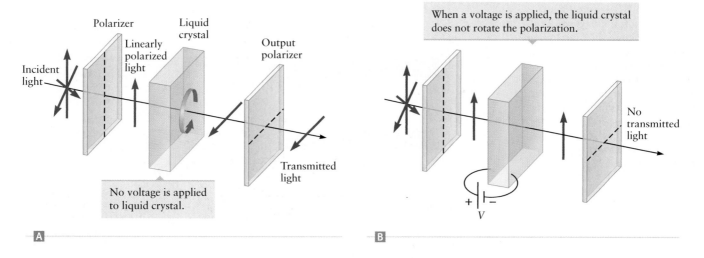

Figure 23.29 Light passing through a liquid crystal display. **A** When no voltage is applied to the liquid crystal layer, it rotates the plane of polarization by 90°, allowing light to pass through the output polarizer. **B** When a voltage is applied to the liquid crystal, it no longer rotates the plane of polarization and no light is transmitted through the output polarizer.

This effect is used in studies of sugar molecules in solution. The rotation direction gives information on the type of sugar present, while the amount of rotation is used to deduce the molecular concentration.

Applications of Polarized Light

The displays in wristwatches, cell phones, and many other devices make use of polarizers and optical activity. These displays are called LCDs ("liquid crystal displays") and work as follows. The incident light is linearly polarized with a polarizer sheet and then passes through a film containing an optically active material called a liquid crystal (Fig. 23.29A). The long, screw-shaped molecules in the liquid crystal rotate the plane of polarization by 90° so that the outgoing light can pass through a second "output" polarizer. To "turn off" the display, a voltage is applied to the liquid crystal, aligning the molecules in such a way that they no longer rotate the direction of polarization (Fig. 23.29B). The light that comes out of the liquid crystal is then polarized at 90° relative to the output polarizer, so no light is transmitted and the display is dark. By applying different voltages to different areas of the liquid crystal, a pattern of light and dark regions corresponding to different letters or numbers can be formed, resulting in the letters and numbers you see in the display of your wristwatch, cell phone, or MP3 player, or even the displays at gasoline station pumps. You can tell that the light from a display is polarized by placing another polarizer such as the polarizers in your sunglasses in front as shown in Figure 23.30. Try it.

Figure 23.30 A liquid crystal display on a wristwatch as viewed through a polarizer. The output light is polarized (Fig. 23.29A), so **A** when the plane of polarization is at a right angle with the axis of a polarizer placed in front of the display, no light is transmitted and the display is dark. **B** When the polarizer is rotated, light from the display passes through.

23.7 | DOPPLER EFFECT

In Section 13.6, we discussed the Doppler effect, a phenomenon in which the frequency of a wave is affected by the relative motion of the source and observer. The Doppler effect was used by the astronomer Edwin Hubble in his studies of the motion of stars and galaxies in the universe in the early 1900s. Light is emitted and absorbed by atoms of different elements at the surface of the star. From the work of physicists in the 1800s, Hubble knew that each atom emits and absorbs light of certain characteristic colors. Figure 23.31 (top) shows some of the colors absorbed by hydrogen, magnesium, and sodium on the surface of our Sun. Each vertical dark

Figure 23.31 Spectral lines showing absorption by hydrogen (H), magnesium (Mg), and sodium (Na) atoms. (Emission lines occur at the same wavelengths.) Top: Spectrum found when the atoms are at rest relative to the observer. Bottom: Spectral lines of the same element as we observe from a distant galaxy. Each line is shifted toward longer wavelengths (i.e., lower frequencies).

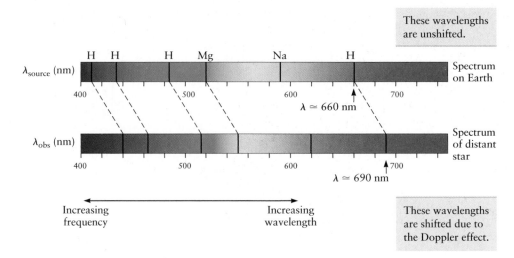

line in this plot corresponds to a color absorbed by these atoms, and the location of each line along the horizontal axis corresponds to a certain wavelength of light. This type of plot is called a *spectrum* of the starlight, and each vertical line (each color) is called a *spectral line*. A similar set of spectral lines is found if one measures the frequencies emitted by an atom; one finds that an atom absorbs and emits radiation at the same frequencies. The origin of these spectral lines was not understood until the discovery of quantum mechanics in the 1920s and will be explained in Chapter 29.

Hubble was using measurements of spectra to identify the elemental composition of distant galaxies when he observed cases in which the spectra did not seem to correspond to any known elements. One example is shown at the bottom of Figure 23.31. Hubble showed that spectra like this one are "shifted" versions of the expected spectra from known elements. In Figure 23.31, the spectral lines below can all be derived from those above by shifting each line to a longer wavelength (lower frequency), and the shift factor is the same for each line. In this example and in all others found by Hubble, the spectral lines of distant galaxies were always shifted to longer wavelengths relative to the wavelength of the same spectral lines emitted by atoms on Earth. Because red light is at the long-wavelength end of the visible part of the electromagnetic spectrum (Fig. 23.8), this effect is called a *red shift*.

To explain this result, Hubble proposed that the galaxy emitting the red-shifted light was moving away from the Earth. The Doppler effect then causes the frequency of the light as measured by an observer on Earth to be shifted to lower frequencies.[2] This shift is similar to the Doppler shift for the sound emitted by an ambulance siren when the ambulance is going away from you. The size of the frequency shift can be used to determine the velocity of the emitting galaxy. After many galaxies were studied in this way, it was found that most galaxies in the observable universe are moving away from the Earth.

At first glance, you might think that if most or all of the galaxies are moving away from us, the Earth must be at the "center" of the universe, but Hubble found that this picture is *not* correct. He showed that the farther a galaxy is from the Earth, the faster it is receding from the Earth, leading to the picture in Figure 23.32 of an expanding universe at two different times. The galaxy at point *B* might, for example, contain the Earth. As time passes, moving from early in the universe's history to later, the distance from *B* to galaxy *A* and from *B* to all other galaxies increases. Hence, *as viewed from the Earth*, all other galaxies are moving away from us. At the same time, observers in other locations find that other galaxies are also moving away from them. For example, the distances from galaxy *F* to galaxies *B*, *D*, and *E* also increase as time passes, so observers at *F* would also say that all other galaxies are moving away from them.

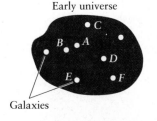

Early universe

Galaxies

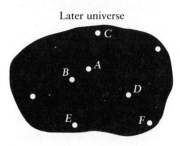

Later universe

Figure 23.32 Schematic of an expanding universe. As the universe expands, the distance between any two points (i.e., between any two galaxies) increases with time. The relative velocity between two nearby points is less than the relative velocity between two widely separated points.

---

[2]We'll say more about Hubble's discovery in Chapter 31, where we'll explain how this red shift is similar to, but not quite the same as, an ordinary Doppler shift.

Measurements of the red shift are now used to determine the motion of many different objects in the universe and are the basis of much of modern astronomy and astrophysics. The analysis of red shift experiments is based on how the Doppler effect causes the frequency seen by an observer f_{obs} to be shifted relative to the frequency of the source f_{source}. In Chapter 13, we derived relations for the Doppler shift for sound waves. Due to the effect of relativity (Chapter 27), however, the Doppler shift relations for light are different. For Doppler shifts of light and other electromagnetic waves, the frequency measured by an observer is given by

$$f_{obs} = f_{source} \frac{\sqrt{1 - v_{rel}/c}}{\sqrt{1 + v_{rel}/c}} \tag{23.19}$$

where v_{rel} is the velocity of the source relative to the observer. Here a positive value of v_{rel} corresponds to a source moving away from the observer.

EXAMPLE 23.7 | Ⓑ Doppler Shift of a Distant Galaxy

Figure 23.31 shows how several spectral lines from a distant galaxy are Doppler-shifted due to the galaxy's motion relative to the Earth. Use the data in this figure to estimate how fast this galaxy is moving relative to the Earth.

RECOGNIZE THE PRINCIPLE

When a light source is moving away from an observer, the wavelength of the light as measured by the observer is longer than the wavelength emitted by the source. Hence, the frequency seen by the observer f_{obs} is lower than the source frequency f_{source}.

SKETCH THE PROBLEM

Figure 23.31 shows the wavelengths at the observer λ_{obs} and source λ_{source}. From the data given in the figure, we can measure the ratio $\lambda_{obs}/\lambda_{source}$.

IDENTIFY THE RELATIONSHIPS AND SOLVE

If the velocity of the source galaxy relative to an observer on the Earth is v_{rel}, the frequencies at the observer and source are related by (Eq. 23.19)

$$f_{obs} = f_{source} \frac{\sqrt{1 - v_{rel}/c}}{\sqrt{1 + v_{rel}/c}}$$

Rearranging leads to

$$\frac{\sqrt{1 - v_{rel}/c}}{\sqrt{1 + v_{rel}/c}} = \frac{f_{obs}}{f_{source}}$$

$$\frac{1 - v_{rel}/c}{1 + v_{rel}/c} = \left(\frac{f_{obs}}{f_{source}}\right)^2$$

Because $f = c/\lambda$, we also have

$$\frac{1 - v_{rel}/c}{1 + v_{rel}/c} = \left(\frac{\lambda_{source}}{\lambda_{obs}}\right)^2 \tag{1}$$

Defining $y = (\lambda_{source}/\lambda_{obs})^2$, we get

$$1 - v_{rel}/c = y(1 + v_{rel}/c)$$

Next we solve for v_{rel}/c,

$$(v_{rel}/c)(1 + y) = 1 - y$$

$$v_{rel}/c = \frac{1 - y}{1 + y} = \frac{1 - 0.91}{1 + 0.91} = \boxed{0.05}$$

where we have inserted the value of $y = (\lambda_{source}/\lambda_{obs})^2 \approx 0.91$ derived from the red shift data in Figure 23.31. We thus get

$$v_{rel} \approx \boxed{2 \times 10^7 \text{ m/s}}$$

The positive value of v_{rel} indicates that the source is moving away from the observer.

What does it mean?
The value of v_{rel} is about 5% of the speed of light. Hubble discovered that the speed of a galaxy is related to its distance from the Earth by what is now known as Hubble's law. This law can be written as

$$d = \frac{v_{rel}}{H_0}$$

where H_0 is called Hubble's constant, which has the value $H_0 = 2.2 \times 10^{-18} \text{ s}^{-1}$. Inserting our value of v_{rel} into Hubble's law gives

$$d = \frac{v_{rel}}{H_0} = \frac{2 \times 10^7 \text{ m/s}}{2.2 \times 10^{-18} \text{ s}^{-1}} = 1 \times 10^{25} \text{ m}$$

This distance is incredibly large; it can be appreciated better if it is expressed in light-years, the distance that light travels in one year. We find

$$d = 1 \times 10^9 \text{ light-years}$$

The spectrum in Figure 23.31 thus involves light that spent about one billion years traveling to the Earth!

23.8 | DEEP CONCEPTS AND PUZZLES CONNECTED WITH ELECTROMAGNETIC WAVES

Maxwell's work and the subsequent experimental studies of electromagnetic waves by Hertz, Marconi, and others led to profound changes in the way we think about electromagnetism and the laws of physics.

Fields Are Real

We first introduced the notion of a *field* in our discussion of gravitation. When analyzing the force of gravity, lines of force led naturally to the concept of gravitational field lines. The gravitational field is certainly a convenient way to visualize and think about gravity, but, in the opinion of Newton and most of his contemporaries, forces were "real" and field lines were just a helpful picture.

We took a similar approach in our work with electromagnetism. There we started with electric forces given by Coulomb's law and then noticed that these forces could be conveniently described using the notion of electric field lines. Magnetic forces and magnetic fields could be understood in a similar way, but the question remains, "Are these fields real, or are they just a useful mathematical tool?"

The existence of electromagnetic waves means that electric and magnetic fields *are indeed real*. There is no other way to explain how an electromagnetic wave can propagate through a vacuum, carrying energy and momentum. An analogy with the gravitational field also suggests the existence of **gravitational waves**. Physicists are now working to detect the gravitational waves believed to come to us from distant galaxies.

How Can Electromagnetic Waves Travel through a Vacuum?

Although Maxwell's theoretical work (based on the equations of electricity and magnetism) showed that electromagnetic waves can travel through a vacuum, physicists at the time had difficulty accepting this result. All mechanical waves travel through a material medium, and most physicists believed that there must be a similar medium called the *ether* to support electromagnetic waves. It was believed that the ether permeated all space, including material substances and the vacuum, too. The ether was assumed to have some amazing properties; for example, an object moving through a vacuum would experience no frictional force due to the ether. By postulating the existence of the ether, it was claimed that electromagnetic waves were mechanical waves through the ether. There were many experimental attempts to study the ether and its properties, but eventually (around 1900) the existence of the ether was disproved. Since there is no ether, electromagnetic waves need no medium for travel. The electric and magnetic fields associated with an electromagnetic wave can carry energy even through empty space (a vacuum). This finding opened the way to Einstein's work and his theory of relativity (Chapter 27), which completely changed the way physicists view the universe. Maxwell's work was thus a catalyst for much of 20th-century physics.

Electromagnetic Waves and Quantum Theory

What is light? This question was debated by scientists for many centuries. Newton and others proposed that light is made up of particles, whereas other physicists suggested that light is a wave. We'll discuss the experimental evidence for both points of view in the next several chapters. Maxwell's work seemed to demonstrate conclusively that light is indeed a wave, but that is not the end of the story. We mentioned earlier that Newton's laws break down when applied to very small objects such as electrons and protons and that such objects must be described using quantum theory (Chapters 28 and 29). Maxwell's equations must also be modified to describe light and other electromagnetic waves at the quantum level. Physicists now know that light has *both* wavelike *and* particle-like properties. The "particles" of light are called *photons*.

SUMMARY | Chapter 23

KEY CONCEPTS AND PRINCIPLES

A time-varying magnetic field produces an electric field
According to Faraday's law, a changing magnetic flux Φ_B produces an induced emf $\mathcal{E}$ with $\mathcal{E} = -\Delta\Phi_B/\Delta t$. An emf $\mathcal{E}$ is just an electric potential difference (voltage) ΔV, and there is always an electric field associated with a voltage. Hence, Faraday's law means that *a changing magnetic field produces an electric field*. Maxwell showed that *a changing electric field produces a magnetic field*. This interdependency of E and B leads to the existence of *electromagnetic waves*.

(Continued)

Properties of electromagnetic waves

Electromagnetic waves are *transverse waves*: their electric and magnetic fields are perpendicular to each other, and both $\vec{E}$ and $\vec{B}$ are perpendicular to the propagation direction. These waves can travel through a vacuum and through many material substances. In a vacuum, they travel at a speed c:

$$c = \frac{1}{\sqrt{\varepsilon_0 \mu_0}} = 3.00 \times 10^8 \text{ m/s} \qquad \textbf{(23.1)} \text{ and } \textbf{(23.2)} \text{ (page 764)}$$

The *intensity* of an electromagnetic wave (I) is related to the amplitude of the electric field E_0 by

$$I = \tfrac{1}{2} \varepsilon_0 c E_0^2 \qquad \textbf{(23.9)} \text{ (page 767)}$$

The electric and magnetic field amplitudes are related by

$$E_0 = c B_0 \qquad \textbf{(23.6)} \text{ (page 766)}$$

In addition to carrying energy, electromagnetic waves carry momentum, leading to a *radiation pressure*. When light (or any other electromagnetic wave) is absorbed by an object, the radiation pressure on the object is

$$P_{\text{radiation}} = \frac{I}{c} \qquad \textbf{(23.13)} \text{ (page 769)}$$

Maxwell's equations

The equations of electromagnetism—the laws of Gauss, Coulomb, Ampère, and Faraday—are called *Maxwell's equations*.

APPLICATIONS

The electromagnetic spectrum

Electromagnetic waves are classified according to their frequency. This frequency ranges from a few Hz for *radio waves* to 10^{22} Hz and above for *gamma rays*. In between we have *microwaves*, *infrared radiation*, *visible light*, *ultraviolet (UV) light*, and *X-rays* (Fig. 23.8).

Polarization

Electromagnetic waves can be *polarized*. If the electric field vector points along a single direction, the wave is *linearly polarized*, whereas light from many sources (such as the Sun) is *unpolarized*. Light can be polarized by transmission through a polarizer film or by scattering and reflection.

Doppler effect

The frequency of an electromagnetic wave depends on the relative motion of the source and observer, the *Doppler effect* we studied in connection with sound in Chapter 13. This effect is used by astronomers to study the motion of distant galaxies.

1. A general rule of thumb when listening to and watching a thunderstorm is that the distance from a listener to the lightning strike can be estimated in the following way: First, count the number of seconds between the time you first see the lightning and the time you first hear the thunder. Then multiply this time (in seconds) by 1000 ft to get the distance to the lightning strike. Explain the origin of this rule of thumb.

2. Most vehicles have a radio antenna that is oriented along the vertical direction. Why? Is having an antenna mounted at a slanting angle an advantage or disadvantage? Explain.

3. Electromagnetic waves do not easily penetrate far into metal surfaces. Discuss why that is the case in terms of the known properties of metals, atoms, and electrons.

4. Microwave ovens heat food items by bathing them in electromagnetic waves, yet you can watch your food cook through a window in the door (Fig. Q23.4). How is it possible for the door, with only a thin layer of metal with a grid of holes, to contain the powerful electromagnetic radiation?

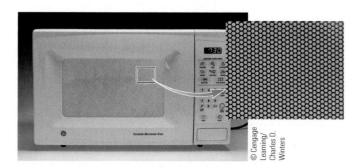

© Cengage Learning/Charles D. Winters

Figure Q23.4

5. A method for spacecraft propulsion called a "solar sail" has been proposed. (See Example 23.3 and Fig. 23.6.) It would use a sail exposed to the Sun in much the same way a sailboat's sail uses wind. The sail on such a spacecraft might reflect light, or it might absorb light. Which type of sail would yield the largest force on the sail? Explain your answer.

6. Consider the radiation pressure on an object in outer space. Explain why, all else being similar, this pressure has a larger effect on the motion of a small particle than on a large particle.

7. Explain how an AC circuit can generate electromagnetic radiation.

8. Consider the *RC* circuit in Figure Q23.8. The switch is closed at *t* = 0 and then kept closed for a very long time. Will this circuit emit electromagnetic radiation (a) just after the switch is closed? (b) After a very long time? Explain your answers. Can you think of ways you might have observed radiation when you opened or closed a switch?

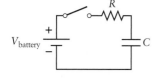

Figure Q23.8

9. In the United Kingdom (and many other European countries), residents are not allowed to use a TV without purchasing a yearly license. Fines of 1000 British pounds are levied against those who are caught using a TV without a license. The law is enforced through devices that can detect a TV in use from 20 m away or more. How might this device, such as the handheld unit shown in Figure Q23.9, work? *Hint*: All TVs have *LRC* circuits in them.

© Touhig Sion/Corbis Sygma

Figure Q23.9

10. SSM The glare off horizontal surfaces that you see while driving a car on a sunny day can be minimized if you wear sunglasses with polarizer films built into the lenses so that light polarized in a certain direction is not allowed to reach your eyes. Discuss how such sunglasses can reduce glare. Is the polarization axis of the glasses vertical or horizontal?

11. Suppose you are given a polarizer, but the axis of polarization is not marked. Devise an experiment you could do to determine the polarization axis.

12. Photographers often use a "haze filter" to remove (or at least minimize) the effects of haze on their photographs. Explain how these filters work.

13. It is a very dusty day, and you are standing by the side of the road looking directly across the road as a car is approaching (Fig. Q23.13). Due to the dust particles in the air, you can see light from the car's headlights as it scatters from the particles. Will this scattered light be (partially) polarized? If so, will the polarization axis be horizontal or vertical?

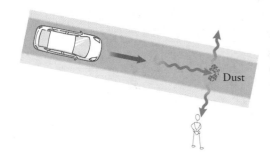

Dust

Figure Q23.13

14. Light from bulbs and other forms of indoor lighting is typically unpolarized, whereas radio waves picked up by radios or cell phones are mostly polarized. Explain why these two types of electromagnetic radiation are different in this way.

15. Can sound waves be polarized? What about ocean waves or waves propagating on a string? What fundamental characteristic allows waves to have a polarization?

16. Consider a bright sunny day at the beach. Why is it that polarizing sunglasses work well at reducing the glare off the sand and water except when you lie on your side (or tilt your head toward a shoulder)?

17. In Figure 23.5, we analyzed the force exerted by the electric and magnetic fields of an electromagnetic wave on a positive charge at the surface of a material when the electric field has the direction shown in Figure Q23.17A. (a) Assume the charge is instead negative (i.e., an electron). Show that the combination of electric and magnetic fields shown in Figure Q23.17A leads to a force on the electron with a component to the right. (b) After a short time, the electric field of the wave at the location of the electron changes direction and is pointing down as shown in Figure Q23.17B. Consider again the combined electric and magnetic force on the electron and show that it again has a component to the right. The conclusion of this analysis is that although the electric and magnetic fields oscillate in direction, there is over time always a component of the force along the direction of propagation (except at instants when the force is zero). This force is connected with the wave's momentum.

Figure Q23.19

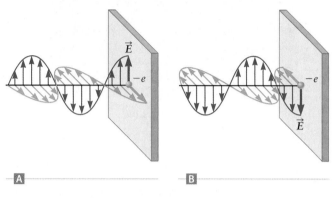

Figure Q23.17

18. Figure 23.18 shows the direction of the electric field produced by an antenna. Sketch the direction of the corresponding magnetic field near the antenna for parts B through H of the figure.

19. Fluorescent lights will glow faintly if brought under a high-tension power line as shown in Figure Q23.19. The lights glow even though they are not plugged into any fixture. Give an explanation of why they glow under the power lines.

20. Figure Q23.20 shows linearly polarized light striking a polarizer, with different angles between the incoming electric field $\vec{E}_{in}$ and the polarizer axis. Is there a force on the polarizer due to the incoming light? If so, what is the direction of the force and in which case is the magnitude of the force largest? Explain.

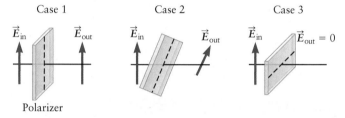

Figure Q23.20

21. Air traffic controllers at an airport use radar to monitor the location of nearby planes. Their radar units deduce this distance from the time it takes a radio wave to travel from their radar antenna to the airplane and back. The radar signal also gives information about the size and type of the plane. How do they do it?

PROBLEMS

SSM = solution in Student Companion & Problem-Solving Guide
★ = intermediate   ✪ = challenging

⊗ = life science application
Ⓡ = reasoning and relationships problem

23.2 PROPERTIES OF ELECTROMAGNETIC WAVES

23.3 ELECTROMAGNETIC WAVES CARRY ENERGY AND MOMENTUM

1. Ⓡ Approximately how long does it take a light wave to travel from your head to your toes? Assume you are standing straight up.

2. Figure P23.2 shows the directions of the electric and magnetic fields associated with several different electromagnetic waves. What are the directions of propagation of these waves?

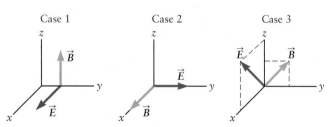

Figure P23.2

3. ✪ ⊗ The human eye can detect light intensities as small as about 1×10^{-23} W/m². (a) What is the corresponding electric field amplitude? (b) Suppose the eye senses this light by detecting the force on electrons inside a photoreceptor cell within the eye. What is the electric force exerted on an electron by the electric field found in part (a)? Is it a large force or a small force? Explain your answer.

4. ★ The electric field of an electromagnetic wave lies along the x axis, whereas the magnetic field is parallel to the z axis. What is the direction of the wave? *Note*: There are two correct answers.

5. SSM ★ The light from a red laser pointer has a wavelength of about 600 nm. If this laser has a power of 0.10 mW, what is the momentum carried by a 3.0-s pulse of this radiation?

6. ★ Ⓡ How long does it take light to travel from (a) the Earth to the Moon? (b) The Sun to the Earth? (c) Jupiter to the Earth, assuming Jupiter is at its closest distance to the Earth? (d) Jupiter to the Earth, assuming Jupiter is at its farthest distance to the Earth? The difference between the times in parts (c) and (d) was used by Danish astronomer Ole Romer to give the first experimental measurement of the speed of light. (Romer did this work in the 1670s, about the time Newton was doing his work on mechanics.)

7. ★ Ⓡ The spacecraft that have landed on Mars send their information to the Earth via radio waves. How long do these waves take to reach the Earth when (a) Mars is at its closest to the Earth? (b) Mars is farthest from the Earth? This time delay is important for NASA when it sends a spacecraft to Mars.

8. ✪ Ⓡ Galileo devised an experiment to measure the speed of light. He and an assistant went to the top of nearby mountains

and did the following. (1) Galileo switched on a lantern that was visible to his assistant. (2) Immediately after seeing the light from Galileo's lantern, the assistant switched on his lantern so that Galileo could see it. (3) Galileo then measured the interval between the time that he (Galileo) switched on his lantern and the time that he first saw the light from his assistant's lantern. Suppose the distance between Galileo and his assistant was 5 km. (a) What is the round-trip travel time for the light in this experiment? (b) Do you think that this experiment was successful? *Hint:* The typical human reaction time is about 0.2 s. (c) Now include the typical human reaction time of 0.2 s in your analysis and answer the following question. How far apart should Galileo and his assistant be in order to observe the contribution to the total time delay from the speed of light?

9. ✪ ® In 2005, the spacecraft named *Huygens* landed on one of the moons of Saturn. The data gathered by *Huygens* were relayed by radio to the Earth by a second satellite in orbit about the moon. Estimate the time required for this radio signal to reach the Earth.

10. ✪ ® A radio station called WWV broadcasts from Fort Collins, Colorado. This station broadcasts the time and has served as a "standard clock" for a many years. The radio waves transmitted by this station travel through the atmosphere at approximately the speed of light, so they reach different listeners at different times. Suppose the signal announcing 9:00 AM reaches a person in Chicago at time $t = 0$ (on her local stopwatch). At what time (on this stopwatch) will this signal reach (a) a listener in New York City? (b) A listener in Honolulu?

11. The intensity of the electromagnetic radiation from the Sun when it reaches the Earth's surface is approximately 1000 W/m^2. What is the amplitude of the associated magnetic field?

12. The intensity of moonlight when it reaches the Earth's surface is approximately 0.02 W/m^2 (for a full moon). What are the amplitudes of the associated electric and magnetic fields?

13. An electromagnetic wave has an electric field amplitude of 20 V/m. What is the magnetic field amplitude associated with this wave?

14. SSM ✪ The magnetic field associated with an electromagnetic wave has an amplitude of $1.5 \ \mu\text{T}$. (a) What is the amplitude of the electric field associated with this wave? (b) What is the intensity of this wave? (c) What is the radiation pressure associated with this wave? (d) How much momentum is carried by a 2.0-s pulse of this wave over an area of 20 m^2?

15. ® What is the electric field amplitude associated with the light from a laser pointer? Assume a power of 0.10 mW with a wavelength of 600 nm.

16. ✪ What is the electric field amplitude at a distance of 0.75 m from a 100-W lightbulb? Assume all the power of the bulb goes into light of a single color with $\lambda = 500$ nm and assume the bulb produces a spherical wave.

17. ✪ ® You are initially standing at rest on the surface of a frozen lake where the ice is perfectly frictionless. At $t = 0$, you turn on a laser pointer ($P = 1.0$ mW) and shine it in a horizontal direction. After a period of time, you notice that you are moving in a direction opposite to the light from the laser pointer. Explain why that is so and estimate your speed after 10 min.

18. SSM ✪ Compare the electric field amplitude for the laser pointer in Problem 15 with the electric field from a point charge. Consider a point charge Q and suppose the magnitude of the electric field at a distance of 1.0 cm from Q is equal to the electric field from the laser pointer. Find Q. Express your answer in coulombs and as an equivalent number of electrons.

19. Radar is used to scan for airplanes near an airport. If an airplane is 30 km from the airport, what is the round-trip travel time for the electromagnetic wave used by the radar system?

20. ✪ ® The satellite TV system used by the author employs a satellite in geosynchronous orbit about the Earth. This satellite transmits an electromagnetic wave with a total power of approximately 40 W. Assume all this power is at a single frequency and is broadcast to an area of just the continental United States. Estimate the following quantities. (a) The author's satellite dish has a diameter of about $d = 50$ cm. What is the total power received by this antenna? (b) What is the amplitude of the electric field received by the antenna? (c) Estimate the magnitude of the voltage that is induced at the antenna. For simplicity, assume $|V| \approx Ed$.

21. ® **Global Positioning System.** Global Positioning System (GPS) satellites transmit radio signals used by receivers on the Earth to determine position with very high precision. The operating principle is that a receiver gets signals simultaneously from several different satellites (typically at least four) and uses them to "triangulate" its position. This method relies on precise knowledge of the time and also the time it takes for a satellite signal to reach the receiving position. Until a few years ago, the U.S. government intentionally degraded the timing information so that it had an uncertainty of about 350 ns. Based on this timing error, estimate the precision in meters with which the position can be determined.

22. ✪ ® The GPS satellites described in Problem 21 are placed in orbits for which the period is approximately 12 h. (They are thus not in geosynchronous orbits.) Estimate the time it takes for a GPS signal to travel to the Earth's surface from (a) a GPS satellite that is directly overhead and (b) a GPS satellite that is on the horizon.

23.4 TYPES OF ELECTROMAGNETIC RADIATION: THE ELECTROMAGNETIC SPECTRUM

23. The light used in a CD player has a frequency of about 5.0×10^{14} Hz. What is its wavelength?

24. ✪ ® What is the approximate frequency at which the Sun's radiation has the highest intensity?

25. ✪ The human eye is most sensitive to light with a wavelength of about 550 nm. What is this light's frequency?

26. The dial on an FM radio contains numbers ranging from about 88 to about 108. These numbers correspond to the frequency of the radio station as measured in megahertz. What is the corresponding range of wavelengths?

27. SSM X-rays are electromagnetic waves with very short wavelengths. Suppose an X-ray has a wavelength of 0.10 nm. What is its frequency?

28. ✪ The electromagnetic waves used in a microwave oven have a wavelength of 12.24 cm. (a) What is the frequency of this radiation? (b) Electromagnetic waves in a microwave can exhibit standing waves just as we discussed for other types of waves in Chapters 12 and 13. If there is a standing wave in your microwave oven, what is the distance between nodes? (c) Based on your answer in part (b), why do you think microwave ovens contain rotating carousels?

29. Some cordless telephones use radio waves with a frequency near 2.4 GHz to transmit to their "base station." What is the wavelength of these waves?

30. AM radio stations use frequencies near 1 MHz. What is the corresponding wavelength?

23.5 GENERATION AND PROPAGATION OF ELECTROMAGNETIC WAVES

31. The U.S. Navy uses extremely low frequency electromagnetic waves to communicate with submarines. Suppose the frequency of these waves is 50 Hz. What is the wavelength? Typically, the size of an antenna is equal to about half of the wavelength. Would this antenna fit in your backyard?

32. ⭐ Ⓡ Consider an antenna used to transmit your favorite FM radio station ($f = 103$ MHz). (a) How long would a dipole antenna be for this frequency? (b) Compare your result in part (a) to the height of a typical radio station tower. Does your answer to part (a) make sense?

33. ⭐ Ⓡ (a) What is the approximate size of the antenna in your cell phone? (b) Based on the length of this antenna, estimate the wavelength and frequency of the radiation used by the cell phone.

34. The "wireless" connection used by your laptop computer employs electromagnetic waves with a frequency near 2.4 GHz. (a) What is the wavelength of this radiation? (b) What is the approximate size of a dipole antenna for this frequency? (c) Will such an antenna fit easily into the case of your laptop computer?

35. A dipole antenna is oriented vertically as shown in Figure P23.35 and is used to generate an electromagnetic wave. A second dipole antenna is used to receive this radiation. How should the second antenna be oriented? (a) vertically along z, (b) horizontally along x, (c) horizontally along y, or (d) it doesn't matter. Explain your answer.

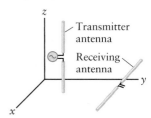

Figure P23.35

36. ⭐ The intensity of light from a lightbulb is 10 W/m² at a distance of 0.50 m from the bulb. What is the intensity across the room at a distance of 3.5 m?

37. SSM ⭐ The distance from the Sun to Jupiter is 5.2 times longer than the Earth–Sun distance. If the intensity of sunlight at the Earth is 1000 W/m², find (a) the intensity of sunlight at Jupiter and (b) the amplitude of the associated electric field at Jupiter.

38. ⭐ Light from the star Vega has an intensity of about 2×10^{-8} W/m². If Vega emits radiation with the same power as our Sun, how far is Vega from the Earth?

23.6 POLARIZATION

39. SSM Unpolarized light with an electric field amplitude of 0.25 V/m is incident on a polarizer. What is the electric field amplitude of the transmitted light?

40. Linearly polarized light propagating along the y direction is incident on a polarizer whose axis is parallel to the z direction. If the intensity of the transmitted light is equal to 35% of the incident intensity, what is the angle of polarization of the incident light?

41. ✪ Unpolarized light with intensity I_0 is incident on two polarizers as sketched in Figure P23.41. Suppose the angle between

the axes of the two polarizers is $\theta = 60°$. What is the intensity of the transmitted light? Express your answer as a fraction of I_0.

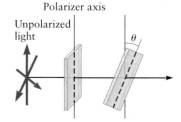

Figure P23.41 Problems 41 and 42.

42. ✪ For the polarizers in Figure P23.41, suppose the incident light is linearly polarized, the transmitted intensity (through both polarizers) is $0.15 \times I_0$, and the axis of second polarizer makes an angle $\theta = 60°$ with the axis of the first polarizer. What is the angle the initial direction of polarization makes with the first polarizer?

23.7 DOPPLER EFFECT

43. SSM ⭐ When viewed by an observer on the Earth, light from a distant galaxy is Doppler shifted to the red such that the ratio of the unshifted frequency to the observed frequency is 2.5. What is the galaxy's speed relative to the Earth?

44. We know from experience that the Doppler shift of sound from a moving car is easily detectable. Assuming light with a frequency of 6.0×10^{14} Hz, estimate the Doppler shift for *light* from a car moving toward you at a speed of 50 m/s (about 100 mi/h). Will this light be shifted to the blue (a higher frequency) or to the red (lower frequency)?

45. ⭐ A galactic accident causes a red giant star to be moving on a collision course with our solar system. This star is moving very fast, and its light is Doppler shifted. Which of the following statements might be true?
(a) Light from the star is observed as blue.
(b) Light from the star is observed as X-rays.
(c) Light from the star is observed in the infrared.
(d) Light from the star interferes with cell phone reception.

46. ⭐ Estimate the Doppler shift for a radio wave reflected from an airplane moving directly toward you at a speed of 250 m/s. Assume the radio wave has a frequency of 1.0×10^7 Hz.

47. ✪ How fast would a star have to travel to make violet light ($\lambda = 400$ nm) appear to be red ($\lambda = 600$ nm)? Assume the star is moving along the line that connects the star to the Earth. Is this star moving toward you or away from you?

48. ⭐ Consider a Doppler radar system that monitors the wind speed of a tornado by measuring the shift in frequency of radio waves reflected from water droplets or other particles in the air. If this wind speed is 45 m/s (about 100 mi/h) and the transmitted radio wave has a frequency of 1.0 MHz (10^6 Hz), what is the frequency of the reflected wave? Assume the droplets reflecting the wave signal are moving toward the antenna. *Hint*: Take all given numbers as "exact" and keep a large number of significant figures in your final calculation.

ADDITIONAL PROBLEMS

49. ⭐ A powerful AM radio station has a radiated power of 100,000 W. Suppose this energy is emitted into a region shaped as a half-sphere (Fig. P23.49). If you are a distance of 100 km from the station, what is the electric field amplitude there?

50. ✪ Ⓡ (a) Estimate the radiation pressure exerted by sunlight on the Earth. (b) What is the total force on the Earth due to this radiation pressure? (c) How does this force compare with the force of gravity exerted by the Sun on the Earth?

51. SSM ⭐ The intensity of sunlight near the Earth is approximately 1000 W/m². (a) If a spacecraft near the Earth uses a

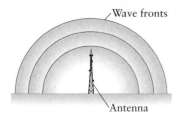

Figure P23.49

square solar sail 200 m on a side and absorbs all light that strikes it, what is the force on the spacecraft? (b) This spacecraft eventually reaches Jupiter, where its distance from the Sun is 5.2 times larger than the distance from the Sun to the Earth. What is the force on the sail now?

52. ★ The momentum carried by sunlight can be used by a solar sail to propel a spacecraft (Example 23.3). (a) Is the force on a sail directed toward or away from the Sun? (b) Suppose a solar sail is used for a spacecraft traveling from the Earth to Pluto. What is the ratio of the force on the sail when it is near Pluto to the force when it is near the Earth?

53. ✪ ✖ **LASIK eye surgery.** Lasers are used extensively in eye surgery to remove unwanted "material" from the eye or reshape the eye's lens. These lasers use pulses of very intense radiation to literally "blast" away material. These pulses are typically 10 ns long (10^{-8} s) and carry an energy of 2.0 mJ each. (a) What is the power during one of these pulses? (b) If this energy is delivered to a spot on the eye that is 1.0 mm in diameter, what is the intensity during a pulse? (c) If the radiation has a wavelength of 350 nm, what is the electric field amplitude during a pulse?

54. ✪ Consider the force of an electromagnetic wave on the electrons in an antenna. Assume the amplitude of the electric field associated with the wave is $E_0 = 200$ V/m. (a) What is the magnitude of the electric force on an electron? Express your answer in terms of e and E_0, and in newtons using the given value of E_0. (b) What is the magnitude of the magnetic force on an electron in the antenna? Express your answer in terms of e, B_0 (the magnetic field amplitude), and the speed v of the electron. (c) Suppose v is the drift velocity of an electron. In Chapter 19 (Section 19.3), we found that for typical values of the current the drift velocity is about 0.01 m/s. What is the magnetic force on the electron? What is the ratio of the magnetic force to the electric force found in part (a)? (d) At what speed v does the magnetic force on an electron equal the electric force?

55. ★ Ⓡ A college football stadium happens to be in your neighborhood, and you are watching the game at home on your TV. The team makes a touchdown, and the capacity crowd in the stadium cheers. Like an echo, 3.5 s after you see the touchdown and hear the cheering on your home TV, you then hear the cheer of the crowd through your window. Approximately how far away, in kilometers, is the stadium from your house? Assume there are no additional delays at the TV station.

56. ✪ **Lightning Detection and Ranging (LDAR).** Lightning is a rapid transfer of charge from a cloud to the ground (or other parts of the cloud) and constitutes a very large, varying current. When a lightning strike occurs, a large pulse of electromagnetic radiation propagates outward in a wide range of frequencies from light waves to radio waves. NASA has developed a system called LDAR that is implemented near Cape Canaveral to monitor the sometimes sudden and dangerous lightning storms that pass through Florida. The difference in time of arrival of these radio waves is used in trilateration ("triangulation") to measure the position of a lightning strike. Consider three antennas arranged in a right triangle formation as shown in Figure P23.56. Each antenna can pick up a radio pulse from as far away as 100 km. Placing antenna 1 at the origin, find the

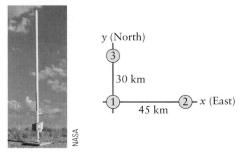

Figure P23.56 An image of an LDAR antenna station and a geographic arrangement of three such stations.

x and y coordinates of a lightning strike given the following travel times to antennas 1, 2, and 3: 93 ns, 131 ns, and 51 ns, respectively.

57. ✪ Consider three polarizing filters oriented as shown in Figure P23.57. A beam of unpolarized light is directed through the three filters. The first filter has a polarization axis along the vertical, the second has its polarization axis rotated at an angle of 45° from the vertical, and the third has its axis along the horizontal. (a) What fraction of the light intensity comes out of the third filter? (b) What fraction makes it through the system if the middle filter is removed? (c) Explain how removing a filter can result in less light transmitted. What if the first or third filter were removed instead of the middle filter? Does the result change? Explain.

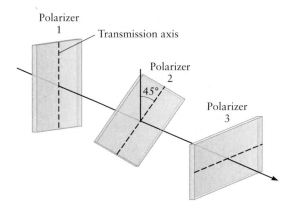

Figure P23.57

58. ✪ **Speed limit enforced by radar.** The use of Doppler radar in the enforcement of traffic law has become common in the United States and many parts of the world. A typical "radar gun" (Fig. P23.58) emits a 24-GHz beam. A highway patrol officer aims the beam at a bus on the highway. (a) If the reflected wave is 4.5 kHz higher in frequency than the emitted beam, how fast is the bus traveling in units of meters per second and miles per hour? (b) Is the bus approaching or moving away? *Hint:* Here there are two shifts to the emitted beam: one because the wave incident on the moving bus is Doppler shifted (as if the bus were stationary with the radar gun approaching) and a second shift from the beam reflected off the moving bus (equivalent to the bus sending out the frequency it sees).

Figure P23.58

59. SSM ★ Ⓡ How fast would a car need to travel to make a yellow light appear to be green as the car approached an intersection? Would a judge accept the excuse of such a Doppler shift for running a yellow light?

60. ✪ The radiation pressure from the Sun has a negligible effect on the orbit of a satellite, but it does pose a problem in terms of keeping a spacecraft pointing in the appropriate orientation. Consider the GOES-N geosynchronous weather satellite shown in Figure P23.60. The satellite body is roughly a cube 2.0 m on

a side with a 2.0-m by 4.0-m solar panel attached to the center of one face of the satellite's body. Assume the center of mass of the spacecraft is at the center of the cube. (a) Find the maximum force on the solar panel from the radiation pressure when the panel is perpendicular to the Sun's rays. (b) Calculate the torque on the spacecraft about its center of mass when the Sun's rays are perpendicular to the solar panel. (c) Estimate the moment of inertia about the center of mass of the satellite, assuming uniform density and assuming the solar panel and other projections are negligible compared with its overall mass of 900 kg. (d) What maximum rotational velocity, in degrees per hour, will the torque produce from a 5-min exposure to full sunlight? Assume the initial angular speed is zero.

Figure P23.60 GOES-N geosynchronous weather satellite.

61. Shooting the Moon. The *Apollo* astronauts placed special reflector arrays on the Moon (Fig. P23.61A). The array was comprised of "corner reflector" prisms that have the property of reflecting a beam of light exactly back along its incoming path. Using the Apache Point Observatory Lunar Laser-Ranging Operation (Fig. P23.61B), powerful laser pulses are directed at the Moon, and the time it takes them to reflect back is measured. (a) If the time of reflection can be measured to an accuracy of one hundredth of a nanosecond, what is the

uncertainty in the measured Earth–Moon distance? (b) A 2.3-W laser is used to get a very faint reflected pulse. When the laser's beam reaches the lunar surface, it is approximately 2.0 km in diameter. What is the intensity of the laser light at the Moon's surface, and how does it compare to sunlight there? Is shooting the Moon with this laser destructive to the lunar surface?

Figure P23.61 ◭ An *Apollo* reflector array on the lunar surface. ◉ The Apache Point Observatory Lunar Laser-Ranging Operation. The bright object in the sky in this photo is the Moon. The photo is overexposed so as to show the laser beam used in the experiments (visible as it approaches the Moon from the right).

62. ✪ ℝ Consider a cell phone that emits electromagnetic radiation with a frequency of 1.0 GHz (1.0×10^9 Hz) with a power of 1.0 W. (a) What is the approximate electric field amplitude a distance 1000 m from the phone? Assume it radiates as a point source. (b) What is the approximate electric field amplitude at your ear as you use the phone? Again assume it radiates as a point source. (c) If the phone can be turned on and transmitting for 3 h before discharging the battery, how much energy can be stored in the battery?

Geometrical Optics

The optical images formed by mirrors and lenses play an important role in everyday life. Here two women play with their images in a funhouse mirror. (© Charles Gupton/Stone/Getty Images)

The study of light is called **optics**. This area of physics dates to at least the third century BC. Eyeglasses were invented around AD 1300, and microscopes and telescopes were first developed around 1600. These applications are based on the ability of lenses and mirrors to focus light, and that is the main topic of this chapter. Light is an electromagnetic wave, and our discussions of optics in this chapter and in Chapters 25 and 26 will need to account for the wave nature of light using concepts such as wave fronts and rays. So, now would be a good time to review these and other basic properties of waves discussed in Chapter 12.

In this chapter, we consider **geometrical optics**, the regime in which light travels in straight-line paths and effects involving wave interference are not important. In general, geometrical optics describes cases in which the wavelength of light is much smaller than the size of objects in the light's path. The wavelength of visible light is less than 1 μm (10^{-6} m) and is about one hundred times smaller

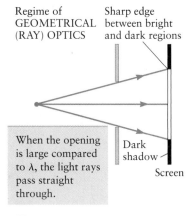

Regime of GEOMETRICAL (RAY) OPTICS

Sharp edge between bright and dark regions

When the opening is large compared to λ, the light rays pass straight through.

Dark shadow

Screen

A

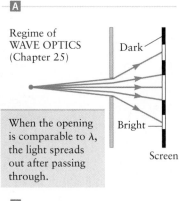

Regime of WAVE OPTICS (Chapter 25)

Dark

When the opening is comparable to λ, the light spreads out after passing through.

Bright

Screen

B

Figure 24.1 Rays of light passing through an opening and onto a screen. **A** A large opening casts an ordinary shadow on the screen. This is the regime of geometrical (also called ray) optics. **B** If the size of the opening is comparable to or smaller than the wavelength λ, the light "spreads out" as it passes through the opening. Instead of just a bright spot and a dark shadow, there is a pattern of bright and dark regions that can only be explained by taking into account the wave nature of light (Chapter 25).

than the diameter of a human hair. Most objects we work with in everyday life are much larger than that, so geometrical optics describes many everyday applications of optics, including the behavior of lenses and mirrors. In this chapter, we'll study how lenses and mirrors work, laying the foundation for our discussion of optical devices such as microscopes and telescopes in Chapter 26. The physics of **wave optics** is the subject of Chapter 25. When the wavelength of light is not small compared to the size of objects in its path, the wave nature of light leads to effects described by wave optics. Such effects give the colors of a thin soap film and are used to read information from CDs and DVDs.

24.1 | RAY (GEOMETRICAL) OPTICS

Figure 24.1A shows a light wave as it passes through an opening that is large compared to the wavelength of the light. This figure also shows **rays**, indicating the path and direction of propagation. We first introduced rays in Chapter 12 (see Fig. 12.12) when describing water waves and sound, but rays are useful for describing the propagation of any type of wave. Strictly speaking, rays indicate the path followed by a wave, but for convenience we will often say that a ray "travels" or "propagates" from one place to another, with the understanding that it is really a light wave that propagates. In Figure 24.1A, rays pass through a wide opening and make an ordinary shadow on the blocked regions of the screen; to a very good approximation, the rays follow straight lines that pass through the opening. If, however, the width of the opening is made very small (Fig. 24.1B), one would now observe that the light "spreads out" after passing through the opening. This spreading is due to the wave nature of light (including interference). We'll come back to this and related situations in Chapter 25 when we discuss wave optics.

The behavior seen in Figure 24.1B is found whenever a wave passes through an opening whose width is about the same size or smaller than the wavelength λ. The wavelength of visible light is less than 1 μm (10^{-6} m), so this behavior is found only with very narrow openings. In most familiar cases, such as light passing through a window, the rays are straight lines as in Figure 24.1A. This straight-line propagation is characteristic of geometrical optics, the regime we'll be discussing throughout the rest of this chapter.

Section 12.4 introduced the concept of a **wave front**. Two examples of wave fronts are shown in Figure 24.2. Wave front surfaces are determined by the crests and troughs of the wave, and are always perpendicular to the associated rays. The shape of a wave front thus depends on how the wave is generated and the distance from the source; wave fronts can be simple planes as in Figure 24.2A or spherical as in Figure 24.2B. When a point source is very far away, the wave fronts form very large spheres, which appear as nearly flat to an observer and can be used to form approximate plane waves (Fig. 24.2C).

Although the behavior of light rays and wave fronts can be derived mathematically from Maxwell's equations, two important properties of light were discovered through experiments several centuries before Maxwell. First, the motion of light along a light ray is *reversible*. That is, if light can travel in one direction along a ray that connects point A to point B, light can also propagate in the reverse direction from point B to point A. Second, the *perpendicular distance* between two wave fronts is proportional to the speed of light because of the way wave fronts are related to the crests and troughs of a wave (see parts A and B of Fig. 24.2). In Section 24.3, we'll build on the proportionality between wave front separation and wave speed to predict the shapes of wave fronts and the directions of rays in more general cases.

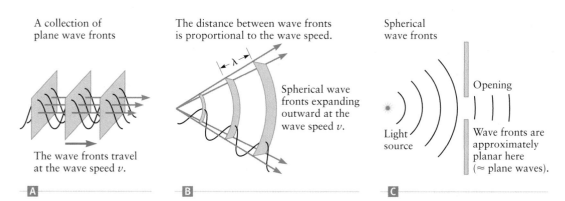

A collection of plane wave fronts

The distance between wave fronts is proportional to the wave speed.

Spherical wave fronts

The wave fronts travel at the wave speed v.

Spherical wave fronts expanding outward at the wave speed v.

Light source

Opening

Wave fronts are approximately planar here ($\approx$ plane waves).

A

B

C

Figure 24.2 The shape of a wave front is determined by the crests and troughs of the wave. **A** A plane wave traveling from left to right. The blue arrows denote the corresponding rays; the wave fronts are planes perpendicular to the rays. **B** When a wave is emitted from a point source, the wave fronts are spherical. **C** When a point source is very far away, the wave fronts over any given (small) region are approximately planar and can be used to make an approximate plane wave.

Forming an Image: Ray Tracing

Light that emanates from an object is used by your eye to form an *image* of the object. Figure 24.3A shows how rays originating from two different points on an object reach your eye.[1] When your eye combines these rays to form an image, your brain extrapolates the rays back to their point of origin. This notion of extrapolating rays back to a common point is an essential part of all image formation. The method of following individual rays as they travel from an object to your eye or to some other point is called *ray tracing*. Ray tracing involves the extensive use of geometry; hence, ray optics is also called "geometrical optics." (Now might be a good time to study the quick review of geometry in Chapter 1 and Appendix B.)

Figure 24.3A shows rays from only two points on the object and only a few rays from each point. It would be more realistic to draw a very large ("infinite") number of rays originating from each point on an object. Figure 24.3B shows how rays from another point on the object reach your eye. The light waves associated with *all* these rays contribute to the image formed by your eye. In most of our ray diagrams, we draw only a few rays from the top or bottom of the object, but the complete image is formed by light that originates from all points on the object.

To understand how images are formed, we must consider two basic problems: (1) what happens to light rays when they reflect from a surface such as a mirror or a piece of glass, and (2) what happens to light rays when they pass across a surface from one material to another such as when they pass from air into a piece of glass. In each case, we must distinguish between a flat surface and a curved surface. We'll first consider flat surfaces in Sections 24.2 and 24.3, and then build on that work to deal with curved surfaces in later sections.

24.2 | REFLECTION FROM A PLANE MIRROR: THE LAW OF REFLECTION

Light rays travel in straight lines until they strike something. What happens next depends on what the rays strike. One possibility is that the rays are reflected as sketched in Figure 24.4A, showing the reflection of rays from a *plane mirror*, a flat surface that reflects all or nearly all the light that strikes it.

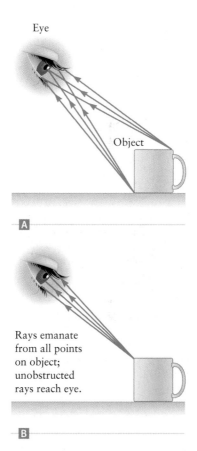

Eye

Object

A

Rays emanate from all points on object; unobstructed rays reach eye.

B

Figure 24.3 **A** An optical image is formed in your eye by rays of light that emanate from an object. **B** Many rays (an infinite number) emanate from each point on an object.

---

[1]We note again that rays themselves do not "travel" but rather show the direction of propagation of the corresponding light wave.

Figure 24.4 **A** Rays reflecting
from a flat surface. A surface can
be considered flat if its roughness
is smaller than λ. This mirror is
called a plane mirror. **B** The angle
of reflection is equal to the angle of
incidence.

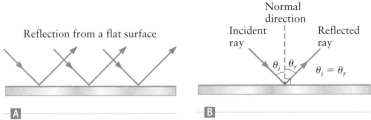

Reflection from a flat surface

Normal direction

Incident ray   Reflected ray

θ_i θ_r $\theta_i = \theta_r$

A   **B**

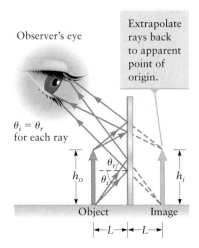

Rough surface

The reflected rays will not all be
parallel to one another. Compare
with Figure 24.4.

Figure 24.5 Rays are reflected
in different directions by different
parts of a rough surface, forming a
diffuse reflection.

Observer's eye

Extrapolate
rays back
to apparent
point of
origin.

$\theta_i = \theta_r$
for each ray

h_o θ_r θ_i h_i

Object   Image

$|\leftarrow L \rightarrow|\leftarrow L \rightarrow|$

Figure 24.6 Forming an image
with a plane mirror. Rays from
the object (the arrow on the left)
reflect from the mirror and reach
your eye. These rays appear to
have come from the image, on the
right of the mirror. It is a virtual
image because light does not actually pass through the image point.
This sketch shows only rays from
the top and bottom of the object,
but rays from other points on the
object also contribute to the image.

Figure 24.4A describes the case of an incoming plane wave, so the rays are parallel and strike the surface at many different points. We can characterize this reflection by considering only a single ray as shown in Figure 24.4B. (For a plane wave,
the other rays are parallel as in Fig. 24.4A and reflect at the same angle.) The vertical dashed line in Figure 24.4B is drawn perpendicular (also called "normal") to
the mirror, and we can measure the directions of the incoming and outgoing rays
relative to this perpendicular direction. The incoming ray is called the **incident**
ray, and the angle it makes with the surface normal is the angle of incidence θ_i. The
outgoing ray is called the **reflected** ray, and its angle with the normal is the angle
of reflection θ_r. For reflection from a flat surface, the angle of incidence is equal to
the angle of reflection,

$$\theta_i = \theta_r \tag{24.1}$$

This equation, called the **law of reflection**, can be derived from our general principle concerning the reversibility of the propagation of light. Suppose the angle of
reflection is smaller than θ_i. If we were to reverse the outgoing ray, its angle of reflection would be even smaller and the new reflected ray would not retrace the original
incoming ray. The only way the reflection process can be reversible is for the angle
of reflection to equal the angle of incidence (Eq. 24.1).

Reflections from a perfectly flat mirror (Fig. 24.4) are called **specular** reflections. If the reflecting surface is rough, we must consider reflections from all the
individual pieces of the surface. In that case, an incident plane wave will give rise to
many reflected rays propagating outward in many different directions (Fig. 24.5).
This reflection is called a **diffuse** reflection.

Image Formation by a Plane Mirror

When you view an object through its reflection in a mirror, you are viewing an
image of the object. An image formed by a plane mirror is shown in Figure 24.6.
Here we show a particularly simple object (an arrow), but the object could be a tree
or a person. We will usually use an arrow as our example object because we'll often
be interested in the orientation of the image relative to the object. As in Figure 24.3,
an infinite number light rays emanate from each point on the object, although we
only show two rays emanating from the top and bottom here. Some of these rays
strike the mirror and are reflected so as to reach your eye, which then uses the rays
to form an image. To your eye, the location of this image—that is, the location of
the arrow—is the point from which these rays appear to emanate. This point can
be found by extrapolating the rays back in a straight-line fashion to their apparent
point of origin as shown by the dashed lines in Figure 24.6. This is our first example
of the ray-tracing procedure. Since each ray obeys the law of reflection (Eq. 24.1),
we can use that together with some geometry to determine the image location. A
few rays are shown in Figure 24.6 for an object located a distance L in front of the
mirror. When the rays that reach the observer's eye are extrapolated back to their
apparent common point of origin, the image is found to be a distance L behind the
mirror.

In addition to locating the image, Figure 24.6 gives several other results. First,
the size of an image formed by a plane mirror is equal to the size of the object; that
is, the height of the image h_i is equal to the height of the object h_o. Second, because

the image point is located behind the mirror, light does *not* actually pass through the image. For this reason, the image is called a ***virtual image***.

| EXAMPLE 24.1 | Reflection from a Corner Cube |

Figure 24.7A shows a combination of two plane mirrors connected to form a right angle. In three dimensions, one can form a similar structure called a ***corner cube*** by combining mirrors to make three faces of a cube. The light ray in Figure 24.7A is incident at an angle θ_i relative to the direction normal (perpendicular) to mirror 1. Find the angle of the outgoing light ray θ_{out} relative to this normal direction.

RECOGNIZE THE PRINCIPLE

To analyze reflection from a single plane mirror, we must first determine the angle of incidence, the angle the incident ray makes with the direction normal to the mirror. According to the law of reflection, the angle of reflection is then equal to the angle of incidence. Here we must use this procedure first for the reflection from mirror 1 and then a second time for the reflection from mirror 2.

SKETCH THE PROBLEM

Figure 24.7A shows the incident ray along with the normal direction for mirror 1 (shown as the dashed line). Reflection from mirror 1 gives an outgoing ray making an angle θ_r with the normal to mirror 1 as sketched in Figure 24.7B. This ray is then the incident ray for reflection from the bottom mirror. We must apply the law of reflection (Eq. 24.1) for each reflection in Figure 24.7B.

IDENTIFY THE RELATIONSHIPS

The angle θ_i in Figure 24.7A is the angle the incoming ray makes with the normal to its reflecting surface, so it is also the angle of incidence in the law of reflection, Equation 24.1. The angle of reflection from this first surface is thus $\theta_r = \theta_i$. The two reflecting surfaces (the two mirrors) are at right angles, so the dashed lines showing the directions normal to each mirror in Figure 24.7B also form a right angle. Because the interior angles of a triangle add up to 180°, the angle of incidence for the second reflection is

$$\theta_{i2} = 90° - \theta_r = 90° - \theta_i$$

Applying the law of reflection to the ray as it reflects from mirror 2, we find

$$\theta_{r2} = \theta_{i2} = 90° - \theta_i \qquad (1)$$

SOLVE

From the geometry in Figure 24.7B, the angle the outgoing ray makes with the horizontal is

$$\theta_{out} = 90° - \theta_{r2}$$

Combining this with Equation (1) gives

$$\theta_{out} = 90° - \theta_{r2} = 90° - (90° - \theta_i)$$

$$\theta_{out} = \boxed{\theta_i}$$

The outgoing ray is thus parallel to the incident ray.

What does it mean?

An interesting feature of this result is that the outgoing ray is *always* parallel to the incident ray for *any* value of θ_i. A combination of mirrors called a corner cube, the three-dimensional version of Figure 24.7, has a similar property: its reflected rays are always parallel to the incident rays. Such devices, called ***retroreflectors***,

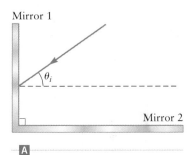

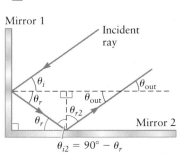

Figure 24.7 Example 24.1. Reflection from two mirrors that form a right angle. ⒜ The incident ray. ⒝ Reflected rays and the corresponding angles of incidence and reflection.

Figure 24.8 Reflective clothing contains tiny corner cubes, each of which reflects light back parallel to the incident rays as in Figure 24.7B. This cyclist's vest seems to "glow" in the dark because it reflects light (such as from a car's headlights) back toward its source.

© Paul Ridsdale/Alamy

are useful in many applications. For example, the Apollo astronauts who visited the Moon left a number of corner cube reflectors on its surface. When laser light from the Earth strikes one of these corner cubes, the reflected light travels in a path parallel to the incident ray and back to the Earth. The time delay between incident and reflected light pulses can be used to measure the Earth–Moon distance with high precision. (See Problem 23.61, page 794.) In principle, a single, flat mirror would reflect light in the same way, but the mirror would have to be aligned very precisely; otherwise, the reflected light would miss the Earth. With a retroreflector, the reflected light is always parallel to the incident ray, no matter what the alignment. A more down-to-Earth application of retroreflectors is found on many roads. Most roadside reflectors, and even some paints, contain tiny crystals that act as corner cubes to reflect the rays from your headlights back to you, thus giving a noticeable indication of the edges of the road. That is also how reflective clothing works (Fig. 24.8).

CONCEPT CHECK 24.1 | Reflection from Two Mirrors

Consider two plane mirrors that make an angle of 60° with each other (Fig. 24.9). An incident ray is parallel to the bottom mirror. This ray reflects from the mirror at the left and then strikes the bottom mirror. What angle of incidence θ_{i2} does the reflected ray make with the bottom mirror, (a) 30°, (b) 45°, or (c) 60°?

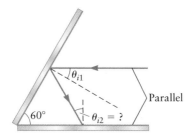

Figure 24.9 Concept Check 24.1.

24.3 | REFRACTION

When a light ray strikes a transparent material like a windowpane, some of the light is reflected (according to the law of reflection described in Section 24.2), and the rest passes into the material. Figure 24.10 shows a light ray incident on a piece of glass with two flat, parallel sides. The ray that travels into the glass is called a **refracted** ray.

The Index of Refraction and Snell's Law

The direction of the refracted ray is measured using the angle θ_2 this ray makes with the surface normal. The value of θ_2 depends on the incident angle (which we now denote as θ_1) and also on the speed of light within the material. The speed of light in vacuum is $c = 3.00 \times 10^8$ m/s. When light travels through a material substance, however, the associated electric and magnetic fields interact with the atoms of the substance, and this interaction affects the speed of the wave. As a result, the speed of light inside a substance such as glass, water, or air is less than the speed of light in vacuum. Table 24.1 lists the speed of light in several substances.

Figure 24.11A shows a plane wave initially traveling in a vacuum and then striking the flat surface of a piece of glass; this sketch illustrates how the change in the speed of light in going from vacuum to glass affects the direction of the rays inside the glass. Here we show several rays along with some associated wave fronts in both the vacuum and the glass. The incoming rays make an angle θ_1 relative to the surface normal, and the incident wave fronts are perpendicular to these rays. The incident light travels at speed c, the speed of light in a vacuum. The light inside the glass, however, moves at speed v, which is less than c.

From Section 24.1 (Fig. 24.2), the perpendicular distance between wave fronts is proportional to the wave speed. The wave fronts in Figure 24.11A are drawn at successive, equally spaced moments in time, so the spacing between wave fronts is ct on the vacuum side and vt in the glass. Figure 24.11B shows an expanded view of two adjacent wave fronts, and two right triangles are indicated. On the vacuum side, the right triangle shaded in red has sides L and ct, with angle θ_1 adjacent to side L. From the geometry of this triangle, we have

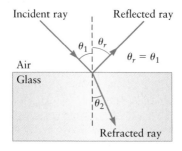

Figure 24.10 An incident light ray is both reflected and refracted when it strikes the surface of a transparent object such as a piece of glass. A portion of the light energy is reflected with $\theta_r = \theta_1$, whereas the rest is refracted as it passes into the glass.

Table 24.1 Speed of Light and Index of Refraction for Several Substances for Room Temperature and Yellow Light ($\lambda = 550$ nm)

SUBSTANCE	SPEED OF LIGHT (m/s)	INDEX OF REFRACTION
Vacuum	2.998×10^8	1.0000 (exactly)
Air	2.997×10^8	1.0003
Water (liquid)	2.26×10^8	1.33
Water (ice)	2.29×10^8	1.31
Benzene	2.00×10^8	1.50
Quartz crystal	2.05×10^8	1.46
Diamond	1.24×10^8	2.42
Pyrex glass	2.04×10^8	1.47
Plexiglas (plastic)	2.01×10^8	1.49
Regions in the eye:		
vitreous humor	2.23×10^8	1.34
aqueous humor	2.26×10^8	1.33
cornea	2.17×10^8	1.38
lens	2.13×10^8	1.41

$$\sin \theta_1 = \frac{ct}{L} \qquad (24.2)$$

There is a corresponding triangle shaded in yellow on the glass side with edge lengths L and vt and an angle θ_2 adjacent to side L, which leads to

$$\sin \theta_2 = \frac{vt}{L} \qquad (24.3)$$

We can rearrange each of these relations to solve for L, resulting in $L = ct/\sin \theta_1$ and $L = vt/\sin \theta_2$. Setting them equal, we find

$$\frac{ct}{\sin \theta_1} = \frac{vt}{\sin \theta_2}$$

The factor of t can be canceled, giving

$$\sin \theta_1 = \frac{c}{v} \sin \theta_2 \qquad (24.4)$$

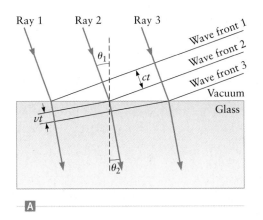

A

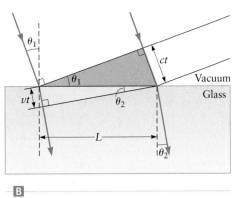

B

Figure 24.11 **A** Refraction occurs due to a difference in the wave speeds in two regions, which causes the rays to change direction at the interface. **B** Derivation of Snell's law. The speed of light in a vacuum is c, whereas the speed of light in glass is v.

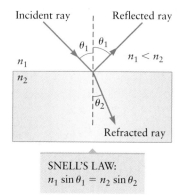

SNELL'S LAW:
$n_1 \sin \theta_1 = n_2 \sin \theta_2$

Figure 24.12 Snell's law gives the relationship between the angle of incidence and the angle of refraction.

The ratio c/v is called the ***index of refraction*** of the glass and is denoted by

$$n = \frac{c}{v} \tag{24.5}$$

We can thus rewrite Equation 24.4 as

$$\sin \theta_1 = n \sin \theta_2 \tag{24.6}$$

In Figure 24.11, we assumed there is vacuum on the incident side and glass on the refracted side. A common situation involves light passing between two substances such as air and glass or air and water. If the two substances have indices of refraction n_1 and n_2 (Fig. 24.12), the incident and refracted angles are related by

$$n_1 \sin \theta_1 = n_2 \sin \theta_2 \tag{24.7}$$

This relation between the incident and refracted angles is called ***Snell's law***. Table 24.1 lists the indices of refraction for some common substances. Since n is the ratio of two speeds (Eq. 24.5), the index of refraction is a dimensionless number.

EXAMPLE 24.2 | **Refraction on a Pier**

A lobster fisherman looks just over the edge of a pier and spots a lobster resting at the bottom. This fishing spot is $d = 3.5$ m deep. Judging from the angle at which he spots the lobster, the fisherman thinks the lobster is a horizontal distance $L_{app} = 6.0$ m from the shore, but when he drops his trap at that location, he does not catch the lobster. What is the true horizontal distance of the lobster from the shore?

RECOGNIZE THE PRINCIPLE

The fisherman determines the lobster's apparent position by extrapolating the ray that meets his eye back to its apparent point of origin at the bottom of the harbor. This extrapolation is shown by the dashed line in Figure 24.13. However, because the ray from the lobster is refracted when it passes from the water into the air, the actual position of the lobster is much closer to the shore. We can use Snell's law to find the angle θ_1 that the true ray from the lobster makes with the surface normal, and from that we can locate the true position of the lobster.

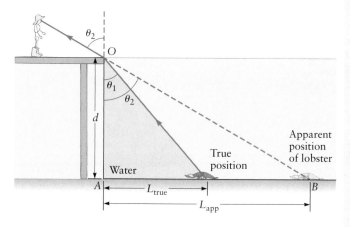

Figure 24.13 Example 24.2.

SKETCH THE PROBLEM

Figure 24.13 shows the problem, including the apparent position of the lobster.

IDENTIFY THE RELATIONSHIPS

To apply Snell's law, we denote the angle of incidence (on the water side of the water–air interface) as θ_1 and the angle of refraction (on the air side of the interface) as θ_2. The angle θ_2 is related to L_{app}, the apparent distance of the lobster from the shore, by

$$\tan \theta_2 = \frac{L_{app}}{d}$$

because L_{app} and d are two sides of the right triangle OAB in Figure 24.13. Inserting the given values $d = 3.5$ m and $L_{app} = 6.0$ m, we find

$$\theta_2 = \tan^{-1}\left(\frac{6.0 \text{ m}}{3.5 \text{ m}}\right) = 60°$$

which is also the angle of refraction for the ray that travels from the lobster to the fisherman. From Snell's law, we thus have $n_1 \sin \theta_1 = n_2 \sin \theta_2$, where n_1 is the index

Insight 24.2

REVERSIBILITY OF LIGHT RAYS AND REFRACTION

In Figure 24.12, light is incident at an angle θ_1 from side 1, leading to a refracted angle θ_2 given by Snell's law (Eq. 24.7). If the light had instead been incident from side 2 (from the bottom of the figure) at an incident angle θ_2, applying Snell's law would give a refracted angle θ_1. Hence, in accord with the general principle of reversibility of propagation direction stated in Section 24.1, the refraction of light is reversible.

of refraction of water and n_2 is the index of refraction for air. Using the values of n_1 and n_2 from Table 24.1 along with our value of the angle θ_2, we have

$$\sin \theta_1 = \frac{n_2}{n_1} \sin \theta_2 = \frac{1.00}{1.33} \sin(60°)$$

$$\theta_1 = 41°$$

SOLVE

Now that we have the value of θ_1, we can find the lobster's true distance from the shore, L_{true}. Using the right triangle shaded in green in Figure 24.13, we have

$$\tan \theta_1 = \frac{L_{\text{true}}}{d}$$

$$L_{\text{true}} = d \tan \theta_1 = (3.5 \text{ m})\tan(41°) = \boxed{3.0 \text{ m}}$$

What does it mean?

The lobster is thus much closer to shore than it appears to the naive fisherman, which would affect where he drops his trap to have a chance of catching the lobster. This effect is illustrated in Figure 24.14, which shows a pencil inserted into a glass of water. The pencil appears to bend where it enters the water due to refraction of light rays as they leave the water.

Figure 24.14 This pencil is partially immersed in water. It appears to "bend" under the water's surface due to refraction of rays that pass from the water into the air.

© Cengage Learning/Charles D. Winters

Applying Snell's Law

When a light ray strikes a plane surface, the angle of the reflected ray is given by the law of reflection (Eq. 24.1) and the angle of the refracted ray is given by Snell's law (Eq. 24.7). Figure 24.12 shows the case in which the incident ray comes from the side with the smaller index of refraction, whereas in Figure 24.15 the light is incident from the side with the larger index. For instance, substance 1 (the lower substance in Fig. 24.15) might be water or glass and substance 2 could be air.

Snell's law reads the same whether light begins in the substance with the larger or smaller index of refraction,

$$n_1 \sin \theta_1 = n_2 \sin \theta_2 \tag{24.8}$$

Possible angles of incidence always lie between zero and 90°, so $0° \leq \theta_1 \leq 90°$ and $0° \leq \theta_2 \leq 90°$. For these angles, the function $\sin \theta$ increases as θ is increased. Hence, according to Equation 24.8, the side with the *larger* index always has the *smaller* angle. In words, we say that light is refracted *toward* the normal direction when moving into the substance with the larger index of refraction (Fig. 24.12). Light is refracted *away* from the normal direction when moving into the substance with the smaller index of refraction (Fig. 24.15).

| CONCEPT CHECK 24.2 | Snell's Law

Light travels from a vacuum ($n_1 = 1.00$) into a plate of glass with $n_2 = 1.55$, with an angle of incidence of 30°. Is the angle of refraction (a) larger than 30° or (b) smaller than 30°? Also, use Snell's law to calculate the angle of refraction.

Total Internal Reflection

Let's now consider light incident from the side with the larger index of refraction in Figure 24.15 (the bottom side) a little more carefully. Because the index n_1 is greater than n_2, the angle of refraction θ_2 is greater than θ_1 and light is refracted *away* from the normal direction. As the incident angle θ_1 is made larger and larger, the refracted angle θ_2 increases more and more. Eventually, θ_2 will reach 90° and the refracted ray will emerge *parallel* to the surface. The value of the incident angle at which that occurs is called the ***critical angle***. If the incident angle θ_1 in Figure 24.15

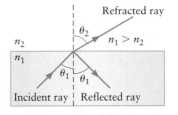

Figure 24.15 Reflection and refraction when light travels from a region of higher to a region of lower index of refraction. Here $n_1 > n_2$; region 1 at the bottom might be glass, and region 2 at the top might be air. Compare with Figure 24.12, in which $n_1 < n_2$.

Figure 24.16 An example of total internal reflection. Light from a laser at the lower left is reflected at the air–water interface. There is no refracted ray.

Courtesy of The Harvard Science Center. © The President & Fellows of Harvard College.

Total internal reflection

is increased beyond the critical angle, Snell's law has no solution for the refracted angle θ_2. Physically, there is no refracted ray, and all the energy from the incident light ray is reflected. This behavior, called *total internal reflection*, is possible only when light is incident from the side with the larger index of refraction.

Figure 24.16 shows an example of total internal reflection. Here light from a laser travels into the side of a tank of water. The light is reflected at the interface between the water and the air, but the angle of incidence exceeds the critical angle, so there is no refracted ray. Hence, no light gets through this otherwise "transparent" interface!

We can find the value of the critical angle using Snell's law. When the angle of incidence equals the critical angle, the angle of refraction in Figure 24.15 is $\theta_2 = 90°$ (because the direction of the refracted ray would be parallel to the interface). Inserting this angle into Snell's law (Eq. 24.8) and using $\sin(90°) = 1$, we get

$$n_1 \sin \theta_1 = n_2 \sin \theta_2$$

$$n_1 \sin \theta_1 = n_2 \sin \theta_2 = n_2 \sin(90°) = n_2$$

$$\sin \theta_1 = \frac{n_2}{n_1}$$

Denoting the critical angle as θ_{crit} $(= \theta_1)$ gives

$$\sin \theta_{\text{crit}} = \frac{n_2}{n_1}$$

or

$$\theta_{\text{crit}} = \sin^{-1}\left(\frac{n_2}{n_1}\right) \tag{24.9}$$

The value of θ_{crit} thus depends on the ratio of the indices of refraction of the substances on the two sides of the interface. When the angle of incidence is equal to or greater than the critical angle, light is reflected completely at the interface. Amazingly, this phenomenon can occur even when the interface between the two substances would seem to be completely transparent (such as between water and air). Look for a reflection from such an interface the next time you go swimming or gaze through a window.

Total internal reflection is used in fiber optics. Optical fibers are composed of specially made glass and are used to carry telephone and computer communications signals. These signals are sent as light waves, using total internal reflection as sketched in Figure 24.17 to keep the light from "leaking" out the sides of the fiber. Optical fibers are designed to carry sig-

Total internal reflection

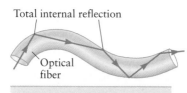

Optical fiber

Light is confined to an optical fiber by total internal reflection.

A

Figure 24.17 **A** Total internal reflection of light traveling inside an optical fiber. **B** Light escapes from an optical fiber only through the end.

© Tetra Images/Jupiterimages

B

nals over very long distances, so to minimize signal loss it is important that these reflections be as perfect as possible. (The model of an optical fiber shown here is highly simplified; we'll describe a more realistic picture in Chapter 26.) A mirage is another example of total internal reflection (Fig. 24.18). A mirage is formed by total reflection of light between layers of cool air and warm air.

CONCEPT CHECK 24.3 | When Is Total Internal Reflection Possible?

In which of the following cases is it possible to have total internal reflection?
  (a) Light traveling from water into air
  (b) Light traveling from air into water
  (c) Light traveling from glass into water
  (d) Light traveling from water into diamond
  (e) Light traveling from benzene into quartz (*Hint:* See Table 24.1.)

Figure 24.18 A mirage forms when there is a layer of warm air next to the ground. Light from this car undergoes total internal reflection from this layer.

| EXAMPLE 24.3 | Total Internal Reflection between Water and Air |

Find the critical angle for total internal reflection between water and air.

RECOGNIZE THE PRINCIPLE

When we used Snell's law to derive Equation 24.9, we saw how to calculate the critical angle. The purpose of this problem is to get a feel for typical values of the critical angle.

SKETCH THE PROBLEM

Figure 24.19A shows light incident on the interface between water and air. The incident ray is at the critical angle, so the refracted ray (if it existed) would be parallel to the interface and $\theta_2 = 90°$.

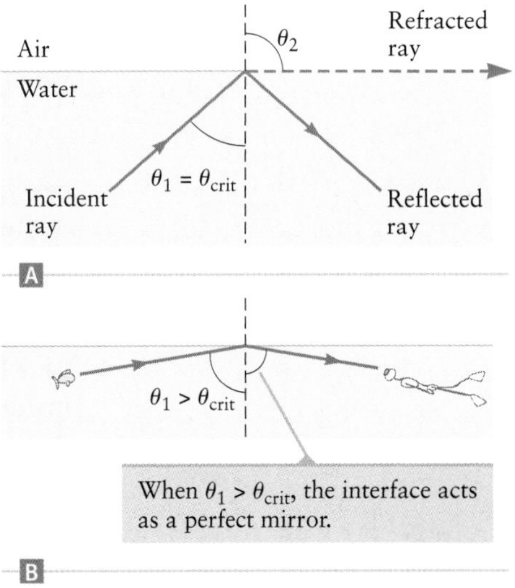

IDENTIFY THE RELATIONSHIPS

We can find the critical angle using the result for θ_{crit} in Equation 24.9:

$$\theta_{crit} = \sin^{-1}\left(\frac{n_2}{n_1}\right) = \sin^{-1}\left(\frac{n_{air}}{n_{water}}\right)$$

We can get values for the indices of refraction for water (n_{water}) and air (n_{air}) from Table 24.1.

SOLVE

Inserting the values $n_{air} = 1.0003 \approx 1.00$, and $n_{water} = 1.33$, we get

$$\theta_{crit} = \sin^{-1}\left(\frac{n_{air}}{n_{water}}\right) = \sin^{-1}\left(\frac{1.00}{1.33}\right) = \sin^{-1}(0.75) = \boxed{49°}$$

What does it mean?
The incident ray in Figure 24.19A is drawn at $\theta_{crit} = 49°$, showing that the incident ray does not have to be "close" to the interface to achieve total internal reflection. You can test this result the next time you go swimming. If you are underwater and look in a direction for which the angle of incidence is greater than the critical angle (Fig. 24.19B), the water–air interface will appear as a perfect mirror. The swimmer in Figure 24.19B will see the fish reflected in this interface mirror. Try it.

Figure 24.19 Example 24.3. **A** Total internal reflection at an air–water interface. **B** When a swimmer (on the right) looks at light coming from the air–water interface, all light with an angle of incidence greater than the critical angle is reflected. There are no rays incident from the air that give rays at this angle; the interface thus acts as a perfect mirror when viewed at this angle.

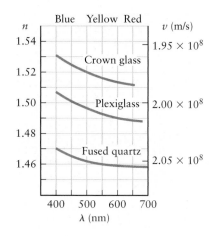

Figure 24.20 Variation of the speed of light (right axis) and the index of refraction (left axis) as a function of wavelength for several materials.

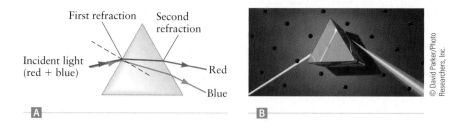

Figure 24.21 **A** The angle of refraction depends on the wavelength (i.e., the color) of the light. Blue light and red light incident from the left will therefore refract at different angles and emerge from the prism propagating in different directions. **B** This enables a prism to separate different colors in an incident beam of white light.

Dispersion

The direction of a refracted ray depends on the indices of refraction of the materials on the two sides of an interface, and these indices of refraction depend on the speed of light in each material (Eq. 24.5). The speed of light in a vacuum has the constant value c for all wavelengths, that is, for all colors. When light travels in a material, however, the speed depends on the color of the light. This dependence of wave speed on color is called *dispersion*, and the variation of the speed of light with wavelength for several materials is shown in Figure 24.20.

For example, the index of refraction for red light in quartz is $n_{red} = 1.46$ and that for blue light is $n_{blue} = 1.47$. The difference $n_{blue} - n_{red}$ is much smaller than the difference between the (average) index of refraction for quartz and water (Table 24.1). This difference, however, does mean that the angle of refraction is slightly different for different colors in quartz (and in other materials, too). This effect is used by a *prism* to separate a beam of light into its component colors.

Prisms are typically composed of glass, with a triangular cross section (Fig. 24.21). In Figure 24.21A, a beam of light is incident on one surface of a prism; we suppose this light is composed of two colors (red and blue), and we use two rays with different colors to denote the incident light. Each ray is refracted twice: once when it enters the prism and again when it exits. The first refraction occurs at the left-hand face of the prism. Because the refractive index of glass depends on the color of the light, incident beams of different color have slightly different angles of refraction. Blue light has a larger index of refraction (Fig. 24.20) than red light, so the blue ray inside the glass makes a smaller angle with the normal direction on the left than does the red ray. Hence, inside the glass the red and blue components travel in different directions. This difference in propagation direction is increased at the second refraction, when the light leaves the prism. The calculation of these outgoing angles involves two applications of Snell's law, one for each surface of the prism. (We leave this calculation for the end-of-chapter problems; see Problems 29–31.) The result is shown in Figure 24.21B. In words, we say that the prism has "dispersed" the light into different directions according to color. That is why the variation of the index of refraction with wavelength (Fig. 24.20) is called dispersion.

CONCEPT CHECK 24.4 | Refraction of Rays with Different Colors

A mixture of red light and blue light is incident from vacuum onto a thick plexiglass slab. If $n_{red} = 1.49$ and $n_{blue} = 1.51$, which color will have the larger angle of refraction, (a) the red light or (b) the blue light? Or, will they (c) have the same angle of refraction? *Hint*: You should be able to answer this question without working out values of the angles of refraction.

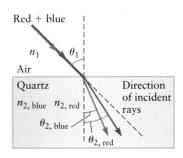

Figure 24.22 Example 24.4.

EXAMPLE 24.4 | Dispersing Light by Refraction

Figure 24.22 shows light incident from air onto a flat quartz slab. This light is a mixture of red and blue, denoted in the figure by the two incident rays with these colors. Both rays have an angle of incidence $\theta_1 = 45.0°$. Find the angles of refraction for the two rays in the quartz plate. Assume the indices of refraction for red and blue light in quartz are $n_{quartz,\, red} = n_{2,\, red} = 1.459$ and $n_{quartz,\, blue} = n_{2,\, blue} = 1.467$. Give your answers to three significant figures.

RECOGNIZE THE PRINCIPLE

We can find the angle of refraction for each color using Snell's law and the values of the index of refraction for each color. The point of this example is to get a feeling for how much the two rays are "dispersed." How large is the difference between the directions of the two rays?

SKETCH THE PROBLEM

Figure 24.22 shows the problem. The index of refraction in quartz is greater for blue light than for red light, so the blue ray is refracted closer to the normal as shown in the figure.

IDENTIFY THE RELATIONSHIPS

The indices of refraction for red and blue light in quartz are given. In air, the indices of refraction for both colors are the same to four significant figures (Table 24.1): $n_{air,\, red} = n_{air,\, blue} = n_1 = 1.000$.

SOLVE

We can find the angles of refraction using Snell's law (Eq. 24.7). For the red light ray, we have

$$n_1 \sin \theta_1 = n_{2,\, red} \sin \theta_{2,\, red}$$

Solving for $\theta_{2,\, red}$, we get

$$\sin \theta_{2,\, red} = \frac{n_1}{n_{2,\, red}} \sin \theta_1 \qquad (1)$$

Inserting the values of the indices of refraction and θ_1 gives

$$\sin \theta_{2,\, red} = \frac{n_1}{n_{2,\, red}} \sin \theta_1 = \frac{1.000}{1.459} \sin(45.0°)$$

$$\boxed{\theta_{2,\, red} = 29.0°} \qquad (2)$$

We have kept three significant figures because, as we'll see below, the difference between the refracted angles for the two rays will be small.

We compute the angle of refraction for the blue light ray in the same way. Equation (1) leads to

$$\sin \theta_{2,\, blue} = \frac{n_1}{n_{2,\, blue}} \sin \theta_1 = \frac{1.000}{1.467} \sin(45°)$$

$$\boxed{\theta_{2,\, blue} = 28.8°} \qquad (3)$$

What does it mean?

The angles of refraction of the two rays differ by only a small amount, just 0.2°, but that is enough to give the dispersion of light by a prism in Figure 24.21B. (To find the angle of the outgoing ray for a prism, we must also include the refraction of the light rays at the second surface of the prism.) Also, the angle of refraction is *smaller* for blue light (see Concept Check 24.4), so the blue ray is *closer* to the normal direction. Thus, refraction causes blue light to be deflected *more* (with respect to the incoming direction) than red light.

24.4 | REFLECTIONS AND IMAGES PRODUCED BY CURVED MIRRORS

In Section 24.2, we saw that a plane mirror can produce an image of an object (Fig. 24.6). That image and all others formed by plane mirrors are the same size as the original object. A curved mirror, however, can achieve a **magnified** image, an image that appears larger or smaller than the original object. Magnified images are used in many applications, ranging from telescopes to a car's rearview mirror, so let's now consider how to produce one.

Ray Tracing for Spherical Concave Mirrors

Consider the spherical mirror in Figure 24.23A in which the surface of the mirror forms a section of a spherical shell (like a beach ball). The radius R of this sphere is called the radius of curvature of the mirror. The mirror's **principal axis** (parts B and C of Fig. 24.23) is the line that extends from the center of curvature C to the center of the mirror. The mirror shown in Figure 24.23 is **concave**, curving toward objects placed in front of it.

Concave spherical mirrors have two important properties. First, consider Figure 24.23B, which shows incoming light rays directed parallel to the principal axis of the mirror. If these rays are close to the principal axis, then after reflecting at the surface of the mirror they all pass through the single point F indicated in Figure 24.23B. (We'll discuss the behavior of rays far from the axis in Section 24.8.) This point F, called the **focal point** of the mirror, is located a certain distance f from the mirror surface along the principal axis. The distance f is called the **focal length** of the mirror. Second, applying the general principle that the propagation of light is reversible, rays that originate at the focal point reflect from the mirror and propagate outward, parallel to the principal axis as in Figure 24.23C. The mirror thus "works" both ways: either focusing parallel light rays at the focal point F or generating a set of parallel rays from light that originates at point F and strikes the mirror.

When an object is placed in front of a spherical mirror, light from the object is reflected from the mirror and forms an **image**. This is illustrated in Figure 24.24, which shows photos of objects placed in front of a concave mirror. Accompanying each photo is a schematic showing the object, the mirror, and the "observer" (the camera that took the photo). In Figure 24.24A, the object—a candle—is placed fairly close to the mirror. The observer can see both the original candle and its image formed by the mirror. Here the image appears larger than the true candle. In Figure 24.24B, the object—a person's head—is somewhat farther from the mirror. The observer now sees the back of the person's head along with an image of the person's face. Note that the face is upside down; we call this an **inverted image**. Figure 24.24 shows two examples of images formed by curved mirrors. Let's now apply ray tracing to analyze some of the image properties.

Figure 24.25 is a ray diagram for an object placed in front of a curved mirror; here the object is the upward-pointing arrow. The figure shows rays that emanate from the tip of the arrow on the left. Although these rays reflect from different parts of the mirror, they all intersect at a single point, which is the tip of the arrow's image. A similar result is found for rays from other points on the object.

We can use the following ray-tracing procedure to find the location of the image formed by a spherical mirror. Figure 24.26 shows an object far from the mirror,

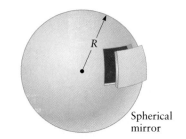
Spherical mirror

A

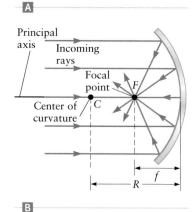

B

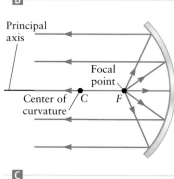

C

Figure 24.23 **A** A spherical mirror matches the shape of a sphere. The radius R of this sphere is called the radius of curvature of the mirror. **B** If the incoming rays are parallel to the principal axis of the mirror, all the reflected rays converge at the focal point of the mirror. **C** If a light source is placed at the focal point, the outgoing rays reflected from the mirror are parallel to the axis.

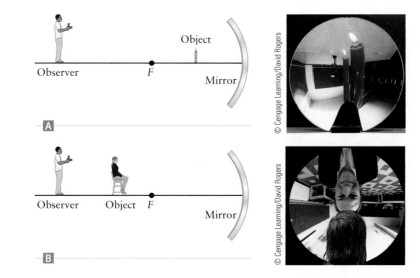

Figure 24.24 Images formed by a concave mirror. The sketches next to each image show the relative positions of the object, mirror, and observer (i.e., the camera that took the photo).

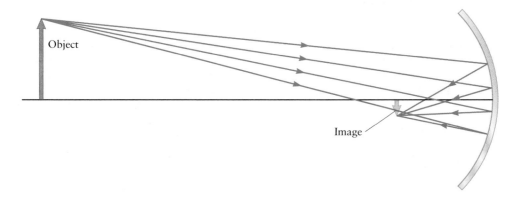

Figure 24.25 Ray diagram for a curved mirror. Rays from each point on the object are reflected by the mirror and intersect at a corresponding image point. Here we show only rays from the tip of the object.

meaning that the center of curvature and the focal point are between the object and the mirror (corresponding to the case in Fig. 24.24B.) The upward-pointing arrow on the left again represents the object, and for simplicity we place the bottom of the object arrow on the principal axis. Although many rays (an infinite number) emanate from each point on the object, three ray directions are particularly easy to draw. Figure 24.26 shows these three rays as they emanate from the tip of the object and intersect at the tip of the image.

One of these rays—called the *focal ray*—begins at the object's tip and passes through the focal point *F*. After reflection from the mirror, this ray travels parallel to the principal axis (compare with Fig. 24.23C). A second ray—called the *parallel ray*—begins at the object's tip and is initially parallel to the principal axis, so after reflection it passes through the focal point (compare with Fig. 24.23B). The third ray—called the *central ray*—begins at the object's tip and passes through the center of curvature *C* of the mirror. The central ray is directed along the radius of the spherical mirror and is therefore perpendicular to the mirror's surface. By the law of reflection, this ray reflects back on itself and passes back through *C*. These three rays all intersect (are "focused") at the tip of the arrow's image. Other rays that emanate from the tip of the object also intersect at the corresponding point on the image (see Fig. 24.25), but the focal, parallel, and central rays drawn from the object's tip provide the easiest way to locate the tip of the image and reveal its nature. These three rays are very useful in many applications of ray tracing, so we have drawn them with different colors here and in other figures. These ray colors are *not* the color of the light, but rather are intended to help you identify these rays.

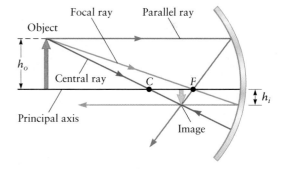

Figure 24.26 Using ray tracing to find the image formed by a concave spherical mirror. For an object outside the focal length, the image is real, inverted, and reduced. Rays emanating from the tip of the object intersect at the tip of the image. *Important note:* The colors of these rays do *not* indicate the color of the light.

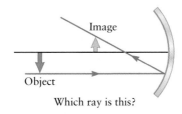

Image

Object

Which ray is this?

Figure 24.27 Concept Check 24.5.

CONCEPT CHECK 24.5 | Which Ray Is It?

Figure 24.27 shows an object in front of a spherical mirror along with the image and one ray that travels from the tip of the object to the tip of this image. Which ray is it, (a) the focal ray, (b) the parallel ray, or (c) the central ray?

Properties of an Image

The image in Figure 24.26 is smaller than the object and is inverted (upside down). The ratio of the height of the image h_i to the height of the object h_o is called the *magnification*, *m*:

Magnification

$$m = \frac{h_i}{h_o} \qquad (24.10)$$

By convention, the image height h_i in Figure 24.26 is taken to be negative; the magnification in this example is thus also negative. Images smaller than the corresponding objects are said to be *reduced* ($|m| < 1$).

Real image

The rays that form the image in Figure 24.26 all pass through a point on the image. We call the image *real*. A real image differs from the virtual image in Figure 24.6 in two ways:

1. Light rays only *appear* to emanate from a virtual image; they do not actually pass through the image. For a real image, the rays *do* pass through the image.
2. An object and its real image are both on the same (front) side of the mirror. A virtual image is located behind the mirror, while the object is in front.

Forming a Virtual Image

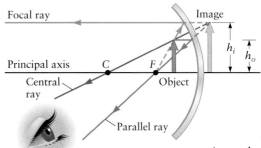

Focal ray

Image

Principal axis

C

F

Central ray

Object

Parallel ray

Observer

h_i h_o

Figure 24.28 If an object is close to a concave mirror (inside the focal point), the outgoing light rays appear to emanate from behind the mirror, producing a virtual, upright image similar to the image in Figure 24.24A.

We can use ray tracing to find the image of an object placed close to a spherical mirror—that is, closer to the mirror than the focal point and the center of curvature—as shown in Figure 24.28. Here we draw just the focal ray, the parallel ray, and the central ray as they emanate from the object's tip (the arrow to the left of the mirror). You should remember, however, that there are actually many rays (an infinite number) that emanate from the tip of the arrow, reflect from the mirror, and then converge at the image. In Figure 24.28, the focal, parallel, and central rays do not intersect at any point on the left (the front) of the mirror, but if we extrapolate the rays back behind the mirror, we find that these extrapolations all intersect at a single "image point." If we were to observe the light reflected by the mirror as shown at the lower left in Figure 24.28, the rays would appear to emanate from this image point behind the mirror. This image is virtual because light does not actually pass through any point on the image. The object and its image are on different sides of the mirror; the image is upright (h_i is positive) and enlarged.

CONCEPT CHECK 24.6 | Magnification of an Image

Consider the virtual image formed by the mirror in Figure 24.28. Which of the following statements is correct?
(a) The magnification is positive and less than 1.
(b) The magnification is positive and greater than 1.
(c) The magnification is negative with an absolute value less than 1.
(d) The magnification is negative and with an absolute value greater than 1.

Rules for Ray Tracing with Mirrors

Figures 24.26 and 24.28 show two examples of how to use ray tracing to construct the image produced by a spherical concave mirror. The general approach can be summarized as follows.

1. Construct a figure showing the mirror and its principal axis. The figure should also show the focal point and the center of curvature.

2. Draw the object at the appropriate point. One end of the object will often lie on the principal axis.

3. Draw three rays that emanate from the tip of the object:

   (a) The *focal ray* (or its extrapolation) passes through the focal point. After reflection, this ray will be parallel to the axis.

   (b) The *parallel ray* is initially parallel to the axis. After reflection, this ray (or its extrapolation) passes through the focal point.

   (c) The *central ray* (or its extrapolation) passes through the center of curvature of the mirror. After reflection, this ray passes back through the tip of the object.

4. The point where the focal, parallel, and central rays (or their extrapolations) intersect is the image point. This point may be in front of the mirror giving a *real image*, or it may be necessary to extrapolate the rays back behind the mirror to locate a *virtual image*.

5. This ray-tracing procedure can be repeated for any desired point on the object, thus locating other points on the image. It is usually sufficient to consider just the tip of the image, but additional points can be used if needed.

In many of our ray diagrams, we use different colors to clearly denote the focal, parallel, and central rays. This use of color is to help you distinguish the different rays; these ray colors are *not* the color of the light used to form the image!

Ray Tracing for Spherical Convex Mirrors

A mirror that curves away from the object as in Figure 24.29 is called a *convex* mirror. For a convex mirror, both C and the focal point lie *behind* the mirror. The location of the focal point is illustrated in Figure 24.29A, which shows a collection of parallel rays incident on the mirror. After striking the convex surface, the reflected rays diverge from the mirror axis. These rays, however, appear to come from a virtual image point behind the mirror, the focal point F. That is, all these rays extrapolate back to the focal point F, which lies behind the mirror.

Figure 24.29B shows an application of ray tracing to a convex mirror following the suggested ray tracing rules listed above. This sketch shows the principal axis of the mirror along with the center of curvature C and the focal point F. We also show the focal ray, the parallel ray, and the central ray. Notice that the focal ray is initially directed toward the focal point, but because F is behind the mirror, this ray is reflected before reaching the focal point. These three rays extrapolate to a point of intersection that lies *behind* the mirror; hence, we have a virtual image. This image is upright, so the image height h_i and magnification m are both positive.

Finding the Location of the Image: The Mirror Equation

While ray tracing is an instructive way to understand image formation, it is not a convenient way to find the image location quantitatively. Suppose we know the radius of curvature of a mirror and the distance from the object to the mirror. How can we calculate the location of the image? To answer this question, we must consider the geometry of the reflected rays. Figure 24.30A shows this geometry for a concave mirror, concentrating on two particular rays: the central ray that passes through C and a ray that reflects from the point where the principal axis meets the mirror. (This second ray is *not* a focal or parallel ray.) These two rays intersect at the image point, giving the height of the image h_i. The image here is inverted, so

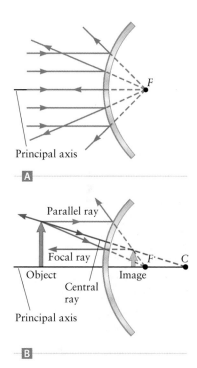

Figure 24.29 Ray diagram for a convex mirror. **A** If the incoming light is parallel to the axis of a convex mirror, the reflected rays appear to emanate from a focal point F behind the mirror. **B** Ray tracing with a convex mirror. The image is behind the mirror (a virtual image).

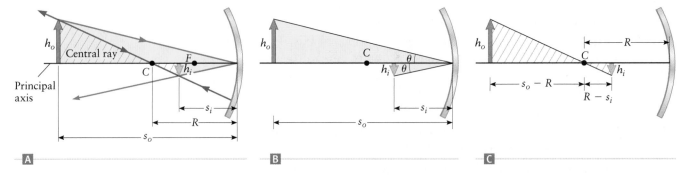

Figure 24.30 Geometry for deriving the mirror equation. ▲ Finding the location of the image formed by a concave mirror. ▣ Two similar triangles from part A. ◗ Two other similar triangles from part A.

by our sign convention h_i is negative and h_o is positive. The figure also defines the distance s_o from the object to the mirror and the distance s_i between the image and the mirror.

The two triangles shaded in yellow in Figure 24.30A are redrawn in part B. They are both right triangles with the same values for the interior angle θ, so the ratios of their sides are equal, giving

$$\frac{|h_o|}{s_o} = \frac{|h_i|}{s_i} \qquad (24.11)$$

We have used absolute values of the object and image heights because h_o and h_i can be negative if they lie below the principal axis, and Equation 24.11 holds for the magnitudes of the sides of the similar triangles shaded in yellow in Figure 24.30B. We next consider the pair of similar triangles crosshatched in red in Figure 24.30A. These triangles are redrawn in Figure 24.30C; the ratios of their sides are equal, giving

$$\frac{|h_o|}{s_o - R} = \frac{|h_i|}{R - s_i} \qquad (24.12)$$

We now take the ratios of Equations 24.11 and 24.12:

$$\frac{|h_o|/s_o}{|h_o|/(s_o - R)} = \frac{|h_i|/s_i}{|h_i|/(R - s_i)}$$

Canceling the common factors of $|h_o|$ and $|h_i|$ and doing some rearranging leads to

$$\frac{R - s_o}{s_o} = \frac{s_i - R}{s_i}$$

$$s_i(R - s_o) = s_o(s_i - R)$$

$$s_iR - s_is_o = s_os_i - s_oR$$

$$s_oR + s_iR = 2s_is_o$$

Dividing both sides by the factor s_is_oR, we finally get

$$\frac{1}{s_o} + \frac{1}{s_i} = \frac{2}{R} \qquad (24.13)$$

which is called the ***mirror equation***. If we know a mirror's shape (its radius of curvature R) and are given the location of an object s_o, we can use the mirror equation to calculate s_i, the location of the image.

The Mirror Equation and Focal Length

Figure 24.31 shows rays for an object very far from a mirror so that the rays that strike the mirror are all approximately parallel to the axis. This object is a distance

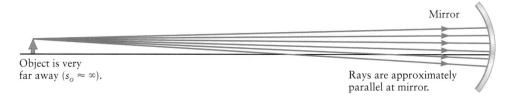

Mirror

Object is very
far away ($s_o \approx \infty$).

Rays are approximately
parallel at mirror.

$s_o \approx \infty$ from the mirror, so we call it an object "at infinity." Inserting an infinite object distance into Equation 24.13, we get

$$\frac{1}{\infty} + \frac{1}{s_i} = \frac{2}{R}$$

Because $1/\infty$ is zero, the image distance for an object at infinity is $s_i = R/2$. A collection of parallel rays would thus reflect from the mirror and intersect at the point a distance $R/2$ from the mirror. By definition, however, rays that are parallel to the principal axis all reflect and pass through the focal point F (compare with Fig. 24.23B), so we have just shown that the focal point is a distance $R/2$ from the mirror. This distance is called the *focal length*, f, of the mirror:

$$f = \frac{R}{2} \qquad (24.14)$$

Focal length for a spherical mirror

Equation 24.14 enables us to rewrite the mirror equation in terms of the mirror's focal length:

$$\frac{1}{s_o} + \frac{1}{s_i} = \frac{1}{f} \qquad (24.15)$$

Mirror equation in terms of focal length

Our analysis of the rays in Figure 24.30 has involved considerable geometry, so let's recap what we have found. Our main results are the mirror equation in its two forms, one using the radius of curvature of the mirror R (Eq. 24.13) and another written in terms of the focal length f (Eq. 24.15). These relations give us a way to calculate the location of an image produced by a spherical curved mirror. We have also shown that the focal length of such a mirror is related to its radius of curvature by $f = R/2$ (Eq. 24.14). So, if we know the curvature of a spherical mirror, we can now give a complete analysis of the images it produces.

Sign Conventions

We derived the mirror equation for a concave mirror, but that relation also applies to a convex mirror provided we use the following sign conventions for the object and image distances s_o and s_i, the focal length f, and the object and image heights h_o and h_i.

> *Sign conventions for the mirror equation*
>
> 1. **All diagrams with mirrors should be drawn with the light rays incident on the mirror from the left.**
> 2. **The object distance s_o is positive when the object is to the left of the mirror and negative if the object is to the right (behind the mirror[2]).**
> 3. **The image distance s_i is positive when the image is to the left of the mirror and negative if the image is to the right (behind the mirror). Hence, $s_i > 0$ for a real image and $s_i < 0$ for a virtual image.**
> 4. **The focal length is positive for a concave mirror and negative for a convex mirror. Hence, for a spherical mirror we have**
>
> *(Continued)*

[2]So far, all our objects have been in front of the mirror, but when two or more mirrors and lenses are combined, it is possible for an object to be behind a mirror. We'll see such examples in Chapter 26.

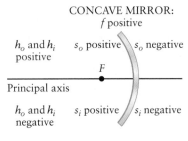

CONCAVE MIRROR:
f positive

h_o and h_i s_o positive s_o negative
positive

F

Principal axis

h_o and h_i s_i positive s_i negative
negative

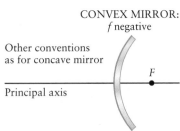

CONVEX MIRROR:
f negative

Other conventions
as for concave mirror

F

Principal axis

Figure 24.32 Sign conventions for application of the mirror equation.

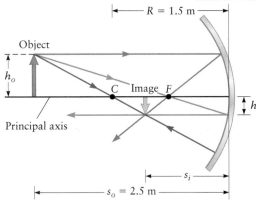

Figure 24.33 Using ray tracing together with the mirror equation to find the location of an image.

$$\text{concave mirror:} \quad f = \frac{R}{2}$$

$$\text{convex mirror:} \quad f = -\frac{R}{2} \qquad (24.16)$$

5. **The object and image heights h_o and h_i are positive if the object/image is upright and negative if it is inverted.**

We'll follow these sign conventions (summarized in Fig. 24.32) throughout this book. Other books may follow different sign conventions, however, so you should always check the sign conventions before applying relations involving s_o, s_i, and so forth from other books.

CONCEPT CHECK 24.7 | Object and Image Distances

Figure 24.27 shows an object and its image as formed by a curved mirror. What are the signs (positive or negative) of (a) the object distance s_o, (b) the image distance s_i, and (c) the focal length f?

Applications of the Mirror Equation

To see the mirror equation in action, consider the problem sketched in Figure 24.33 in which an object is placed a distance $s_o = 2.5$ m in front (to the left) of a concave mirror with $R = 1.5$ m. Let's find the location of the image and also determine if the image is real or virtual. The distances s_o and s_i along with the three rays in our standard ray-tracing diagram are shown in Figure 24.33. (Compare with Fig. 24.26.) While this diagram gives information about the image, the mirror equation provides quantitative answers.

We first use the given value of R to find the focal length $f = R/2 = 0.75$ m and show the position of the focal point F in the ray diagram. (We could use the form of the mirror equation in Eq. 24.13 to find the image location s_i without first finding f, but it is usually a good idea to include the focal point F in all ray diagrams.) The mirror equation can then be used to find the image location. Solving Equation 24.15 for $1/s_i$, we get

$$\frac{1}{s_o} + \frac{1}{s_i} = \frac{1}{f}$$

$$\frac{1}{s_i} = \frac{1}{f} - \frac{1}{s_o}$$

Substituting values for the focal length and object distance gives

$$\frac{1}{s_i} = \frac{1}{0.75 \text{ m}} - \frac{1}{2.5 \text{ m}} = 0.933 \text{ m}^{-1}$$

$$s_i = 1.07 \text{ m}$$

Here we keep three significant figures to avoid roundoff problems later in this example. Since s_i is positive, we expect to find a real image on the left (in front) of the mirror, which is confirmed by the ray diagram. The ray diagram also shows that the image is inverted, so according to our sign conventions the image height h_i is negative. To find the value of h_i, we recall the triangles in Figure 24.30B, which led to Equation 24.11; now we must be careful about the signs of h_o and h_i. We see from Figure 24.33 that the object is above the principal axis, so according to our sign conventions h_o is positive. Since h_i is below the principal axis, h_i is negative. We thus have

$$\frac{-h_i}{s_i} = \frac{h_o}{s_o} \qquad (24.17)$$

Solving for the magnification $m = h_i/h_o$ (Eq. 24.10), we get

$$m = \frac{h_i}{h_o} = -\frac{s_i}{s_o} \qquad (24.18)$$

$$m = -\frac{1.07 \text{ m}}{2.5 \text{ m}} = -0.43$$

The negative value of m means that we have an inverted image (as already noted). Because $|m| < 1$, the image is reduced as shown in Figure 24.33.

| EXAMPLE 24.5 | Applying the Mirror Equation to a Convex Mirror |

Consider again the problem sketched in Figure 24.33, but now assume the mirror is convex. (a) Construct the ray diagram. (b) Is the image inverted or upright? (c) Find the image location s_i. (d) Find the size of the image. Assume the same values of R and s_o as in Figure 24.33 and assume the object height is $h_o = 1.0$ cm.

RECOGNIZE THE PRINCIPLE

A ray diagram should always be your first step in problems involving image formation. We first construct a ray diagram, similar to the ray diagram we used in dealing with a concave mirror. We next use our ray diagram to apply the sign conventions for the image and object distances and heights and for the focal length. The mirror equation can then be applied to find the image's location and properties.

SKETCH THE PROBLEM

(a) We follow our recommended procedure for ray tracing applied to spherical mirrors. Step 1: Figure 24.34A shows the mirror with its principal axis, center of curvature C, and focal point F. According to our sign conventions, the focal length for a convex mirror is negative, so F is behind the mirror. The focal length is (from Eq. 24.16) $f = -R/2 = -0.75$ m. Step 2: We add the object to our picture. Step 3: The focal ray, parallel ray, and central ray are added (Fig. 24.34B). The focal ray is initially directed toward the focal point; after reflecting from the mirror, it travels parallel to the axis. The parallel ray initially travels parallel to the principal axis and then reflects from the mirror so that it appears to emanate from the focal point. The central ray is initially directed toward the center of curvature of the mirror and reflects back toward the object. Step 4: Extrapolations of the rays in Figure 24.34B intersect at the image point, which is behind the mirror.

IDENTIFY THE RELATIONSHIPS AND SOLVE

(b) The ray diagram in Figure 24.34B shows that the image is behind the mirror (a virtual image), is upright, and is smaller than the object.

(c) We next use our sign conventions for image location. The object is in front of the mirror, so s_o is positive with $s_o = +2.5$ m. Using this value in the mirror equation (Eq. 24.15) along with the focal length determined above, we get

$$\frac{1}{s_i} = \frac{1}{f} - \frac{1}{s_o} = \frac{1}{-0.75 \text{ m}} - \frac{1}{2.5 \text{ m}}$$

$$s_i = \boxed{-0.58 \text{ m}}$$

The negative value of s_i indicates that the image is to the right of the mirror (behind), so it is a virtual image as we noted in connection with the ray diagram in Figure 24.34B.

(d) We can find the image height h_i using Equation 24.17; we get

$$-\frac{h_i}{s_i} = \frac{h_o}{s_o}$$

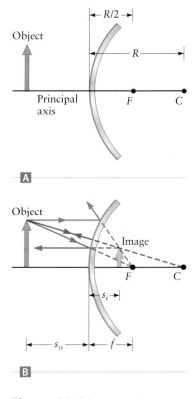

Figure 24.34 Example 24.5. Ray diagram for a convex mirror. ◳ A sketch of just the object, mirror, focal point, and center of curvature, in preparation for ray tracing. ◳ Same as part A, but now with the parallel, focal, and central rays included.

Note that in this example, h_i is positive while s_i is negative. Solving for h_i and using our result for s_i gives

$$h_i = -h_o\left(\frac{s_i}{s_o}\right) = -h_o\left(\frac{-0.58 \text{ m}}{2.5 \text{ m}}\right) = +0.23h_o = \boxed{0.23 \text{ cm}} \quad (1)$$

where we have used the given value of $h_o = 1.0$ cm.

According to our sign conventions, a positive value of h_o corresponds to an upright object as confirmed in Figure 24.34B. A positive value of h_o in Equation (1) gives a positive value for h_i, so we again find that the image is $\boxed{\text{upright}}$.

What does it mean?

When an object is placed in front of a convex mirror, the solution of the mirror equation for s_i is always negative, so the image will *always* be virtual. In addition, if the object is upright, the image will always be upright. This property is handy in many applications, one of which is discussed in Example 24.6.

EXAMPLE 24.6 | ℝ Designing a Car Rearview Side Mirror

The external rearview side mirrors on most cars carry the phrase "objects in the mirror are closer than they appear." **(a)** What does this phrase tell us about the magnification of the image? **(b)** Is this mirror convex or concave? **(c)** Estimate the radius of curvature of a typical rearview mirror.

RECOGNIZE THE PRINCIPLE

(a) The "apparent size" of an object gets smaller as the object is moved farther and farther away. When viewed by its reflection from a mirror, this apparent size is also just the image size. When judging the distance from your eye to an object, your brain compares the known size of the object (as stored in your memory) with the size of the image (the apparent size). So, when you judge an object to be farther away that it actually is, the image must be *smaller* than expected for that particular object distance. Hence, the image formed by the car's rearview side mirror must have a $\boxed{\text{magnification } |m| < 1}$.

SKETCH THE PROBLEM

(b) The next step is to draw a ray diagram, but should this diagram be for a concave mirror or a convex mirror? A car's external rearview mirror *always* produces an upright (noninverted) image, for any object distance. (You should check for yourself.) We have seen that the image produced by a concave mirror can be upright or inverted, depending on the object distance (see Figs. 24.26 and 24.28), so a rearview mirror cannot be concave. We have also seen that a convex mirror always produces a virtual image, as can be seen from the ray diagrams in Figure 24.29. Or, if we insert a negative value of f (as appropriate for a convex mirror) into the mirror equation together with a positive s_o, we find that s_i is always negative, which means that the image is behind the mirror and is thus a virtual image. (See also Example 24.5.) So, a rearview mirror must be $\boxed{\text{convex}}$, and the ray diagram will look like Figure 24.34B.

IDENTIFY THE RELATIONSHIPS

(c) Because $f = -R/2$ (Eq. 24.16 for a convex mirror), we can find the radius of curvature of a rearview mirror if we can estimate the focal length. If we know something about the object and image distances, the value of f can be found from the mirror equation. We have already argued that the magnification is less than 1, but what is its value? A value of $m = 0.1$ would be a very small image, whereas (based on the author's experience) the image is only a little smaller than the true object size. We therefore estimate a magnification of about $m \approx \frac{1}{3}$ for a car that is $s_o = 30$ m from the mirror. These estimates are only rough, but they should not be off by more than a factor of three. (You should think about how you could check these values for yourself.) The magnification is related to the object and image distances by Equation 24.18, giving $m = -s_i/s_o$ or $s_i = -ms_o$. Inserting this information into the mirror equation, we find

$$\frac{1}{s_o} + \frac{1}{s_i} = \frac{1}{f} = \frac{1}{s_o} - \frac{1}{ms_o}$$

and, solving for f, we get

$$f = s_o\left(\frac{m}{m-1}\right)$$

SOLVE

Using our estimates for m and s_o gives

$$f = s_o\left(\frac{m}{m-1}\right) = (30\text{ m})\left(\frac{\frac{1}{3}}{\frac{1}{3}-1}\right) = -15\text{ m}$$

The radius of curvature of the mirror is thus

$$R = -2f = \boxed{30\text{ m}}$$

What does it mean?

This large value of R means that the mirror has just a very slight curvature, which is why it is easily mistaken for a plane mirror. But what is the purpose for using a curved mirror in the first place? Why not just use a flat mirror? The reason is that a curved mirror gives a larger "field of view," allowing the driver to see objects over a wider angle and thus spot cars and other things that might be missed with a plane mirror (Fig. 24.35).

24.5 | LENSES

We have seen how a mirror uses reflection to form an image. A *lens* uses refraction to accomplish the same thing. Typical lenses are composed of glass or plastic, and the refraction of light rays as they pass from the air into the lens and then back into the air causes the rays to be redirected as shown Figure 24.36. Refraction occurs at the two separate surfaces as shown in Figure 24.36A, but for simplicity our ray diagrams will usually be drawn as in parts B and C of the figure, showing a single deflection of each ray as it passes through the lens.

In the simplest case, the lens surfaces are spherical (Fig. 24.37). A careful treatment of the refraction of rays that are close to the principal axis shows that if the incident rays are parallel to the principal axis of the lens, then after passing through the lens the rays all pass through a point on the axis called the *focal point*, F. This is true for parallel rays incident from either side of the lens (parts B and C of Fig. 24.36). Hence, there are two focal points, and they are at identical distances from the lens. This symmetry is related to the reversibility principle of light propagation (Section 24.1).

Lenses can be classified according to the curvatures of their two surfaces (Fig. 24.38). Given the radius of curvature of each surface (Fig. 24.37), it is possible to use Snell's law to calculate the focal length. (We'll see how that works in Section 24.6.) For most applications, we only need to know two things about a lens: the value of the focal length f and whether the lens is *converging* or *diverging*. The lenses in Figure 24.36 are converging lenses: all the incoming rays intersect at the focal point on the opposite side. A ray diagram for a diverging lens is shown in Figure 24.39. In that case, parallel rays from the left diverge after passing through the lens, but they all extrapolate back to the focal point on the left (the same side as the incident rays).

Figure 24.35 Example 24.6. View through a car's rearview side mirror.

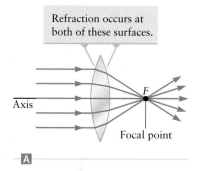

Refraction occurs at both of these surfaces.

Axis

F

Focal point

A

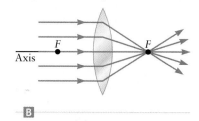

Simplified ray diagram showing a single deflection of each ray.

Axis

F F

B

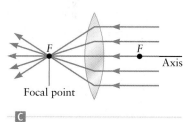

F F

Axis

Focal point

C

Figure 24.36 **A** Light rays are refracted at *both* surfaces of a lens. When parallel light rays strike a converging lens, the refracted rays converge at the focal point F of the lens. **B** For simplicity, most ray diagrams with lenses show only a single "deflection" of a ray as it passes through the lens. **C** If light strikes the lens from the opposite side, the rays converge at a second focal point F on the left side of the lens.

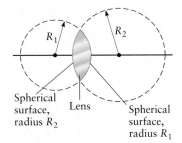

Figure 24.37 A spherical lens: both surfaces follow the shape of a sphere. The radii of curvature of the lens are the radii R_1 and R_2 of the corresponding spheres.

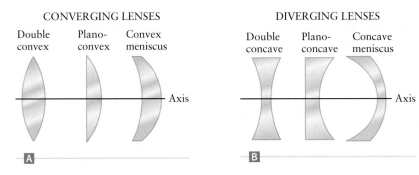

Figure 24.38 Some examples of **A** converging lenses and **B** diverging lenses.

Figure 24.39 With a diverging lens, parallel incident rays from the left are refracted away from the axis. The rays on the right appear to emanate from a point F on the left side of the lens. This point is one of the focal points.

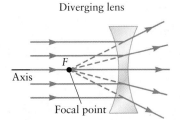

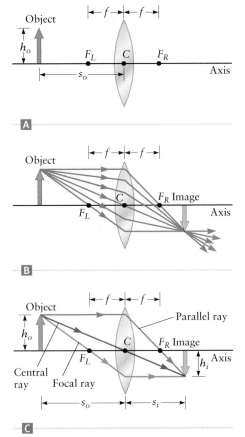

Figure 24.40 Ray tracing with a converging lens when the object is outside the focal point F_L. **A** Sketch showing the object, lens, and focal points. **B** Some typical rays from the tip of the object, intersecting at the tip of the image. **C** Ray diagram showing just the parallel ray, focal ray, and central ray, all emanating from the tip of the object. In this case, the image is real. (When the object is inside the focal point and $s_o < f$, the image is virtual instead.)

Forming an Image with a Lens: Ray Tracing

Our next job is to analyze the images produced by a lens. We'll start with a ray-tracing analysis and then consider how to calculate quantitatively the image properties. Let's begin with the case of a converging lens. Figure 24.40A shows the object along with the focal points on the left (F_L) and right (F_R) sides of the lens. In a typical case, we know the distance s_o from the object to the lens and the object's height h_o, and we want to find the location of the image and its size. Figure 24.40B shows many rays emanating from the object; these rays are refracted by the lens and then intersect to form the image. For simplicity, this diagram shows only rays that emanate from the tip of the object, but rays also emanate from other points on the object and intersect to form other points on the image. In Figure 24.40C, we show only three rays that emanate from the tip of the object; these three rays are particularly useful in a ray tracing analysis. The *parallel ray* is initially parallel to the principal axis of the lens. The lens refracts this ray so that it passes through the focal point F_R on the right. The *focal ray* passes through the focal point on the left. When the focal ray strikes the lens, the ray is refracted to be parallel to the axis on the right. The *central ray* passes straight through the center of the lens (denoted as C). If the lens is very thin (a good approximation for many lenses), this ray is not deflected by the lens. These three rays come together at the tip of the image on the right of the lens. In this case, the image is inverted, and because the rays pass through the image, the image is real.

This approach to ray tracing with lenses is described by the following rules.

Ray tracing applied to lenses

1. Construct a figure showing the lens and the principal axis of the lens. The figure should also show the focal points on both sides of the lens.

2. Draw the object at the appropriate point. One end of the object will generally lie on the axis.

3. Draw three rays that emanate from the tip of the object:

(a) The *parallel ray* is initially parallel to the axis. After passing through the lens, this ray or its extrapolation passes through one of the focal points.

(b) The *focal ray* is directed at the other focal point. After passing through the lens, this ray is parallel to the axis.

(c) The *central ray* passes straight through the center of the lens and is not deflected.

4. These three rays or their extrapolations intersect at the image. If the rays actually pass through the image, the image is *real*; if they do not, the image is *virtual* (just as in the case of images formed by a mirror). When a lens forms a real image, the object and image are on opposite sides of the lens (Fig. 24.40C). In Example 24.7, we'll show that when a lens forms a virtual image, the object and image are on the same side of the lens.

This approach to ray tracing is very similar to the method we used for mirrors; the three rays in Figure 24.40C are similar to the rays used in ray diagrams for mirrors (Fig. 24.26). Our ray diagrams have emphasized the parallel, focal, and central rays because these rays are the simplest ones to draw. However, all other rays that pass through the lens will also pass through the image point as illustrated in Figure 24.40B.

EXAMPLE 24.7 | Ray Tracing for a Diverging Lens

Consider a diverging lens with an object to the left of the focal point on the left (incident) side of the lens. **(a)** Use ray tracing to construct the resulting image. **(b)** Is the image real or virtual?

RECOGNIZE THE PRINCIPLE

These questions can both be answered with a ray diagram, which we can construct using the rules for ray tracing applied to lenses.

SKETCH THE PROBLEM

Figure 24.41 shows the problem, including a full ray diagram.

IDENTIFY THE RELATIONSHIPS AND SOLVE

(a) We follow the suggested steps for making a ray diagram with lenses. Step 1: Figure 24.41A shows the lens, its axis, and the focal points. Step 2: We add the object to the drawing. Step 3: Figure 24.41B shows the parallel ray, the focal ray, and the central ray, drawn according to our ray-tracing rules. Step 4: These rays do not intersect at any point to the right of the lens. If the rays on the right are extrapolated back to the left side of the lens, however, the extrapolations do intersect. The point of intersection is the image point at the tip of the image.

(b) The image is | virtual | because the rays do not pass through the image point, and the image is on the same side of the lens as the object.

What does it mean?

For all problems involving the image formed by a lens, you should always begin with a ray diagram. Our ray tracing rules apply to both converging and diverging lenses.

Finding the Image Location for a Lens

Ray tracing enables us to describe qualitatively the image formed by a lens. We now want to derive a quantitative approach for calculating the image location. As with our work with the mirror equation, we must adopt sign conventions for dealing with the object and image distances, heights, and so forth. The conventions we'll follow are shown in Figure 24.42 and are listed below.

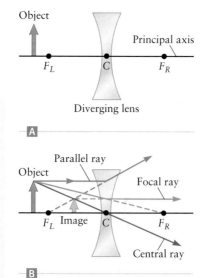

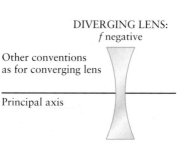

Figure 24.41 Example 24.7. Ray tracing with a diverging lens.

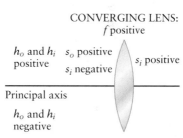

Figure 24.42 Sign conventions for application of the thin-lens equation.

Once again, remember that there are several different sign conventions for working with lenses, so be careful when comparing our equations with those from other books.

CONCEPT CHECK 24.8 | Object and Image Distances

Figure 24.41B shows an object and its image as formed by a diverging lens. What are the signs (positive or negative) of (a) the object distance s_o, (b) the image distance s_i, and (c) the focal length f?

The Thin-Lens Equation

Figure 24.43A shows a ray diagram for a converging lens. We can use the geometry of this sketch to find a mathematical relation for locating the image produced by a converging lens. Within this figure, we identify two pairs of similar right triangles. One pair is shaded in yellow in Figure 24.43A; one yellow triangle has side lengths h_o and s_o, and the other yellow triangle has sides of length h_i and s_i. The similarity of these triangles leads to

$$\frac{h_o}{s_o} = \frac{-h_i}{s_i} \qquad (24.19)$$

The negative sign here is due to our convention that h_i is negative because the image in this case points downward (extends below the axis). Another pair of similar triangles is shaded in red in Figure 24.43B and leads to

$$\frac{h_o}{f} = \frac{-h_i}{s_i - f} \qquad (24.20)$$

with the same sign convention for the image height as in Equation 24.19. Taking the ratio of Equations 24.19 and 24.20 gives

$$\frac{h_o/s_o}{h_o/f} = \frac{h_i/s_i}{h_i/(s_i - f)}$$

Canceling the factors of h_o and h_i, we have

$$\frac{f}{s_o} = \frac{s_i - f}{s_i}$$

and cross multiplying, we get

$$s_i f = s_o(s_i - f)$$

We can now rearrange to obtain

$$s_o f + s_i f = s_o s_i$$

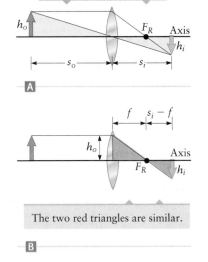

The two yellow triangles are similar.

The two red triangles are similar.

Figure 24.43 Geometry for deriving the thin-lens equation.

[3]Notice, however, that when light passes through two or more lenses (as in a telescope or microscope), it is possible for the object distance s_o to be negative. We'll see examples of that in Chapter 26.

Dividing both sides by a factor of fs_os_i leads to our final result,

$$\frac{1}{s_o} + \frac{1}{s_i} = \frac{1}{f}$$ (24.21) Thin-lens equation

Equation 24.21 is called the **thin-lens equation**. It can be used to calculate the image properties of both converging lenses and diverging lenses, provided we use the sign conventions described above.[4] Notice that the thin-lens equation is actually identical to the mirror equation (Eq. 24.15).

The thin-lens equation (Eq. 24.21) gives a relation between the object and image distances s_o and s_i, and can thus be used to find the image location. Equation 24.21 can also be used to obtain the magnification. As with mirrors, the magnification is the ratio of the image height to the object height,

$$m = \frac{h_i}{h_o} = -\frac{s_i}{s_o}$$ (24.22)

Applying the Thin-Lens Equation

Let's now see how to use the thin-lens equation together with a ray diagram to locate and describe the image formed by a lens. Consider a converging lens with $f = +0.50$ m and an object at $s_o = +0.30$ m. We have included the positive signs here to emphasize that f is positive (a converging lens) and that s_o is positive because the object is on the left side of the lens. Figure 24.44, which also shows our standard ray diagram, is sketched to scale. We can find the image distance from the thin-lens equation,

$$\frac{1}{s_o} + \frac{1}{s_i} = \frac{1}{f}$$

Solving for $1/s_i$ first, we get

$$\frac{1}{s_i} = \frac{1}{f} - \frac{1}{s_o} = \frac{s_o - f}{fs_o}$$

Rearranging to find s_i gives

$$s_i = \frac{fs_o}{s_o - f}$$ (24.23)

Inserting the given values of f and s_o, we obtain

$$s_i = \frac{fs_o}{s_o - f} = \frac{(0.50 \text{ m})(0.30 \text{ m})}{0.30 \text{ m} - 0.50 \text{ m}} = -0.75 \text{ m}$$

According to our sign convention, this negative value for s_i means that the image is on the left side of the lens, so it is a virtual image. This result is confirmed by the ray diagram in Figure 24.44.

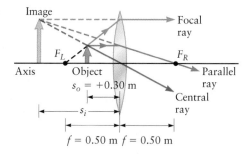

Figure 24.44 Applying the thin-lens equation to find the location of an image formed by a converging lens.

| CONCEPT CHECK 24.9 | Image Formed by a Converging Lens

Is the image in Figure 24.44 (a) real and upright, (b) virtual and upright, (c) real and inverted, or (d) virtual and inverted?

| **EXAMPLE 24.8** | Image Formed by a Diverging Lens

Suppose the converging lens in Figure 24.44 is replaced by a diverging lens with a focal length $f = -0.50$ m. **(a)** If the object is again at $s_o = +0.30$ m (i.e., on the left side of the lens), what is the location of the image? **(b)** Is this image real or virtual?

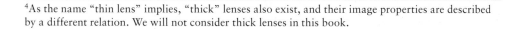

[4]As the name "thin lens" implies, "thick" lenses also exist, and their image properties are described by a different relation. We will not consider thick lenses in this book.

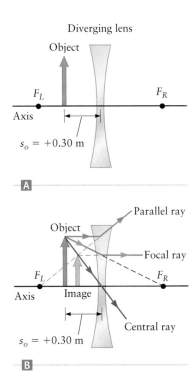

Diverging lens

A

B

Figure 24.45 Example 24.8.

We begin with a ray diagram, which will give us a qualitative understanding of the image. We then proceed to apply the thin-lens equation to find the actual value of s_i.

SKETCH THE PROBLEM

Figure 24.45 shows a ray diagram constructed by following our suggested procedures for ray tracing with lenses. Step 1: Figure 24.45A shows the lens and the focal points on the left and right. Step 2: We add the object to our drawing. Step 3: In Figure 24.45B, we add three rays. The parallel ray is initially parallel to the principal axis; after passing through the lens, this ray is refracted away from the principal axis since we have a diverging lens. Extrapolating this ray back to the left-hand side, we see that it passes through the focal point on the left, F_L. The focal ray is directed from the object toward the focal point to the right of the lens, F_R; this ray is refracted by the lens and like the parallel ray is deflected away from the principal axis. On the right side of the lens, the focal ray travels parallel to the principal axis. The central ray passes through the center of the lens and is not deflected by the lens. Step 4: These rays do not intersect at any point on the right side of the lens. Extrapolating them back, however, we find intersection at a common image point to the left of the lens. This ray diagram tells us much about the image and provides a check on results obtained with the thin-lens equation.

IDENTIFY THE RELATIONSHIPS AND SOLVE

(a) We apply the thin-lens equation (Eq. 24.21) and solve for s_i just as we did in deriving Equation 24.23:

$$s_i = \frac{fs_o}{s_o - f}$$

Substituting the given values of the focal length ($f = -0.50$ m) and object distance ($s_o = +0.30$ m), we get

$$s_i = \frac{fs_o}{s_o - f} = \frac{(-0.50\text{ m})(0.30\text{ m})}{0.30\text{ m} - (-0.50\text{ m})} = \boxed{-0.19\text{ m}}$$

(b) According to the ray diagram, light does not pass through this image point, so the image is virtual. This conclusion is confirmed by our analysis; since the image distance s_i is negative, the image is indeed on the left side of the lens and is $\boxed{\text{virtual}}$.

What does it mean?

Because the focal length f for a diverging lens is negative, when s_o is positive (an object on the left side of the lens), the thin-lens equation will always give a negative value for s_i and thus a virtual image. This result is different from a converging lens, which can give either a real or a virtual image, depending on how far the object is from the lens.

Applications of the thin-lens equation are mathematically straightforward, provided you carefully follow the sign conventions in Figure 24.42. Constructing the image using ray tracing is *always* advisable, however, because it provides a check on results from the thin-lens equation and gives an intuitive understanding of the image properties.

24.6 | ✪ HOW THE EYE WORKS

Lenses and mirrors are used in many different kinds of optical instruments, including microscopes and telescopes, which we'll discuss in Chapter 26. In a sense, the oldest optical instrument is the eye. A sketch of a vertebrate eye is shown in Figure 24.46A; it is a complicated structure with several different regions, but its basic

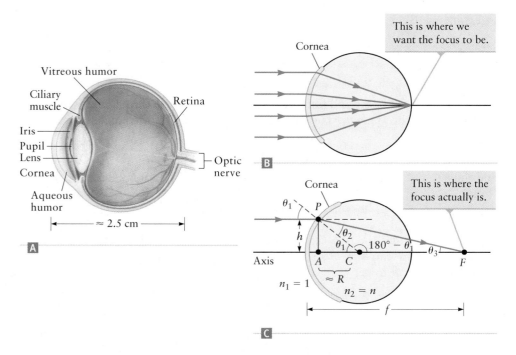

Figure 24.46 A Structure of the eye. B Corneal refraction model of the eye. The curved surface of the cornea refracts and thus focuses light near the retina (the back surface of the eye). C The refraction at the cornea can be analyzed using Snell's law. Here we ignore refraction when the light leaves the right surface of the spherical eye.

function can be understood using the principles of refraction and the properties of lenses. Light enters through the cornea, passes through a liquid region called the aqueous humor, through a lens, into a gel-like region called the vitreous humor, and then to the retina at the back surface of the eye. The retina contains light-sensitive cells that convert light into electrical signals carried to the brain by the optic nerve.

A Simple Model of the Eye: Corneal Refraction

Different regions of the eye have different indices of refraction, so light rays are refracted and change direction as they pass from the air into the cornea and then from region to region within the eye. The indices of refraction of the different regions inside the eye are all in the range of approximately 1.33 to 1.40 (Table 24.1), so the biggest index change and largest refraction occur when light first enters the cornea. In the simplified model in Figure 24.46B, we approximate the eye as spherical, and we also treat all the different parts of the eye as a single region with a single index of refraction. In this model, the only refraction occurs at the surface of the cornea, and we want to calculate how this refraction focuses light. We are especially interested in the focal length of this corneal "lens" and how close (or how far away) the focal point is to the retina. Although this illustration is certainly a simplified version of the real eye, it will still help us understand how the eye works. We'll then consider improvements to the model that make it closer to the real thing.

Figure 24.46C shows a ray diagram for refraction by the cornea. The incident angle is θ_1, the angle the incoming ray makes with the normal to the surface. Here we consider a single incident ray that initially travels parallel to the principal axis; we want to find where its refracted ray intersects the axis since that will give the location of the focal point F. Because we have assumed the surface of the eye is spherical, the normal direction (the dashed line through points P and C) passes through the center of the eye at C. For simplicity, we assume the incoming ray is near the axis, so the height h and the angle θ_1 are small. The refracted angle is given by Snell's law (Eq. 24.7), $n_1 \sin \theta_1 = n_2 \sin \theta_2$. The outside region is air, so $n_1 = n_{air} \approx 1.00$, and we denote the index of refraction inside the eye as simply n. We also assume the angles are small so that $\sin \theta_1 \approx \theta_1$ (and so forth for θ_2). We now have

$$n_1 \sin \theta_1 = n_{air} \sin \theta_1 \approx (1.00)\theta_1 = \theta_1$$

$$n_2 \sin \theta_2 = n \sin \theta_2 \approx n\theta_2$$

Inserting into Snell's law gives

$$\theta_1 = n\theta_2$$

$$\theta_2 = \frac{\theta_1}{n} \tag{24.24}$$

To locate the focal point and find the focal length f, we use the geometry in Figure 24.46C to find the angle θ_3. The sum of the interior angles of triangle PFC is 180°, so we have

$$180° - \theta_1 + \theta_2 + \theta_3 = 180°$$

and thus $\theta_3 = \theta_1 - \theta_2$. Inserting the result for θ_2 from Equation 24.24 gives

$$\theta_3 = \theta_1 - \theta_2 = \theta_1 - \frac{\theta_1}{n} = \theta_1\left(\frac{n-1}{n}\right) \tag{24.25}$$

We next consider the two right triangles PFA and PCA. From the first of these triangles, we find[5] $\tan\theta_3 = h/f$. The angle θ_3 is small, so we can also use the approximation $\tan\theta_3 \approx \theta_3$ to get

$$\theta_3 \approx \tan\theta_3 = \frac{h}{f}$$

$$h \approx f\theta_3 \tag{24.26}$$

Applying the same approach to triangle PCA leads to

$$\theta_1 \approx \tan\theta_1 = \frac{h}{R}$$

$$h \approx R\theta_1 \tag{24.27}$$

Setting Equations 24.26 and 24.27 equal gives $f\theta_3 = R\theta_1$. The focal length of the corneal "lens" is then

$$f = R\frac{\theta_1}{\theta_3}$$

and inserting the result for θ_3 from Equation 24.25, we find

$$f = R\frac{\theta_1}{\theta_1(n-1)/n}$$

$$f = \frac{n}{n-1}R \tag{24.28}$$

Properties of the Spherical Eye

Equation 24.28 gives the distance from the refracting surface (the cornea) to the focal point for our simplified model of the eye in Figure 24.46C. Where does this focal point lie? Is it inside the eye? We can estimate this focal length using the average index of refraction of the eye, which is about $n = 1.4$; we find

$$f = \frac{n}{n-1}R = \frac{1.4}{1.4-1}R = \frac{1.4}{0.4}R = 3.5R \tag{24.29}$$

In our simple model of a spherical eye, the retina is a distance $2R$ from the front surface of the cornea, so the focal point found in Equation 24.29 is about $1.5R$ *behind* the retina.

We have worked through this simplified model of the eye for two reasons. First, it shows how to deal with refraction from a spherical surface. Second, the result for the focal length in Equation 24.29 shows that this simple spherical eye would not function very well because it does not form an image at the retina; indeed, that

---

[5]We assume point A is very close to the spherical refracting surface. This approximation is good when the rays are close to the axis (as we have assumed) and h is small.

explains why the eye actually deviates somewhat from a spherical shape. In the more realistic picture of Figure 24.46A, you will notice that the cornea protrudes slightly from the otherwise spherical profile of the eye. This protrusion makes the radius of curvature of the cornea smaller than the radius of the main spherical portion of the eye. The quantity R in Equation 24.29 is the radius of curvature of the *cornea*, so allowing for this more realistic shape will bring the focal point closer to the retina. Hence, our simple model has given some insight into why the cornea is shaped the way it is.

Refining the Model: Adding the Lens

Let's now add the eye's lens to our model. In a real eye, the lens is inside the eye (Fig. 24.46A), and light passes through this lens after being refracted by the cornea. We will instead consider a slightly simpler model in which the lens is in front of the cornea and hence just outside the eye. Our model will thus not apply strictly to a real eye, but it will yield some useful insights and can be a good way to understand the combination of an eye with a contact lens. So, in our simple model, the incoming light first passes through and is partially focused by the lens and then enters the cornea, which finishes the focusing job. By placing the lens in front of the cornea, we can more easily apply our results for lenses, including the thin-lens equation. We now ask two questions: "What kind of lens is needed for the eye to function properly (i.e., converging or diverging)?" and "What is the shape of the lens (i.e., its radius of curvature)?"

We begin with the ray diagrams for two different cases. In Figure 24.47A, the eye is focused on an object at infinity (i.e., on an object that is very far away). The incoming rays are parallel to the axis, and we assume the cornea has the correct radius of curvature so that these rays are focused on the retina. Hence, we assume the focusing for distant objects is done solely by the cornea using the refraction focusing of our first model in Figure 24.46B. The ray diagram in Figure 24.47B shows why we need the added lens. Here the object has been moved very close to the eye; the incoming rays are now diverging from the axis, and the refracted rays produced by the cornea alone intersect at a point behind the retina. We need the extra focusing of the added lens in Figure 24.47C to focus the image on the retina.

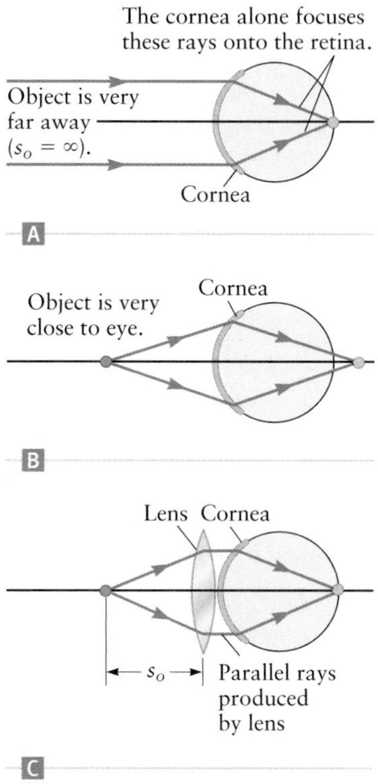

Now let's consider what properties this lens must have to make the ray diagram in Figure 24.47C work. We place the object a distance s_o in front of the lens and suppose the lens produces parallel, outgoing rays (i.e., an image at infinity) as shown in the figure. We know from Figure 24.47A that the cornea by itself can focus these parallel rays on the retina. Hence, the lens should be designed such that an object at s_o focuses at infinity (so that $s_i = \infty$), producing parallel rays for the cornea. Using the thin-lens equation (Eq. 24.21) with an assumed focal length of f_{lens}, we have

$$\frac{1}{f_{lens}} = \frac{1}{s_o} + \frac{1}{s_i} = \frac{1}{s_o} + \frac{1}{\infty} = \frac{1}{s_o} \tag{24.30}$$

and thus

$$f_{lens} = s_o \tag{24.31}$$

The eyes of a person with normal vision are able to focus on objects as close as $s_o = 25$ cm from each eye. The lens in Figure 24.47C must therefore have a focal length $f_{lens} = 25$ cm. Since f_{lens} is positive, the lens is a converging lens.

How must this lens be shaped to give this value for the focal length? The focal length is related to the radii of curvature of the two surfaces of a lens by a relationship[6] called the *lens maker's formula*. For a converging lens with two convex surfaces with radii of curvature R_1 and R_2, this formula reads

$$\frac{1}{f} = (n - 1)\left(\frac{1}{|R_1|} + \frac{1}{|R_2|}\right) \tag{24.32}$$

Lens maker's formula

---

[6]The lens maker's formula can be derived by considering the refraction from the two surfaces of the lens in succession, similar to the calculation we did for our model eye in Figure 24.46C.

The cornea alone focuses these rays onto the retina.

Object is very far away ($s_o = \infty$).

Cornea

A

Object is very close to eye.

Cornea

B

Lens Cornea

$\leftarrow s_o \rightarrow$ Parallel rays produced by lens

C

Figure 24.47 Function of the eye's lens. **A** In this model, light from a very distant object is focused by the cornea alone on the retina. **B** When an object is very close to the eye, the refracted rays do not fall on the retina, but instead intersect behind it. **C** If we place a lens in front of the eye, light from a nearby object is focused by the combination of the eye's lens and the cornea so that an image is formed at the retina.

where n is the index of refraction of the lens. Equation 24.32 applies only to the double convex lens shape in Figure 24.37 and assumes the lens is in air. More general versions of the lens maker's formula apply to other lens shapes and to situations such as a lens under water. The absolute value terms in Equation 24.32 emphasize that both these terms are positive for the double convex lens considered here. For other lens shapes, one or both of these terms may be negative and hence describe a diverging lens with a negative value for f. A few such cases are discussed in the end-of-chapter problems. (See Problems 81 and 82.)

Applying the Lens Maker's Formula

Let's now apply the lens maker's formula to estimate the radius of curvature of the lens in our model of the eye in Figure 24.47C. The required focal length of this lens was estimated using Equation 24.31, where we found $f_{lens} \approx 25$ cm. Although the lens maker's formula involves the radii of curvature for both sides of the lens separately, in the spirit of our approximate treatment we assume these radii are equal and $R_1 = R_2 = R$. Using this assumption in Equation 24.32 gives

$$\frac{1}{f_{lens}} = (n - 1)\left(\frac{1}{|R_1|} + \frac{1}{|R_2|}\right) = \frac{2(n - 1)}{R}$$

or

$$f_{lens} = \frac{R}{2(n - 1)}$$

where we have now dropped the absolute value signs and assume R is positive. Solving for R, we find

$$R = 2(n - 1)f_{lens} \tag{24.33}$$

Inserting the value of $f_{lens} = 25$ cm given above and using the value of the index of refraction for the eye's lens $n = 1.41$ (see Table 24.1), we have

$$R = 2(n - 1)f_{lens} = 2(1.41 - 1)(0.25 \text{ m}) = 0.21 \text{ m} \tag{24.34}$$

The radius of curvature of the lens is thus several times larger than the diameter of the eye. (Recall that the diameter of the eye is about 2.5 cm = 0.025 m.) This result for the radius of curvature is in accord with the realistic drawing of the eye in Figure 24.46A. The front surface of the lens within the eye, which is responsible for most of the focusing, has a rather flat surface, so its radius of curvature is much larger than the radius of the eye.

In addition to allowing the eye to focus on a nearby object, the eye's lens plays another crucial role. When the object is at a particular location, the focal length of the eye's lens has a value such that the final image from the cornea is properly focused on the retina. If the object is moved closer to or farther away from the eye, the focal length of the lens must change to keep the final image in focus. This change of shape is accomplished using muscles attached to the edges of the lens (Fig. 24.46A) to pull on the edges and change the lens' radius of curvature. The lens must be sufficiently flexible to allow its shape to change accordingly. As a person grows older, the lens becomes less flexible and the eye can no longer focus properly for objects that are both far away and very close. That is why the author wears glasses.

24.7 | OPTICS IN THE ATMOSPHERE

Rainbows

In Section 24.3, we showed that refraction of light by a prism causes rays to change direction and that the amount light is deflected depends on the color of the light.

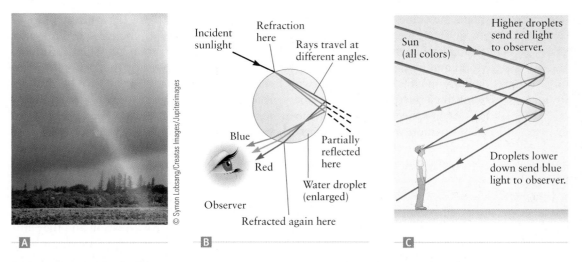

Figure 24.48 ◧ A typical rainbow. Rainbows are caused by refraction, dispersion, and reflection of light from water droplets in the atmosphere. ◪ When light enters a droplet, refraction and dispersion separate different colors just as in a prism. After reflecting from the back of the droplet (on the right), these rays are separated further by refraction and dispersion when they leave the droplet. They finally emerge from the droplet at different angles. ◫ Different colors of the rainbow come from different parts of the sky. The colors of these rays *do* represent the color of the light.

A similar effect is responsible for rainbows (Fig. 24.48A), but in this case the refracting object is a water droplet. Water droplets in the atmosphere are illuminated by light from the Sun, and these droplets disperse and redirect light to an observer according to the ray diagram in Figure 24.48B. An incident ray from the Sun is refracted when it enters a droplet; the refracted angles depend on the color of the light, so rays for different colors travel at different angles inside the droplet. When light reaches the back surface of the droplet, a portion is reflected, giving a set of reflected rays. (The rest is refracted and propagates out the back of the droplet.) The reflected rays are refracted when they leave the droplet; the refracted angles again depend on the color, so the final rays leave the droplet at different angles. The rays in Figure 24.48B enter droplets at one particular angle, but other rays will enter at different angles, so the outgoing rays will emerge over a range of angles. However, for each color there is one particular outgoing angle for which the outgoing rays have the greatest intensity, and this "best" outgoing angle depends on the color due to dispersion. As a result, the different colors of a rainbow appear at different (angular) positions in the sky (Fig. 24.48C).

Occasionally, a secondary rainbow is also visible, outside the primary brighter rainbow shown in Figure 24.48A. We'll consider the origin of the secondary rainbow in Problem 77.

Why Stars Twinkle

The index of refraction of a substance can depend on many factors. For example, the index of refraction of air depends on the pressure and temperature of the air. Wind and other effects cause the air temperature and pressure, and hence the index of refraction, to fluctuate from place to place in the atmosphere. These fluctuations cause each small region of air to act as a small lens or prism, deflecting light. As an example, Figure 24.49 shows starlight as it travels through the atmosphere and through one pocket of cool air on its way to an observer on the Earth. This pocket of air has a different index of refraction than the air around it, so rays from the star are refracted as they enter and leave this region, just as light is refracted by a lens. This "lens" of cool air thus deflects rays from the star and affects the apparent location of a star as seen by an observer on the Earth. These "lenses" change

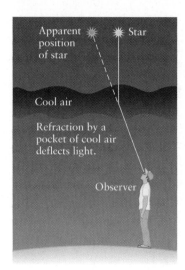

Figure 24.49 Pockets of cool air or warm air refract starlight as it travels to an observer. These pockets shift with time, so the apparent position of the star changes with time and the stars appear to "twinkle." The amount a ray is deflected is much smaller than drawn here.

shape and come and go as air—both warm air and cool air—is blown from place to place. Thus, the deflection of the rays from a particular star changes from moment to moment, causing a star to "twinkle." A similar effect is found with light from the Moon and planets; these sources, however, appear much larger than stars (because they are much closer), so the twinkling effect is not as apparent as for stars. This twinkling effect is why astronomers prefer to have telescopes located on mountaintops or in space, with less of the atmosphere between the telescope and the star. We'll say more about this subject when we discuss telescopes in Chapter 26.

24.8 | ABERRATIONS

In our simple ray diagrams and our derivations of the mirror equation and the thin-lens equation, we found that rays emanating from a particular point on an object intersect (focus) at a single image point. That result relied on several assumptions, and although these assumptions are usually good approximations, they are still just approximations. The images formed by most real mirrors and lenses suffer from effects called *aberrations*. Aberrations are deviations from the ideal images we have encountered so far; the most common is when all the light rays from a point on an object do not quite focus at a single image point.

Chromatic Aberration

Chromatic aberrations are related to the dispersion of light into different colors as in a prism. Because the index of refraction of a lens depends on the color of the light, different colors are refracted by different amounts. For a plane surface such as the prism in Figure 24.21, the way the rays of different colors disperse can be calculated using Snell's law (Example 24.4). For a lens, we can use the lens maker's formula to estimate the effect of chromatic aberration. According to this formula (Eq. 24.32), the focal length f depends on the index of refraction of the lens. Hence, f is different for different colors and the location of an image is different for each color as shown in Figure 24.50. Many optical instruments (as we'll describe in Chapter 26) employ multiple lenses, and it is sometimes possible to compensate for chromatic aberration by arranging the aberrations of separate lenses to cancel each other. Even with this approach, however, most optical systems have at least a small amount of chromatic aberration.

Spherical Aberration

Since it is relatively simple to make an accurate spherical shape, spherical surfaces for mirrors and lenses are convenient. In this chapter, we have assumed the surfaces of curved mirrors and lenses are spherical and have showed that the resulting focal lengths depend on the radii of curvature. Ray diagrams such as Figures 24.26 and 24.41 rely on the assumption that the rays drawn are very close to the axis of the mirror or lens. Mathematically, this assumption allows us to make the approximations $\sin \theta \approx \theta$ and $\tan \theta \approx \theta$, and if those approximations are accurate, the rays will focus at a single image point. (Those approximations were used in the derivations of the mirror equation and the thin-lens equation.) They are, however, only approximations; in any real situation, rays that are not exactly on the axis will not focus exactly at the ideal image points we found for mirrors or lenses. Such deviations are called *spherical aberrations* because they are due to the spherical shape of the mirror or lens.

Certain curved surfaces with nonspherical shapes give better focusing properties than those of spherical shapes. For this reason, high-precision optical components often employ parabolically shaped surfaces. Components of this shape cost more to manufacture, but they give better image properties for off-axis rays.

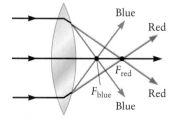

Figure 24.50 Chromatic aberration produced by a converging lens. Rays of different wavelengths (colors) focus at different points.

Ray optics (also called geometrical optics)

In the *ray optics* approximation, light travels along straight lines called *rays*. These rays change direction when light is reflected or refracted. Ray optics gives an accurate description of light propagation when light passes through openings or strikes objects that are large compared to its wavelength.

Reflection and refraction

For reflection and refraction, we measure the direction of a ray using the angle θ made by the ray with the direction normal (perpendicular) to the surface of interest.

According to the *law of reflection*, the angle of incidence is equal to the angle of reflection:

$$\theta_i = \theta_r \qquad \textbf{(24.1)} \text{ (page 798)}$$

Refraction is described by *Snell's law*,

$$n_1 \sin \theta_1 = n_2 \sin \theta_2 \qquad \textbf{(24.7)} \text{ (page 802)}$$

where the *index of refraction* $n = c/v$ is the ratio of the speed of light in a vacuum (c) to the speed of light in the refracting material (v). By applying the law of reflection and Snell's law, we can calculate how light interacts with plane mirrors, curved mirrors, lenses, and other refracting objects such as flat plates and water droplets.

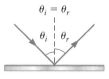

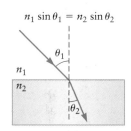

Ray tracing

Ray tracing is the process of using light rays that emanate from an object to derive the location and other properties of an image. Important properties of an image include the following:

- *Real image*: Light from the object passes through the image point.
- *Virtual image*: Light rays do not pass through the image point. The image point is found by extrapolating rays back to a common point behind a mirror or in front of a lens.
- *Upright image*: The image has the same orientation as the object.
- *Inverted image*: The image is upside down compared with the object.
- *Enlarged image*: The image is larger than the object (the magnification is $|m| > 1$).
- *Reduced image*: The image is smaller than the object ($|m| < 1$).

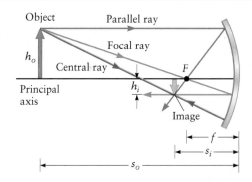

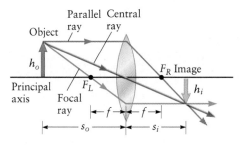

(Continued)

Magnification

Magnification is the ratio of the image height h_i to the object height h_o,

$$m = \frac{h_i}{h_o} \qquad \textbf{(24.10)} \text{ (page 810)}$$

According to our *sign conventions* for h_i and h_o, the magnification can be either positive or negative. See pages 813–814 and 820, and Figures 24.32 and 24.42, for a description of these sign conventions.

Mirror equation

The *mirror equation* relates the object distance, the image distance, and the *focal length* for a curved mirror:

$$\frac{1}{s_o} + \frac{1}{s_i} = \frac{1}{f} \qquad \textbf{(24.15)} \text{ (page 813)}$$

For a concave spherical mirror, f is given by

$$f = \frac{R}{2} \qquad \textbf{(24.14)} \text{ (page 813)}$$

where R is the radius of curvature. For a convex spherical mirror, f has the same magnitude, but is negative. See Figure 24.32 for a description of the mirror sign conventions.

Thin-lens equation

The images formed by a lens are described by the *thin-lens equation*,

$$\frac{1}{s_o} + \frac{1}{s_i} = \frac{1}{f} \qquad \textbf{(24.21)} \text{ (page 821)}$$

where f is the focal length of the lens. For a converging lens, f is positive; for a diverging lens, it is negative. See Figure 24.42 for a description of the sign conventions for lenses.

Aberrations

Real lenses do not form perfect images. *Chromatic aberrations* are deviations due to the dependence of the refractive index on color. *Spherical aberrations* result from rays that do not propagate close to the principal axis of a lens or mirror with spherical surfaces.

QUESTIONS

SSM = answer in Student Companion & Problem-Solving Guide   ⊗ = life science application

1. An angler observes a fish whose image is 10 m from the shore. Is this fish closer to or farther from the shore than its image? Explain. *Hint*: Draw a ray diagram.

2. Where is the image formed for an object in front of a plane mirror?

3. Does a plane mirror produce a real image or a virtual image? Explain.

4. What is the focal length of a plane mirror? Explain.

5. Explain why reflection from a window consisting of a single pane of glass produces two images.

6. The author's satellite TV dish receives signals from a satellite in geosynchronous orbit about the Earth. (For this orbit, the period is precisely one day, so the satellite is always "at" the same point in the sky.) The signal is picked up with a receiver

located 40 cm from the dish's center as shown Figure Q24.6. Explain why this "dish" is concave and estimate its radius of curvature.

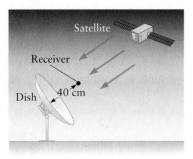

Figure Q24.6

7. When finding the image produced by a lens via ray tracing, it was suggested that you should always draw a ray that goes through the focal point (the focal ray). (In some cases, an extrapolation of the focal ray passes through the focal point.) Which of the following statements gives the most accurate description of this ray?
    (a) The focal ray passes through the center of the lens.
    (b) The focal ray always leaves the lens parallel to the axis of the lens.
    (c) The focal ray always leaves the object parallel to the axis of the lens.
    (d) The focal ray always passes through a virtual image.
    (e) There are two focal rays, one for each focal point.

8. In the ray diagram in Figure Q24.8, a combination of blue ($n = 1.60$) and red ($n = 1.50$) light strikes the surface of a flat glass plate. Because of dispersion, there are two different refracted rays inside the glass. Which of the following statements is correct?
    (a) Ray 1 is the blue ray, and ray 2 is the red ray.
    (b) Ray 1 is the red ray, and ray 2 is the blue ray.
    (c) Different rays reflect at different angles due to their color.
    (d) This diagram shows an example of total internal reflection.
    (e) The index of refraction in the glass is larger for red light.

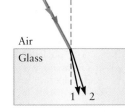

Figure Q24.8

9. Consider a diverging lens with an object in front of the lens. Prove that the image is always virtual.

10. Explain how a refraction experiment could be used to measure the index of refraction of a material as a function of the frequency of the light.

11. Imagine a light ray as it passes from one material to another (as in Fig. 24.11) and consider the electric and magnetic fields associated with these rays at the boundary between the materials. At (i.e., very near) the boundary, the component of the electric field directed parallel to the boundary must be the same on the two sides of the boundary. Use this fact to argue how the frequency of the light wave must change (or not change) when you cross the boundary.

12. Light travels from air into plastic. Which of the following statements is correct? The results from Question 11 will be useful here.
    (a) The frequency of the light is greater in air than in plastic.
    (b) The frequency of the light is smaller in air than in plastic.
    (c) The frequency of the light is the same in both air and plastic.
    (d) The wavelength of the light is greater in air than in plastic.
    (e) The wavelength of the light is smaller in air than in plastic.
    (f) The wavelength of the light is the same in both air and plastic.

13. **SSM** The image formed by a plane mirror appears to be a "reverse" of the object, with left and right interchanged, as illustrated in Figure Q24.13. It is actually more accurate to think of this reversal as front to back. (a) Use a ray diagram to explain why this front-to-back reversal occurs. (b) Use your ray diagram to explain why a top-to-bottom reversal does *not* also occur.

Figure Q24.13

14. When you look at yourself at the center of two plane mirrors forming a 90° corner, your image is not reversed from right to left. In other words, you see yourself like others see you. Using a ray diagram, explain this observation.

15. In Figure Q24.15, water is streaming out of a hole in a can, and it appears that light is also "pouring" out of the can. Explain what is actually happening.

Figure Q24.15

16. ✖ Under water, the human eye cannot focus. Why?

17. In many convenience stores, a mirror is placed in a top corner of the store to allow the clerk to have a full view of the store. Is this mirror spherical concave or convex? Explain your answer.

18. Figure Q24.18 is a picture of a mirror one might find in a bathroom. The mirror is actually three mirrors with different curvatures so that you can view your image in different ways. The largest of the mirrors produces a normal image, whereas the smaller, two-sided mirror is used to magnify your image by two different amounts. Discuss the curvatures (plane, concave, convex) of the three mirrors required to produce these three images.

Figure Q24.18

19. When you observe a fish in a spherical fishbowl, the fish appears larger than it actually is. Why?

20. Funhouse mirrors distort your image so that you look shorter, taller, fatter, or thinner than you actually are. For the four images shown in Figure Q24.20, describe the curvature of the mirror.

Figure Q24.20

21. A magnifying glass produces an enlarged image that is upright and virtual as shown in Figure Q24.21. Does it use a converging or a diverging lens? How close must you hold the magnifying glass to the object to view the image in this way?

Figure Q24.21

22. [SSM] A plastic rod becomes invisible when inserted in a container of vegetable oil as shown in Figure Q24.22. What can you conclude about the plastic rod and the vegetable oil?

23. Explain how you could measure the index of refraction of a liquid.

24. Explain why dispersion is essential for producing a rainbow.

25. Explain how you could use Snell's law to measure the speed of light in a piece of plastic.

26. Explain how a person swimming under water could observe total internal reflection and how she would know when she saw it.

Figure Q24.22

24.1 RAY (GEOMETRICAL) OPTICS

1. [SSM] Light from a very tiny lightbulb passes through a hole of diameter 0.15 m in a dark screen that is 2.5 m away (Fig. P24.1). Assume the light travels as ideal rays as it passes from the bulb, through the hole, and onto a second screen on the far right in Figure P24.1. If the light rays illuminate a spot of diameter 0.65 m on the second screen, what is the spacing between the two screens?

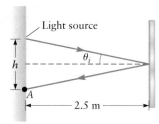

Dark screen
Lightbulb
Hole
0.65 m
|← 2.5 m →|← L = ? →|
Screen

Figure P24.1

24.2 REFLECTION FROM A PLANE MIRROR: THE LAW OF REFLECTION

2. The light ray in Figure P24.2 emanates from a point on one wall of a room and is incident on a plane mirror that lies flat on the opposite wall. The reflected ray strikes the left-hand wall at point A, which is a distance $h = 1.4$ m below the light source. What is the angle of incidence θ_i?

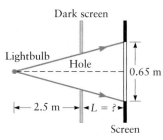

Light source
θ_i
h
A
|← 2.5 m →|

Figure P24.2

3. ✪ Two plane mirrors meet at a right angle (Fig. P24.3). An incident ray strikes the bottom mirror at point B with an angle of incidence θ_i; this ray reflects from each mirror and strikes the screen on top at point A. (The screen is parallel to the bottom mirror.) Given the dimensions in Figure P24.3, what is θ_i?

A Screen
|← 3.0 m →|
θ_i Mirrors 2.0 m
B
|← 2.5 m →|

Figure P24.3

4. A pencil is placed in front of a plane mirror and is oriented parallel to the plane of the mirror (Fig. P24.4). Construct a careful ray diagram and use it to find the image of the pencil. Be sure to draw several rays from at least two different points on the pencil (not just the tip).

Pencil

Figure P24.4
Problems 4 and 5.

5. What is the magnification of the pencil in Problem 4 and Figure P24.4? Is the image upright or inverted?

6. ★ A light ray reflects from a mirror hanging vertically on a wall in your bathroom, and the ray strikes a particular point on the opposite wall (Fig. P24.6). The bottom end of the mirror is then tilted slightly away from the wall so that the mirror makes an angle of 5° with the wall. The spot where the reflected ray hits is found to move upward a distance of 0.25 m. What is the length L of the room?

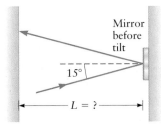

Figure P24.6

7. ✪ The two plane mirrors in Figure P24.7 come together as shown. Find two images of point *A* as viewed by an observer at *B*. Do your analysis graphically and assume the images are made by just a single reflection.

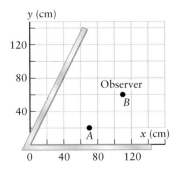

Figure P24.7

8. ✪ A marble is placed near a plane mirror as shown in Figure P24.8. (a) Where will the image be located? (b) How does the image location change as the observer moves closer to the mirror? (c) How does the image location change as the marble is moved away from the mirror?

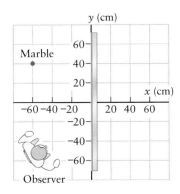

Figure P24.8

9. SSM ✪ Two plane mirrors are each *L* = 3.5 m long. They are parallel and are placed 0.25 m apart. If a light ray enters from one end as shown in Figure P24.9, how many reflections occur before it leaves the region between mirrors?

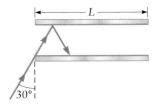

Figure P24.9

10. ✪ A plane mirror made of a very thin piece of glass lies flat on the ground. As shown in Figure P24.10, one end of the mirror is 2.0 m from you and the other end is 30 m from a nearby tree. You are 1.8 m tall, and the mirror has a length *L*. The mirror is arranged so that you can just see the image of the top of the tree at one edge of the mirror. (a) How tall is the tree? (b) If you can just see a point halfway up the tree at the other edge of the mirror, what is *L*?

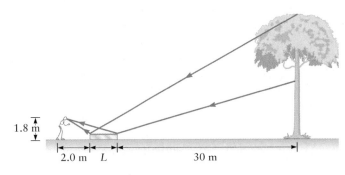

Figure P24.10 Not to scale.

11. ✪ With an ideal corner cube, the reflected ray is precisely parallel to the incident ray. No corner cube will ever be perfect, however. The astronauts who visited the Moon left corner cubes so that scientists can do experiments in which light is reflected from the Moon and back to the Earth. Suppose one of these corner cubes has an angle of θ = 90.00001° (Fig. P24.11). By what distance will the reflected ray miss its target on the Earth?

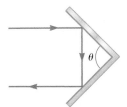

Figure P24.11

12. ✪ In a typical changing room at a clothing store, you may find two mirrors on opposite walls, a setup that allows you to see how you look from the front and the back (Fig. P24.12). A consequence of this configuration is that you see a seemingly endless number of images of yourself getting farther and farther away. Assuming you're standing in the middle of the changing room and the mirrors are separated by a distance *d*, derive a general expression for the positions of the images you see.

Figure P24.12

24.3 REFRACTION

13. A light ray is incident from air onto the surface of a very quiet lake. If the angle of incidence is 35° and the angle of refraction is 25°, what is the index of refraction of the water?

14. The index of refraction of beer is about 1.35. What is the speed of light in beer?

15. Light is refracted through diamond from air. If the angle of incidence is 30° and the angle of refraction is 12°, what is the speed of light in diamond?

16. Suppose light with a wavelength of 560 nm in a vacuum is incident upon fused quartz, which has an index of refraction of 1.46. What will the wavelength of this light be inside the quartz?

17. A particular type of glass has an index of refraction of 1.45. What is the frequency of light with a wavelength of 450 nm inside this glass? Recall from Chapter 12 that the speed of a wave is equal to the product $f\lambda$.

18. Light from a helium–neon laser has a wavelength of 630 nm in a vacuum. What is its wavelength in water ($n = 1.33$)?

19. ⭐ Light is incident on a plane mirror with an angle of incidence of $\theta_i = 65°$. If this experiment takes place under water ($n = 1.33$), what is the angle of reflection?

20. What is the critical angle for light to stay inside a thin glass rod ($n = 1.40$) surrounded by air? What if the rod is surrounded by water?

21. A ray of light is incident upon a still water surface from above. What is the maximum possible angle of refraction?

22. A light beam containing several different colors is incident on a flat piece of glass with $\theta_i = 25°$. Which of the following statements is correct?

   (a) The angle of refraction will be greater for red light than for blue light.

   (b) The angle of refraction will be smaller for red light than for blue light.

   (c) The angle of refraction will be the same for red light as for blue light.

23. For Problem 22, if the angle of refraction for green light is 14°, what is the refractive index for this wavelength (i.e., color)?

24. ⭐ A thin film of water sits on top of a flat piece of glass, and light is incident from below (Fig. P24.24). If the angle of incidence is 30°, what is the angle of refraction of the final outgoing ray?

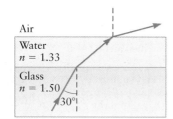

Figure P24.24

25. What is the critical angle for total internal reflection for light traveling between glass ($n = 1.65$) and water ($n = 1.33$)? Does the light start in the glass or the water?

26. ⭐ A combination of red light and blue light is incident on a flat piece of quartz (Fig. P24.26). The two colors refract at different angles. (a) Which refracted ray in Figure P24.26 is red, and which is blue? (b) What is the angle $\Delta\theta$ between the two rays? *Hint*: Use the data in Figure 24.20 and the discussion on page 806.

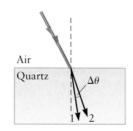

Figure P24.26

27. ✪ A light ray is incident on a plate of plastic ($n = 1.50$) that is $d = 2.5$ cm thick (Fig. P24.27). (a) Show that the outgoing ray is parallel to the incident ray. (b) If the angle of incidence is 40°, what is the displacement Δh of the outgoing ray relative to the incident ray?

28. ✪ Figure 24.14 shows a pencil that penetrates into a body of water. The pencil appears to be bent due to refraction. Explain this effect and calculate the angle that the pencil appears to be bent at the point it enters the water.

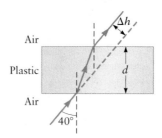

Figure P24.27

29. [SSM] ⭐ Monochromatic light (i.e., light of a single color) is incident on the symmetric triangular prism in Figure P24.29 as shown with $\theta = 55°$. If the index of refraction of the prism is $n = 1.60$, what angle does the outgoing ray make with the horizontal?

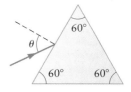

Figure P24.29
Problems 29 and 30.

30. ✪ Consider again the prism in Figure P24.29. White light is now incident on the prism. Calculate the angular separation of red light and blue light that is refracted by the prism. Assume the prism is composed of quartz and use the values for the indices of refraction in Figure 24.20.

31. ⭐ The prism shown in Figure P24.31 has an index of refraction $n = 1.75$ for a particular color of light. Four different incident rays are shown in the figure. Construct a ray diagram showing the outgoing rays in all four cases and also calculate the angles these outgoing rays make with a horizontal direction.

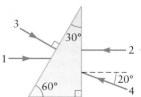

Figure P24.31

32. ⭐ The prism in Figure P24.32 is symmetric (i.e., forms an isosceles triangle). If the ray shown is perpendicular to the bottom surface of the prism and just barely undergoes total internal reflection at its first reflection (i.e., it strikes the interface at the critical angle) at the left face of the prism, what is the index of refraction of the prism?

Figure P24.32

33. ✪ A rectangular slab of glass has an index of refraction $n = 1.80$. Light is incident at 45°, striking the upper surface very close to the right edge as shown in Figure P24.33, and is refracted as it enters the glass forming ray 2. When ray 2 strikes the right side of the slab, will any light emerge from the glass?

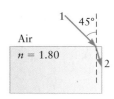

Figure P24.33
Problems 33 and 34.

34. ✪ Consider again the rectangular slab of glass in Figure P24.33 with light incident on the top surface with an angle of incidence of 45°. The refracted ray (denoted ray 2) will undergo partial reflection at the right side of the slab and again at the bottom surface, then pass back through the top surface. What angle will the ray that exits the top surface make with the direction normal to that surface?

35. ⭐ When an observer looks into the empty swimming pool in Figure P24.35 and peers over the top edge, she finds that she can just see the far corner of the bottom of the pool. When the pool is filled with water, she can see a much bigger portion of the pool's bottom; that is, she can now see a point on the bottom a distance L from the far edge. Find L.

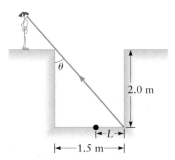

Figure P24.35

36. ✪ Figure P24.36 shows light incident at an angle θ on a prism with index of refraction $n = 1.55$. It is found that light reflects from the upper right face of the prism and travels back along the path followed by the incident ray. Find θ.

Figure P24.36

24.4 REFLECTIONS AND IMAGES PRODUCED BY CURVED MIRRORS

37. An object 5.0 cm tall is 45 cm in front of a concave mirror with a radius of curvature of 15 cm. Construct a ray diagram to scale following the steps outlined in the procedure for "Ray Tracing Applied to Spherical Mirrors." (a) Indicate the parallel, focal, and central rays in your diagram. (b) Where is the image? (c) Is the image real or virtual? How big is it?

38. A cat is located 4.5 m in front of a concave mirror. The mirror has a radius of 1.5 m. Construct a ray diagram to scale following the steps outlined in the procedure for Ray Tracing Applied to Spherical Mirrors. (a) Where is the image of the cat? (b) Is this image real or virtual? (c) If the image of the cat is 5.0 cm tall, how tall is the cat?

39. If the cat in Problem 38 moves away from the mirror, does the cat's image get larger or smaller in magnitude?

40. Where should the cat in Problem 39 sit so as to appear as large as possible (i.e., to be as scary as possible) to a dog?

41. A concave mirror has a focal length of 10 cm. What is the magnification of an object placed 100 cm in front of this mirror?

42. A ball whose diameter is 0.25 cm is placed 50 cm in front of a convex spherical mirror. If the focal length of the mirror is −40 cm, what is the diameter of the ball as viewed in the mirror?

43. An object is placed 75 cm in front of a concave mirror. If the magnification is −0.75, what is the focal length of the mirror?

44. An object 22 mm tall is located 1.5 m in front of a convex mirror. If the image is 1.2 m behind the mirror, what is the height of the image?

45. An object 5.0 cm tall is 25 cm in front of a convex mirror with a radius of curvature of 25 cm. Construct a ray diagram to scale following the steps outlined in the procedure for "Ray Tracing Applied to Spherical Mirrors." (a) Indicate the parallel, focal, and central rays in your diagram. (b) Where is the image? (c) Is the image real or virtual? How big is it?

46. ✪ A convex mirror has a focal length of −20 cm. Where is the image if the image is upright and one-fourth the size of the object?

47. A pencil is placed 45 cm in front of a convex mirror. Is the image in front of or behind the mirror?

48. ✪ When an object is placed 12 cm in front of a mirror, an image is formed 25 cm behind the mirror. Is the mirror concave or convex? What is the mirror's focal length?

49. [SSM] ✪ An object is 35 cm in front of a concave mirror and produces an upright image that is 2.5 times larger than the object. What is the focal length of the mirror?

50. A convex mirror has a radius of curvature of 85 cm. An image is produced 30 cm in front of the mirror. Where is the object?

51. ✪ Suppose an object is somehow located a distance 35 cm *behind* a concave mirror with a radius of curvature of 1.50 m. Such a "virtual object" can be produced using the image of a second mirror. (a) Use the mirror equation to find the location of the image. (b) Is the image real or virtual? (c) What is magnification of the object?

52. A monster is 10 m in front of a convex mirror and appears with a magnification of 0.50. What is the radius of curvature of the mirror?

53. ✪ Suppose the rearview mirror on a car is convex with a radius of curvature of 25 m. (a) If a motorcycle is 40 m behind the car, what is the magnification of the motorcycle? (b) How does your answer change when the motorcycle is 15 m behind the car?

54. An object 3.5 cm high is placed 20 cm in front of a convex mirror with a radius of 12 cm. (a) Make an accurate ray diagram showing the location and size of the image. Construct your ray diagram following the steps outlined in the procedures for "Ray Tracing Applied to Spherical Mirrors." (b) Check your results using the mirror equation.

55. An object is placed in front of a concave mirror with a focal length of 10 cm. What is the position of the resulting image if the image is upright and twice the size of the object?

56. ✪ A meterstick lies along the optical axis of a convex mirror of focal length −40 cm, with its nearest end 60 cm from the mirror. How long is the image of the meterstick?

24.5 LENSES

57. An object 1.5 cm high is placed a distance of 10 cm in front of a converging lens with a focal length of 5.0 cm. (a) Follow the steps outlined in the procedures for "Ray Tracing Applied to Lenses" and make an accurate ray diagram to scale showing the location and size of the image. (b) Check your results using the thin-lens equation.

58. An object is a distance $2f$ from a converging lens. (a) Following the steps outlined in the procedures for "Ray Tracing Applied to Lenses," construct a ray diagram showing the location and orientation of the image. (b) Where is the image located? (c) Is it a real image or a virtual image?

59. Repeat Problem 58 for an object that is a distance $f/3$ in front of the lens.

60. An object is 45 cm to the left of a lens, and the image is 25 cm to the left of the lens. What is the magnification?

61. An object is placed in front of a lens. The image is upright, 35 cm to the left of the lens, and is half as tall as the object. What is the focal length of the lens?

62. An object is a distance $2|f|$ in front of a diverging lens. (a) Following the steps outlined in the procedures for "Ray Tracing Applied to Lenses," construct a ray diagram showing the location and orientation of the image. (b) Where is the image located? (c) Is it a real image or a virtual image?

63. An object 5.0 cm tall is 25 cm in front (on the left) of a converging lens with a focal length of 40 cm. Where is the image?

64. ✪ A fly appears to be upside down and twice its actual size when viewed through a converging lens with a focal length of 5.5 cm. How far is the fly from the lens?

65. ✪ If the fly in Problem 64 moves farther from the lens, will the image become larger or smaller?

66. ✪ Two converging lenses with $f_1 = 25$ cm and $f_2 = 35$ cm are located 1.50 m apart as shown in Figure P24.66. An object is placed 50 cm in front of the lens on the left. (a) Construct a ray diagram following the steps outlined in the procedures for "Ray Tracing Applied to Lenses." Find the location and size of the image produced by the first lens. (b) Construct a ray diagram for the second lens, using the image from part (a) as the object. Where is the image produced by the second lens? How big is it?

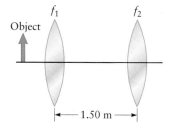

Figure P24.66

67. ✪ A converging glass lens ($n = 1.50$) is immersed in water. Does the focal length get longer or shorter and by what factor? Explain why. *Hint*: In the lens maker's formula in Equation 24.32, the term $n - 1$ is actually the difference between the index of refraction of the lens and the surrounding air.

68. A converging lens with a focal length of 25 cm is used to form an image on a screen that is 3.0 m from the lens. Where must the object be placed?

69. A converging lens has a focal length of 10 cm. How far from the lens should an object be placed to get (a) a real image with a magnification of 2.5 and (b) a virtual image with a magnification of 2.5?

70. SSM ✪ For a converging lens of focal length f, determine the minimum object-to-image distance so that the image is (a) real and (b) virtual.

24.6 HOW THE EYE WORKS

71. ✖ A young child is just able to read a newspaper that is only 10 cm from her eyes. What is the focal length of the child's eye? Assume her retina is 1.5 cm in diameter.

72. ✪ ✖ Your eye (diameter 2.5 cm) is able to focus the image of a tree on your retina. The tree is 25 m tall and 75 m away. What is the size of the image of the tree on your retina?

73. ✪ ✖ A cell phone 7.0 cm tall is 40 cm in front of your eye (diameter 2.5 cm). (a) Find the size of the image of the cell phone on your retina. (b) The cell phone is now moved to a spot 60 cm from your eye. Is the image on the retina larger or smaller than the image in part (a)? By what factor?

74. ✪ ✖ ⑱ **Horse's eye.** Have you ever noticed that as you approach a horse, it appears to raise its head in a way that seems aggressive? In fact, the horse is not usually showing aggression, but rather is raising its head to see you. The lens in the eye of a horse does not have the ability to change shape as in the human eye. Instead, the back of the eye (the retina) is curved in an elliptical shape so that the straight-line distance from the front of the eye to the retinal surface depends on where the light is hitting the retinal surface. The result is that when a horse is trying to view objects up close it must lift its head in such a way that it uses the upper portion of its retina. More distant objects are

viewed using the central portion of the retina in much the same way as a person who is presbyopic uses bifocal lenses to view both objects near and far. (a) Given the information provided in Figure P24.74, determine the near-point distance for the horse. The *near-point distance* is the shortest object distance for which the eye is able to focus. (b) For a typical horse, estimate the distance from its eye to the tip of its nose. Using this estimate, does your answer to part (a) seem reasonable? Explain.

24.7 OPTICS IN THE ATMOSPHERE

75. SSM A rainbow is formed by refraction of light as it passes through a water droplet. In the ray diagram in Figure 24.48B, which of the following statements best explains the origin of a rainbow? Explain your answers.
   (a) The water droplet must be exactly spherical for a rainbow to occur.
   (b) Different rays refract at different angles due to their color.
   (c) Different rays reflect at different angles due to their color.
   (d) The water droplet acts like a lens and focuses the light.
   (e) Light from the water droplet is polarized.

76. ✪ Figure 24.48B shows a ray diagram for light from the Sun as it passes through a water droplet and produces a rainbow. Construct a careful ray diagram of this type for light rays of different colors (e.g., red and blue) and use it to explain the order (sequence) of colors in a rainbow.

77. ✪ On rare occasions, it is possible to observe a "double" rainbow (Fig. P24.77A). (The secondary rainbow is always present, but is usually too faint to be observed.) The ray diagram in Figure P24.77B shows how the light rays that form the secondary rainbow are refracted and reflected by a water droplet. Use this ray diagram to explain why the order of colors in the secondary rainbow is opposite to the order of colors in the primary rainbow.

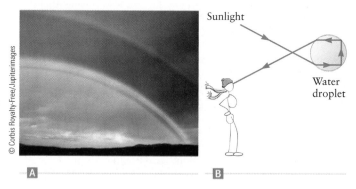

Figure P24.77

78. ✪ ⑱ **Stellar illusion.** As we discussed in the chapter, the position of a star in the sky as viewed from the surface of the Earth is not where it actually is because the light from the star refracts as it enters the Earth's atmosphere. Approximating the Earth's atmosphere to be homogenous with a thickness of 100 km and having an index of refraction of 1.0003, determine the angular separation between the observed position of a star just barely above the horizon and the actual position of the star.

24.8 ABERRATIONS

79. ✪ A converging lens has a focal length of $f = 55$ cm for blue light ($n = 1.65$). The index of refraction of this lens for red light is $n = 1.55$. What is the focal length for red light?

80. ✪ Consider the lens in Problem 79. An object is placed 95 cm in front of the lens and is illuminated with white light. Chromatic aberration causes the image formed with blue light to be at a different position than the red image. How far apart are the two images?

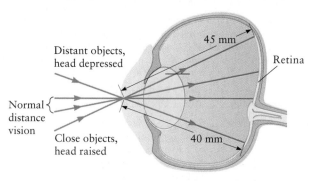

Figure P24.74

81. **Lens maker's formula for a double concave lens.** Equation 24.32 gives the focal length of a double convex lens (i.e., a lens for which both surfaces are convex) in terms of the radii of curvature of the two surfaces. For a double concave lens (Fig. P24.81), the lens maker's formula has a slightly different form:

$$\frac{1}{f} = (n-1)\left(-\frac{1}{|R_1|} - \frac{1}{|R_2|}\right)$$

This relation always gives a negative focal length f as expected for a double concave lens (because it is a diverging lens). Consider such a lens composed of quartz with $R_1 = R_2 = 50$ cm and $n = 1.56$ for blue light. What is its focal length for blue light?

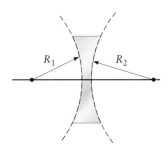

Figure P24.81

82. ✪ Compare the forms of the lens maker's formulas in Equation 24.32 and Problem 81 and devise a formula for a lens that has one concave surface and one convex surface.

83. ✪ In Chapter 23, we discussed the polarization of light by reflection. Figure P24.83 shows light that is both reflected and refracted as it passes from air into glass with $n = 1.60$. The incident light is unpolarized, containing components with the electric field perpendicular to the plane of the drawing and parallel to this plane. When the reflected and refracted rays are precisely perpendicular, the reflected ray will be purely linearly polarized, with $\vec{E}$ perpendicular to the plane of the drawing as indicated. Find the angle of the incident ray for which the refracted and reflected rays are perpendicular. It is called **Brewster's angle**.

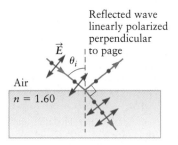

Figure P24.83

84. SSM ✪ ⓡ In Problem 83, we saw how reflection can lead to polarization of reflected light. Is this effect important for reflection of sunlight from a car's windshield? Investigate this question by working through the following steps. (a) Calculate Brewster's angle for glass. (b) Draw a ray diagram of the problem. Show the Sun, windshield, and driver along with the incident and refracted rays. The angle of incidence should equal Brewster's angle found in part (a).

85. Brewster's angle (see Problem 83) for a particular prism is found to be 65°. What is the index of refraction of the prism?

86. ✶ A slide projector uses a converging lens to project images on a screen 4.0 m away. If the distance between the slide (image) and the lens is 5.0 cm, what are (a) the focal length of the lens and (b) the magnification of the image?

87. SSM ✶ Prior to the development of the digital camera, the 35-mm film camera was the norm. The designation "35 mm" refers to the size of the film (screen) on which the image is formed. Suppose you wish to use one of these cameras to photograph a 10-m-tall tree that is 30 m away and have the tree's image fill the entire 35-mm film. What must the focal length of the camera lens be?

88. ✪ Suppose a ray of light is incident on three stacked layers of glass with different indices of refraction as shown in Figure P24.88. If the ray enters from air, what is the angle θ_3?

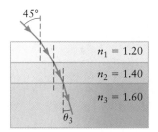

Figure P24.88

89. ✪ A ray of light is incident on the surface of a block of clear ice ($n = 1.309$) at an angle of 30° to the normal (Fig. P24.89). A portion of the light is reflected from the surface, and a portion is refracted into the ice. Determine the angle between the reflected and refracted light.

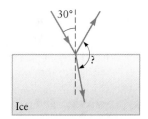

Figure P24.89

90. ✪ A laser beam strikes in the center of one end of a long, thin glass plate that has an index of refraction of 1.5 as shown in Figure P24.90. Determine the number of internal reflections the beam undergoes before it exits the plate at the other end.

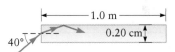

Figure P24.90 Not to scale.

91. ✪ **Refracting surfaces.** A more general analysis of refracting surfaces would include the case of refraction of light at a spherical interface such as a fishbowl. Similar to the thin-lens equation, the expression relating the object distance, image distance, indices of refraction, and the radius of curvature of the surface is given by

$$\frac{n_1}{s_o} + \frac{n_2}{s_i} = \frac{n_2 - n_1}{R}$$

where n_1 and n_2 are indices of refraction where the object and image are located, respectively. For the configuration shown in Figure P24.91, the radius of curvature of the bowl is $R = -15$ cm and

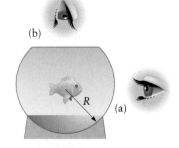

Figure P24.91

the fish is 10 cm from the curved surface and 8.0 cm down from the top. We can neglect the glass bowl and only consider the refraction due to the water. Determine the image distance of the fish as viewed from (a) the side and (b) the top. *Hint*: You'll need to determine the radius of curvature of a flat surface to answer part (b).

92. A rock is dropped in a large tank of benzene and sinks to the bottom of the tank. If the apparent depth of the rock as viewed directly from above is 35 cm, what is the actual depth of the rock? *Hint*: See Problem 91.

93. A swimming pool constructed below ground level is 3.0 m square and 2.0 m deep (Fig. P24.93). A drain is at the bottom center of the pool. How far from the edge of the pool can a person whose eyes are 1.2 m above the ground stand and still see the drain?

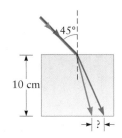

Figure P24.93   **Figure P24.94**

94. Suppose a beam of white light is incident at a 45° angle upon a glass plate 10 cm thick. The index of refraction of the glass for blue light is 1.65, and for red light it is 1.60 (Fig. P24.94). Determine the distance between the blue light and red light as they exit the glass plate.

95. A thin, converging lens whose focal length is 80 cm is placed 5.0 m from a convex spherical mirror whose focal length is −50 cm (Fig. P24.95). For an object placed 1.0 m to the left of the lens, determine (a) the location of the image, (b) the magnification of the image, and (c) the nature of the image (real/virtual, upright/inverted, enlarged/reduced). *Hint*: Use the image formed by the lens as the object for the mirror.

Lens   Mirror

Figure P24.95

96. A material with an index of refraction of 1.80 is machined into the shape of a quarter of a cylinder. The radius of the cylinder is 10 cm; it rests on a flat, horizontal surface; and it is surrounded by a vacuum. A beam of light is incident upon the cylinder as shown in Figure

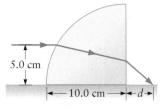

Figure P24.96

P24.96. Determine the distance d from the quarter-cylinder at which the exiting beam strikes the horizontal surface.

97. A birdwatcher observes a hummingbird at a distance of 10 m through a simple eyepiece (converging lens) of focal

length 25 cm (Fig. P24.97). As the hummingbird flies away at a speed of 2.5 m/s, how fast does the image move? Does the image move toward or away from the lens?

Figure P24.97

98. **Renaissance technology.** The Hockney–Falco theory claims that early Renaissance painters used concave mirrors to aid their depth perception (e.g., Jan van Eyck's *Portrait of Giovanni Arnolfini and his Wife*; Fig. P24.98A). Reportedly, van Eyck traced the reflected image from his subjects according to the schematic shown in Figure P24.98B. Calculate the diameter of the blown-glass sphere from which the reflective mirror was cut and comment on the likelihood of such a theory.

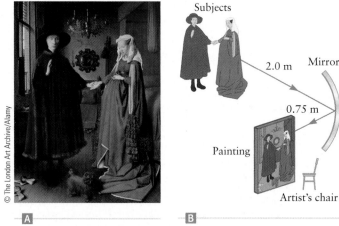

Figure P24.98

99. An overhead projector produces an image of an object (transparency) on a distant screen by using a light source, a plane mirror, and a converging lens ($f = 50$ cm). The distance between the lens and the mirror is 15 cm. (a) If you wish to project a focused image on a screen 3.0 m away from the lens, what distance above the transparency must you set the center of the mirror–lens combination? (b) Describe how you must place the transparency on the projector so that the image is oriented properly.

100. Sometimes on hot summer days the road appears to be wet as shown in Figure P24.100. The black surface becomes

Figure P24.100

very hot, heating up the air just above the road. This increase in temperature decreases the density of the air and in turn decreases its index of refraction. The result is that light from a distant object can be totally internally reflected, producing a mirage. (a) If your eye level while driving your car is 2.0 m above the surface of the road and you observe a mirage 100 m ahead, what is the index of refraction of air just above the ground? (b) Describe how you might use a mirage to estimate the temperature of the air just above the ground.

101. ✪ **Camera obscura.** Prior to the development of the photographic camera, artists would use a camera obscura like the one shown in Figure P24.101 to draw the image of a subject. In this figure, light from a subject passes through a small hole or lens in the box, is reflected from a plane mirror, and is projected onto a translucent sheet of paper. The artist would then trace the image on the paper. Room-sized camera obscuras were also used. Suppose the distance from the pinhole to the

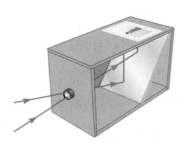

Figure P24.101

mirror is 1.0 m and the mirror-to-screen distance is 50 cm. (a) How far away should a 1.5-m-tall subject stand so that she fills a 25-cm-square screen? (b) Describe how adjusting the size of the pinhole will affect the clarity and brightness of projected image.

102. ✪ **Pinhole camera.** The earliest cameras were simply light-tight boxes with a very small hole in one side of the box. By uncovering the pinhole for a brief period of time, light from an object could pass through the hole, exposing a photographic plate (film) on the inside of the box. Later versions of this type of camera replaced the hole with a lens. For a pinhole camera, the image appears upside down on the photographic plate in relation to the object. (a) Draw a ray diagram to explain how a pinhole camera is able to form a clear image without the use of a lens. (b) If the distance between the photographic plate and the pinhole is 30 cm, what is the size of the image for an object 10 m away and 2.0 m tall?

103. ✪ ℝ **Solar eclipse.** Suppose you use the pinhole camera described in Problem 102 to photograph the Sun during an eclipse. Determine the size of the image of the Sun. Repeat the calculation for the Moon.

104. ✪ **Underwater stargazers.** Some creative amateur astronomers have been known to improve their view of the night sky by wearing scuba gear and observing the sky from the bottom of a swimming pool. Looking up from the bottom of a swimming pool, you are able to see a wider angle of the sky with a pool full of water than when the pool is empty. For a swimming pool 10 m wide and 5.0 m deep, determine the ratio of the angular field of view one sees when the pool is filled with water to that when the pool is empty.

Chapter 25

Wave Optics

This photograph shows two water waves undergoing interference with each other, producing regions of high and low wave intensity. Interference is the central subject of this chapter. (© Illustration-Nature/Alamy)

Is light a particle, or is it a wave? This question was debated for many centuries. Newton believed that light is a particle; he reasoned that since light travels in straight-line paths (as we found in our ray diagrams in Chapter 24), it must consist of particles moving with constant velocity. Newton (and others) developed a particle theory of light that was able to explain reflection, refraction, and all the other ray properties of geometrical optics discussed in Chapter 24. Over time, however, experiments revealed several properties of light that can only be explained with a wave theory. Maxwell's theory of electromagnetism then convinced physicists that light is an electromagnetic wave. In this chapter, we discuss properties of light that depend on its wave nature, the field known as **wave optics**. We also show

how geometrical optics—ray tracing and the focusing of light by mirrors and lenses—fits together with the wave properties of light.

The wavelength of light λ plays a key role in determining when we can or cannot use geometrical optics. When discussing image formation and the properties of mirrors and lenses over distances much larger than λ, geometrical optics is extremely accurate, but for all discussions of interference effects and the propagation of light through small openings (openings whose sizes are comparable to or smaller than λ), wave optics is required. Hence, in their appropriate regimes, geometrical optics and wave optics both describe light.

However, there is still more to this story: new properties of light were discovered only a few decades after Maxwell. These properties could not be understood with Maxwell's theory and led to the development of the quantum theory of light in which light has properties of both particles *and* waves! We'll say a little more about this quantum theory at the end of this chapter and a lot more about it in Chapter 28. The quantum theory of light builds on Maxwell's theory, so everything we learn in this chapter about the wave properties of light is true also in the quantum theory.

25.1 | COHERENCE AND CONDITIONS FOR INTERFERENCE

How is a wave different from a particle? How can we tell one from the other? In some cases (such as water waves), the repeating oscillations of wave motion are obvious, but in other cases (such as light and sound), the wave nature of a phenomenon is not as apparent. How can we really prove that light and sound are wave phenomena? One property is unique to waves: only waves can exhibit *interference*. We described the interference of waves on a string and water waves in Chapter 12 and of sound in Chapter 13. In this chapter, we study the interference of light waves.

Let's first consider when light from two or more sources can and cannot exhibit interference. In particular, can any two light waves exhibit interference effects? If not, how can we tell in advance if light from two or more sources is capable of exhibiting interference? For example, can light from the Sun exhibit interference with light from a car's headlight?

We can answer these questions if we first consider the interference of sound waves emitted by two speakers as sketched in Figure 25.1A. The speakers emit sound waves of wavelength λ that come together and combine at a listener's location. If the sources emit identical waves and if the distances from the sources to the listener are equal ($L_1 = L_2$), the two waves undergo the same number of oscillations on their way to the listener. In this case, we say that the waves are *in phase* (Fig. 25.1B). When two waves are in phase, their maxima occur at the same time at a given point in space. The total wave displacement at the listener's location is the sum of the displacements of the two individual waves. If two waves are in phase, the sum of their displacements is large and the waves interfere constructively to produce a large amplitude and a large intensity.

Now suppose the listener moves relative to the speakers so that the distances L_1 and L_2 traveled by the two waves are no longer equal. If these distances differ by an amount $\lambda/2$, one of the waves travels an extra one-half wavelength on the way

Figure 25.1 🄰 Sound waves emitted by two separate sources (two speakers) combine when they reach a listener. These waves travel distances L_1 and L_2 on their way to the listener, where L_1 and L_2 may or may not be equal. 🄱 Here the two waves arrive in phase, and their sum is larger than either of the individual waves. 🄲 If the two waves are 180° out of phase when they reach the listener, their sum is zero, and no sound (and no energy) arrives at the listener.

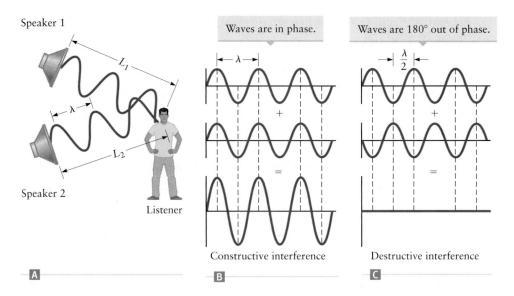

to the listener and the maximum of one wave will coincide with the minimum of the other (Fig. 25.1C). We say that these waves are[1] *out of phase*; in this example, the sum of the two wave displacements at the listener's location is zero and there is *destructive* interference.

Figure 25.1 illustrates several crucial aspects common to all interference experiments, including those involving light.

General conditions for interference. Two waves can interfere if all the following conditions are met.

1. **Two or more interfering waves travel through different regions of space over at least part of their propagation from source to destination.**

2. **The waves are brought together at a common point.**

3. **The waves must have the same frequency and must also have a *fixed phase relationship*. This means that over a given distance or time interval the phase difference between the waves remains constant. Such waves are called *coherent*.**

Coherence

Interference conditions 1 and 2 are illustrated in Figure 25.1, but condition 3 and the notion of coherence warrants more discussion. When the ear detects a sound wave or the eye detects a light wave, it is actually measuring the wave intensity averaged over many cycles. For example, the frequency of blue light is about 7×10^{14} Hz. The eye cannot follow the variation of every cycle; instead, the eye averages the light intensity over time intervals of about 0.1 s, which corresponds to approximately 7×10^{13} cycles, and then sends this average signal to the brain. Suppose the waves in Figure 25.1 are two interfering light waves. In order for the waves to interfere constructively and give a large total intensity, they must stay in phase during the time the eye is averaging the intensity. Likewise, for the two waves in Figure 25.1C to interfere destructively, they must stay out of phase during this averaging time. Each scenario is possible only if the two waves have precisely the same frequency. If their frequencies differ, two waves that are initially in phase with each other (and interfere constructively) will be out of phase at a later time (and interfere destructively).

---

[1]Mathematically, we would say that the waves in Figure 25.1C are 180° out of phase, corresponding to a shift of one wave relative to another by $\lambda/2$. Figure 25.1 considers waves that are either perfectly in phase or completely (180°) out of phase, but other relationships (specified by other phase angles) are also possible.

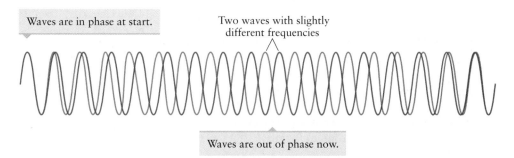

Waves are in phase at start.

Two waves with slightly different frequencies

Waves are out of phase now.

Figure 25.2 Two waves with slightly different frequencies. Initially, they are in phase, but as time passes they become out of phase. The in-phase and out-of-phase portions average out over many cycles, so on average there is no interference. *Important note*: The different colors here do *not* indicate different colors of light, but rather denote the amplitudes of the different waves.

Figure 25.2 shows two waves with slightly different frequencies. The waves are initially in phase, as their maxima and minima occur at about the same times. After many cycles, however, one of the waves "gets ahead" of the other; eventually, the wave with the higher frequency has completed an extra one-half cycle, and the two waves are out of phase. The two waves thus go in and out of phase as time passes. When averaged over a very large number of cycles, the two waves display neither constructive nor destructive interference; on average, there is no interference at all. So, to exhibit interference, two light waves must have exactly the same frequency.

Most visible light is emitted by atoms; you might expect that light waves produced by two atoms of the same element would be "identical" and have the same frequency and could therefore exhibit interference. To exhibit interference, however, two waves must satisfy an additional condition. Consider atoms in a gas that are emitting light. These atoms undergo collisions with other atoms in the gas and with the walls of the container, and each collision causes the phase of the emitted light to jump abruptly to a new value. In other words, each collision "resets" the light wave, starting its oscillation over again. These collisions are different for each atom and spoil interference effects because they cause waves that may have originally been in phase to suddenly become out of phase and vice versa. Experiments show that phase jumps in typical light waves occur about every 10^{-8} s; hence, light from two different atoms of a given element are not coherent. (Certain special cases are an exception, such as when the atoms are all in the same laser. See Chapter 29.) One way to eliminate the effect of phase jumps is to derive both waves from the *same* source. In Section 25.2, we'll explain how to do that.

CONCEPT CHECK 25.1 | Conditions for Interference

Consider an (attempted) experiment to observe interference of light from two different helium–neon lasers. The light from these lasers is emitted by neon atoms. Is it possible to observe interference with these two light sources? Explain why or why not.

EXAMPLE 25.1 | Coherence Time for a Light Wave

The phase of light emitted by an atom changes (jumps) after a typical time interval $\Delta t \approx 10^{-8}$ s, called the *coherence time* of the light. How many cycles of a light wave occur during this coherence time? Assume the light is blue, with $f = 7.0 \times 10^{14}$ Hz.

RECOGNIZE THE PRINCIPLE

A wave's frequency is inversely related to its period T, with $f = 1/T$, and the period is the time it takes to undergo one cycle of oscillation. Given the frequency, we can find the period and from that calculate how many cycles are completed in the coherence time Δt.

SKETCH THE PROBLEM

No sketch is necessary.

Recall some basic facts about oscillations and waves from Chapters 11 and 12. The period of this light wave is (using Eq. 11.1)

$$T = \frac{1}{f} = \frac{1}{7.0 \times 10^{14} \text{ Hz}} = 1.4 \times 10^{-15} \text{ s}$$

which is the time required to complete one cycle of the wave. The number of cycles N contained in the coherence time Δt is thus

$$N = \frac{\Delta t}{T}$$

SOLVE

Inserting our values of Δt and T gives

$$N = \frac{\Delta t}{T} = \frac{1 \times 10^{-8} \text{ s}}{1.4 \times 10^{-15} \text{ s}} = \boxed{7 \times 10^6 \text{ cycles}}$$

What does it mean?

Although the coherence time Δt is very short, the frequency of visible light is very high, so there are more than one million oscillations—quite a large number—during this coherence time. The precise value of Δt depends on the type of atom and its environment; the value 10^{-8} s is only a typical estimate.

25.2 | THE MICHELSON INTERFEROMETER

An optical instrument called a *Michelson interferometer* is based on interference of reflected waves (Fig. 25.3). This clever device, invented by Albert Michelson in the late 1800s, played an important role in the discovery of relativity (Chapter 27). The device contains two reflecting mirrors mounted at right angles. At least one of these mirrors is movable; here the mirror on the far right can be moved along the horizontal axis. A third partially reflecting mirror called a "beam splitter" is mounted at a 45° angle relative to the other two. The beam splitter reflects half the light incident on it and lets the other half pass through.

Light incident from the left in Figure 25.3 strikes the beam splitter and is divided into two waves. One of these waves (denoted 1) travels to the mirror at the top where it is reflected and then returns to the beam splitter. When it reaches the beam splitter, it is again split into two waves, and one of these waves travels downward in Figure 25.3 to the detector as wave 1. The other wave from the beam splitter (wave 2) travels to the mirror on the right, where it is also reflected back to the beam splitter. When wave 2 returns to the beam splitter, a portion is reflected to the detector at the bottom of the figure. Waves 1 and 2 thus combine at the detector; they are the two waves that interfere, and they follow very similar paths as they travel through the interferometer. First, both waves are reflected once by the beam splitter. Second, both waves pass through the beam splitter once. Third, both waves are reflected once by a mirror (at the far right or top). The *only* difference between the two waves is that they travel different distances between their respective mirrors and the beam splitter. After the waves are created by the beam splitter, wave 1 travels a round-trip distance $2L_1$ and wave 2 travels a round-trip distance $2L_2$ before they recombine. The implications for interference are shown in Figure 25.4. The *path length difference* is

$$\Delta L = 2L_2 - 2L_1 \tag{25.1}$$

If we assume $L_2 > L_1$, wave 2 travels an extra distance ΔL and undergoes a number of extra oscillations before it reaches the detector. Each oscillation occurs over a

MICHELSON INTERFEROMETER

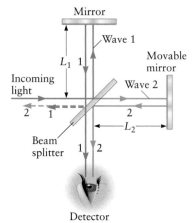

Figure 25.3 Schematic diagram of a Michelson interferometer. Light is incident from the left and split into two separate waves (also called beams) by partial reflection at the beam splitter in the center. These waves reflect from the mirrors at the top and right and arrive back at the beam splitter, where they recombine and travel together to the detector at the bottom. (Waves 1 and 2 also recombine and travel back to the left, but these beams are not used in the experiment.) The colors of the rays do *not* indicate the color of the light.

distance equal to the wavelength λ, so the number of extra oscillations made by wave 2 is

$$N = \frac{\Delta L}{\lambda} \tag{25.2}$$

If N is an integer, then N complete oscillations fit into the extra path length and the two waves are in phase when they recombine, producing constructive interference. On the other hand, if N is half integral (i.e., if $N = \frac{1}{2}, \frac{3}{2}, \ldots, m + \frac{1}{2}$, where m is an integer), wave 2 travels an "extra" one-half wavelength relative to wave 1 and the maxima of one wave will coincide with the minima of the other as in Figure 25.1C. That is the condition for destructive interference. The interference conditions for a Michelson interferometer are thus

$$\Delta L = m\lambda \qquad \text{(constructive interference)}$$

$$\Delta L = (m + \tfrac{1}{2})\lambda \quad \text{(destructive interference)} \tag{25.3}$$

where m is an integer. If the interference is constructive, the light intensity at the detector is large, whereas if the interference is destructive, the detector intensity is zero. Figure 25.5 shows how the light intensity at the detector varies as a function of ΔL. The places on the intensity curve where the interference intensity is greatest are called bright "fringes," and the places where the intensity is zero are called dark "fringes."[2]

The Michelson interferometer uses reflection to satisfy the general requirements for interference from Section 25.1. First, reflection at the central beam splitter produces two separate waves that travel through different regions of space. Second, these waves are brought back together by the beam splitter so that they recombine at the detector. Third, because they are produced from a common incident wave, the two interfering waves are coherent.

Using a Michelson Interferometer to Measure Length

The wavelengths of light emitted by various sources are known. For example, a helium–neon (He–Ne) laser emits light with a wavelength of approximately $\lambda_{\text{He–Ne}} = 633$ nm. Recall that 1 nm = 1×10^{-9} m and that the wavelength of visible light is in the range of about 400 nm to 750 nm. As a rough comparison, the wavelength of visible light is about one thousand times less than the thickness of a human fingernail.

Knowing the value of $\lambda_{\text{He–Ne}}$, an experimenter can use a Michelson interferometer to measure a displacement in the following way. Using light from the laser, the mirror on the far right in Figure 25.3 is adjusted to give constructive interference and its position is noted; this corresponds to one of the bright fringes in Figure 25.5. The mirror on the far right is then moved horizontally, changing the path length difference ΔL, and the intensity at the detector moves along the curve in Figure 25.5. The intensity changes from a high value to zero and back to a high value every time the path length difference ΔL changes by one wavelength of the light. According to Equation 25.3, moving the mirror from one bright fringe to the next bright fringe corresponds to starting at the condition for constructive interference with $\Delta L = m\lambda_{\text{He–Ne}}$ and increasing the path length difference to $\Delta L = (m + 1)\lambda_{\text{He–Ne}}$. If the interferometer mirror is moved so as to pass through N cycles from bright to dark and back to bright, the change in the path length difference is

$$(\Delta L)_{\text{change}} = N\lambda_{\text{He–Ne}} \tag{25.4}$$

When the mirror moves a distance d, the distance traveled by the light changes by $2d$ because the light travels back and forth between the beam splitter and the mirror

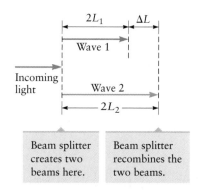

Figure 25.4 A Michelson interferometer with the waves redrawn. The two waves travel distances $2L_1$ and $2L_2$, respectively, as they travel between the beam splitter and the mirrors. The path length difference ΔL leads to interference; compare with Figure 25.1A.

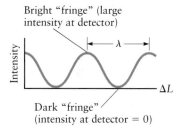

Figure 25.5 The light intensity at the detector of a Michelson interferometer oscillates as a function of the path length difference $\Delta L = 2L_2 - 2L_1$, which is varied by moving one or both mirrors.

---

[2]We'll see why they are called fringes in Section 25.3, when we consider some examples of how interfering light waves appear when viewed by the eye.

(Fig. 25.3). Hence, if the mirror moves through N bright fringes, the distance d traveled by the mirror satisfies

$$(\Delta L)_{change} = 2d = N\lambda_{He-Ne}$$

so that

$$d = \frac{N}{2}\lambda_{He-Ne} \qquad (25.5)$$

The accuracy of this measurement of d depends on the accuracy with which the wavelength λ_{He-Ne} is known. For this reason, physicists have devoted a lot of effort to measuring the wavelength of light from certain light sources. In fact, the wavelength produced by a specially constructed helium–neon laser is known to be

$$\lambda_{He-Ne} = 632.99139822 \text{ nm} \qquad (25.6)$$

Suppose this light source is used in a Michelson interferometer and one of the mirrors is moved a distance d such that exactly $N = 1,000,000$ bright fringes are counted. According to Equation 25.5, we have $d = N\lambda_{He-Ne}/2$. Since N can be counted directly (and is thus accurately known) and the wavelength in Equation 25.6 is also known with high accuracy, this approach provides a very precise way to measure length. Moreover, people in different laboratories can use helium and neon to construct lasers that produce the wavelength given in Equation 25.6, so they can compare their separate length measurements with high accuracy. The Michelson interferometer thus makes possible a very convenient standard for the measurement of length. A similar type of interference effect is used to read information from CDs and DVDs.

CONCEPT CHECK 25.2 | Analyzing the Michelson Interferometer

A Michelson interferometer using light from a helium–neon laser ($\lambda_{He-Ne}=$ 633 nm) is adjusted to give an intensity maximum (a bright fringe) at the detector. One of the mirrors is then moved a very short distance so that the intensity changes to zero. How far was the mirror moved, (a) $\lambda_{He-Ne}/4$, (b) $\lambda_{He-Ne}/2$, (c) λ_{He-Ne}, or (d) $2\lambda_{He-Ne}$?

EXAMPLE 25.2 | Applications of a Michelson Interferometer: Detecting Gravitational Waves

An experiment called the LIGO (for *l*aser *i*nterferometer *g*ravitational wave *o*bservatory) is designed to detect very small vibrations associated with gravitational waves that arrive at the Earth from distant galaxies. The LIGO experiment involves several Michelson interferometers; in one of these interferometers, the mirrors are placed a distance $L_2 = 4$ km (kilometers) from the beam splitter. Suppose changes in the interference pattern corresponding to 1% ($\frac{1}{100}$) of a cycle in Figure 25.5 can be detected. What is the change in the mirror–beam splitter distance that can be detected by the LIGO experiment? Assume the light used in LIGO has a wavelength $\lambda = 500$ nm.

RECOGNIZE THE PRINCIPLE

For a general interference experiment, the path length difference ΔL must change by one wavelength λ to go from a bright fringe to a dark fringe and then back to a bright fringe. A 1% change of a fringe thus corresponds to a $\lambda/100$ change in ΔL. When a mirror of a Michelson interferometer moves a distance d, the path length difference changes by $2d$. We can first find the value of d that gives a 1% fringe change and from that get the change of the mirror–beam splitter distance.

SKETCH THE PROBLEM

Figures 25.3 through 25.5 show a Michelson interferometer and the interference conditions.

IDENTIFY THE RELATIONSHIPS

The change in the path length difference equals $2d$, so each complete intensity cycle from bright to dark to bright in Figure 25.5 corresponds to moving the mirror a distance $\lambda/2$. If we can detect $\frac{1}{100}$ of this change, the corresponding displacement of the mirror is

$$d = \frac{\lambda/2}{100} = \frac{\lambda}{200} \qquad (1)$$

SOLVE

Inserting the given value of the wavelength into Equation (1), we get

$$d = \frac{\lambda}{200} = \frac{(500 \times 10^{-9}\ \text{m})}{200} = \boxed{2.5 \times 10^{-9}\ \text{m}}$$

What does it mean?

The value of d found here is 2.5 nm—approximately 10 times the diameter of an atom—which is quite impressive. This result also shows that when adjusting the mirrors of a Michelson interferometer to actually observe interference, an experimenter must be able to move the mirrors with a precision of a few nanometers. Such precision can be accomplished with carefully designed screw adjustments (a low-tech approach, but there are other ways, too).

25.3 | THIN-FILM INTERFERENCE

Soapy water is normally transparent and colorless, but the photo of a thin soap film in Figure 25.6 shows very bright colors resulting from interference of reflected waves from the film's two surfaces. The colorful "bands" in Figure 25.6 are called "fringes" and correspond to the locations of constructive and destructive interference for light waves with different wavelengths (colors) in the incident light.

To understand where these colors come from, we start with the case shown in Figure 25.7A, in which a thin soap film rests on a flat glass surface. For simplicity, we also assume the incident light is monochromatic, that is, that it has a single wavelength. (Later we'll consider the behavior with white light, which is light containing many different colors.) The upper surface of the soap film in Figure 25.7A is similar to the beam splitter in the center of the Michelson interferometer, reflecting part of the incoming light (ray 2) and allowing the rest of the incident light to be transmitted into the soap layer after refraction at the air–soap interface. This transmitted light is partially reflected at the bottom surface, producing the wave that travels back into the air as ray 3. The two outgoing rays denoted as 2 and 3 meet the conditions for interference: they travel through different regions (one travels the extra distance through the soap film), they recombine when they leave the film as parallel rays, and they are coherent because they originated from the same light source and initial wave, ray 1.

To determine whether these interfering waves are in phase or out of phase with each other, we must consider what happens in the extra distance traveled by ray 3. For simplicity, let's assume the incident and reflected rays are all approximately normal to the film (Fig. 25.7B), which corresponds to looking straight down at the soap film. For a film of thickness d, the extra distance traveled by ray 3 is just $2d$. We must next account for the index of refraction of the soap film. Recall that light propagates in a vacuum at a speed $c = 3.00 \times 10^8$ m/s. The frequency f and wavelength $\lambda = \lambda_{\text{vac}}$ are related by

$$\lambda_{\text{vac}} f = c \qquad (25.7)$$

© Terry Oakley/Alamy

Figure 25.6 When a soap film is viewed with white light (as from the Sun), one observes colored interference fringes (colored bands). These colors are due to constructive interference; different colors (different wavelengths) give constructive interference for different values of the film thickness and for different viewing angles.

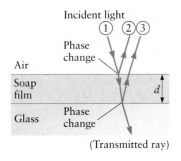

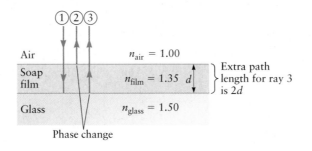

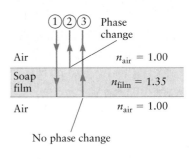

A

B

C

Figure 25.7 **A** Thin-film interference. Incident light (beam 1) is partially reflected at the air–soap film interface. The refracted light enters the film and is partially reflected at the bottom surface, in this case a soap film–glass interface. Beams 2 and 3 have traveled different distances and can interfere either constructively or destructively, depending on the thickness d of the soap film. Beams 2 and 3 both undergo a 180° phase change when they are reflected. (See ahead to Fig. 25.9.) Some light is also transmitted through the bottom surface of the glass. **B** Same as part A but with the incident and reflected rays normal to the film. **C** When the bottom layer is air, there is again interference of the two reflected beams, but now there is no phase change of beam 3 when it reflects at the bottom surface.

When light travels through a substance, its speed is reduced to $v = c/n$, where n is the index of refraction of the substance (Chapter 24). A soap film is mostly water and its index of refraction is about $n_{film} = 1.35$, so light slows down inside the soap film by 35% relative to c. The product of the wavelength inside the film λ_{film} and the frequency inside the film f_{film} equals the wave speed. (Compare with the case in vacuum, Eq. 25.7.) We then have

$$\lambda_{film} f_{film} = \frac{c}{n_{film}} \tag{25.8}$$

Comparing with Equation 25.7, we see that either the wavelength or the frequency (or both) must change when light travels from a vacuum into the film. It turns out that the wavelength changes but the frequency does *not*. This can be seen from Figure 25.8, which sketches how the electric field associated with a light wave varies as the wave travels from air into a film. The fields just outside and just inside the film must stay in phase; otherwise, the wave fronts would "pile up" at the boundary. This is similar to the propagation of a wave on a string that travels between two strings of different diameters; the knot between the two strings then corresponds to the interface between air and the soap film in Figure 25.8. The strings on both sides of the knot must move with the same frequency; otherwise, the knot would come apart. Hence, the wave must have the same frequency in both strings. The same kind of argument shows that the frequency of light (or any other wave) does not change as it travels from one medium into another.

Although the frequency of light does not change when it enters the soap film in Figure 25.7, its wavelength does change. Inserting $f_{film} = f$ (the frequency in vacuum) into Equation 25.8, the wavelength in the film is

$$\lambda_{film} = \frac{v}{f} = \frac{c/n_{film}}{f}$$

Using Equation 25.7, this leads to

$$\lambda_{film} = \frac{\lambda_{vac}}{n_{film}} \tag{25.9}$$

and the wavelength of light inside the film is thus shorter than its wavelength in a vacuum. The same is true for light traveling in air, with $\lambda_{air} = \lambda_{vac}/n_{air}$. Since n_{air} is very close to 1 ($n_{air} = 1.0003$), the wavelength of light in air is very close to its wavelength in a vacuum:

$$\lambda_{air} \approx \lambda_{vac} \tag{25.10}$$

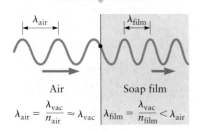

Figure 25.8 When a light wave passes from one medium to another (e.g., from air on the left into a soap film on the right), the waves on the two sides must stay in phase at the interface. That can only happen if the frequency is the same on both sides of the interface.

We can usually ignore the very small difference between the wavelengths of light in air and in a vacuum and will denote the wavelength in a vacuum as simply λ (without the subscript "vac").

Ray 3 in Figure 25.7B is traveling the distance $2d$ inside the soap film, and its wavelength is given by Equation 25.9. The extra number of wavelengths this wave travels is thus

$$N = \frac{2d}{\lambda_{\text{film}}} = \frac{2d}{\lambda/n_{\text{film}}} \qquad (25.11)$$

If N is an integer, ray 3 will travel some number of complete, extra wavelengths compared with ray 2. As with the Michelson interferometer, the value of N will help determine how rays 2 and 3 interfere. Before we can say if the interference is constructive or destructive, however, we must consider how the reflections at the surfaces of the soap film affect the waves.

Effect of Reflection on the Phase of a Light Wave

The result in Equation 25.11 is analogous to what we found for the extra number of cycles in one of the interfering waves in a Michelson interferometer (Eq. 25.2). We might therefore expect similar conditions for interference as for a Michelson interferometer. However, we also must consider one additional piece of physics. In Chapter 12, we found that under certain conditions a reflected wave is inverted relative to the incident wave; this happens for waves on a string (Fig. 12.17) when the reflecting end of the string is tied to a rigid wall. Similarly, when a light wave reflects from a surface it *may* also be inverted, corresponding to a phase change of 180° as illustrated in Figure 25.9A. This phase change on reflection is found whenever the index of refraction on the incident side (n_1) is less than the index of refraction on the opposite side (n_2). On the other hand, if the index of refraction is larger on the incident side (Fig. 25.9B), the reflected light wave is not inverted and there is no phase change.

In a thin-film interference experiment, we must take account of the *total* phase difference between the two interfering light waves; the total phase difference includes the phase difference due to the extra distance traveled by ray 3 plus phase changes that occur due to reflection. In Figure 25.7B, we have a soap film on top of a glass layer; to three significant figures, the indices of refraction are $n_{\text{air}} = 1.00$,

Phase change on reflection

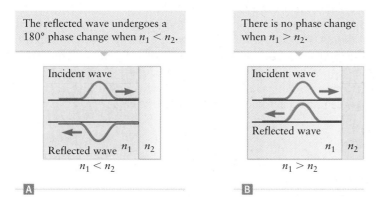

Figure 25.9 Phase change upon reflection. When light strikes an interface at which the index of refraction changes, part of the wave is reflected even when both substances are transparent. A When the index of refraction n_1 of the first substance is less than the index of refraction n_2 of the second substance, the phase of the reflected wave changes by 180° and the reflected wave is inverted, similar to the behavior of a wave on a string when the string is tied to a wall (Fig. 12.17, page 392). B On the other hand, if $n_1 > n_2$, the phase of the reflected wave does not change and the wave is not inverted, similar to the behavior of a wave on a string when the end of the string is able to slide up and down (Fig. 12.20).

$n_{film} \approx 1.35$, and $n_{glass} \approx 1.50$. These values are typical and will vary slightly depending on the color of the light and the composition of the film and glass, but for a soap film on glass we'll always have $n_{air} < n_{film} < n_{glass}$. Hence, there are phase changes for both reflections at the soap film interfaces. The wave reflected at the upper surface of the film (wave 2) undergoes a 180° phase change because $n_{air} < n_{film}$, and the wave reflected at the bottom surface (wave 3) undergoes a 180° phase change because $n_{film} < n_{glass}$. Both interfering waves undergo the same phase change due to reflection, so their *relative* phase is not affected by the reflections and the nature of the interference is determined only by the extra path length $2d$ traveled by wave 3. Hence, we finally arrive at interference conditions similar to those found with the Michelson interferometer, with the only difference being the extra factor of n_{film} due to the change in wavelength inside the film.

Thin-film interference case 1: When both waves reflected by a thin film undergo a phase change. In this case the number of extra cycles traveled by the ray inside the film completely determines the nature of the interference. If the number of extra cycles N equals an integer, there is constructive interference; if N is a half integer, there is destructive interference. Using Equation 25.11, we have

$$2d = \frac{m\lambda}{n_{film}} \qquad \text{(constructive interference)}$$

$$2d = \frac{(m + \frac{1}{2})\lambda}{n_{film}} \qquad \text{(destructive interference)}$$

(25.12)

These interference conditions apply whenever the indices of refraction are related as in Figure 25.7B, with $n_{air} < n_{film} < n$ (substance below film).

Thin-film interference case 2: When only one wave reflected by a thin film undergoes a phase change. A different case, with different interference conditions, is shown in Figure 25.7C. Here a soap film is surrounded by air; that is, there is air above and below the film. That is also the situation in a soap bubble as in Figure 25.6. When ray 1 reflects from the top surface, there is a phase change because $n_{air} < n_{film}$ and the index of refraction is smaller on the incident side of the interface. For the reflection at the bottom surface of the film, however, there is no phase change because now the index is larger on the incident side. One of the interfering waves (2) thus undergoes a phase change due to reflection, and the other (3) does not. Figure 25.1C shows that a 180° phase change is equivalent to a shift of the wave by $\lambda/2$. This is equivalent to inserting an extra half cycle into the wave, which changes the conditions for achieving constructive interference and destructive interference.

The case of only one wave undergoing a phase change occurs in thin-film interference when the index of refraction of the film in the middle is greater than the indices of the regions on either side. Now only one of the interfering waves (wave 2 in Fig. 25.7C) undergoes a 180° phase change due to reflection, and wave 3 must travel an extra half wavelength to get back in phase with wave 2. The interference conditions are thus

$$2d = \frac{(m + \frac{1}{2})\lambda}{n_{film}} \qquad \text{(constructive interference)}$$

$$2d = \frac{m\lambda}{n_{film}} \qquad \text{(destructive interference)}$$

(25.13)

Thin-Film Interference with White Light

Thin-film interference is responsible for the colors often seen in thin layers of water or oil on flat pavement and in free-standing films (Fig. 25.6). This effect also gives the bright colors found on the wings of many insects and birds (Fig. 25.10). Those cases usually involve sunlight (so-called white light), whereas in the situation ana-

Figure 25.10 The colors of many insect wings are due to thin-film interference. For a wing of a particular thickness, constructive interference and hence a bright interference fringe occurs for only one or a few values of the wavelength, giving the wing its color when viewed in reflected light.

lyzed in Figure 25.7 we assumed light with a single wavelength. We can understand what happens with white light using Figure 25.11A, which shows light of just two different colors (red and blue) incident on a thin film. We also assume the condition for constructive interference is satisfied for blue light at normal incidence (on the left in the figure). The wavelength of red light is longer than the wavelength for blue light, so for the same angle of incidence the condition for constructive interference will usually *not* be satisfied for red light at the same time. However, if the angle of incidence θ_i for red light is increased (on the right in Fig. 25.11A), the path length difference between the two light waves increases. It is therefore possible to have constructive interference for red light, but *for a different angle of incidence* than for blue light.

When an observer looks at a soap film, white light from the sky illuminates the film over a range of angles (Fig. 25.11B). Each different color component of the sunlight can exhibit constructive interference for a different angle of incidence, giving a series of interference fringes with different colors and producing brightly colored fringes even with a soap film of constant thickness.

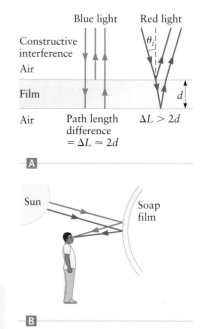

Figure 25.11 ◢ On the right, light is incident on a thin film at an angle away from the normal direction. The path length difference between the reflected waves is greater than for light at normal incidence (on the left). ◣ When you view a soap film (perhaps as part of a soap bubble) with reflected sunlight, different colors give constructive interference at different angles, producing the "colored" interference fringes in Figure 25.6.

EXAMPLE 25.3 | Color and Thickness of a Soap Film

Consider a bubble formed from an *extremely* thin, soapy film that looks blue when viewed at normal incidence. Estimate the thickness of the film. Assume the index of refraction of the soap film is $n_{film} = 1.35$ and blue light has a wavelength $\lambda_{blue} = 400$ nm. Also assume the film is so thin that thinner films are not able to give constructive interference.

RECOGNIZE THE PRINCIPLE

The blue color of the soapy film comes from interference of waves reflected from the two film surfaces as sketched in Figure 25.7C. The wave reflected from the top surface (denoted by ray 2) undergoes a phase change on reflection because the index of refraction on the incident side (air) is less than the index of the film. On the other hand, the wave reflected from the bottom surface (denoted by ray 3) does not undergo a phase change on reflection because the index on the incident side (the film) is greater than the index of the opposite side (air). (See Fig. 25.9.)

SKETCH THE PROBLEM

Figure 25.7C describes thin-film interference for a soap film with air on both sides.

IDENTIFY THE RELATIONSHIPS

The interference conditions that apply with a phase change at only one of the reflections are given by Equations 25.13. Since the film has a blue color, there must be constructive interference for blue light of the given wavelength $\lambda_{blue} = 400$ nm $= 4.0 \times 10^{-7}$ m.

Applying Equation 25.13 for constructive interference, we have

$$2d = \frac{(m + \frac{1}{2})\lambda_{blue}}{n_{film}} \quad (1)$$

Recall that m in Equation (1) is an integer that can have a value 0, 1, 2, ... Equation (1) thus has many solutions for the film thickness d, depending on the value of m. How can we determine the value of m? The problem states that the film is extremely thin and that thinner films do not give constructive interference. The smallest possible value of d corresponds to $m = 0$, so that case must apply for the film.

SOLVE

Using $m = 0$ in Equation (1), we find

$$2d = \frac{(m + \frac{1}{2})\lambda_{blue}}{n_{film}} = \frac{\lambda_{blue}}{2n_{film}}$$

Solving for d and inserting values for the wavelength and index of refraction gives

$$d = \frac{\lambda_{\text{blue}}}{4n_{\text{film}}} = \frac{4.0 \times 10^{-7}\ \text{m}}{4(1.35)} = 7.4 \times 10^{-8}\ \text{m} = \boxed{74\ \text{nm}}$$

What does it mean?

This value is about five hundred times smaller than the diameter of a human hair. It is amazing that a soap film this thin can be strong enough to form a bubble.

EXAMPLE 25.4 Measuring the Diameter of a Thin Fiber

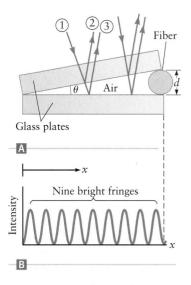

Figure 25.12 Example 25.4. ▲ Two glass plates separated at one end by a thin fiber. ▶ Reflected light intensity as a function of position along the plates.

Figure 25.12 shows how thin-film interference can be used to measure the diameter of a thin fiber such as a strand of hair. Two glass plates are in contact at one end but are held apart at the other end by the fiber. Light incident from above is reflected at the top and bottom of the "film" of air between the plates, producing interference. When viewed at normal incidence and with a small wedge angle θ, the path length difference between the interfering light waves is equal to twice the plate separation. This separation increases from left to right in Figure 25.12A, so the interference alternates between constructive and destructive. The resulting interference pattern is sketched as a function of position x along the plates in Figure 25.12B. The light intensity is zero at the left end (see Concept Check 25.3) and then passes through nine bright fringes (places where the intensity is a maximum), and the intensity at the right end next to the fiber is again zero. Find the diameter of the fiber. Assume the wavelength of the light is $\lambda = 600$ nm.

RECOGNIZE THE PRINCIPLE

The interference involves waves like those represented by ray 2 that reflect at the glass–air interface and waves represented by ray 3 that reflect at the air–glass interface. Ray 2 does not undergo a phase change at this reflection because $n_{\text{air}} < n_{\text{glass}}$, similar to the reflection from the bottom surface in Figure 25.7C. Ray 3 does undergo a 180° phase change on reflection, however, as found for the waves in Figure 25.7B. Hence, Equations 25.13 are the correct interference conditions for this case.

SKETCH THE PROBLEM

Figure 25.12 shows the problem.

IDENTIFY THE RELATIONSHIPS

The condition for destructive interference in this case is (Eq. 25.13)

$$2d = \frac{m\lambda}{n_{\text{film}}} \quad \text{(destructive interference)} \tag{1}$$

To pass from one dark fringe in the interference pattern to the next dark fringe, the path difference between the interfering waves must change by exactly one wavelength. In terms of the condition for destructive interference in Equation (1), this change corresponds to going from m to $m + 1$. We are told that there are nine complete fringes in Figure 25.12B, so if the separation between the plates on the right-hand side is equal to the diameter of the fiber d, the condition for destructive interference becomes

$$2d = \frac{m\lambda}{n_{\text{film}}} = \frac{9\lambda}{n_{\text{air}}}$$

In this case the "film" between the plates is filled with air.

SOLVE

Solving for d, we get

$$d = \frac{9\lambda}{2n_{\text{air}}} = \frac{9(6.0 \times 10^{-7}\ \text{m})}{2(1.00)} = \boxed{2.7 \times 10^{-6}\ \text{m}}$$

Figure 25.13 shows a photograph of the interference fringes produced by an air wedge. Each bright fringe is a "strip" on which the intensity is high, corresponding to the intensity peaks in Figure 25.12B. Similar interference methods are used to measure extremely small distances or heights in many applications, including the reading and writing of information on CDs and DVDs.

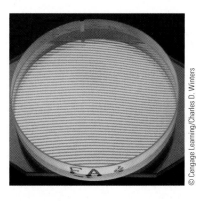

Figure 25.13 Photo of interference fringes (the horizontal bands) produced by two glass plates separated by a small amount at one end.

CONCEPT CHECK 25.3 | Reflection from an Air Wedge

Explain why the waves that reflect from the far left edge of the plates in Figure 25.12A exhibit destructive interference (zero intensity) at $x = 0$. The puzzle: At the far left edge, the plates are in contact, so the path length difference between beams 2 and 3 is zero. Why is there still destructive interference?

Antireflection Coatings

A partially reflecting mirror splits a light wave into two separate waves, which may later be recombined and interfere. Nearly any flat piece of glass will act in this way. You can readily demonstrate this fact by looking through the nearest window; even though the window glass is designed to let light pass through, a small amount (about 4%) of the light is reflected. That is not a problem for an ordinary window, but is bad for some applications. For example, light in a typical camera must pass through several lenses and filters; if 4% of the intensity is reflected at each interface, it would take only a few surfaces to reduce the light intensity by 25% or more, which would be bad for the camera's performance. To avoid this problem, most camera lenses have *antireflection coatings* that make a lens appear slightly dark in color when viewed in reflected light as in Figure 25.14A.

Antireflection coatings are often made from MgF_2, which has an index of refraction of about $n_{MgF_2} = 1.38$. Incident light reflects from the front surface of the MgF_2 film as in Figure 25.14B and again from the MgF_2–glass interface. There is a 180° phase change at both reflections (as with the soap film on glass in Fig. 25.7B), so the interference conditions in Equation 25.12 apply and destructive interference occurs when

$$2d = \frac{(m + \frac{1}{2})\lambda}{n_{MgF_2}}$$

Here, m can have the values $0, 1, 2 \ldots$, so the smallest possible value of d and hence the thinnest MgF_2 film that gives destructive interference corresponds to $m = 0$. We thus get

$$2d = \frac{\lambda}{2n_{MgF_2}}$$

If we want to design an antireflection coating of MgF_2 to operate for red light ($\lambda = 6.0 \times 10^{-7}$ m), we have

$$d = \frac{\lambda}{4n_{MgF_2}} = \frac{6.0 \times 10^{-7}\text{ m}}{4(1.38)} = 1.1 \times 10^{-7}\text{ m} = 110\text{ nm}$$

MgF_2 is a popular material for such coatings in part because it can be made into very uniform films with this thickness. An antireflection coating will only work best at one particular wavelength in the visible range; ideally, it will give "perfect" destructive interference at only one wavelength, but it will also give "partially" destructive interference at nearby wavelengths. Antireflection coatings that function well over the entire range of visible wavelengths are made using multiple layers that give "perfect" destructive interference at different wavelengths.

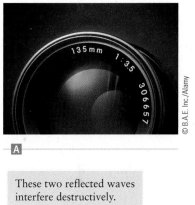

A

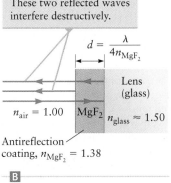

B

Figure 25.14 **A** The lenses on most binoculars and cameras are coated with antireflection coatings. Although these coatings are transparent, they give the lens a dark tint. **B** Thin-film interference in an antireflection coating composed of MgF_2.

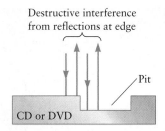

Figure 25.15 Reflection and interference of light from the edge of a "pit" in a CD or DVD.

Interference and the Design of CDs and DVDs

Thin-film interference is used in CDs and DVDs, which encode information as a series of pits in a layer of plastic. The depth of the pit is chosen to give destructive interference between light waves reflecting from the bottom and top surfaces at a pit edge (Fig. 25.15). Because of this destructive interference, the reflected intensity is zero when the incident light is over the edge of a pit. Light striking the plastic surface away from a pit edge, whether fully inside or fully outside the pit, is reflected to the detector with a large intensity. The presence or absence of a pit edge can thus be detected by the absence or presence of a reflected light wave. The design of CDs and DVDs is discussed in more detail in Chapter 26.

25.4 | LIGHT THROUGH A SINGLE SLIT: QUALITATIVE BEHAVIOR

The photo at the start of this chapter shows two interfering circular water waves. The analogous experiment with light can be performed by letting light shine onto a pair of slits in an opaque screen. To understand the resulting double-slit interference pattern, we must first consider carefully what happens when light passes through a single slit. In this section, we discuss the qualitative behavior; in the two following sections, we then give a more detailed analysis.

The basic experiment is sketched in Figure 25.16A. Light passes through a slit and then illuminates a screen. For simplicity, the incident light is assumed to be a plane wave, represented by the parallel beams on the left side of the slit. When the slit is very wide, the light that reaches the screen is a projection of the slit. If we plot the light intensity versus position along the x axis marked on the screen in Figure 25.16A, we find a "step" in intensity approximately equal to the width w of the slit as shown at the top of Figure 25.16B, but a very careful look reveals that the edges of this shadow are slightly "smeared out." As the slit is made narrow, the intensity pattern on the screen spreads out. If we keep reducing the slit width so that it is comparable to or even smaller than λ, the intensity pattern spreads more and more, and some additional intensity maxima become noticeable.

This behavior occurs for all types of waves, including light and sound; it is easy to observe with a water wave because the wavelength of a water wave is large enough for individual waves to be seen easily by eye. In Figure 25.17, a water wave is incident from the left and passes through an opening whose width is approximately equal to λ. The outgoing wave on the right spreads out as a broad beam, analogous to striking a screen over a broad region as in the lowermost pattern in Figure 25.16B. Figures 25.16 and 25.17 illustrate the phenomenon of *single-slit diffraction*. Dif-

Single-slit diffraction

Figure 25.16 **A** Light passing through an opening or slit and onto a screen. **B** If the slit is very wide compared with the wavelength λ, an ordinary shadow (projection of the slit) appears on the screen (top). If the slit width is comparable to or smaller than the wavelength, the wave intensity is spread over a larger region of the screen. As the slit width is made narrower, the amount of spreading increases.

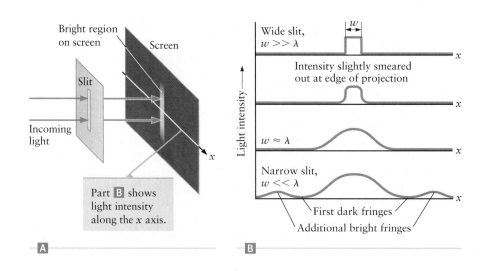

fraction is the bending or spreading of a wave when it passes through an opening. In single-slit diffraction, the opening is just a single slit.

To understand the qualitative behavior of how light passes through a very narrow slit, it is useful to draw the wave fronts and rays for the incident and diffracting waves. Figure 25.18 shows the results for an extremely narrow slit. In this case, the outgoing wave radiates from the slit over a very wide range of angles. This is an example of *Huygens's principle*, which can be stated as follows.

Huygens's principle
All points on a wave front can be thought of as new sources of spherical waves.

When the incoming wave encounters the narrow openings in Figures 25.17 and 25.18, each point inside the opening acts as a new source of spherical waves. In Figure 25.18, the opening is *very much smaller* than the wavelength; in this case, the opening acts as a single point source of waves, and the simple outgoing spherical waves sketched in Figure 25.18 are the result.

We will not attempt to derive or justify Huygens's principle. It was proposed (by Huygens) in about 1680, long before there was a detailed theory of light. This principle gives a qualitative way to understand how a very narrow slit can convert a plane wave into a spherical one. This conversion is key to understanding the effects of interference and diffraction when light passes through a small opening.

25.5 | DOUBLE-SLIT INTERFERENCE: YOUNG'S EXPERIMENT

We next consider the behavior of light passing through two very narrow slits, a problem called *double-slit interference*. We jump to the double-slit case before finishing the discussion of the single slit from Section 25.4 because a full analysis of light passing through a single slit is in some ways more complicated than that for two *very narrow* slits. When the two slits in the double-slit case are both very narrow, each slit acts as a simple point source of new waves, and the outgoing waves from each slit are like the simple spherical waves in Figure 25.18.

The double-slit experiment is famous in the history of physics because it demonstrated conclusively that light is a wave. This experiment, first carried out around 1800 by Thomas Young, is shown in Figure 25.19. Light is incident from the left onto two slits and after passing through them strikes a screen on the right. The incident light here is a plane wave; this important requirement is easy to achieve today using light from a laser. It can also be accomplished with a lens as described in Chapter 24.

The double-slit experiment in Figure 25.19 is very similar to the interference arrangement involving sound waves in Figure 25.1 and satisfies the general conditions for interference from Section 25.1. First, the interfering waves travel through different regions of space as they travel through different slits. Second, the waves come together at a common point on the screen where they interfere. Third, the waves are coherent because they come from the same source, the incoming plane wave on the left. We thus expect interference to determine how the intensity of light on the screen varies with position. For simplicity, we assume the two slits are very narrow, so according to Huygens's principle each slit acts as a simple source with circular wave fronts as viewed from above. If we cover one of the slits in Figure 25.19, a very wide area of the screen is illuminated with light as with the single slit in Figure 25.18, but when we allow light to pass through both slits, we get the very different result shown in Figure 25.19. The light intensity on the screen alternates between bright and dark as we move along the screen, signaling regions of constructive interference and destructive interference.

Figure 25.17 Diffraction of a water wave as it passes from left to right through a single narrow slit, producing an outgoing wave on the right. The outgoing wave front is approximately spherical.

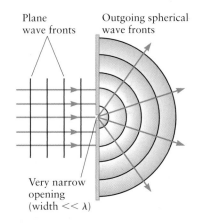

Figure 25.18 A plane wave incident on a single narrow slit. According to Huygens's principle, the slit acts as a new wave source, producing a spherical wave that propagates away from the slit.

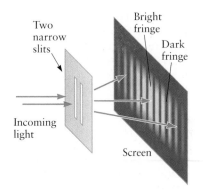

Figure 25.19 Interference of light produced by two slits. The pattern on the screen at right is a series of bright and dark interference fringes.

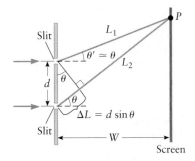

Figure 25.20 Analysis of double-slit interference. Waves that pass through the two slits travel different distances L_1 and L_2 on their way to point P on the screen. (Compare with Fig. 25.1.) Not to scale.

Double-slit interference conditions

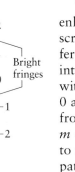

Figure 25.21 Double-slit interference. Values of θ from Equation 25.15 give the locations of the bright fringes on the screen.

Where Are the Fringes?

To analyze the interference pattern in Figure 25.19, we must determine the path length between each slit and the screen. Consider the path lengths for waves that arrive at point P on the screen in Figure 25.20. For simplicity, we assume the screen is very far away from the slits, so W is very large. The distance from the bottom slit to P is greater than the distance between P and the top slit. Also, because W is large, the angles θ and θ' that specify the directions from the slits to point P are approximately equal, and we denote both by θ.

If the slits are separated by a distance d, the extra path length traveled by the wave from the bottom slit is ΔL, which is given by

$$\Delta L = d \sin \theta \qquad (25.14)$$

as shown in Figure 25.20. If this extra path length is equal to an integral number of complete wavelengths, the two waves will be in phase when they strike the screen; the interference is then constructive and the light intensity is large. On the other hand, if the extra path length is equal to $\lambda/2$, $3\lambda/2$, $5\lambda/2$, . . . , the interference is destructive and the intensity at the screen is zero. The condition for constructive interference and a bright fringe in the interference pattern is thus

$$d \sin \theta = m\lambda \quad m = 0, \ \pm 1, \ \pm 2, \ \pm 3, \ldots \qquad (25.15)$$

$$(\text{constructive interference})$$

whereas the condition for destructive interference and a dark fringe is

$$d \sin \theta = (m + \tfrac{1}{2})\lambda \quad m = 0, \ \pm 1, \ \pm 2, \ \pm 3, \ldots \qquad (25.16)$$

$$(\text{destructive interference})$$

The double-slit intensity pattern at the screen is shown greatly enlarged in Figure 25.21. The angle θ varies as we move along the screen. At certain values of θ, the condition for constructive interference is satisfied, producing a maximum in the intensity (a bright interference fringe). Each bright fringe satisfies Equation 25.15 with a different value of the integer m. The value $m = 0$ gives $\theta = 0$ and corresponds to the center of the screen. Moving up or down from this point gives bright fringes with $m = +1, +2, +3, \ldots$ and $m = -1, -2, -3, \ldots$. Negative values of m indicate that the path to those points on the screen from the lower slit is shorter than the path to those points from the upper slit.

Now let's consider what the interference pattern will look like for particular values of the separation d between the slits and the perpendicular distance to the screen W. Suppose the slits are $d = 0.10$ mm apart and the screen is located $W = 50$ cm $= 0.50$ m from the slits. If we use light from a red laser with $\lambda = 630$ nm, what is the separation between the bright fringe at the center of the screen ($\theta = 0$) and the nearest bright fringe on either side in Figure 25.21? Two such bright fringes correspond to $m = 0$ (the central fringe) and $m = 1$ in Equation 25.15. The angle θ measured at the slits between the central fringe and the first fringe above it thus satisfies $d \sin \theta = \lambda$, and we have

$$\sin \theta = \frac{\lambda}{d} \qquad (25.17)$$

The wavelength is much smaller than the spacing d between the slits, so the ratio on the right-hand side of Equation 25.17 is very small. Hence, $\sin \theta$ is also very small, and we can use the approximation $\sin \theta \approx \theta$ (where the angle is measured in radians; see Appendix B) to get

$$\sin \theta \approx \theta \approx \frac{\lambda}{d} \qquad (25.18)$$

To find the separation between the bright fringes on the screen, we use the right triangle in Figure 25.21 with sides of length W and h. The spacing h between the two fringes corresponding to $m = 0$ and $m = 1$ is

$$h = W \tan \theta$$

Using the approximation for small angles $\tan \theta \approx \theta$ (Appendix B), we get

$$h = W\theta$$

Our result for θ from Equation 25.18 then gives

$$h = W\theta = W\frac{\lambda}{d} \tag{25.19}$$

Inserting the values of W, d, and λ given above, we find

$$h = W\frac{\lambda}{d} = (0.50 \text{ m})\frac{6.3 \times 10^{-7} \text{ m}}{1.0 \times 10^{-4} \text{ m}} = 3.2 \times 10^{-3} \text{ m} = 3.2 \text{ mm} \tag{25.20}$$

This fringe spacing is large enough to be seen easily by the naked eye.

Our analysis of Young's double-slit experiment and the result in Equation 25.20 shows that the bright and dark fringes associated with interference are observable, provided the slits are fairly close together. The results for h in Equations 25.19 and 25.20 show that a slit separation of about 0.1 mm (or smaller) is a good choice. If the spacing is larger than a few millimeters or so, the fringes are quite close together and can be difficult to observe.

CONCEPT CHECK 25.4 | Double-Slit Interference

Suppose the distance between the slits in Figure 25.19 is reduced. How would this change affect the interference pattern?
(a) The bright fringes would be more closely spaced on the screen.
(b) The bright fringes would be more widely spaced on the screen.
(c) The fringe spacing on the screen would not change.

EXAMPLE 25.5 | Measuring the Wavelength of Light

Young's double-slit experiment shows that light is indeed a wave, and also gives a way to measure the wavelength. Suppose the double-slit experiment in Figure 25.21 is carried out with a slit spacing $d = 0.40$ mm and the screen at a distance $W = 1.5$ m. If the bright fringes nearest the center of the screen are separated by a distance $h = 1.5$ mm, what is the wavelength of the light?

RECOGNIZE THE PRINCIPLE

In a double-slit experiment, the bright fringes occur when light waves from the two slits interfere constructively. That happens when the path length difference ΔL (Fig. 25.20) is equal to an integral number of wavelengths. We used this condition to derive Equation 25.19, which gives a relation between h, W, d, and λ for the first bright fringe.

SKETCH THE PROBLEM

Figures 25.20 and 25.21 describe the problem.

IDENTIFY THE RELATIONSHIPS

Equation 25.19 relates the slit spacing d and the perpendicular distance W to the screen, along with h and λ. Rearranging to solve for the wavelength gives

$$\lambda = \frac{hd}{W}$$

Insight 25.2
REMEMBER THE APPROXIMATIONS FOR SMALL ANGLES
Because the wavelength of light is small, the conditions for constructive interference and destructive interference often involve small angles. For small values of the angle θ, the sine and tangent functions can be approximated by

$$\sin \theta \approx \theta \quad \text{and} \quad \tan \theta \approx \theta$$

These relations are true only if θ is measured in radians. See Appendix B.

SOLVE

Inserting the given values of h, d, and W, we find

$$\lambda = \frac{hd}{W} = \frac{(1.5 \times 10^{-3} \text{ m})(4.0 \times 10^{-4} \text{ m})}{1.5 \text{ m}} = 4.0 \times 10^{-7} \text{ m} = \boxed{400 \text{ nm}}$$

What does it mean?

Young's double-slit experiment gives a way to measure even the very short wavelength of visible light. The values of h, W, and d can all be measured with an ordinary ruler, but together they enable us to measure the wavelength λ, which is much too small to be measured with a ruler. While technology has certainly advanced since Young's work two hundred years ago, methods involving interference are still widely used in precision measurements of the wavelength of light.

Interference with Monochromatic Light

The conditions for interference listed in Section 25.1 (page 842) state that the interfering waves must have the same frequency and hence the same value of the wavelength λ. Light with a single frequency is called *monochromatic* because it consists of a single color. Light from most sources, such as the Sun or an ordinary lightbulb, contains a distribution of wavelengths. Such light sources are generally not useful in double-slit interference experiments. Suppose a light source emits mainly at two colors, that is, at two wavelengths λ_1 and λ_2. The angles that give bright fringes in the double-slit experiment (Eq. 25.15) depend on the value of λ. Different values of the wavelength thus give bright fringes at different locations on the screen, and places on the screen where wavelength λ_1 gives a bright fringe will often coincide with places where wavelength λ_2 gives a dark fringe. If a light source emits a broad distribution of wavelengths, the total intensity pattern will be a "washed-out" sum of bright and dark regions, and no bright or dark fringes will be visible. For this reason, interference experiments with light usually work best with monochromatic waves.

25.6 | SINGLE-SLIT DIFFRACTION: INTERFERENCE OF LIGHT FROM A SINGLE SLIT

We began our treatment of how light passes through small openings by first discussing the qualitative behavior for a single slit (Section 25.4), including how the intensity pattern changes as we go from a very wide to a very narrow slit (Fig. 25.16). We next discussed the behavior with two slits (Section 25.5), with the simplifying assumption that the slits are very narrow so that each acts as a simple point source of new spherical waves (Huygens's principle). With the insight gained from the double-slit case, we are ready to tackle the behavior of a single slit in a more general way. We'll now see how to deal with slits that are narrow enough to exhibit diffraction but not so narrow that they can be treated as a single point source of waves.

Analysis of Single-Slit Diffraction

Figure 25.22A shows light incident on a single slit of width w. Light is diffracted as it passes through the slit and then propagates to the screen. The intensity pattern at the screen contains a central very bright fringe with dark fringes alongside as well as other fringes where the intensity peaks at values much smaller than the intensity of the central fringe. We qualitatively sketched this pattern at the bottom of Figure 25.16B. We want to explain the origin of these fringes and calculate where bright and dark fringes occur on the screen. The key to this calculation is Huygens's principle, which states that each point on a wave front acts as a new source of waves. All points across the slit in Figure 25.22 act as wave sources, and these different waves interfere at the screen.

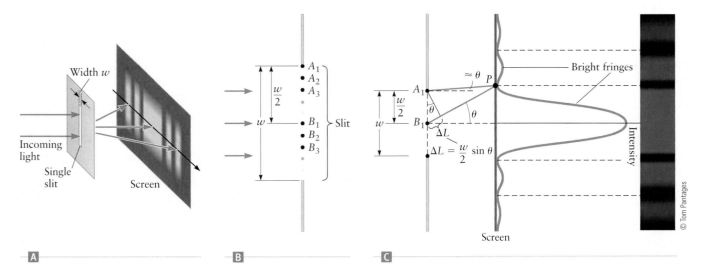

Figure 25.22 **A** Light passing through a single slit and then on to a screen. **B** Each point within the slit $A_1, A_2, \ldots, B_1, B_2, \ldots$ acts as a source of spherical waves. **C** Analysis of the single-slit diffraction pattern. Notice that the slit widths in parts B and C are greatly expanded to show the geometry of the interfering waves. In a scale drawing, the slit widths would be much *smaller* than the width of the central diffraction peak in part C.

Where Are the Dark Fringes?

We divide the slit into two halves and denote points in one half by A and those in the other half by B (Fig. 25.22B). Each of these points is a source of Huygens (i.e., spherical) waves, including point A_1 at one edge of the slit and the corresponding source point B_1 in the other portion of the slit. Figure 25.22C shows the distance from these points to a point P on the screen and also shows the path length difference ΔL. We again assume the screen is very far away from the slit, so all the angles denoted by θ are approximately equal. Suppose the path length difference is $\Delta L = \lambda/2$ so that the waves from A_1 and B_1 interfere destructively when they reach a particular point P on the screen. If this condition for destructive interference is met for points A_1 and B_1, the waves from A_2 and B_2 will also interfere destructively and so on for points A_3 and B_3 and for *all* similar pairs of points within the slit. Destructive interference will thus produce a dark fringe on the screen. This condition can be written as $\Delta L = \lambda/2$. From Figure 25.22C, we see that $\Delta L = (w/2)\sin\theta$, so we have

$$\frac{w}{2}\sin\theta = \Delta L = \frac{\lambda}{2}$$

which gives the angle for the first dark fringe above the center of the screen. There is a first dark fringe below center, which corresponds to negative values of θ with

$$\frac{w}{2}\sin\theta = -\frac{\lambda}{2}$$

The first dark fringes are thus at the angles

$$w\sin\theta = \pm\lambda \quad \text{single-slit diffraction: first dark fringes} \qquad (25.21)$$

$$\text{(destructive interference)}$$

The position of the second dark fringe can be found using Figure 25.23. In Figure 25.23A, we divide the slit into four regions and denote the points in these four regions by A, B, C, and D. Figure 25.23B shows the path lengths from the top point in each region (A_1, B_1, C_1, and D_1) to a point P on the screen. Suppose the path length difference for the Huygens waves from A_1 and B_1 is $\Delta L = \lambda/2$, so these waves interfere destructively. The path length difference for C_1 and D_1 has the same value,

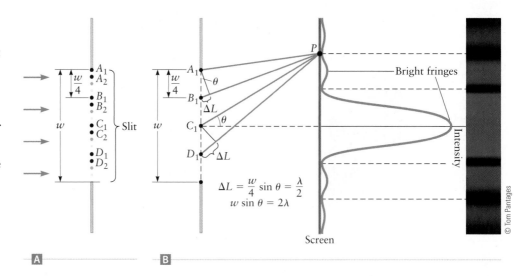

Figure 25.23 **A** To locate the second dark fringe in the intensity pattern, we imagine the slit as split into four regions, A, B, C, and D. Points in these regions can be grouped in pairs, A_1 and B_1, A_2 and B_2, . . . , C_1 and D_1, . . . , similar to the pairing in Figure 25.22B. **B** Analysis of the location of the second dark fringe in a single-slit diffraction pattern. Notice that the slit widths are greatly expanded to show the geometry of the interfering waves. In a scale drawing, the slit widths would be much *smaller* than the width of the central diffraction peak in part B.

$$\Delta L = \frac{w}{4}\sin\theta = \frac{\lambda}{2}$$
$$w\sin\theta = 2\lambda$$

so these waves will also interfere destructively. Also note that because the figure is not to scale, the two path length differences ΔL for the pairs AB and CD appear to have very different lengths. However, for a small slit width w, they will be very close in length. The path length difference will be the same for waves from others pairs of points A_2 and B_2, C_2 and D_2, and so on (Fig. 25.23A), so these waves will also interfere destructively, and the total intensity on the screen is zero. For the case of the second dark fringe, we thus have $\Delta L = (w/4)\sin\theta$. Setting this expression equal to $\lambda/2$, the interference condition in this case is

$$w\sin\theta = \pm 2\lambda \quad \text{single-slit diffraction: second dark fringes}$$

$$\text{(destructive interference)}$$

where the negative sign again corresponds to a fringe below the center of the screen. The same argument can be used to find the third, fourth, . . . dark fringes by dividing the slit into six, eight, . . . regions (following Figs. 25.22 and 25.23). The result for the mth dark fringe is

Single-slit interference conditions: destructive interference

$$w\sin\theta = \pm m\lambda \quad \text{single-slit diffraction: } m\text{th dark fringes} \qquad \textbf{(25.22)}$$

$$\text{(destructive interference)}$$

where $m = 1, 2, 3, \ldots$. The values of θ from Equation 25.22 give the angles of all dark fringes in the diffraction pattern.

Analyzing the Complete Diffraction Pattern

The angles that give the intensity peaks in Figures 25.22 and 25.23 are not as easy to calculate as the angles of the dark fringes (Eqs. 25.21 and 25.22), and there is no simple formula for the angles at which the bright fringes occur. However, the intensity on the screen can be calculated by adding up all the Huygens waves that come from all the source points A_1, B_1, A_2, B_2, and so forth. The resulting intensity curve is sketched on the right in Figures 25.22C and 25.23B. There is a central bright diffraction fringe (the central maximum), with other "bright" fringes much lower in intensity. A full calculation of the intensity curve shows that the central bright fringe is about 20 times more intense than the bright fringes on either side. The width of the central bright fringe is given approximately by the angular separation of the first dark fringes on either side. The angles for these first dark fringes are given by Equation 25.21:

$$\frac{w}{2}\sin\theta = \pm\frac{\lambda}{2}$$

$$\sin\theta = \pm\frac{\lambda}{w}$$

If the angle θ is small (as it usually is), then $\sin \theta \approx \theta$, which gives

$$\theta = \pm \frac{\lambda}{w} \quad \text{(first dark fringes)} \qquad (25.23)$$

These two dark fringes occur on either side of the central bright fringe in Figure 25.22A, so the *full* angular width of the central intensity is 2θ:

$$\text{full angular width of central bright fringe} = 2\frac{\lambda}{w} \qquad (25.24)$$

As a rough rule of thumb, one can say that diffraction of a light wave as it passes through a slit produces a central bright fringe extending over the angular range given by Equation 25.24. If the slit is much wider than the wavelength, this angle is small. For example, if the slit width is equal to the thickness of a typical fingernail ($w \approx 0.5$ mm $= 5 \times 10^{-4}$ m) and we consider red light ($\lambda = 630$ nm), the angle is

$$\text{full angular width} = 2\frac{\lambda}{w} = \frac{2(6.3 \times 10^{-7} \text{ m})}{5 \times 10^{-4} \text{ m}} = 0.0025 \text{ rad} = 0.14° \ \ (25.25)$$

This angle is very small, so the light beam would essentially pass straight through the slit with almost no effect from diffraction. The central diffraction fringe is much broader when the slit width approaches λ. For example, suppose the slit width is only five times larger than the wavelength; for the red light considered in Equation 25.25, we get $w = 5\lambda = 3.2 \times 10^{-6}$ m $= 3.2 \ \mu$m (about 300 times smaller than 1 mm). The central bright diffraction fringe would then correspond to an angle

$$\text{full angular width} = \frac{2\lambda}{w} = \frac{2\lambda}{5\lambda} = \frac{2}{5} = 0.40 \text{ rad} = 23°$$

so the spreading of the light beam due to diffraction is now quite substantial.

CONCEPT CHECK 25.5 | Single-Slit Diffraction

A single-slit diffraction experiment is performed with red light, and the width of the central diffraction fringe is measured. The experiment is then repeated with blue light. Which of the following statements describes the result?
 (a) The width of the central diffraction fringe is greater with blue light.
 (b) The width of the central diffraction fringe is smaller with blue light.
 (c) The width of the central diffraction fringe does not change.

Double-Slit Interference with Wide Slits

In our discussion of Young's double-slit experiment in Section 25.5, we assumed for simplicity that the slits are very narrow (widths much less than the wavelength of light); such slits produce an intensity pattern that varies approximately sinusoidally with position on the screen as shown in Figure 25.21. When the slits are not extremely narrow, the single-slit diffraction pattern produced by each slit is combined with the sinusoidal double-slit interference pattern. A full calculation of the intensity pattern is complicated, but the result can be understood by combining Figure 25.24A, which shows the double-slit pattern with two very narrow slits, and Figure 25.24B, which shows the single-slit pattern for a wide slit. For two double slits that are wide, the intensity pattern is essentially the product of these two patterns (Fig. 25.24C).

25.7 | DIFFRACTION GRATINGS

In the past few sections, we have discussed the interference patterns found when light passes through either a single slit or a double slit. As you should expect, similar interference effects also occur with three or more slits. In fact, physicists have

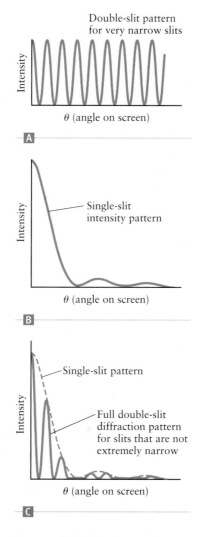

Figure 25.24 The complete interference pattern produced by two slits is a combination of **A** the double-slit pattern and **B** the single-slit pattern. The combination of these two gives the intensity pattern in **C**.

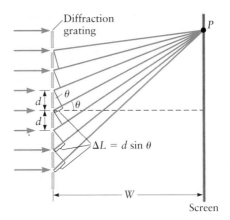

Figure 25.25 Interference experiment with a diffraction grating.

Diffraction from a grating: conditions for constructive interference

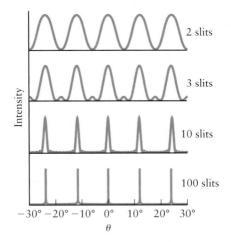

Figure 25.26 Interference patterns with 2, 3, 10, and 100 equally spaced slits. The main intensity peaks (the principal maxima) become narrower as the number of slits is increased. With 100 slits, these peaks are very narrow.

Figure 25.27 ◪ Sketch of a diffraction experiment with a grating using red light and green light. The colors of the rays here *do* indicate the color of the light. The bright fringes are at different angles for the different colors because their wavelengths are different. ◪ Diffraction experiment with both red laser light and green laser light shining on a grating.

found that systems with hundreds or even thousands of slits are quite useful; you probably own one without even knowing it as we'll explain below.

Figure 25.25 shows an arrangement of many slits called a ***diffraction grating***. When light passes through a diffraction grating, a pattern of light and dark fringes is produced on the screen. To analyze this pattern, we'll assume for simplicity that the slits are each very narrow (so that each one produces a single outgoing wave) and the screen is very far away. The rays connecting each slit with a point P on the screen are then approximately parallel, and all make an angle θ with the horizontal axis in Figure 25.25. If the slit-to-slit spacing is d, the path length difference for rays from two adjacent slits is

$$\Delta L = d \sin \theta$$

which is the same as in the case of a double slit (Fig. 25.20). If ΔL is equal to $0, \pm\lambda, \pm2\lambda, \pm3\lambda, \ldots$, two adjacent rays will interfere constructively at the screen. In addition, since the path length difference is the same for every pair of adjacent rays, *all* rays will interfere constructively. The condition for a bright fringe is thus

$$d \sin \theta = m\lambda \quad m = 0, \pm 1, \pm 2, \ldots \tag{25.26}$$

constructive interference for a diffraction grating

The condition for a bright fringe from a diffraction grating is identical to the condition for constructive interference from a double slit (Eq. 25.15). The angles that give constructive interference and the positions of the bright fringes on the screen are thus the same in the two cases. However, the overall intensity pattern depends on the number of slits in the grating. Figure 25.26 shows the intensity pattern found with two slits (just the double-slit pattern from Fig. 25.21), three slits, ten slits, and one hundred slits in the grating. The main intensity peaks are called the ***principal maxima*** of the pattern. These maxima occur at the same angles for 2, 3, or any number of slits. As the number of slits increases, however, smaller peaks occur between the main peaks. Most importantly, the main peaks become narrower as the number of slits increases, and when the number of slits is large, these peaks are *extremely* narrow. Compare the intensity pattern for a double slit (top of Fig. 25.26) with the result for a grating containing one hundred slits (bottom). The intensity peaks are *much* narrower with the grating.

Using a Grating to "Separate" Colors

The narrow intensity peaks for a grating with many slits make the grating a powerful tool for "separating" different light waves according to their wavelengths, that is, according to their colors. Figure 25.27A shows light composed of two different colors, red and green, incident on a grating. The grating produces an intensity pattern on the screen for each color, with the bright fringes given by the interference

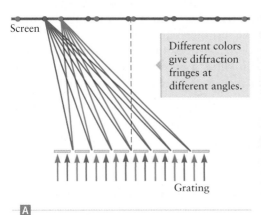

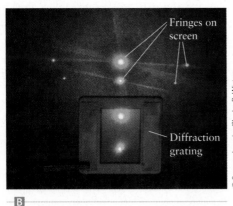

conditions in Equation 25.26. The two colors have different wavelengths, so the fringes for red and green light are at different angles and thus different places on the screen as illustrated in Figure 25.27B. The diffraction fringes appear here as spots on the screen because the laser illuminates only a small spot on the diffraction grating. Gratings are widely used to analyze the colors in a beam of light.

Diffraction and the "Color" of a CD

The diffraction gratings in Figures 25.25 and 25.27 contain straight slits arranged with a constant spacing. Similar diffraction effects occur in other situations in which many light waves interfere, such as with light reflected by a compact disc (CD). A CD contains "pits" in a reflecting layer, with the pits arranged in arcs ("tracks") as sketched in Figure 25.28A. Light reflected from these arcs acts as sources of Huygens waves, just like the slits in a grating. The reflected waves exhibit constructive interference (bright fringes) at certain angles, depending on the color (wavelength) of the light. Light reflected from a CD therefore has the colors "separated" as shown in Figure 25.28B.

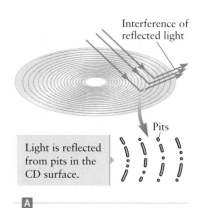

A

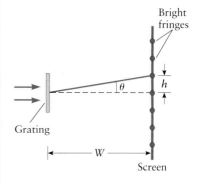

B

| **EXAMPLE 25.6** | Diffraction of Light by a Grating |

A diffraction experiment is carried out with a grating. Using light from a red laser ($\lambda = 630$ nm), the diffraction fringes are separated by $h = 0.15$ m on a screen that is $W = 2.0$ m from the grating. Find the spacing d between slits in the grating.

RECOGNIZE THE PRINCIPLE

The bright fringes in a grating's diffraction pattern occur when the waves from adjacent slits interfere constructively. The condition for constructive interference is given by Equation 25.26.

SKETCH THE PROBLEM

Figure 25.29 shows the problem and indicates the fringe spacing h and the distance W from the grating to the screen.

IDENTIFY THE RELATIONSHIPS

From the geometry of the diffraction pattern in Figure 25.29, we can calculate the angle θ corresponding to the first bright diffraction fringe. We can then use Equation 25.26 with $m = 1$ to find the slit spacing. From Figure 25.29, h and W form two sides of a right triangle containing θ, which leads to

$$\tan \theta = \frac{h}{W}$$

SOLVE

Inserting the given values of h and W and solving for θ gives

$$\tan \theta = \frac{0.15 \text{ m}}{2.0 \text{ m}} = 0.075$$

$$\theta = 4.3°$$

From Equation 25.26, the angle for the first diffraction fringe (corresponding to $m = 1$) is

$$d \sin \theta = m\lambda = \lambda$$

Rearranging to find d and using our value of θ, we get

$$d = \frac{\lambda}{\sin \theta} = \frac{630 \times 10^{-9} \text{ m}}{\sin(4.3°)} = 8.4 \times 10^{-6} \text{ m} = \boxed{8.4 \; \mu\text{m}}$$

Figure 25.28 **A** The pits in a CD have a regular spacing. Reflections from these pits act as separate but coherent sources of light, just like the slits of a grating, and produce diffraction fringes at different angles for different colors. **B** Hence, CDs appear brightly colored when viewed with reflected light.

Figure 25.29 Example 25.6.

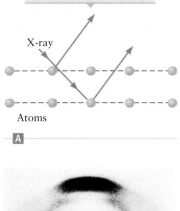

Waves are reflected by each atomic plane. These reflected waves interfere.

X-ray

Atoms

Ⓐ

Omikron/Photo Researchers

Ⓑ

Figure 25.30 Ⓐ The atomic planes in a crystal reflect electromagnetic radiation. These reflected waves interfere in a manner similar to that found with visible light and a grating. Diffraction experiments with a crystal are usually done with X-rays. The angles at which X-rays interfere constructively can be used to deduce the distance between atomic planes and the arrangement of atoms in the crystal. Ⓑ This X-ray diffraction pattern was obtained with a DNA crystal and led to the discovery of the double-helix structure of DNA. [See J. D. Watson and F. H. C. Crick, "Molecular structure of nucleic acid," *Nature* 4356:737 (1953).]

Diffraction of X-rays by a Crystal

In our discussion of gratings, we have so far considered the diffraction of light waves. Similar diffraction effects also occur with other types of electromagnetic waves (and with any other type of wave, such as sound). The atoms in a crystal are arranged in a periodic array, forming planes of atoms (Fig. 25.30A). These planes reflect electromagnetic radiation, leading to interference of the reflected waves, just as the light reflected by a CD can act as a grating (Fig. 25.28).

In a diffraction grating designed to work with light, the spacing between slits is typically 10^{-5} m to 10^{-4} m, which is 100 to 1000 times larger than the wavelength of light. For diffraction from the atomic planes in a crystal, the effective slit spacing is the distance between atomic planes, which is typically 3×10^{-10} m and hence much smaller than the spacing in a diffraction grating. To observe diffraction using the atomic planes, one therefore needs to use wavelengths much shorter than visible light, and X-rays with wavelengths on the order of 10^{-10} m are a convenient choice. Experiments of this type are called *X-ray diffraction*. By measuring the angles that give constructive interference of reflected X-rays, one can determine the spacing between atomic planes, just as the spacing between the slits in a diffraction grating can be determined from the diffraction pattern produced by a grating (Example 25.6). Because the atomic planes consist of atoms instead of simple slits, however, the intensity peaks appear as diffraction "spots" instead of fringes. An example is shown in Figure 25.30B, which shows the intensity pattern produced by diffraction of X-rays from a crystal of DNA. This diffraction pattern is very famous and led to the discovery of the structure of DNA. Because each DNA molecule has a helical structure, a crystal of DNA molecules gives a more complicated array of diffraction spots than most crystals. Through careful analysis of this pattern (along with other evidence), James Watson and Francis Crick deduced the double helix structure of DNA.

25.8 | OPTICAL RESOLUTION AND THE RAYLEIGH CRITERION

Diffraction causes light to spread out after passing through a narrow slit. A similar diffractive spreading is found whenever a beam of light passes through an opening of any shape, including the aperture of a telescope or microscope or the pupil of your eye. The result in Equation 25.23 gives the angle between the central bright fringe and the first dark fringe for the case of a slit. For a circular opening of diameter D, the result is almost the same, but the circular geometry leads to an additional numerical factor of 1.22:

$$\theta = \frac{1.22\lambda}{D} \tag{25.27}$$

where we assume D is much larger than the wavelength. The angle in Equation 25.27 is the angle between the central bright maximum and the first minimum in the diffraction pattern. Figure 25.31A shows the diffraction pattern produced by a circular opening, and Figure 25.31B shows the light intensity as a function of angle on the screen. The first dark fringe occurs at $\theta = 1.22\lambda/D$ away from the central maximum.

This result is important in many situations. For example, suppose you are looking through a telescope at a distant star. Light from the star must pass through a circular opening (an aperture) in the telescope before it reaches a screen or your eye, as sketched in parts A and B of Figure 25.32. Diffraction at the opening produces a circular diffraction spot like the one in Figure 25.31. Now suppose there are actually two light sources (two stars) viewed again through the telescope. If these two stars are at slightly different angles with respect to the central axis of the opening, your eye will see them at slightly different viewing angles. Because they are two separate light sources, their waves are incoherent and do not interfere. Light from each source produces its own diffraction pattern, however, resulting in two diffraction spots that may overlap at your eye.

Figure 25.33 shows schematically how the two light sources in Figure 25.32C would actually appear through a telescope. Both sources produce a circular diffraction spot, and if the two sources are sufficiently far apart, two separate diffraction spots are observable as shown in Figure 25.33A. If the sources are too close together, however, their diffraction spots will overlap so much that they appear as a single spot as in Figure 25.33C. As a rough criterion, two sources will be resolved as two distinct sources of light if their angular separation is greater than the angular spread of a single diffraction spot. This result is called the *Rayleigh criterion*. For a circular opening, the Rayleigh criterion for the angular resolution is (using Eq. 25.27)

$$\theta_{min} = \frac{1.22\lambda}{D} \tag{25.28}$$

In words, Equation 25.28 says that two objects will be resolved when viewed through an opening of diameter D (which might be a telescope or your eye) if the light rays from the two objects are separated by an angle at least as large as θ_{min}.

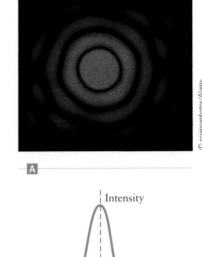
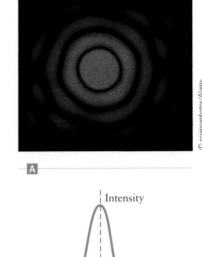

Figure 25.31 A When a diffraction opening is circular, a circular diffraction spot is created. B The intensity curve has a shape similar to that of a single slit, with the first intensity minima at an angle given by Equation 25.27.

Figure 25.32 A The aperture of a telescope diffracts the incoming light. B If light from a single point source passes through an opening, it produces a single diffraction spot like the one in Figure 25.31. C If there are two nearby sources, their diffraction spots may overlap on the screen.

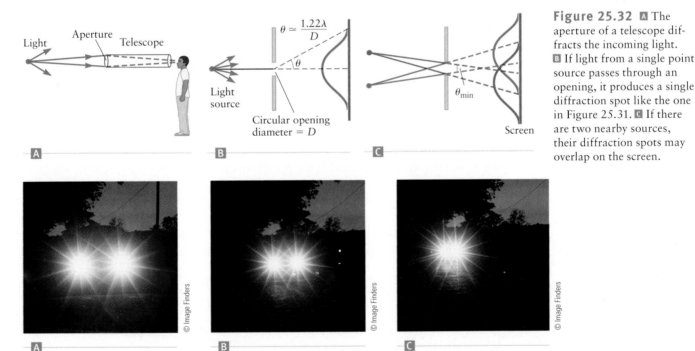

Figure 25.33 Light from two car headlights demonstrates the Rayleigh criterion. A When the headlights are far apart, they can easily be resolved as two separate light sources. B As the headlights are brought close together their images overlap. Here they can still just be resolved as separate sources. This corresponds to approximately the angular separation given by the Rayleigh criterion (Eq. 25.28). C When the headlights are very close together, they cannot be resolved as separate sources.

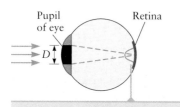

Pupil of eye   Retina

D

Diffraction causes the image from a point object to spread over an angle θ_{min} given by Rayleigh's criterion.

Figure 25.34 Example 25.7. The angular resolution of the eye is determined by the Rayleigh criterion.

EXAMPLE 25.7 | ⊗ ⓡ **Limits of Resolution of the Eye**

The Rayleigh criterion sets a limit on the resolution of the eye. Light enters the eye through a circular opening called the pupil, resulting in a diffraction spot when the light is focused on the retina. **(a)** Estimate the limiting angular resolution of the eye as measured at the retina. Assume green light is incident with $\lambda = 500$ nm. *Note:* The wavelength λ in the Rayleigh criterion is the wavelength after the waves have passed through the opening, inside the eye in this example. We must therefore allow for the index of refraction of the eye and use the wavelength found in the same way we did for light in a film in Equation 25.9. Use the (approximate) value $n_{eye} = 1.3$ for the index of refraction of the eye. **(b)** Using your result for the angular resolution in part (a), suppose two very tiny pieces of dust are separated by 50 μm (approximately the diameter of a human hair) and are located a distance $L = 25$ cm from your pupil. Can you resolve them as two separate dust particles?

(a) Let's consider part (a) of this example before moving on to part (b).

RECOGNIZE THE PRINCIPLE

The pupil is approximately circular and diffracts light, just like the aperture of a telescope (Fig. 25.31A). A true point source of light will be imaged as a "spot" on the retina; the size of this spot determines how well your eye can resolve two tiny objects that are close together. Using the Rayleigh criterion, we can determine how this effect limits the resolution of the eye.

SKETCH THE PROBLEM

Figure 25.34 shows a simplified sketch of the eye. Light is diffracted by the pupil, forming a diffraction spot on the retina.

IDENTIFY THE RELATIONSHIPS

The angular resolution limit from the Rayleigh criterion is

$$\theta_{min} = \frac{1.22\lambda_{eye}}{D}$$

(Eq. 25.28), where λ_{eye} is the wavelength of the light as it travels inside the eye. By analogy with Equation 25.9, we have

$$\lambda_{eye} = \frac{\lambda}{n_{eye}} \quad (1)$$

where λ is the wavelength outside the eye (given as 500 nm). We are given the wavelength λ and n_{eye}, but must estimate D to apply the Rayleigh criterion. Looking at your own eye in a mirror, you can estimate the pupil diameter as about $D = 5$ mm.

SOLVE

Substituting the λ_{eye} from Equation (1) and the estimated pupil diameter gives

$$\theta_{min} = \frac{1.22(\lambda/n_{eye})}{D} = \frac{1.22(5.0 \times 10^{-7} \text{ m})/1.3}{5 \times 10^{-3} \text{ m}} = \boxed{9 \times 10^{-5} \text{ rad}} \quad (2)$$

Notice that the units here are radians because the expression for θ_{min} is a dimensionless ratio.

What does it mean?
The calculated angle is only 9×10^{-5} rad = 0.005°! The angular resolution of the human eye is quite good; two beams of light separated by only 0.005° will be seen as separate spots on the retina.

(b) Now let's apply the result from part (a) to determine how well the eye can resolve two small pieces of dust.

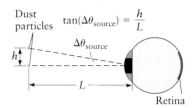

Figure 25.35 Example 25.7. Application of the Rayleigh criterion. Can your eye resolve the dust particles?

SKETCH THE PROBLEM

The geometry is shown in Figure 25.35. The dust particles are a distance L from the eye and are separated by a distance h. Given the values of L and h, we want to know if the particles can be resolved as separate pieces of dust.

IDENTIFY THE RELATIONSHIPS

When you view the dust particles, your retina is detecting light reflected by the particles. The dust particles thus act as separate sources of light, similar to the two sources of light in Figure 25.32C. From Figure 25.35, we see that light waves from the dust have an angular separation at the retina of $\Delta\theta_{source}$, where

$$\tan(\Delta\theta_{source}) \approx \frac{h}{L}$$

Because the dust particles are very close together, $\Delta\theta_{source}$ is very small and we can use the small-angle approximation $\tan(\Delta\theta) \approx \Delta\theta$.

SOLVE

We are left with

$$\tan(\Delta\theta_{source}) \approx \Delta\theta_{source} \approx \frac{h}{L}$$

Inserting the values of h and L, we find

$$\Delta\theta_{source} = \frac{h}{L} = \frac{50 \times 10^{-6} \text{ m}}{0.25 \text{ m}} = 2.0 \times 10^{-4} \text{ rad} \qquad (3)$$

The limiting value of θ_{min} in Equation (2) is smaller than the angular separation of the dust particles $\Delta\theta_{source}$ in Equation (3). Hence, the eye can resolve the pieces of dust as two separate objects.

What does it mean?

Our estimation of the eye's resolution using the Rayleigh criterion is only an approximate treatment, but even though we have ignored factors such as the separation of the detector cells (called rods and cones) in the retina, our estimate matches the actual resolution of the eye to within about a factor of two. You should check for yourself. You can do so by carefully placing two strands of hair side by side and pushing them as close as possible, but not so close that they appear as one strand of hair. Then use a microscope or magnifying glass to measure the distance between the strands.

CONCEPT CHECK 25.6 | The Rayleigh Criterion and the Diameter of a Telescope

The Rayleigh criterion determines the ultimate resolution of a telescope. This resolution is commonly measured in terms of the smallest angular separation θ_{min} between two objects that the telescope can resolve (Fig. 25.32C). If the diameter of the telescope aperture is increased, does θ_{min} (a) get smaller, (b) get larger, or (c) stay the same?

Limits on Focusing

Suppose you are asked to focus a light beam to a very tiny spot using a very high quality (and expensive) lens. Is there any limit to how small you can make this spot? According to the ray optics approximation of Chapter 24, a perfect lens will

Figure 25.36 Focusing light from a distant object. The lens acts as an opening as in Figure 25.32, causing the image to be broadened ("smeared out") by diffraction. This sets limits on focusing and resolution of the lens. *Note:* The colors of these rays do *not* represent the colors of the light waves.

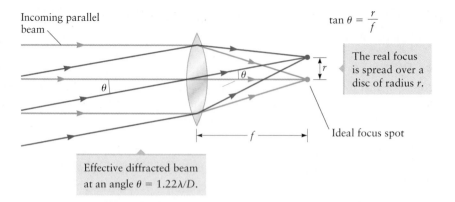

Incoming parallel beam

$\tan \theta = \dfrac{r}{f}$

The real focus is spread over a disc of radius r.

Ideal focus spot

Effective diffracted beam at an angle $\theta = 1.22\lambda/D$.

focus a very narrow parallel beam of light to a precise point at the focal point of the lens. However, the ray optics approximation ignores diffraction. If the lens has a diameter D, it acts just like the openings in Figures 25.31 and 25.32 and according to the Rayleigh criterion produces a diffracted beam spread over a range of angles up to $\theta \approx 1.22\lambda/D$ (Eq. 25.27). This spreading is modeled in Figure 25.36, which shows the original beam (drawn in green) traveling along the principal axis of a lens. The diffraction spreading is modeled by the second incident beam (drawn in red), which makes an angle $\theta \approx 1.22\lambda/D$ relative to the original beam. Without diffraction, the rays would all focus (intersect) at a point on the axis a distance f from the lens, where f is the focal length. Diffraction spreads this point over a disk of radius r as shown near the intersection of the rays drawn in red. From the geometry in Figure 25.36, we find $\tan \theta = r/f$. The angle θ is usually small, so we can use the small-angle approximation to get

$$\tan \theta \approx \theta \approx \frac{r}{f}$$

which can be rearranged to solve for r:

$$r = f\theta \tag{25.29}$$

Inserting θ from Equation 25.27 gives

$$r = f\theta = f\frac{1.22\lambda}{D} \tag{25.30}$$

The lens thus focuses parallel incoming light to a "smeared-out" circular spot of radius r. According to Equation 25.30, the value of r depends on the diameter D of the lens and its focal length f, but these two quantities are not independent of each other. The lens maker's formula from Chapter 24 tells how f is related to the radius of curvature R of the lens. It is not physically possible to make the diameter of the lens opening larger than $2R$. (See Fig. 24.37.) For a lens with an index of refraction n and two convex surfaces both with a radius of curvature R, the focal length is

$$\frac{1}{f} = \frac{2(n-1)}{R}$$

as obtained from the lens maker's formula (Eq. 24.32) with $R_1 = R_2 = R$. We have just argued that the upper limit on the opening of a lens is $D = 2R$, so the limit on focal length is

$$\frac{1}{f} = \frac{2(n-1)}{D/2}$$

which leads to

$$f = \frac{D}{4(n-1)}$$

For a typical lens material such as glass, $n \approx 1.5$ (see Table 25.1), and we get

$$f \approx \frac{D}{2} \tag{25.31}$$

Inserting this into Equation 25.30 gives the radius of the focus spot (also called the focal disk) as

$$r = f\frac{1.22\lambda}{D} \approx \frac{D}{2}\frac{1.22\lambda}{D} = \frac{1.22\lambda}{2} \tag{25.32}$$

Our result for r is only approximate because it depends on our estimate for how the lens opening D depends on its radius of curvature R. In the spirit of an estimation problem, we can thus write

$$r \approx \lambda \tag{25.33}$$

In words, Equation 25.33 says that the wave nature of light limits the focusing qualities of even a perfect lens. *It is not possible to focus a beam of light to a spot smaller than approximately the wavelength.*

This fundamental limit on focusing determines the ultimate performance of microscopes and other optical instruments, as we'll explain in Chapter 26. Equation 25.33 defines yet again the boundary between the regimes of ray optics and wave optics. As a general rule, the ray approximation of geometrical optics can only be applied at size scales much greater than the wavelength. When a slit or a focused beam of light is made so small that its dimensions are comparable to λ, we must expect that diffraction effects will be important. These effects are the essence of wave optics.

25.9 | WHY IS THE SKY BLUE?

The wave nature of light leads to interesting effects when light reflects from very small objects. When light reflects from a plane surface, the angle of reflection is equal to the angle of incidence (the law of reflection, Eq. 24.1), but this result is true only when the size of the reflecting surface is much greater than the wavelength. Figure 25.37 shows the opposite situation, when λ is larger than the reflecting object. The reflected waves now travel away in all directions and are called *scattered* waves. This behavior is found for all types of waves, including sound (Fig. 13.14).

The amplitude of the scattered wave depends on the size of the scattering object relative to the wavelength. For particles much smaller than the wavelength of light, the amplitude of the scattered wave grows rapidly as the wavelength is decreased (Fig. 25.38). Blue light and ultraviolet light are thus scattered much more strongly than red light. This effect is called *Rayleigh scattering*.

Figure 25.38 helps answer the very simple everyday question, "Why is the sky blue?" The Earth's atmosphere is composed mainly of oxygen and nitrogen molecules, which are both colorless, so one might expect the atmosphere to be colorless. We know that the atmosphere is blue on a normal clear day, however, and Figure 25.39 shows why. The light that we see from the sky overhead is sunlight scattered by N_2, O_2, and other molecules in the atmosphere. These molecules are much smaller than the wavelength of visible light, so (according to Fig. 25.38) they scatter blue light more strongly than red light, giving the atmosphere its blue color.

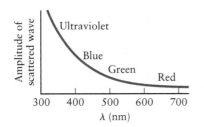

Figure 25.38 When light is scattered by small particles in the atmosphere, the amplitude of the scattered wave in Figure 25.37 increases as the wavelength decreases.

Table 25.1 Index of Refraction of Some Common Substances for Yellow Light

SUBSTANCE	INDEX OF REFRACTION
Air	1.0003
Water	1.33
Glass	1.5[a]
MgF_2	1.38

[a]Typical value. The precise value depends on composition.

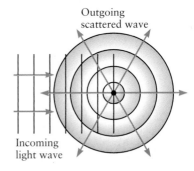

Figure 25.37 When a plane light wave strikes an object that is much smaller than the wavelength, a spherical wave propagates outward in all directions from the object. This outgoing wave is called scattered light. A similar phenomenon occurs for sound waves; see Section 13.5 and Figure 13.14.

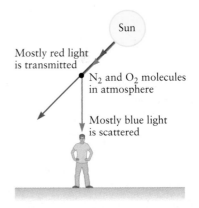

Figure 25.39 Scattering of sunlight by N_2, O_2, and other molecules in the air is responsible for the color of the atmosphere. Since these molecules are smaller than the wavelength of visible light, they scatter short-wavelength light most strongly, giving the atmosphere its blue color.

EXAMPLE 25.8 | Color of the Sun

In Chapter 14, we learned that the Sun emits light over a range of wavelengths (colors) as shown in Figure 14.32. The maximum intensity falls close to the middle of the visible range, and the Sun appears white when it is overhead (Fig. 25.40A), but when the Sun is close to the horizon (at sunrise or sunset), it has a distinctly reddish color (Fig. 25.40B). Explain this effect in terms of Rayleigh scattering.

Figure 25.40 Example 25.8.
A When the Sun is high in the sky, it has a white to pale yellow color due to its blackbody spectrum, which peaks near the middle of the visible range. **B** When the Sun is low on the horizon, it appears more reddish in color.

© Rob Casey/Brand-X Pictures/Jupiterimages

© Comstock Images/Jupiterimages

A

B

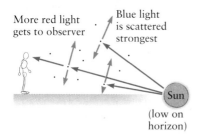

Figure 25.41 The red color is a result of scattering by molecules and small particles in the atmosphere, removing most of the blue light. Compare with Figure 25.39.

SKETCH THE PROBLEM

Figure 25.41 shows light traveling from the Sun to an observer on the Earth.

IDENTIFY THE RELATIONSHIPS AND SOLVE

When the Sun is near the horizon, the sunlight you observe travels a greater distance through the atmosphere than when the Sun is directly overhead. Because of Rayleigh scattering, molecules and other small particles in the air scatter short wavelengths (blue light) more strongly than long wavelengths (red light). By the time the light from a rising or setting Sun reaches an observer, a substantial fraction of the blue component has been scattered away. The remaining light (Fig. 25.41) contains a greater proportion of red light than when the Sun is high in the sky.

What does it mean?
The color of an object depends on both the colors of light the object emits and what happens to the different colors (wavelengths) as they travel to an observer. The scattering of light can have a big effect on "the" color of an object.

CONCEPT CHECK 25.7 | Color of the Atmosphere

Why does a forest fire tend to give the sky a pronounced reddish color?
(a) The small particles of soot and molecules produced by combustion from a fire usually have a red color.
(b) The small particulates from a fire scatter red light from the Sun less than blue light.
(c) Heat from the fire removes water vapor from the air.

25.10 | THE NATURE OF LIGHT: WAVE OR PARTICLE?

The observations of double-slit interference and of diffraction by a single slit show convincingly that light has wave properties. However, certain properties of light and optical phenomena can *only* be explained with the particle theory of light. One very common effect that can only be correctly explained by the particle theory of light is color vision.

Insight 25.3
WHY IS THE SKY BLUE AND NOT VIOLET?
The Rayleigh scattering of light by molecules in the atmosphere gets stronger as the wavelength decreases. We have claimed that this scattered light gives the sky its blue color (Fig. 25.39), but violet light has a shorter wavelength than blue light, so why isn't the sky violet instead? Two factors combine to make the sky blue instead of violet. First, the Sun emits more strongly in the blue than in the violet range (Fig. 14.32). Second, our eyes are more sensitive to blue light than to violet light. Hence, even though violet light is scattered more strongly than blue light, we still perceive the sky as blue.

⊗ Origin of Color Vision

Human and many other vertebrate eyes are sensitive to color; that is, they are able to distinguish light waves according to their wavelength. This is accomplished by cells in the retina called *cones*.[3] There are three types of cones, and Figure 25.42 shows that these different cone cells are sensitive to light of different colors. Some cones are most sensitive to red light—that is, they send a large signal to the brain when exposed to red light and a much smaller signal with either blue or green light—and other cones are sensitive to mainly blue or green light.

How can we explain the behavior of different cone cells? The energy carried by an ordinary wave (such as a sound wave) depends on its amplitude; a large amplitude corresponds to a high intensity and vice versa. So, we should always be able to arrange for two light waves with different colors (say, red and blue) to carry the same total energy just by adjusting their relative intensities, but even when two such waves carry the same total energy into the eye, the red cone cells still respond mainly to red light and the blue cone cells respond mainly to blue light. How can a cone cell be much more sensitive to light of one color (i.e., one wavelength) than another despite such differences in intensity? How can a red cone cell be almost completely insensitive to blue and violet light, or a blue cone cell insensitive to red light? Studies of these and related questions led to the quantum theory of light (Chapter 28). According to that theory, particles of light called *photons* carry an amount of energy that depends on the frequency of the light, with blue photons carrying more energy than red photons. A complete explanation of color vision involves the properties of photons along with the quantum properties of molecules inside cone cells, and we'll come back to this problem when we describe quantum theory in Chapter 28.

So, although Newton did not realize it, the evidence for his particle theory of light was literally right in front of him. No one else realized it either until about 1905, when the beginnings of a quantum theory of light were first developed by Einstein. If light is a particle, though, how do we explain interference and diffraction? After all, aren't they strictly wave phenomena? It would thus seem that we have strong evidence that light is both a particle *and* a wave! This concept is called "particle-wave duality." Reconciling these notions is a major part of quantum theory, as we'll discuss in Chapter 28.

The study of light has been a central thread in the development of physics as a science, connecting Newton and mechanics to the theory of waves, to Maxwell's work on electromagnetism, and all the way to quantum theory. In Chapter 28, we'll see how quantum theory extends the concept of particle-waves to electrons, protons, and other objects that a classical physicist (Newton!) would have classified as particles. The discovery that light has properties of both classical particles and waves thus opened the way to 20th-century physics.

---

[3]Cells in the retina called *rods* also detect light, but rods are not sensitive to color.

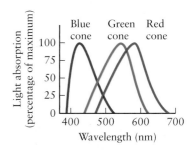

Figure 25.42 Color vision is due to light detectors in the eye called cones. The three types of cones cells are sensitive to light from different regions of the visible spectrum.

SUMMARY | Chapter 25

KEY CONCEPTS AND PRINCIPLES

Conditions for interference
Light is a wave and exhibits interference if certain conditions are met.
1. Two or more waves must travel through different regions of space.
2. The waves must be brought together at a common point.
3. The waves must be *coherent*.

(Continued)

Constructive and destructive interference
Interference of two waves is **constructive** if the waves are **in phase** and is **destructive** if the waves are 180° **out of phase**.

Phase change upon reflection
When light reflects from a surface, the phase of the reflected wave changes by 180° if the index of refraction on the incident side is smaller than the index on the opposite side. This effect, called a **phase change on reflection**, is important in thin-film interference.

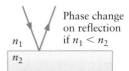

Phase change on reflection if $n_1 < n_2$

n_1
n_2

Huygens's principle
All points on a wave front act as sources of new spherical waves. Huygens's principle explains how light passing through a narrow opening is **diffracted**.

Double-slit interference
Young's double-slit experiment demonstrates that light passing through two narrow slits can interfere. The interference pattern on a distant screen shows constructive or destructive interference, depending on the outgoing angle. Constructive interference (and thus a bright fringe in the interference pattern) is found when

$$d \sin \theta = m\lambda \quad m = 0, \pm 1, \pm 2, \pm 3, \ldots \quad \textbf{(25.15)} \text{ (page 856)}$$

Destructive interference resulting in a dark fringe is found when

$$d \sin \theta = (m + \tfrac{1}{2})\lambda \quad m = 0, \pm 1, \pm 2, \pm 3, \ldots \quad \textbf{(25.16)} \text{ (page 856)}$$

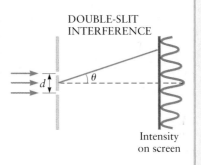

DOUBLE-SLIT INTERFERENCE

θ

d

Intensity on screen

Single-slit diffraction
Single-slit diffraction occurs when light passes though a narrow opening. Dark fringes (destructive interference) occur when

$$w \sin \theta = \pm m\lambda \quad m = \pm 1, \pm 2, \ldots \quad \textbf{(25.22)} \text{ (page 860)}$$

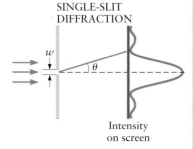

SINGLE-SLIT DIFFRACTION

w

θ

Intensity on screen

APPLICATIONS

Michelson interferometer
The **Michelson interferometer** uses the interference of light waves to measure small displacements. It was first used in fundamental studies of light and tests of relativity (Chapter 27) and is still widely used in studies of the optical properties of matter.

Thin-film interference
Light waves reflected from the two surfaces of a thin film can interfere to produce brightly colored interference fringes. The relative phases of the interfering waves are affected by the different path lengths traveled by the two waves and by any phase changes on reflection.

(Continued)

Rayleigh criterion

When light passes through a circular opening, diffraction causes the light beam to spread. The **Rayleigh criterion** states that two light beams can be resolved only if their directions differ by an amount as large as or greater than

$$\theta_{min} = \frac{1.22\lambda}{D} \qquad \textbf{(25.28)} \text{ (page 865)}$$

This result means that it is impossible to focus a light beam to a spot size much smaller than the wavelength.

Rayleigh scattering

Rayleigh scattering of light by small particles in the atmosphere produces the color of the atmosphere. This scattering is strongest for shorter wavelengths and gives the atmosphere its normal blue color.

QUESTIONS

1. Is it possible for light waves of different color to exhibit constructive or destructive interference? Explain.

2. Why don't you observe interference between light waves produced by two lightbulbs?

3. Figure Q25.3 shows the bands of color that are produced when a thin soap-bubble film is exposed to white light. What mechanism produces the colors seen here? Is the bubble of uniform thickness? If not, where is the bubble thinnest? Thickest? How do you know?

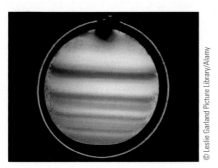

Figure Q25.3

4. If a soap film suspended in air is extremely thin, what color will it be when viewed in reflected light?

5. Explain why two light waves can interfere only if they are coherent.

6. Yellow light is used to make a double-slit interference pattern with two very narrow slits. If the same slits are used with blue light, will the separation between two adjacent bright fringes be larger or smaller?

7. SSM How does a double-slit interference pattern change when the following changes are made? Do the interference fringes become closer together or farther apart? Explain your answers.
(a) The two slits are brought closer together.
(b) The two slits are moved farther apart.

(c) The wavelength is increased.
(d) The wavelength is decreased.
(e) The entire experiment is immersed in water.

8. Green light is used to make a single-slit diffraction pattern with a particular slit. If the same slit is used with red light, will the width of the central bright fringe be wider or narrower?

9. Explain how a thin coating on a lens (called an antireflection coating) can reduce the amount of reflected light. Most expensive lenses have these coatings.

10. Some vivid rainbows, called supernumerary rainbows, have extra inner arcs of color (Fig. Q25.10) that do not follow the normal pattern in Figure 24.48. The many internal reflections that occur within a spherical raindrop can produce rays of different colors leaving the drop along the same path. Given this information, can you suggest a mechanism that could produce these extra bows?

Figure Q25.10

11. Design an experiment in which light from a helium–neon laser ($\lambda = 630$ nm) can be used to measure (a) the width of a single slit and (b) the distance between two nearby slits.

12. In Section 25.2, we assumed the Michelson interferometer was operated with monochromatic light (light of a single wavelength). Suppose a source of white light is used instead. Is it possible to have constructive interference and hence a bright fringe with white light? If your answer is yes, what value(s) of the path length difference ΔL is required for constructive interference and thus a bright fringe?

13. Some small, bright lightbulbs on a strand (Fig. Q25.13) are grouped in pairs according to color and then spaced 5 cm apart. (a) From a long distance, it becomes difficult to tell if each color is coming from one bulb or from a pair. Why? (b) There is a distance from the lights where the pairs of blue bulbs can be resolved but the pairs of red bulbs would look like one point. What feature of the red and blue light would account for this difference?

Figure Q25.13

14. SSM The star in the bottom of the Northern Cross asterism is actually a binary pair. The photograph in Figure Q25.14 came from a telescope with a mirror 10 inches in diameter. Which feature of the telescope is more important in enabling it to resolve the pair when your eye sees them as one point of light, (a) magnification or (b) the size of mirror or primary lens? (You can assume fluctuations in the atmosphere as described in Section 24.7 are negligible in this case.)

© Richard Yandrick (cosmicimage.com)

Figure Q25.14

15. A guitar player is at a rock concert and watches the fingers of the lead guitarist from a few rows back (Fig. Q25.15). She notes that she can see the guitarist's chord positions clearly when a yellow or blue spotlight falls on him through the smoke, but when a bright purple light is used, the scene becomes blurry. Why can she see less detail when the purple light is used?

© Mel Yates/Taxi/Getty Images

Figure Q25.15

16. ⊗ **Eyeshine.** The eyes of cats, foxes, raccoons, and many other mammals brightly reflect light that shines into them back toward the source and makes the animal's eyes seem to glow as seen in Figure Q25.16. The reflection occurs within a thin membrane behind the retina called the *tapetum lucidum*, which allows the animal's eye more sensitivity at night. There is no metallic material within the eye, and membrane does not look metallic under magnification. Describe a mechanism that might produce the bright reflection.

© Mark L. Stephenson/CORBIS

Figure Q25.16

17. You are designing a billboard that will be viewed at night. The letters used on this billboard are made of an array of bright lights like that shown in Figure Q25.17. If you want to make the sign legible from the greatest distance, which color should you choose for the lights? Why?

© Mike Dobel/Alamy

Figure Q25.17

18. Which single-slit diffraction experiment will give the widest central bright fringe, (a) a 900-nm slit with light of wavelength 500 nm or (b) a 600-nm slit with light of wavelength 450 nm?

19. The sky has a blue color due to the scattering of light by molecules in the air (Fig. 25.39). Explain why the sky's color can be a deeper blue when viewed through polarizing sunglasses.

20. Figure 25.21 shows the intensity pattern produced in a double-slit interference pattern with two very narrow slits. Suppose one of the slits is covered so that light can only pass through one slit. Sketch what the new intensity pattern would look like.

21. Constructive interference occurs when the crests of two coherent waves overlap (Fig. 25.1B). What happens when the troughs of two coherent waves overlap? Is the interference constructive or destructive? Explain.

SSM = solution in Student Companion & Problem-Solving Guide
★ = intermediate ✪ = challenging

❌ = life science application
ℝ = reasoning and relationships problem

25.1 COHERENCE AND CONDITIONS FOR INTERFERENCE

1. Light waves are emitted coherently from two sources and detected at a spot equidistant from the two sources. Do these waves interfere constructively or destructively?

2. Sound waves with a wavelength of 0.25 m are emitted coherently (i.e., in phase with each other) from two sources. These waves are found to interfere constructively at a point P that is 1.60 m from one of the sources. What is the distance from P to the other source? *Note:* There is more than one correct answer. Give three of them.

3. Coherent sound waves with $\lambda = 0.75$ m are emitted from points P_1 and P_2 in Figure P25.3. Do these waves interfere constructively or destructively at point P_3?

4. Two sources produce coherent light waves that come together at a detector located 1.50000 mm from one source and 1.50500 mm from the other. If the two waves interfere constructively, what is a possible value for the wavelength? Assume the light is visible light.

5. The detection point in Problem 4 is moved so that it is 1.50000 mm from one source and 1.50150 mm from the other. If the two waves interfere destructively, what is a possible value for the wavelength? Assume the light is visible light and keep three significant figures in your calculation.

6. SSM ★ Coherent sound waves are emitted from points P_1 and P_2 in Figure P25.6. The waves have a wavelength $\lambda = 1.5$ m. (a) Find the point on the x axis nearest the origin at which the waves interfere constructively. (b) Find the point on the x axis nearest the origin at which the waves interfere destructively.

7. Waves from a radio station have a wavelength of 250 m. These waves can travel directly from the antenna to a receiver or can reflect from a nearby mountain cliff and then reach the receiver (Fig. P25.7). If the distance from the receiver to the cliff is $L = 1000$ m, is there constructive or destructive interference at the receiver? Assume there is no phase change when the radio wave reflects from the cliff.

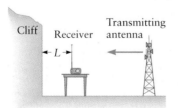

Figure P25.7 Problems 7 and 8.

8. ✪ Consider again the interfering radio waves in Problem 7, but now assume the amplitude of the reflected wave is exactly half the amplitude of the wave that arrives directly. If the amplitude

of the direct wave is A, what is the intensity at the receiver when there is (a) constructive interference or (b) destructive interference?

9. ✪ An omnidirectional speaker that emits sound with a wavelength λ is positioned between two vertical walls spaced 2.00 m apart as shown in Figure P25.9. A listener approaches as shown and finds that there is constructive interference of the waves that reflect from the two walls when he is a distance $L = 5.00$ m from the speaker. Find the longest possible wavelength.

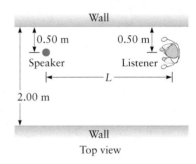

Figure P25.9

25.2 THE MICHELSON INTERFEROMETER

10. One of the mirrors in a Michelson interferometer is moved a distance of 2.0 mm. Meanwhile, the interference intensity moves through 7000 dark fringes. What is the wavelength of the light?

11. Suppose the wavelength in Problem 10 is increased by a factor of 1.2. How many dark fringes would now be found when the mirror moves the same distance?

12. ✪ Both arms of a Michelson interferometer are enclosed in pipes so that the air can be removed from the regions between the mirrors and the beam splitter. There is initially a vacuum in both regions, and the mirrors are the same distance from the beam splitter. Air is then let into one of the pipes, and light that travels to and from one mirror now propagates in air while the other part of the interferometer is still in a vacuum. If the interference pattern shifts to the 4000th bright fringe when the air is added, what is the distance from a mirror to the beam splitter? Assume $\lambda = 600$ nm.

13. ✪ Consider a Michelson interferometer that uses a sodium lamp as its source. A sodium lamp gives a large intensity at two wavelengths, 589.00 nm and 589.59 nm, in the yellow range of the visible spectrum. The mirrors are adjusted so that there is constructive interference with the Nth bright fringe for one of these wavelengths and the (N + 1)st bright fringe for the other wavelength. Find N. Be careful with significant figures in your calculation, and give your answer for N to two significant figures.

14. SSM ★ Consider a Michelson interferometer operated with light from a sodium lamp with $\lambda_{Na} = 589$ nm. Suppose the interferometer is used to check the accuracy of a meterstick by moving one of the mirrors a distance of precisely 1.00 m (as determined by the meterstick) and counting bright fringes as they appear at the detector. How many bright fringes will be observed while the mirror is being moved?

15. ✪ A Michelson interferometer is used to check the length of a "standard" platinum meterstick. The interferometer is operated with light from a specially constructed helium–neon laser, which gives red light with $\lambda = 632.99139822$ nm. (a) The standard meterstick is placed next to one of the interferometer's arms, and the mirror is moved from one end of the meterstick to the other. How many interference fringes N are counted as the mirror is moved? (b) Suppose the wavelength λ is known with an accuracy of ±0.00000002 nm. Also suppose the number of interference fringes N is known to an accuracy of ±1. What determines the uncertainty in the measured length of the meterstick, the uncertainty in the wavelength or the uncertainty in N? (c) In a real experiment, you would not simply count the number of bright fringes but instead would measure "fractions of a fringe" by measuring the intensity and comparing it with the sinusoidal curve in Figure 25.5. If the number of fringes could be measured to ±0.0001, would the uncertainty in N still be a limiting factor?

25.3 THIN-FILM INTERFERENCE

16. An extremely thin film of soapy water ($n = 1.35$) sits on top of a flat glass plate with $n = 1.50$. The soap film has a red color when viewed at normal incidence. What is the thickness of the film? ($\lambda_{red} = 600$ nm.)

17. Repeat Problem 16 assuming the soap film rests on a material having an index of refraction $n = 1.10$.

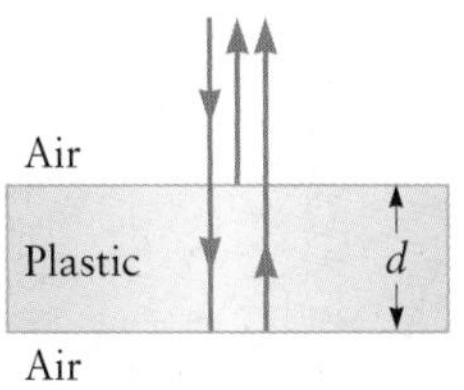

18. ✪ A plastic film of thickness 250 nm appears to be green ($\lambda_{green} = 500$ nm) when viewed in reflection at normal incidence (Fig. P25.18). What is the plastic's index of refraction?

Figure P25.18

19. ✪ A very thin sheet of glass ($n = 1.55$) floats on the surface of water ($n = 1.33$). When illuminated with white light at normal incidence, the reflected light consists predominantly of the wavelengths 560 nm and 400 nm. How thick is the glass?

20. A compact disk uses the interference of reflected waves from the CD to encode information. Light from a red laser ($\lambda = 630$ nm) is reflected from small pits in the CD, and waves from the bottom of a pit and the adjacent top edge interfere (Fig. P25.20). What is the minimum pit depth for which the two waves interfere destructively? Assume this pit is on the surface of the CD so that the region above the pit is air.

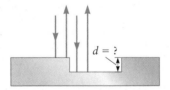

Figure P25.20 Problems 20 and 21.

21. ✪ In a real CD, the pits in Figure P25.20 are covered with a layer of plastic so that surface scratches do not spoil the pits. If the index of refraction of the plastic coating is $n = 1.70$, what minimum depth should a pit have to give destructive interference using light with $\lambda = 630$ nm?

22. SSM ✪ Two glass plates are arranged as in Figure 25.12 so that light reflected from the bottom surface of the top plate interferes with light reflected from top surface of the bottom plate. As one moves along the x axis, the total reflected intensity alternates between bright and dark fringes. If there are a total of 25 bright fringes across the entire width of the plates, what is the plate spacing d at the right edge? Assume the source of illumination is blue light with a wavelength of 420 nm.

23. ✪ A device called a Fabry–Perot interferometer contains two parallel mirrors as shown in Figure P25.23. Multiple reflec-

tions occur between the inner mirror surfaces, and the waves emitted at the top can interfere. Assume the light waves travel nearly perpendicular to the mirrors, the emitted waves interfere constructively, and the spacing between the mirrors is exactly 3.5 μm. What is a possible value for the wavelength in the visible range between 600 and 700 nm?

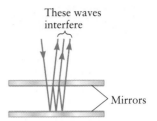

These waves interfere

Mirrors

Figure P25.23

25.4 LIGHT THROUGH A SINGLE SLIT: QUALITATIVE BEHAVIOR

24. SSM Single-slit diffraction can be observed with any type of electromagnetic wave (not just light). Suppose you want to make a diffraction slit whose width is five times larger than the wavelength for the following cases. How wide would the slit be?
(a) A radio wave for your favorite FM station ($f = 100$ MHz)
(b) The waves that carry your cell phone signals ($f = 2.0$ GHz)
(c) Red light ($f = 5.0 \times 10^{14}$ Hz)
(d) The X-rays used in your dentist's office ($f \approx 4 \times 10^{20}$ Hz)

25.5 DOUBLE-SLIT INTERFERENCE: YOUNG'S EXPERIMENT

25. A double-slit interference experiment is performed with two very narrow slits separated by 0.15 mm. The experiment uses red light with a wavelength of 600 nm and projects the interference pattern onto a screen 7.5 m away from the slits. (a) What is the distance between two nearby bright fringes on the screen? (b) What is the distance between two nearby dark fringes on the screen? Assume these fringes are all near $\theta = 0$.

26. Two small stereo speakers are 2.5 m apart and act as a double-slit interference experiment with sound waves. If the opposite wall of the room is 5.5 m away and the sound frequency is 1.5 kHz, what is the distance between the two interference maxima nearest the center of the wall? Assume the speakers act as point sources of sound.

27. ✪ In Problem 26, you were told to assume the speakers could be treated as point sources. Which of the following statements justifies this assumption?
(a) Small speakers could be much smaller than the wavelength of the sound waves.
(b) The speakers could be larger than the wavelength but smaller than the size of the room.
(c) The wavelength is smaller than the size of the room.
(d) The wavelength is smaller than the spacing between the speakers.

28. SSM ✪ Monochromatic light ($\lambda = 500$ nm) passes through two slits and onto a screen 2.5 m away. (a) If two nearby bright fringes are separated by 12 mm, what is the slit spacing? (b) If the slit spacing is reduced by a factor of three, what is the new distance between bright fringes?

29. Two slits are separated by 0.35 mm and are used to perform a double-slit experiment with a screen 1.5 m away from the slits. If the distance between a dark region and the nearest bright spot on the screen is 1.0 mm, what is the wavelength?

30. ✪ Two narrow slits are used to produce a double-slit interference pattern with monochromatic light. The slits are separated by 0.90 mm, and the interference pattern is projected onto a screen 9.5 m away from the slits. The central bright fringe is at a certain spot on the screen. Using a ruler with one end placed at the central fringe, you move along the ruler passing by two more bright fringes and find that the next bright fringe is 20 mm away from the central fringe. What is the wavelength of the light?

31. Two narrow slits are used to produce a double-slit interference pattern with monochromatic light ($\lambda = 450$ nm). The slits are separated by 0.90 mm, and the interference pattern is projected onto a screen 7.5 m away from the slits. (a) What is the distance between two nearby bright fringes on the screen? (b) What is the distance between two nearby dark fringes on the screen? Assume the fringes are near the center of the interference pattern.

32. ✪ A Young's double-slit interference experiment gives bright fringes separated by 2.5 cm on a screen when the experiment is performed in air. The entire apparatus then placed under water ($n = 1.33$). What is the new spacing between bright fringes? Assume the fringes are near the center of the interference pattern.

33. ✪ Two narrow slits are used to produce a double-slit interference pattern with monochromatic light. It is found that the third bright fringe from the center with blue light ($\lambda_1 = 420$ nm) occurs at the same place on the screen as the second bright fringe with light of an unknown wavelength λ_2. Find λ_2. Here the central bright fringe is fringe number "zero."

25.6 SINGLE-SLIT DIFFRACTION: INTERFERENCE OF LIGHT FROM A SINGLE SLIT

34. Red light ($\lambda = 600$ nm) is diffracted by a single slit, and the first intensity minimum (the first dark fringe) occurs at $\theta = 6.9°$. What is the width of the slit?

35. SSM ★ Ⓡ A doorway acts as a single slit for light that passes through it, but a doorway is much larger than the wavelength of visible light, so the diffraction angles are very small. Estimate the angle of the 10th diffraction minimum for a typical doorway.

36. A slit with width 0.15 mm is located 10 m from a screen. If light with $\lambda = 450$ nm is passed through this slit, what is the distance between the central bright fringe and the third dark fringe on the screen?

37. Light with $\lambda = 550$ nm passes through a single slit and then illuminates a screen 2.0 m away. If the distance on the screen from the first dark fringe to the center of the interference pattern is 5.5 mm, what is the width of the slit?

38. A helium–neon laser ($\lambda = 630$ nm) is used in a single-slit experiment with a screen 3.0 m away from the slit. If the slit is 0.25 mm wide, what is the width of the central bright fringe on the screen? Measure this width using the locations where there is destructive interference.

25.7 DIFFRACTION GRATINGS

39. The slits in a diffraction grating are spaced 25 μm apart. What is the diffraction angle for the second bright fringe away from $\theta = 0$ with red light ($\lambda = 600$ nm)?

40. ★ A diffraction grating is a square 1.0 cm on a side and contains 3000 slits. What is the angle of the first intensity maximum in the diffraction pattern with blue light ($\lambda = 450$ nm)?

41. ✪ A sodium lamp produces yellow light with wavelengths λ and $\lambda + \Delta\lambda$, where $\lambda = 589.00$ nm and $\Delta\lambda = 0.59$ nm. This light is incident on a diffraction grating with a slit spacing of 15 μm, and the diffracted light illuminates a screen 2.5 m away. What is the distance on the screen between the fifth bright fringes for these two wavelengths?

42. ★ Light from a helium–neon laser is incident on a diffraction grating and then illuminates a screen that is 10 m away. If the bright diffraction fringes near $\theta = 0$ are spaced 0.50 m apart, what is the spacing between the slits in the grating?

43. SSM ★ The atomic planes in a crystal are spaced $d = 0.50$ nm apart. These planes are used in a diffraction experiment with X-rays with $\lambda = 0.05$ nm. Assume the bright fringes in the diffraction pattern are at the same angles as for a grating with slits with the same value of d. What are the diffraction angles for the first three bright fringes?

44. ★ Consider again diffraction from a crystal in Problem 43, but now assume light with $\lambda = 600$ nm is used. What is the diffraction angle for the first bright fringe? Your answer explains why visible light cannot be used in diffraction experiments with crystals.

45. ✪ You are designing a lecture demonstration on diffraction gratings. You plan to use light from a laser pointer ($\lambda = 630$ nm) and shine the light from the grating on the far wall of the lecture room ($W = 9.0$ m). You want the separation between the central diffraction spot and one of the spots closest to it on the wall to be 1.0 m apart (Fig. 25.27B). The grating store offers gratings with a specified number of slits per millimeter. How many slits per millimeter do you want for your demonstration?

25.8 OPTICAL RESOLUTION AND THE RAYLEIGH CRITERION

46. ★ Ⓡ Is the resolution of an optical instrument greater with blue light or with red light? Explain your answer. By approximately what factor are the resolutions different?

47. ✪ Ⓡ The camera on a spy satellite has a lens with a diameter of 1.5 m. This satellite is in low-Earth orbit about 2.9×10^5 m above the surface of the Earth (see Chapter 5, Eq. 5.28). What is the approximate size of the smallest feature the camera can resolve when taking a picture of something on the Earth's surface? Assume blue light with a wavelength of 400 nm. Ignore the effect of the Earth's atmosphere on the resolution.

48. ★ A terrestrial telescope (a "spyglass") with a diameter of 2.0 cm is attached to a rifle and used for target practice with targets that are 300 m away. What is the smallest feature on the target that can be resolved with this telescope? Assume $\lambda = 400$ nm. Ignore the effect of the Earth's atmosphere on the resolution.

49. SSM ★ Radio telescopes use radio waves (instead of light) to produce images. If radio waves with a frequency of 500 MHz are used together with a radio-wave mirror of diameter 1000 m, what is the best possible angular resolution that can be achieved? Ignore the effect of the Earth's atmosphere on the resolution.

50. A large optical telescope has a mirror with a diameter of 15 m. What is the angular resolution of this telescope when used to form an image with blue light? *Hint:* The diameter of the mirror is the size of the telescope's "opening" (also called its aperture). Ignore the effect of the Earth's atmosphere on the resolution.

51. ★ Ⓡ An amateur astronomer's telescope containing a lens with a diameter of 15 cm is used to study the Moon. What is the smallest object on the Moon that can be resolved with this telescope using blue light? Ignore the effect of the Earth's atmosphere on the resolution.

52. ✪ (a) A large optical telescope with an aperture of diameter 15 m is used to study the Moon. What is the size of the smallest feature on the Moon that this telescope can resolve using blue light? (b) Calculate the size of the smallest object on Mars that this telescope can resolve. Assume Mars is at its closest distance from the Earth. Ignore the effect of the Earth's atmosphere on the resolution.

25.9 WHY IS THE SKY BLUE?

53. ✪ Ⓡ The Sun appears red when it is on the horizon due to scattering of light by molecules in the atmosphere (Fig. 25.41). The Sun does not appear as reddish in color when it is overhead because the thickness of the atmosphere through which sunlight must travel is smaller when the Sun is overhead than when it is near the horizon. (a) Take the approximate thickness of the atmosphere to be 2×10^5 m. What distance through the atmosphere must light travel to reach us when the Sun is at the horizon? (b) Estimate the ratio of the distances that sunlight travels through the atmosphere in these two cases.

54. ✪ You want to design an antireflection coating for a lens. You want this coating to work well at two wavelengths, $\lambda_1 = 450$ nm and $\lambda_2 = 650$ nm, which you accomplish using two separate layers. One layer uses MgF_2 and is designed to remove the reflections at wavelength λ_1 ($n = 1.38$). The second coating is composed of a material with $n = 1.55$ and is designed to remove the reflections at λ_2. If the MgF_2 coating is deposited on a lens whose index of refraction is $n = 1.60$ and the other coating sits on top of the MgF_2, what is the thickness of each layer?

55. SSM ✪ A very thin soap bubble is floating in air. If the bubble strongly reflects green light ($\lambda = 520$ nm), what is the minimum possible thickness of the bubble? Assume the index of refraction of the bubble is $n = 1.35$.

56. ✪ A challenge astronomers face is that any telescope on the ground must look through many kilometers of air to peer into space. The air is not of uniform refractive index and is also turbulent, which limits the angular resolution of any telescope to approximately 1 arc second (1/3600 of a degree). (a) What is the maximum diameter of a ground-based telescope where the resolution limit by diffraction (Rayleigh limit) is equal to that of atmospheric turbulence? Assume a wavelength of 450 nm (blue light) for the calculation. (b) Find the angular separation of Pluto and its moon Charon, which are separated by approximately 2.3×10^4 km. Assume Pluto and the Earth are at their closest. Would this telescope be able to resolve Pluto from Charon?

57. ✪ If a laser is directed at a thin wire or hair, a diffraction pattern will occur similar to that produced by a slit of the same width. If the distance from the thin object to the screen on which the diffraction pattern is cast is known, a measurement of the spacing between peaks in the diffraction pattern can be used to determine the object's width. This method is used in the wire-making industry to monitor the production of wire with a uniform width. A hair is placed a distance $L = 1.2$ m away from a screen as shown in Figure P25.57A and illuminated with a green 532-nm laser pointer. The resulting diffraction pattern is shown in Figure P25.57B, where the distance to the first dark fringe is $d = 1.4$ cm. What is the diameter of the hair?

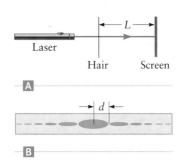

Figure P25.57

58. ✪ An oil tanker has collided with a smaller vessel, resulting in an oil spill in a large, calm-water bay of the ocean. You are investigating the environmental effects of the accident and need to know the area of the spill. The tanker captain informs you that 20,000 liters of oil have escaped and that the oil has an index of refraction of $n = 1.1$. From the deck of your ship, you note that in the sunlight the oil slick appears to be blue. A spectroscope confirms that the dominant wavelength from the surface of the spill is 420 nm. Assuming a uniform thickness, what is the total area of the oil slick?

59. ✪ An astronomical interferometer is an apparatus that combines a number of telescopes into a much larger equivalent light-gathering instrument. The very first extrasolar planet to be discovered was imaged in 2005 (Fig. P25.59) with the interferometer in the mountains of northern Chile known as the Very Large Telescope (VLT) project, which consists of four separate optical telescopes each 8.3 m in diameter. The interferometer operates with infrared wavelengths (1.4 μm), which are much less affected by atmospheric turbulence than other wavelengths. (a) The star system with the extrasolar planet was found to be at a distance of 200 light-years, and the planet orbits the star at a distant of 55 astronomical units (55 times the distance between the Earth and the Sun). What is the angular separation of the planet and star? Give your answer in units of arc seconds. (b) The interferometer has an angular resolution of 0.0020 arc second. What is its effective aperture? That is, a telescope of what diameter would have such resolution?

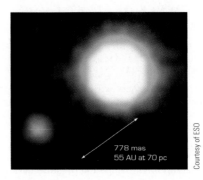

Figure P25.59 The first optical image of an extrasolar planet found in April 2005 with the ESO's Very Large Telescope Array in the mountains of northern Chile. This star is the brown dwarf 2M1207.

60. ⓧ ✪ **Visual acuity.** Optometrists quantify the resolution of a patient's vision using a concept called *visual acuity*, which is calculated as the inverse of the angular resolution of the eye measured in arc minutes (where 60 arc minutes = 1°) under standard illumination. (a) What is the visual acuity of a patient who can resolve two periods separated by 1.0 mm at a distance of 35 cm from the page? (b) Another patient has a visual acuity of 0.25. Calculate the farthest distance from which she could read 12-point lettering. Assume a needed resolution of one point to be able to discern letter shapes. (In 12-point type, a period has a diameter of about 0.35 mm.)

61. ⓧ ✪ **Resolution of the eye.** The angular resolution of the eye is determined by two factors, diffraction (the Rayleigh criterion) and the density of photoreceptor cells within the retina. Here we investigate which one is more important in limiting resolution. The center circular area of the retina, roughly 0.3 mm in diameter, is called the *fovea centralis* and is the most sensitive part. The fovea has photoreceptor cells spaced about 1 μm apart center to center, compared with a spacing of 3 μm outside this region. The diagrams in Figure P25.61 illustrate how the angular resolution is limited by receptor density. Notice that for two objects to be resolved, there must be an unstimulated cell between the image of the objects on the retina. In the upper diagram of Figure P25.61, the objects A and B are seen as one; in the lower diagram, there is an unstimulated receptor between, so the objects are seen as two distinct objects. The distance from the lens to the retina is about 16 mm. (a) What is the angular resolution of the eye due to the spacing of the photoreceptors in the fovea? (b) Under bright light conditions, the pupil is about 2.0 mm in diameter. What is the angular resolution of an eye for blue light (450 nm) for a pupil of this diameter? Assume n(eye) = 1.3. (c) What is the resolution for blue light under dim light conditions where the pupil's diameter is 6.0 mm? (d) Do bright light conditions lead to better resolution? Why or why not? (e) At what pupil size is the angular resolution due to diffraction less than that due to receptor spacing?

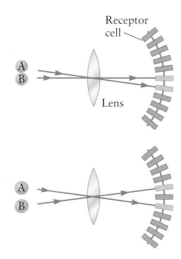

Figure P25.61 Resolution of two objects and receptor cell density.

62. ✪ **CD or DVD?** A simple experiment can tell the difference between a CD and a higher-data-density DVD. The tracks of pits on an optical data disc can act like a grating when laser light is reflected from its surface. Figure P25.62 shows a simple experimental configuration to produce an interference pattern from a data disc using a red laser pointer (λ = 650 nm). In this particular experiment, the point on the disc struck by the laser is positioned 10 cm above the tabletop. When the disc is oriented at a 45° angle with respect to the vertical, the reflected laser light shows an interference pattern where the first maximum is 4.5 cm beyond the reflected beam (directly below the disc). (a) What is the spacing of the tracks on the CD? (b) If the CD is replaced with a DVD, will the first maximum in the interference pattern move closer to or farther from the reflected beam? Explain. (c) The standard track spacing on a DVD is 0.74 μm. What is the distance from the reflected beam to the first interference maximum using this experimental configuration?

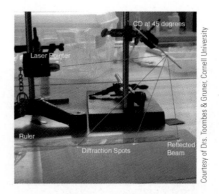

Figure P25.62

63. Ⓡ ☆ To make transistors and other circuit elements that are very small, modern integrated circuits (like those used in a computer) are produced using very high resolution optical patterning. The process uses a very expensive lens to produce the pattern. If the smallest circuit feature size is 200 nm, what does the Rayleigh criterion predict for the approximate wavelength of the light used to make the pattern? How does that compare with the wavelength of blue light?

Chapter 26

Applications of Optics

The Hubble Space Telescope is one of the most advanced optical instruments ever constructed. In this chapter, you'll find out how it works. (NASA)

It is difficult to imagine a world without cameras, microscopes, CDs and DVDs, and (especially) eyeglasses and contact lenses. These and many other devices are based on the principles of optics developed in the last two chapters. In this chapter, we discuss a few of these optics-based inventions and analyze how they work.

Eyeglasses were invented in the late 1200s and are perhaps the oldest optical instruments. Microscopes and telescopes were developed around 1600, CDs and DVDs were developed much more recently, and new optical devices are constantly being invented. In addition, there have been recent improvements to telescopes and microscopes. All these devices are discussed in this chapter.

26.1 | ⊗ APPLICATIONS OF A SINGLE LENS: CONTACT LENSES, EYEGLASSES, AND THE MAGNIFYING GLASS

In Section 24.6, we described how light from an object is refracted by the cornea and other regions in the eye to form an image on the retina. For simplicity, we can model the entire eye as a single lens with a focal length f_{eye}. Figure 26.1A shows a ray diagram for a "normal" eye; light emanating from a point on the object is focused to a corresponding point on the retina. The object is a distance s_o from the front of the eye, and the image is formed on the retina at an image distance $s_i = d_{eye}$, where d_{eye} is the diameter of the eye ($d_{eye} \approx 2.5$ cm in a typical case). To understand the eye, we need to know the value of the focal length f_{eye}. We can find it using the thin-lens equation (Eq. 24.21),

$$\frac{1}{s_o} + \frac{1}{s_i} = \frac{1}{f} \tag{26.1}$$

Thin-lens equation

Substituting $s_i = d_{eye}$ and $f = f_{eye}$ gives

$$\frac{1}{s_o} + \frac{1}{d_{eye}} = \frac{1}{f_{eye}} \tag{26.2}$$

and solving for the focal length of the eye leads to

$$f_{eye} = \frac{s_o d_{eye}}{s_o + d_{eye}} \tag{26.3}$$

The object distance s_o in Equation 26.3 varies depending on the location of the object. For a young adult, the eye can typically focus on objects as near as about 25 cm. This distance, called the **near-point** distance, is denoted as s_N. Objects closer to the eye than s_N cannot be focused on the retina; rays from such an object extrapolate to an intersection point behind the retina as shown in Figure 26.1B. This ray diagram corresponds to placing an object, such as this book, very close to your eyes; the page then appears "out of focus" on your retinas.

A typical young adult can also focus on objects that are very far away, that is, with $s_o \approx \infty$ (at "infinity"). Hence, the value of s_o in Equation 26.3 can vary from s_N to infinity. To satisfy Equation 26.3 at these two values of the object distance and at all values in between, the eye must adjust its focal length, which it does by using muscles that deform and change the shape of the eye's lens (Fig. 24.46).

Let's estimate the focal length of the eye when an object is at the typical near-point distance $s_N = s_o = 25$ cm. Inserting this value along with $d_{eye} \approx 2.5$ cm (as mentioned above) into Equation 26.3 gives

$$f_{eye, near} = \frac{s_N d_{eye}}{s_N + d_{eye}} = \frac{(25 \text{ cm})(2.5 \text{ cm})}{25 \text{ cm} + 2.5 \text{ cm}} = 2.3 \text{ cm} \tag{26.4}$$

On the other hand, for at object at infinity we have $s_o \approx \infty$, which when inserted into Equation 26.2[1] leads to

$$\frac{1}{f_{eye, distant}} = \frac{1}{s_o} + \frac{1}{d_{eye}} = \frac{1}{\infty} + \frac{1}{d_{eye}} = \frac{1}{d_{eye}}$$

$$f_{eye, distant} = d_{eye} = 2.5 \text{ cm} \tag{26.5}$$

Hence, as an object moves from the near point to a point very far away, the eye must adjust its focal length from $f_{eye} = 2.3$ cm to 2.5 cm, about a 10% change.

Eyeglasses and Contact Lenses: Adding a Lens in Front of the Eye

Figure 26.1 and our applications of the thin-lens equation in Equations 26.2 through 26.5 describe the optics of a "normal" eye. In many cases, such as when a person is nearsighted or farsighted, the eye is aided by eyeglasses or contact lenses, which are simply lenses placed in front of the eye. To understand how eyeglasses and contact lenses work, we must deal with optical "systems" that contain two or more lenses; here one of the lenses is the lens of the eye and the other is the lens in the eyeglass or

Near point of the eye

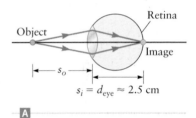

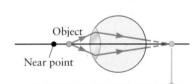

Figure 26.1 **A** When properly focused, the eye forms an image on the retina. **B** If an object is located closer than the near point, the eye is unable to produce an image at the retina. Instead, the refracted rays extrapolate to a point behind the retina.

---

[1] We could get the same result by inserting $s_o = \infty$ into Equation 26.3 and carefully canceling terms in the numerator and denominator.

Figure 26.2 Sign conventions for working with the thin-lens equation.

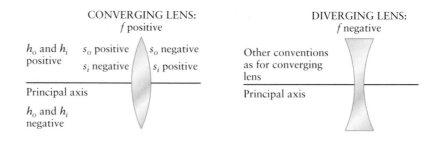

CONVERGING LENS:
f positive

| h_o and h_i positive | s_o positive | s_o negative |
| | s_i negative | s_i positive |

Principal axis

h_o and h_i negative

DIVERGING LENS:
f negative

Other conventions as for converging lens

Principal axis

RAY DIAGRAM FOR A FAR-SIGHTED PERSON

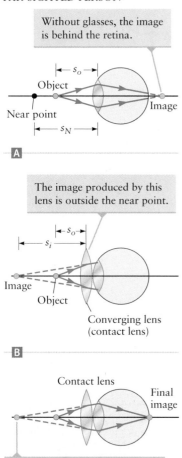

Without glasses, the image is behind the retina.

Object

Near point

s_o

s_N

Image

A

The image produced by this lens is outside the near point.

s_o

s_i

Image

Object

Converging lens (contact lens)

B

Contact lens

Final image

Image produced by contact lens is the object for the eye.

C

Figure 26.3 **A** For a farsighted person, the near point is farther away than is convenient for reading or other close work. Rays from an object within the near point focus behind the retina. **B** When a converging lens (such as a contact lens) is placed in front of the eye, a virtual image is produced by the lens. **C** The image produced by the contact lens acts as the object for the eye. If this image is at or outside the near-point distance, the eye can form an image on the retina.

contact lens. Learning how to deal with systems containing multiple lenses is very useful because telescopes, microscopes, and many other optical devices contain two or more lenses.

In Chapter 24, we learned how to use ray tracing along with the mirror equation and the thin-lens equation to analyze the images produced by curved mirrors and lenses. In a typical case, an object is placed in front of a lens or mirror and the location of the image is then calculated. For a system containing two lenses, the general approach is as follows.

> *Analyzing an optical system with two (or more) lenses*
> 1. **Draw a picture showing the object of interest and the lenses in the problem.**
> 2. **Use the rules for ray tracing along with the thin-lens equation from Chapter 24 to find the location and magnification of the image produced by the first lens in the system.**
> 3. **The *image produced by the first lens* then acts *as the object for the second lens* in the system.**
> 4. **Use ray tracing and the thin-lens equation a second time to find the image produced by the second lens in the system.**

We'll illustrate this general approach in a number of examples throughout this chapter. Most of these examples involve lenses, so Figure 26.2 shows our sign conventions (from Fig. 24.42) for applying the thin-lens equation to a single lens. A positive image distance s_i corresponds to an image on the right of the lens, whereas a negative value of s_i describes an image to the left of the lens (a "virtual image"). Notice also that a positive object distance s_o corresponds to an object to the left of the lens. In all examples in Chapter 24, the object distance was positive. Our sign conventions also allow for a negative value of s_o, which means that the object is to the right of the lens (a "virtual object"). We'll encounter such negative values of s_o in systems with two lenses, when the image produced the first lens is "beyond" the second lens.

Rather than dwell on the rules for analyzing systems of multiple lenses, we'll illustrate the key concepts by working through a series of examples.

Designing a Pair of Contact Lenses

As you age, the lenses in your eyes usually become less flexible and are not able to change focal length over the range found in Equations 26.4 and 26.5. The focal length is instead confined to a narrower range of values, reducing your ability to focus. In some cases, the near-point distance increases and is greater than the near-point distance for a normal eye; objects located closer than this near-point distance cannot be focused on the retina (Fig. 26.3A), and you must move things (such as a newspaper or this book) to a greater distance to see them clearly. You are then *farsighted*. To compensate, you can place a lens (a contact lens) immediately in front each eye. Let's now see how the required focal length of this lens depends on a person's near-point distance.

Suppose the near-point distance for a certain individual is $s_N = 75$ cm, which is much greater than the "normal" near-point distance of 25 cm. We wish to design a contact lens that enables this person to focus on objects at the normal near-point distance of 25 cm. The contact lens is placed just in front of the eye as in Figure 26.3B; it forms the first lens of our "system," and the eye acts as the second lens. We want this system of two lenses to be able to take an object at $s_o = 25$ cm from the contact lens and produce a final image on the retina. We follow our general rules for analyzing optical systems with two lenses and analyze the problem one lens at a time. The contact lens produces an image (Fig. 26.3B) that then acts as the object for the second "lens," the eye, but where should the image formed by the contact lens be located? We are given that the near-point distance for the eye is 75 cm, which means that the eye alone can focus an object that is 75 cm away to produce an image on the retina. So, if the image produced by the contact lens is at this distance (or greater, Fig. 26.3C), the eye can produce a final image at the retina. We thus want the contact lens to take an object at $s_o = 25$ cm and produce an image that is 75 cm to the left of the eye. Since the contact lens is touching the eye, the image distance is $s_i = -75$ cm; this image distance is negative because it is to the left of the contact lens (and is thus a virtual image; see the sign conventions in Fig. 26.2).

We now apply the thin-lens equation (Eq. 26.1) to the contact lens; we substitute $s_o = 25$ cm and $s_i = -75$ cm to find

$$\frac{1}{f_{\text{lens}}} = \frac{1}{s_o} + \frac{1}{s_i} = \frac{1}{+25 \text{ cm}} + \frac{1}{-75 \text{ cm}} \qquad (26.6)$$

and solve for f_{lens}:

$$f_{\text{lens}} = 38 \text{ cm} \qquad (26.7)$$

which is the focal length required for the contact lens. Since f_{lens} is positive, this lens is a converging lens.

If the person's near-point distance is larger than 75 cm, Equation 26.6 shows that a lens with a shorter focal length is needed. In words, we say that a person with a large near-point distance needs a "stronger" lens than a person with a short s_N. The strength of a lens is sometimes measured in terms of its **refractive power**, defined as

$$\text{refractive power} = \frac{1}{f_{\text{lens}}} \qquad (26.8)$$

The unit of refractive power is m^{-1}, which is called a **diopter**, and

$$1 \text{ diopter} = 1 \text{ m}^{-1}$$

The contact lens with the focal length in Equation 26.7 thus has a refractive power of $1/f_{\text{lens}} = 1/(0.37 \text{ m}) = 2.7$ diopters.

Let's now consider what type of contact lens is needed by a **nearsighted** person. Such a person is unable to focus light from distant objects on the retina as shown in the ray diagram in Figure 26.4A. The incoming rays from an object very far away are approximately parallel to the axis (an object at "infinity"); a nearsighted eye produces an image in front of the retina.

A nearsighted person is able to properly focus objects that are within a certain distance. Suppose a person can focus objects within 2.0 m on the retina. What kind of contact lens does this person need? We follow the approach we took for a farsighted person in Figure 26.3. We want an object that is very far away ($s_o = \infty$) to lead to a final image at the retina. It is also given that the eye alone can focus an object that is 2.0 m away on the retina. So, the contact lens must take an object at $s_o = \infty$ and give an image at $s_i = -2.0$ m (Fig. 26.4B).

Applying the thin-lens equation to the contact lens, we have $s_o = \infty$ and $s_i = -2.0$ m, and we get

$$\frac{1}{f_{\text{lens}}} = \frac{1}{s_o} + \frac{1}{s_i} = \frac{1}{\infty} + \frac{1}{-2.0 \text{ m}}$$

RAY DIAGRAM FOR A
NEAR-SIGHTED PERSON

Without glasses, the image is in front of the retina.

Distant object

$s_o = \infty$

Image

A

Virtual image produced by diverging lens

s_i

B

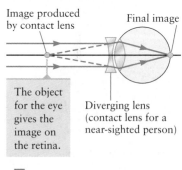

Image produced by contact lens

Final image

The object for the eye gives the image on the retina.

Diverging lens (contact lens for a near-sighted person)

C

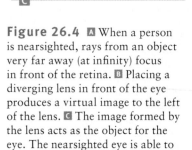

Figure 26.4 **A** When a person is nearsighted, rays from an object very far away (at infinity) focus in front of the retina. **B** Placing a diverging lens in front of the eye produces a virtual image to the left of the lens. **C** The image formed by the lens acts as the object for the eye. The nearsighted eye is able to focus this image on the retina.

Solving for f_{lens}, we have

$$f_{lens} = -2.0 \text{ m}$$

which is the focal length required for the contact lens. Since f_{lens} is negative, this lens is a diverging lens.

CONCEPT CHECK 26.1 | What Kind of Image Is It?

Figures 26.3 and 26.4 show ray diagrams for different types of contact lenses. Which of the following statements correctly describes the images formed by these contact lenses?
  (a) A contact lens always forms a real image.
  (b) A contact lens always forms a virtual image.
  (c) A contact lens can form a real or virtual image, depending on the type of lens.

CONCEPT CHECK 26.2 | ⊗

A farsighted person has a very long near-point distance. What should the focal length of this person's contact lenses be so that he can read a newspaper at a normal distance of (a) 25 cm, (b) 400 cm, or (c) 4.0 diopters? *Hint*: Assume the near-point distance is *extremely* long so that $s_N \approx \infty$.

Designing a Pair of Eyeglasses

The problem of selecting the proper lens for a set of eyeglasses is very similar to the contact lens problems considered in Figures 26.3 and 26.4. The main difference is that the eyeglass lens is a short distance in front of the eye, rather than touching it (Fig. 26.5). This distance must be accounted for when relating the image distance from the eyeglass lens and the object distance for the eye. The rest of the analysis is the same as with the contact lens problem and is illustrated in Example 26.1.

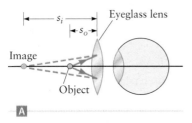

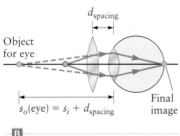

Figure 26.5 Image produced by the combination of an eyeglass lens plus the eye. Ⓐ The eyeglass lens produces an image that then acts as the object for the eye. Ⓑ When calculating the object distance for the eye, we must account for the distance $d_{spacing}$ between the eyeglass lens and the eye.

EXAMPLE 26.1 | ⊗ What Kind of Eyeglass Lenses Do You Need?

Consider a farsighted person with a near-point distance of 75 cm. In Figure 26.3 (and Eqs. 26.6 and 26.7), we designed a contact lens that would enable this person to read a book located at the "normal" near-point distance of 25 cm. The same person now wants a set of eyeglasses that he can use to read this book at 25 cm. What is the required focal length of the eyeglass lenses? Assume the glasses are positioned about $d_{spacing} = 2$ cm in front of the person's eye.

RECOGNIZE THE PRINCIPLE

With an object 25 cm in front of the eye, we want the eyeglass lens to produce an image 75 cm in front of the eye. This image then forms the object for the eye of our farsighted person, and we know that for this distance he will then be able to focus the final image on his retina.

SKETCH THE PROBLEM

The problem is described by the ray diagrams in Figure 26.5.

IDENTIFY THE RELATIONSHIPS

Since the eyeglass lens is $d_{spacing} = 2$ cm in front of the eye and the object is 25 cm in front of the eye, the object and image distances *for the eyeglass lens* are $s_o = 25 - d_{spacing} = 23$ cm and $s_i = -73$ cm (Fig. 26.5A). This image is a virtual image (s_i is negative) because light does not actually pass through this point.

Using the thin-lens equation with $s_o = 23$ cm and $s_i = -73$ cm, we get

$$\frac{1}{f_{\text{lens}}} = \frac{1}{s_o} + \frac{1}{s_i} = \frac{1}{23 \text{ cm}} + \frac{1}{-73 \text{ cm}}$$

Solving for f_{lens} gives

$$f_{\text{lens}} = \boxed{34 \text{ cm}}$$

Since f_{lens} is positive, this lens is a converging lens.

What does it mean?

Comparing this result for f_{lens} with the result in Equation 26.7 for a contact lens, we see that the eyeglass lenses have a focal length about 10% shorter. That is why your prescription for contact lenses is slightly different than for your eyeglasses.

The Magnifying Glass

The magnifying glass is one of the most basic of all optical devices and is a component in microscopes, telescopes, and other instruments. The simplest magnifying glass is a single lens, used to view things that are too small to examine with the naked eye (Fig. 26.6). To analyze this optical "instrument," we must again consider a system consisting of two lenses—the magnifying glass lens and the eye behind it—so the problem is similar to our analysis of contact lenses and eyeglasses. In those applications, the goal was to choose a lens that would allow the eye to produce an image at the retina. With a magnifying glass, the goal is to produce a greatly magnified image at the retina. We thus want the image on the retina to be as large as possible.

Consider first the case without a magnifying glass. To examine a small object, the natural thing to do is to place the object as close as possible to your eye. Figure 26.7A shows how the position of an object in front of the eye affects the size of its image on the retina. As the object is brought closer and closer to the eye from point *A* to *B*, the image on the retina grows larger and the apparent size of the object increases, but there is a limit to how close you can bring the object and still keep it in focus. The largest clearly focused image for the unaided eye results when the object is at the near point. The object's apparent size when it is located at the near point can be measured using the angle θ defined in Figure 26.7B; we'll call it the image angle.

To make the object even larger on the retina than in Figure 26.7B, we need to move it closer to the eye and hence inside the near point. The eye by itself cannot then produce an image on the retina, but it can do so with the help of the magnifying glass lens.

Figure 26.6 A magnifying glass produces an upright virtual image.

© Thinkstock Images/Jupiterimages

Image Properties with a Magnifying Glass

A magnifying glass is a single converging lens placed in front of the eye as shown in Figure 26.7C. The object is positioned inside the focal length of this lens, that is, so that the focal point *F* of the lens is to the left of the object in Figure 26.7C. We showed in Chapter 24 (Fig. 24.44) that this position leads to an upright virtual image at a point farther from the eye, on the far left in Figure 26.7C. As far as the eye is concerned, light from the object emanates from this virtual image, and if this image produced by the magnifying glass is at the near point, the eye is able to focus this light onto the retina. Comparing the ray diagrams in Figures 26.7B and C shows that the image angle with the magnifying glass θ_M (Fig. 26.7C) is greater than the image angle for the eye alone (θ in Fig. 26.7B), so the image *on the retina* is enlarged by the magnifying glass.

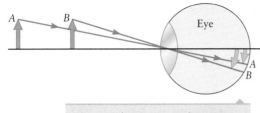

The size of an image on the retina determines the object's apparent size.

A

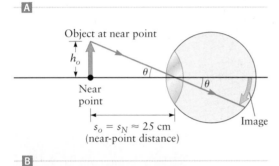

B

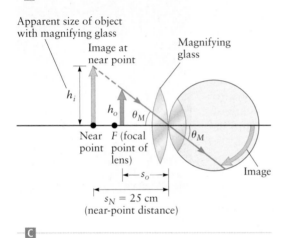

C

Figure 26.7 **A** The size of an object as perceived by the eye depends on the size of the image on the retina. If an object is brought closer to the eye (from A to B), the rays from its tip make a larger angle with the axis and the image on the retina becomes larger. **B** The closest object distance at which the eye can focus the image on the retina is the near point, which for a typical young person is approximately 25 cm from the eye. **C** A magnifying glass produces a magnified (enlarged) image at the near point of the eye. The eye focuses the light from this image on the retina.

The enlargement of the image on the retina is given by the *angular magnification* m_θ, defined by

$$m_\theta = \frac{\theta_M}{\theta} \tag{26.9}$$

The angular magnification is related to (but slightly different from) the linear magnification m defined in Chapter 24. The linear magnification involves the height of the object and image, whereas the angular magnification involves the angle the image makes on the retina.

We can calculate the angular magnification m_θ by applying the thin-lens equation (Eq. 26.1) to the lens of the magnifying glass (Fig. 26.7C):

$$\frac{1}{s_o} + \frac{1}{s_i} = \frac{1}{f} \tag{26.10}$$

As always, we must be careful to follow the proper sign conventions. Because the magnifying glass is a converging lens, its focal length f is positive. The object is on the left side of the lens, so the object distance s_o is positive; the image is also on the left side of the lens, so the image distance s_i is negative. For simplicity, let's assume the magnifying glass is very close to the eye; the image is at the near point and hence $s_i = -s_N$ in Equation 26.10. Inserting all this information leads to

$$\frac{1}{s_o} - \frac{1}{s_N} = \frac{1}{f}$$

$$\frac{1}{s_o} = \frac{1}{f} + \frac{1}{s_N} = \frac{s_N + f}{f s_N}$$

Solving for s_o, we get

$$s_o = \frac{s_N f}{s_N + f} \tag{26.11}$$

We can now find the image angle with and without the magnifying glass in place. From the geometry in parts B and C of Figure 26.7, we have

$$\tan \theta = \frac{h_o}{s_N} \quad \text{and} \quad \tan \theta_M = \frac{h_i}{s_N}$$

These angles are usually small, so we can use the approximations $\tan \theta \approx \theta$ and $\tan \theta_M \approx \theta_M$, leading to

$$\theta = \frac{h_o}{s_N} \quad \text{and} \quad \theta_M = \frac{h_i}{s_N} \tag{26.12}$$

Inserting these angles into the definition for the angular magnification (Eq. 26.9) gives

$$m_\theta = \frac{\theta_M}{\theta} = \frac{h_i}{h_o}$$

There are two right triangles in Figure 26.7C, one involving h_o and s_o and another involving h_i and s_N. These triangles are similar (having the same interior angles), so $h_i/s_N = h_o/s_o$. The angular magnification is thus

$$m_\theta = \frac{h_i}{h_o} = \frac{s_N}{s_o}$$

Inserting the result for s_o from Equation 26.11 gives our final result,

$$m_\theta = \frac{s_N}{s_o} = \frac{s_N}{(s_N f)/(s_N + f)}$$

$$m_\theta = \frac{s_N + f}{f} = \frac{s_N}{f} + 1 \tag{26.13}$$

Practical magnifying glass lenses have focal lengths that are smaller than the near-point distance by a factor of 10 or more, so the ratio s_N/f in Equation 26.13 is much greater than 1. Hence, to a good approximation we can write

$$m_\theta = \frac{s_N}{f} \qquad (26.14)$$

Magnification of a magnifying glass

For a typical magnifying glass, this result is accurate to 10% or better (and we can always use the result for m_θ in Eq. 26.13 if greater accuracy is needed). This expression for m_θ is also found when a magnifying glass is used as the eyepiece of a microscope or telescope, which we'll explore in Example 26.2.

The near-point distance s_N varies from person to person, so the magnification actually obtained with a particular magnifying glass (having a particular value of f) will depend on the person. Because the "typical" near-point distance for a young adult is 25 cm, this value is used as a convention when specifying magnifying glasses, so

$$m_\theta = \frac{s_N}{f} = \frac{25 \text{ cm}}{f}$$

For a magnifying glass specified by the manufacturer as providing $m_\theta = 10$, the focal length is thus

$$f = \frac{25 \text{ cm}}{m_\theta} = \frac{25 \text{ cm}}{10} = 2.5 \text{ cm}$$

The angular magnification of a typical magnifying glass or eyepiece is usually 10 or 20.

EXAMPLE 26.2 | **Angular Magnification for an Object at the Focal Point**

Figure 26.8A shows an object at the focal point of a magnifying glass. This arrangement is slightly different from the one we considered in Figure 26.7C, where the object was inside the focal point. Now the image from the magnifying glass is at infinity instead of at the near point of the eye (Fig. 26.7C). Show that the angular magnification in this case is given by $m_\theta = s_N/f$ (Eq. 26.14).

RECOGNIZE THE PRINCIPLE

To find the angular magnification, we must compare the angle θ that the object makes with the principal axis of the eye when the object is at the near point (without the magnifying glass; Fig. 26.7B), with the angle θ_f when the object is at the focal point of the magnifying glass (Fig. 26.8B). The magnification is the ratio of θ_f to θ. (Compare with Eq. 26.9.)

$$m_\theta = \frac{\theta_f}{\theta} \qquad (1)$$

SKETCH THE PROBLEM

Figure 26.8B shows a ray diagram. The object is at the focal point of a converging lens. From Chapter 24, we know that the lens will form an image at infinity ($s_i = \infty$); hence, the rays between the magnifying glass and the eye are parallel as shown in Figure 26.8B.

IDENTIFY THE RELATIONSHIPS AND SOLVE

To find θ_f, we consider the right triangle shaded in red in Figure 26.8B. We assume the magnifying glass is very close to the eye, so the red triangle has sides f and h_o and contains the angle θ_f. From the geometry of this triangle, $\tan \theta_f = h_o/f$. Using the small-angle approximation ($\tan \theta \approx \theta$) gives $\theta_f = h_o/f$. From Figure 26.7B, an object at

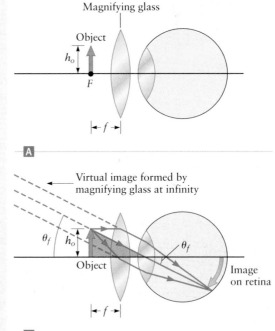

Figure 26.8 Example 26.2. **A** An object at the focal point of a magnifying glass. **B** The lens forms a virtual image an infinite distance (at "infinity") in front of the eye. The light from this virtual image forms a parallel bundle of rays that are then focused by the eye to form an image on the retina.

the near point viewed with the unaided eye has an angular size θ with $\tan \theta = h_o/s_N$. Using the small-angle approximation here leads to $\theta = h_o/s_N$. Inserting these results for θ_f and θ into the expression for m_θ in Equation (1), we get

$$m_\theta = \frac{\theta_f}{\theta} = \frac{h_o/f}{h_o/s_N} = \boxed{\frac{s_N}{f}}$$

which is Equation 26.14.

What does it mean?

As viewed by the eye, the object in Figure 26.8 appears to be at a location very far away (at "infinity") because the magnifying glass produces a collection of parallel rays that enter the eye. The magnifying glass is a component in many other optical instruments, including microscopes and telescopes, where it also produces a similar set of parallel rays that are viewed by an observer. This result for the magnification will help us analyze the role of the magnifying glass in those instruments.

CONCEPT CHECK 26.3

Does a magnifying glass form a real image or a virtual image? Explain.

26.2 | ✪ MICROSCOPES

A magnifying glass is very useful for viewing small objects, but there is a practical limit to the magnification that can be obtained with a single lens. Lenses with a focal length of less than a few millimeters are difficult to make, which leads to a limit on the angular magnification (Eq. 26.14) of around 100 or so for a typical near-point distance. Most practical magnifying glasses have angular magnifications of only 10 or 20. A more useful way to achieve higher magnification is by using two lenses arranged as a *compound microscope*. The image produced by one lens is used as the object for the second lens, and the image produced by the second lens is then viewed by the eye of an observer. The total magnification is the product of the magnifications of the two lenses; an angular magnification as high as $m_\theta = 1000$ is readily attained in this way.

Figure 26.9A shows a compound microscope and the usual vertical path followed by the light, and Figure 26.9B shows a ray diagram with the light traveling from left to right according to our usual convention. The two lenses in a compound microscope are called the *objective* and the *eyepiece*. The objective is a converging lens, so according to our sign conventions (Fig. 26.2) it has a positive focal length that we denote by f_{obj}. The eyepiece is simply a magnifying glass; it also has a positive focal length that we denote by $f_{eyepiece}$. To analyze the image produced by a compound microscope, we first apply ray tracing and the thin-lens equation to find the image produced by the objective lens. This image acts as the object for the eyepiece; the image produced by the eyepiece is then viewed by the eye.

The object to be studied is placed just outside (i.e., just to left) of the focal point of the objective lens (Fig. 26.9B). In Chapter 24, we showed that an object located outside the focal length of a converging lens gives an image that is real and inverted (Fig. 24.40C). The image produced by the objective in Figure 26.9B is therefore drawn a distance s_i to the right of the objective lens. The image produced by the objective lens forms the object for the eyepiece. The eyepiece then behaves in the same way as the magnifying glass discussed in Section 26.1. The distance between the objective lens and the original object is adjusted so that the image produced by the objective falls at the focal point of the eyepiece, which then gives a final virtual image for the observer. (Compare the rays produced by the eyepiece in Fig. 26.9B with the corresponding rays in Fig. 26.8B.)

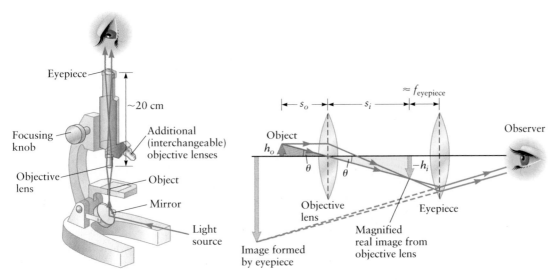

Figure 26.9 ◨ A compound microscope contains two lenses. The object of interest is placed close to the objective lens, and the observer's eye is close to the eyepiece. ◧ Ray diagram for a compound microscope. The object is placed just outside the focal point of the objective lens, producing a magnified real image. The image from the objective lens is then the object for the eyepiece, which produces a final virtual image much larger than the original object.

To find the total magnification of the microscope, we must first calculate s_i. Figure 26.9B shows two similar triangles, one shaded in red with sides h_o and s_o and another shaded in yellow with sides $-h_i$ and s_i. (The negative sign in front of h_i is due to our sign convention for lenses, which says an inverted image has a negative value for its height.) Both of these right triangles have an interior angle θ, so we have

$$\frac{h_o}{s_o} = \tan \theta = \frac{-h_i}{s_i}$$

The object is placed very close to the focal point of the objective; hence, $s_o \approx f_{\text{obj}}$. In Figure 26.9B, the image produced by the objective is larger than the original object. The ratio of the image and object heights is called the *linear magnification*, which we discussed in Chapter 24. For the objective lens in Figure 26.9B, we have a linear magnification $m_{\text{obj}} = h_i/h_o$. Putting it all together, we find

$$m_{\text{obj}} = \frac{h_i}{h_o} = -\frac{s_i}{f_{\text{obj}}} \tag{26.15}$$

The final virtual image produced by the eyepiece is a greatly magnified version of the original object. The total magnification of the microscope is the product of the linear magnification of the objective and the angular magnification of the eyepiece. If the eyepiece has a focal length f_{eyepiece}, its angular magnification according to Equation 26.14 is $m_{\theta, \text{eyepiece}} = s_N/f_{\text{eyepiece}}$, where $s_N = 25$ cm is the value (by convention) of the near-point distance. Combining this expression with Equation 26.15, we have

$$m_{\text{total}} = m_{\text{obj}} m_{\theta, \text{eyepiece}}$$

$$m_{\text{total}} = -\frac{s_i}{f_{\text{obj}}} \frac{s_N}{f_{\text{eyepiece}}} \tag{26.16}$$

Magnification of a compound microscope

Notice that m_{total} is negative, which means that the image is inverted.

Using a compound microscope is a good way to attain a large magnification. For a typical student microscope like the one in Figure 26.9A, the focal lengths might be $f_{\text{obj}} = 4$ mm and $f_{\text{eyepiece}} = 2.5$ cm (a so-called 10× eyepiece), and the distance between the objective and the eyepiece is about 20 cm. Inserting these values into Equation 26.16 gives

$$m_{total} = -\frac{s_i}{f_{obj}}\frac{s_N}{f_{eyepiece}} = -\frac{(0.20\text{ m})(25\text{ cm})}{(0.004\text{ m})(2.5\text{ cm})} = -500$$

which is quite a useful value.

Advances in Microscope Design

The first compound microscopes, which were invented around the time of Galileo, followed the simple design in Figure 26.9 with two lenses, but this design has the following problem. The index of refraction of the glass used to make lenses is slightly different for light of different colors, which, according to the lens maker's formula (Eq. 24.32), makes the focal length f slightly different for different colors and affects the focusing properties of a microscope. When light of a particular color is in focus, light of a different color will not quite be in focus. This chromatic aberration can be minimized by making the objective lens out of several separate lenses, composed of different types of glass with different indices of refraction. Clever optical design approximately cancels the aberrations from these individual lenses. Such lenses are called *achromatic* and are used in most modern high-quality microscopes.

Although most microscopes are thus not quite as simple as the one sketched in Figure 26.9A, the basic result for the total magnification in Equation 26.16 still applies, but now f_{obj} is the focal length of the objective lens system and $f_{eyepiece}$ is the focal length of the complete eyepiece. Hence, to obtain a high magnification, these two focal lengths should be as short as possible. This aspect of microscope design is explored in Example 26.3.

EXAMPLE 26.3 | ⊗ ℝ Designing a Compound Microscope

You are given the task of designing a compound microscope like the one in Figure 26.9A. The eyepiece is to have an angular magnification $m_{\theta,\text{ eyepiece}} = 10$, and you want the total magnification to be $m_{total} = -250$. To achieve this magnification, what should the focal length of the objective lens f_{obj} be?

RECOGNIZE THE PRINCIPLE

The total magnification of a compound microscope depends on several factors, including the focal lengths of the objective and eyepiece and the distance between the two lenses. We are not given this distance, but we can estimate a typical value from the information in Figure 26.9A. We can also find the focal length of the eyepiece from its magnification (which is given).

SKETCH THE PROBLEM

Figure 26.9 shows the geometry of a compound microscope.

IDENTIFY THE RELATIONSHIPS

The total magnification of a compound microscope is (Eq. 26.16)

$$m_{total} = -\frac{s_i}{f_{obj}}\frac{s_N}{f_{eyepiece}} \quad (1)$$

The desired magnification is given, and by convention the near-point distance is $s_N = 25$ cm. The angular magnification of the eyepiece is also given, and from Equation 26.14 we have $m_{\theta,\text{ eyepiece}} = s_N/f_{eyepiece}$. Hence, to use Equation (1) to find f_{obj} we need to estimate s_i in Figure 26.9B. This distance is nearly equal to the distance between the objective and the eyepiece (because s_i is usually much larger than the focal length of the eyepiece); from Figure 26.9A, we see that this distance is also the approximate length of the body of the microscope. From our experience with student microscopes (and Fig. 26.9A), we estimate $s_i = 20$ cm. Rearranging to solve for f_{obj} leads to

$$f_{obj} = -\frac{s_i}{m_{total}}\frac{s_N}{f_{eyepiece}}$$

Using $m_{\theta,\,eyepiece} = s_N/f_{eyepiece}$ this equation can also be written as

$$f_{obj} = -\frac{s_i}{m_{total}}m_{\theta,\,eyepiece} \qquad (2)$$

SOLVE

Inserting the values of the various quantities gives

$$f_{obj} = -\frac{s_i}{m_{total}}m_{\theta,\,eyepiece} = -\left(\frac{20\text{ cm}}{-250}\right)(10) = 0.80\text{ cm} = \boxed{8.0\text{ mm}}$$

What does it mean?

This value for f_{obj} and the general result in Equation (2) lead to some useful insights. If we want a higher total magnification, we need an objective with an even smaller focal length. Because the object must be placed a distance f_{obj} from the approximate *center* of the objective lens (see Fig. 26.9B), however, a practical limit arises. A real lens is at least several millimeters thick, so we cannot make f_{obj} smaller than that.

Insight 26.1

FOCUSING A COMPOUND MICROSCOPE

A typical compound microscope like the one in Figure 26.9A has several interchangeable objective lenses, allowing the user to obtain different magnifications. According to the result for the total magnification in Equation 26.16, if we wish to increase the magnification, we must make the focal length of the objective very short. The object is positioned just outside the focal point of the objective (Fig. 26.9B), and since f_{obj} is small, the object must be very close to the objective lens. For an objective lens with a magnification of -100, the distance between the object and the surface of the objective lens can be less than 1 mm. This distance is adjusted with the microscope's focusing knob.

Ultimate Resolution of a Microscope

Example 26.3 mentioned some practical limits on the total magnification of a compound microscope. Might a very clever physicist devise a different microscope design that can achieve much greater magnifications, or is there a fundamental limit to the magnification that any microscope can achieve? One can indeed use additional lenses, more powerful eyepieces, and so forth and make the magnitude of the total magnification much greater than the value of 500 to 1000 that is typical for compound microscopes. Even with such a high magnification, however, there is still a fundamental limit to the **resolution** that can be achieved with *any* microscope that relies on focusing.[2] This limit is due to the diffraction of light passing through the aperture of the microscope (i.e., through the lens openings). In Section 25.8, we described the Rayleigh criterion and showed that even when a perfect lens is used to focus light, diffraction prevents the size of the focused spot from being less than a value approximately equal to the wavelength of the light (Fig. 25.36 and Eq. 25.33).

The same diffraction effect limits the resolution achievable with a microscope. Suppose a microscope is used to observe two very closely spaced features on an object. Figure 26.10 shows light from these two features as it travels through the objective lens. The aperture of the objective leads to a slight spreading of the outgoing light rays from the two features, just the "inverse" of the spreading found when focusing light to a spot. An analysis similar to the one given in Section 25.8 shows that it is possible to resolve the outgoing light from the two features only if they are separated by a distance greater than approximately the wavelength λ. If they are closer, it is not possible to tell that there are two separate features. Hence, the ultimate optical resolution of a microscope is set by diffraction and is approximately equal to the wavelength of the light that is used. Applications requiring the best possible resolution use blue or even ultraviolet light because these colors have the shortest wavelengths compared with other colors of visible light.

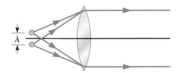

Objects are indistinguishable in the outgoing beam.

Figure 26.10 When two objects are close together, their diffraction spots overlap. If they are too close together (separated by about λ or less), they appear as a single object instead of two separate objects when viewed through a lens.

---

[2] A microscope that does not rely on focusing is not restricted by this limit; one such microscope is described in Section 26.7.

Figure 26.11 ◢ A confocal microscope is similar in design to the compound microscope in Figure 26.9, but with the addition of a small pinhole (aperture) in front of the observer. This pinhole selects light from a specific distance below the objective lens. By scanning the objective lens back and forth and also up and down, one can construct a three-dimensional image. ◩ Confocal light micrograph of the retina. The tissue has been fluorescently stained and the colors shown do not correlate with the colors of the rays in part A. Optic nerve fibers are shown in red and blood vessels are shown in blue. The region shown here is about 0.6 mm on a side.

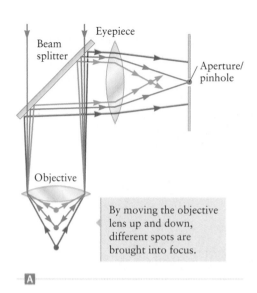

Beam splitter

Eyepiece

Aperture/pinhole

Objective

By moving the objective lens up and down, different spots are brought into focus.

A

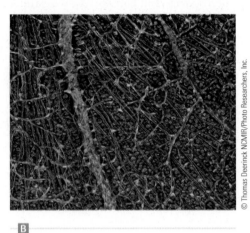

B

© Thomas Deerinck NCMIR/Photo Researchers, Inc.

The Confocal Microscope

Biologists examining samples of cells or tissue often want to resolve two nearby features according to their depth within a semitransparent object. Figure 26.11A shows a schematic diagram of a *confocal microscope*. This instrument is designed so that features at only one particular depth form the final image, which is accomplished by placing a very small opening, called a pinhole or aperture, in front of the observer. This pinhole allows mainly light from the selected depth to reach the observer and blocks most of the light from other depths. By combining images made at different depths, a confocal microscope yields a three-dimensional image of a sample. Confocal microscopes provide an extremely powerful way to examine features inside transparent objects such as a living cell (Fig. 26.11B). The depth resolution of a confocal microscope is again limited by diffraction effects. Separate features can be resolved only if their depths differ by more than approximately λ.

Although the optical principles of a confocal microscope have been understood since the time of Galileo, modern computers and lasers make confocal microscopes practical. The pinhole in front of the observer blocks much of the light from the object, so a very intense light source is needed. For this reason, most confocal microscopes use lasers, and lasers have been available for this kind of application only since the 1960s. Second, collecting many images at different depths requires that the objective lens be scanned side to side and also up and down in Figure 26.11A. Computers are essential for automating the scanning and for analyzing the large amount of image data needed for a three-dimensional image.

26.3 | TELESCOPES

When you observe two stars with your naked eyes or with a telescope, light rays extending from the stars to you are nearly parallel to one another. Figure 26.12 shows light rays emanating from two stars; here rays from one star (star 1) make a very small angle θ with the axis of a telescope, whereas light from star 2 travels parallel to the axis. The purpose of the telescope is to increase the angular separation of rays from these two stars so that your eye can distinguish one star from the other.

Refracting Telescopes

The simplest telescopes employ two lenses called an objective and an eyepiece as sketched in Figure 26.12. They are called *refracting* telescopes because they use light refracted by the objective lens. This type of telescope was invented around 1600 and then refined by several physicists, including Galileo who used it in his studies of the solar system. In a way, this arrangement of two lenses is similar to that used in a compound microscope (Fig. 26.9), but now the object to be studied is very far from the objective. The objective lens forms an image of a star or whatever is being studied. This image then acts as the object for a second lens (the eyepiece), which forms the final image viewed by an observer.

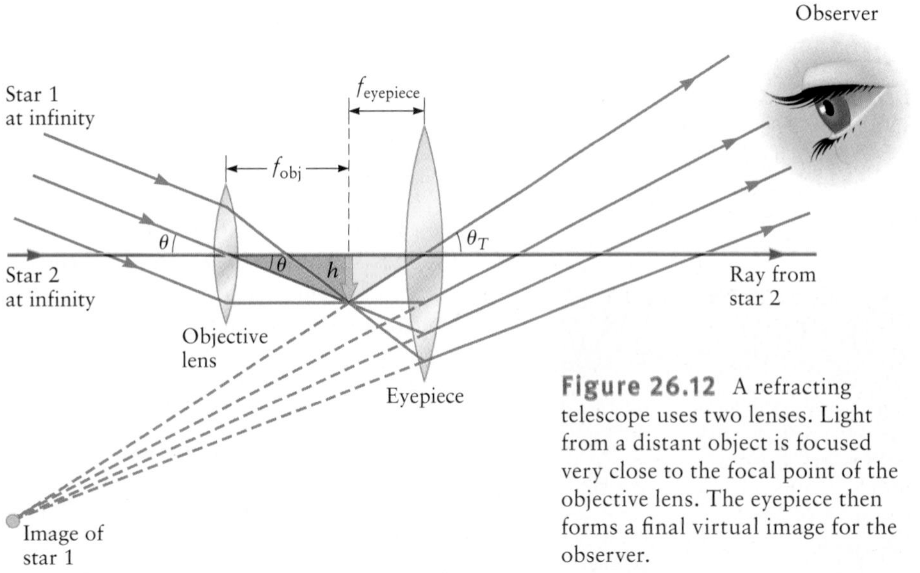

Let's consider light from star 1 as it moves through the telescope in Figure 26.12. This star is very far away (at "infinity") and the objective is a converging lens, so a real image is formed. For an object distance $s_o = \infty$, the thin-lens equation (Eq. 26.1) gives an image distance $s_i = f_{obj}$, where f_{obj} is the focal length of the objective lens. The rays from star 1 make a small angle θ with the axis, so the image of star 1 is a small distance h off the principal axis. The telescope's eyepiece is located such that the image formed by the objective is very close to the focal point of the eyepiece. The image from the objective is now the object for the eyepiece; since this new object is a distance $f_{eyepiece}$ from the eyepiece lens, the rays emanating from this object are refracted by the eyepiece to form a "bundle" of nearly parallel rays that are perceived by the observer.

The telescope's magnification is determined by the angles that rays from star 1 make with the axis; before reaching the telescope (on the far left in Fig. 26.12), this angle is θ; after passing through the eyepiece, it is θ_T. The total magnification of the telescope is the ratio of these two angles,

$$m_\theta = \frac{\theta_T}{\theta} \qquad (26.17)$$

This is actually an angular magnification because it involves the two angles, but it is usually called just the "magnification." A large magnification m_θ means that rays from the two stars in Figure 26.12 are separated by a large amount and hence an observer can more easily resolve that there are actually two separate stars present.

We can calculate the value of m_θ from Figure 26.12. The red right triangle in Figure 26.12 has sides of length h and f_{obj} and contains the angle θ; it leads to $\tan \theta = h/f_{obj}$. Using the small-angle approximation ($\tan \theta \approx \theta$), we get

$$\theta = \frac{h}{f_{obj}} \qquad (26.18)$$

The height h is also one side of the yellow right triangle in Figure 26.12 containing the angle θ_T and leads to

$$\tan \theta_T = \frac{h}{f_{eyepiece}}$$

The angle θ_T is greatly exaggerated in Figure 26.12. In practice, this angle is still small, so we can again use the small-angle approximation ($\tan \theta_T \approx \theta_T$), yielding

$$\tan \theta_T \approx \theta_T = \frac{h}{f_{eyepiece}} \qquad (26.19)$$

Figure 26.12 A refracting telescope uses two lenses. Light from a distant object is focused very close to the focal point of the objective lens. The eyepiece then forms a final virtual image for the observer.

Inserting these results for the angles θ and θ_T into Equation 26.17, we have

$$m_\theta = \frac{\theta_T}{\theta} = \frac{h/f_{\text{eyepiece}}}{h/f_{\text{obj}}}$$

$$m_\theta = \frac{f_{\text{obj}}}{f_{\text{eyepiece}}} \tag{26.20}$$

Magnification of a refracting telescope

Since one usually wants the magnification to be as large as possible, Equation 26.20 suggests that the focal length of the objective lens should be as large as possible. That is why long refracting telescopes are usually preferred, and such telescopes with lengths of 10 m or more were common in the 1700s and 1800s (Fig. 26.13).

EXAMPLE 26.4 | ® Finding the Magnification of a Refracting Telescope

The refracting telescope in Figure 26.13 has a total length of approximately 9.0 m. Suppose this telescope has an eyepiece with a magnification of 20. Estimate the total angular magnification of this telescope. *Hint*: The eyepiece is just a magnifying glass.

RECOGNIZE THE PRINCIPLE

The total magnification of a refracting telescope equals the ratio of the focal length of the objective lens f_{obj} to the focal length of the eyepiece f_{eyepiece} (Eq. 26.20),

$$m_\theta = \frac{f_{\text{obj}}}{f_{\text{eyepiece}}} \tag{1}$$

The eyepiece is a magnifying glass, and given its magnification we can find f_{eyepiece}. From our discussion of the refracting telescope, the sum of the two focal lengths is approximately equal to the telescope's total length, which gives us a way to find f_{obj}.

SKETCH THE PROBLEM

Figures 26.12 and 26.13 describe the problem.

IDENTIFY THE RELATIONSHIPS

The magnification of the eyepiece is given by (Eq. 26.14) $m_\theta = s_N/f_{\text{eyepiece}}$, where $s_N = 25$ cm is the normal value of the near-point distance. Rearranging to solve for f_{eyepiece} leads to

$$f_{\text{eyepiece}} = \frac{s_N}{m_\theta}$$

Inserting the given value $m_\theta = 20$ gives

$$f_{\text{eyepiece}} = \frac{s_N}{m_\theta} = \frac{25 \text{ cm}}{20} = 1.25 \text{ cm}$$

where we keep an extra significant figure to avoid problems with roundoff errors below. According to the ray diagram for a refracting telescope in Figure 26.12, the sum of f_{obj} and f_{eyepiece} is equal to distance between the lenses, and this distance is approximately the total length of the telescope, which is given as 9.0 m. We thus have $f_{\text{eyepiece}} + f_{\text{obj}} = 9.0$ m, which leads to

$$f_{\text{obj}} = (9.0 \text{ m}) - f_{\text{eyepiece}} = (9.0 \text{ m}) - (1.25 \text{ cm}) \approx 9.0 \text{ m}$$

SOLVE

Inserting our values of the focal lengths in Equation (1), the total magnification of the telescope is

$$m_\theta = \frac{f_{\text{obj}}}{f_{\text{eyepiece}}} = \frac{9.0 \text{ m}}{0.0125 \text{ m}} = \boxed{720}$$

Figure 26.13 The magnification of a refracting telescope is proportional to the focal length of the objective lens, which motivated people to construct some very long telescopes. This one, at Lick Observatory in California, is more than 9 m long.

© UC Regents/Lick Observatory

Reflecting Telescopes

The refracting telescope was an important experimental tool in the history of physics and astronomy. Galileo and others used it to observe the motion of planets and moons in the solar system, and their findings refuted the Earth-centered theory of the universe. Newton was also a serious astronomer, and during his work he noticed the chromatic aberration produced by a single objective lens (Section 24.8). This aberration led him to devise another telescope design, shown in Figure 26.14A; it is called a *reflecting telescope* because it replaces the objective lens in Figure 26.12 with a mirror. The particular telescope design in Figure 26.14A is called a Newtonian reflector.

Newton realized that, unlike lenses, a mirror would not have any chromatic aberration, and that is a big advantage for reflecting telescopes. Other advantages are even more important today. For instance, it is easier to make a large, high-quality mirror than an equally large lens of the same quality because the lens requires two carefully made surfaces and the glass must be absolutely transparent. Also, for a given diameter, a mirror is lighter and easier to support than a comparable lens, which is important because modern telescopes have mirrors that are 10 m in diameter or even larger. A lens of this size would, if supported by its edges, sag considerably under its own weight and be useless for a refracting telescope.

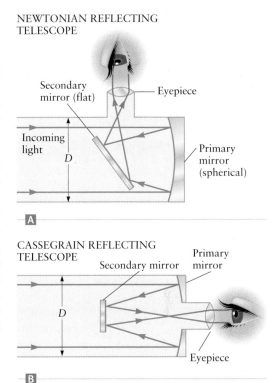

NEWTONIAN REFLECTING TELESCOPE

A

CASSEGRAIN REFLECTING TELESCOPE

B

Magnification of a Reflecting Telescope

Figure 26.14A shows how light travels through a Newtonian reflecting telescope. By itself, the concave mirror at the far right would form a real image of a distant object at a point very close to the focal point of the mirror (see Fig. 24.23B). A second mirror is positioned in front of the focal point of the concave mirror, reflecting the light to an eyepiece and on to an observer. The Newtonian reflector's magnifying elements are similar to those in the refracting telescope, so the total angular magnification is given by a result similar to Equation 26.20. If the concave mirror has a focal length f_M, the total magnification is

$$m_\theta = \frac{f_M}{f_{eyepiece}} \qquad (26.21)$$

Resolution of a Telescope

In addition to magnification, another important issue in the performance of a telescope is its resolution. Resolution determines how close together in angle two stars can be and yet still be seen as two separate stars (see Fig. 26.12). The resolution of a telescope is limited by two main factors: diffraction at the telescope's aperture (the "opening") and atmospheric turbulence. As in a compound microscope (Section 26.2), diffraction spreads the light rays when they enter a telescope. The size of a telescope aperture is normally equal to the diameter D of the objective lens or

Figure 26.14 A reflecting telescope uses a concave mirror (called the primary mirror) in place of the objective lens. **A** The design shown here was invented by Newton and also uses a flat mirror (called the secondary mirror). **B** A Cassegrain reflecting telescope uses a slightly different design.

Insight 26.2

THE CASSEGRAIN REFLECTING TELESCOPE

Nowadays, many reflecting telescopes (especially those used by hobbyists) follow the Cassegrain design (Fig. 26.14B). Light from a large concave mirror (the primary mirror) strikes a second mirror and is reflected along the axis and through a small hole in the center of the primary mirror. The light then travels through an eyepiece and to the observer. The Hubble Space Telescope is a Cassegrain reflecting telescope.

Figure 26.15 This reflecting telescope on Mount Palomar has a primary mirror with a diameter of 5.0 m.

Insight 26.3

SIZE AND WEIGHT OF THE HALE REFLECTING TELESCOPE

The Hale Observatory on Mount Palomar contains a typical reflecting telescope used in astronomical research (Fig. 26.15). The mirror of the Hale telescope has a diameter of 5.0 m and a thickness of about 0.5 m, with a total mass of approximately $m = 25{,}000$ kg. The mirror's weight is thus $mg \approx 2.5 \times 10^5$ N, which is the weight of about 25 average-sized cars! The mirror's weight is so great that the supporting structure must be specially designed to prevent the mirror from sagging or bending as the telescope is rotated to view different parts of the sky.

primary mirror, and the Rayleigh criterion for the limiting angular resolution set by diffraction is (Eq. 25.28)

$$\theta_{min} = \frac{1.22\lambda}{D} \qquad (26.22)$$

As an example, one of the largest reflecting telescopes in the world, with a mirror diameter $D = 10$ m, is at the Keck Observatory in Hawaii. Inserting this diameter in Equation 26.22 and assuming blue light ($\lambda = 400$ nm) gives

$$\theta_{min} = \frac{1.22\lambda}{D} = \frac{1.22(4 \times 10^{-7}\,\text{m})}{10\,\text{m}} = 5 \times 10^{-8}\,\text{rad} \qquad (26.23)$$

which is a very small angle. To get an idea of how small this angle θ_{min} is, let's convert it to degrees. For historical reasons, astronomers work in degrees, and fractions of a degree called arc minutes and arc seconds, with

$$1° = 60 \text{ arc minutes} = 60'$$

$$1 \text{ arc minute} = 60 \text{ arc seconds} = 60''$$

Converting from radians in Equation 26.23 to arc seconds, we find

$$\theta_{min} = 5 \times 10^{-8}\,\text{rad} \times \frac{360°}{2\pi\,\text{rad}} \times \frac{60'}{1°} \times \frac{60''}{1'} = 0.01'' \qquad (26.24)$$

Unfortunately, most telescopes do not attain the resolution limit θ_{min} set by diffraction. Starlight must pass through many kilometers of air (the atmosphere) before reaching an observer on the Earth, and turbulent motion of the air causes fluctuations in the refractive index from place to place. These fluctuations act like lenses and refract the incoming light from the star (Fig. 24.49). These "lenses" are constantly changing, so the direction of the starlight changes too, making the star "twinkle." Twinkling is a random process, causing light rays from a star to fluctuate through an angle that depends on the wind speed and other meteorological factors. For a location at the Earth's surface, the angular spread caused by these atmospheric fluctuations is typically

$$\Delta\theta \approx 1 \text{ arc second} = 1'' \qquad (26.25)$$

This is usually much larger than the resolution limit due to diffraction (Eq. 26.24), so most ground-based telescopes are limited more by atmospheric fluctuations than by diffraction.

The value of $\Delta\theta$ due to the atmosphere becomes smaller at high altitudes because there is less air between the Earth's surface at high altitudes and outer space, which is why most observatories are placed on mountaintops. Even at a good observatory, however, the resolution limit due to the atmosphere is rarely smaller than $\Delta\theta = 0.5''$. This value is still much larger than the typical diffraction limit, so, for example, the resolution of the 10-m Keck telescope considered in Equation 26.24 is limited by the atmosphere and not by diffraction at the telescope aperture. Why, then, did astronomers bother making such a large telescope? The answer is that the amount of light a telescope collects for an image depends on the telescope's size. Although the resolution of a 10-m telescope is not any better than the resolution of a 5-m telescope, the larger telescope collects light from a larger area and can thus observe fainter stars.

It is interesting to calculate the telescope diameter at which the diffraction limit for $\Delta\theta$ is equal to the atmospheric limit. Converting the atmospheric limit of about $\Delta\theta = 1$ arc second to radians gives

$$\Delta\theta = 1'' \times \frac{1°}{3600''} \times \frac{2\pi\,\text{rad}}{360°} = 5 \times 10^{-6}\,\text{rad}$$

Setting $\Delta\theta$ equal to the diffraction limit (Eq. 26.22) and solving for the diameter of the telescope, we find

$$\frac{1.22\lambda}{D} = 5 \times 10^{-6} \text{ rad}$$

$$D = \frac{1.22\lambda}{5 \times 10^{-6} \text{ rad}}$$

Assuming $\lambda = 400$ nm (blue light) gives a diameter of

$$D = \frac{1.22\lambda}{5 \times 10^{-6} \text{ rad}} = \frac{1.22(4 \times 10^{-7} \text{ m})}{5 \times 10^{-6} \text{ rad}}$$

$$D = 0.1 \text{ m} = 10 \text{ cm} \tag{26.26}$$

So, if you have a telescope with this diameter (or larger), your resolution is as good as *any* telescope located near sea level. Many amateur astronomers have telescopes that are at least this large.

This discussion shows that the resolution of a telescope located on the Earth's surface is severely limited by the atmosphere, which has prompted astronomers to move their telescopes to orbits around the Earth. Several such telescopes are now in operation and many more are in the planning stages, but the best known is the Hubble Space Telescope (see the photo at the beginning of this chapter). The Hubble telescope's resolution is limited by diffraction, as considered in Example 26.5.

EXAMPLE 26.5 | Resolution of the Hubble Space Telescope

The Hubble Space Telescope is a Cassegrain reflector (Fig. 26.14B), with a mirror diameter $D = 2.4$ m. How does its resolution compare with that of a ground-based telescope of the same size? Assume the observer is using blue light with $\lambda = 400$ nm.

RECOGNIZE THE PRINCIPLE

Because there is no atmosphere around the Hubble telescope, its resolution is limited only by diffraction and is given by Equation 26.22. Given the diameter of the Hubble's mirror, we can find its resolution limit. A ground-based telescope of the same size would be limited to about one arc second of resolution by the atmosphere (Eq. 26.25).

SKETCH THE PROBLEM

Figure 26.14B shows the design of a Cassegrain reflecting telescope. The resolution is determined by the diameter D of the primary mirror.

IDENTIFY THE RELATIONSHIPS AND SOLVE

The resolution limit of the Hubble telescope is given by $\theta_{min} = 1.22\lambda/D$ (Eq. 26.22). Inserting the given values $D = 2.4$ m and $\lambda = 400$ nm $= 4.0 \times 10^{-7}$ m, we get

$$\theta_{min} = \frac{1.22\lambda}{D} = \frac{1.22(4.0 \times 10^{-7} \text{ m})}{2.4 \text{ m}}$$

$$\theta_{min} = 2.0 \times 10^{-7} \text{rad} \times \frac{360°}{2\pi \text{ rad}} \times \frac{3600''}{1°} = 0.041''$$

This value is the Hubble telescope's resolution limit as set by diffraction. The atmospheric limit for a telescope on the Earth is about 1 arc second $= 1''$, so the Hubble telescope's resolution is better than ground-based resolution by about

$$\frac{1''}{0.041''} = \boxed{24 \text{ times}}$$

What does it mean?

The superior resolution of the Hubble Space Telescope means that it can resolve much more detail in photos of distant galaxies as illustrated in Figure 26.16,

Figure 26.16 Example 26.5. Because it is above the Earth's atmosphere, the Hubble Space Telescope has much better resolution than a ground-based telescope. These images are of the same regions of space; much more fine detail is visible in the Hubble telescope's images when compared to images obtained with the Subaru Telescope, an 8.2-m telescope in Hawaii.

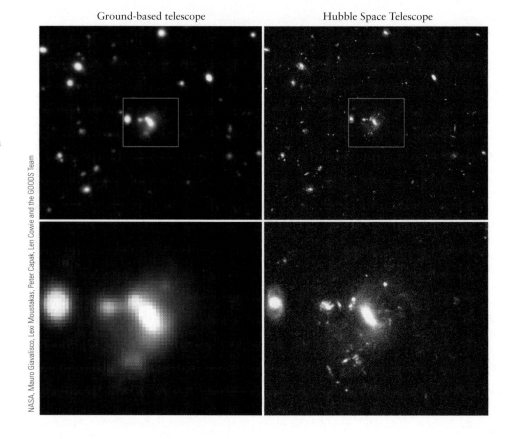

Ground-based telescope · Hubble Space Telescope

NASA, Mauro Giavalisco, Lexi Moustakas, Peter Capak, Len Cowie and the GOODS Team

which compares photos of a particular region of space taken using the Hubble telescope and a ground-based telescope. Telescopes in space thus have a natural advantage over conventional telescopes on the Earth's surface. We'll next see how an "unconventional" ground-based telescope can overcome the resolution limit caused by atmospheric fluctuations.

Adaptive Optics

The distortion of starlight by the atmosphere is due to the refraction of light by small pockets of the atmosphere that have a different refractive index than the neighboring air. This refraction is illustrated very schematically in Figure 26.17. Figure 26.17A shows a hypothetical case with no atmospheric distortion, and the light is focused perfectly by the telescope's mirror. Figure 26.17B shows a pocket of air acting as a lens, refracting some of the light rays in a different direction. Any focusing done by one lens can always be undone by a second lens, so a very clever astronomer could place a second lens in front of the telescope that would completely correct for distortion caused by the atmosphere (Fig. 26.17C). Atmospheric distortion changes as the winds and so forth change, however, so the corrective lens must change too. In practice, this corrective procedure is not done with a lens, but rather through very small changes in the shape of the mirror, made by pushing and pulling on the mirror's mounting platform (Fig. 26.17D). The mirror shape is adjusted many times each second to match fluctuations in the atmosphere by monitoring the image of a "reference star." The technology of building telescopes with adjustable mirrors to compensate for atmospheric distortion is called *adaptive optics*.

A "reference star" is an object known to appear as a point source at the telescope (so the object cannot be a planet or galaxy). If there were no distortion from the atmosphere, light from the reference star would focus to an image point limited only by diffraction. As the atmosphere causes the image of the reference star to be smeared out, the telescope mirror is adjusted to make this image as perfect as pos-

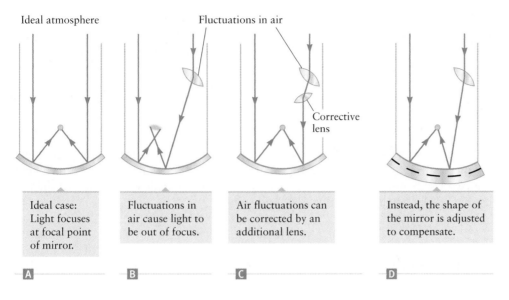

Ideal atmosphere   Fluctuations in air

Corrective lens

A	B	C	D
Ideal case: Light focuses at focal point of mirror.	Fluctuations in air cause light to be out of focus.	Air fluctuations can be corrected by an additional lens.	Instead, the shape of the mirror is adjusted to compensate.

Figure 26.17 Adaptive optics. **A** Under ideal conditions, a telescope focuses light from a star onto a spot that is limited only by diffraction. **B** Under real conditions, fluctuations in the atmosphere refract some of the incoming light, producing a larger focus "spot." **C** These fluctuations can in theory be corrected by using a lens to "undo" the refraction of atmosphere. Since the atmosphere fluctuates rapidly with time, the focal length of such a (hypothetical) lens must be rapidly adjusted to match the atmospheric fluctuations. **D** A practical approach is to adjust the shape of the mirror instead of adding an extra lens.

sible. Rapid and accurate control of the mirror shape requires modern computers. The technology of adaptive optics is fairly new and is still being improved, but in some cases it is currently possible to reduce atmospheric distortion by a factor of three or more.

26.4 | CAMERAS

Cameras are extremely common optical devices. In its simplest version (Fig. 26.18), a camera consists of a single lens positioned in front of a light-sensitive material. In an old-fashioned camera, this material is called *film*. The chemistry of photographic film is surprisingly complicated, but roughly speaking, the regions of film exposed to light change their chemical properties so as to record how much light energy has been absorbed and the color of this light. Digital cameras replace film with an electronic detector that also records information on the light intensity and color, which can then be transferred to a computer. In both types of camera, the lens forms an image on the light-sensitive detector, and an aperture (the shutter) is opened for a short time to allow sufficient light energy to enter.

The simplest cameras use lenses consisting of a single piece of glass, but many camera lenses contain multiple lens elements. In either case, their overall behavior is like a single converging lens, and a camera functions in many ways like the eye. However, the eye adjusts its focus by changing the shape of its lens, while the focus of a camera is adjusted by moving the lens relative to the film (or detector) as indicated in Figure 26.18.

Design and Operation of a Film Camera

The distance between the camera's lens and the film determines which objects are in focus and which are not. For a typical film camera, the film has dimensions of approximately 24×35 mm² and is known as a "35-mm camera." The standard lens for such a camera has a focal length of about $f = 40$ mm, but other lenses can be purchased with both larger and smaller values of f. To understand how these different focal lengths affect the resulting photographs, consider the ray diagram in Figure 26.19, which shows the image formed by an object that is very far from the camera. The thin-lens equation (Eq. 26.1) gives a relation between the object distance s_o, the image distance s_i, and the focal length f. We have

$$\frac{1}{s_o} + \frac{1}{s_i} = \frac{1}{f} \qquad (26.27)$$

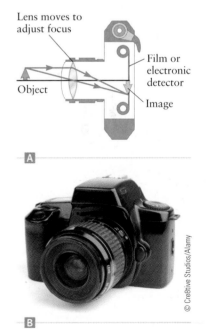

Lens moves to adjust focus

Film or electronic detector

Object

Image

A

B

© Cre8tive Studios/Alamy

Figure 26.18 **A** A camera is focused by moving the lens relative to the film or an electronic (CCD) detector. **B** Typical camera used by an amateur photographer.

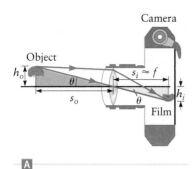

Camera

Object

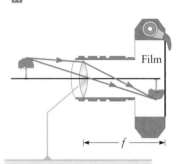

A telephoto lens has a large focal length f. This lens must be farther from the film to focus the image properly.

B

Figure 26.19 A The size of the image on the film depends on the focal length of the lens. B Increasing the focal length by using a telephoto lens produces a larger image. These drawing are not to scale. The object (the tree) is typically far from the camera.

The object (the tree) in Figure 26.19A is very far from the camera, so s_o is very large, much larger than both f and s_i. The term $1/s_o$ in Equation 26.27 is thus much smaller than the other terms, and to a very good approximation we have

$$\frac{1}{s_i} \approx \frac{1}{f}$$

and

$$s_i \approx f \qquad (26.28)$$

This result tells us that the lens should be positioned a distance of approximately f from the film. It also enables us to calculate the size of the image on the film. There are two right triangles in Figure 26.19A involving the angle θ. One side of the red triangle is the height of the object h_o and yields

$$\tan \theta = \frac{h_o}{s_o} \qquad (26.29)$$

A similar analysis of the right triangle containing the image distance s_i and the image height h_i (shaded in yellow in Fig. 26.19A) leads to

$$\tan \theta = \frac{-h_i}{s_i} \qquad (26.30)$$

where the negative sign comes from our sign convention for geometric optics (Fig. 26.2) that a downward-pointing image has a negative height. Inserting $s_i = f$ from Equation 26.28 and setting the results for $\tan \theta$ in Equations 26.29 and 26.30 equal gives

$$\frac{h_o}{s_o} = -\frac{h_i}{f}$$

The linear magnification of the image is the ratio of the image height to the object height, $m = h_i/h_o$, and we find

$$m = \frac{h_i}{h_o} = -\frac{f}{s_o} \qquad (26.31)$$

The magnification is negative since the image is inverted. The object distance s_o is much larger than the focal length, so the magnification is much less than one and the image of a tree or other large object can thus fit on a piece of 35-mm film.

This result for the magnification tells us when a lens with a large value of the focal length (i.e., a telephoto lens) would be handy. Telephoto lenses usually contain several component lenses, and the overall focal length can be adjusted by changing the spacing between these lenses. Suppose a normal lens with $f = 40$ mm is used in Figure 26.19A, while Figure 26.19B shows the corresponding diagram using a telephoto lens with $f = 200$ mm. With a longer focal length, the lens must be positioned farther from the film (because $s_i \approx f$, Eq. 26.28), giving a larger magnification for a given value of the object distance s_o (Eq. 26.31). A telephoto lens is thus convenient for photographs of distant objects.

CONCEPT CHECK 26.4 | The Image Produced by a Camera

Is the image formed by the camera in Figure 26.19 a real image or a virtual image? Explain why a camera could not work if it formed a virtual image.

EXAMPLE 26.6 Designing a Camera

You are asked to design a film camera using a lens with $f = 40$ mm and want to determine how much the lens will have to move to change the focus. Your camera must focus on objects very far away and also on objects that are as close as 1.00 m. Calculate how much the distance between the lens and the film must change to adjust the focusing from one case to the other.

RECOGNIZE THE PRINCIPLE

The camera's lens must form an image on the film for two different object distances, $s_o = \infty$ and $s_o = 1.00$ m. The focal length f is given, so we must find the image distances that correspond to these two object distances. We can do so using the thin-lens equation.

SKETCH THE PROBLEM

Figure 26.19A shows the problem.

IDENTIFY THE RELATIONSHIPS

We already analyzed this problem for the case of a very distant object and in that case found $s_i = f$ (Eq. 26.28). We thus only need to consider the case with $s_o = 1.00$ m and find the corresponding image distance. The thin-lens equation reads

$$\frac{1}{s_o} + \frac{1}{s_i} = \frac{1}{f}$$

Rearranging to solve for the image distance yields

$$\frac{1}{s_i} = \frac{1}{f} - \frac{1}{s_o}$$

and doing some arithmetic gives

$$\frac{1}{s_i} = \frac{1}{f} - \frac{1}{s_o} = \frac{s_o}{s_o f} - \frac{f}{s_o f} = \frac{s_o - f}{s_o f}$$

SOLVE

We can now solve for s_i:

$$s_i = \frac{s_o f}{s_o - f}$$

Inserting the given value of f and $s_o = 1.00$ m for the smallest object distance, we find

$$s_i = \frac{s_o f}{s_o - f} = \frac{(1.00 \text{ m})(0.040 \text{ m})}{1.00 \text{ m} - 0.040 \text{ m}} = 0.042 \text{ m} = 42 \text{ mm}$$

This is the image distance when the object is very close; it is also the distance between the lens and the film. When the object is very far away, we had $s_i = f = 40$ mm. So, we only need to move the lens $(42 - 40)$ mm = $\boxed{2 \text{ mm}}$ farther from the film to focus properly.

What does it mean?
The focusing knob or the auto-focus mechanism of a camera moves the lens relative to the film. It only has to move the lens a few millimeters.

Digital Cameras

The digital camera has developed rapidly in recent years and is now the camera of choice for most amateurs and many professionals. Digital cameras are made possible by a light-sensitive detector called a charge-coupled device (CCD), which replaces the film in a film camera. A CCD uses a type of capacitor to detect light and record its intensity. The optical system (lens and aperture) of a digital camera is basically the same as that of a film camera, but the cameras differ in a few important ways.

Detecting Light with a CCD: What Is a Pixel?

A CCD is fabricated in an integrated circuit chip; current CCD chips are about the size of your thumbnail (Fig. 26.20A). A CCD chip contains many capacitors

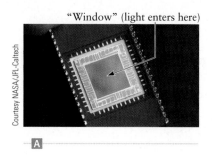

"Window" (light enters here)

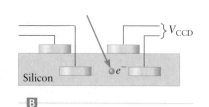

V_{CCD}

Silicon e^-

A   B   C

Figure 26.20 **A** A CCD integrated circuit chip. Light enters through the "window" on the chip. **B** The CCD chip contains many capacitors constructed on a piece of silicon. When light is absorbed, an electron is "released" from one of the covalent bonds in the silicon, leading to a charge (and hence a voltage) on the nearest capacitor. **C** The CCD chip contains an array of millions of these capacitors.

arranged in a grid over the surface of the chip as sketched in parts B and C of Figure 26.20. The structure of the capacitors in Figure 26.20B is somewhat simplified, showing only the two capacitor plates and a dielectric material in between. When light strikes the chip, it is absorbed in the dielectric layer and ejects some electrons from their normal chemical bonds. These electrons then move to the capacitor plates, leading to a voltage across the capacitor that is closest to where the light was absorbed. This voltage, denoted V_{CCD} in Figure 26.20B, is then detected by additional circuitry, and its value is stored in a computer memory in the camera. The magnitude of V_{CCD} depends on the light intensity because a greater intensity ejects more electrons and gives a larger voltage. The pattern of voltages on the capacitors in the grid (Fig. 26.20C) thus gives the light intensity as a function of position on the surface of the CCD, just as ordinary film records the intensity as a function of position on the surface of a piece of film.

The capacitors in Figure 26.20B do not distinguish the color of the light; a charge on the capacitors indicates that light was absorbed, but does not tell its color. One way to measure the color is to combine the information from four adjacent capacitors as sketched in Figure 26.21. A filter is placed in front of each of the capacitors; one of the filters allows red light to pass through to the CCD capacitor behind it, and other filters allow blue light and green light to pass. From the voltages on the four capacitors, the camera's computer can estimate the average color over this region. The CCD contains many such regions, called *pixels* (an abbreviation for "picture element"). The image produced by the CCD chip is thus stored by the camera as a set of intensity and color values for each pixel.

An important specification for a digital camera is the number of pixels in each photograph. A larger number of pixels indicates a finer level of detail in the photograph, so generally speaking, better cameras produce photos with larger numbers of pixels.

Optics of a Digital Camera

The optics and ray diagram for a digital camera are very similar to those for a film camera. Figure 26.19 thus still applies, but there is one important difference between the two types of cameras. The film for a 35-mm camera is (as noted above) approximately 24 × 35 mm². The CCD detector is much smaller, typically about 6 × 8 mm². Our derivation of the magnification for a film camera applies just as well to a digital camera, so we can again apply Equation 26.31:

$$m = \frac{h_i}{h_o} = -\frac{f}{s_o}$$

The CCD detector is smaller than a piece of 35-mm film; hence, the image height h_i must be smaller so that the image still fits on the detector. For a given value of the object distance, the focal length f of the lens must therefore be smaller for a digital

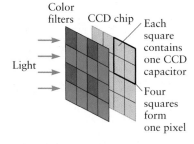

Color filters   CCD chip   Each square contains one CCD capacitor

Light

Four squares form one pixel

Figure 26.21 A CCD detector detects colors by using filters in front of the CCD capacitors. Each filter (red, green, or blue) lets only one color of light pass to the capacitor behind it, so each capacitor is sensitive to one particular color. The voltages from four adjacent capacitors are analyzed by the camera's computer to find the overall color of the light striking that region. These four adjacent capacitors form a pixel.

camera than for a film camera. The CCD detector is about four times smaller on each side, so the focal length is also about four times smaller.

In our analysis of the camera in Figure 26.19, we also showed that the distance s_i between the lens and the film is approximately equal to the focal length f (Eq. 26.28). Since it has a shorter focal length, the lens in a digital camera must be closer to the detector, which is why a digital camera can be much thinner and more compact than a conventional film camera (Fig. 26.22).

As with a film camera, the magnification of a digital camera can be changed by moving the lens relative to the CCD detector. (Compare again with Fig. 26.19.) For a digital camera, this magnification is called "optical zoom." It is also possible to use a "digital zoom" to enlarge an image. The digital zoom process constructs the entire photo using just the image data from near the center of the CCD grid. Hence, a photo taken with the digital zoom uses fewer pixels and has poorer resolution than a photo obtained with the digital zoom turned off.

Figure 26.22 Digital cameras can be very compact.

f-Number

Two important camera settings that apply to both film and digital cameras are the shutter speed and the *f-number*. The shutter "speed" is the amount of time the film is exposed to light from the object, and the *f*-number is associated with the camera's *aperture*, an opening that controls the open area of the lens (Fig. 26.23). A large aperture admits light from a large area of the lens, whereas a small aperture admits only a small amount of light (passing through the center of the lens) to the detector (the film or CCD). There are thus two ways to change the total amount of light energy that reaches the film: changing the shutter speed or changing the diameter of the aperture. The *f*-number is the ratio of the focal length to the aperture diameter D,

$$f\text{-number} = \frac{f}{D} \tag{26.32}$$

A larger aperture (a larger value of D) lets more light reach the film or CCD and gives a smaller *f*-number. A large *f*-number (small D) allows less light to get to the film or CCD. To have a properly exposed photograph (not too bright and not too dark), the total light energy must lie within a certain range. If you reduce the shutter speed and thus keep the shutter open longer, you must compensate by using a smaller aperture and hence a larger *f*-number.

> **CONCEPT CHECK 26.5 | Aperture and Shutter Speed for a Camera**
>
> A photographer is taking photos using an *f*-number of 4 and shutter speed setting that keeps the aperture open for $1/100 = 0.010$ s. The photographer then increases the *f*-number to 16. How long should the shutter now be open so as to have the same amount of light energy reach the detector, (a) the same amount of time, (b) 0.0025 s, (c) 0.04 s, or (d) 0.16 s?

Depth of Focus

Because of the trade-off between shutter speed and *f*-number, a photographer is often faced with the question of how to choose these two settings.[3] Using a short shutter speed reduces problems with vibrations when holding a camera and also makes it easier to photograph moving objects. To get enough light to the detector when using a short shutter speed, you need a small *f*-number (large aperture). The *f*-number, however, also affects a property called the *depth of focus* illustrated in Figure 26.24, which shows two photographs of the same scene using the same camera. Figure 26.24A was taken with a large *f*-number, and all the objects are fairly well focused. Figure 26.24B was taken with a smaller *f*-number, and only objects in the center of the scene are in focus, whereas those farthest from and closest to the

Large aperture/small *f*-number

Small aperture/large *f*-number

Figure 26.23 The adjustable opening in front of a camera lens is called the aperture.

---

[3]Some inexpensive cameras make this choice for you; others allow you to make choices such as "indoor" or "action" and then set the shutter speed and *f*-number accordingly.

Figure 26.24 Changing the aperture of a camera changes the *f*-number. **A** A larger *f*-number gives a larger depth of focus. Here objects from a range of distances in front and behind the best focus distance are all well focused on the film. **B** With a small *f*-number, only objects a particular distance from the camera are in focus. Objects farther away or closer than this distance are now far out of focus.

Figure 26.25 Ray diagrams for **A** a small aperture (large *f*-number) and **B** a large aperture (small *f*-number). With the small aperture the rays diverge more slowly from the central ray, making the depth of focus larger.

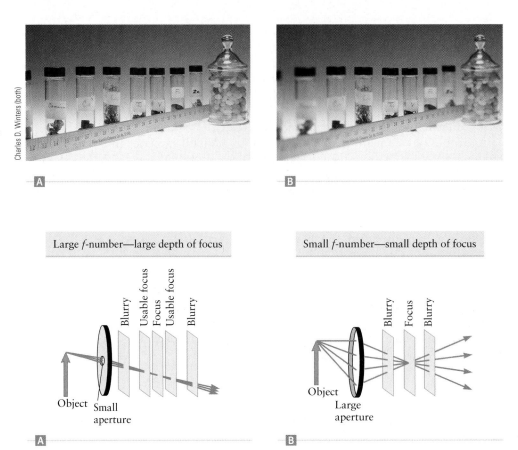

camera are badly out of focus. A photographer would say that the depth of focus is much larger in Figure 26.24A. Having a large depth of focus means that objects that are not at the best focusing point will produce images that are still close to ideal.

The *f*-number and depth of focus are related as shown by the ray diagrams in Figure 26.25. When the aperture is large (a small *f*-number; Fig. 26.25B), some rays make a large angle with the central ray, so they diverge more quickly as one moves away from the image point. With a small aperture (large *f*-number; Fig. 26.25A), the diverging angle is smaller, and the blurring of images away from the best focus is smaller than with a large aperture.

The Pinhole Camera

The ray diagrams in Figure 26.25 show that the "sharpness" of a photograph improves as the aperture is made smaller and smaller. In fact, if we make the aperture extremely tiny, we can make an extremely good image even without a lens! The *pinhole camera* uses this principle. This device, invented in the 1500s and sometimes called[4] the *camera obscura*, is sketched in Figure 26.26. Light from the object passes through the aperture (a tiny hole) in a screen and forms an image when it strikes a piece of film that lies behind the screen (most pinhole cameras use film rather than a CCD). Since the aperture is very small, the intensity at the film is very low and the film must be exposed to the image for a long time. Pinhole cameras are easy to make and are used mainly by hobbyists. They can also be used to view very intense light sources safely; for example, they are a safe way to view a solar eclipse and to search for sunspots (Fig. 26.27).

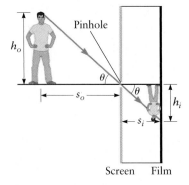

Figure 26.26 A pinhole camera forms an image without a lens. The lens of a conventional camera (Fig. 26.19) is replaced by a small pinhole. The magnification is determined by the ratio of the image distance s_i and the object distance s_o.

---

[4]Film and CCDs had not yet been invented at that time, so the first pinhole cameras were used to form images on a screen or wall behind the pinhole.

EXAMPLE 26.7 | Magnification of a Pinhole Camera

A pinhole camera can be made from a shoebox. Assume the distance from the pinhole to the film is $s_i = 20$ cm (about the size of a shoebox). If the object is $s_o = 20$ m from the camera, what is the magnification of the camera?

RECOGNIZE THE PRINCIPLE

There is no lens in a pinhole camera, so the object and image distances do not depend on any focal lengths and the thin-lens equation does not apply. We must rely on the basic definition of magnification as the ratio of the image height to the object height.

SKETCH THE PROBLEM

Figure 26.26 shows the problem and defines the image and object distances as well as the image and object heights.

IDENTIFY THE RELATIONSHIPS

Consider the two right triangles with interior angle θ in Figure 26.26, one on the object side of the pinhole with sides s_o and h_o, and the other on the image side of the pinhole with sides s_i and h_i. These triangles have the same interior angles, so the ratios of their corresponding sides are equal and we have

$$\frac{h_o}{s_o} = \frac{-h_i}{s_i}$$

The negative sign here comes from our sign convention that an inverted image is assigned a negative height (Fig. 26.2).

SOLVE

The magnification is $m = h_i / h_o$; hence,

$$m = \frac{h_i}{h_o} = -\frac{s_i}{s_o}$$

Inserting the given values for s_o and s_i gives

$$m = -\frac{s_i}{s_o} = -\frac{0.20 \text{ m}}{20 \text{ m}} = \boxed{-0.010}$$

What does it mean?
The negative value for the magnification shows that the image is inverted, as we knew from Figure 26.26. The image is also greatly reduced ($|m| \ll 1$).

Figure 26.27 A Landscape photograph taken with a pinhole camera. B Viewing the Sun during an eclipse through a pinhole.

26.5 | CDS AND DVDS

Most optical devices, including eyeglasses, microscopes, telescopes, and cameras, form images using lenses and mirrors and can therefore be analyzed using the principles of geometrical optics. Some of their performance limits, however, can only be understood by applying wave optics as in the Rayleigh criterion. We now consider several applications based on the wave nature of light that can only be understood using the ideas of wave optics from Chapter 25. The CD and DVD are applications of this type. CDs and DVDs operate through similar principles, so we'll focus mainly on the CD.

Figure 26.28 shows the structure of a CD; it is like a sandwich with the thickest layer composed of a plastic similar to Plexiglas. The bottom surface of the plastic in Figure 26.28A is smooth, but the top surface contains a pattern of pits used to

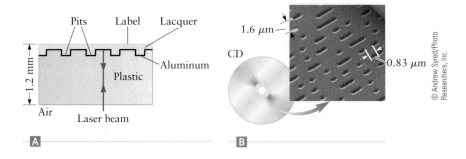

Figure 26.28 The information on a compact disc is encoded in a series of pits buried within the CD. **A** Light from a laser is incident from below and reflects from the aluminum layer. **B** The pits are arranged in a long spiral track. The presence or absence of a pit edge represents a binary 1 or 0.

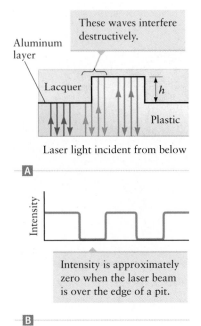

Figure 26.29 **A** Light from a laser comes in from below a CD; see Figure 26.28. *Note:* The colors of these rays do *not* represent the color of the light. Light waves reflected from the top and bottom of a pit edge interfere destructively, so there is no reflected intensity at the edge of a pit (where the interfering waves are denoted in green). Reflections from elsewhere on the CD (denoted by the red and blue rays) give a large intensity. **B** The reflected intensity thus encodes the presence or absence of a pit edge, representing a binary 1 or 0.

encode information in the CD. This top surface is coated with a thin layer of aluminum to make it reflecting and then covered with a protective layer of lacquer (a varnishlike coating), and the label of the CD is placed on top of the lacquer.

The pits are arranged in a long spiral "track" with sequential pits as close as 0.83 μm and neighboring paths on the spiral about 1.6 μm apart (Fig. 26.28B). Information encoded in the pits is read by reflecting a laser beam from the aluminum surface. Laser light passes in and out through the bottom surface of the plastic in Figure 26.28A, so that surface must be kept clean. The technology of focusing a laser beam onto a rapidly rotating CD is quite impressive. This beam must focus to a spot size smaller than a single pit and the CD player must control the spot's position on the spiral, all in the presence of vibrations and dirt. These challenging problems were solved by clever engineers. Here we consider just the optical problem of how the laser light actually detects the presence or absence of a pit.

The thin layer of aluminum in the CD acts as a mirror, reflecting the laser light. The pits influence this reflection through the thin-film interference effect illustrated in Figure 26.29. When the laser beam is positioned directly over the edge of a pit (the green rays in Fig. 26.29A), some of the light reflects from the bottom of the pit, and the rest reflects from the level outside the pit. These reflected beams are very similar to reflections from a soap film (Section 25.3). Destructive interference occurs if the difference in path lengths of the two reflected beams is $\lambda/2$, $3\lambda/2$, $5\lambda/2$, and so on.[5] The pit depth is designed to produce destructive interference, so there is *no* reflected light when the laser beam is over a pit edge, whereas the intensity is large when the laser beam is over the center of a pit or is outside a pit. As the laser beam travels along a track, the reflected light intensity varies between zero and a large value depending on the presence or absence of pit edges as sketched in Figure 26.29B. These high and low values of the intensity correspond to ones and zeros in a binary encoding of information on the CD.

EXAMPLE 26.8 | Designing a CD

The CD drive in your music player uses laser light with $\lambda_{air} = 780$ nm in air (in the infrared part of the spectrum). What is the smallest pit depth h in Figure 26.29 that will result in destructive interference for a laser beam positioned at the edge of a pit? The index of refraction of the plastic portion of the CD is $n_{plastic} = 1.55$.

RECOGNIZE THE PRINCIPLE

To have destructive interference with the smallest possible pit depth, the path lengths of the two reflected light beams must differ by

$$\Delta L = \frac{\lambda_{plastic}}{2}$$

---

[5]There is also a phase change of the light wave on reflection from the aluminum surface. This phase change occurs for both reflected light beams in Figure 26.29A and hence does not affect their *relative* phases.

where $\lambda_{plastic}$ is the wavelength of the light when it travels in the plastic. Recall that the wavelength of light in a substance (such as plastic) depends on the index of refraction of the substance (Eq. 25.9). The index of refraction of air is $n_{air} = 1.00$, so for light traveling within the CD we have

$$\lambda_{plastic} = \frac{\lambda_{air}}{n_{plastic}} \tag{1}$$

SKETCH THE PROBLEM

Figure 26.29 shows the problem. The two interfering waves are represented by the green rays near the pit edge on the left.

IDENTIFY THE RELATIONSHIPS

Using the given values of λ_{air} and $n_{plastic}$ in Equation (1) gives

$$\lambda_{plastic} = \frac{\lambda_{air}}{n_{plastic}} = \frac{780 \text{ nm}}{1.55} = 503 \text{ nm}$$

The path length difference is equal to *twice* the depth of a pit h (Fig. 26.29A), so to get destructive interference we must have

$$h = \frac{\Delta L}{2} = \frac{\lambda_{plastic}}{4}$$

SOLVE

Inserting our value for $\lambda_{plastic}$, we find

$$h = \frac{\lambda_{plastic}}{4} = \frac{503 \text{ nm}}{4} = 126 \text{ nm} = \boxed{1.26 \times 10^{-7} \text{ m}}$$

What does it mean?
The pit depth in a CD is very small, about five hundred times smaller than the thickness of a typical human hair. Covering the reflecting side of the pits with a layer of plastic protects them from scratches that would otherwise ruin the CD.

To store as much information as possible on a CD, the area of a pit must be as small as possible. Here again we encounter a limit set by wave optics. The minimum size of a focused spot of light is approximately equal to the wavelength (Eq. 25.33). CDs use the light from a laser with $\lambda_{air} = 780$ nm and $\lambda_{plastic} = 503$ nm (see Example 26.8), so the CD pits cannot be smaller than about $\lambda_{plastic}$ in width and length, as you can confirm from Figure 26.28B. DVDs can store information more densely than CDs because of several design changes. Most importantly, DVD players today have a green laser with $\lambda_{plastic} = 420$ nm. The shorter wavelength allows the focus spot to be smaller, which in turn allows the pits in a DVD to be more closely spaced than on a CD. New DVD designs have been developed that can store even more information by using multiple aluminum layers in different layers of pits that are probed with laser beams from both sides instead of in a single layer as in current DVDs. These new-style DVDs can store at least 30 times more information than a CD.

CONCEPT CHECK 26.6 | Designing a DVD

The depth of a pit in a CD or DVD is based on the wavelength of the light used to read the information. If a CD uses light with $\lambda_{plastic} = 503$ nm and a DVD uses $\lambda_{plastic} = 420$ nm, how does the depth of a pit in a DVD compare to the depth in a CD?

    (a) The pits in a DVD are deeper by a factor of 503/420 = 1.20.
    (b) The pits are the same depth.
    (c) The pits in a DVD are shallower by a factor of 503/420 = 1.20.

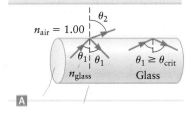

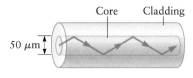

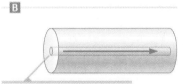

The diameter of the core of a modern optical fiber is comparable to λ.

Figure 26.30 Ⓐ Light traveling inside a glass fiber is reflected at the fiber's surface. The angle of incidence is always greater than the critical angle for total internal reflection, so the light does not leave the fiber. Ⓑ Real optical fibers consist of two (or more) types of glass with different indices of refraction. The light is confined to the core of the fiber. Ⓒ When the diameter of the core is comparable to the wavelength (as in current fibers), the light wave travels directly along the axis of the fiber.

26.6 | OPTICAL FIBERS

Optical fibers are flexible strands of glass that conduct and transmit light using total internal reflection. We described them briefly in Section 24.3, but provide more details here. Recall that light undergoes partial reflection and refraction when it encounters an interface between materials with different indices of refraction. Figure 26.30A shows a simplified model of an optical fiber. Although the fiber is cylindrical, we assume its diameter is large compared to the wavelength of light so that reflection of light at the glass–air interface is just like reflection from a flat interface. If a light ray traveling inside the fiber meets the sidewall of the fiber with an angle of incidence θ_1, there is a reflected ray with an angle of reflection equal to θ_1. If it exists, the refracted ray in the air travels at an angle θ_2. This angle of refraction is given by Snell's law (Eq. 24.7),

$$n_{\text{glass}} \sin \theta_1 = n_{\text{air}} \sin \theta_2 \qquad (26.33)$$

which can be arranged to get

$$\sin \theta_2 = \frac{n_{\text{glass}}}{n_{\text{air}}} \sin \theta_1 \qquad (26.34)$$

The value of n_{glass} depends on the composition of the glass, but is typically around $n_{\text{glass}} \approx 1.5$, while $n_{\text{air}} \approx 1.00$. Hence, the ratio $n_{\text{glass}}/n_{\text{air}}$ is greater than 1, so if $\sin \theta_1$ is large enough, the right-hand side of Equation 26.34 is greater than 1. Since the sine of an angle cannot be larger than 1, Snell's law does not have any solution for the angle of refraction θ_2 in this case. This is the phenomenon of total internal reflection described in Section 24.3, where we showed that there is no refracted ray when the angle of incidence is greater than a certain critical angle θ_{crit}. Whenever the angle of incidence is greater than the critical angle, all the light is reflected at the surface, which allows optical fibers to carry light over long distances without any leakage of light energy out of the fiber.

The optical fiber in Figure 26.30A is only a simplified design. All practical optical fibers consist of at least two different types of glass as sketched in Figure 26.30B. The central core is surrounded by an outer layer called the cladding. The core and the cladding are both made of glass, but with different compositions and different indices of refraction. The index of refraction of the cladding is smaller than the index of the core, $n_{\text{cladding}} < n_{\text{core}}$, enabling total internal reflection of light within the core. In an application, light from a laser enters the fiber at one end, and the entry angle is arranged so that the light always undergoes total internal reflection at the boundary between the core and the cladding. Signals (such as a telephone signal) are carried from one end of the fiber to the other by pulses of laser light within the core. These fibers can be very long (many kilometers), and a single fiber can carry many more simultaneous signals than is possible with a conventional metal wire.

Some early optical fibers were designed as in Figure 26.30B, and although this approach works, it has the following flaw. The laser light carrying the signal consists of pulses that encode information in a binary fashion (ones and zeros, corresponding to "on" and "off" values of the light intensity). There is a certain delay between the time a pulse enters the fiber and the time it exits at the other end. For the optical fiber in Figure 26.30B, a pulse is carried by a collection of light waves that reflect at different angles as they move along the fiber. Since they propagate at different angles, some waves travel a greater total path length than others; all travel at the same speed, so some reach the end of the fiber before others, causing the pulses to be smeared out in time, thus limiting the amount of information the fiber can carry.[6]

---

[6]Shorter pulses transmit information at a higher rate because more of them can be sent in a given amount of time, but the time between pulses must be longer than the smearing in time to avoid losing information; hence, pulse smearing limits the transmission rate.

This problem is overcome using the design sketched in Figure 26.30C. This fiber is similar to the one in Figure 26.30B except that the diameter of the core is much smaller. In fact, the core diameter is so small that ray optics can no longer be used to describe how light travels along the fiber. An analysis using wave optics shows that a light wave travels directly along the axis of the core, with no side-to-side reflections as for the larger core in Figure 26.30B. Fibers with very small diameter cores are called "single-mode" fibers because there is only one way for light to propagate in the fiber. These fibers greatly reduce the smearing of light pulses, allowing the fibers to carry more information than the "multimode" fiber in Figure 26.30B.

CONCEPT CHECK 26.7 | **What Happens When You Bend an Optical Fiber?**

The simplest optical fibers consist of a single strand of glass. If such a fiber is straight, the angle of incidence θ_i for a ray inside the glass is large enough that there is total internal reflection and no light leaves the fiber (Fig. 26.31A). However, if the fiber is bent through a large enough angle, the angle of incidence can become small enough that some light does "escape" from the fiber (Fig. 26.31B). If, starting from a straight fiber, the bending angle of the fiber is gradually increased, which color of light escapes first?
  (a) The longest wavelengths (red light) escape first.
  (b) The shortest wavelengths (blue and violet light) escape first.
  (c) All wavelengths escape at the same time.

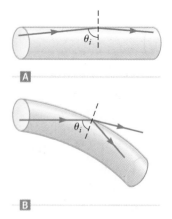

Figure 26.31 Concept Check 26.7.

EXAMPLE 26.9 | Analysis of an Optical Fiber

Consider an optical fiber like the one in Figure 26.30B, with length $L = 10$ km and core diameter $d = 40$ μm. Suppose the index of refraction of core and cladding are $n_{core} = 1.80$ and $n_{cladding} = 1.40$. Assuming light travels along the fiber as rays as sketched in Figure 26.32, estimate (a) the minimum and (b) the maximum travel times of light rays along the fiber.

RECOGNIZE THE PRINCIPLE

The minimum travel time will be for light rays parallel to the axis of the fiber, which can be calculated from the length of the fiber and the speed of light. The maximum travel time occurs for rays at the critical angle for total internal reflection (Fig. 26.32), and we can find that angle using Snell's law. Using geometry, we can compare the minimum and maximum travel distances and therefore the minimum and maximum travel times.

SKETCH THE PROBLEM

Figure 26.32 shows the problem. One ray (shown in blue) lies along the axis of the fiber, and the other ray (shown in green) makes an angle θ_{crit} (the critical angle) with the direction normal to the wall of the fiber.

IDENTIFY THE RELATIONSHIPS AND SOLVE

(a) We first find the minimum travel time. The speed of light in a vacuum is $c = 3.00 \times 10^8$ m/s, but in the core this speed is reduced to $v_{core} = c/n_{core}$. The time for a light ray to travel the length of the fiber L is

$$t_{min} = \frac{L}{v_{core}} = \frac{L}{c/n_{core}}$$

Inserting the given values, we find

$$t_{min} = \frac{L}{c/n_{core}} = \frac{10 \times 10^3 \text{ m}}{(3.00 \times 10^8 \text{ m/s})/1.80} = 6.0 \times 10^{-5} \text{ s} = \boxed{60 \ \mu s} \qquad (1)$$

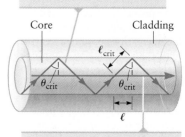

Light on this path travels the maximum distance and time.

Core   Cladding

ℓ_{crit}

θ_{crit} θ_{crit}

ℓ

Light traveling straight through the fiber travels the minimum distance and time.

Figure 26.32 Example 26.9.

(b) To find the maximum travel time, we first find the maximum travel distance, which depends on the critical angle θ_{crit} (Fig. 26.32). The critical angle for reflection at the interface between the core and the cladding can be found by applying Snell's law (Eq. 26.33) to the green ray in Figure 26.32. At the critical angle, we have an angle of refraction $\theta_2 = 90°$ in the cladding. We get

$$n_{cladding} \sin \theta_2 = n_{core} \sin \theta_{crit}$$

Inserting our value of θ_2 and the indices of refraction, we can solve for θ_{crit}:

$$n_{cladding} \sin \theta_2 = n_{cladding}(1) = n_{core} \sin \theta_{crit}$$

$$\sin \theta_{crit} = \frac{n_{cladding}}{n_{core}} = \frac{1.40}{1.80} = 0.78$$

$$\theta_{crit} = 0.89 \text{ rad} = 51°$$

The path length traveled along a ray that makes an angle θ_{crit} with the normal to the interface is longer than the path parallel to the axis of the fiber. The critical-angle path in Figure 26.32 has a length ℓ_{crit}, whereas the path along the axis has a length ℓ. From the geometry in Figure 26.32,

$$\ell_{crit} = \frac{\ell}{\sin \theta_{crit}}$$

The corresponding travel time for light along this path is thus longer than the minimum (direct) travel time by a factor

$$\frac{\ell_{crit}}{\ell} = \frac{1}{\sin \theta_{crit}}$$

Light traveling along the axis takes a time t_{min} as found in Equation (1), so light at the critical angle takes a longer time:

$$t_{max} = t_{min}\left(\frac{1}{\sin \theta_{crit}}\right)$$

Using our results above for t_{min} and $\sin \theta_{crit}$, we find

$$t_{max} = \frac{t_{min}}{\sin \theta_{crit}} = \frac{60 \ \mu s}{0.78} = 7.7 \times 10^{-5} \text{ s} = \boxed{77 \ \mu s}$$

The difference in travel times is thus

$$\Delta t = t_{max} - t_{min} = 77 \ \mu s - 60 \ \mu s = \boxed{17 \ \mu s}$$

What does it mean?
The value of Δt is nearly one-third of the minimum travel time, so the laser pulses will be quite smeared out. That is why single-mode fibers were invented.

26.7 | ⊗ MICROSCOPY WITH OPTICAL FIBERS

The wavelength of light sets a limit on the resolution of a compound microscope because light cannot be focused to a spot size smaller than approximately λ. However, light can still be used to perceive details smaller than λ using an approach called ***near-field scanning optical microscopy*** (NSOM). The NSOM technique uses an optical fiber to illuminate a very tiny region at the end of the fiber. The light is most intense at the opening of the tip, which can be much smaller than the wavelength. The tip is positioned very close to the object to be studied and is then is scanned over one of its surfaces (Fig. 26.33A). Because the tip is so close to the object, only a very small area near the tip is strongly illuminated; the resolution is then determined by the spacing from the surface and the tip diameter, both of which can be much smaller than the wavelength. By measuring the scattered or reflected

Core — Cladding

A

400 nm

B

Figure 26.33 **A** The diameter of an optical fiber tip can be much smaller than the wavelength of light, producing a very tiny region just outside the tip where the light intensity is high. The tiny region at the fiber's tip is used in a near-field optical microscope by scanning it across the surface of the object of interest. **B** Image of a specially patterned layer of gold on the surface of silicon taken with a near-field scanning optical microscope. The scale bar at the lower left is 400 nm, the approximate wavelength of blue light. The small features resolved in this image are thus much smaller than the wavelength of visible light.

light, one can construct an image of the surface based on the scattered intensity as a function of the tip's position.

Current NSOM fibers have openings of 25 nm or even smaller, much less than the wavelength of visible light. (Blue light has a wavelength of about 400 nm.) The resolution of the NSOM microscope is constantly improving, and the method is now being used to create images of objects as small as individual molecules! Conventional microscopes are limited by the ability of a lens to focus, so their resolution is set by the Rayleigh criterion (Fig. 26.10) and is approximately equal to the wavelength of the light that is used. An NSOM microscope can thus resolve much smaller objects and features than is possible with a conventional microscope.

SUMMARY | Chapter 26

KEY CONCEPTS AND PRINCIPLES

Optics and the eye

The eye is an essential part of many optical instruments, including magnifying glasses, microscopes, and telescopes. The purpose of these devices is to produce an image on the retina.

A *magnifying glass*, also called an *eyepiece*, is a converging lens that produces an upright virtual image when an object is placed near its focal point. This image appears at the *near point* of the eye, which is approximately $s_N = 25$ cm from the front of the eye.

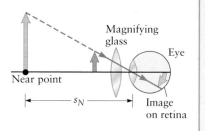

Diffraction and the resolution of microscopes and telescopes

The ultimate resolution of compound microscopes and telescopes is determined by diffraction. A compound microscope can resolve two nearby objects only when they are separated by a distance greater than approximately the wavelength of light.

APPLICATIONS

Optical instruments

Microscopes and refracting *telescopes* employ (at a minimum) an *objective* lens together with an *eyepiece*. In a reflecting telescope, the objective lens is replaced by a mirror; this design has some practical advantages when the telescope is very large.

A *camera* produces an image on a light-sensitive detector, either film or a CCD. Focusing of a camera is done by moving the lens relative to the detector.

Although microscopes, telescopes, and cameras were all invented long ago, optical technology is still improving. CDs, adaptive optics, optical fibers, and near-field scanning optical microscopes have all been developed recently (during the author's lifetime!), and all can be understood using the principles of geometrical and wave optics.

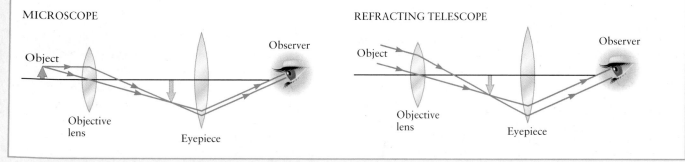

1. Explain why a magnifying glass uses a convex (converging) lens.

2. A magnifying glass uses a lens with a focal length of magnitude $|f| = 300$ cm. Is f positive or negative?

3. Will a particular magnifying glass produce the same magnification for any user? Why or why not? If a magnifying glass is marked 5×, what does that imply?

4. In Section 26.1, we showed that the angular magnification of a magnifying glass is equal to the ratio of the near-point distance to the focal length of the lens ($m_\theta = s_N/f$, Eq. 26.14). Hence, by making the focal length sufficiently short, a very large magnification can be achieved in principle. However, other practical considerations limit the maximum useable value of m_θ. Explain why it is hard to find magnifying glasses with magnifications greater than about 20.

5. ⊗ Your friend has eyeglasses that make his eyes appear larger than they really are. Is your friend nearsighted or farsighted? Explain.

6. ⊗ If you are nearsighted, should your eyeglasses contain converging lenses or diverging lenses? What if you are farsighted?

7. ⊗ Bifocals are eyeglasses that contain two different lenses for each eye (Fig. Q26.7). Each of these lenses is made from one piece of glass. One lens (lens 1) is used when looking at things far away, and the other (lens 2) is used for reading. The author uses bifocals because, without glasses, his eyes can only focus correctly on objects that are about 1 m away. (a) Is lens 1 a converging lens or a diverging lens? (b) Is lens 2 a converging lens or a diverging lens?

Figure Q26.7

8. ⊗ Surgery can be used to correct the vision of a person who is nearsighted. Should the surgeon increase or decrease the curvature of the cornea? What should the surgeon do if the person is farsighted?

9. SSM For typical cameras, the f-number can be selected from a set of fixed values such as 2, 2.8, 4, 5.6, 8, 11, and 16. Explain why this combination of f-numbers is used. *Hint*: The diameter of the aperture is inversely proportional to the f-number (Eq. 26.32), while the total intensity of the light that reaches the film is proportional to the area of the aperture.

10. ⊗ You want to increase the magnification of a particular compound microscope. Which of the following changes would achieve that goal?
(a) Increase the focal length of the objective lens.
(b) Decrease the focal length of the objective lens.
(c) Increase the focal length of the eyepiece.
(d) Decrease the focal length of the eyepiece.
(e) Increase the spacing between the objective lens and the eyepiece.

11. Is the final image produced by a refracting telescope real or virtual? Explain.

12. Many of the refracting telescopes designed and built in the 18th and 19th centuries were extremely long (Fig. 26.13). Explain why they were made in this way; that is, what desirable property of the telescope becomes enhanced when it is made very long? What are the disadvantages of this design approach?

13. What kind of telescope did Galileo build and use in his studies? Do some investigating and try to find answers to the following questions. (See also Problem 30.)
(a) How long was his telescope?
(b) What was the total magnification?
(c) What was the focal length of the objective lens?
(d) What was the magnification of the eyepiece?

14. ⊗ Will the image produced by the contact lenses designed in Section 26.1 (and Eq. 26.7) be upright or inverted? What about the image produced by glasses for a nearsighted person? Explain.

15. ⊗ When an optometrist determines your eyeglass prescription, she always places a set of "test" lenses in front of your eyes as if they were your real glasses. The optometrist is interested in learning what lenses work best for you, but she is also measuring several other properties of your eyes. What are they?

16. ⊗ Why does squinting often allow you to read at a distance? For example, even if you forget to wear corrective lenses, forming small slits by squinting your eyes often allows you to read a clock across a room or a sign across a street. Why does squinting help?

17. Some recent water fountain landscape installations include arcs of water exhibiting laminar flow like those shown in Figure Q26.17. The arc of water is illuminated at the base, and the light appears to follow along with the flow of the stream into the pond. Why does the light stay mostly within the arcing stream of water?

Figure Q26.17 Question 17 and Problem 58.

18. The largest telescopes are all the reflector variety. Describe and explain two main design considerations in the construction of large-diameter telescopes that favor a reflector over a refractor.

19. Consider two telescopes, one a reflector and the other a refractor. Assume both telescopes are of equal diameter, are of equal magnification, and have the same focal length. In what ways is the reflector superior to the refractor and vice versa?

20. ⊗ The pupil of the eye (Fig. Q26.20) defines the aperture of the eye and thus determines how much light enters the eye and strikes the retina. If the pupil "opens up," does the f-number of the eye increase or decrease?

Figure Q26.20

21. Does a camera use a diverging lens or a converging lens? Does a camera produce a real image or a virtual image?

22. [SSM] A photographer intends to take a photograph of a country landscape with a large depth of field such that the flowers in the foreground of the picture and a barn in the distance are both in sharp focus. What conditions are more suitable to such a photo, a bright and sunny day or a dim and overcast day? Explain your choice.

23. DVD players use a laser with a higher frequency than that of CD players. Why can DVD players can read CDs, but CD players cannot read DVDs?

24. You are asked to design an optical fiber. Should the refractive index of the core material be larger or smaller than that of the surrounding cladding? For the fastest transmission times, do you want a high or a low refractive index for the core material?

25. You have two unlabeled laser discs, one a CD and the other a DVD. Describe how you might use a laser pointer to tell which is which.

Figure Q26.26

26. Small binoculars or opera glasses (Fig. Q26.26) are useful in seeing the stage in detail. No prisms or mirrors are used, and yet the magnified image in these glasses is not inverted. Are the eyepieces converging lenses or diverging lenses?

26.1 APPLICATIONS OF A SINGLE LENS: CONTACT LENSES, EYEGLASSES, AND THE MAGNIFYING GLASS

1. (X) You are nearsighted and can only focus clearly on things that are closer than 2.5 m. What should the focal length be for contact lenses that correct this problem?

2. (X) Your friend is farsighted with a near-point distance of 500 cm. What should the focal length be for the lenses in his reading glasses? Assume the glasses are placed 2.0 cm in front of his eyes.

3. [SSM] ⊠ (X) A person is nearsighted and can clearly focus on objects that are no farther than 3.0 m away from her eyes. She borrows a friend's glasses and finds that she can now focus on objects as far away as 4.5 m. What is the focal length of the glasses? Do they use converging lenses or diverging lenses? Assume the glasses are placed 2.0 cm in front of her eyes.

4. ⊠ (X) Consider again the nearsighted person in Problem 3, but now assume the borrowed glasses make things worse; that is, the person can now focus only on things that are within 1.5 m away. Do these glasses contain converging lenses or diverging lenses? What is the focal length of the glasses?

5. (X) The distance from the front of the eye to the retina is about 2.5 cm. If this eye is focused on a newspaper that is 40 cm in front of the eye, what is the focal length of the eye?

6. ⊠ (X) A person uses eyeglasses that have a focal length $f = -25$ cm and are mounted 1.5 cm in front of his eyes. (a) If he switches to contact lenses that are mounted in contact with the eye, what focal length should the contact lenses have? (b) Make a qualitative sketch of the shape of the lenses. Are both lens surfaces concave, convex, or . . .?

7. ⊕ (X) (R) The author is nearsighted and (at present) is unable to focus clearly on objects that are farther than about 2.0 m from his eyes. (a) What type of lens (converging or diverging) should his eyeglasses contain so that he can focus on more distant objects? (b) What should be the focal length of his eyeglass lenses so that he can focus on a very distant object? Assume the glasses are placed 2.0 cm in front of the eyes. (c) What is the refractive power of the lenses in part (b) in units of diopters?

8. ⊠ (X) (R) The author's near-point distance is about 40 cm. What is the focal length of his reading glasses if using them allows him to bring his newspaper to within 10 cm of his eyes? Assume the glasses are placed 2.0 cm in front of his eyes.

9. (X) ⊕ **Helping your optometrist.** In Section 26.1, we designed a set of contact lenses. For a person with a particular value of

the near-point distance, we found that lenses with focal length $f = 38$ cm are required (Eq. 26.7). An optometrist needs instructions on how to make these lenses; specifically, he needs to know the radii of curvature R_1 and R_2 of the two lens surfaces. In Chapter 24, we used the lens maker's formula to relate the focal length of a lens to R_1 and R_2. For a double convex lens, the lens maker's formula reads

$$\frac{1}{f} = (n - 1)\left(\frac{1}{|R_1|} + \frac{1}{|R_2|}\right)$$

where n is the index of refraction of the glass used to make the lens. To apply the lens maker's formula to our contact lens, assume one side of the lens is flat (which is not very realistic) so that its radius of curvature is $R_2 = \infty$. Find the radius of curvature R_1 of the other surface of the lens. Assume the refractive index is $n = 1.55$. Does your value for R_1 seem reasonable? How does it compare with your experience with real contact lenses?

10. ⊕ (X) **Designing an eyeglass lens.** In Problem 9, we calculated the shape (i.e., the radius of curvature) of a contact lens. In this problem, you will do the same for an eyeglass lens. One side of a contact lens must be flat (or nearly flat) so that it can rest against the surface of the eye, but with an eyeglass lens both surfaces can be convex or concave. Your job is to design a double convex lens with a focal length $f = 38$ cm (the same as in Problem 9). For simplicity, assume both surfaces have the same radius of curvature R and the lens is made from glass with $n = 1.55$. Find R. *Hint*: Use the lens maker's formula given in Problem 9.

11. What is the refractive power of the lens whose focal length is $f = 37$ cm? Give your answer in diopters.

12. (X) The cornea of an average human eye has a refractive power of about 40 diopters. To what focal length does this value correspond? How does this focal length compare to the size of the eye?

13. ⊠ (X) Design a set of eyeglasses for a nearsighted person. Assume the person is able to focus only on objects closer than 2.0 m. What type of lens (converging or diverging) with what focal length is required so that the person can view an object at infinity? Assume the glasses are placed 2.0 cm in front of the eyes.

14. ⊕ (X) A pair of glasses is designed for a person with a far-point distance of 3.0 m so that she can read street signs 20 m away. (The far-point distance is the distance from the eye at which you are just able to properly focus a distant object.) (a) If the glasses are to be worn 1.0 cm from her eyes, what is the needed focal length? (b) Compare this focal length (find the percent difference) to the focal length she would need if she chooses a style of glasses that fit on her face so that the lenses are instead 2.0 cm away from her eyes. For a nearsighted person, is the position of the glasses from the eyes as important as that for a farsighted person?

15. ☆ ⊗ **Accommodation in the eye.** A relaxed human eye has a focal length of about 2.5 cm. When looking at something very close, however, the refractive power of the eye can change by up to 16 diopters for a young adult. The shaping of the lens of the eye through muscles that stretch and compress is known as the process of *accommodation*. (a) Does the refractive power increase or does it decrease by this amount when viewing an object close up? (b) What is the focal length of the eye when focusing on something that is very close?

16. A magnifying glass gives an actual magnification of 10 for a person with a near-point distance of $s_N = 35$ cm. What is the focal length of the lens?

17. SSM ☆ ⊗ An entomologist with a near-point distance of 30 cm has two lenses that she uses to examine small beetles. One lens has a focal length $f_1 = 2.3$ cm, and the second has a focal length $f_2 = 7.5$ cm. (a) Which of the two lenses would allow her to see the smallest detail? (b) Calculate the maximum magnification available to her for each lens. (c) She examines a beetle that is 3.7 mm in length. What is the apparent length of the beetle under the highest magnification available to her?

18. ⊗ A person with a near-point distance of 25 cm is able to get a usable magnification of 20 with a particular magnifying glass. Her friend is only able to get a usable magnification of 15. What is the friend's near-point distance?

19. ⊗ Will you achieve a larger magnification with a particular magnifying glass if you have a near-point distance of 25 cm or of 15 cm? If the focal length of the magnifying glass is 7.5 cm, what is the ratio of the magnifications in the two cases?

20. ☆ ⊗ Ⓡ Suppose you have a near-point distance of 25 cm. The angular resolution of the eye is determined by diffraction and the Rayleigh criterion (Chapter 25) and is approximately 1 arcmin = 1′. What is the smallest object your eye can resolve without a magnifying glass?

21. ☆ ⊗ Ⓡ Consider again Problem 20. With a near-point distance of 25 cm, what is the smallest object a person can resolve using a magnifying glass with a magnification of 10?

26.2 MICROSCOPES

22. ⊗ A compound microscope in which the objective and eyepiece are 20 cm apart uses an objective lens with $f_{obj} = 1.0$ cm and an eyepiece with $f_{eyepiece} = 5.0$ cm. What is the magnification of the microscope?

23. ⊗ For the microscope in Problem 22, how would the magnification change if the focal length of the objective lens were decreased by a factor of three?

24. ☆ ⊗ The distance between the objective lens and the eyepiece of a microscope is 30 cm. If the eyepiece has a magnification of 20 and the total magnification of the microscope is $|m| = 250$, what is the focal length of the objective lens?

25. SSM ☆ ⊗ A compound microscope has an objective lens that by itself produces a magnification of −15. The total magnification of the microscope is designed to be −150. (a) What is the focal length of the eyepiece? (b) Why is the value of the magnification negative?

26. ⊗ A compound microscope has a total magnification $|m| = 500$, using an eyepiece whose magnification is 20. What is the focal length of the objective lens?

27. ☆ ⊗ Figure P26.27 shows a typical student microscope with a rotating platform that allows the user to select from three objective lenses. A common model has objective focal lengths of 18 mm, 4.0 mm, and 1.6 mm. Each objective lens

© Image Source Black/Jupiterimages

Figure P26.27

forms an image 10 cm beyond its focal point. If the eyepiece has an angular magnification of 10×, calculate the highest and lowest overall angular magnifications available.

26.3 TELESCOPES

28. Consider a refracting telescope like that described in Figure 26.12. If it has an objective lens with $f_{obj} = 1.5$ m and an eyepiece with $f_{eyepiece} = 2.0$ cm, what is the angular magnification of this telescope?

29. ☆ You wish to design a telescope using an objective lens with $f_{obj} = 0.75$ m. If you use an eyepiece with a magnification of 20, what are (a) the total magnification of the telescope and (b) the length of the telescope?

30. ✪ Galileo's telescope consisted of two lenses, an objective and an eyepiece. Unlike the telescope in Figure 26.12, however, Galileo used a diverging lens as his eyepiece (Fig. P26.30). This telescope was designed for the object (i.e., the Moon) to be very far away (at infinity) and produced an image that was also at infinity. In contrast to the case with a converging lens for an eyepiece, Galileo's telescope produced an *upright* image. (a) Construct a ray diagram using Figure P26.30, showing how an object very far to the left will produce an upright image that is also very far to the left. *Hint*: Consider the situation in which the image produced by the objective lens is to the right of the eyepiece. The image produced by the objective then acts as the object for the eyepiece lens. (b) Suppose $f_{obj} = 1.5$ m and $f_{eyepiece} = -1.0$ cm. What is the approximate total length of the telescope? (c) What is the total magnification?

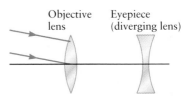

Objective lens   Eyepiece (diverging lens)

Figure P26.30

31. ☆ Ⓡ Galileo was famous for his use of a telescope to study the solar system. Among his many discoveries, he showed that there are mountains on the surface of the Moon. Estimate the size of the smallest object that Galileo's telescope could resolve on the Moon. *Hint*: Galileo used a refracting telescope with an aperture of approximately 37 mm.

32. SSM ☆ Ⓡ The Hubble Space Telescope is a Cassegrain reflecting telescope (Fig. 26.14B) with mirror of diameter of 2.4 m. What is the size of the smallest object that this telescope can resolve (a) on Mars, (b) on the Moon, and (c) on the surface of one of the moons of Saturn?

33. ☆ One of the largest refracting telescopes made to date has an objective lens with a focal length of approximately 20 m. When this telescope is used to look at the Moon, what is the angle subtended by the edges of the Moon in the image formed by the objective? If the eyepiece has a focal length of 5.0 cm, what is the total magnification?

34. Consider a reflecting telescope described by the ray diagram in Figure 26.14A. If f_M is the focal length of the mirror and $f_{eyepiece}$ is the focal length of the eyepiece, the total magnification of the telescope is (Eq. 26.21)

$$m_\theta = \frac{f_M}{f_{eyepiece}}$$

This telescope has a main mirror with $f_M = 550$ mm. If an eyepiece with a magnification of 25 is used, what is the total magnification of the telescope?

26.4 CAMERAS

35. A film camera uses a single lens with a focal length $f = 60$ cm. (a) Calculate the image distance for an object at infinity. (b) Calculate the image distance for an object that is 2.5 m away. (c) How far must the lens be moved with respect to the film when an object moves from infinity to a point 2.0 m from the camera?

36. The distance between a camera lens and the film in a particular camera is 5.0 cm. If this lens is focused on an object that is very far away, what is the focal length of the lens?

37. ☆ The *f*-number on a camera is changed from 4 to 8. If the photographer wants to have the same amount of light energy strike the film, should she increase or decrease the shutter speed, and by what factor?

38. ☆ The shutter speed on a camera is increased from 1/60 s to 1/500 s. If the *f*-number is originally 2.8, what new setting would you recommend?

39. ℝ For a camera with a shutter speed set to "250" as read on the camera, how long is the film exposed to light?

40. A photograph of a particular scene is nicely exposed with an *f*-number of 5.6 and an exposure time of 0.010 s. If the exposure time is changed to 0.020 s, what *f*-number should be used?

41. A camera has an aperture of 2.0 cm and a lens with $f = 4.2$ cm. (a) What is the *f*-number? (b) If the *f*-number is reduced by a factor of three, by what factor should the exposure time be changed to have the same total light energy strike the film?

42. SSM ☆ A 35-mm camera uses film that is 24 mm tall and 35 mm wide. If it is used to take a photograph of a car that is 2.5 m long and 11 m away from the camera and the car just fills the photo, what is the focal length of the lens?

43. ✪ A digital camera uses a CCD detector that is 7.2 mm wide and 5.3 mm tall. (a) If the detector is rated for "4.0 megapixels" and the pixels are equally spaced, what is the distance between the centers of adjacent pixels on the CCD chip? (Use Figure 26.21 to model the arrangement of CCD pixels and ignore any space between the edges of adjacent pixels.) (b) When an object is at infinity, the lens is found to be 1.5 cm from the CCD detector. What is the focal length of the lens? (c) If an object is placed 30 cm from the lens, what spacing between the lens and the CCD detector is required for the image to be at the detector?

44. ✪ The author has an inexpensive digital camera that has the following specifications: "5× optical zoom, 7× digital zoom." (a) The "lens" is actually a combination of several lenses, and the value of the optical zoom is adjusted by changing the focal length *f* of the lens system. If the magnification is changed by a factor of five (the maximum optical zoom), by what factor does *f* change? (b) The digital zoom is adjusted by using the image from only a portion of the CCD detector. When the digital zoom is "off," all the CCD's pixels are used. What fraction of the pixels is used when the digital zoom is set for its maximum value (7×)?

26.5 CDS AND DVDS

45. SSM A hypothetical CD player uses a laser with $\lambda = 880$ nm in air. Inside the plastic coating of a hypothetical CD the wavelength is 550 nm. What is the index of refraction of the plastic?

46. ✪ ℝ A real CD can hold approximately 600 Mbytes of information. Each byte is an 8-bit binary number, so it takes eight digits of zeros and ones to make 1 byte. In order that each pit edge can be optically resolved by the laser beam that reflects from the CD (Fig. 26.28), the length of a pit must be no smaller than the wavelength. Assume each pit on a CD stores 2 bits of information (two edges), the pits are 3λ long, and $\lambda = 503$ nm in the plastic of the CD. Approximately how many megabytes of information can be stored on this model of CD? Explain why your number agrees or does not agree with the storage capacity of a real CD.

47. ℝ The light used in a CD has a wavelength of 503 nm (inside the plastic coating), whereas a Blu-ray disc uses a wavelength of 260 nm (again, inside the plastic coating). Based solely on the effects of diffraction, a Blu-ray disc can contain more pits and thus hold more information than a CD. Estimate the ratio of the number of pits that can be on a Blu-ray disc to the number on a CD. How does this answer compare with what you know (or can look up) about the amount of information that can be stored on a CD and a Blu-ray disc?

26.6 OPTICAL FIBERS

48. SSM ☆ The optical fiber in Figure P26.48 has a core of diameter 50 μm and a total diameter of 60 μm. The indices of refraction are $n_{core} = 1.65$ and $n_{cladding} = 1.55$. What is the maximum value of θ in Figure P26.48 that gives total internal reflection? This is called the "acceptance" angle of the fiber.

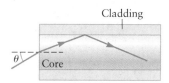

Figure P26.48 Problems 48 and 49.

49. ✪ Consider the simple optical fiber in Figure 26.30A that does not have a cladding and suppose the fiber has an index of refraction $n = 1.75$. The acceptance angle θ is the maximum angle that an incoming ray can have with the end of the fiber and still achieve total internal reflection inside the fiber. (See Fig. P26.48.) (a) What is the acceptance angle of this fiber? (b) If the fiber is immersed in water (perhaps as part of an undersea cable that leaks), what is the new acceptance angle? Assume that the ends of the fiber are in air while the sides are immersed in water.

50. ✪ In Example 26.9, we discussed how light pulses in a multimode optical fiber are "smeared out" in time due to the different angles of travel that are possible within the fiber. This problem does not occur in single-mode fibers because all light travels along the axis of the fiber, but there is another smearing effect in single-mode fibers due to dispersion. The laser light traveling in the fiber has a small spread in wavelength; hence, a single pulse consists of components of different colors. Since the index of refraction depends on the wavelength, different components of a pulse travel at different speeds, and eventually the faster components of one pulse will "overtake" the slower components of the pulse ahead of it. Consider a light pulse with just two components with wavelengths (in the fiber) of $\lambda_1 = 503.0$ nm and $\lambda_2 = 504.0$ nm, with corresponding indices of refraction $n_1 = 1.5500$ and $n_2 = 1.5510$, traveling in a fiber of length 20 km. (a) Which wavelength travels fastest? (b) How long will it take each component to travel the entire length of the fiber? (c) Two very short pulses, one with wavelength λ_1 and another with λ_2, are started simultaneously at one end of the fiber. When the faster component reaches the end of the fiber, how far behind is the slower component?

ADDITIONAL PROBLEMS

51. ✪ ✖ ℝ An objective lens has a focal length of 0.50 cm. Use it to design a compound microscope with a total magnification of 500. First identify the quantities whose values you must choose for the design and then find a set of appropriate values.

52. ✪ ℝ The author has a Cassegrain reflecting telescope (Fig. 26.14B) with a mirror diameter of 10 cm that he is using to study the many rings of Saturn. Figure P26.52 shows a photo taken with the Hubble Space Telescope showing various features of the rings. The most prominent (brightest) are the A and B rings. Can the author's telescope resolve the gap between Saturn and the inner edge of the B ring? Estimate the data you need from the figure. The distance from the Earth to Saturn is approximately 1.3×10^9 km.

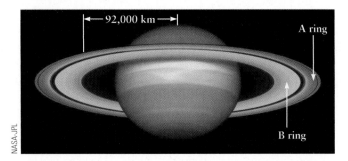

Figure P26.52

53. ✪ You have lenses with focal lengths $f_1 = 4.0$ cm and $f_2 = 12$ cm. Assume your near point distance is 25 cm and consider the following configurations of these lenses. (a) For magnifying glasses, what magnification could you achieve with each lens? (b) What is the maximum magnification you could achieve if you used these lenses in a compound microscope of length 25 cm? (c) What maximum magnification could you achieve if you made them into a telescope? What is the separation distance between the lenses for the telescope configuration?

54. ✪ Two lenses are mounted in a tube such that the distance between them can be varied. This device can be used as either a telescope or a compound microscope. One lens has a focal length of 27 cm. (a) Find the focal length of the second lens such that a telescope made from the two lenses would have the same magnification as a microscope made from the same two lenses. *Hint*: It will be a low-magnification microscope, so you will need to substitute Equation 26.13 for $m_{\theta, \text{eye}}$ in deriving a result to replace Equation 26.16. (b) What is the magnification achieved?

55. Completed in 1895, the Yerkes telescope operated by the University of Chicago remains the biggest (largest-diameter) refracting telescope ever built. Figure P26.55 shows a photograph of the

Figure P26.55 Albert Einstein and observatory staff stand in front of the Yerkes 40-in. refracting telescope in 1921.

telescope taken in 1921, when it was the most important astronomical instrument in the world. The refractor has an *f*-number of 19 and an objective lens that measures 1.02 m across. What are the focal length and approximate length of this telescope? Do your calculations agree with an estimate made from the photograph?

56. ✪ Ⓡ A digital camera has an 8.3-megapixel CCD array with the pixels evenly spaced over a rectangular region, 7.2 mm $\times$ 5.3 mm. The camera has a lens 2.0 cm in diameter with a focal length of 3.5 cm. (a) What is the resolution limit due to pixel size? (b) What is the approximate resolution limit due to aperture size? (c) If an object 1.5 m high is 200 m away, how many pixels high will the image be? (d) What is the maximum distance from the camera that this page could be photographed in order to read this paragraph from the digital photo? (e) In a movie theater, the screen is 4.0 m wide and 2.2 m tall. How far back from the screen would one have to sit to capture an image of the whole screen with this camera?

57. ✪ Ⓡ **Blu-ray disc system data storage.** As the name implies, the Blu-ray disc has the same diameter and refractive index as a DVD or CD, but the laser light used is in the blue–violet region with a wavelength in air of 405 nm. The index of refraction of the plastic coating on the disc is $n = 1.55$. (a) What is the wavelength of this light within the disc? (b) How deep are the pits in a Blu-ray disc? Are they deeper or shallower than the pits in a CD or DVD? (The laser light used by a CD has a wavelength of 780 nm while for a DVD it is 650 nm.) (c) Tracks (adjacent rows of pits) are separated by 1.6 μm in a CD, 0.74 μm in a DVD, and 0.32 μm in a Blu-ray disc. Calculate the approximate amount of memory in megabytes (1 byte = 8 bits, and 1 Mb = 10^6 bytes) per square centimeter of disc area for each format. Note that each pit edge, going from the bottom of a pit to the upper surface or vice versa, counts as 1 bit.

58. ✪ Consider the water fountain in Figure Q26.17, where the stream of water acts like an optical fiber. What is the minimum angle of total internal reflection in the stream of water for blue light with $\lambda = 450$ nm?

59. SSM ✪ The constellation Orion is approximately 800 light-years from Earth. (One light-year is the distance traveled by light in a vacuum in one year.) Assume that one of the stars in this constellation has a planet orbiting at the same distance as the Earth orbits the Sun. How large should the aperture of a telescope be so that it can resolve the star from its planet using light with $\lambda = 450$ nm?

Relativity

The propagation of light through the universe has some surprising features that are explained by Einstein's theories of relativity. This is a photo of the Sombrero Galaxy—it is 28 million light-years from the Earth and is believed to have a black hole at its center. (NASA/JPL-Caltech/University of Arizona/STScI)

Newton discovered his laws of mechanics around 1700. Over the next two centuries, these laws were applied to a wide variety of problems and seemed to work perfectly. With the development of Maxwell's equations in the late 1800s, the physics of electromagnetism and light fell into place. By 1900, many physicists thought that physics was completely understood with nothing more to be done, but how wrong they were! Two completely new and unanticipated subfields of physics—relativity and quantum theory—were discovered in the first decades of the 1900s. For historical reasons, these topics are called "modern physics," although it seems peculiar to use the term *modern* to describe work nearly one hundred years old. Relativity was developed mainly by Albert Einstein (although other physicists had important roles), and his theory of relativistic mechanics has profoundly changed the way we think about space and time.

There are actually two types of relativity theory. **Special relativity** is concerned with objects and observers moving with constant velocity, the case we will consider through most of this chapter. **General relativity** applies to situations in which an object or observer is accelerated, and it has some surprising implications for the understanding of gravitation.

27.1 I NEWTON'S MECHANICS AND RELATIVITY

In physics, the term *relativity* arises when we describe a situation from two different points of view. Figure 27.1 shows a person (Ted) standing on an open railroad car. In Figure 27.1A, the car is at rest while Ted throws a ball straight up into the air and then catches it when it falls back down to him. Figure 27.1B shows a similar event, but now the railroad car is moving with constant speed v to the right and a person on the ground next to the car (Alice) is watching. According to Ted—that is, *relative to Ted*—the ball travels straight up and then back down, just as in Figure 27.1A. Relative to Alice, the ball has a horizontal component of velocity equal to that of the railroad car and follows a parabolic trajectory. The ball's motion can be analyzed by both Ted and Alice using the results for motion with constant acceleration (projectile motion) from Chapter 4.

These two observers are in different **reference frames**. You can think of a reference frame as a set of coordinate axes. In this example, Alice is at rest relative to the x–y axes, whereas Ted would describe the motion of the ball using the coordinate axes x'–y' that travel along with his railroad car.

Reference frames that move with a constant velocity, like the reference frames in Figure 27.1, are called **inertial reference frames**. The idea that the laws of motion should be the same in all inertial frames can be traced to Galileo and is called the **principle of Galilean relativity**. Newton certainly understood that his laws of mechanics obey Galilean relativity; that is, if Newton's laws of mechanics are obeyed in one inertial reference frame, they are obeyed in all inertial frames. For example, Newton's second law reads

$$\sum \vec{F} = m\vec{a} \tag{27.1}$$

where $\vec{a}$ is the acceleration of a particle of mass m and $\sum \vec{F}$ is the total force on the particle. Adding or subtracting a constant velocity does not change the acceleration of a particle. In Figure 27.1, Ted would say that the horizontal component of the ball's velocity (the component along x') is zero, but Alice would disagree; in her reference frame, the horizontal (x) component of the ball's velocity is v. Both Ted and Alice would agree, however, that the ball's acceleration is directed downward (along either y' or y) with magnitude g, and both would agree that the ball's horizontal acceleration is zero. They would also agree that the only force on the ball is the force of gravity, so both would say that Newton's second law (Eq. 27.1) is obeyed.

Inertial reference frame: a set of coordinate axes that move at constant velocity relative to other inertial reference frames

Figure 27.1 **A** When Ted throws a ball upward or drops a ball downward, he observes that the ball's motion is purely along the vertical (y') direction. **B** When Ted's railroad car has a speed v relative to a second observer (Alice), the ball undergoes projectile motion with a nonzero displacement along both x and y in Alice's reference frame. However, as viewed by Ted using his reference frame and coordinates x' and y', the ball's motion is still purely vertical as in part A.

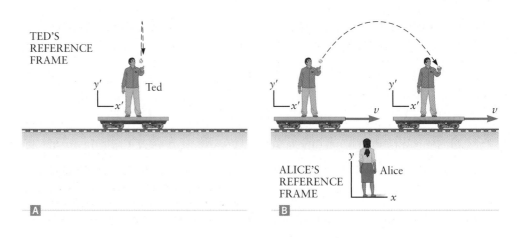

Newton's second law is thus obeyed in both of these inertial reference frames. This result is similar to what we found in our analysis of the airplane accelerometer in Chapter 4; there again, observers in different inertial reference frames both found that Newton's second law is obeyed.

The principle of Galilean relativity is certainly reasonable. It agrees with our everyday intuition and the experiment in Figure 27.1, and was believed to be an exact law of physics for more than two hundred years. The first hint of a problem came from Maxwell's work in electromagnetism. According to Maxwell's equations, the speed of light is (Eq. 23.1)

$$c = \frac{1}{\sqrt{\varepsilon_0 \mu_0}} \qquad (27.2)$$

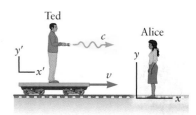

Figure 27.2 Ted moves at speed v relative to Alice when his flashlight emits a light pulse that moves at speed c relative to him. What is the speed of the light pulse relative to Alice?

Maxwell also showed that this result is independent of the motion of both the source of the light and the observer. To see why this finding is surprising, suppose Ted is again moving with a speed v relative to Alice when he turns on a flashlight that generates a light pulse traveling to the right in Figure 27.2. This is a one-dimensional situation, so we need consider only the component of velocity along the horizontal (x) direction. The light wave in Figure 27.2 has a speed c relative to Ted and his flashlight. Since Ted's speed relative to Alice is v, Newton's mechanics predicts that the speed of the light wave relative to Alice should be $c + v$. According to Maxwell's theory, however, Ted and Alice should *both* observe the light wave to move with speed c. Maxwell's theory of electromagnetism is thus not consistent with the predictions of Galilean relativity for observers in different inertial frames.

Which prediction is correct? Initially, Maxwell and other physicists thought that there must be a problem with Maxwell's theory, but an experimental test is difficult. For example, if Ted in Figure 27.2 is moving at 100 m/s (about 200 mi/h), Galilean relativity predicts that Alice will observe a speed of

$$v(\text{Alice}) = c + v(\text{Ted}) = [(3.00 \times 10^8) + 100] \, \text{m/s} \qquad (27.3)$$

This differs from c by less than one part in one million, and such a small difference would be very hard to measure using the technology available in Maxwell's time. Successful experiments were carried out only well after Maxwell's death and showed that Maxwell's theory has it right: the speed of light in a vacuum is always c, and the prediction of Galilean relativity for how the speed of light depends on the motion of the source is wrong. In 1905, Einstein worked out the correct theory, which had profound implications for all physics.

27.2 | THE POSTULATES OF SPECIAL RELATIVITY

The theory Einstein developed to analyze the predicament of Ted and Alice in Figure 27.2 is called *special relativity.* According to Einstein, his work on this theory was not motivated by any particular experiment; indeed, at the time he developed his theory (1905), there were not yet any clear experimental results to show that Galilean relativity was wrong. Those experiments came only after Einstein's theory. Instead, Einstein suspected that Maxwell's result—that the speed of light is the same in all reference frames—is correct, and he then worked out what that implies for the other laws of physics such as Newton's laws. In a sense, Einstein's work was similar to that of Newton: both proposed basic laws of physics and then worked out the consequences. For Newton, these basic laws were his three laws of motion. For Einstein, the basic laws are now known as two "postulates" about the laws of physics. We'll first state the postulates and then spend most of the rest of this chapter working out the consequences. Experiments performed after 1905 showed that Newton's theory breaks down for fast-moving objects (such as light) and that Einstein's theory gives a correct description of motion in such a regime.

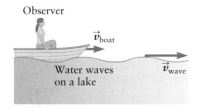

Figure 27.3 According to our intuition, water waves on a lake obey Galilean relativity. The velocity of these waves relative to the water (the wave medium) is $\vec{v}_{\text{wave}}$, whereas the velocity of a boat relative to the water is $\vec{v}_{\text{boat}}$. In this example, the velocity of the waves relative to the boat is $\vec{v}_{\text{wave}} - \vec{v}_{\text{boat}}$.

Postulates of Special Relativity

1. **All the laws of physics are the same in all inertial reference frames.**
2. **The speed of light in a vacuum is a constant, independent of the motion of the light source and all observers.**

The first postulate can be traced to the ideas of Galileo and Newton on relativity, but this postulate goes further than Galileo because it applies to *all* physical laws, not just mechanics. The second postulate of special relativity is motivated by Maxwell's theory of light, which we have seen is *not* consistent with Newton's mechanics. Newton would have predicted that the speed of the light pulse relative to Alice in Figure 27.2 is $c + v$ (Eq. 27.3), whereas Ted (who is in a different inertial reference frame) would measure a speed c. According to the second postulate of special relativity, both observers will measure the *same* speed c for light.

The postulates of special relativity will lead us to a new theory of mechanics that corrects and extends Newton's laws. Postulate 2 also tells us something very special and unique about light. Light is a wave, and all the other waves we have studied travel in a mechanical medium. For example, Figure 27.3 shows an observer traveling in a boat with velocity $\vec{v}_{\text{boat}}$ relative to still water. If there are waves traveling at velocity $\vec{v}_{\text{wave}}$ relative to still water and if the boat is traveling along the wave propagation direction, Newton's mechanics (and Galilean relativity) predict that the velocity of the waves *relative to this observer* is $\vec{v}_{\text{wave}} - \vec{v}_{\text{boat}}$. An observer who is stationary relative to the water would thus measure a different wave speed than the observer in the boat. This is *not* what the two observers in Figure 27.2 find; they measure the *same* speed for the light wave.

The conclusion from Figures 27.2 and 27.3 is that our everyday experience with conventional waves *cannot* be applied to light. According to postulate 2 of special relativity, the speed of a light wave is independent of the velocity of the observer. What role does the medium have in this case? Light does not depend on having a conventional material medium in which to travel. For a light wave, the role of the medium is played by the electric and magnetic fields, so a light wave essentially carries its medium with it as it propagates (which is why light can travel through a vacuum). The lack of a conventional medium is surprising and difficult to reconcile with one's intuition, and Maxwell's results were therefore slow to be accepted. Experiments, however, show that nature does work this way; the speed of light *is* independent of the motion of the observer. This example is just one of many for which conventional intuition fails. We'll come to more such failures as we study special relativity.

Finding an Inertial Reference Frame

Inertial reference frames played a role in our work with Newton's mechanics, and they play a crucial role in special relativity. We have stated that an inertial frame is one that moves with constant velocity, but relative to what? Newton believed that the heavens and stars formed a "fixed" and absolute reference frame to which all other reference frames could be compared. We now know that stars are also in motion, so we need a better definition of what it means to be "inertial." Nowadays, we define an inertial reference frame as one in which Newton's first law holds. Recall that Newton's first law states that if the total force on a particle is zero, the particle will move with a constant velocity, in a straight line with constant speed. So, we can test for an inertial reference frame by observing the motion of a particle for which the total force is zero. If the particle moves with a constant velocity, the reference frame in question—the reference frame we use to make the observation—is an inertial frame.

Alternate definition of an inertial reference frame

This definition of how to find an inertial reference frame is thus tied to Newton's first law and the concept of inertia, so the definition may seem a bit circular. However, the notion of Galilean relativity also asserts that Newton's other laws of mechanics are valid in all inertial frames. Hence, Newton's second law ($\sum \vec{F} = m\vec{a}$) and third law (the action–reaction principle) should apply in all inertial frames, too.

The Earth as a Reference Frame

Inertial reference frames move with constant velocity; hence, their acceleration is zero. Since the Earth spins about its axis as it orbits the Sun, all points on the Earth's surface have a nonzero acceleration. Strictly speaking, then, a person standing on the surface of the Earth is not in an inertial reference frame. However, the Earth's acceleration is small enough that it can be ignored in most cases, so in most situations we can consider the Earth to be an inertial reference frame.

27.3 | TIME DILATION

Einstein's two postulates seem quite "innocent." The first postulate—that the laws of physics must be the same in all inertial reference frames—is in accord with Newton's laws, so it does not seem that this postulate can lead to anything new for mechanics. The second postulate concerns the speed of light, and it is not obvious what it will mean for objects other than light. Einstein, however, showed that these two postulates together lead to a surprising result concerning the very nature of time. He did so by considering in a very careful way how time can be measured.

Let's use the postulates of special relativity to analyze the operation of the simple clock in Figure 27.4. This clock keeps time using a pulse of light that travels back and forth between two mirrors. The mirrors are separated by a distance ℓ, and light travels between them at speed c. The time required for a light pulse to make one round trip through the clock is thus $2\ell/c$. That is the time required for the clock to "tick" once.

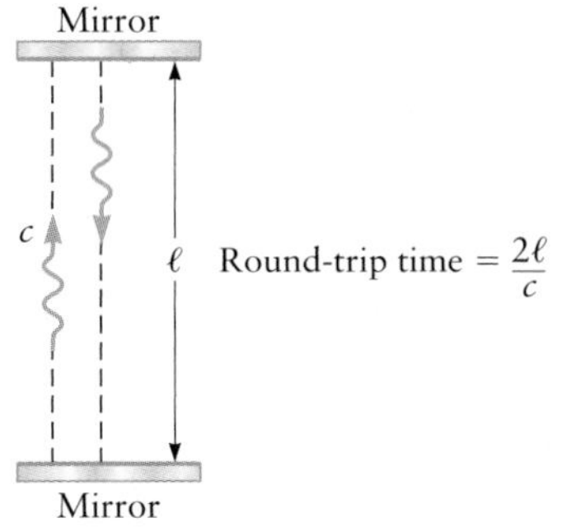

Analysis of a Moving Light Clock

We now place our light clock on Ted's railroad car in Figure 27.5A, so the clock is moving at constant velocity with speed v relative to the ground. How does that affect the operation of the clock? Let's first analyze the clock from Ted's viewpoint—that is, in Ted's reference frame—as we ride along on the railroad car in Figure 27.5B. In this reference frame, the operation of the clock is identical to that shown in Figure 27.4; the light pulse simply travels up and down between the two mirrors. The separation of the mirrors is still ℓ, so the round-trip time is still $2\ell/c$.

Figure 27.4 A light clock. Each round-trip motion of the light pulse between the two mirrors corresponds to one tick of the clock.

Round-trip time $= \dfrac{2\ell}{c}$

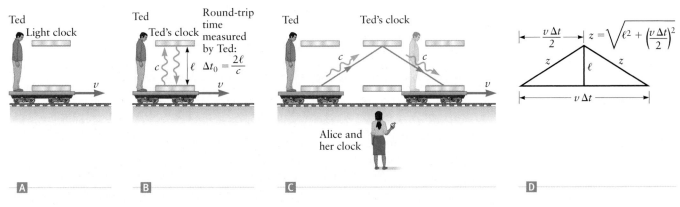

Figure 27.5 **A** A light clock traveling with Ted on his railroad car. **B** According to Ted, light pulses travel back and forth in the clock just as in Figure 27.4. Each tick of the clock takes a time $\Delta t_0 = 2\ell/c$. According to Ted, the operation of the clock is the same whether or not his railroad car is moving. **C** Motion of the light pulses in Ted's clock as viewed by Alice, who is at rest on the ground. **D** According to Alice, the round-trip travel distance for a light pulse is $2z$, where $z = \sqrt{\ell^2 + (v\,\Delta t/2)^2}$, which is longer than the round-trip distance 2ℓ seen by Ted in part B.

If Δt_0 is the time required for the clock to make one "tick" as measured by Ted, we have the result

$$\Delta t_0 = \frac{2\ell}{c} \tag{27.4}$$

A second observer, Alice, is standing on the ground watching Ted's railroad car travel by and sees things differently. In her reference frame, Ted's clock is moving horizontally, so from her point of view the light pulse does not simply travel up and down between the mirrors, but must travel a longer distance as shown in Figure 27.5C. According to postulate 2 of special relativity, the speed of light is the same for Alice as it is for Ted. Because light travels a longer distance in Figure 27.5C than in Figure 27.5B, according to Alice the light will take longer to travel between the mirrors. Let's use Δt to denote the round-trip time as observed by Alice; that is the time it takes for the clock to complete one "tick" in Alice's reference frame. We can find Δt by using a little geometry. In a time Δt (one "tick"), Alice sees the clock move a horizontal distance $v\, \Delta t$. The path of the light pulse forms each hypotenuse z of the back-to-back right triangles in Figure 27.5D. Using the Pythagorean theorem,

$$z^2 = \ell^2 + \left(\frac{v\, \Delta t}{2}\right)^2 \tag{27.5}$$

Since z is half the total round-trip distance, Alice finds

$$z = \frac{c\, \Delta t}{2}$$

or

$$z^2 = \frac{c^2 (\Delta t)^2}{4} \tag{27.6}$$

Combining Equations 27.5 and 27.6 gives

$$\frac{c^2 (\Delta t)^2}{4} = \ell^2 + \frac{v^2 (\Delta t)^2}{4}$$

We next solve for Δt:

$$(\Delta t)^2 = \frac{4\ell^2}{c^2} + \frac{v^2}{c^2}(\Delta t)^2$$

$$(\Delta t)^2 \left(1 - \frac{v^2}{c^2}\right) = \frac{4\ell^2}{c^2}$$

$$(\Delta t)^2 = \frac{4\ell^2/c^2}{1 - v^2/c^2}$$

Taking the square root of both sides and using Equation 27.4 finally leads to

$$\Delta t = \frac{\Delta t_0}{\sqrt{1 - v^2/c^2}} \tag{27.7}$$

Recall that Δt and Δt_0 are the times required for one tick of the light clock as observed by Alice and Ted, respectively. In words, Equation 27.7 thus says that these times are *different*! The operation of this clock *depends on the motion of the observer*. Let's now consider the implications of Equation 27.7 in more detail.

Moving Clocks Run Slow

The clock in Figure 27.5 is at rest relative to Ted, and he measures a time Δt_0 for each tick. The same clock is moving relative to Alice, and according to Equation 27.7 she measures a longer time Δt for each tick. This result is not limited to light clocks. Postulate 1 of special relativity states that *all* the laws of physics must be the same in *all* inertial reference frames. We could use a light clock to monitor or time any process in any reference frame. Since Equation 27.7 holds for light clocks, it must therefore apply to *any* type of clock or process, including biological ones.

According to Equation 27.7, the ratio of Δt (the time measured by Alice) to the time Δt_0 (measured by Ted) is

$$\frac{\Delta t}{\Delta t_0} = \frac{1}{\sqrt{1 - v^2/c^2}} \qquad (27.8)$$

Time dilation: moving clocks run slow.

Assuming v is less than the speed of light c (discussed further below), the factor on the right-hand side is always greater than 1. Hence, the ratio $\Delta t/\Delta t_0$ is larger than 1, and Alice measures a longer time than Ted does. In words, a moving clock will, according to Alice, take longer for each tick. Hence, special relativity predicts that *moving clocks run slow*. This effect is called **time dilation**.

This result seems very strange; our everyday experience does not suggest that a clock (such as your wristwatch) traveling in a car gives different results than an identical clock at rest. If Equation 27.8 is true (and experiments definitely show that it *is* correct), why haven't you noticed it before now? Figure 27.6 shows a plot of the ratio $\Delta t/\Delta t_0 = 1/\sqrt{1 - v^2/c^2}$ as a function of the speed v of the clock. At ordinary terrestrial speeds, v is much smaller than the speed of light c and the ratio $\Delta t/\Delta t_0$ is very close to 1. For example, when $v = 50$ m/s (about 100 mi/h), the ratio is

$$\frac{\Delta t}{\Delta t_0} = \frac{1}{\sqrt{1 - v^2/c^2}} = \frac{1}{\sqrt{1 - (50 \text{ m/s})^2/(3.00 \times 10^8 \text{ m/s})^2}}$$

$$\frac{\Delta t}{\Delta t_0} = 1.000000000000014 \qquad (27.9)$$

The result in Equation 27.9 is extremely close to 1, so for all practical purposes the times measured by Ted and by Alice are the same if Ted moves at 50 m/s relative to Alice. For typical terrestrial speeds, the difference between Δt and Δt_0 is completely negligible.

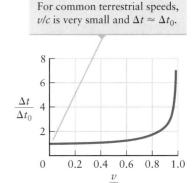

For common terrestrial speeds, v/c is very small and $\Delta t \approx \Delta t_0$.

Figure 27.6 Time dilation factor $\Delta t/\Delta t_0$ as a function of v/c.

Nature's Speed Limit

A curious feature of the time-dilation factor in Equation 27.8 is that its value is imaginary when v is greater than the speed of light. For example, if we insert $v = 2c$ into Equation 27.8, we get

$$\frac{\Delta t}{\Delta t_0} = \frac{1}{\sqrt{1 - (2c)^2/c^2}} = \frac{1}{\sqrt{-3}}$$

This result is an imaginary number! Does it mean that special relativity predicts that some time intervals Δt can be imaginary numbers? No, it does not. Speeds greater than c have never been observed in nature. We'll come back to this issue in Section 27.9 when we discuss work and energy in special relativity, and we'll see why it is not possible for an object to travel faster than the speed of light.

| EXAMPLE 27.1 | Time Dilation for an Astronaut |

The astronauts who traveled to the Moon in the Apollo missions hold the record for the highest speed traveled by people, with $v = 11,000$ m/s. What is the ratio $\Delta t/\Delta t_0$ for the Apollo astronauts' clock?

RECOGNIZE THE PRINCIPLE

The Apollo astronauts have the role of Ted in Figure 27.5 because we are interested in a clock that travels with them, while an observer on the Earth has the role of Alice. The time measured by the astronaut's clock thus reads the time interval Δt_0, and an observer on the Earth measures a longer time interval Δt.

Insight 27.1

RELATIVISTIC CALCULATIONS WHEN V IS SMALL

The factor $\sqrt{1 - v^2/c^2}$ arises often in special relativity. When v is small compared with c, this factor is very close to 1. In fact, the difference between it and 1 can be so small that your calculator may have trouble dealing with it. In such cases, the approximations

$$\sqrt{1 - v^2/c^2} \approx 1 - \frac{v^2}{2c^2}$$

and

$$\frac{1}{\sqrt{1 - v^2/c^2}} \approx 1 + \frac{v^2}{2c^2}$$

are very handy and are quite accurate at terrestrial speeds. In practice, they can be used whenever v is less than about $0.1c$. (See Figs. 27.6 and 27.15.) We sometimes also have expressions like $1/(1 + A)$, where A is very small. In such cases, we can use the approximation

$$\frac{1}{1 + A} \approx 1 - A$$

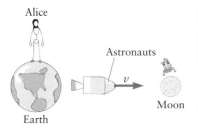

Figure 27.7 Example 27.1.

This problem is described by Figure 27.7. We assume the astronauts are carrying a light clock with them to the Moon, just as Ted carried a clock in his railroad car in Figure 27.5.

IDENTIFY THE RELATIONSHIPS

We can find $\Delta t/\Delta t_0$ using our analysis of Figure 27.5 (and Eq. 27.8), substituting $v = 1.1 \times 10^4$ m/s and the known speed of light.

SOLVE

Inserting the given values, we have

$$\frac{\Delta t}{\Delta t_0} = \frac{1}{\sqrt{1 - v^2/c^2}}$$

$$= \frac{1}{\sqrt{1 - (1.1 \times 10^4 \text{ m/s})^2/(3.00 \times 10^8 \text{ m/s})^2}} = \boxed{1.00000000067}$$

When v is such a small fraction of the speed of light, you may be limited by the number of significant figures your calculator can display. In such cases, you can use one of the approximations given in Insight 27.1. The second approximation gives

$$\frac{\Delta t}{\Delta t_0} \approx 1 + \frac{v^2}{2c^2} = 1 + \frac{(1.1 \times 10^4 \text{ m/s})^2}{2(3.00 \times 10^8 \text{ m/s})^2} \approx \boxed{1 + 6.7 \times 10^{-10}} \tag{1}$$

What does it mean?
Time dilation is a very small effect, even at this (relatively) high speed, yet it is possible to make clocks that are accurate enough to observe the small amount of slowing down in Equation (1). Experiments have shown that the time dilation predicted by special relativity is indeed correct. This result for Δt applies to all clocks, including the biological clocks of the Apollo astronauts. Hence, these astronauts aged slightly less than other people who stayed behind!

Proper Time

We derived Equation 27.8 from an analysis of a light clock, but the result applies to all time intervals measured with any type of clock. The time interval Δt_0 for a particular clock or process is measured by an observer *at rest relative to the clock* (Ted in Fig. 27.5). The quantity Δt_0 is called the ***proper time***. The proper time is always measured by an observer at rest relative to the clock or process that is being studied. So, while Ted is moving on his railroad car in Figure 27.5, the clock is moving along with Ted. Hence, Ted is at rest relative to this clock and he measures the proper time. On the other hand, Alice in Figure 27.5 is moving *relative to the clock*, so she does not measure the proper time. The time interval Δt measured by a moving observer (Alice) for the same process is always longer than the proper time.

When an observer is at rest relative to a clock or process, the start and end of the process occur at the same location for this observer. For the light clock in Figure 27.5B, Ted might be standing next to the bottom mirror, so from his viewpoint the light pulse starts and ends at the same location. By comparison, for Alice in Figure 27.5C, the light pulse begins at the bottom mirror when the clock is at the left; the pulse returns to this mirror when the clock is in a different location (relative to Alice), and Alice measures a longer time interval Δt. The proper time is always the shortest possible time that can be measured for a process, by any observer.

CONCEPT CHECK 27.1 | Measuring Proper Time

Ted is traveling in his railroad car with speed v relative to Alice, who is standing on the ground nearby (Fig. 27.8). Ted is playing with his yo-yo and uses a clock

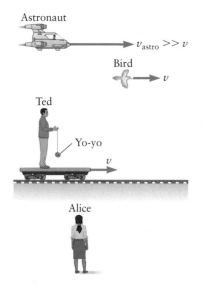

Figure 27.8 Concept Check 27.1.

on the railroad car to measure the time it takes for the yo-yo to complete one up-and-down oscillation. The yo-yo is also observed by Alice, by a bird flying nearby (also with speed v), and by an astronaut who is cruising by at a very high speed v_{astro}. Which observer measures the proper time for the yo-yo's period, (a) Ted, (b) Alice, (c) the bird, or (d) the astronaut?

| EXAMPLE 27.2 | Time Dilation for a Muon |

Subatomic particles called muons are created when cosmic rays collide with atoms in the Earth's atmosphere. Muons created in this way have a typical speed $v = 0.99c$, very close to that of light. Muons are unstable, with an average lifetime of about $\tau = 2.2 \times 10^{-6}$ s before they decay into other particles. That is, physicist 1 at rest relative to the muon measures this lifetime τ. Another (physicist 2) studies the decay of muons that are moving through the atmosphere with a speed of $0.99c$ relative to her laboratory (Fig. 27.9). What lifetime would physicist 2 measure?

RECOGNIZE THE PRINCIPLE

The muon acts as a sort of "clock" in which its lifetime corresponds to one "tick." Our results for a light clock, including Equation 27.8, apply to this muon "clock" since the results of special relativity apply to *all* physical processes. A clock moving along with the muon measures the proper time Δt_0, just as Ted in Figure 27.5B measures the proper time of a clock that travels along with him in his railroad car. The muon is moving relative to physicist 2, so that physicist is just like Alice in Figure 27.5C. Hence, that physicist measures a longer time Δt for the muon's lifetime.

SKETCH THE PROBLEM

Figure 27.9 shows the problem schematically.

IDENTIFY THE RELATIONSHIPS

Applying the time dilation result from Equation 27.7, we have

$$\Delta t = \frac{\Delta t_0}{\sqrt{1 - v^2/c^2}}$$

The lifetime for the muon at rest (i.e., measured by a clock at rest relative to the muon) is $\Delta t_0 = \tau$. We are given $v = 0.99c$.

SOLVE

The lifetime of the moving muon is

$$\Delta t = \frac{\tau}{\sqrt{1 - v^2/c^2}} = \frac{\tau}{\sqrt{1 - (0.99c)^2/c^2}}$$
$$\Delta t \approx \boxed{7.1 \times \tau}$$

What does it mean?
According to physicist 2, the moving muon exists for a much longer time than a muon at rest. Experiments with muons show that this result is correct: moving muons do indeed "live" longer before decaying than muons at rest in the laboratory. This is another surprising and counterintuitive result of special relativity.

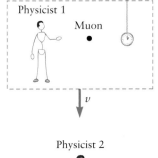
A clock moving with the muon measures the proper time for the muon's lifetime.

Physicist 1 — Muon

v

Physicist 2

Clock at rest on the Earth

Figure 27.9 Example 27.2.

The Twin Paradox

Example 27.2 describes the effect of time dilation on the lifetime of a muon, but a similar result applies in other cases, including the lifetime of a person. Consider an astronaut (Ted) who is on a mission to travel to the nearby star named Sirius[1] and

---

[1]Sirius is actually a double star, but that does not affect the mission; Ted gets to visit both stars.

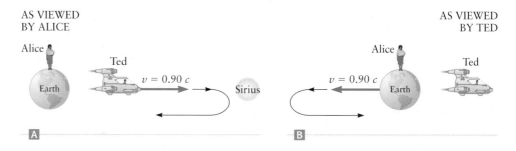

then return to the Earth (Fig. 27.10A). Sirius is 8.6 light-years (ly) from the Earth, which means that light takes 8.6 years to travel from the Earth to Sirius. Ted must therefore travel in a very fast spaceship to complete his mission before he is too old to be an astronaut, so NASA has given him a ship that travels at a speed of $0.90c$. Ted's mission is being tracked by a second astronaut, Alice, who stays behind on Earth. Suppose Alice is Ted's twin, and one of her jobs is to monitor Ted's health and study how he ages during the trip. According to Alice, the round-trip distance is $2 \times 8.6 = 17.2$ ly, so the trip takes a time Δt with

$$\Delta t = \frac{17.2 \text{ ly}}{0.90c} = 19 \text{ years} \qquad (27.10)$$

Alice also knows about time dilation and realizes that Ted's body acts as a clock (just like the muon in Example 27.2). Ted's body clock measures the proper time Δt_0 since it travels along with him. From our work on time dilation, we know that the proper time is shorter than Alice's time in Equation 27.10, with (from Eq. 27.7)

$$\Delta t_0 = (\Delta t)\sqrt{1 - v^2/c^2} \qquad (27.11)$$

Inserting $v = 0.90c$ and our value of Δt gives

$$\Delta t_0 = (\Delta t)\sqrt{1 - v^2/c^2} = (19 \text{ years})\sqrt{1 - (0.90c)^2/c^2} = 8.3 \text{ years}$$

which is the time the trip takes according to Ted's body clock; in other words, Ted ages only 8.3 years, whereas his twin, Alice, ages by 19 years during the trip. When they compare notes at the end of the journey, Ted will be younger than his twin! This result can be understood in simple terms from the basic statement about time dilation: moving clocks (in this case, Ted himself) run slow.

It is interesting to ask now how Ted views the trip. According to Ted, Alice and the Earth both travel at a speed $0.90c$ *relative to his spaceship*, returning to him at the same speed at the end of his journey as sketched in Figure 27.10B. Ted then reasons that he can apply the results for time dilation described above to calculate that, according to his clock, Alice's trip will take 19 years while her body clock will age by the proper time $\Delta t_0 = 8.3$ years. Ted thus concludes that when they get back together, Alice will be younger than he is!

This problem is called the twin "paradox", as it appears that time dilation has led to contradictory results. Alice and Ted cannot both be right; only one can be the younger twin at the end of the journey. Alice's analysis is the correct one: Ted ages less than she does during the trip. The mistake in Ted's analysis is that special relativity applies only to inertial reference frames. Alice stays on the Earth, so she is always in an inertial reference frame and she can apply the results of special rela-tivity. On the other hand, Ted spends some of his time in an accelerating reference frame when his spaceship turns around at Sirius to return to the Earth, and during this time he cannot use special relativity to analyze how Alice ages. That is why Ted's conclusion is wrong, and is the resolution of this apparent paradox.

How Do We Know That Time Dilation Really Happens?

Our applications of time dilation to analyze the decay of a muon in Example 27.2 and the aging of two twins in the twin paradox (Fig. 27.10) are good examples of special relativity, but they don't seem very relevant to everyday life. A similar application of time dilation, however, does have important practical applications. The Global Positioning System (GPS) consists of about 30 satellites that orbit the Earth twice each day (Fig. 27.11). These satellites send signals to receivers on the Earth, and the receivers use the signals to "triangulate" their position with an accuracy of about 10 m. Each satellite contains a very accurate clock, and the GPS receivers compare their clocks with the time signal from each satellite to do this triangulation. The GPS satellite clocks are moving in orbit, so they run slow by the factor $\sqrt{1 - v^2/c^2}$ according to Equation 27.11. The GPS satellites move at a speed of about 4000 m/s, which is much less than the speed of light. Even so, because of time dilation the GPS clocks run slow by about 7×10^{-6} s per day. The satellite signals travel at the speed of light, so the corresponding distance is more than 200 m, which is much larger than the theoretical accuracy of the GPS system. Only by accounting for the effect of time dilation on the satellite clocks can the GPS system successfully determine a position with an uncertainty of only 10 m. (We'll apply time dilation to analyze the GPS performance in Problem 70 at the end of the chapter.)

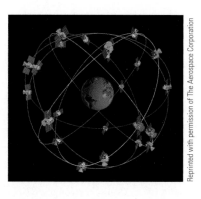

Figure 27.11 The Global Positioning System (GPS). Approximately 30 GPS satellites orbit the Earth, sending signals to the Earth. These signals are used by GPS receivers to determine their location with an accuracy of better than 10 m.

27.4 | SIMULTANEITY IS NOT ABSOLUTE

Two events are simultaneous if they occur at the same time. Our everyday experiences and intuition suggest that the notion of simultaneity is "absolute"; that is, two events are either simultaneous or they are not, for all observers. However, to determine if two events are simultaneous (or not), involves the measurement of time, and our studies of time dilation show that different observers do not always agree on measurements involving clocks and time intervals. So, let's examine what special relativity implies for the notion of simultaneity. If two events are judged simultaneous by one observer, will other observers also find them to be simultaneous? We can answer this question by analyzing the situation in Figure 27.12; Ted is standing at the middle of his railroad car, moving with a speed v relative to Alice, when two lightning bolts strike the ends of the car. The lightning bolts leave burn marks on the ground (points A and B), which indicate the locations of the two ends of the car when the bolts struck. We now ask, "Did the two lightning bolts strike simultaneously?"

We first ask this question of Alice, who notices that she is midway between the burn marks at A and B. Alice also observes that the light pulses from the lightning bolts reach her at the same time (Fig. 27.12C). Since she is midway between points A and B and the light pulses reach her at the same time, Alice concludes that the lightning bolts struck Ted's railroad car at the same time. *As viewed by Alice*, the bolts are simultaneous.

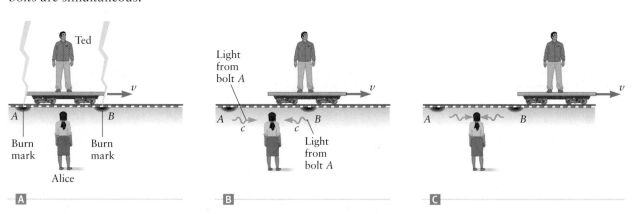

Figure 27.12 An experiment to study simultaneity. **A** Two lightning bolts strike at the ends of Ted's moving railroad car, leaving burn marks on the ground. **B** According to Alice, the lightning bolts are simultaneous. She comes to this conclusion because she is midway between the two burn marks and **C** the light pulses from the two bolts reach her at the same time.

What does Ted have to say? Ted stands at the middle of his railroad car, so (like Alice) he is also midway between the places where the bolts strike. Hence, if the two events are simultaneous *as viewed by Ted*, the light pulses should reach him at the same time. Do they? Alice can also answer this question! Although the light flashes are traveling to Alice, Ted's railroad car is moving to the right. Alice realizes that since Ted is moving, the flash emitted from B will reach him *before* the flash from point A in Figure 27.12B reaches him. Two observers must always agree on the order of two events that occur at the same point in space. (See Insight 27.2.) Hence, Ted will agree that the light pulse from B reaches him first. Ted therefore concludes that the lightning bolt at B struck *before* the bolt at A. In Ted's reference frame, the two lightning bolts are *not* simultaneous. The two lightning bolts in Figure 27.12 are therefore simultaneous for one observer (Alice) but not for another observer (Ted). The question of simultaneity is thus "relative" and can be different in different reference frames.

Time dilation and the relative nature of simultaneity mean that special relativity conflicts with many of our intuitive notions about time. Measurements of time intervals and judgments about simultaneity depend on the motion of the observer. That is very different from Newton's picture, in which "time" is an absolute, objective quantity, the same for all observers.

27.5 | LENGTH CONTRACTION

In the past few sections, we have seen that special relativity forces us to give up the notion of absolute time. Measurements of time and simultaneity are "relative"; that is, they can be different for different observers. Time is just one aspect of a reference frame; reference frames also involve measurements of position and length. What does relativity have to say about these quantities?

Let's consider how Ted and Alice might work together to measure a length or distance. Suppose Alice marks two spots A and B on the ground and measures these spots to be a distance L_0 apart on the x axis (Fig. 27.13). Ted travels along the x direction at constant speed v, and as he passes point A he reads his clock. Ted reads his clock again when he passes point B and calls the difference between the two readings Δt_0. This is a proper time interval because Ted measures the start and finish times at the same location (the center of his railroad car) with the same clock. From Section 27.3 (and Eq. 27.7), we know that when Alice measures the time it takes for Ted to travel from A to B with her clock, she will find a value

$$\text{time measured by Alice} = \Delta t = \frac{\Delta t_0}{\sqrt{1 - v^2/c^2}} \qquad (27.12)$$

Since Ted is traveling relative to Alice at speed v and points A and B are a distance L_0 apart, Alice concludes that

$$v = \frac{L_0}{\Delta t}$$

which can be rearranged as

$$\text{distance measured by Alice} = L_0 = v\,\Delta t \qquad (27.13)$$

Ted will calculate the distance from A to B in the same way, but he will use Δt_0, the time interval measured with his clock, so

$$\text{distance measured by Ted} = L = v\,\Delta t_0 \qquad (27.14)$$

Comparing Equations 27.13 and 27.14, we see that since Δt is different from Δt_0 due to time dilation, the lengths measured by Alice and Ted will also be different. Using Equation 27.11, we get

$$L = v\,\Delta t_0 = v(\Delta t\,\sqrt{1 - v^2/c^2})$$
$$L = L_0\sqrt{1 - v^2/c^2} \qquad (27.15)$$

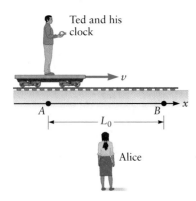

Figure 27.13 Ted can measure the distance between points A and B by using a clock on his railroad car to measure the time Δt_0 it takes for him to travel from A to B, together with his known speed v.

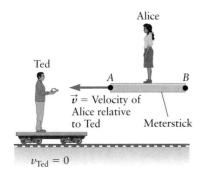

Figure 27.14 A hypothetical experiment in which Ted is at rest and Alice travels on a meterstick with speed v relative to Ted. Ted finds that the moving meterstick is shorter than the length measured by Alice. This situation is very similar to Figure 27.13 because the meterstick (whose ends represent the points A and B) is at rest relative Alice, while it has speed v relative to Ted.

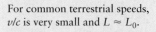

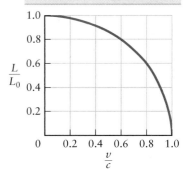

Figure 27.15 Length contraction factor $L/L_0 = \sqrt{1 - v^2/c^2}$ as a function of v/c.

Hence, the length L measured by Ted is *shorter* than Alice's length L_0. This effect is called **length contraction**.

Proper Length

Length contraction also plays a role in the situation in Figure 27.14. Points A and B are now the two ends of a meterstick, and Ted is measuring the length of the meterstick as it moves past him. The meterstick is at rest relative to Alice, who might be standing on it, just as she is at rest relative to the marks A and B on the ground in Figure 27.13; she measures the meterstick in Figure 27.14 to have a length of exactly $L_0 = 1$ m. Ted measures a length L (Eq. 27.15), which is shorter than L_0. We can thus say that *moving metersticks are shortened*. The quantity L_0 is called the **proper length** because it is measured by an observer (Alice) *who is at rest relative to the meterstick.*

Length contraction is described by the factor

$$\frac{L}{L_0} = \sqrt{1 - v^2/c^2} \tag{27.16}$$

Length contraction: moving metersticks are contracted.

This factor is plotted in Figure 27.15; it is very close to 1 when the speed is small and approaches zero as v approaches c.

Notice that we denote the proper time by Δt_0 and the proper length by L_0, in both cases using a subscript zero. In both cases, they are measurements made by an observer who is at rest relative to the "thing" being measured. Proper time is measured by an observer (Ted in Fig. 27.5B) who is at rest relative to the clock used for the measurement. Proper length is measured by an observer (Alice in Fig. 27.14) who is at rest relative to the object whose length is being measured.

Proper time and proper length

| EXAMPLE 27.3 | Length Contraction of a Moving Car |

Consider a race car measured by Ted to be 4.0 m long when moving past him at a speed of 100 m/s (about 200 mi/h). How long is the car when the race is finished and the car is stopped?

RECOGNIZE THE PRINCIPLE

This problem is an example of length contraction, with the race car playing the role of the meterstick in Figure 27.14. The length measured during the race (when $v = 100$ m/s) by an observer (Ted) watching from trackside is the contracted length $L = 4.0$ m (Fig. 27.16A). The length of the race car at the end of the race (when it is not moving) is the proper length L_0; that is the length that would be measured by an observer at rest relative to the car, such as Ted in Figure 27.16B.

SKETCH THE PROBLEM

Figure 27.16 describes the problem.

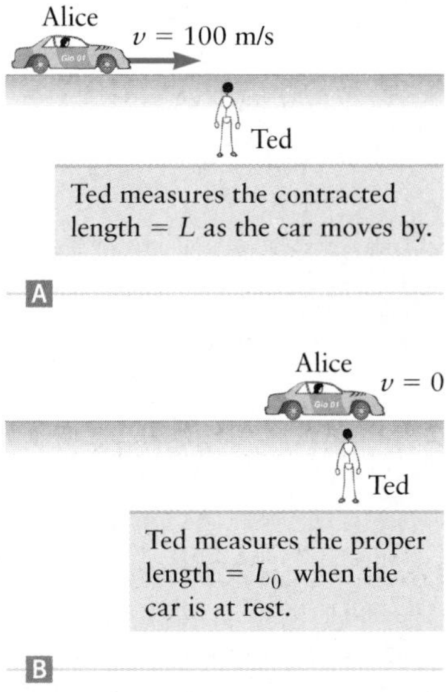

Alice $v = 100$ m/s

Ted

Ted measures the contracted length $= L$ as the car moves by.

A

Alice $v = 0$

Ted

Ted measures the proper length $= L_0$ when the car is at rest.

B

Figure 27.16 Example 27.3.

IDENTIFY THE RELATIONSHIPS

Length contraction is described by Equation 27.15; we have

$$L = L_0\sqrt{1 - v^2/c^2}$$

Rearranging to solve for the proper length gives

$$L_0 = \frac{L}{\sqrt{1 - v^2/c^2}}$$

SOLVE

Inserting the given values of L and v, we find

$$L_0 = \frac{L}{\sqrt{1 - v^2/c^2}} = \frac{4.0\ \text{m}}{\sqrt{1 - (100\ \text{m/s})^2/(3.00 \times 10^8\ \text{m/s})^2}}$$

$$L_0 = \boxed{4.000000000000022\ \text{m}} \tag{1}$$

Your calculator may not be able to show this many zeros to the right of the decimal point. If that is the case, we can again use the results in Insight 27.1 (as shown in Example 27.1) and write

$$L_0 = \frac{L}{\sqrt{1 - v^2/c^2}} \approx L\left(1 + \frac{v^2}{2c^2}\right)$$

Inserting the values of v and c gives

$$L_0 \approx L\left(1 + \frac{v^2}{2c^2}\right) = (4.0\ \text{m})\left[1 + \frac{(100\ \text{m/s})^2}{2(3.00 \times 10^8\ \text{m/s})^2}\right] \approx \boxed{4.0 + 2.2 \times 10^{-13}\ \text{m}}$$

What does it mean?

We had to include many places to the right of the decimal point in Equation (1) to show the effect of length contraction in this example. Length contraction is quite negligible in this case and is always extremely small when v is a typical terrestrial speed. Length contraction is usually only important and measurable when v approaches the speed of light c. (See Fig. 27.15.)

| EXAMPLE 27.4 | Length Contraction and a Moving Muon |

Length contraction can be important for subatomic particles such as muons because these particles often move at very high speeds. Example 27.2 discussed muons created by cosmic rays in the Earth's atmosphere. Consider a muon created at point A in Figure 27.17 moving downward toward the Earth at speed $v = 0.99c$, decaying after a time $\tau = 2.2 \times 10^{-6}$ s. How far will the muon travel relative to the Earth before it decays? Give the answer from (a) the point of view of the muon and (b) from the point of view of an observer, Alice, standing on the ground.

RECOGNIZE THE PRINCIPLE

The muon travels a distance $= v \times$ its lifetime before it decays. Different observers measure different values for this lifetime and hence find different values for the distance traveled. For part (a), consider an observer, Ted, riding along with the muon so that the muon is at rest relative to Ted. The muon exists for a time $= \Delta t_0 = \tau$, which is a proper time because Ted records both the muon's creation and its decay on his clock at the same location (Ted's location). According to Ted, the Earth moves toward him and the muon through a distance $v\,\Delta t_0 = v\tau$ during the muon's lifetime. For part (b), let Alice be an observer on the ground. She measures a longer muon lifetime than τ due to time dilation (Example 27.2), so she will measure a longer travel distance than Ted. Notice that Alice measures the proper length between the muon's starting point at A and ground level at B because she is at rest relative to the points where the muon is created and decays.

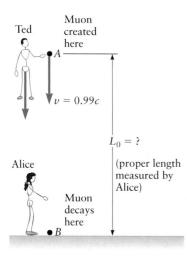

Figure 27.17 Example 27.4.

SKETCH THE PROBLEM

Figure 27.17 shows the problem.

IDENTIFY THE RELATIONSHIPS AND SOLVE

(a) Riding along with the muon, Ted measures the travel distance as

$$L = v \, \Delta t_0$$

Inserting the given values of v and $\Delta t_0 = \tau$, we get

$$L = v \, \Delta t_0 = 0.99c \, \Delta t_0 = 0.99(3.00 \times 10^8 \text{ m/s})(2.2 \times 10^{-6} \text{ s})$$

$$L = \boxed{650 \text{ m}} \tag{1}$$

(b) From Example 27.2, Alice observes the lifetime of the moving muon as

$$\Delta t = \frac{\Delta t_0}{\sqrt{1 - v^2/c^2}} = \frac{\tau}{\sqrt{1 - (0.99c)^2/c^2}} = 7.1\tau \tag{2}$$

During this time, the muon travels a distance

$$L_0 = v \, \Delta t$$

Inserting the values of v and Δt leads to

$$L_0 = v \, \Delta t = (0.99c)(7.1\tau) = 0.99(3.00 \times 10^8 \text{ m/s})(7.1)(2.2 \times 10^{-6} \text{ s})$$

$$L_0 = \boxed{4600 \text{ m}} \tag{3}$$

What does it mean?

Suppose Alice finds that the muon just reaches the ground when it decays (point *B* in Fig. 27.17). Ted will agree with her finding because he is traveling with the muon and will find himself on the ground when it decays. Comparing the answers for parts (a) and (b), we see that the distance traveled by the muon is different for Ted and Alice. How can they agree that the muon reaches the ground? Ted and the muon observe the *contracted* length between *A* and *B*, which is (Eq. 27.15)

$$L = L_0\sqrt{1 - v^2/c^2} \tag{4}$$

Inserting the values of L from Equation (1) and L_0 from Equation (3) along with the speed of the muon gives

$$\text{left side of Equation (4)} = 650 \text{ m}$$

$$\text{right side of Equation (4)} = (4600 \text{ m})\sqrt{1 - (0.99c)^2/c^2} = 650 \text{ m}$$

Equation (4) is thus satisfied by the results in parts (a) and (b) due to the effect of length contraction. Hence, Ted and Alice agree that the muon reaches the ground when it decays. Ted explains it in terms of length contraction and claims that the muon started only 650 m above the Earth's surface at point *A*. Alice explains the result in terms of time dilation and Equation (2).

CONCEPT CHECK 27.2 | Length Contraction and Muons

Suppose the speed of the muon in Example 27.4 is increased to $0.995c$. As viewed by Ted (and the muon), how far does the muon travel before it decays, (a) 6600 m, (b) 4600 m, or (c) 660 m?

What Is "Relative" and What Is Not?

The postulates of special relativity seem quite "innocent," but they have forced us to give up the notions of absolute time and space. These notions were dear to Newton and are ingrained in our intuition, so you should ask, "Are these predictions of special relativity really true?" They are indeed true; a large number of experiments have shown that time dilation and length contraction actually do occur. However, we have seen in Examples 27.1 and 27.3 that at ordinary terrestrial speeds these effects are negligibly small. The reason can be traced to the factor $\sqrt{1 - v^2/c^2}$ that appears in calculations of both time dilation and length contraction. At the typical terrestrial speed of 100 m/s (about 200 mph), this factor is approximately

$$\sqrt{1 - v^2/c^2} = \sqrt{1 - (100 \text{ m/s})^2/(3.00 \times 10^8 \text{ m/s})^2} \approx 1 - 6 \times 10^{-14}$$

The effects of relativity are proportional to how much this factor differs from 1, so relativity is completely negligible in nearly all everyday situations. We can thus continue to use our normal intuition at such "low" speeds, but must realize that these intuitive notions of time and length break down for objects moving at speeds approaching the speed of light.

27.6 | ADDITION OF VELOCITIES

Special relativity has forced us to give up some of our intuitive notions about space and time. Other quantities in mechanics, such as velocity and energy, involve displacement and time, so we must also revisit the behavior of these quantities.

The second postulate of special relativity—that the speed of light is the same for all observers—involves speed, so let's consider how the speed and velocity of an object appear to observers in different reference frames. The "object" could be a particle, a conventional wave (such as a water wave), or a light pulse. In Figure 27.18, Ted is traveling on his railroad car at constant speed v_{TA} relative to Alice when he throws an object, which might be a baseball. We now use the subscript *TA* to indicate that it is the velocity of Ted as measured by Alice, that is, the velocity of Ted relative to Alice. In this section, we will be concerned mainly with the component of the velocity along the *x* axis. Because this component can be positive or negative, we'll refer to it as the *velocity* rather than the speed. For simplicity, we ignore gravity and assume the object (the baseball in Fig. 27.18) travels horizontally, parallel to the velocity of the railroad car. If the velocity of the baseball *relative to Ted* is v_{OT}, what is the velocity v_{OA} of the ball relative to Alice who is at rest on the ground? (Notice how we again use subscripts to indicate which velocity we are referring to and who is measuring it.)

Newton's answer to this question was discussed in Section 27.2 and also in Chapter 4. For the baseball in Figure 27.18, Newton would predict that

$$v_{OA} = v_{OT} + v_{TA} \tag{27.17}$$

or, in words,

velocity of the object (the ball) relative to Alice (v_{OA}) =

velocity of the object relative to Ted (v_{OT}) + velocity of Ted relative to Alice (v_{TA})

This result agrees with our everyday (classical) intuition, and it would be quite an accurate description for a person throwing a real baseball from a real railroad car. Equation 27.17, however, is inconsistent with the postulates of special relativity when the speeds are very high. To see the problem, suppose an object's speed relative to Ted is $v_{OT} = 0.9c$ and that Ted's railroad car is traveling very fast with $v_{TA} = 0.9c$. Inserting these values into Equation 27.17 gives a velocity $v_{OA} = v_{OT} + v_{TA} = 1.8c$. Newton's theory thus gives a velocity greater than the speed of light. Indeed, nothing in Newton's mechanics prevents Ted or a baseball or any other object from traveling at a speed greater than *c*. For an object or reference frame traveling at speed *v*, the factor that appears in time dilation (Eq. 27.7) and length contraction (Eq. 27.15) is $\sqrt{1 - v^2/c^2}$. If $v > c$, the result is an imaginary number, suggesting that Newton's formula for the addition of velocities (Eq. 27.17) cannot be correct for speeds close to the speed of light.

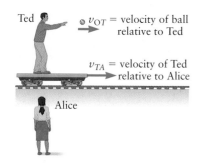

Figure 27.18 While he is moving with velocity v_{TA} relative to Alice, Ted throws a baseball with velocity v_{OT} relative to his railroad car. According to Galilean relativity and Newton's mechanics, the velocity of the ball relative to Alice is $v_{TA} + v_{OT}$. Special relativity gives a different result. Compare this situation with the light from Ted's flashlight in Figure 27.2.

Relativistic Addition of Velocities

It is possible to analyze the situation in Figure 27.18 using the postulates of special relativity. We will not give the derivation, but simply state the result of special relativity for the addition of velocities:

$$v_{OA} = \frac{v_{OT} + v_{TA}}{1 + \dfrac{v_{OT}\,v_{TA}}{c^2}} \qquad\qquad (27.18) \qquad\qquad \text{Relativistic addition of velocities}$$

This relation involves the component along a particular direction (x) of three different velocities, and it is easy to confuse them.

1. v_{OT} is the velocity of an object relative to observer an (e.g., Ted).
2. v_{TA} is the velocity of one observer (Ted) relative to a second observer (Alice).
3. v_{OA} is the velocity of the object relative to the second observer (Alice).

When the velocities v_{OT} and v_{TA} are much less than the speed of light, the factor $v_{OT}v_{TA}/c^2$ in the denominator in Equation 27.18 is much less than 1. In such cases, the relativistic addition of velocities formula (Eq. 27.18) gives nearly the same result as Newton's mechanics (Eq. 27.17). For example, suppose Ted is able to throw the baseball in Figure 27.18 with velocity $v_{OT} = 0.10c$, and his railroad car is moving at velocity $v_{TA} = 0.10c$ relative to Alice. Inserting these values into Equation 27.18 gives

$$v_{OA} = \frac{v_{OT} + v_{TA}}{1 + \dfrac{v_{OT}\,v_{TA}}{c^2}} = \frac{0.10c + 0.10c}{1 + \dfrac{(0.10c)(0.10c)}{c^2}} = 0.198c \qquad\qquad (27.19)$$

We have kept an extra significant figure here to emphasize that the result is very close but not quite equal to the prediction of Newton's formula (Eq. 27.17), which gives

$$v_{OA} = v_{OT} + v_{TA} = 0.10c + 0.10c = 0.200c$$

This answer differs from the relativistic result (Eq. 27.19) by only a small amount (approximately 1%). Hence, for speeds less than about 10% of the speed of light, the Newtonian velocity addition formula works very well (although it is not exact!).

The results are quite different for objects traveling near the speed of light. Suppose $v_{OT} = 0.90c$ and $v_{TA} = 0.90c$. Inserting these values into Equation 27.18, we find

$$v_{OA} = \frac{v_{OT} + v_{TA}}{1 + \dfrac{v_{OT}\,v_{TA}}{c^2}} = \frac{0.90c + 0.90c}{1 + \dfrac{(0.90c)(0.90c)}{c^2}} = 0.994c \qquad\qquad (27.20)$$

We have given the result to three significant figures to emphasize that v_{OA} is just slightly smaller than c, which is quite different from Newton's prediction for this case, which is $v_{OA} = v_{OT} + v_{TA} = 1.8c$. Experiments with particles moving at very high speeds show that the relativistic result is correct.

Relativistic Velocities and the Speed of Light as a "Speed Limit"

The relativistic formula for the addition of velocities (Eq. 27.18) applies when the velocity of the object is parallel (or antiparallel) to the relative velocity of the two observers $\vec{v}_{TA}$ as is the case for Ted, Alice, and the baseball in Figure 27.18. A slightly different result applies when the object is moving perpendicular to $\vec{v}_{TA}$. It is again found that if v_{OT} and v_{TA} are both less than c, then v_{OA} is also less than c. In general, *if an object has a speed less than c for one observer, its speed is less than c for all other observers.* Since no experiment has ever observed an object with a speed greater than the speed of light, c is again a sort of universal "speed limit."

As a final application of Equation 27.18, suppose the object leaving Ted's hand in Figure 27.18 is not a baseball but a pulse of light. Its velocity is then $v_{OT} = c$. What speed will Alice find? Inserting $v_{OT} = c$ into Equation 27.18 gives

$$v_{OA} = \frac{v_{OT} + v_{TA}}{1 + \dfrac{v_{OT} v_{TA}}{c^2}} = \frac{c + v}{1 + \dfrac{cv}{c^2}} = \frac{c + v}{1 + \dfrac{v}{c}}$$

Multiplying the numerator and denominator by c leads to

$$v_{OA} = \frac{c(c + v)}{c\left(1 + \dfrac{v}{c}\right)} = \frac{c(c + v)}{c + v} = c \qquad (27.21)$$

Alice thus finds a speed $v_{OA} = c$ regardless of Ted's speed v. In words, if an object moves at the speed of light for one observer, it moves at the speed of light for all observers. The only known "object" that can move at this speed is light; hence, this result is just a restatement of postulate 2 of special relativity.

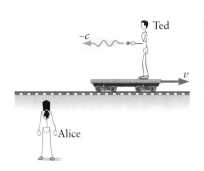

Figure 27.19 Example 27.5.

EXAMPLE 27.5 | Addition of Velocities

Ted is moving at a velocity $+v$ in his railroad car when he aims his flashlight backward as shown in Figure 27.19. What velocity will Alice find for the light from Ted's flashlight?

RECOGNIZE THE PRINCIPLE

According to the second postulate of special relativity, the speed of light must be the same for all observers, so Alice must measure a speed c for the light from Ted's flashlight. This result can also be obtained by applying the addition of velocities relation of special relativity.

SKETCH THE PROBLEM

The problem is described by Figure 27.19.

IDENTIFY THE RELATIONSHIPS

Light from Ted's flashlight moves at a velocity $v_{OT} = -c$ relative to Ted, whereas Ted moves at a velocity $v_{TA} = v$ relative to Alice. We wish to find the velocity of v_{OA} of the light relative to Alice.

SOLVE

Inserting these velocities into Equation 27.18 gives

$$v_{OA} = \frac{v_{OT} + v_{TA}}{1 + \dfrac{v_{OT} v_{TA}}{c^2}} = \frac{-c + v}{1 - \dfrac{cv}{c^2}} = \frac{-c + v}{1 - \dfrac{v}{c}}$$

Multiplying the numerator and denominator by c leads to

$$v_{OA} = \frac{c(-c + v)}{c\left(1 - \dfrac{v}{c}\right)} = \frac{-c(c - v)}{c - v} = \boxed{-c}$$

What does it mean?
The speed measured by Alice is again equal to the speed of light. The negative sign means that the light pulse is moving in the $-x$ direction in Alice's reference frame.

CONCEPT CHECK 27.3 | Riding on a Light Beam

Einstein said that some of the ideas for relativity theory occurred to him when he tried to imagine how the universe would look to a person "riding on a beam of light." Consider an experiment in which Ted rides on a light pulse with both Ted and the light traveling at speed c (Fig. 27.20). Ted is carrying his own flashlight and uses it to send a second light pulse to the right. According to postulate 2 of special relativity, the second light pulse travels at speed c relative to Ted. Is the speed of the second pulse as measured by Alice (a) $2c$, (b) c, or (c) 0?

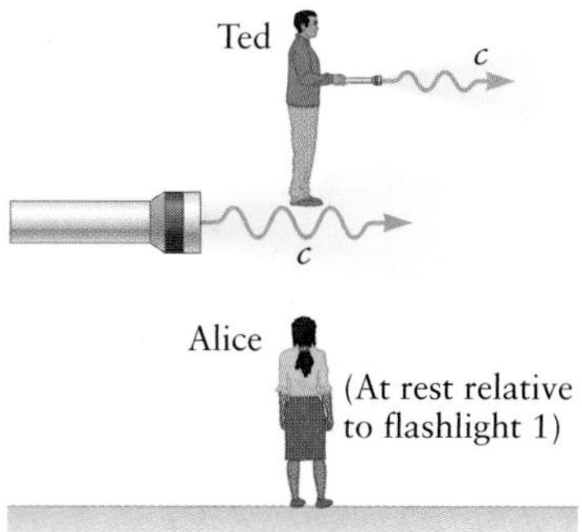

Alice

(At rest relative to flashlight 1)

27.7 | RELATIVISTIC MOMENTUM

Figure 27.20 Concept Check 27.3. What is the speed of the light beam from Ted's flashlight relative to Alice?

According to Newton's mechanics, a particle of mass m_0 moving with a speed v has a momentum

$$\text{Newton's mechanics: } p = m_0 v \tag{27.22}$$

In Chapter 7, we showed that the total momentum of two colliding particles is conserved. We also found that when there are no external forces on a system of particles, the total momentum of the system is conserved. The principle of conservation of momentum is one of the fundamental conservation rules in physics and is believed to be satisfied by all the laws of physics, including the theory of special relativity. In Equation 27.22, however, there is a problem with the expression for momentum.

We can write Newton's result for the momentum of a single particle as

$$p = m_0 \frac{\Delta x}{\Delta t} \tag{27.23}$$

From our analyses of time dilation and length contraction, we know that measurements of both time Δt and length Δx can be different for observers in different inertial reference frames. Should we use the proper time Δt_0 or the proper length Δx_0 to calculate the momentum? Einstein showed that we should use the proper time Δt_0 to calculate the momentum, which amounts to using a clock that travels along with the particle. At the same time, we should use the measurement of length Δx taken by an observer who watches the particle move by with speed v. Because $\Delta t_0 = \Delta t \sqrt{1 - v^2/c^2}$ (from Eq. 27.7), the result is

$$\text{special relativity: } p = m_0 \frac{\Delta x}{\Delta t_0} = m_0 \frac{\Delta x}{\Delta t \sqrt{1 - v^2/c^2}}$$

and since $v = \Delta x / \Delta t$, we get

$$\text{special relativity: } p = \frac{m_0 v}{\sqrt{1 - v^2/c^2}} \tag{27.24}$$

Relativistic momentum

How do we know that this method is the correct way to calculate momentum? Einstein showed that when the momentum is calculated using Equation 27.24, the principle of conservation of momentum is obeyed exactly. Equation 27.24 is indeed the correct expression for momentum and applies even for particles moving at high speeds (i.e., close to the speed of light).

Comparing Equations 27.22 and 27.24, we see that the mass m_0 in Newton's expression is replaced by the factor $m_0/\sqrt{1 - v^2/c^2}$, which suggests that special relativity is trying to tell us something about the concept of mass. Before we discuss that, however, let's analyze the relativistic result for momentum in a little more detail.

When a particle's speed is small compared with the speed of light, the relativistic momentum (Eq. 27.24) becomes

$$(\text{relativistic momentum when } v \ll c) = \frac{m_0 v}{\sqrt{1 - v^2/c^2}} \approx \frac{m_0 v}{\sqrt{1 - 0}} = m_0 v$$

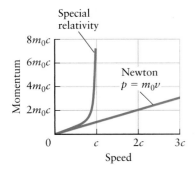

Figure 27.21 Momentum of a particle of mass m_0 as a function of the particle's speed v.

The relativistic momentum is thus equal to Newton's result for the momentum for "slowly" moving objects. As v approaches the speed of light, however, the relativistic result for momentum is very different from Newton's result as shown in Figure 27.21, which compares the two expressions for momentum (Eqs. 27.22 and 27.24).

Newton's result gives a linear relation between p and v. In contrast, the result from special relativity shows that p becomes extremely large as the particle's speed approaches c. A particle's momentum can be increased by the application of forces, which could occur in a collision or in some other way. By applying a very large force or arranging for a very violent collision, the particle's momentum can be made very large, and there is no limit to how large the momentum can be. However, even when the momentum is very large, the particle's speed never quite reaches the speed of light. Hence, when the speed is large, a particle responds to forces and impulses *as if its mass had increased* relative to m_0.

| EXAMPLE 27.6 | Relativistic Momentum of an Electron |

In experiments to study the properties of subatomic particles, physicists routinely accelerate electrons to speeds quite close to the speed of light. For an electron moving with speed $v = 0.99c$, compare its actual momentum to the momentum predicted by Newton's mechanics. Express your answer as the ratio p(special relativity)/p(Newton).

RECOGNIZE THE PRINCIPLE

Since the electron's speed is very close to the speed of light, special relativity will have a very large effect. According to Figure 27.21, the true momentum (as predicted by special relativity) is greater than the momentum predicted by Newton's mechanics.

SKETCH THE PROBLEM

Figure 27.21 shows the difference between the classical (Newton's) momentum and the true momentum (from special relativity).

IDENTIFY THE RELATIONSHIPS

This example is a direct application of the relativistic formula for momentum for an object (an electron) with speed $v = 0.99c$. We take the ratio of the relativistic momentum in Equation 27.24 to p in Equation 27.22. The mass of the electron cancels, leaving only terms that depend on v and c.

SOLVE

Taking the ratio of Equations 27.24 and 27.22 gives

$$\frac{p(\text{special relativity})}{p(\text{Newton})} = \frac{m_0 v / \sqrt{1 - v^2/c^2}}{m_0 v} = \frac{1}{\sqrt{1 - v^2/c^2}}$$

Inserting the given value of v we find

$$\frac{p(\text{special relativity})}{p(\text{Newton})} = \frac{1}{\sqrt{1 - v^2/c^2}} = \frac{1}{\sqrt{1 - (0.99c)^2/c^2}} = \boxed{7.1}$$

What does it mean?
When an electron moves at 99% of the speed of light relative to an observer (physicists in the laboratory), the true (relativistic) momentum is more than seven times larger than that predicted by Newton. When a particle's speed approaches the speed of light, the effects of special relativity become very significant.

CONCEPT CHECK 27.4 | Relativistic Momentum of a Proton

Consider again Example 27.6, but now the particle is a proton instead of an electron. The mass of a proton is about 1800 times greater than the mass of an electron. If the proton is moving at a speed $v = 0.99c$, is the ratio of p(special relativity)/p(Newton) for the proton

    (a) the same as found for an electron with the same speed,

    (b) greater than that found for an the electron with the same speed, or

    (c) less than that found for an electron with the same speed?

27.8 | WHAT IS "MASS"?

The concept of "mass" is basic to all physics. When we first discussed mass in Chapter 1, we appealed to your intuition and suggested in very rough terms that mass is the "amount of matter" carried by a particle. This definition is admittedly imprecise, but it is surprisingly difficult to give a better one. Indeed, our intuitive notion of mass can be traced to Newton; in words, his laws of mechanics say that the mass of a particle determines how it will move (i.e., accelerate) in response to applied forces. According to Newton's second law, the acceleration of a particle of mass m_0 is given by

$$a = \frac{\sum F}{m_0} \tag{27.25}$$

so m_0 is just the constant of proportionality that relates acceleration and force. Newton's laws implicitly assume this constant of proportionality is really a constant,[2] but how do we know that this is true?

The only way to prove that Newton's second law is correct is through experiments, and experiments show that it does work with very high accuracy *provided the object's speed is small compared with the speed of light.* In the early part of the 1900s, however, experiments began to probe the regime in which a particle's speed approaches c, and these experiments showed that Newton's second law breaks down at such high speeds. When the postulates of special relativity are applied to this problem, we find that acceleration and force are related instead by

$$a = \frac{\sum F}{m_0/(1 - v^2/c^2)^{3/2}} \tag{27.26}$$

Comparing this relation with Newton's second law (Eq. 27.25), we have replaced m_0 by the factor $m_0/(1 - v^2/c^2)^{3/2}$. That is,

$$m_0 \quad \rightarrow \quad \frac{m_0}{(1 - v^2/c^2)^{3/2}} \tag{27.27}$$

At low speeds ($v \ll c$), the quantity on the right approaches m_0, so the two acceleration equations (Eqs. 27.25 and 27.26) are indistinguishable. However, as v approaches the speed of light, the denominator in Equation 27.27 becomes very small. When $v \approx c$, the acceleration in Equation 27.26 is thus very small, even when the force is very large. Hence, the particle responds to a force *as if it had a mass larger than* m_0. What's more, this "enhancement of the mass" depends on the particle's speed.

We encountered a similar effect when we considered relativistic momentum (Eq. 27.24). There we noticed that the relativistic result for p can be obtained by replacing the mass m_0 in Newton's expression by the term $m_0/\sqrt{1 - v^2/c^2}$. Then, in our interpretation of the relativistic formula, we noted that at high speeds v the particle responds to impulses and forces "as if" its mass were larger than m_0.

So, when dealing with both acceleration and momentum, a particle behaves "as if" it had a mass larger than expected from its behavior at low speeds. For this reason, physicists have introduced the term **rest mass**, denoted by m_0, for the mass measured by an observer who is moving very slowly relative to the particle (so that

---

[2]Assuming, of course, the particle does not break apart, for example.

$v \approx 0$ in Eqs. 27.24 and 27.26). Some authors call it the "proper mass" in analogy with the proper length of a meterstick at rest, but the term *rest mass* is much more widely used in physics, so we'll use it in this book.

When dealing with momentum and acceleration for particles moving at very high speeds, we can solve any problem we might encounter by simply using the results from special relativity in Equations 27.24 and 27.26, but we need to understand what these results mean: that Newton's simple notions about mass break down for particles moving at high speeds. The best way to describe the "mass" of a particle is through its rest mass m_0, which describes the motion of a particle for an observer who is moving very slowly relative to the particle.

27.9 | MASS AND ENERGY

As we saw in the last two sections, the concepts of mass and momentum are not as simple as Newton had assumed. You should therefore not be surprised that we must also reexamine our concepts of work and energy. That will lead us to what is probably the most famous equation in physics.

In Chapter 6, we derived a relation between mechanical work and kinetic energy and found that the kinetic energy of a particle of mass m_0 moving at speed v is

$$\text{Newton's kinetic energy: } KE = \tfrac{1}{2}m_0 v^2 \tag{27.28}$$

We can derive the corresponding relativistic expression for kinetic energy by applying work–energy ideas. According to the work–energy theorem (Chapter 6), the work done on a particle is equal to the change in the particle's kinetic energy. The work done by an applied force F is

$$W = F\,\Delta x$$

where Δx is the particle's displacement. Inserting the relation between F and acceleration according to special relativity (Eq. 27.26), we get

$$W = F\,\Delta x = \frac{m_0 a}{(1 - v^2/c^2)^{3/2}}\,(\Delta x) \tag{27.29}$$

which is similar to our derivation of Newton's kinetic energy in Chapter 6 (Eqs. 6.6 through 6.10). Simplifying the right-hand side of Equation 27.29 to find W involves more mathematics than we can include here, so we'll simply state the answer for a particle that starts at rest and comes to a final speed v. In this case, the work done equals the particle's relativistic kinetic energy KE, and the result is

Relativistic kinetic energy

$$KE = \frac{m_0 c^2}{\sqrt{1 - v^2/c^2}} - m_0 c^2 \tag{27.30}$$

Equation 27.30 looks quite different from Newton's result ($KE = \tfrac{1}{2}m_0 v^2$), but the two are actually closely related. To see the connection, we first rewrite Equation 27.30 as

$$KE = \frac{m_0 c^2}{\sqrt{1 - v^2/c^2}} - m_0 c^2 = m_0 c^2 \left(\frac{1}{\sqrt{1 - v^2/c^2}} - 1 \right)$$

According to Insight 27.1, for small v ($v \ll c$) we have

$$\frac{1}{\sqrt{1 - v^2/c^2}} \approx 1 + \frac{v^2}{2c^2}$$

Inserting this result into our expression for the kinetic energy gives

$$KE \approx m_0 c^2 \left(1 + \frac{v^2}{2c^2} - 1 \right)$$

$$KE \approx \tfrac{1}{2}m_0 v^2 \tag{27.31}$$

which is identical to Newton's kinetic energy. Hence, for small v (particles moving at speeds much less than the speed of light), we get our familiar result for KE as we should have expected. As v approaches c, however, the relativistic result for the kinetic energy has a different behavior than does Newton's expression. Figure 27.22 shows the kinetic energy as a function of the particle speed v (Eq. 27.30). At high speeds, v runs into the "speed limit" c. Although the kinetic energy can be made very large, the particle's speed never quite reaches the speed of light. As the speed increases, the particle responds as if it had a larger mass than at low speeds.

The expression for the kinetic energy in Equation 27.30 is the change in energy of a particle due to work done on the particle. We can also think of it as the difference between the final energy and the initial energy,

$$KE = \overbrace{\frac{m_0 c^2}{\sqrt{1 - v^2/c^2}}}^{\text{final energy}} - \overbrace{m_0 c^2}^{\text{initial energy}}$$

The last term, $m_0 c^2$, is a constant, however. What is the meaning of this initial energy term? Einstein proposed that this term is the energy of the particle, even when it is at rest. We now call it the **rest energy** of the particle:

$$\text{rest energy} = m_0 c^2 \qquad (27.32)$$

In words, Equation 27.32 asserts that a particle has an amount of energy $m_0 c^2$ even when its speed v is zero. When a particle is in motion, it also has kinetic energy; to get the total energy of the particle, we must then add the rest energy to the kinetic energy to get the total relativistic energy TE:

$$TE = \frac{m_0 c^2}{\sqrt{1 - v^2/c^2}} \qquad (27.33)$$

The result for the rest energy in Equation 27.32 forces us to rethink our notion of mass. The rest energy relation of special relativity implies that **mass is a form of energy**. This suggests that it is possible to convert an amount of energy $m_0 c^2$ into a particle of rest mass m_0 or to convert a particle of rest mass m_0 into an amount of energy $m_0 c^2$. So, the principle of conservation of energy must be extended to include this type of energy.

Since the speed of light is a very large number, the magnitude of the rest energy can be very large even when the rest mass m_0 is small. A rest mass of only 1 kg corresponds to a rest energy of

$$m_0 c^2 = (1 \text{ kg})(3.00 \times 10^8 \text{ m/s})^2 = 9 \times 10^{16} \text{ J} \qquad (27.34)$$

which is a tremendous amount of energy. Approximately 3×10^{17} J of energy is consumed each day in the United States. According to Equation 27.34, this energy could be obtained by converting a rest mass of just 3 kg into energy. Special relativity tells us that this much energy is available in principle, but it does not tell us how to actually convert mass into energy. Nuclear reactions (Chapter 30) are one way to do so.

CONCEPT CHECK 27.5 | Relativistic Kinetic Energy

A particle has a kinetic energy that is twice its rest energy. What is the speed of the particle, (a) $c/2$, (b) $c/4$, or (c) $(2\sqrt{2}\,c)/3$?

EXAMPLE 27.7 | Mass–Energy and Nuclear Weapons

The nuclear bomb dropped on Hiroshima, Japan, in 1945 released an amount of energy equivalent to that found in an explosion of approximately 20 kilotons of trinitrotoluene (TNT), a chemical explosive that releases about 4.2 MJ/kg when it explodes. How much rest mass was converted to energy at Hiroshima?

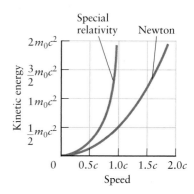

Figure 27.22 Kinetic energy of a particle as a function of its speed v. Even if the particle has a very large kinetic energy, its speed never quite reaches the speed of light.

Rest energy

Total relativistic energy

RECOGNIZE THE PRINCIPLE

This problem involves the relativistic relation between mass and energy. Given the properties of TNT as an explosive, we can find how much energy was released at Hiroshima. We can then use the relation (rest energy) = $m_0 c^2$ to find the corresponding amount of rest mass that was converted to energy in the explosion.

SKETCH THE PROBLEM

No sketch is necessary.

IDENTIFY THE RELATIONSHIPS

We first compute the mass of 20 tons of TNT, arriving at a value in kilograms. Then we can find the amount of energy released by using the fact that 1 kg of TNT yields an energy of 4.2 MJ. The expression for the rest energy (Eq. 27.32) then gives the amount of mass required (i.e., consumed in the explosion).

SOLVE

A weight of 1 ton is equal to 2000 lb (U.S. customary units). We can obtain the mass of TNT in kilograms using the fact that a force of 1 lb equals 4.45 N. This corresponds to a mass m in kilograms given by 1 lb = 4.45 N = mg = m × (9.8 m/s^2). We can thus write:

$$\text{mass (in kilograms)} = \text{weight (in pounds)} \times \left(\frac{4.45}{9.8}\right) = \text{weight (in pounds)} \times 0.45$$

$$m_{\text{TNT}} = (20 \times 10^3 \text{ ton})\left(\frac{2000 \text{ lb}}{1 \text{ ton}}\right)(0.45) = 1.8 \times 10^7 \text{ kg}$$

An explosion with this amount of TNT releases an energy

$$E = m_{\text{TNT}}\left(\frac{4.2 \text{ MJ}}{1 \text{ kg}}\right) = (1.8 \times 10^7)(4.2 \times 10^6) \text{ J} = 7.6 \times 10^{13} \text{ J}$$

This result equals the amount of energy released by the uranium fission bomb that exploded at Hiroshima, so it is the amount of rest energy released. We thus have

$$E = \text{rest energy} = m_0 c^2 = 7.6 \times 10^{13} \text{ J}$$

Solving for m_0 gives

$$m_0 = \frac{7.6 \times 10^{13} \text{ J}}{c^2} = \frac{7.6 \times 10^{13} \text{ J}}{(3.00 \times 10^8 \text{ m/s})^2} = \boxed{8.4 \times 10^{-4} \text{ kg}}$$

What does it mean?

Less than 1 g of matter was converted to energy in this explosion! Even a small amount of rest mass contains a very large rest energy.

Mass–Energy Conversion and Chemical Reactions

The conversion of mass into energy is an important effect in nuclear reactions (Example 27.7), but it also occurs in other cases. Consider a chemical reaction in which a hydrogen atom is dissociated. The ionization energy of a hydrogen atom is 13.6 eV (electron-volts), meaning that when a proton and electron are bound together in a hydrogen atom their total energy is 13.6 eV lower than when they are separated. The principle of conservation of energy then implies that

total energy of hydrogen atom =

total energy of electron plus total energy of proton − 13.6 eV

If all these particles are observed at rest, the total energies of the hydrogen atom and of the electron and proton are just the rest energies. We thus have

$$m_0(\text{hydrogen atom})c^2 = m_0(\text{electron})c^2 + m_0(\text{proton})c^2 - 13.6 \text{ eV} \qquad \text{(27.35)}$$

The mass of a hydrogen atom must therefore be less than the sum of the masses of an electron and proton. So, mass is *not conserved* when a hydrogen atom dissociates. In this case and in other chemical reactions, the change in the total rest mass Δm_0 in the reaction is small. For hydrogen dissociation (Eq. 27.35), we get

$$\Delta m_0 c^2 = 13.6 \text{ eV}$$

Expressing Δm_0 in SI units, we find

$$\Delta m_0 = \frac{13.6 \text{ eV}}{c^2} = \frac{(13.6 \text{ eV})(1.60 \times 10^{-19} \text{ J/eV})}{(3.00 \times 10^8 \text{ m/s})^2} = 2.4 \times 10^{-35} \text{ kg}$$

The mass of a proton is 1.7×10^{-27} kg, so Δm_0 is only about 0.000001% of the mass of the proton.

EXAMPLE 27.8 | Energy Released in Electron–Positron Annihilation

A positron is a subatomic particle with precisely the same mass as an electron, but with a positive charge $+e$; that is, its charge is opposite that of the electron. A positron is an example of **antimatter** (Chapter 31), and it is the antimatter "twin" of an electron. When an electron comes into close contact with a positron, they can annihilate each other, converting all their rest mass into energy. Typically, this annihilation reaction results in electromagnetic radiation. How much energy is released when an electron and a positron annihilate each other?

RECOGNIZE THE PRINCIPLE

In electron and positron annihilation, all the mass of both the electron and the positron are converted into energy. Both of these particles have a rest mass $m_0 = m_e$ (the mass of the electron), so we can use the known mass of the electron to find the amount of energy released from the relation (rest energy) $= m_0 c^2$.

SKETCH THE PROBLEM

Figure 27.23 describes the problem.

IDENTIFY THE RELATIONSHIPS

The (rest) mass for an electron is $m_e = 9.11 \times 10^{-31}$ kg (see Appendix A, Table A.1), and the positron's rest mass also equals m_e. We can use Equation 27.32 to get the total rest energy of the two particles:

$$E_{\text{total}} = \text{rest energy} = 2(m_e c^2)$$

SOLVE

This rest energy is the energy released in the annihilation process. Inserting the values of m_e and c gives

$$E_{\text{total}} = 2(m_e c^2) = 2(9.11 \times 10^{-31} \text{ kg})(3.00 \times 10^8 \text{ m/s})^2 = \boxed{1.6 \times 10^{-13} \text{ J}}$$

What does it mean?

When an electron and positron annihilate, their combined electric charge also disappears. Charge is conserved because the total charge of the two particles is $+e - e = 0$ before the annihilation. Although rest mass is not conserved, all reactions must conserve both electric charge and total energy.

CONCEPT CHECK 27.6 | Relativistic Mass and Energy

Ted takes a walk on a very dark night. It is so dark that he uses his flashlight to find his way home. Is the total mass of the flashlight (including its batteries) when Ted gets home (a) less than, (b) greater than, or (c) exactly the same as its mass just before he left?

BEFORE ANNIHILATION

 AFTER ANNIHILATION

Electromagnetic radiation is emitted.

Figure 27.23 Example 27.8. In electron–positron annihilation, all the rest mass of the two particles is converted to energy.

Insight 27.4
CONSERVATION PRINCIPLES

In Newton's mechanics, mass is a conserved quantity, so the total mass of a closed system cannot change. Special relativity tells us that mass is in fact not conserved. The principle of conservation of energy must instead be extended to include the rest energy (Eq. 27.32). What about our other conservation rules? Momentum is still conserved in collisions, but we must use the relativistic expression for momentum in Equation 27.24. Electric charge is also conserved, but it is possible to create or annihilate charges (as in Example 27.8) as long as the total charge does not change.

27.10 | THE EQUIVALENCE PRINCIPLE AND GENERAL RELATIVITY

Special relativity is concerned with the laws of physics in inertial reference frames; these are reference frames that move at constant velocity. A *noninertial* reference frame is one that has a nonzero acceleration. Physics in noninertial frames is described by *general relativity*. Most of the results of special relativity, such as time dilation and length contraction, can be worked out using algebra, but general relativity involves much more complicated mathematics. We will therefore only describe the main ideas and results of the general theory.

General relativity is based on a postulate known as the *equivalence principle*:

Equivalence principle

The effects of a uniform gravitational field are identical to motion with constant acceleration.

An example of this equivalence is shown in Figure 27.24. In Figure 27.24A, Ted stands in an elevator compartment near the Earth's surface, with the elevator at rest. Ted feels the normal force exerted by the floor on his feet; he has mass m_0, so this force has magnitude $m_0 g$. From his study of physics, Ted deduces that g is a result of the Earth's gravity and hence that he is situated in a gravitational field.

Figure 27.24B shows the same elevator, but now it is located in distant space, far from any planets or stars, and the elevator compartment has an acceleration $a = g$. Since Ted's acceleration is a, Newton's second law tells us that a force $F = m_0 a = m_0 g$ is exerted by the elevator floor on Ted's feet.[3] The force in Figure 27.24B is thus not due to a gravitational force. According to the equivalence principle, however, there is no way for Ted to tell the difference between the effects of the gravitational field in Figure 27.24A and the accelerated motion in Figure 27.24B; they are completely equivalent.

This equivalence argument seems simple and innocent, but it has the following profound consequences.

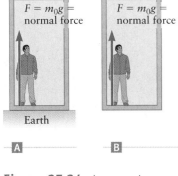

Figure 27.24 A person in an elevator feels a normal force $m_0 g$ on the bottoms of his feet. He cannot tell if he is **A** at rest in a gravitational field so that the normal force is the result of a gravitational force or if **B** there is no gravitational field and the elevator has an acceleration $a = g$. According to the equivalence principle, there is no way to tell the difference between these two situations.

1. *Equivalence of inertial and gravitational mass.* The concept of mass enters Newton's mechanics in two places. The mass in Newton's second law is called the *inertial mass* $m_{inertial}$, while the masses in the law of gravitation (Eq. 5.16) are called *gravitational mass* m_{grav}. Newton assumed these two types of mass are the same, but why should that be? In the elevator in a gravitational field (Fig. 27.24A), the force exerted on Ted is due to gravity, so $F = m_{grav} g$. In the accelerated elevator in Figure 27.24B, we have $F = m_{inertial} a = m_{inertial} g$. According to the equivalence principle, these two forces are precisely the same, so *the inertial mass must equal the gravitational mass*. We can also turn this argument around: the observation (by Newton and many others) that the inertial and gravitational masses are the same leads us to the equivalence principle.

2. *Deflection of light by gravity.* Figure 27.25 shows another experiment with Ted's elevator. A light pulse travels into the elevator car through a window on the left. In Figure 27.25A, the elevator is in distant space and has zero acceleration ($a = 0$). An observer (Ted) in the elevator then sees the light pulse travel in a straight-line path across the car. In Figure 27.25B, the elevator is accelerating upward; Ted now finds that *relative to him* the light beam travels on a downward arc. A nonaccelerated observer outside the elevator would say that the light pulse travels in a straight line, but *relative to the elevator* the light travels along a curved path. In Figure 27.25C, the experiment is repeated in a gravitational field and with no acceleration. According to the equivalence principle, the results must be the same as in the accelerated case in Figure 27.25B. So, Ted

---

[3]We assume the elevator's speed is much smaller than the speed of light, so the difference between Newton's second law (Eq. 27.25) and the relativistic result in Equation 27.26 is negligible.

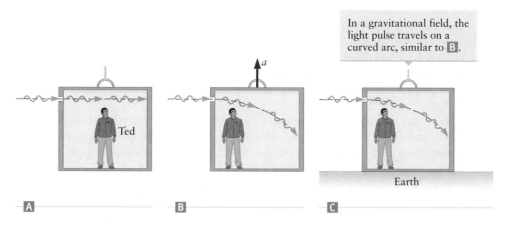

In a gravitational field, the light pulse travels on a curved arc, similar to B.

a

Ted

Earth

A   B   C

Figure 27.25 A light ray moving through the elevator compartment in Figure 27.24. A When the acceleration of the car is $a = 0$, the light travels in a straight line through the elevator. In B, the elevator is accelerating upward. According to an observer in the elevator, the light beam travels in a downward arc. C According to the equivalence principle, the situation in part B is identical to the effect of a gravitational field. Hence, a gravitational field must also deflect a light beam.

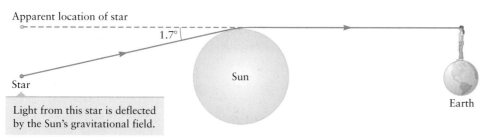

Apparent location of star

1.7°

Sun

Star

Earth

Light from this star is deflected by the Sun's gravitational field.

Figure 27.26 Deflection of light by the Sun's gravitational field. The deflection is largest for light beams that travel nearest the Sun. Light that just grazes the surface of the Sun is deflected through an angle of about 1.7°. This effect is easiest to measure when the Sun's light is blocked by the Moon during a solar eclipse.

must again observe that the light follows a curved path, but now this curvature is produced by the gravitational field. In words, we say that light is "bent" by gravity. This conclusion is a straightforward result of the equivalence principle, but when Einstein proposed that light is bent by a gravitational field his prediction was met with considerable skepticism. Einstein's ideas were confirmed in experiments in 1919, in which light passing near the Sun during an eclipse (Fig. 27.26) was found to be deflected by the predicted amount. This result made Einstein an international celebrity.

Black Holes

The most massive object in our solar system is the Sun, but even its large gravitational field only deflects light a small amount (Fig. 27.26). However, the universe contains other, much more massive objects. *Black holes* contain so much mass that light is not able to escape from their gravitational attraction. They are called "black" because any light that passes nearby is attracted to the black hole and never escapes. Hence, a region of space containing a black hole appears dark to an observer on the Earth. How, then, do we "see" a black hole? One way to find a black hole is through its effect on the motion of nearby objects. Stars near a black hole move along curved trajectories due to the gravitational attraction of the black hole. Astronomical observations of these curved trajectories give the mass and location of the black hole.

EXAMPLE 27.9 | Size of a Black Hole

One way to measure the size of a black hole is to study the motion of light emitted a distance R from the center of the black hole. If R is small, the gravitational force of the black hole is strong enough that all light is attracted to the black hole and cannot escape. On the other hand, if R is large, the light can escape. This problem is similar to finding the escape velocity for a rocket launched from the surface of a planet into

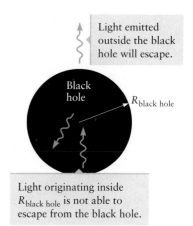

Light emitted outside the black hole will escape.

Black hole

$R_{black\ hole}$

Light originating inside $R_{black\ hole}$ is not able to escape from the black hole.

Figure 27.27 Example 27.9.

outer space (Chapter 6), with light now playing the role of the rocket. We calculated the velocity required for a rocket to escape from a planet of radius R in Equation 6.24. Use that result to find the value of R at which light can just barely escape from a black hole of mass M.

RECOGNIZE THE PRINCIPLE

In Chapter 6, we considered the motion of a particle with speed v that is "fired" into space at a distance R from the Earth's center. Using the principle of conservation of energy, we found that such a particle could escape to an infinite distance if its speed exceeds the escape velocity, which is given by

$$v_{escape} = \sqrt{\frac{2GM}{R}} \tag{1}$$

where M is the mass of the Earth and R is the initial distance from the Earth's center. The same result for the escape velocity also applies to light near a black hole. Since light moves at speed c, it can escape from a black hole if the escape velocity is less than c.

SKETCH THE PROBLEM

Figure 27.27 describes the problem.

IDENTIFY THE RELATIONSHIPS AND SOLVE

Setting $v_{escape} = c$ in Equation (1) and solving for R leads to

$$v_{escape} = c = \sqrt{\frac{2GM}{R}}$$

$$c^2 = \frac{2GM}{R}$$

$$\boxed{R = \frac{2GM}{c^2}} \tag{2}$$

What does it mean?

In Chapter 30, we'll discuss how stars use nuclear reactions to generate their energy. It is believed that black holes are produced by stars that collapse to a very small size when they exhaust their nuclear fuel. Let's calculate the value of R for a black hole with the mass of the Sun ($M \approx 2.0 \times 10^{30}$ kg). Inserting this value into Equation (2) gives

$$R = \frac{2GM}{c^2} = \frac{2(6.67 \times 10^{-11} \text{ N} \cdot \text{m}^2/\text{kg}^2)(2.0 \times 10^{30} \text{ kg})}{(3.00 \times 10^8 \text{ m/s})^2} \approx 3000 \text{ m}$$

Considering the mass and (current) radius of the Sun, this size is incredibly small. For comparison, a typical airport runway is also about 3000 m in length. A black hole is thus an extremely dense object.

Detecting a Black Hole: Gravitational Lensing

Another way to detect a black hole is through *gravitational lensing*. This effect is illustrated in Figure 27.28A, which shows light from a star as it passes by a black hole on its way to an observer on the Earth. If the black hole is between the star and the Earth, light from the star can pass by either side of the black hole and still be bent by gravity so as to reach the Earth. This figure shows a view in only one plane; light from the star can also pass above and below the plane of the drawing. Hence, light from a single star can produce multiple images and even a set of images along a ring or arc. Astronomers have identified several such gravitational lenses; one example is shown in Figure 27.28B. The positions and shapes of the images can be

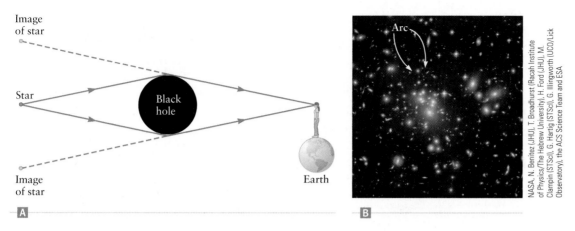

A   B

Figure 27.28 ◰ A very massive object such as a black hole acts as a gravitational "lens." Here light rays from a star pass on both sides of a black hole and arrive at an observer, forming two separate images of the star. There will also be rays that come out of the plane of the drawing, producing additional images or a ring of images. ◰ Example of gravitational lensing observed with the Hubble Space Telescope. Each circular arc of light (one is indicated by the arrows) comes from a single star. These arcs are due to light rays that travel out of the plane in part A.

used to deduce the mass of the black hole. From this and other types of astronomical observations, a number of black holes have been found. In fact, a black hole resides at the center of the Milky Way, our galaxy.

CONCEPT CHECK 27.7 | Size of the Black Hole at the Center of the Milky Way

Astronomers have found that the black hole at the center of the Milky Way has a mass about three million times larger than our Sun. Hence, this black hole has a mass $M = (3 \times 10^6) \, M_{Sun} = 6 \times 10^{36}$ kg. Is the radius of this black hole (a) 3000 m, (b) 9×10^9 m, (c) 3×10^9 m, or (d) 9.5×10^6 m? Is it larger or smaller than our Sun? *Hint*: Use the result in Example 27.9.

27.11 | RELATIVITY AND ELECTROMAGNETISM

Special relativity is all about how the physical world looks to inertial observers moving at different speeds. So far, our two observers, Ted and Alice, have focused on several problems in mechanics. Let's now consider a problem in electromagnetism and give them the job of calculating the force between an infinite line of charge with charge per unit length λ and a point charge $+q$ located a distance r away. Figure 27.29A shows Alice's analysis; she is at rest relative to both the charged line and the point charge. Alice applies results from Chapter 17 and finds that the line of charge produces an electric field $E = \lambda/(2\pi\varepsilon_0 r)$ directed away from the line, so the force on the point charge is

$$F_E(\text{Alice}) = \frac{q\lambda}{2\pi\varepsilon_0 r} \qquad (27.36)$$

Let's now consider how Ted would analyze this situation from his moving railroad car. Ted is moving to the left at speed v in Figure 27.29B, so from his point of view the charged line and the point charge are both moving to the right with speed v as shown in Figure 27.29C. In Ted's reference frame, the moving charged line acts as a current, so according to him this line produces both an electric field *and* a magnetic field. Since the test charge $+q$ is moving through this magnetic field, Ted concludes that there is also a magnetic force on this test charge. The current is to the right in Figure 27.29D, so the magnetic field at the test charge is directed into the page; applying right-hand rule 2 (Chapter 20) gives a magnetic force F_B (Ted) directed toward the line.

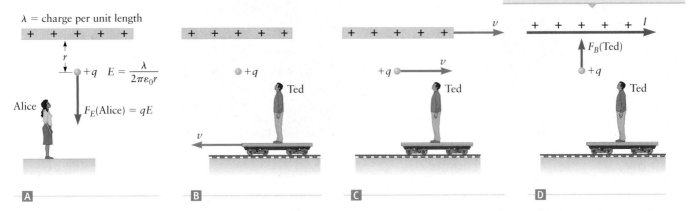

λ = charge per unit length

$E = \dfrac{\lambda}{2\pi\varepsilon_0 r}$

$F_E(\text{Alice}) = qE$

Alice

Ted

Ted

Ted

$F_B(\text{Ted})$

Figure 27.29 ◻ An infinite line of charge at rest relative to Alice. This line produces an electric field $E = \lambda/(2\pi\varepsilon_0 r)$ and hence an electric force on the test charge $+q$. ◻ The same situation viewed by Ted as he moves past on his railroad car. ◻ In Ted's frame of reference, both the charged line and the test charge move with speed v. ◻ According to Ted, the moving charged line is equivalent to an electric current. This current produces a magnetic field and thus also a magnetic force on the test charge.

Ted will thus say that there is an electric force *and* a magnetic force on the particle. On the other hand, according to Alice there is *only* an electric force. Who is correct? The answer is that they are *both* correct!

How can Ted possibly agree with Alice? According to Ted, the moving charged line is contracted (length contraction) as discussed in Section 27.5. Squeezing this charge into a smaller length *increases* the charge per unit length and thus *increases* the electric force calculated by Ted when he uses Equation 27.36. Thus, Ted observes a *larger* electric field and a *larger* electric force directed away from the line than that found by Alice. We have also seen, however, that Ted observes a magnetic force directed *toward* the line. A careful analysis shows that this magnetic force precisely cancels the extra electric force, so Ted and Alice agree on the total force on the particle. Alice will attribute it completely to an electric force, whereas Ted will attribute it to a combination of electric and magnetic forces.

This example shows that the values of the electric and magnetic fields in a particular situation depend on the motion of the observer; here Ted finds a nonzero magnetic field, while for Alice the magnetic field is zero. It is interesting that Maxwell's equations already contain this effect; his equations of electromagnetism were already consistent with the theory of special relativity, but Maxwell (and others) didn't fully appreciate it until Einstein discovered special relativity.

27.12 WHY RELATIVITY IS IMPORTANT

Generally speaking, relativistic effects become important for objects or observers whose speed approaches the speed of light. Although there are a few everyday situations, such as the Global Positioning System (GPS) discussed in Section 27.3, where the effect of special relativity is important, relativity plays no significant role in most terrestrial situations. Even so, relativity is an extremely important part of physics for the following reasons.

1. The relation between mass and energy and the possibility that mass can be converted to energy (and vice versa) mean that mass is not conserved. Instead, we have a more general view of energy and its conservation. The three hallmark

conservation principles of physics are thus (a) conservation of energy, (b) conservation of momentum, and (c) conservation of charge. It is believed that *all* the laws of physics must obey these three principles.

2. The rest energy of a particle is huge, which has important consequences for the amount of energy available in processes such as nuclear reactions.

3. Relativity changes our notions of space and time. At the start of this book, we mentioned that time and position are two primary quantities in physics and that it is not possible to give precise definitions of such quantities. The best we can do is use our intuition to understand the meaning of space and time. Now we have found that our everyday intuition breaks down when applied to special relativity. While this may seem surprising, there is no reason why an intuition developed from terrestrial experience should be reliable when applied to stars and black holes.

4. Relativity plays a key role in understanding how the universe was formed and how it is evolving. Black holes can't be understood without relativity.

5. Relativity shows that Newton's mechanics is not an exact description of the physical world. Instead, Newton's laws are only an approximation that works very well in some cases, but not in others. So, we shouldn't discard Newton's mechanics, but we should understand its limits.

SUMMARY | Chapter 27

KEY CONCEPTS AND PRINCIPLES

The postulates of special relativity
Special relativity is based on two postulates:

1. **The laws of physics are the same in all inertial reference frames.**
2. **The speed of light in a vacuum is constant, independent of the motions of the source and observer.**

Time dilation and length contraction are consequences of these two postulates.

General relativity
General relativity is based on the **equivalence principle**: the effects of a uniform gravitational field are identical to motion with constant acceleration.

One consequence of the equivalence principle is that light rays are deflected by a gravitational field.

APPLICATIONS

Time dilation
The **proper time** Δt_0 is the time interval between two events as measured by an observer for which the two events occur at the same location (i.e., an observer at rest relative to the events). A second observer moving at a relative speed v measures a longer time,

$$\Delta t = \frac{\Delta t_0}{\sqrt{1 - v^2/c^2}}$$

(27.7) (page 922)

In words, we say that *moving clocks run slow.*

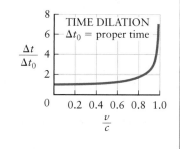

(Continued)

Length contraction

The *proper length* of an object is the length L_0 measured by a person at rest relative to the object. If a second observer is moving with a speed v relative to the object, that observer will measure a shorter length,

$$L = L_0\sqrt{1 - v^2/c^2} \qquad \textbf{(27.15)} \text{ (page 928)}$$

An equivalent situation occurs when the object is moving with speed v relative to an observer. In words, we say that *moving metersticks are contracted*.

Relativistic addition of velocities

Observer Alice is at rest while Ted moves with a velocity v_{TA} relative to Alice. Ted observes an object moving with a velocity v_{OT} relative to him. When $v_{TA} \ll c$ and $v_{OT} \ll c$, the velocity of the object relative to Alice is given by the Galilean addition of velocities relation (the result from Newton's mechanics),

$$v_{OA} = v_{OT} + v_{TA} \qquad \textbf{(27.17)} \text{ (page 932)}$$

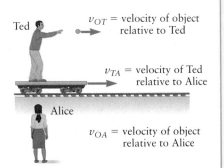

v_{OT} = velocity of object relative to Ted

v_{TA} = velocity of Ted relative to Alice

v_{OA} = velocity of object relative to Alice

For speeds that are not small compared with c, we must use the relativistic addition of velocities relation,

$$v_{OA} = \frac{v_{OT} + v_{TA}}{1 + \dfrac{v_{OT}v_{TA}}{c^2}} \qquad \textbf{(27.18)} \text{ (page 933)}$$

If both $v_{TA} < c$ and $v_{OT} < c$, then $v_{OA} < c$. No object can move faster than c.

Relativistic momentum

In Newton's mechanics, the momentum of a particle of mass m_0 moving with a velocity v is

$$\text{Newton's mechanics: } p = m_0 v \qquad \textbf{(27.22)} \text{ (page 935)}$$

Special relativity gives a different result:

$$\text{special relativity: } p = \frac{m_0 v}{\sqrt{1 - v^2/c^2}} \qquad \textbf{(27.24)} \text{ (page 935)}$$

where m_0 is called the *rest mass* of the particle.

Relativistic energy

In Newton's mechanics, the kinetic energy of a particle of mass m_0 is

$$\text{Newton's kinetic energy: } KE = \tfrac{1}{2}m_0 v^2 \qquad \textbf{(27.28)} \text{ (page 938)}$$

In special relativity, one considers instead the total energy of a particle:

$$\text{total relativistic energy: } TE = \frac{m_0 c^2}{\sqrt{1 - v^2/c^2}} \qquad \textbf{(27.33)} \text{ (page 939)}$$

When $v = 0$, the total energy is $TE = m_0 c^2$, which is called the *rest energy* of the particle. This result implies that rest mass is a form of energy. When $v > 0$, the total energy is larger than the rest energy; the energy in excess of the rest energy is the *relativistic kinetic energy*,

$$KE = \frac{m_0 c^2}{\sqrt{1 - v^2/c^2}} - m_0 c^2 \qquad \textbf{(27.30)} \text{ (page 938)}$$

SSM = answer in Student Companion & Problem-Solving Guide   ⊗ = life science application

1. It is impossible for a particle of rest mass m_0 to travel at a speed greater than c. Is there an upper limit to the momentum or kinetic energy of the particle? Explain why or why not.

2. You are traveling in a windowless spacecraft, far from any planets or stars. Describe an experiment you could do to tell whether you are in an inertial or a noninertial frame.

3. SSM A constant force F is applied to a spacecraft of mass m. The spacecraft is initially at rest. Sketch the speed of the spacecraft as a function of time according to Newton's laws (a) without allowing for relativity and (b) allowing for relativity. Compare the two results and explain why they are different.

4. Give an example of an inertial frame of reference.

5. Give four examples of noninertial reference frames.

6. The speed of light inside a substance depends on the index of refraction n. For water, $n \approx 1.33$. (a) What is the speed of light in water? (b) Is it possible for a particle to travel faster than your result in part (a)? Explain why or why not.

7. Length contraction applies to the length of an object measured in the direction of motion. Consider the length of an object measured in a direction perpendicular to its velocity. Give an argument that explains why there is no contraction along the perpendicular direction.

8. A particle is "extremely" relativistic if its speed is very close to c. What is the ratio of the total energy to the momentum for such a particle?

9. Two observers move at different speeds. Which of the following quantities will they agree on? In each case, give a reason for your answer.
(a) The length of an object
(b) The time interval between two events
(c) The speed of light in a vacuum
(d) The speed of a moving object
(e) The relative speed between the observers

10. At a typical everyday speed such as 65 mi/h, does length contract, time dilate, and mass increase by amounts that could be easily measured? Explain.

11. A wind-up toy has a coil spring that stores spring potential energy. As the toy is wound, does its mass increase, decrease, or stay the same? Explain.

12. SSM Travelers onboard a spaceship are moving at $0.9c$ from one star to another. The stars are at rest relative to each other, and this is the speed of the spaceship relative to the stars. Which of the following statements are true?
(a) Observers at rest with respect to the stars measure a shorter distance between the stars than the travelers do.
(b) The observers see the travelers moving in slow motion relative to themselves.
(c) The observers see the travelers age more slowly than they themselves do.
(d) The travelers measure a shorter distance between the stars than the observers do.
(e) The travelers believe they are aging more slowly than the observers are.

13. Two spaceships, each traveling at $0.5c$ relative to an observer on the Earth, approach each other head on. One ship fires its laser beam weapon at the other ship. What does the other ship measure for the speed of the laser light?

14. Suppose an object moves away from a plane mirror at a speed of $0.9c$. Does the image recede from the source at a speed of $1.8c$? Explain.

15. You are standing on the Earth and observe a spherical spacecraft fly past at very high speed. Sketch the shape of the ship as observed by you.

16. Two identical clocks are synchronized. One stays on the Earth while the other goes in orbit around the Earth for 1 year as measured by the clock on the Earth. Which of the following statements are true after both clocks are again on the Earth?
(a) The clocks are still synchronized.
(b) The clock that made the trip runs slower after it returns.
(c) The clock that made the trip does not have the same time as the clock that stayed on the Earth.
(d) The clock that stayed on the Earth has the wrong time.
(e) The clock that orbited the Earth has the wrong time.

17. If you are traveling very close to the speed of light and hold a mirror in front of your face, what will you see?

18. If the Earth were compressed to the density of a black hole, would it be about the size of (a) an atom, (b) a grape, (c) an orange, or (d) a basketball?

SSM = solution in Student Companion & Problem-Solving Guide   ⊗ = life science application
⭐ = intermediate   ✪ = challenging   ® = reasoning and relationships problem

27.1 NEWTON'S MECHANICS AND RELATIVITY

1. SSM Ted travels in a railroad car at constant velocity while his motion is watched by Alice, who is at rest on the ground (Fig. P27.1). Ted's speed v is much less than the speed of light. Ted releases a ball from his hand and observes that in his reference frame the ball falls directly downward. Hence, according to Ted, the component of the ball's velocity along the horizontal direction is zero. (a) According to Alice, what is the ball's velocity along x just after the ball is released? (b) According to Alice, what is the ball's velocity along x just before the ball lands at Ted's feet? (c) According to Alice, what is the acceleration of the ball along x? (d) According to Ted, what is the acceleration of the ball along x? (e) What is the force F_x on the ball along x? Do

Ted and Alice agree on the value of F_x? Explain how this answer is expected from the principle of Galilean relativity.

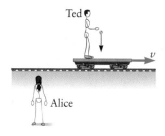

Figure P27.1

27.2 THE POSTULATES OF SPECIAL RELATIVITY

2. A person travels in a car with a speed of 75 m/s. To get the driver's attention, a stationary police officer behind the car shines his spotlight at the car. If the officer is 300 m behind the car when he turns on his spotlight, how long does it take for the light to reach the car?

3. The spotlight in Problem 2 emits light at a wavelength $\lambda = 450$ nm as viewed by the police officer. The wavelength seen by the driver is λ'. What is the percentage difference between these two wavelengths? *Hint:* Use the results for the Doppler effect in Chapter 23.

4. [SSM] Identify inertial and noninertial reference frames in the following list.
   (a) A spinning ice skater
   (b) A car turning a corner
   (c) A spacecraft moving at constant speed in a straight line
   (d) A spacecraft moving with constant velocity
   (e) A spacecraft landing on the Moon
   (f) A spacecraft traveling with its engines off, far from any stars or planets

27.3 TIME DILATION

5. Ted is traveling on his railroad car (Fig. P27.5) with speed $0.85c$ relative to Alice. Ted travels for 30 s as measured on his watch. (a) Who measures the proper time, Ted or Alice? (b) How much time elapses on Alice's watch during this motion?

Figure P27.5 Problems 5 and 6.

6. ⊗ Consider again the time intervals measured by Ted and Alice in Figure P27.5, but now suppose Alice measures the time required for her heart to beat 75 times and measures an elapsed time of 80 s with her watch. (a) Who measures the proper time, Ted or Alice? (b) How much time elapses on Ted's watch?

7. Ted is traveling in his railroad car at speed v relative to Alice. He is also carrying a light clock in his luggage. Alice compares notes with Ted and finds that each tick of the light clock takes 1.0 s in Ted's reference frame, whereas Alice measures 3.0 s per tick. Find v.

8. Consider the motion of a muon as it moves through the Earth's atmosphere (see Example 27.4). A particular muon is created 15,000 m above the Earth's surface and just reaches the ground before it decays. What is the speed of the muon?

9. ⊗ An astronaut travels at a speed of $0.98c$ from the Earth to a distant star that is 4.5×10^{17} m away. If the astronaut is 25 years old when she begins her trip, how old is she when she arrives at the star?

10. A subatomic particle has an unknown lifetime τ. The particle is created at $t = 0$ and travels around the Earth's equator at speed $0.999c$. If the particle decays at the moment it completes one trip around the Earth, what is τ?

11. [SSM] ✮ An astronaut is traveling to the Moon at a speed of $0.85c$. When his spacecraft is 2.0×10^8 m from the Moon, there is an explosion on the Moon. How long does it take for light from the explosion to reach the astronaut as measured on the astronaut's clock?

12. ✮ An astronaut travels away from the Earth at a speed of $0.95c$ and sends a light signal back to the Earth every 1.0 s as measured by his clock. An observer on the Earth finds that the arrival time between consecutive signals is Δt. Find Δt.

27.4 SIMULTANEITY IS NOT ABSOLUTE

13. [SSM] Ted is traveling on his railroad car (length 25 m as measured by Ted) at a speed of $0.95c$. Alice arranges for two small explosions to occur on the ground next to the ends of the railroad car (Fig. P27.13). According to Alice, the two explosions occur simultaneously, and she uses the burn marks on the ground to measure the length of Ted's railroad car. (a) According to Ted, do the two explosions occur simultaneously? (b) If not, then according to Ted which explosion occurs first?

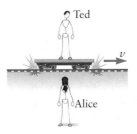

Figure P27.13

14. Ted is traveling on his railroad car (length 25 m, as measured by Ted) at a speed of $0.95c$. Ted arranges for two small explosions to occur on the ends of the railroad car (Fig. P27.14). According to Ted, the two explosions occur simultaneously, and he asks Alice to use the burn marks on the ground to measure the length of the railroad car. (a) According to Alice, do the two explosions occur simultaneously? (b) If not, then according to Alice which explosion occurs first?

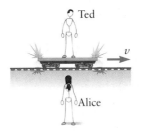

Figure P27.14

27.5 LENGTH CONTRACTION

15. A spaceship moves past you at speed v. You measure the ship to be 300 m long, whereas an astronaut on the ship measures a length of 400 m. Find v.

16. A muon is created at the top of Mount Everest (height 8900 m) and travels at constant speed to sea level, where it decays. According to a person traveling on the muon, what is the height of Mount Everest? Assume the muon travels vertically downward.

17. A meterstick moves toward you at speed $0.93c$. What length do you measure for the meterstick?

18. Ted is traveling on his railroad car (length 25 m) at speed $v = 0.95c$ (Fig. P27.18). As measured on Alice's clock, how long does it take for Ted's railroad car to pass Alice?

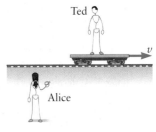

Figure P27.18

19. A car of rest length 2.5 m moves past you at speed v, and you measure its length to be 2.2 m. What is the car's speed?

20. An astronaut travels from a distant star to the Earth at speed $0.999c$. The astronaut ages 4.0 years during his journey. What is the distance from the star to the Earth as measured by an observer on the Earth?

21. Parking in a hurry. Your car is 5.0 m long, but you need to fit into a parking space that is only 4.5 m long. How fast must your car be moving so that you will not get a parking ticket from the police officer observing your predicament?

22. SSM ★ A military jet that is $L_0 = 35$ m long (as measured on its runway prior to taking off) travels at a speed of 600 m/s. What is the length L of the jet as measured by an observer on the ground? Keep extra significant figures so that you can obtain the *difference* in length $\Delta L = L_0 - L$ to two significant figures. Compare ΔL to the diameter of the hydrogen atom (approximately 5×10^{-11} m).

23. In studies of elementary particles such as electrons, physicists use machines called "accelerators" in which the particles move at very high speeds. Suppose an electron in one of these machines has a speed of $0.9999c$ and it travels along a straight line for a distance of 500 m as measured by physicists doing an experiment. How long is the accelerator as measured by a person riding along with the electron?

27.6 ADDITION OF VELOCITIES

24. Traveling on a light beam. When he was first thinking about relativity (and before he developed his theory), Einstein considered how the universe would look to a person who traveled on a light beam. Suppose you are traveling on a light beam when you encounter a friend who is traveling on a second light beam in the opposite direction. What is your friend's speed relative to your reference frame?

25. An electron (electron 1) moves to the left with speed $0.95c$ while a second electron (electron 2) moves to the left with speed $0.70c$. What is the speed of electron 2 as measured by an observer sitting on electron 1?

26. A spacecraft travels at a speed of $0.75c$ relative to the Earth. The spacecraft then launches a probe at speed v relative to the spacecraft (Fig. P27.26). If the probe has a speed of $0.90c$ relative to the Earth, what is v?

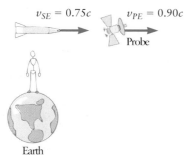

$$v_{SE} = 0.75c \qquad v_{PE} = 0.90c$$

Probe

Earth

Figure P27.26

27. SSM Two electrons are moving in a collision course, each with speed $0.99c$ relative to a stationary observer. What is the speed of one electron as measured by an observer traveling on the other? Express your answer to five significant figures.

28. Spacecraft 1 is moving at speed $0.85c$ and is catching up with spacecraft 2, which is moving at speed $0.75c$ (both measured by an observer on the Earth). What is the speed of spacecraft 2 as measured by the pilot of spacecraft 1?

29. If the two spacecrafts in Problem 28 are initially 5.0×10^{10} m apart according to an observer on Earth, how long does it take for them to meet as measured by the Earthly observer?

30. Two asteroids are traveling along a common line and are on course for a head-on collision. Astronomers on the Earth observe that one asteroid has a speed of $0.75c$ and the other has a speed of $0.95c$. What is the relative speed of the two asteroids as measured by an observer on one of the asteroids?

31. An ice skater travels on a frozen lake at a speed of 30 m/s relative to the shore. The skater then throws a baseball in the direction parallel to her velocity with a speed of 20 m/s relative to the skater. What is the speed v of the baseball relative to the shore? Keep enough significant figures in your calculation so that you can calculate the difference $v - 50$ m/s to two significant figures.

32. A distant galaxy is moving at a speed of $0.93c$ away from the Earth. The galaxy ejects some material toward the Earth with speed $0.85c$ relative to the Earth. What is the speed of the material relative to the galaxy?

27.7 RELATIVISTIC MOMENTUM

33. An electron moves with speed $0.95c$. (a) What is the momentum of the electron according to Newton's mechanics? (b) What is the correct value of the electron's momentum according to special relativity? (c) Explain in words why the answer to part (b) is larger than the answer to part (a).

34. A baseball has a mass of about 0.14 kg; if thrown by a major league pitcher, it has a speed of about 50 m/s. Now suppose an electron is given a very large speed v so that it has the same momentum as the baseball. Find v. Express your answer as $c - v$.

35. SSM A proton has a relativistic momentum 10 times larger than its classical (Newton's law) momentum. What is the speed of the proton? Express your answer with three significant figures.

36. At what value of v is the relativistic momentum of a particle twice its classical (Newton's law) momentum?

37. The factor $m_0/\sqrt{1 - v^2/c^2}$ for an electron (see Eq. 27.24) is equal to five times its rest mass. What is the speed of the electron?

38. The factor $m_0/\sqrt{1 - v^2/c^2}$ for an electron (see Eq. 27.24) is equal to the rest mass of a neutron. Find the speed of the electron.

39. The relativistic momentum of a particle of rest mass m_0 and speed v is equal to $5m_0 v$. What is the speed of the particle?

27.8 WHAT IS "MASS"?

27.9 MASS AND ENERGY

40. The kinetic energy of a particle is equal to the rest energy of the particle. What is the speed of the particle?

41. ★ An electron is accelerated through an electric potential difference ΔV such that its kinetic energy is equal to its rest energy. Find ΔV.

42. Annihilate this. The antiproton is a type of antimatter and is the antimatter "cousin" of the proton; the two particles have the same rest mass. It is possible for a proton and an antiproton to annihilate each other, producing pure energy in the form of electromagnetic radiation (Chapter 31). How much energy is released when a proton and an antiproton annihilate each other? Assume the two particles are rest just prior to the annihilation.

43. An asteroid of mass 2500 kg has a (relativistic) kinetic energy of 1.5×10^{20} J. What is the speed of the asteroid?

44. ★ An electron in a television picture tube has a classical kinetic energy of 30 keV, which is the kinetic energy that Newton would calculate using the measured speed and rest mass of the electron. What is the actual kinetic energy of the electron; that is, what is the value found using the relativistic result for the kinetic energy?

45. Cosmic ray protons have been observed with kinetic energies claimed to be as high as 10^{20} eV. What is speed of these protons? Express your answer as $c - v$ where v is the speed of the protons.

46. A proton has a kinetic energy of 1.5×10^7 eV. What is its speed?

47. SSM ✴ A chemical reaction that produces water molecules is

$$2H_2 + O_2 \ \rightarrow \ 2H_2O$$

This reaction releases 570 kJ for each mole of oxygen molecules that is consumed. The relativistic relation between mass and energy then implies that the mass of two hydrogen molecules plus the mass of one oxygen molecule is slightly larger than the mass of two water molecules. Find this mass difference.

48. ✪ The most common helium nucleus in nature contains two protons and two neutrons. The mass of this nucleus is 6.64466×10^{-27} kg, while the mass of a proton is 1.67262×10^{-27} kg and the mass of a neutron is 1.67493×10^{-27} kg. Find the binding energy of this helium nucleus.

49. ✴ One kg of the chemical explosive TNT releases approximately 4.2×10^6 J of energy when it explodes. How much of the initial mass of TNT is converted to energy in such an explosion?

50. In 2005, the total energy "consumed" by people on Earth was about 5×10^{20} J. About 85% of this was from the burning of fossil fuels. How much fossil fuel mass was converted to energy in 2005?

27.10 THE EQUIVALENCE PRINCIPLE AND GENERAL RELATIVITY

51. ✪ In Chapter 23, we learned that an accelerating electric charge produces electromagnetic waves. In the same way, the general theory of relativity predicts that an accelerated mass produces gravitational waves. As the Earth moves in orbit around the Sun, it undergoes accelerated motion (i.e., uniform circular motion), and it has been predicted that the power carried away by the resulting gravitational radiation is about 0.001 W. How long will it take for the Earth to fall into the Sun? That is, how long will it take for the Earth's orbit to have the same radius as the Sun's radius?

ADDITIONAL PROBLEMS

52. The muon in Example 27.4 has a relativistic momentum of 4.0×10^{-19} kg · m/s. The rest mass of a muon is $m_0 = 1.9 \times 10^{-28}$ kg. Find the kinetic energy of the muon.

53. What is the relativistic momentum of an electron whose relativistic kinetic energy is 4.0×10^{-22} J?

54. ✪ The Tevatron is an accelerator at Fermi National Laboratory near Chicago. It uses the collisions between protons and antiprotons to generate and study quarks and other elementary particles. If the kinetic energy of a proton in the Tevatron is 1.0 TeV (= 1.0×10^{12} eV), what are (a) the speed of the proton and (b) the momentum of the proton? (c) Compare your result to the momentum of a mosquito ($m = 1$ mg) flying at a speed of 1 m/s.

55. ✴ The kinetic energy of a particle of rest mass m_0 is equal to three times its rest energy. What is the momentum of the particle? Express your answer in terms of m_0.

56. ✪ The space shuttle has a length of 37 m. (a) In its normal orbital motion around the Earth, the shuttle has a speed of about 8000 m/s and an orbital period of about 90 minutes. Allowing for length contraction, what is the length of the shuttle as viewed by an observer at rest on the ground? (b) Suppose the shuttle increases its speed so that it has a length of 25 m when viewed by a stationary observer. How long would it take to complete one orbit?

57. An electron has a momentum equal to that of a baseball ($m = 0.22$ kg) moving at 45 m/s (about 100 mi/h). What is the kinetic energy of the electron?

58. ✪ Ⓡ Suppose two highly precise, identical clocks are synchronized and one clock is placed on the North Pole and the other on the equator. After 100 years, how much will the clocks differ in time? Assume the Earth is a perfect sphere.

59. SSM ✴ Ⓡ Like all stars, the Sun converts mass into energy that radiates out in all directions. The average rate at which this radiant energy reaches the Earth is approximately 1.4×10^3 W/m². (a) Calculate the rate at which the Sun is losing mass. (b) Assuming this rate remains constant, estimate the lifetime of the Sun.

60. Consider a proton moving at relativistic speed. (a) Determine the proton's rest energy in electron volts (1 eV = 1.60×10^{-19} J). (b) Suppose the total energy of the proton is three times its rest energy. With what speed is the proton moving? (c) At this speed, determine the kinetic energy of the proton in electron-volts. (d) What is the magnitude of the proton's momentum?

61. ✪ Ⓡ As we'll discuss in Chapter 30, radioactive decay involves the nucleus of an atom decaying by emitting either a particle or energy or both. The ^{216}Po nucleus decays to ^{212}Pb by emitting an alpha particle, which is a helium nucleus, ^{4}He. Using the mass data given in Appendix A, find (a) the mass change in this decay and (b) the energy that this mass represents. (c) Given that no other particles or energy are emitted in this process, where must the energy go?

62. ✪ **Falling light.** According to general relativity and the equivalence principle, light is bent by gravity. Suppose you stand two tall, perfectly reflecting mirrors exactly 1 m apart and facing each other. A beam of light is directed horizontally through a hole in one of the mirrors 10 m above the ground. (a) Determine the time it takes for the light to strike the ground. (b) The light will undergo N reflections (i.e., $N/2$ reflections from each mirror) before it strikes the ground. Find N.

63. SSM ✴ Suppose an electron is accelerated from rest through a potential difference of 100,000 volts. Determine the electron's final kinetic energy, speed, and momentum (a) ignoring relativistic effects and (b) including relativistic effects.

64. ✪ Suppose a rocket ship leaves the Earth in the year 2020. One of a set of twins born in 2000 remains on the Earth while the other rides in the rocket. The rocket ship travels at $0.90c$ in a straight line path for 10 years as measured by its own clock, turns around, and travels straight back at $0.90c$ for another 10 years as measured by its own clock before landing back on the Earth. (a) What year is it on the Earth? (b) How old is each twin? (c) How far away from the Earth did the rocket ship travel as measured by each twin?

65. ✪ One consequence of general relativity is that a clock in a gravitational field runs slower. This time dilation is distinct from the time dilation from relative motion (special relativity) and is given by $\Delta t = \Delta t_0 / \sqrt{1 - 2GM/(Rc^2)}$, where Δt is the dilated time, Δt_0 is the time when no gravitational field is present, G is the universal gravitational constant, M is the mass of the object producing the gravitational field, R is the distance from the center of the mass, and c is the speed of light. In 1976, the Smithsonian Astrophysical Observatory sent aloft a Scout rocket to a height of 10,000 km to confirm this effect. Show that at this height a clock should run faster than an identical clock on the Earth by a factor of 4.5×10^{-10}.

66. ✪ Thomas Jefferson National Accelerator Facility in Newport News, Virginia, is host to a continuous electron beam. Within the facility, electrons travel around a $\frac{7}{8}$-mi-long tunnel five times

---

in 23.5 millionths of a second. (Actually, the facility is capable of producing even higher speeds.) (a) What is the speed of the electrons? (b) Determine the length around the tunnel as measured in the reference frame of the electrons. (c) Determine the time it takes for the electrons to complete one lap around the tunnel in the reference frame of the electrons. (d) What are the momentum and kinetic energy of the electrons?

67. ✪ **Gravitational red shift.** An important consequence of Einstein's general relativity is that gravity must affect a light wave's frequency and wavelength. As light moves upward from the Earth's surface, the wavelength of the light gets longer and the frequency gets lower as gravity "drains" the light of some energy. In a famous experiment in 1960, Robert Pound and Glen Rebka of Harvard University successfully tested this effect to within 10% of the predicted value. Later, in 1964, they improved the agreement to within 1%. The experimental procedure involved placing a detector 22.6 m above a radioactive source placed on the ground. Quantized "particles" of light called photons (see Chapter 28) are given off by the source with an energy of 14.4 keV. Using a highly sensitive technique called the Mössbauer effect, they were able to measure a small shift in the energies of the photons as they moved upward through the Earth's gravitational field. Although photons are massless, we can treat their energy as rest energy and find a corresponding "effective mass" for the photons. Using this approach, determine the shift in the photons' energy as they move the 22.6 m up to the detector. This shift is called a *gravitational red shift* because it results in a shift toward a lower frequency and longer wavelength.

68. ✪ Ⓡ By what amount is the diameter of the Moon shortened (as measured by a stationary observer on the Earth) due to its orbital motion around the Earth?

69. ✪ Suppose you take a trip to Mars, which is 80 million km from the Earth. You head directly toward Mars onboard your spaceship, traveling at $0.80c$. Unbeknownst to you, your mortal enemy placed a bomb on your ship that has a 4-minute timer that was initiated upon your departure. Will you live long enough to reach Mars? Answer the problem first without relativity and then using relativity from both your point of view and that of your Earth-bound enemy.

70. ✪ Ⓡ **GPS and relativity.** The Global Positioning System (GPS) consists of a network of about 30 satellites in orbit, each carrying atomic clocks on board. The orbital radius of the satellites is about four Earth radii (26,600 km). The orbits are nearly circular, with a typical eccentricity of less than 1%. The onboard atomic clocks keep highly accurate time, and oscillators produce signals at several different frequencies. The frequency used by nonmilitary GPS receivers is 1575.42 MHz. Two relativistic corrections must be made to this signal so that accurate position readings on the Earth are possible: first, special relativity predicts that moving clocks will appear to tick slower than nonmoving ones, and second, general relativity predicts that clocks in a stronger gravitational field will tick at a slower rate. Determine (a) the orbital period and (b) the orbital speed of these satellites. (c) Apply special relativity to determine the correction to the frequency of the signal received on the Earth. (d) Applying general relativity and following the method described in Problem 67, determine the correction to the frequency of the signal received on Earth. *Hint*: The energy of a photon is proportional to its frequency.

Chapter 28

Quantum Theory

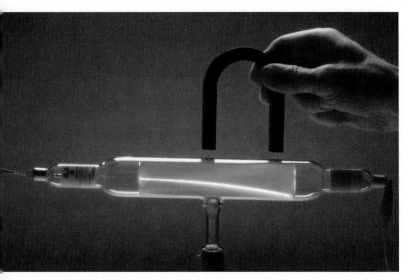

A modern reconstruction of the apparatus used in experiments by J. J. Thomson in his discovery of the electron in 1897. The bright horizontal beam is produced by electrons as they travel from left to right inside the glass tube. Thomson's discovery and other work described in this chapter led to the development of quantum theory. (© Charles D. Winters/Photo Researchers, Inc.)

In this text, our work on mechanics, electricity, and magnetism has been based on two pillars of **classical** physics: Newton's laws and Maxwell's equations. Classical physics deals with the macroscopic world, including baseballs, automobiles, and planets. We have already mentioned that Newton's laws fail when applied to the atomic-scale world of electrons and atoms. Similarly, Maxwell's equations correctly describe electromagnetic phenomena in the macroworld, but fail when applied at the atomic scale. This atomic-scale world is called the **quantum** regime. In this chapter, we describe how quantum behavior is different from anything we have seen before. "Quantum" refers to a very small increment or parcel of energy. These parcels, called **quanta**, are a central aspect of the quantum world.

The discovery and development of quantum theory began in the late 1800s and continued during the first few decades of the twentieth century. As we introduce the key ideas of quantum theory, we'll also mention some of the very interesting history of this work.

28.1 | PARTICLES, WAVES, AND "PARTICLE-WAVES"

In the macroworld of Newton and Maxwell, energy can be carried from one point to another by only two types of "objects": particles (such as baseballs or bullets) and waves (such as sound or light). We have an intuitive understanding of particles and waves through our everyday experiences, and it is natural to use this macroworld intuition when we consider the behavior of things in the microworld such as electrons and atoms. We'll see, however, that this intuition is completely incapable of describing the quantum regime!

How Do Waves and Particles Differ?

In Chapter 25, we studied the interference and diffraction of light. Let's now consider the double-slit experiment in Figure 28.1A in which light is incident on an opaque barrier containing two very narrow openings. A double-slit interference pattern, consisting of a series of bright and dark fringes, is formed on the screen on the far right. The bright fringes are produced by constructive interference between waves that pass through the two slits, and the dark fringes are at locations where there is destructive interference. A similar result would be found with other types of waves, including sound and water waves.

Figure 28.1B shows another double-slit experiment, this time using bullets instead of light. Our macroworld intuition correctly predicts what will happen in this case: only bullets that pass through one or the other of the two slits will reach the screen on the far right; the other bullets (the ones that strike the barrier) are stopped by the barrier. The pattern of bullets (or holes!) on the screen is quite different from the pattern found with light waves; the bullet pattern corresponds only to the "shadows" of the slits. There is no constructive or destructive interference with bullets or other classical particles such as baseballs or rocks.

The behaviors sketched in Figure 28.1 illustrate some important classical differences between particles and waves.

Properties of waves and particles in the classical regime:

- Waves exhibit interference; particles do not.

- Particles often deliver their energy in discrete amounts. For example, when a bullet in Figure 28.1B strikes the screen, all its energy is deposited on the screen (assuming the bullet does not bounce off). Energy arrives at the screen in discrete parcels, with each parcel corresponding to the kinetic energy carried by a single bullet.

- The energy carried or delivered by a wave is not discrete, but varies in a continuous manner. Recall from Chapter 23 that the energy carried by a wave is described by its intensity, which equals the amount of energy the wave transports per unit time across a surface of unit area. Hence, for the light wave in Figure 28.1A, the amount of energy absorbed by the screen depends on the intensity of the wave and the absorption time. The amount of absorbed energy can thus take on any nonnegative value. Classically, wave energy is not delivered in discrete parcels.

An Interference Experiment with Electrons

In the world of classical physics, the experiment in Figure 28.1 can be used to distinguish between particles and waves. According to classical physics, only these two types of behavior are possible: waves exhibit interference; particles do not.

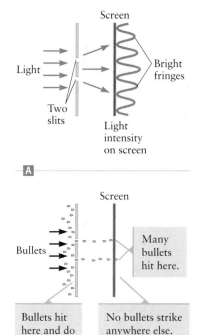

Figure 28.1 🄰 When light passes through a double slit, constructive and destructive interference produce bright and dark fringes on the screen, evidence for the wave nature of light. 🄱 Classical particles such as bullets do not undergo interference when they pass through a double slit.

However, this tidy separation of particles versus waves is *not* found in the quantum world. Figure 28.2 shows a hypothetical double-slit experiment performed with electrons. A beam of electrons, all with the same speed, are incident from the left and pass through a pair of slits. The electrons then travel on to a screen on the right, which records where each electron strikes. Case 1 in Figure 28.2 shows the result after 20 electrons have passed through the slits. Each dot shows where a particular electron arrived at the screen, and the arrival points seem to be distributed randomly. If this experiment is repeated with another 20 electrons, the precise arrival points would be different, but the general appearance would be the same. If we proceed with the experiment and wait until 100 electrons have arrived, we find the result shown in case 2. The arrival points are still spread out, but we can now see that the electrons are more likely to strike at certain points than at others. By the time 300 electrons have reached the screen in case 3, it is clear that electrons are much more likely to hit certain points on the screen. Case 4 on the far right shows the results after a very large number of electrons have passed through the slits. This sketch shows the probability that electrons will arrive at different points. This probability curve has precisely the same form as the variation of light intensity in the double-slit interference experiment in Figure 28.1A. The experiment shows that electrons undergo constructive interference at certain locations on the screen, giving a large probability for electrons to arrive at those locations. At other places, the electrons undergo destructive interference, and the probability for an electron to reach those locations is very small or zero.

The results in Figure 28.2 show that electrons can exhibit interference, a property that classical theory says is possible only for waves. This experiment also shows aspects of particle-like behavior since the electrons arrive one at a time at the screen, with each dot in cases 1 through 3 of the figure corresponding to the arrival of a single electron as it deposits its parcel of energy on the screen.

The behavior in Figure 28.2 is characteristic of the quantum regime and shows that electrons behave in some ways as *both a classical particle and a classical wave*. In fact, *all* objects in the quantum world behave in this way. Electrons, protons, atoms, and molecules have all been found to give the results shown in Figure 28.2. Moreover, even light waves exhibit particle-like behavior. The clear-cut distinction between particles and waves thus breaks down in the quantum regime. It may be more helpful for your classical intuition to call these things "particle-waves," all of which exhibit the following properties.

Properties of waves and particles in the quantum regime:

- *All* objects, including light and electrons, can exhibit interference.
- *All* objects, including light and electrons, carry energy in discrete amounts. These discrete "parcels" are called quanta.

We explore the implications of these ideas for light and for electrons in the next two sections, beginning with an experiment showing that light energy is quantized.

Figure 28.2 An interference experiment with electrons. Each dot in cases 1 through 3 indicates where an electron strikes the screen. Case 1: pattern formed by the first 20 electrons. Case 2: pattern formed by the first 100 electrons. Case 3: After 300 electrons, it is clear that electrons tend to arrive at certain places on the screen and not at others. Case 4: When the experiment is carried out with a very large number of electrons, the probability for an electron to reach the screen forms a double-slit interference pattern, similar to the pattern for light in Figure 28.1A.

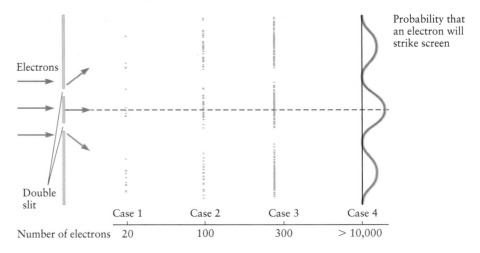

Around 1800, Thomas Young performed his double-slit interference experiment, which provided the first clear evidence that light is a wave. Maxwell's theory of electromagnetic waves was worked out about 60 years later, so physicists then had a detailed theory of light as an electromagnetic wave and believed that the nature of light was well understood. In the 1880s, however, studies of what happens when light is shined onto a metal gave some very puzzling results that could not be understood with the wave theory of light. To appreciate these experiments, we must first recall that a metal contains electrons that are free to move around within the metal. However, these electrons are still bound as a whole to the metal because of their attraction to the positive charges of the metal atom nuclei; that is why these electrons do not just "fall out of" the metal. The minimum energy required to remove a single electron from a piece of metal is called the *work function*, W_c. The value of the work function is different for different metals; it can be measured by applying an electric potential difference between the metal of interest and a second metal plate (all in vacuum) and observing the value of the potential at which electrons first leave the metal (Fig. 28.3). If V is the electric potential at which electrons begin to jump across the vacuum gap in Figure 28.3, the work function is $W_c = eV$, where e is the magnitude of the electron's charge ($e = 1.60 \times 10^{-19}$ C).

The work function has units of energy, so it can be measured in joules. Since the experiment in Figure 28.3 involves electrons, it is sometimes convenient to measure W_c in units of electron-volts (eV). The conversion factor 1.60×10^{-19} J/eV is listed on the page facing the inside front cover; see also the definition of the electron-volt in Eq. 18.19. Values of the work function in both units are listed in Table 28.1.

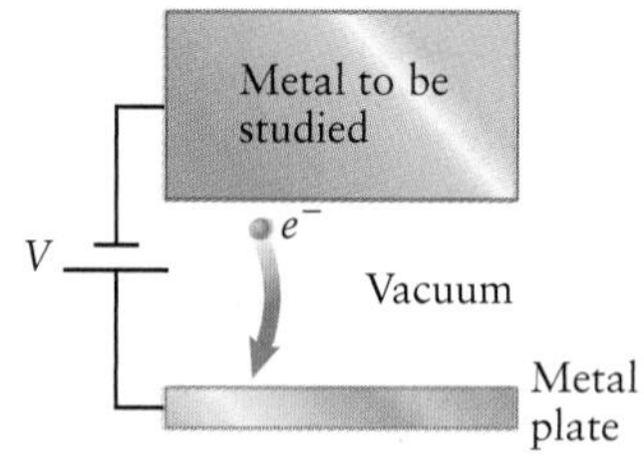

Figure 28.3 Measuring the work function W_c of a metal by applying an electric potential. When $eV > W_c$, electrons are ejected from the metal on top and jump to the metal plate below. Measuring the smallest V able to eject electrons gives the value of W_c.

The Photoelectric Effect

Another way to extract electrons from a metal is by shining light onto it. Light striking a metal is absorbed by the electrons, and if an electron absorbs an amount of light energy greater than W_c, it is ejected from the metal as sketched in Figure 28.4A. This phenomenon is called the *photoelectric effect*.

Experimental studies of the photoelectric effect carried out around 1900 found that no electrons are emitted unless the light's frequency is greater than a critical value f_c. When the frequency is above f_c, the kinetic energy of the emitted electrons varies linearly with frequency f as shown in Figure 28.4B. Physicists initially tried to explain these results using the classical wave theory of light, but there were two difficulties with the classical explanations. First, experiments show that the critical

Photoelectric effect: ejection of electrons from a metal after absorbing energy from light

Table 28.1	Work Function of Several Metals	
METAL	**WORK FUNCTION (J)**	**WORK FUNCTION (eV)**
Na (sodium)	3.8×10^{-19}	2.4
Al (aluminum)	6.9×10^{-19}	4.3
Ca (calcium)	4.5×10^{-19}	2.8
Fe (iron)	6.9×10^{-19}	4.3
Cu (copper)	7.0×10^{-19}	4.4
Mo (molybdenum)	6.7×10^{-19}	4.2
Ag (silver)	6.9×10^{-19}	4.3
Pt (platinum)	1.0×10^{-18}	6.4
Pb (lead)	6.4×10^{-19}	4.0

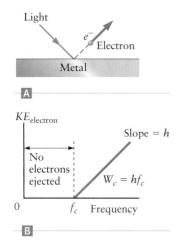

Light

e^-

Electron

Metal

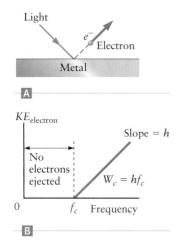

A

KE_{electron}

Slope = h

No electrons ejected

$W_c = hf_c$

0   f_c Frequency

B

Figure 28.4 A The photoelectric effect: electrons are ejected when light strikes a metal. B Experiments show that the kinetic energy of the ejected electrons (KE_{electron}) depends on the frequency of the light. When the frequency is below a certain critical value f_c, no electrons are ejected. The work function is related to the critical frequency by $W_c = hf_c$, where h is Planck's constant defined in Equation 28.2.

frequency f_c is *independent* of the intensity of the light. According to classical wave theory, the energy carried by a light wave is proportional to the intensity, so it should always be possible to eject electrons by increasing the intensity to a sufficiently high value. Experiments found that when the frequency is below f_c, however, there are no ejected electrons no matter how great the light intensity. Second, the kinetic energy of an ejected electron is independent of the light intensity. Classical theory predicts that increasing the intensity will cause the ejected electrons to have a higher kinetic energy, but the experiments show no such result. In fact, the experiments show that the electron kinetic energy depends on the light's frequency (Fig. 28.4B) instead of the intensity.

The classical wave theory of light is thus not able to explain the photoelectric effect experiments. Albert Einstein surprised the physics world in 1905 when he offered the following explanation.[1] He proposed that light carries energy in discrete quanta, now called *photons*. According to Einstein, each photon carries a parcel of energy

$$E_{\text{photon}} = hf \tag{28.1}$$

where h is a constant of nature called *Planck's constant*, which has the value

$$h = 6.626 \times 10^{-34} \, \text{J} \cdot \text{s} \tag{28.2}$$

and f is the frequency of the light.

Planck's constant had been introduced a few years earlier by Max Planck to explain another unexpected property of electromagnetic radiation (the blackbody radiation spectrum, page 962). Let's now see how Einstein's photon theory explains the photoelectric effect; we'll also see why the slope of the kinetic energy curve in Figure 28.4B is equal to Planck's constant h.

Einstein suggested that a beam of light should be thought of as a collection of particles (photons), each of which has an energy that depends on frequency according to Equation 28.1. If the intensity of a monochromatic (single-frequency) light beam is increased, the number of photons is increased but the energy carried by each photon does not change. This theory explains the two puzzles associated with photoelectric experiments. First, the absorption of light by an electron is just like a collision between two particles, a photon and an electron. The photon (according to Eq. 28.1) carries an energy hf that is absorbed by the electron. If this energy is less than the work function, the electron is not able to escape from the metal. For monochromatic light, increasing the light intensity increases the number of photons that arrive each second, but if the photon energy hf is less than the work function, even a high intensity will not eject electrons. The energy of a single photon—and hence the energy gained by any particular electron—depends on frequency f but not on the light intensity. Second, Einstein's theory also explains why the kinetic energy of ejected electrons depends on light frequency but not intensity. The critical frequency in the photoelectric effect (Fig. 28.4B) corresponds to photons whose energy is equal to W_c:

$$hf_c = W_c$$

Such a photon has barely enough energy to eject an electron from the metal, but the ejected electron then has no kinetic energy. If a photon has a higher frequency and thus a greater energy, the extra energy above the work function goes into the kinetic energy of the electron. We have

$$KE_{\text{electron}} = hf - hf_c = hf - W_c \tag{28.3}$$

which is the equation of a straight line; hence, the kinetic energy of an ejected electron should be linearly proportional to f. This linear behavior is precisely what is

---

[1]Einstein made a number of other monumental discoveries in the same year. He was awarded the Nobel Prize in Physics in 1921 for his theory of photons and the photoelectric effect.

found in experiments as shown in Figure 28.4B. The slope of this line is the factor multiplying f in Equation 28.3, which is just Planck's constant h. Thus, photoelectric experiments give a way to measure h, and the values found agreed with the value known prior to Einstein's theory.

Photons Carry Energy and Momentum

Einstein's photon theory asserts that light energy can only be absorbed or emitted in discrete parcels, that is, as single photons. Each photon carries an energy $E_{\text{photon}} = hf$. From Chapter 23, a light wave with an energy E also carries a certain amount of momentum $p = E/c$ (Eq. 23.12). Identifying the energy E with E_{photon}, Einstein's theory thus predicts that the momentum carried by a single photon is

$$p_{\text{photon}} = \frac{hf}{c} \qquad (28.4)$$

Momentum of a photon

Wavelength is related to frequency by $f\lambda = c$, so Equation 28.4 can also be written as

$$p_{\text{photon}} = \frac{h}{\lambda} \qquad (28.5)$$

The quantum theory of light thus predicts that "particles" of light called photons carry a discrete amount (a quantum) of both energy and momentum. Notice that we put the term *particles* in quotation marks. We did so because photons have two important properties that are quite different from classical particles: photons do not have any mass (!), and they exhibit interference effects (as in the double-slit interference experiment in Figure 28.1A). Photons are thus unlike anything we have encountered in classical physics.

Let's use Equation 28.1 to calculate the energy carried by a single photon. The result depends on the frequency of the light, so consider the light from a green laser pointer. This light has a wavelength of about $\lambda = 530$ nm, so its frequency is

$$f = \frac{c}{\lambda} = \frac{3.00 \times 10^8 \text{ m/s}}{530 \times 10^{-9} \text{ m}} = 5.7 \times 10^{14} \text{ Hz}$$

Inserting this frequency in Equation 28.1 gives

$$E_{\text{photon}} = hf = (6.63 \times 10^{-34} \text{ J} \cdot \text{s})(5.7 \times 10^{14} \text{ Hz}) = 3.8 \times 10^{-19} \text{ J} \quad (28.6)$$

which is a very small amount of energy, much smaller than is normally encountered in the macroworld (as shown in Example 28.1 and Problem 24). In most applications, one detects the presence or absence of light (the presence or absence of one or more photons) through the energy carried by the light. Hence, to detect a single photon of green light it is necessary to detect amounts of energy as small as 3.8×10^{-19} J. In Section 28.7, we describe how a cell in the human retina can respond to just a single photon! The eye is thus an exquisitely sensitive detector of light.

| EXAMPLE 28.1 | Number of Photons Emitted by a Lightbulb |

The light emitted by an ordinary lightbulb consists of photons, each of which has an energy given by Equation 28.1. Consider a 60-W lightbulb and for simplicity assume it emits only green light with $\lambda = 530$ nm as considered in Equation 28.6. How many photons does the lightbulb emit in 1.0 s?

RECOGNIZE THE PRINCIPLE

This example is an application of the relation between photon energy and frequency. The power rating P of the lightbulb gives the rate at which it emits energy. Power is

Each photon carries an energy $E_{photon} = hf$

Figure 28.5 Example 28.1.

the energy emitted per unit time, so from the value of P we can get the energy emitted in 1.0 s. We know the energy of a single photon from Equation 28.6, so we can then find the total number of photons emitted in this time.

SKETCH THE PROBLEM

Figure 28.5 shows a lightbulb emitting photons. The energy emitted each second equals the energy of a single photon times the number of photons emitted in 1 second.

IDENTIFY THE RELATIONSHIPS

The lightbulb has a power rating of $P = 60$ W $= 60$ J/s, so the total energy emitted in $t = 1.0$ s is

$$E_{total} = Pt = (60 \text{ J/s})(1.0 \text{ s}) = 60 \text{ J}$$

This energy is carried by N photons, each of which has the energy E_{photon} found in Equation 28.6. We thus have

$$E_{total} = NE_{photon}$$

SOLVE

Solving for N, we get

$$N = \frac{E_{total}}{E_{photon}} = \frac{60 \text{ J}}{3.8 \times 10^{-19} \text{ J}} = \boxed{1.6 \times 10^{20} \text{ photons}}$$

What does it mean?
In this example as in most ordinary situations, the number of photons involved is extremely large, so the typical observer would not notice that the energy in a light beam actually does arrive as discrete quanta.

CONCEPT CHECK 28.1 | Energy of a Radio Frequency Photon

Your favorite AM radio station has a frequency near $f_{AM} \approx 1$ MHz, whereas the author's favorite FM station has $f_{FM} \approx 100$ MHz. Radio waves are electromagnetic radiation, just like visible light, so both AM and FM signals are carried by photons. Which of the following statements is true?
  (a) The AM photons have a higher energy than the FM photons.
  (b) The FM photons have a higher energy than the AM photons.
  (c) They are both radio waves, so they have the same photon energy.

EXAMPLE 28.2 | The Photoelectric Effect with Aluminum

A photoelectric experiment is planned using aluminum (Al). **(a)** What is the minimum photon frequency that will eject electrons? **(b)** If the experiment is performed using blue light with $\lambda = 450$ nm, will there be any ejected electrons? If so, find the kinetic energy of an ejected electron.

RECOGNIZE THE PRINCIPLE

This example is a direct application of the photoelectric effect. **(a)** The minimum photon frequency f_c required to eject an electron is proportional to the work function. **(b)** We can determine if a photon will eject an electron by comparing its frequency with f_c.

SKETCH THE PROBLEM

No sketch is necessary.

The energy of a photon of frequency f is $E_{photon} = hf$ (Eq. 28.1), so we can use the value of the work function for Al from Table 28.1 to find the minimum photon frequency. We can then compare this frequency with the energy of a photon of blue light to decide if such a photon will eject an electron. (The variation of the frequency and wavelength across the electromagnetic spectrum is also shown in Fig. 23.8.) From Table 28.1, the work function for Al is $W_c = 4.3$ eV (electron-volts), or, expressed in units of joules, $W_c = (4.3$ eV$)(1.60 \times 10^{-19}$ J/eV$) = 6.9 \times 10^{-19}$ J. The work function W_c equals the energy of a photon that will just barely eject an electron, so

$$hf_c = W_c$$

SOLVE

(a) Solving for f_c gives

$$f_c = \frac{W_c}{h} = \frac{6.9 \times 10^{-19} \text{ J}}{6.63 \times 10^{-34} \text{ J} \cdot \text{s}} = \boxed{1.0 \times 10^{15} \text{ Hz}}$$

(b) The frequency of a photon with $\lambda = 450$ nm $= 4.5 \times 10^{-7}$ m is

$$f = \frac{c}{\lambda} = \frac{3.00 \times 10^8 \text{ m/s}}{4.5 \times 10^{-7} \text{ m}} = 6.7 \times 10^{14} \text{ Hz}$$

This frequency is smaller than f_c from part (a), so this photon will not have enough energy to eject an electron.

What does it mean?
To eject an electron requires a photon with a higher frequency (and hence shorter wavelength) than the blue photon considered here. So, only ultraviolet light can eject electrons from aluminum.

CONCEPT CHECK 28.2 | The Photoelectric Effect

A photoelectric effect experiment finds that the minimum photon frequency required to eject an electron from a metal is half that found for copper. What is the work function of the metal, (a) 4.4 eV, (b) 3.5 eV, (c) 2.2 eV, or (d) 8.8 eV?

| EXAMPLE 28.3 | Photons and Chemical Reactions |

Light plays a role in chemical processes called *photochemical reactions*. For example, the absorption of light can cause a hydrogen atom to dissociate, a process in which the atom absorbs a single photon and the electron is ejected (i.e., the hydrogen atom literally breaks apart). If the energy required for dissociation is 13.6 eV, what is the smallest photon frequency that can cause this reaction?

RECOGNIZE THE PRINCIPLE

This example is another application of the relation between photon frequency and energy. The minimum energy required for dissociation is called the ionization energy, which for hydrogen is $E_{ionization} = 13.6$ eV. To dissociate a hydrogen atom, the energy of an incoming photon must be equal to or greater than this value.

SKETCH THE PROBLEM

Figure 28.6 shows the problem.

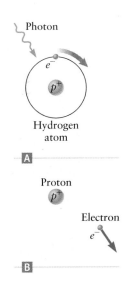

Figure 28.6 Example 28.3. Ionization of a (highly schematic) hydrogen atom. See Chapter 29 for a more accurate description of atomic structure. A Incoming photon. B The atom's proton and electron have dissociated.

The photon frequency must satisfy the condition

$$E_{photon} = hf \geq E_{ionization} \tag{1}$$

We know $E_{ionization}$ and Planck's constant h, so we can solve for the minimum photon frequency.

SOLVE

The lowest photon frequency that satisfies Equation (1) is

$$f = \frac{E_{ionization}}{h}$$

We are given the value $E_{ionization} = 13.6$ eV. Converting to units of joules, we use the conversion factor listed on the page facing the inside front cover and get $E_{ionization} = (13.6 \text{ eV})(1.60 \times 10^{-19} \text{ J/eV}) = 2.2 \times 10^{-18}$ J, which leads to

$$f = \frac{E_{ionization}}{h} = \frac{2.2 \times 10^{-18} \text{ J}}{6.63 \times 10^{-34} \text{ J} \cdot \text{s}} = \boxed{3.3 \times 10^{15} \text{ Hz}}$$

This frequency lies in the ultraviolet range and is not visible to the eye.

What does it mean?
Photochemical reactions are quite common in nature. In photosynthesis, for example, the absorption of a photon initiates a chain of reactions that leads to the storage of a portion of the photon's energy within a cell.

Insight 28.1
DETECTING SINGLE PHOTONS
Although the energy carried by a photon of visible light is small (see Eq. 28.6), it is nevertheless possible to detect single photons. Since the energy carried by a photon increases as the frequency increases, it is easiest to detect single photons of high-frequency radiation such as visible light, X-rays, and gamma rays. It is much more challenging to detect individual infrared, microwave, and radio frequency photons because they carry much less energy.

Blackbody Radiation

Planck's constant has an important role in Equation 28.1 and the photon theory of light. This constant was introduced by Max Planck in 1901 when he was studying the problem of *blackbody radiation*. Section 14.8 introduced the concept of a blackbody and described how the absorption and emission of light by a blackbody depends on its temperature. The specific problem that puzzled Planck is typified by the glowing oven in Figure 28.7A. This oven emits blackbody radiation over a range of wavelengths (λ) and frequencies as sketched in Figure 28.7B. To the eye, the color of the cavity is determined by the wavelength at which the radiation intensity is

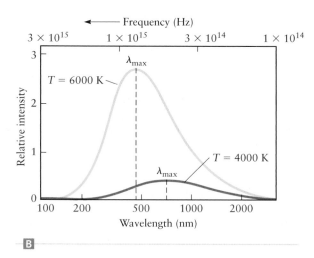

Figure 28.7 ◩ This oven is an approximate blackbody. Light emitted from an ideal blackbody follows the blackbody spectrum in ◪. The wavelength λ_{max} at which the blackbody radiation intensity is largest depends on the temperature of the blackbody.

largest. This wavelength is denoted as λ_{max} in Figure 28.7B and is determined by the temperature T of the blackbody through Wien's law (Eq. 14.14),

$$\lambda_{max} = \frac{2.90 \times 10^{-3} \, \text{m} \cdot \text{K}}{T} \tag{28.7}$$

Here T is measured in kelvins and the numerical constant in Equation 28.7 is a combination of several fundamental constants. According to Wien's law, the value of λ_{max} and hence the color of the blackbody depend on the temperature. A hotter object (higher T) has a smaller value of λ_{max}, and the entire blackbody curve in Figure 28.7B shifts to shorter wavelengths when the temperature is increased. A flame that appears blue is therefore hotter than one that is red.

Experiments prior to Planck's work showed that the intensity curve in Figure 28.7B has the same shape for a wide variety of objects: the blackbody intensity falls to zero at both long and short wavelengths, corresponding to low and high frequencies, with a peak in the middle at the location given by Wien's law (Eq. 28.7). Planck tried to explain this behavior. At that time, physicists knew that electromagnetic waves form standing waves as they reflect back and forth inside an oven's cavity. Such standing waves are just like the standing waves we discussed in Chapter 12, where we found that standing waves on a string have frequencies following the pattern $f_n = nf_1$, where f_1 is the fundamental frequency and $n = 1, 2, 3, \ldots$. Standing electromagnetic waves in a cavity follow the same mathematical pattern. There is no limit to the value of n (as long as it is an integer), so the frequency f_n of a standing electromagnetic wave in a blackbody cavity can be infinitely large. According to classical physics, each of these standing waves carries energy, and as their frequency increases, so does their energy. As a result, the classical theory predicts that the blackbody intensity should become *infinite* as the frequency approaches infinite values!

Planck's Hypothesis and Quanta of Electromagnetic Radiation

It was obvious to Planck and other physicists of the time that any theory predicting that the intensity of blackbody radiation is infinite could not possibly be correct. Such a theory would also be in conflict with the experimental intensity curves in Figure 28.7B because the true blackbody intensity falls to zero at high frequencies. (It would also violate "common sense"; nearly all objects act as approximate blackbodies, and it is very hard to imagine how all objects could emit an infinite amount of energy and still be consistent with our ideas about conservation of energy.) Despite much effort, however, physicists were not able reconcile classical theory with the observed blackbody behavior. This disagreement was called the "ultraviolet catastrophe" since the predicted infinite intensity is found at high frequencies and high-frequency visible light falls at the ultraviolet end of the electromagnetic spectrum.

Planck proposed to resolve this catastrophe by assuming the energy in a blackbody cavity must come in discrete parcels (quanta), with each parcel having an energy $E = hf_n$, where f_n is one of the standing wave frequencies and h is the constant given in Equation 28.2. Planck showed that this assumption fixes the theory of blackbody radiation so that it correctly produces the blackbody spectrum in Figure 28.7B, but he could give no reason or justification for his assumption about standing wave quanta. His theory could fit the experiments perfectly, but no one (including Planck) knew *why* it worked. That question was answered in part by Einstein: the standing electromagnetic waves in a blackbody cavity consist of photons with quantized energies given by Equation 28.1. While Einstein's photon theory supported Planck's result, it would still be several decades before the photon concept was fully worked out and understood.

Planck's theory of blackbody radiation was developed in 1900 and is generally cited as the beginning of quantum theory, giving Planck an important place in the history of physics. Blackbody radiation also has several practical uses. Figure 14.33B (page 458) shows an "ear thermometer," which detects the blackbody radiation from inside your ear and uses this measurement to calculate your body temperature. According to Wien's law (Eq. 28.7), the wavelength at which the blackbody radiation intensity is largest depends on temperature. This thermometer measures λ_{max} and then uses Wien's law to find your body temperature.

| EXAMPLE 28.4 | Temperature of the Sun

The Sun emits radiation over a wide range of wavelengths, following an approximate blackbody curve. To the eye, the Sun appears yellow, and yellow light has a wavelength near $\lambda_{max} = 5.5 \times 10^{-7}$ m (550 nm). What is the approximate temperature of the Sun?

RECOGNIZE THE PRINCIPLE

The color of a blackbody, such as the Sun, depends on its temperature. The color is determined by the wavelength at which the intensity has its maximum, which is the wavelength λ_{max} given by Wien's law.

SKETCH THE PROBLEM

Figure 28.7B gives the blackbody spectrum at two temperatures, and shows how this spectrum shifts to shorter wavelengths as the temperature of the blackbody increases.

IDENTIFY THE RELATIONSHIPS

According to Wien's law (Eq. 28.7), the location of the intensity peak is related to the temperature of the blackbody by

$$\lambda_{max} = \frac{2.90 \times 10^{-3} \text{ m} \cdot \text{K}}{T}$$

We know λ_{max} for the Sun, so we can find its temperature T.

SOLVE

Rearranging Wien's law to solve for the temperature gives

$$T = \frac{2.90 \times 10^{-3} \text{ m} \cdot \text{K}}{\lambda_{max}}$$

Inserting the observed value for λ_{max}, we find

$$T = \frac{2.90 \times 10^{-3} \text{ m} \cdot \text{K}}{5.5 \times 10^{-7} \text{ m}} = 5.3 \times 10^{3} \text{ K} = \boxed{5300 \text{ K}}$$

What does it mean?
The temperature of any star can be determined from its blackbody spectrum (and value of λ_{max}). Stars with a reddish color are thus cooler than the Sun, whereas stars that have a blue color are hotter.

CONCEPT CHECK 28.3 | ⊗ Blackbody Radiation from the Body
The human ear emits blackbody radiation. At what wavelength is the intensity of blackbody radiation from the ear largest, (a) 0.93 m, (b) 3.2×10^{-5} m, (c) 9.1×10^{-6} m, or (d) zero? *Hint*: The temperature of the human body is approximately 320 K.

Particle-Wave Nature of Light

We have discussed two phenomena, the photoelectric effect and blackbody radiation, that can only be understood in terms of the particle nature of light. While light thus has some properties like those of a classical particle, it also has wave properties *at the same time.* Electromagnetic radiation still exhibits some of the characteristic properties of classical waves, including interference. Hence, light has both wave-like and particle-like properties. In Figure 28.2, we described a double-slit experiment with electrons and described how the electrons arrive one at a time at the screen. Precisely the same behavior is found with light. Consider a double-slit experiment performed with light at very low intensity and suppose the screen responds to the arrival of individual photons by emitting light from the spot where the photon strikes. (The properties of this *fluorescent* screen are explained in Chapter 29.) The results would then appear just as in Figure 28.2, with the full interference pattern at the far right becoming visible only after many photons have reached the screen, thus demonstrating both the particle nature of light (the arrival of individual photons) and the wave nature (interference) in the same experiment.

28.3 | WAVELIKE PROPERTIES OF CLASSICAL PARTICLES

In Section 28.1, we described in a general way the wavelike properties of electrons and introduced the concept of "particle-waves." The notion that the properties of both classical waves and classical particles are present at the same time is also called *wave–particle duality* and is essential for understanding the microscale world of electrons, atoms, and molecules. Let's now discuss what wave–particle duality means in more quantitative terms. By the early 1920s, the photon theory of light was well established, but physicists were still struggling with how to describe particles such as electrons in the quantum regime. In particular, physicists did not yet realize that electrons are capable of the interference behavior sketched in Figure 28.2. This possibility was first proposed by Louis de Broglie in 1924, who suggested that *all* classical particles have wavelike properties. At the time, this idea must have seemed crazy because the experimental evidence for interference with electrons (similar to Fig. 28.2) had not yet been discovered. That did not stop de Broglie, however, who developed his theory in analogy with the behavior of photons. We have seen that a photon has a momentum given by (Eq. 28.5)

$$p = \frac{h}{\lambda}$$

De Broglie turned this result around and suggested that if a particle has a momentum p, its wavelength is

$$\lambda = \frac{h}{p} \tag{28.8}$$

If a particle has a wavelength, it should be able to exhibit interference just as waves do, so the test of de Broglie's hypothesis was to look for interference involving classical particles. Such an experiment is easiest if the wavelength is long, and according to Equation 28.8, a long wavelength implies a small value of momentum p. For a classical particle, $p = mv$, so a particle with a very small mass is required, and the lightest known particle (at that time) was an electron. It is thus not surprising that the first observation of wavelike behavior with a classical particle came from an experiment done with electrons by Clinton Davisson and Lester Germer (Fig. 28.8). A beam of electrons was aimed at a crystal target whose atoms formed a regular array (as is typical of many solids), acting as a series of slits for the electrons. Hence, it is really a "multislit" interference experiment similar to the diffraction gratings for light discussed in Chapter 25. Just as with a diffraction grating, the Davisson–Germer experiment exhibits interference when the wavelength of the electrons is comparable to the spacing between the slits (i.e., the spacing between atoms). In Figure 28.8, we sketch the interference of electrons that pass through a crystal. Davisson and Germer actually looked at the electrons reflected by a crystal,

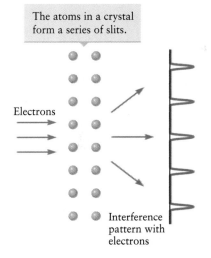

Figure 28.8 Schematic of an interference experiment with electrons, similar to the one first done by Davisson and Germer in 1927. When electrons pass through a thin crystal, the atoms in the crystal act as series of slits, producing interference and demonstrating the wave nature of electrons. Davisson and Germer's studies actually employed reflected electrons, which also form an interference pattern.

but the basic idea is the same. There is interference in both cases, but the reflection experiment is simpler to carry out.

Davisson and Germer's demonstration of interference with electrons showed conclusively that electrons have wavelike properties. Careful measurements of the interference pattern allowed them to deduce the electron wavelength, with a result in good agreement with de Broglie's theory (Eq. 28.8).

EXAMPLE 28.5 | **Wavelength of an Electron**

In their studies of interference with electrons, Davisson and Germer used electrons with a kinetic energy of about $KE = 50$ eV $= 8.0 \times 10^{-18}$ J. **(a)** What was the wavelength of these electrons? Express your answer in nanometers (1 nm $= 10^{-9}$ m). **(b)** The spacing between atoms in a typical crystal is about 0.3 nm. How does this spacing compare with the wavelength of the electrons used by Davisson and Germer?

RECOGNIZE THE PRINCIPLE

Given the kinetic energy, we can find the speed of the electrons, and from that, we can get their momentum ($p = mv$). We then use the de Broglie relation (Eq. 28.8) for the wavelength of a particle-wave to calculate the electron wavelength.

SKETCH THE PROBLEM

No sketch is necessary.

IDENTIFY THE RELATIONSHIPS

The kinetic energy of an electron of mass m and speed v is

$$KE = \tfrac{1}{2}mv^2$$

Rearranging to solve for v gives

$$v = \sqrt{\frac{2(KE)}{m}}$$

SOLVE

(a) Inserting the mass of an electron and the given value for the kinetic energy, we find

$$v = \sqrt{\frac{2(KE)}{m}} = \sqrt{\frac{2(8.0 \times 10^{-18} \text{ J})}{9.11 \times 10^{-31} \text{ kg}}} = 4.2 \times 10^6 \text{ m/s} \qquad (1)$$

The wavelength of an electron with this speed is

$$\lambda = \frac{h}{p} = \frac{h}{mv}$$

Inserting our result for v gives

$$\lambda = \frac{h}{mv} = \frac{6.63 \times 10^{-34} \text{ J} \cdot \text{s}}{(9.11 \times 10^{-31} \text{ kg})(4.2 \times 10^6 \text{ m/s})} = 1.7 \times 10^{-10} \text{ m} = \boxed{0.17 \text{ nm}}$$

(b) This wavelength is similar to the spacing between atoms in a solid (which is typically 0.3 nm). Hence, the conditions for observing interference are indeed satisfied, which is why the Davisson–Germer experiment was successful.

What does it mean?

The natural "slits" formed by atoms in a crystal (Fig. 28.8) are very useful for studying interference effects with particles such as electrons, protons, neutrons, and even atoms. Notice also that for electrons the speed found in Equation (1) is low enough that the classical (Newton's law) relations for the kinetic energy and momentum are accurate, and the relativistic results from Chapter 27 are not required.

Why Don't You See Wave Interference with Baseballs?

De Broglie's ideas were first confirmed with electrons, but he proposed that they apply to all classical particles. Equation 28.8 does indeed hold for all particles, including electrons, protons, and even baseballs, all of which exhibit wavelike properties under suitable conditions. Davisson and Germer demonstrated this with electrons, but the experiment is more difficult with other particles for the following reason. Consider a particle (such as an electron or proton) with mass m, speed v, and kinetic energy KE. Using the classical expression for the kinetic energy, we have

$$KE = \tfrac{1}{2}mv^2$$

Rearranging, we can solve for v:

$$v = \sqrt{\frac{2(KE)}{m}}$$

The momentum of the particle is then

$$p = mv = m\sqrt{\frac{2(KE)}{m}} = \sqrt{2m(KE)}$$

We can use this momentum in de Broglie's relation (Eq. 28.8) to get

$$\lambda = \frac{h}{p} = \frac{h}{\sqrt{2m(KE)}} \tag{28.9}$$

This result shows that for a fixed value of the kinetic energy, the wavelength becomes smaller as the mass of the particle is increased. The mass of a proton is about 1830 times greater than the mass of an electron, so assuming the same kinetic energy, the proton's wavelength will be shorter than that of an electron by a factor of approximately $\sqrt{1830} \approx 43$.

Will an interference experiment work with the much shorter wavelength of such protons? For a diffraction grating (Chapter 25), the condition for constructive interference is

$$d \sin \theta = n\lambda$$

where d is the slit spacing, θ is the diffraction angle, and $n = 0, 1, 2, \ldots$ is an integer. (Here we use n to denote an integer to avoid confusion with the particle mass m.) Had Davisson and Germer used protons, λ would have been much smaller (by a factor of 43); this small size could be compensated for by increasing n by the same factor, which corresponds to the 43rd maximum in an interference pattern such as Figure 28.2. In other words, with protons there would be 43 diffraction maxima squeezed in the same angle as one of the diffraction peaks found with electrons. For this reason, it would be much more difficult, although not completely impossible, to observe interference with protons.

In general, it becomes more and more difficult to observe interference as the mass of the particle is increased because if all else is kept fixed, a larger mass leads to a shorter wavelength (Eq. 28.9). In principle, one could observe interference with baseballs, but this experiment has not yet been done. (See Example 28.6.) This should remind you of the difference between geometrical optics and wave optics. Geometrical optics (Chapter 24) applies when the wavelength is very short, and in this regime light moves in straight lines just like classical particles. On the other hand, wave optics (Chapter 25) applies when the wavelength is relatively long and comparable to the size of any slits or openings, and in that case interference effects are important.

CONCEPT CHECK 28.4 | The de Broglie Wavelength
and the Mass of a Particle

An electron and a proton have the same de Broglie wavelength. Which of the following statements is true? *Note*: The mass of a proton is about 1800 times larger than the mass of an electron.

(a) The electron and proton have the same kinetic energy.
(b) The electron's kinetic energy must be larger than the proton's kinetic energy.
(c) The electron's kinetic energy must be smaller than the proton's kinetic energy.

| EXAMPLE 28.6 | De Broglie's Theory and the Wavelength of a Baseball |

What is the wavelength for a baseball traveling at a speed of 100 mi/h (approximately 45 m/s)? This speed is attained by some major league pitchers. The mass of a baseball is approximately 0.14 kg.

RECOGNIZE THE PRINCIPLE

De Broglie's theory connects the momentum and wavelength of a particle-wave. The momentum of a baseball is $p = mv$, so given the mass and speed we can find its momentum. The wavelength of the baseball is then given by the de Broglie relation.

SKETCH THE PROBLEM

Figure 28.9 describes the problem. While a baseball is a particle-wave, its wavelength is *very* short.

IDENTIFY THE RELATIONSHIPS AND SOLVE

Combining de Broglie's relation for the wavelength of a particle-wave (Eq. 28.8) with $p = mv$ leads to

$$\lambda = \frac{h}{p} = \frac{h}{mv}$$

Using the given values of m and v for the baseball, we find

$$\lambda = \frac{h}{mv} = \frac{6.63 \times 10^{-34} \text{ J} \cdot \text{s}}{(0.14 \text{ kg})(45 \text{ m/s})} = \boxed{1.0 \times 10^{-34} \text{ m}}$$

What does it mean?

This wavelength is *extremely* short. For comparison, the diameter of a proton is about 10^{-15} m, so the wavelength of a baseball is smaller than the size of a proton by the enormous factor of 10^{19}. Extremely small values of the wavelength are found for other macroscopic objects such as hockey pucks, planets, and people. For this reason, the wave properties of macroscopic objects are almost always completely negligible and the particle description of Newton's mechanics works with very high accuracy. The wavelike behavior of a baseball and of most other macroscopic objects will probably never be observed.

28.4 | ELECTRON SPIN

We have seen that electrons behave in some ways like classical particles and in other ways like classical waves. Are there any "new" or "extra" properties that show up in the quantum regime? That is, do electrons have any properties not found in either classical particles or waves? The answer is yes: electrons have another quantum property that involves their magnetic behavior. In Section 20.8 (page 667), we mentioned that an electron has a magnetic moment, a property associated with the phenomenon of **electron spin**. In a classical model, the electron's magnetic moment can be understood by picturing the electron as a spinning ball of charge (Fig. 28.10A). This spinning ball of charge acts as a collection of current loops, producing a magnetic field and thus acting like a small bar magnet as shown in Figure 28.10B.

$v = 45$ m/s

Figure 28.9 Example 28.6. A baseball is a particle-wave, but its wavelength is *very* short as given by the de Broglie relation.

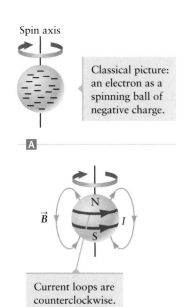

Spin axis

Classical picture: an electron as a spinning ball of negative charge.

Ⓐ

$\vec{B}$ I
N
S

Current loops are counterclockwise.

Ⓑ

Figure 28.10 Ⓐ Classical picture of the electron's spin. Ⓑ A spinning ball of charge (the electron) acts like a collection of current loops, producing a magnetic field.

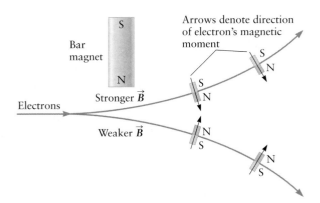

Bar magnet

Electrons

Stronger $\vec{B}$

Weaker $\vec{B}$

Arrows denote direction of electron's magnetic moment

Figure 28.11 Electrons have an intrinsic magnetism called the "spin magnetic moment." When a beam of electrons passes near one end of a bar magnet, it is found that the spin magnetic moment is either "up" or "down," so the electrons are either attracted to or repelled from the bar magnet. According to quantum theory, the electron's magnetic moment can only point in these two directions, up or down, but *not* "sideways" due to the quantization of the electron's spin.

Since an electron behaves as a small bar magnet, it is attracted to or repelled from the poles of other magnets. The magnetic properties of an electron were revealed in an experiment first performed by Otto Stern and Walther Gerlach. A simplified version of their experiment is sketched in Figure 28.11, showing a beam of electrons passing near one pole of a large (macrosize) magnet. (Stern and Gerlach used a slightly different arrangement of macrosize magnets, but the result is the same.) The electrons are attracted to or repelled from the large magnet according to the direction of the electron's north and south "poles." Two examples are shown in Figure 28.11; one electron is attracted to the large magnet's north pole and is thus deflected upward, and the other electron is repelled and deflected downward. The surprise in this experiment is that the outgoing electrons form just two beams. If an electron behaves like a bar magnet, a line connecting the north and south poles defines the axis of the magnet. These axes are indicated by arrows in Figure 28.11, with the arrowheads at the north ends of the magnets. These arrows also denote the direction of the magnetic moment of the electron, which is similar to the magnetic moment of a bar magnet or current loop as discussed in Chapter 20.

Classical theory predicts that the magnetic moment of an electron may point in any direction (up, down, sideways, or at any angle), and if that is true, the electrons in a Stern–Gerlach experiment should be deflected not just as two beams, but over a range of angles. However, only the two possibilities in Figure 28.11 actually occur; experiments always give just two outgoing beams of electrons. Hence two and only two orientations for the electron magnetic moment with respect to the direction of the applied magnetic field are possible. In words, the electron magnetic moment is *quantized*, with only two possible values.

The Stern–Gerlach experiment in Figure 28.11 shows that the quantization of the electron's magnetic moment applies to both the *direction* and *magnitude* of its magnetic moment. Hence, all electrons under all circumstances act as identical bar magnets.

Quantization of Electron Spin

To understand the implications of the Stern–Gerlach experiment, we consider again the picture in Figure 28.10 showing how an electron can act as a small bar magnet. Classically, we can picture an electron as a spinning ball of electric charge and acting as a collection of current loops. The electron in Figure 28.10A spins clockwise as viewed from the top of the figure, so the circulating charge acts as a counterclockwise current loop as viewed from above (because the electron is negatively charged). These current loops then produce a magnetic field in the same way that the current loops in Section 20.1 produce a magnetic field, and the result is called the **spin magnetic moment** of the electron. We also say that the electron has **spin angular momentum**. This angular momentum is similar to the angular momentum of a macroscopic spinning ball.

This classical picture explains how the electron can have a magnetic moment, but it does *not* explain the Stern–Gerlach experiment. The rotation axis of a classical particle such as a spinning baseball can point in any direction, so the classical picture in Figure 28.10 suggests that an electron's magnetic moment can be oriented in any direction. This classical picture also predicts that the electron's magnetic moment can be large, small, or even zero, depending on the angular speed of the electron since a classical ball can spin fast, slow, or not at all. However, the Stern–Gerlach result indicates that there can be only two values for the electron magnetic moment! Physicists now understand that this property is a result of quantum theory: the electron's spin axis and the magnitude of its magnetic moment are both quantized, with only two possible values. These two states are called spin "up" and spin "down." In the Stern–Gerlach experiment, one of the outgoing beams of electrons in Figure 28.11 corresponds to electrons with spin up, whereas electrons in the other beam have spin down.

Other quantum particles, such as protons and neutrons, also have a spin angular momentum and a resulting magnetic moment. In all cases, they are quantized, although the details can be more complicated than for an electron.

The quantization of spin angular momentum and the spin magnetic moment is a purely quantum effect, and the simple classical model in Figure 28.10 gives only an intuitive picture. There is no classical explanation for why spin is quantized (i.e., why there are only two outgoing beams in the Stern–Gerlach experiment). All we can say is that is the way electrons work. Recognizing the existence of electron spin is essential for understanding the structure of atoms and the organization of the periodic table of the elements (Chapter 29).

28.5 | THE MEANING OF THE WAVE FUNCTION

Newton's laws of mechanics are based on the idea that the motion of a particle can be described completely in terms of the particle's position, velocity, and acceleration. After all our work with Newton's mechanics, it is easy to take this notion for granted, but it is only an assumption about how the laws of physics are structured. This assumption works well in the classical world as proven by the outstanding success of Newton's laws in describing the mechanics of macroscale objects, but it fails in the quantum world of particle-waves. In a quantum description, the motion of a particle-wave is described by its *wave function*. The full mathematics of the wave function is beyond the scope of this book, but we can describe a few of its properties.

The wave function can be calculated using an equation developed by Erwin Schrödinger. He was one of the inventors of quantum theory and in 1933 received the Nobel Prize in Physics for this work. Schrödinger's equation plays a role similar to Newton's laws of motion. In quantum theory, one uses the Schrödinger equation to find the wave function and how it varies with time, in much the same way that one uses Newton's laws of mechanics to find the position, velocity, and acceleration of a particle.

In many situations, the solutions of the Schrödinger equation are mathematically similar to standing waves. As an example, consider an electron confined to a particular region of space as sketched in Figure 28.12A. In classical terms, you can think of this region as a very deep canyon or box from which the electron cannot escape. A classical particle would travel back and forth inside the box, bouncing from the walls. The wave function for a particle-wave (such as an electron) in such a box is described by standing waves, similar to the standing waves on a string. Figure 28.12B shows two possible wave function solutions corresponding to electrons with different kinetic energies. The wavelengths of these standing waves are different, as predicted by the de Broglie picture, since the wavelength of an electron depends on its kinetic energy (Eq. 28.9). In a classical wave-on-a-string picture, these two solutions correspond to two standing waves with different wavelengths. (See Fig. 12.23.)

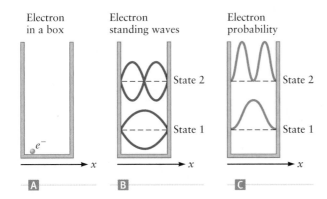

Electron in a box   Electron standing waves   Electron probability

State 2

State 1

e^-

x   x   x

A   B   C

Figure 28.12 ◩ An electron trapped in a box. ◩ The electron wave function forms a standing particle-wave similar to the standing waves on a string fastened to the walls of the box. The electron wavelength must therefore "fit" into the box as it would for a standing wave. ◪ Quantum mechanical probabilities of finding the electron at different locations in the box, corresponding to each of the two wave functions in part B.

After finding the wave function for a particular situation such as for the electron in Figure 28.12B, one can calculate the position and velocity of the electron. However, the results do not give a simple single value for x. Instead, the wave function allows the calculation of the *probability* of finding the electron at different locations in space. This probability for each of the wave functions in Figure 28.12B is shown in Figure 28.12C for our electron in a box. The probability of finding the electron at certain values of x is large in some regions and small in others, corresponding to the antinodes and nodes of the standing wave. The probability distribution (how the probability varies with position in the box) is different for each wave function.

CONCEPT CHECK 28.5 | Energy of an Electron in a Box

Figure 28.12B shows the wave functions for two different electron standing waves. Which of these electrons has the higher kinetic energy?

The Heisenberg Uncertainty Principle

In classical physics, we are used to calculating or measuring the position and velocity of a particle with great precision, but for a particle-wave, quantum effects place fundamental limits on this precision. For example, the electron in Figure 28.12 is described by a standing wave. The associated probability of finding the electron at different places in the box can be calculated from its wave function, yet what is the position of this electron? In a quantum sense, the standing waves in Figure 28.12B *are* the electron, so there is an inherent uncertainty in its position. There is some probability for finding the electron at virtually any spot in the box, and the uncertainty Δx in the electron's position is approximately the size of the box. This uncertainty is due to the wave nature of the electron.

We can gain insight into the uncertainty Δx from the experiment described in Figure 28.13 in which electrons are incident from the left on a narrow slit. This is very similar to the case of single-slit diffraction we studied with light waves in Section 25.6, and the electron wave is diffracted as it passes through the slit. If we carry out this experiment with many electrons, we can observe the associated interference pattern by measuring where the electrons strike the screen to the right of the slit. In quantum terms, the interference pattern gives a measure of how the wave function of the electron is distributed throughout space after it passes through the slit. The electrons coming in from the left in Figure 28.13 are all traveling parallel to the y axis, so their initial momentum is purely along the y direction. However, after being diffracted by the slit, an electron is likely to strike the screen some distance away from the y axis. Hence, a diffracted electron acquires a nonzero momentum along x.

Figure 28.13 also shows how the width Δx of the slit affects the interference pattern. The wide slit in Figure 28.13A produces a narrow electron distribution along the screen (the x direction in the figure). The narrow slit in Figure 28.13B produces a broader distribution as the wave function is more spread out along x.

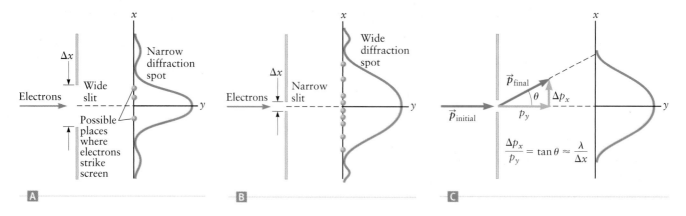

Figure 28.13 Single-slit diffraction with electrons. ▲ If an electron passes through a wide slit, the diffraction spot on the screen is narrow. ᴮ If the slit is made narrower, the diffraction spot becomes wider. ᴄ The spreading of an electron wave as it passes through a slit gives a way to relate the uncertainty Δx in the electron's position to the uncertainty Δp in its momentum.

Relating the Uncertainties in Position (Δx) and Momentum (Δp)

The electrons pass through the slit, so the position of the electron at the moment it passes through the slit is known with an uncertainty Δx equal to the width of the slit. Since an electron may strike the screen over a range of angles defined approximately by the central diffraction maximum, the outgoing electrons have a spread in their momentum along x and we can say that there is some uncertainty Δp_x in the x component of the momentum. Most of the electrons arrive somewhere within the width of the central diffraction maximum, so the final momentum along the x direction is uncertain by an amount approximately equal to the x component of the momentum of an electron at the edge of the diffraction spot. The uncertainty in momentum Δp_x is related to the uncertainty in position Δx. A wide slit (Fig. 28.13A) has a large value of Δx and a smaller value of Δp_x (because the diffraction "spot" is relatively narrow). On the other hand, a narrow slit (Fig. 28.13B) has a small uncertainty in position Δx but a larger uncertainty in momentum Δp_x.

We can derive a quantitative relationship between Δx and Δp_x from Figure 28.13C. The diffraction angle θ is contained in a right triangle with sides p_y and Δp_x, where p_y is the momentum along y. We have

$$\frac{\Delta p_x}{p_y} = \tan \theta$$

If the angle is small, we can use the approximation $\tan \theta \approx \theta$, resulting in

$$\frac{\Delta p_x}{p_y} = \theta \qquad (28.10)$$

The analysis of single-slit diffraction in Chapter 25 (Eq. 25.23) showed that for light of wavelength λ passing through a slit of width w, the first minimum in the diffracted intensity occurs at an angle

$$\theta = \frac{\lambda}{w}$$

This angle θ also describes the spread of the electron diffraction spot in Figure 28.13C, where the slit width w is equal to the uncertainty in the electron's position Δx. In the electron case, we have

$$\theta = \frac{\lambda}{\Delta x} \qquad (28.11)$$

Setting the results in Equations 28.10 and 28.11 equal leads to

$$\frac{\Delta p_x}{p_y} = \frac{\lambda}{\Delta x}$$

or

$$\Delta x \, \Delta p_x = \lambda p_y \qquad (28.12)$$

If we assume the uncertainty in the momentum Δp_x is small, then p_y is approximately equal to the total momentum, which we can denote by simply p. We will also follow standard notation and drop the subscript on Δp_x, referring to it as just the spread (i.e., uncertainty) in the electron's momentum. According to the de Broglie relation (Eq. 28.8), $p = h/\lambda$. Inserting that into Equation 28.12 gives

$$\Delta x \, \Delta p = \lambda p = \lambda \frac{h}{\lambda} = h$$

Thus, the uncertainty Δx in the electron's position is connected with the uncertainty Δp in its momentum in the x direction. Our analysis was only approximate; a more careful treatment gives nearly the same result,

$$\Delta x \, \Delta p = \frac{h}{4\pi} \qquad (28.13)$$

The uncertainties Δx and Δp in Equation 28.13 are the absolute limits set by quantum theory. In many situations, there are additional contributions to these uncertainties resulting from experimental or measurement error. Hence, Equation 28.13 gives a lower limit on the product of Δx and Δp. We can allow for this lower limit by writing

$$\Delta x \, \Delta p \geq \frac{h}{4\pi} \qquad (28.14)$$

Heisenberg uncertainty relation for position and momentum

This is called the *Heisenberg uncertainty principle*, in this case expressed as a relation between position and momentum. We derived this relation between Δx and Δp for the special case in Figure 28.13, but it holds for *any* quantum situation and for any particle-wave (not just electrons).

Position, Momentum, and Energy of a Particle: How Accurately Can We Know Them?

In classical physics, we often wish to calculate the position and momentum of a particle, as when finding the trajectory of a projectile using Newton's laws. Assuming there are no mistakes in our mathematics, we can find x and p to any desired *classical* accuracy. The Heisenberg uncertainty principle, however, dictates that in the quantum regime the uncertainties in x and p are connected. Under the very best of circumstances, the product of Δx and Δp is a constant, proportional to Planck's constant h. So, if you measure a particle-wave's position with great accuracy (small Δx), you must accept a large uncertainty Δp in its momentum. On the other hand, if you know the momentum very accurately (small Δp), you must accept a large position uncertainty Δx. You cannot make both uncertainties small at the same time. This key point is shown in Figure 28.13, where a wide slit (large Δx) leads to a small diffraction width (Δp) as in Figure 28.13A, whereas a narrow slit (small Δx) leads to a larger uncertainty Δp and a wider diffraction width as in Figure 28.13B.

The same ideas that lead to Equation 28.14 can also be used to derive a relation between the uncertainties in the energy ΔE of a particle and the time interval Δt over which this energy is measured or generated. The Heisenberg energy–time uncertainty principle is

$$\Delta E \, \Delta t \geq \frac{h}{4\pi} \qquad (28.15)$$

Heisenberg uncertainty relation for energy and time

If the energy of a particle or system is measured over a time period Δt, Equation 28.15 leads to a minimum uncertainty in the measured energy. This minimum uncertainty is negligibly small for a macroscale object, but it can be important in atomic and nuclear reactions.

Philosophical and Practical Implications of the Uncertainty Principle

The Heisenberg uncertainty principle has profound implications for our physical description of the universe. According to Newton's mechanics, it is possible (if not always easy) to know the position and momentum of a particle with any desired precision. However, quantum theory and Heisenberg's uncertainty relation in Equation 28.14 mean that there is always a trade-off between the uncertainties (i.e., precision) in x and p. It is not possible, *even in principle*, to have perfect knowledge of both x and p, which suggests that there is always some inherent "uncertainty" in our knowledge of the physical universe. The best we can do is compute the probability for finding a particle (e.g., an electron) at a particular position or the probability that a particle will have a certain momentum. Quantum theory thus says that the world is inherently *unpredictable*, in sharp contrast to the perfect predictability of classical physics.

We should also ask what effect the uncertainty principle has on the use of Newton's mechanics. The value of Planck's constant is very small ($h = 6.63 \times 10^{-34}$ J · s). For any macroscale object such as a baseball, a planet, or even an amoeba, the uncertainties in the real measurement will always be much larger than the inherent uncertainties due to the Heisenberg uncertainty relation. (For a numerical example, see Example 28.7.) As a general rule, the Heisenberg uncertainty relations only become important on the scale of electrons, atoms, molecules, and other microscale objects.

EXAMPLE 28.7 | Applying the Heisenberg Uncertainty Principle to a Brick

Consider a typical macroscale object such as a brick with $m = 1.0$ kg sitting on a table. A very careful measurement of the brick's location has an uncertainty of $\Delta x = 0.1$ nm (approximately the diameter of an atom). What is the corresponding uncertainty in the speed of the brick?

RECOGNIZE THE PRINCIPLE

This example involves a direct application of the Heisenberg uncertainty relation between the uncertainty in position and momentum. A given uncertainty Δx in the brick's position leads to a minimum uncertainty Δp in the brick's momentum. Since $p = mv$, there is a corresponding uncertainty in the brick's speed.

SKETCH THE PROBLEM

Figure 28.14 shows the problem.

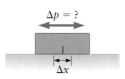

Figure 28.14 Example 28.7. Applying the Heisenberg uncertainty principle to a brick.

IDENTIFY THE RELATIONSHIPS

According to the Heisenberg uncertainty principle (Eq. 28.14),

$$\Delta x \, \Delta p \geq \frac{h}{4\pi}$$

Because we know Δx, we can solve for Δp. The momentum is $p = mv$ and we are given the brick's mass, so we can then find the uncertainty Δv in its speed.

SOLVE

Rearranging Equation 28.14 to solve for Δp gives

$$\Delta p \geq \frac{h}{4\pi(\Delta x)}$$

An uncertainty in the momentum corresponds to an uncertainty in the speed according to

$$\Delta p = m\,\Delta v \geq \frac{h}{4\pi(\Delta x)}$$

Solving for Δv and inserting the given values of m and Δx gives

$$\Delta v \geq \frac{h}{4\pi m\,\Delta x} = \frac{6.63 \times 10^{-34}\ \text{J} \cdot \text{s}}{4\pi(1.0\ \text{kg})(0.1 \times 10^{-9}\ \text{m})}$$

$$\Delta v \geq \boxed{5 \times 10^{-25}\ \text{m/s}}$$

What does it mean?

This uncertainty is *extremely* small. For all practical purposes, this quantum limit on the uncertainty is so small that it could never be measured. For bricks and other macroscale objects, the limits set by the Heisenberg uncertainty principle are unobservable, but they can be important in atomic and nuclear physics experiments, as we explore in Example 28.8.

| EXAMPLE 28.8 | Applying the Uncertainty Principle to a Hydrogen Atom |

A simple picture of the hydrogen atom has an electron moving around a proton as a sort of "electron cloud," with the density of the cloud proportional to the electron probability (Fig. 28.15). We'll see how this cloud picture is related to the electron's wave function in Chapter 29. Even without detailed knowledge of the wave function, however, we can still apply the Heisenberg uncertainty principle (Eq. 28.14), taking the uncertainty Δx in the electron's position equal to the diameter of the cloud. **(a)** If the diameter of a hydrogen atom is 1.0×10^{-10} m ($= 0.10$ nm), what is the minimum uncertainty Δp in the electron's momentum? **(b)** The momentum of the electron must be at least comparable to the uncertainty Δp, so we can use Δp as an estimate for the total momentum ($p \approx \Delta p$). With this approximation, use your result for Δp to find the electron's kinetic energy and compare it with the known ionization energy of a hydrogen atom, 13.6 eV.

RECOGNIZE THE PRINCIPLE

Part (a) is a direct application of the Heisenberg uncertainty relation between the uncertainty in position and momentum, with Δx equal to the given diameter of a hydrogen atom. For part (b), we take p to be equal to the uncertainty Δp in the momentum, and from that we can get the kinetic energy of the electron.

SKETCH THE PROBLEM

Figure 28.15 describes the problem.

IDENTIFY THE RELATIONSHIPS AND SOLVE

(a) Heisenberg's uncertainty relation (Eq. 28.14) gives

$$\Delta x\,\Delta p \geq \frac{h}{4\pi}$$

Solving for Δp, we find

$$\Delta p \geq \frac{h}{4\pi\,\Delta x} = \frac{6.63 \times 10^{-34}\ \text{J} \cdot \text{s}}{4\pi(1.0 \times 10^{-10}\ \text{m})} = \boxed{5.3 \times 10^{-25}\ \text{kg} \cdot \text{m/s}}$$

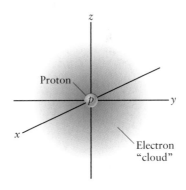

Figure 28.15 Example 28.8. Simple model of a hydrogen atom. The "cloud" represents the probability for finding the electron as a function of position near the proton.

(b) The momentum of an electron is $p = mv$, while the kinetic energy is $KE = \frac{1}{2}mv^2$. Combining these two relations gives

$$KE = \frac{p^2}{2m}$$

Using $p \approx \Delta p$ from part (a) and the known mass of an electron, we find

$$KE = \frac{p^2}{2m} = \frac{(5.3 \times 10^{-25}\ \text{kg} \cdot \text{m/s})^2}{2(9.11 \times 10^{-31}\ \text{kg})} = \boxed{2.9 \times 10^{-19}\ \text{J}}$$

Converting this energy to units of electron-volts (eV) gives

$$KE = (2.9 \times 10^{-19}\ \text{J})\left(\frac{1\ \text{eV}}{1.60 \times 10^{-19}\ \text{J}}\right) = \boxed{1.8\ \text{eV}}$$

What does it mean?
This kinetic energy is about a factor of three smaller than the ionization energy of the hydrogen atom (13.6 eV). The fact that these energies are the same order of magnitude suggests that an accurate theory of the hydrogen atom must be based on quantum theory (the main topic of Chapter 29).

28.6 | TUNNELING

According to classical physics, the electron in Figure 28.12 is trapped in the box and cannot escape. However, a quantum effect called *tunneling* allows such an electron to escape under certain circumstances. Figure 28.16A shows the wave function for this electron, with the region near the wall of the box shown on an expanded scale. Quantum theory allows the electron's wave function to penetrate a short distance *into* the wall. We say that the wave function extends a short distance into the *classically forbidden* region because according to Newton's mechanics the electron must stay completely inside the box and cannot go into the wall. In most cases, the quantum penetration distance is short and of little consequence, but if two boxes are very close together so that the wall between them is very thin, the wave function can extend from the inside of one box and through the wall to the box on the other side (Fig. 28.16B). Since the wave function extends through the wall, the electron has some probability for passing through the wall. If bullets were quantum objects, it would be like a bullet passing through a wall without leaving a hole!

Using Quantum Mechanics to Make a New Kind of Microscope

Tunneling is important in certain nuclear decay processes (Chapter 30) and in the motion of certain molecules. It is also used in the operation of a *scanning tunnel-*

Figure 28.16 ◮ An electron wave function (standing wave) such as for the electron in a box in Figure 28.12B. The electron wave function actually penetrates a short distance into the walls of the box, although the wave function is very small in this region. ◮ If the walls of two boxes are brought close together, the wave functions from the two boxes can overlap, allowing an electron to tunnel from one box to the other.

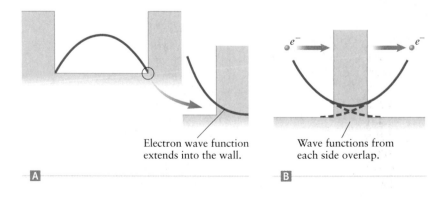

Electron wave function extends into the wall.

Wave functions from each side overlap.

A

B

ing microscope (STM). The basic design of an STM is shown in Figure 28.17A. A very sharp metal tip is positioned near a conducting surface, and an electric potential difference is applied between the tip and the surface. If the separation is large, the space (usually a vacuum) between the tip and the surface acts as a barrier for electron flow. This barrier is similar to the wall in Figure 28.16 since it prevents electrons from leaving the metal (i.e., the box). However, if the tip is brought very close to the surface, electrons may tunnel between them, producing a tunneling current. By measuring this current as the tip is scanned over the surface, it is possible to construct an image of how atoms are arranged on the surface. As the tip is scanned, it moves from regions where it is directly over an atom to other regions where the nearest atoms are much farther away as in Figure 28.17B. The tunneling current is largest when the tip is closest to an atom because the wave function overlap needed for tunneling is largest in this case. An STM image is based on variations in the size of the tunneling current as the tip is scanned across the surface, with high image intensity when the current is large and vice versa. A typical STM image is shown in Figure 28.18, with each "bump" corresponding to an individual iron atom on the surface of a piece of copper. The electric field between the tip and the surface atoms can also be used to push atoms around on the surface. In this experiment the STM tip was used to arrange iron atoms that were initially scattered around the surface (upper left) into a circle (lower right).

Tunneling plays a dual role in the operation of an STM. First, the detector current is produced by tunneling, so without tunneling there would be no image. Second, tunneling is essential for obtaining such high resolution. The STM tip is specially made to be very sharp, but even the best tips are rounded at the end. This rounding typically corresponds to a radius of curvature of 10 nm (10^{-8} m) or so, which is somewhat larger than the size of an atom. (An atom has a diameter of about 0.1 nm.) The effect of tip rounding on tunneling is shown in Figure 28.17C. Electrons can tunnel across the gap along many different paths, with the tunneling probability for each path depending on the size of the wave function in the tunneling gap. The wave function in the gap region falls off rapidly with distance (see Fig. 28.16B), so the shortest tunneling paths have the largest wave functions near the tip. Therefore, the vast majority of electrons that tunnel follow the shortest path, reaching a very small area on the surface, an area much smaller than the tip radius. The STM can then form images of individual atoms, even though the tip is larger than they are.

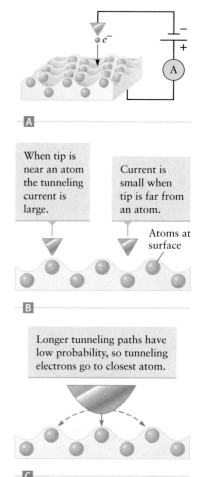

Figure 28.17 **A** In a scanning tunneling microscope (STM), a very sharp tip is brought very close to a surface, detecting the probability for an electron to tunnel between the tip and the surface by measuring the tunneling current. **B** The tunneling probability is largest for the shortest tunneling paths, so there is a large tunneling current when the tip is directly over an atom. **C** The STM is most sensitive to atoms that are directly below the tip.

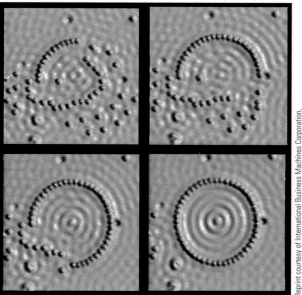

Figure 28.18 STM images of iron atoms on the surface of copper. The electric field from an STM tip can be used to push atoms around on the surface of a metal. In the lower right photo, the iron atoms are arranged to form a circle.

Human vision depends on the refractive properties of the eye as discussed in Chapters 24 and 26, treating light with Maxwell's wave theory, but a complete understanding of human vision also depends on the particle theory of light. In particular, the wave theory of light cannot explain color vision. Light is detected at the back surface of the eye, in a region called the retina. (Review Fig. 24.46.) The retina contains two kinds of light-sensitive cells, called *rods* and *cones*. When these cells absorb light, they generate an electrical signal that travels through the optic nerve to the brain. The rod cells are more sensitive to low light intensities and are used predominantly at night, while the cone cells are responsible for color vision.

Can the Eye Detect Single Photons?

Detection of low-intensity light is the job of the rod cells. When light enters your eye, only about 10% of it actually reaches the retina. The other 90% is reflected or absorbed by the cornea and other parts of the eye. Hence, for every ten photons that enter the eye, only one (on average) strikes the retina. Experiments show that the absorption of only a single photon by a rod cell causes the cell to generate a small electrical signal. The signal of an individual rod cell is not sent directly to the brain, however. Instead, the eye combines the signals from many rod cells before passing that combination signal along the optic nerve. Measurements show that for the combination signal to register at the brain, approximately five photons must be absorbed by rod cells within a period of about 0.1 s. So, although an individual rod cell is sensitive to single photons of visible light, your eye must receive approximately 50 photons within about 0.1 s for the brain to know that light has actually arrived at the eye.

Color Vision

The retina contains three types of cone cells, which respond to light of different colors (Fig. 28.19). The brain deduces the color of light by combining the signals from all three types of cones. For example, blue light triggers a large response from the blue cone cells and much smaller responses from the green and the red cone cells. A large signal from the blue cones together with small signals from the green and red cones tell the brain that the incident light lies in the blue part of the spectrum.[2]

The key property of the cones is that each type of cone cell is most sensitive to light with a particular color—that is, frequency—independent of the light intensity. For example, the green cone cells are most sensitive to green light and less sensitive to both red light (a lower frequency) and blue light (a higher frequency), even if the red or the blue light is very intense.

The correct explanation of color vision relies on two aspects of quantum theory. First, light arrives at the eye as photons whose energy depends on the frequency of the light. According to Equation 28.1, a photon of blue light carries more energy than a photon of green light, which in turn carries more energy than a photon of red light. When an individual photon is absorbed by a cone cell, the energy of the photon is taken up by a pigment molecule within the cell. Second, the energy of a pigment molecule is quantized, just as the energy of a photon is quantized. Figure 28.20A shows a simplified sketch of the quantized energies of blue, green, and red pigments. When a photon of the correct frequency is absorbed by a green pigment molecule, the molecule ends up in the upper energy level in Figure 28.20B, initiating a series of chemical reactions that eventually send an electrical signal to the brain. This photon absorption is possible because the difference in energy levels in the

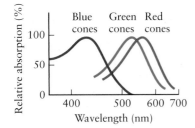

Figure 28.19 Pigment molecules inside the three types of cone cells absorb light preferentially at certain wavelengths (i.e., at certain frequencies).

---

[2]The precise way the brain combines these three signals is surprisingly complex. Some of this complexity was first revealed through clever experiments by Edwin Land, who also invented polarizing film (Chapter 23).

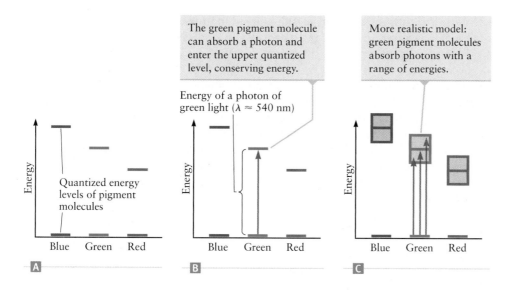

The green pigment molecule can absorb a photon and enter the upper quantized level, conserving energy.

Energy of a photon of green light ($\lambda \approx 540$ nm)

More realistic model: green pigment molecules absorb photons with a range of energies.

Quantized energy levels of pigment molecules

Blue Green Red

Blue Green Red

Blue Green Red

A

B

C

Figure 28.20 🅐 Pigment molecules have quantized energies. 🅑 The molecules preferentially absorb photons whose energy matches the difference between two energy levels. For this reason, a green pigment molecule readily absorbs photons of green light, but not photons of red light or blue light. 🅒 More realistic energy-level diagram. Each pigment molecule responds to a range of photon energies. These ranges overlap somewhat and give the broad absorption curves in Figure 28.19.

green pigment matches the energy of the photon. The quantized energy levels of a pigment molecule thus allow only photons with a certain energy to be absorbed.

According to the simplified energy-level diagram in Figure 28.20A, a pigment molecule can absorb a photon only if the photon energy precisely matches the pigment energy level. A more realistic energy diagram is shown in Figure 28.20C. The pigment molecule sits among other molecules in the retina, and interactions between the pigment and these molecules enable the absorption of photons over a range of energies, not just at the single value of the energy indicated in Figure 28.20A. The pigment molecule still responds most strongly to frequencies near the center of its preferred range, but it also gives a nonzero (although weaker) response to photons with both higher and lower energies, corresponding to nearby colors. The other pigments are sensitive to different energy ranges, so, by combining the response of all three types of pigments, the brain is still able to determine the color of the incident light. Hence, quantum theory and the existence of quantized energies for both photons and pigment molecules lead to color vision.

28.8 | THE NATURE OF QUANTA: A FEW PUZZLES

Quantum theory forces us to rethink our classical view of the world. Newton's laws cannot be applied at the atomic scale, and we must learn to deal with particle-waves. This rethinking leads to many new concepts, but a few of the central foundations of classical physics remain. The principles of conservation of energy, momentum, and charge are believed to hold true under *all* circumstances, but we must now allow for the existence of quanta. According to our understanding of the photon (Eq. 28.1) and the de Broglie theory (Eq. 28.8), its energy and momentum come in discrete quantized units. Electric charge also comes in quantized units, a result we have accepted throughout our study of electricity and magnetism. Indeed, the notion that charge is quantized in units of $\pm e$ is so familiar that it is easy to forget that the electron's existence was not known to Ampère, Faraday, Maxwell, and the other physicists who developed the theory of electricity and magnetism.

Does the discovery of quantum theory and the true nature of electrons and photons as particle-waves mean that we have now arrived at the "final" theory of nature? The answer is a definite *no*. Many puzzles remain, including (but not limited to) the following.

1. The relation between gravity and quantum theory is a major unsolved problem. No one really knows how Planck's constant enters the theory of gravitation or what a quantum theory of gravity looks like.

2. We know that electric charge is quantized in units of $\pm e$, but why are there two kinds of charge, positive and negative? In addition, why do the positive and negative charges come in the same quantized units? Physicists currently have no answer.

3. What new things happen in the regime where the micro- and macroworlds meet? According to quantum theory, all objects—even baseballs and bricks—behave as particle-waves. The quantum behavior of these macroscale objects is for all practical purposes unobservable, however, as we have seen in Examples 28.6 and 28.7. What about objects that are much smaller than baseballs and bricks? The quantum behavior of objects that are intermediate in size between the micro- and macroscales is now being intensively studied. There are also some interesting and unanswered questions about how quantum theory and the uncertainty principle apply to living things. For example, does the unpredictability associated with the Heisenberg uncertainty principle have any connection with consciousness and human free will? Some physicists think so, but the answer to this question is far from settled.

SUMMARY | Chapter 28

KEY CONCEPTS AND PRINCIPLES

Photons
Light energy is carried in quantized parcels called *photons*. For light with frequency f, the energy of a photon is

$$E_{\text{photon}} = hf \qquad \textbf{(28.1)} \text{ (page 958)}$$

The quantity h is *Planck's constant*, with the value

$$h = 6.626 \times 10^{-34} \text{ J} \cdot \text{s} \qquad \textbf{(28.2)} \text{ (page 958)}$$

Wave–particle duality
Electrons and other particles have a dual nature. They behave in some ways like classical particles and in other ways like classical waves. We call them *particle-waves*. The de Broglie wavelength of a particle-wave with momentum p is

$$\lambda = \frac{h}{p} \qquad \textbf{(28.8)} \text{ (page 965)}$$

This relation applies to electrons and all other particle-waves.

Electron spin
Electrons have **spin angular momentum** and behave as small bar magnets, with small magnetic moments. The direction of an electron magnetic moment is *quantized*; it can only point "up" or "down" (i.e., parallel or antiparallel to a particular direction). The magnitude of this magnetic moment is also quantized.

(Continued)

Wave functions

Schrödinger's equation is used to calculate the behavior of objects such as electrons in the quantum regime. The solution of Schrödinger's equation gives the *wave function* of an object. The probability of finding the object at different points in space can be calculated from the wave function.

Heisenberg uncertainty principle

The *Heisenberg uncertainty relations* give fundamental lower limits on the accuracy with which the position, momentum, and energy of an object can be determined, with

$$\Delta x \, \Delta p \geq \frac{h}{4\pi} \qquad \text{(28.14) (page 973)}$$

and

$$\Delta E \, \Delta t \geq \frac{h}{4\pi} \qquad \text{(28.15) (page 973)}$$

APPLICATIONS

Photoelectric effect

The *photoelectric effect* is a process in which light is absorbed by a metal. An electron is ejected from the metal when the photon energy is equal to or greater than the work function of the metal. This experiment can only be explained by the photon theory of light.

Tunneling

Tunneling is a process in which an electron (or any other type of particle-wave) passes through a barrier such as a wall that is classically impenetrable. Tunneling cannot be explained by Newton's mechanics, but is accounted for by quantum theory.

QUESTIONS

SSM = answer in Student Companion & Problem-Solving Guide   (X) = life science application

1. Estimate the de Broglie wavelength of a car that (a) has a speed of 100 mi/h, (b) has a speed of 10,000 mi/h, and (c) is at rest.

2. You have two lightbulbs: one gives off green light with a very low intensity, and the other emits red light with a very high intensity. Which one emits photons with a higher energy? If these two lightbulbs emit the same intensity, which one emits more photons each second?

3. (X) Consider a microscope that uses electrons instead of photons. Under what conditions will this microscope have better resolution than a microscope that uses visible light? *Hint*: Consider the electron's energy and wavelength.

4. What has greater energy, an ultraviolet photon or an X-ray photon?

5. Physicists often conduct experiments in which photons are counted one by one as they arrive at a detector. Explain why these experiments are easiest for photons of visible light and X-rays and are hardest for radio waves.

6. No electrons are ejected when a dim source of red light is directed onto a metallic surface. The intensity of the red light source is increased by a factor of 100. Is it now possible for electrons to be ejected via the photoelectric effect? Why or why not?

7. The (old-fashioned) film used for black-and-white photography is not affected by infrared light, is somewhat sensitive to red light, and exposes rapidly to blue light to the extent that blue filters are sometimes used to obtain appropriate contrast. Give a possible explanation for why the film has this color dependency.

8. (X) **Sunscreen.** Blocking ultraviolet radiation is effective in preventing sunburn. Using the concept of photons, describe why human skin is sensitive to ultraviolet frequencies, but much less so to those of visible light.

9. (X) Color vision can only be understood using a particle (photon) model of light. Compare color vision to the photoelectric effect, explaining how they are similar and how they differ.

10. Apply the Heisenberg uncertainty principle to a car. If you are asked to measure the position of the car, estimate the best accuracy you could expect to achieve with a ruler. Then calculate the minimum possible uncertainty in the car's momentum. Do you think it is feasible to measure the momentum with this accuracy?

11. [SSM] Consider the following types of radiation: visible light, infrared radiation, gamma rays, X-rays, radio waves, and ultraviolet radiation. From smallest to largest, order them according to (a) their frequency, (b) their wavelength, and (c) their photon energy.

12. The peak (maximum-intensity) wavelength emitted by a glowing piece of metal is found to be decreasing as time passes. Is the metal being heated, or is it being cooled?

13. A blacksmith heats a piece of metal in a furnace. The metal initially glows red and later yellow, and then it gets white hot while glowing brighter as more heat is applied. Is it possible for the metal to get "violet hot" and glow purple if it is hot enough? Why or why not?

14. (X) Astronomers classify the color of stars as red, yellow, white, and blue. Why are green stars never seen? *Hint*: Consider how the eye perceives a mixture of different colors of light.

15. Which of the following experiments or phenomena are evidence for the wave nature of light, and which are evidence for the existence of photons?
    (a) Single-slit diffraction
    (b) Photoelectric effect
    (c) Double-slit interference
    (d) Color vision

16. Explain why the wave nature of baseballs and cars is not apparent in everyday life.

17. Explain why the particle nature of light is not apparent in everyday life.

18. In Section 28.4, we gave a classical explanation of the electron's spin magnetic moment using the picture of an electron as a spinning ball of electric charge. The neutron has zero net charge, yet it also has a magnetic moment. Explain how a spinning ball with zero net charge can still have a magnetic moment due to its spin.

19. An electron and a proton have the same kinetic energy. Which one has the larger momentum?

20. [SSM] (X) Human color vision is based on the absorption of light by three different types of cone cells, which are most sensitive to red light, green light, and blue light. (a) Explain how the output of these cells can be used to determine the color of the incident light. (b) Could an eye with just two types of cone cells be used to give color vision? Explain why or why not. If your answer is yes, are there any examples in nature?

21. (X) Suppose the human eye was so sensitive that it could detect a single photon. With input of just a single photon, could our system of rods and cones detect the color of this light?

22. Explain why the existence of a cutoff frequency in the photoelectric effect cannot be explained with the wave theory of light.

23. Explain why a classical picture cannot account for the Stern–Gerlach experiment with electrons (Fig. 28.11). The Stern–Gerlach experiment can also be done with nuclei, and it is found that certain nuclei give rise to three outgoing beams. What does that imply for the number of possible directions for the magnetic moment of that particular nucleus?

24. Planck's constant was first used to explain what phenomenon?

PROBLEMS

[SSM] = solution in Student Companion & Problem-Solving Guide   (X) = life science application
[★] = intermediate   (✿) = challenging   (R) = reasoning and relationships problem

28.1 PARTICLES, WAVES, AND "PARTICLE-WAVES"

28.2 PHOTONS

1. (R) Find the approximate photon energy for (a) an FM radio signal, (b) the radiation in your microwave oven, (c) your cell phone signal, and (d) the light emitted by a burning match.

2. A helium–neon laser emits red light with $\lambda = 632.8$ nm. Find (a) the energy of a single photon and (b) the momentum carried by a single photon.

3. A helium–neon laser ($\lambda = 632.8$ nm) emits radiation with a power of 100 mW. How many photons does it emit in 1 s?

4. (X) X-rays used by your dentist have a wavelength near 0.070 nm. (a) What is the frequency of this radiation? (b) What is the energy of a single X-ray photon with this wavelength?

5. The highest-energy photons emitted by a hydrogen atom have an energy of 13.6 eV. Find the energy, in joules, of one of these photons.

6. [★] (R) The color of a blackbody is related to its temperature through Wien's law (Eq. 28.7). Estimate the energy of a typical photon emitted by (a) an ice cube, (b) a human being, and (c) an oven baking a cake.

7. A photon has an energy of 6.0 eV. What is the frequency of the radiation?

8. Suppose a blackbody emits photons most strongly with a frequency near 2.0 GHz. (a) What is the photon energy? (b) What

is the approximate temperature of the blackbody? (c) In what region of the electromagnetic spectrum does this radiation fall?

9. [★] (X) (R) The intensity of the Sun at the Earth's surface is approximately 1000 W/m². Estimate the number of photons from the Sun that strike an area of 1 m² each second.

10. How many photons of red light ($\lambda = 600$ nm) does it take to have a total energy of 1 J?

11. Radiation from outer space that reaches the Earth follows a blackbody spectrum with a temperature of about 2.7 K. (a) What is the energy of a typical photon from outer space? (b) Where does this photon fall in the electromagnetic spectrum?

12. [★] An FM radio station transmits at a frequency of 95 MHz and a total power of 100 kW. How many photons does this station emit each second?

13. The energy carried by 300 photons is 2.5×10^{-16} J. What is the frequency of one of these photons? Where do these photons fall in the electromagnetic spectrum?

14. A hydrogen atom that is initially at rest absorbs a photon with an energy of 2.0 eV. What is the momentum of the atom after it absorbs the photon?

15. A sodium lamp emits light with a wavelength of 590 nm. If the light radiates with a power of 30 W, how many photons does it emit each second?

16. How many photons does a green laser ($\lambda = 510$ nm) with a power rating of 2.0 W emit in 1 microsecond?

17. The work function for a metal is 5.0 eV. What is the minimum photon frequency that can just eject an electron from the metal?

18. [SSM] ⭐ The work function of gold is 4.6 eV. Photons of a certain energy are found to eject electrons with a kinetic energy of 2.3 eV. (a) What is the speed of the ejected electrons? (b) What is the photon frequency? (c) If the photon energy is increased by a factor of two, what is the kinetic energy of the ejected electrons?

19. ℝ Crystals can be used to do interference experiments with photons (as for electrons in Fig. 28.8). Consider a crystal with a 0.40-nm spacing between atoms. (a) If the photon wavelength must be approximately this size to observe interference, what is the energy of this photon? (b) Where is this radiation in the electromagnetic spectrum?

20. A device called a *photocell* detects light by letting it fall onto a metal and then measuring the current from the ejected electrons. You are designing a photocell to work with visible light and are considering the use of either aluminum (work function = 4.3 eV) or cesium (work function = 2.1 eV). (a) Which one is a better choice? (b) What is the lowest photon frequency that can be measured with your photocell? (c) Where does this frequency fall in the electromagnetic spectrum?

21. Molybdenum has a work function of 4.2 eV. If light of frequency 2.0×10^{15} Hz strikes the surface of molybdenum, what is the energy of the ejected electrons?

22. A nuclear fission bomb reaches a temperature of about 10^8 K when it explodes. What is the approximate photon energy at which the blackbody radiation has its highest intensity?

23. ℝ What is the approximate surface temperature of stars with the color (a) blue and (b) red?

24. ⭐ A "mosquito push-up" is the work done by a mosquito when it lifts its body ($m \approx 1$ mg) through a distance of 1 mm (the approximate length of a mosquito's "arms"). This amount of mechanical energy is very small. (a) How many mosquito push-ups does it take to equal the energy in a baseball ($m = 0.14$ kg) moving at 45 m/s (about 100 mi/h)? (b) How many photons of yellow light ($\lambda = 580$ nm) does it take to equal one mosquito push-up?

25. ✪ The ionization energy of a hydrogen atom is 13.6 eV. When a hydrogen atom absorbs a photon with this energy, the electron is ejected from the atom. (a) What are the frequency and wavelength of a photon with this energy? Where does this photon lie in the electromagnetic spectrum? Could you see light with this frequency? (b) When a photon has an energy greater than 13.6 eV, the "extra" energy (the energy in excess of 13.6 eV) goes into kinetic energy of the ejected electron. If a hydrogen atom absorbs a photon with an energy of 15.0 eV, what are the kinetic energy and speed of the ejected electron? (c) When the light intensity is very high, it is possible for a hydrogen atom to absorb two photons simultaneously. If a hydrogen atom absorbs two photons of equal energy and the atom is just barely ionized, what are the frequency and wavelength of these photons? Could you see this light?

26. Two photons have different frequencies f_1 and f_2. If the ratio of their momenta is 3.0, what is the ratio of their frequencies?

27. [SSM] ⭐ A hydrogen atom emits a photon of energy 3.4 eV. If the atom is at rest before emitting the photon, what is the speed of the atom after emission?

28. ⭐ A red laser that operates at a wavelength of 650 nm is directed at a black surface, which completely absorbs the beam. If the laser exerts a force of 10 nN (1.0×10^{-8} N) on the surface, how many photons are striking the surface every second?

29. ✪ A red laser emits photons with a wavelength of 620 nm. (a) If this laser has a power of 30 W, how many photons does it emit in 1.0 s? (b) If the light emitted by the laser is absorbed by a black sail, what is the force on the sail? (c) If the light from this laser is reflected from a mirror, what is the force on the mirror? You may recall that in Chapter 23 (Ex. 23.3) we discussed the use of such a sail to propel a spacecraft.

28.3 WAVELIKE PROPERTIES OF CLASSICAL PARTICLES

30. The kinetic energy of an electron is increased by a factor of two. By what factor does the wavelength change?

31. Consider two particles with masses that differ by a factor of 210. If they have the same wavelengths, what is the ratio of their kinetic energies?

32. ⭐ The kinetic energy of an electron in an atom such as hydrogen is typically around 10 eV. (a) What is the wavelength of an electron with this energy? (b) How does this wavelength compare with the size of an atom? (c) If the energy were decreased by a factor of 10, how much would the wavelength change?

33. ⭐ An electron has the same energy as a photon of blue light ($\lambda = 400$ nm). (a) What is the momentum of the electron? (b) What is the ratio of the momentum of the electron to the momentum of the photon?

34. ✪ ✗ ℝ An electron microscope uses electrons instead of light to form images. There are two general types of electron microscopes. A *scanning electron microscope* (SEM) forms an image of the surface of an object by reflecting electrons from it. (a) If the electron energy is 100 keV, what is the wavelength? (b) Based on our discussion of light microscopes in Chapter 26, what resolution do you expect to achieve with this SEM? That is, what is the approximate size of the smallest object that you expect to resolve? *Note:* Due to other limitations, SEMs do not achieve this resolution.

35. ✪ ✗ In addition to the scanning electron microscope described briefly in Problem 34, a second type of electron microscope, called a *transmission electron microscope* (TEM), operates by firing electrons through an object. Suppose the electrons in a TEM have energies of 5 MeV. (a) Find the electron wavelength. (b) Do you expect this TEM to have better resolution than the SEM in Problem 34? Explain why or why not. (c) How does your result from part (a) compare with the spacing between atoms in a solid (about 0.3 nm)? With the size of an atom (about 0.05 nm)?

36. An electron and a neutron have the same wavelength. What is the ratio of (a) their kinetic energies and (b) their momenta? Assume the speeds are low enough that you can ignore relativity.

37. (a) What is the wavelength of a photon whose momentum is equal to that of an electron with a speed of 2000 m/s? (b) What is the ratio of their energies?

38. If the wavelength of an electron is 150 nm, what is its speed?

39. A "thermal neutron" is a neutron whose kinetic energy is equal to $k_B T$, with T = room temperature ≈ 300 K. (a) What is the speed of a thermal neutron? (b) What is the wavelength of a thermal neutron? (c) How does the wavelength compare with the spacing between atoms in a solid (about 0.3 nm)? *Note:* To study the structure of solids, thermal neutrons are used in experiments similar to those of Davisson and Germer.

40. ✪ ℝ Oxygen molecules (O_2) in the atmosphere are described by kinetic theory (Chapter 15). What is the average de Broglie wavelength of these molecules at room temperature?

41. [SSM] ⭐ Photons with a frequency of 3.0×10^{15} Hz shine on a piece of aluminum (work function = 4.3 eV). What are the energy and the de Broglie wavelength of the ejected electrons?

42. ⭐ X-rays, electrons, and neutrons can all be used in diffraction experiments with crystals. In all three cases, the wavelength must be smaller than the spacing between atoms. Assuming this spacing is $d = 0.30$ nm, calculate the energies of an X-ray photon, an electron, and a neutron with wavelengths equal to $d/5$.

43. The mass of an electron is approximately 1800 times smaller than the mass of a proton. If an electron and proton have the same wavelength, what is the ratio of their energies?

44. ⭐ The de Broglie relation for the wavelength of a particle-wave (Eq. 28.8) is valid for relativistic particles (i.e., particles moving at speeds approaching the speed of light c) provided the relativistic expression for momentum (Eq. 27.24) is used to compute p. What is the wavelength of an electron with a kinetic energy of 5.0 MeV?

45. ⭐ Ⓡ In principle, a baseball could be used to do a diffraction experiment. If this experiment were to use a doorway as the diffraction slit, estimate the approximate speed that would be required for the baseball. *Hint:* First estimate the necessary wavelength for the baseball.

46. Electrons are used in a single-slit diffraction experiment. We saw in Chapter 25 that such an experiment works well if the slit has a width of 10 times the wavelength. If the slit in an experiment with electrons has a width of 10 nm, what is the electron energy?

47. |SSM| ⭐ Consider a double-slit interference experiment employing electrons. If the separation between two very narrow slits is 150 nm and the electron energy is 10 eV, what is the angle at which the first bright interference fringe is found?

48. Crystals can be used to do interference experiments with electrons. Consider a crystal with a spacing between atoms of 0.30 nm. If the electron wavelength must be approximately this size to observe interference, what is the electron energy?

28.4 ELECTRON SPIN

49. |SSM| ✪ The spin s of an electron is a type of angular momentum. It is quantized and is equal to $s = \pm\frac{1}{2}\hbar$, where the symbol h with a slash through the top is the *reduced* Planck's constant $\hbar = h/(2\pi)$, a quantity often found in equations concerning spin. As discussed in Section 28.4, a positive value of s denotes "spin up" orientation and a negative value means "spin down." This intrinsic property determines the behavior of an electron under the influence of the magnetic field in a Stern–Gerlach experiment. Spin is related to the ***magnetic moment***, which is given by the expression

$$\mu = -\frac{e}{m}s$$

where e and m are the charge and mass of the electron, respectively. (a) Find the value of the magnetic moment of the electron. It is a fundamental quantity called the ***Bohr magneton***. (b) Show that the units of the Bohr magneton can be expressed as J/T (joules per tesla).

28.5 THE MEANING OF THE WAVE FUNCTION

50. If a measurement of an electron's energy takes 3.0 ns (3.0×10^{-9} s), what is the minimum possible uncertainty in the energy?

51. If an electron is confined to a region of size 1.0 nm, what is the minimum possible uncertainty in its momentum?

52. ✪ An atom has a size of approximately 0.1 nm. Take this size as Δx, the uncertainty in the position of an electron in the atom. (a) Use the Heisenberg uncertainty relation to compute the uncertainty Δp in the electron's momentum. (b) If the electron's total momentum is equal to Δp, what is its kinetic energy? (c) How does your answer to part (b) compare with the binding energy of a typical atom? (d) Suppose the atomic size is reduced to 0.05 nm. This size might correspond to the extent of the wave function of the most tightly bound electrons in an atom such as argon. What is the estimated binding energy in this case? Does your result imply that these electrons are more "tightly bound" than the electron in part (c)?

53. ⭐ A proton in an atom is confined to the nucleus, which has a diameter of approximately 1 fm (1×10^{-15} m). This diameter is also the uncertainty Δx in the proton's position. Compute the uncertainty in the proton's momentum. What is the corresponding kinetic energy of the proton? We'll use this result in Chapter 30 when we study nuclear physics.

54. ⭐ Use the Heisenberg uncertainty relation to find the approximate kinetic energy of an electron that is trapped in a region the size of an atomic nucleus (1 fm = 1×10^{-15} m). Is this electron moving relativistically?

55. ✪ Suppose an electron is confined to a region of width 10 nm. The electron wave function forms a standing wave with nodes at the ends of the region. (a) What is the lowest-energy electron that can form such a standing wave? (b) What is the energy of an electron that forms a standing wave with three nodes (one at each end of the region and one in the middle)?

56. |SSM| ⭐ A beam of electrons is directed along the x axis and through a slit that is parallel to the y axis and 10 μm wide. The beam then hits a screen 1.5 m away. The electrons in the beam have a kinetic energy of 70 eV. (a) What is the de Broglie wavelength of the electrons? (b) What is the uncertainty in the y component of their momentum after passing through the slit? (c) How long does it take the electrons to reach the screen? (d) What is the uncertainty in the y position when they hit the screen?

28.6 TUNNELING

57. ✪ When a particle such as an electron in an STM tunnels through a barrier, the Heisenberg uncertainty relation can be applied to get an approximate upper limit on the tunneling distance. Suppose an electron with energy E_1 attempts to tunnel through the barrier sketched in Figure P28.57. For an STM, the width of the barrier is the distance between the STM tip and the object being studied, which is typically 0.5 nm. (a) Take this value for the barrier width as Δx and use the Heisenberg uncertainty relation to find the uncertainty Δp in the momentum of the electron. (b) Assuming Δp is the electron's total momentum, compute the speed v of the electron. (c) How long does it take the electron to travel through the barrier? (d) Use the time found in part (c) in the Heisenberg uncertainty relation (Eq. 28.15) to find the uncertainty ΔE in the energy of the electron. If ΔE is the total energy of the electron, compare it with the kinetic energy calculated with the value for the speed found in part (b). The tunneling probability will be significant if the energy uncertainty ΔE is comparable to or larger than the barrier "height" (measured in terms of the electron's potential energy).

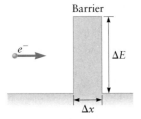

Figure P28.57

58. |SSM| ✪ Consider a hypothetical STM in which the tunneling particles are protons instead of electrons. Assuming an electron STM has a tunneling gap of 0.5 nm, estimate the size of the gap required for a proton STM. Assume all other parameters (including the kinetic energy) are the same.

28.7 ✖ DETECTION OF PHOTONS BY THE EYE

59. ✖ Estimate the photon energies at which the blue, green, and red cone cells in Figures 28.19 and 28.20 are most sensitive.

60. |SSM| ⭐ ✖ Given that the threshold for a human eye to sense the presence of light is 50 photons within 0.1 s, what is the minimum power output of a red ($\lambda = 635$ nm) laser diode that would make it visible in a darkened room?

61. A photoelectric experiment uses light with a wavelength $\lambda = 550$ nm, and the ejected electrons are found to have a kinetic energy of 0.25 eV. What is the work function of the metal?

62. SSM ✪ Ⓡ (a) What is the de Broglie wavelength of a helium atom with kinetic energy 2.0 eV? (b) Could you observe diffraction of these helium atoms using the atoms in a crystal to form the diffraction slits? (c) If your answer to part (b) is no, what is the approximate maximum speed for the helium atoms for which this diffraction experiment would work?

63. The mass of a mosquito is about 1 mg. If its speed is 2 m/s, what is its wavelength?

64. ✪ ✖ Ⓡ The threshold intensity for a human eye to sense the presence of light is 50 photons falling on the retina within a period of 0.1 s. The Sun radiates an output of 2.0×10^{26} W in the visible wavelengths, with a peak at $\lambda = 550$ nm. Estimate how far away a star like the Sun can be and still be visible to the unaided eye on a dark night.

65. Consider the following two quantum processes, each of which always emits a photon of a specific wavelength: (1) a 1.45-nm gamma ray from the nuclear process of cobalt-56 radioactive decay and (2) the 546-nm photon from the de-excitation of mercury in a street lamp (in the regime of chemical binding energies). (a) Determine the energy of the emitted photon (in units of joules and electron-volts) for each process and (b) calculate the ratio of the photon energies. How does the photon energy released from nuclear processes compare with that of chemical processes?

66. ✪ Ⓡ In the mid 1960s, the Stanford Linear Accelerator was the first to attain the ability to accelerate electrons to a kinetic energy of 20 GeV (giga-electron-volts). A series of experiments involved directing the beam of electrons at protons. By examining the angular distribution of the electrons after the interaction, it was discovered that protons are not fundamental particles, but instead are made of smaller pointlike entities, later named quarks. What is the approximate diameter of the smallest object this beam of electrons could be used to detect, and how does it compare with the approximate size of a proton? *Hint*: Consider the de Broglie wavelength.

67. ✖ (a) Calculate the electric potential difference needed to accelerate a proton so that it has a wavelength of 1.0×10^{-12} m. (b) Repeat the calculation for an electron. *Hint*: Here you must use the relativistic expressions for energy and momentum. (c) Compare the energies of these two particles (ratio of proton energy to electron energy) when they have this same wavelength.

68. ✖ ✪ Ⓡ Microwave ovens operate at a wavelength of 12.24 cm and have a typical output of 700 W. (a) How many photons are emitted by the oven's microwave generator per second? (b) Approximately what is the minimum number of photons needed to heat 0.15 L of coffee (mostly water) from 25°C to 90°C?

69. ✪ **Heavy light.** In 1983, a hypothesized heavy force-carrying particle called the Z boson was discovered by studying the collisions of protons at the European Particle Physics laboratory, CERN. Just as a photon transmits the electromagnetic force, the Z boson carries the weak nuclear force (Section 31.5). Consider an experiment in which many Z bosons are produced, and the mass of each can be calculated by looking at the kinetic energies and trajectories of the particles into which it decays. The mass of each Z boson is then plotted in a histogram like that in Figure P28.69, which gives an average value of approximately 91 proton masses! We find that the actual mass of the particle can be quite different from one Z particle to another, and no matter how great the precision of the measurement, the distribution of the mass is spread out with an uncertainty of about 2.5 proton masses. What is the approximate lifetime of this particle? *Hint*: Consider the mass–energy relationship from Chapter 27. Can short-lived particles have a well-defined mass? What is the relationship?

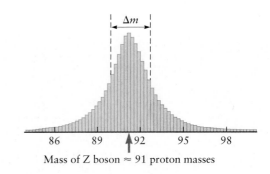

Mass of Z boson ≈ 91 proton masses

Figure P28.69

70. ✪ Using the Apache Point Observatory Lunar Laser-Ranging Operation (Fig. P23.61, page 794), the distance to the Moon is determined by how long it takes for a laser pulse to travel from the Earth to the Moon and then reflect back to the Earth from special reflector arrays placed there by the Apollo astronauts. A 2.3-W laser sends 20 pulses per second (each only 100 picoseconds long) at a wavelength of 532 nm. Due to atmospheric absorption and dispersion of the beam, only a very faint reflected pulse arrives back at the observatory with a power of 3.0×10^{-17} W. (a) How many photons does the observatory send in an outgoing pulse? (b) How many photons are received in a returning pulse? (c) How many photons are sent for every one that is detected upon return?

71. ✪ Electrons are accelerated through a potential difference of 20 kV in a vacuum tube to create a beam directed along the y axis. The beam passes through a 1.0-μm-wide slit and hits a fluorescent screen 25 cm away from the slit (as in Fig. 28.13C). (a) What is the de Broglie wavelength of the electrons in the beam as the beam enters the slit? (b) Consider single-slit diffraction and calculate the expected width along the x direction of the central bright spot on the screen. (Consider the distance to the first minimum in the diffraction pattern.) (c) What is the uncertainty in the x component of an electron's momentum in the beam after passing through the slit? (d) How long does it take for the electrons to reach the screen? (e) What is the uncertainty in the x position when the electrons hit the screen? (f) How does your answer in part (e) compare with that of part (b)? *Hint*: Consider the derivation in Section 28.5.

Chapter 29

Atomic Theory

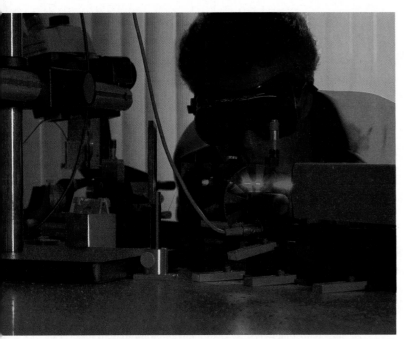

The operation of a laser is based on the quantum properties of the atom. This laser employs helium and neon atoms, producing red light. (© Stockbyte/Alamy)

Matter is composed of atoms, and atoms are themselves assembled from electrons, protons, and neutrons. These facts are such a fundamental part of elementary science courses that it is hard to imagine a world without atoms. You should realize, however, that atoms were discovered *after* Galileo, Newton, Maxwell, and most of the other physicists mentioned so far had passed from the scene. Indeed, electrons were discovered barely one hundred years ago, and the discovery of protons and neutrons came even later. In this chapter, we review some of the history and development of atomic theory and emphasize how key aspects of quantum theory explain the way atoms are put together. The central goal of atomic theory is to understand why different elements have different properties. With this understanding, we can explain the organization of the periodic table and provide a basis for most of chemistry.

29.1 | STRUCTURE OF THE ATOM: WHAT'S INSIDE?

By about 1890 or so, it was apparent to most physicists and chemists that matter is composed of atoms, and it was also widely believed that atoms are indivisible. Perhaps the strongest evidence for this "atomic picture" of matter came from the gas laws (Chapter 15) and chemistry's "law of definite proportions" (Insight 29.1). At the end of the 1800s, about 90 elements were known, suggesting there were the same number of fundamental building blocks of matter. We now know that picture is not quite correct. At last count, there were more than 110 elements but each is composed of just three different types of particles: *electrons*, *protons*, and *neutrons*.

We encountered electrons and protons in our work on electricity (Chapter 17); they are subatomic particles that carry electric charges of $-e$ and $+e$, respectively. Neutrons are a third type of subatomic particle; they have no net electric charge and have a mass approximately the same as that of a proton. Two key questions are at the center of atomic theory: (1) What are the basic properties of these atomic building blocks; for example, what are the mass, charge, and size of an electron, a proton, and a neutron? (2) How do just these three building blocks combine to make so many different kinds of atoms?

The properties of these building blocks were determined by experiments between about 1890 and 1932, which also showed that the behavior of these particles cannot be understood in terms of Newton's mechanics. An understanding of how electrons, protons, and neutrons combine to form atoms required the invention of a new type of mechanics, called *quantum mechanics*. Some key aspects of quantum mechanics were described in Chapter 28, where the notion of a particle-wave was introduced. In this chapter, we apply the ideas of quantum theory to understand the structure of the atom.

Plum-Pudding Model of the Atom

Electrons were the first of the building-block particles to be discovered. In about 1897, experiments involving electricity showed that electrons carry a negative charge and that they are contained within atoms. These experiments also showed that electrons have a very small mass (compared with an atom). Since atoms are electrically neutral—that is, they have zero net charge—there must be some source of positive charge inside them. It was initially suggested that the positive charge is distributed as a sort of "pudding," with electrons suspended within positively charged pudding throughout the atom as sketched in Figure 29.1A. This analogy is known as the "plum-pudding" model of the atom, a name chosen by English physicists who proposed the model and named it after a popular dessert of the time.

The plum-pudding model of the atom raises an interesting conceptual question. A neutral atom has a total electric charge of zero, so the negative charge of the electrons must be exactly canceled by the positive charge of the pudding. Hence, an atom must contain a precisely measured amount of positive pudding, but how that could be accomplished was not clear. To learn more about the positive charge in the atom, physicists studied how atoms collide with other atomic-scale particles. The most famous of these experiments was carried out by Ernest Rutherford and his students Hans Geiger[1] and Ernst Marsden. They arranged for alpha particles to collide with a thin sheet of gold atoms. At that time, it was known that alpha particles are emitted spontaneously from certain radioactive elements, but the details of radioactivity were not understood. We now know that an alpha particle is actually the nucleus of a helium atom (Chapter 30).

Even though the true nature of an alpha particle was not understood at the time of Rutherford's work, prior experiments had shown that alpha particles behave as simple

---

[1]Inventor of the Geiger counter. (See Insight 30.3.)

Insight 29.1
CHEMISTRY, ATOMS, AND THE LAW OF DEFINITE PROPORTIONS
In the early 1800s, chemists discovered that when a compound is completely broken down into its constituent elements, the masses of the constituents always have the same *proportions*, no matter what the quantity of the original substance. This similarity can be explained by assuming the constituents are atoms that combine in definite proportions to make molecules. For example, when water is broken down into hydrogen and oxygen, the number of hydrogen atoms released is always twice the number of oxygen atoms, the mass of hydrogen released is always one-ninth of the mass of the original water, and the mass of oxygen gives the remaining eight-ninths of the mass. This is an example of the *law of definite proportions*.

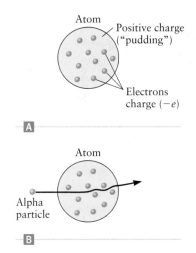

Figure 29.1 ◻ According to the plum-pudding model, the positive charge in an atom is distributed continuously, and the electrons sit in this "pudding." ◻ The pudding was thought to have a very low density, so the plum-pudding model predicts that an alpha particle should be deflected very little when it collides with (passes through) an atom.

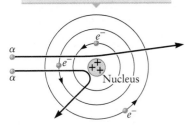

Figure 29.2 Planetary model of the atom, with all the positive charge located in the nucleus. When alpha particles are fired at an atom, they occasionally collide with the nucleus and are deflected through very large angles, but since the alpha particle and the nucleus are both very small, most alpha particles pass right through the atom.

Planetary model of the atom

Atomic number = number of protons in the nucleus

atomic-scale "bullets" when they collide with an atom. Rutherford's idea was to fire these "bullets" at an atom and then infer how the atom is put together by observing what comes out. Electrons have a very small mass, and Rutherford expected that the positively charged pudding would have a low density since (according to the plum-pudding model) this charge is spread throughout the entire atom. Using his knowledge of collisions and conservation of momentum (Chapter 7), Rutherford expected that the relatively massive alpha particles would pass freely through the plum-pudding atom as sketched in Figure 29.1B. For example, in a collision between a car (the alpha particle) and a Ping-Pong ball (an electron), the car is deflected only a very small amount because the momentum of a massive particle is much larger than the momentum of a low-mass particle with the same velocity. Likewise, an alpha particle should pass right through a wall of pudding because the pudding supposedly had a very small density compared with an alpha particle. When he carried out the experiment, Rutherford found that most of the alpha particles did indeed pass freely through the atom, but he also found that a small number of alpha particles were deflected a very *large* amount, some even bouncing directly "backward" toward the source of alpha particles. This surprising result could not be explained by the plum-pudding model. After much study, Rutherford realized that all the positive charge in an atom must be concentrated in a very small volume, with a mass and density about the same as an alpha particle. Most alpha particles in a collision experiment miss this dense region and pass right through the atom. Occasionally, though, an alpha particle collides with the dense region, giving a large deflection. Hence, Rutherford concluded that atoms contain a ***nucleus*** that is positively charged and has a large mass, much greater that the mass of an electron.

The Atomic Nucleus

After his discovery of the nucleus, Rutherford suggested that the atom is a sort of miniature solar system, with electrons orbiting the nucleus just as planets orbit the Sun (Fig. 29.2). This is called the ***planetary model of the atom***. In contrast to the plum-pudding model, the electrons in the planetary model are not stationary. Indeed, the electrons must move in orbits to avoid "falling" into the nucleus as a result of the electric force.

We now know that the atomic nucleus contains protons and that the charge on a proton is precisely equal to $+e$. The charge carried by an electron is $-e$, so the total charge carried by a hydrogen atom (symbol H), which contains one electron along with a nucleus consisting of a single proton, is precisely zero. The total charge of neutral atoms of all other elements is also zero, so the number of protons in the nucleus is equal to the number of electrons in the neutral atom. This number, called the ***atomic number*** of the element, is denoted by Z.

Except for the H nucleus, which is a single proton, all other nuclei contain neutrons. As its name implies, the neutron is a neutral particle, carrying zero net electric charge. The neutron was not discovered until the 1930s, and we'll discuss some of its properties in Chapter 30. Even without a detailed understanding of the neutron, however, it is easy to understand why all nuclei (except H) contain both protons and neutrons. Protons are positively charged particles, repelling one another according to Coulomb's law, so a hypothetical nucleus containing just two or more protons would fly apart spontaneously. Protons are attracted to neutrons by an additional force that overcomes this Coulomb repulsion and thereby holds the nucleus together (described in Chapter 30).

Rutherford's discovery of the nucleus led to the planetary model of the atom. In the solar system, the gravitational force exerted by the Sun on the planets keeps the planets in their orbits; in the atom, the Coulomb force exerted by the nucleus on the electrons keeps the electrons in orbit. Both of these forces (gravity and Coulomb's law) follow the inverse square law, so physicists immediately tried to apply Newton's mechanics to the motion of electrons as they orbit within the atom. Their attempts were not successful as these "planetary" properties of electrons do not correctly predict atomic behavior. Even so, this approach provides some useful insights into the correct quantum theory that was developed later. With that in mind, let's estimate some atomic properties using this planetary model.

The Planetary Model: Energy of an Orbiting Electron

Figure 29.3 shows the planetary model as applied to hydrogen. It is the simplest atom, containing just[2] one proton and one electron. For simplicity, let's assume the electron's orbit is circular with a radius r. To estimate the properties of this orbit, we note that r is the approximate size of the hydrogen atom. Chemists have found that the typical distance between hydrogen atoms in various molecules suggests a radius of about $r \approx 1.0 \times 10^{-10}$ m. Since the electron in Figure 29.3 is moving in a circle, there must be a force directed toward the center of the circle. For an electron of mass m moving with speed v, this attractive force has a magnitude (Eq. 5.6)

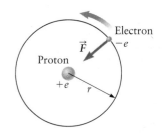

Figure 29.3 Planetary model of the hydrogen atom. A single electron is in orbit around a proton.

$$F = \frac{mv^2}{r} \tag{29.1}$$

The force in Figure 29.3 is the electric force (from Coulomb's law) exerted by the proton on the electron, which is given by (Eq. 17.3)

$$F = \frac{kq_{\text{proton}}q_{\text{electron}}}{r^2} \tag{29.2}$$

The charge on an electron is $q_{\text{electron}} = -e$, while the charge on a proton is $+e$. Inserting into Equation 29.2 gives $F = k(+e)(-e)/r^2 = -ke^2/r^2$. The negative sign indicates an attractive force, and we can drop it when considering just the magnitude of F. Inserting this result for F in Equation 29.1 leads to

$$\frac{ke^2}{r^2} = \frac{mv^2}{r}$$

The charge and mass of the electron are known and we have an estimate for the orbital radius r, so we can now solve for the electron's speed v. We get

$$v^2 = \frac{ke^2}{mr} \tag{29.3}$$

$$v = \sqrt{\frac{ke^2}{mr}} \tag{29.4}$$

Inserting values for the various quantities, including our estimate for r, we find

$$v = \sqrt{\frac{ke^2}{mr}} = \sqrt{\frac{(8.99 \times 10^9 \text{ N} \cdot \text{m}^2/\text{C}^2)(1.60 \times 10^{-19} \text{ C})^2}{(9.11 \times 10^{-31} \text{ kg})(1.0 \times 10^{-10} \text{ m})}}$$

$$v = 1.6 \times 10^6 \text{ m/s} \tag{29.5}$$

This orbiting electron has quite a high speed (approximately 4×10^6 mi/h!). Given this value of the speed, the associated kinetic energy is

$$KE = \tfrac{1}{2}mv^2 = \tfrac{1}{2}(9.11 \times 10^{-31} \text{ kg})(1.6 \times 10^6 \text{ m/s})^2 = 1.2 \times 10^{-18} \text{ J}$$

We'd like to compare this roughly estimated value of the electron's orbital kinetic energy to some energy value associated with a hydrogen atom in a chemical reaction. Chemists measure the **ionization energy**, the energy required to remove an electron from an atom in the gas phase. In mechanical terms, the ionization energy is the energy needed to completely separate the two charges; that is, to take the electron in hydrogen from its original orbit to an infinite distance from the proton. For

---

[2]It is possible to form atoms called deuterium and tritium, which each contain one proton and one electron. A deuterium nucleus also contains one neutron, and a tritium nucleus contains two neutrons.

ease of comparison, let's convert the kinetic energy found above from joules to units of electron-volts (eV). From Equation 18.19, 1 eV = 1.60×10^{-19} J, which leads to

$$KE = (1.2 \times 10^{-18} \text{ J}) \left(\frac{1 \text{ eV}}{1.60 \times 10^{-19} \text{ J}} \right) = 7.5 \text{ eV} \qquad (29.6)$$

In fact, the measured ionization energy of a hydrogen atom is 13.6 eV, so the electron kinetic energy found with the planetary model is indeed of the same order of magnitude as the ionization energy. For this reason, Rutherford's planetary model received strong initial support. In Example 29.1, we'll consider another aspect of the planetary model.

EXAMPLE 29.1 | **Ionization Energy of a Hydrogen Atom**

In our analysis of the planetary model of the hydrogen atom leading to Equation 29.6, we considered only the kinetic energy of the electron. To calculate the total energy required to remove the electron from the proton, we must also consider the potential energy. Assuming again the electron's orbit has a radius $r = 1.0 \times 10^{-10}$ m, find the electric potential energy in this model of the hydrogen atom. What is the change in this potential energy when the atom is ionized?

RECOGNIZE THE PRINCIPLE

The electric potential energy of two charged particles depends on the value of each charge and their separation. In the planetary model, the electron's orbit is circular, so the separation between the charges for the atom is just the radius of the orbit. When the atom is ionized, the electron is very (infinitely) far from the proton.

SKETCH THE PROBLEM

Figure 29.4 shows the problem. The electron is initially in orbit at a distance r_i from the proton (Fig. 29.4A), with a speed found in Equation 29.4. When the atom is ionized, the electron is at a distance $r_f = \infty$ from the proton (Fig. 29.4B).

IDENTIFY THE RELATIONSHIPS

The electron is initially in orbit at $r_i = 1.0 \times 10^{-10}$ m. The electric potential energy is given by (Eq. 18.6)

$$PE_{\text{elec}, i} = \frac{k q_{\text{proton}} q_{\text{electron}}}{r_i}$$

The charge on the proton is $q_{\text{proton}} = +e$ and the charge on the electron is $-e$, so

$$PE_{\text{elec}, i} = \frac{k q_{\text{proton}} q_{\text{electron}}}{r_i} = \frac{-ke^2}{r_i} \qquad (1)$$

SOLVE

Using our estimate for r_i, we find

$$PE_{\text{elec}, i} = \frac{-ke^2}{r_i} = \frac{-(8.99 \times 10^9 \text{ N} \cdot \text{m}^2/\text{C}^2)(1.60 \times 10^{-19} \text{ C})^2}{1.0 \times 10^{-10} \text{ m}}$$

$$PE_{\text{elec}, i} = -2.3 \times 10^{-18} \text{ J}$$

Converting to units of electron-volts as in Eq. 29.6 gives

$$PE_{\text{elec}, i} = (-2.3 \times 10^{-18} \text{ J}) \left(\frac{1 \text{ eV}}{1.60 \times 10^{-19} \text{ J}} \right) = \boxed{-14 \text{ eV}} \qquad (3)$$

When the atom is ionized, the separation between the electron and proton is infinite. Using Equation (1), the corresponding potential energy is $PE_{\text{elec}, f} = 0$. Hence, the change in potential energy when the atom is ionized is

$$PE_{\text{elec}, f} - PE_{\text{elec}, i} = \boxed{+14 \text{ eV}}$$

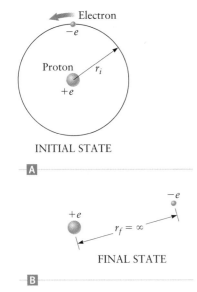

Electron

$-e$

Proton $\quad r_i$

$+e$

INITIAL STATE

A

$-e$

$+e$

$r_f = \infty$

FINAL STATE

B

Figure 29.4 Example 29.1. Calculating the potential energy of an electron in the planetary model of the hydrogen atom.

Problems with the Planetary Model of the Atom

Calculations with Rutherford's planetary model (like the one in Example 29.1) initially suggested that Newton's mechanics could be used to describe the atom, but some fundamental problems with this model soon became apparent. The biggest problem concerns the stability of an electron orbit. The orbital motion of a planet such as the Earth's motion around the Sun is extremely stable. Astronomers believe that the Earth has been traveling in its present orbit around the Sun for billions of years, with relatively little change in the orbital radius and period during that time. Likewise, atoms such as hydrogen can also be extremely stable. (If chemical reactions are avoided, an individual atom can "last" indefinitely.) The electrons in Rutherford's planetary model, however, are undergoing accelerated motion, and in Chapter 23 we saw that accelerated charges emit electromagnetic radiation that carries away energy. If the electron in a hydrogen atom loses energy in this way, it will spiral inward to the nucleus (Fig. 29.5). Hence, according to classical physics (i.e., according to Newton and Maxwell), Rutherford's atom is inherently unstable!

A careful analysis in terms of Newton's laws shows that an electron in the planetary model must spiral into the nucleus in a very small fraction of a second. Hence, according to classical physics, such planetary model atoms cannot exist. Despite much trying, physicists found no way to fix the planetary model to make the atoms in this model stable. This problem was not resolved until the development of quantum theory.

Quantum theory avoids the problem of unstable electron orbits by replacing them with discrete energy levels, just like the discrete energy levels of an electron in a box (Fig. 28.12) and the discrete energies of the pigment molecules involved in color vision (Section 28.7). Quantum theory rejects the notion of the electron as a simple particle that obeys Newton's laws (and spirals into the nucleus in Fig. 29.5). Instead, an electron is a particle-wave described by a wave function with discrete energy levels. Electrons lose or gain energy only when they undergo a transition between energy levels.

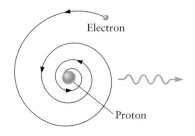

Figure 29.5 Maxwell's theory of electromagnetic waves predicts that an orbiting electron emits radiation, causing it to lose energy and spiral into the nucleus. If this picture is correct, all atoms would collapse, which is clearly *not* the case.

29.2 | ATOMIC SPECTRA

We have just claimed that an electron in an atom can exist only in discrete energy levels. The best evidence for this claim comes from the radiation an atom emits or absorbs when an electron undergoes a transition from one energy level to another. This radiation is key to another question that was studied intensely in the late 1800s: what gives an object its color? Physicists of that time knew about blackbody radiation (Chapters 14 and 28), including the relationship between the temperature of an object and its color. This relationship is described by Wien's law (Eqs. 14.14 and 28.7). The approximate blackbody spectrum of the Sun shown in Figure 29.6A is a continuous curve with a smooth distribution of intensity over a wide range of wavelengths and frequencies.

The spectrum in Figure 29.6A describes the radiation emitted by the Sun in a general way. Careful observations, however, show that the Sun's spectrum also contains sharp dips superimposed on the otherwise smooth blackbody curve as shown schematically in Figure 29.6B. These dips are called spectral "lines" because of their appearance when the spectrum is dispersed by a prism or diffraction grating (Chapters 24 and 25). Examination of the Sun's spectrum with a prism gives a band

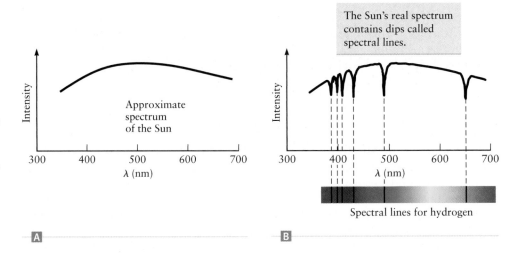

of colors extending from red (long wavelength) to blue (short wavelength) as shown below the graph in Figure 29.6B. The dips in the graph of the figure show up as dark lines in the band of colors, indicating wavelengths at which the light intensity is much lower than the expected blackbody value.

The origin of the dark spectral lines is illustrated in Figure 29.7A; when light from a pure blackbody source passes through a gas, atoms in the gas absorb light at certain wavelengths, producing dips in the spectrum at those wavelengths. Some of the dark lines in the Sun's spectrum are produced when hydrogen atoms in the Sun's atmosphere absorb light. This can be confirmed by an experiment using a blackbody source and hydrogen atoms on the Earth, which gives the spectral lines shown in the top band in Figure 29.7B; these lines have precisely the same wavelengths as those found in the Sun's spectrum.

The dark spectral lines are called ***absorption lines*** because they result from the absorption of light, in this case by hydrogen atoms. These same atoms can also be made to *emit* light. When the spectrum of light emitted by atoms is analyzed, it is found that the emission occurs only at certain wavelengths. The ***emission lines*** for hydrogen are shown in the lower band in Figure 29.7B, demonstrating that these emission and absorption lines occur at the same wavelengths.

Another example of spectral lines is given in Figure 29.8, which shows the emission and absorption spectra for sodium (Na) atoms. Here again the emission and absorption lines occur at the same wavelengths. The pattern of spectral lines is different for each element. In fact, by analyzing the wavelengths at which these lines occur, physicists in the 1800s determined the composition of the Sun's atmosphere.

These observations of atomic spectra lead to several questions. Why do absorption and emission occur only at certain special wavelengths? Why do the absorption and emission lines for a particular element occur at the same wavelengths? What determines the pattern of the absorption and emission wavelengths, and why are they different for different elements?

Many physicists attempted to answer these questions by applying Newton's mechanics to Rutherford's planetary model of the atom, but all ran into the fol-

Atoms emit and absorb light at discrete frequencies.

Figure 29.7 A When the Sun's blackbody radiation passes through its atmosphere, atoms in the atmosphere absorb at certain discrete frequencies, producing dark spectral absorption lines and causing the dips in the spectrum in Figure 29.6B. B Absorption and emission spectra for hydrogen. The absorption and emission lines occur at the same wavelengths. This plot shows just the spectral lines for hydrogen that lie between 400 nm and 700 nm. Figure 29.6B shows these and some other lines with wavelengths just below 400 nm.

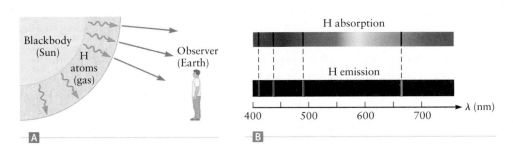

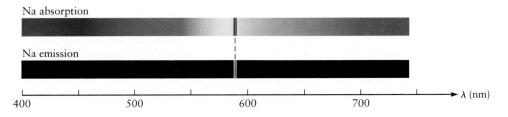

Na absorption

Na emission

λ (nm)

400   500   600   700

Figure 29.8 Sodium (Na) has two closely spaced emission lines in the yellow part of the visible spectrum. Sodium atoms also absorb strongly at the same wavelengths as the emission lines.

lowing problem. When a photon is emitted by an atom, the photon carries away a certain amount of energy; for a photon with frequency f, we have (Eq. 28.1)

$$E_{photon} = hf \qquad (29.7)$$

Total energy must be conserved, so the final energy of the atom is lower than the initial energy by an amount E_{photon}. Atomic emission occurs only at certain discrete wavelengths (i.e., discrete frequencies), which suggests that the energy of the orbiting electron can have only certain discrete values. The kinetic and potential energies of an orbiting electron depend on the radius r of the orbit as we already derived in Equation 29.6 and Example 29.1. According to Newton's mechanics, the orbital radius can be very large or very small, so the total energy of an orbiting electron can have a continuous range of values. Based on Newton's mechanics, there is no way for this planetary orbit picture to give discrete electron energies and no way to explain the existence of discrete spectral lines. This problem is resolved in quantum theory through the description of the electron's state in terms of a wave function instead of an orbit.

Atoms Have Quantized Energy Levels

Spectra like those in Figures 29.7 and 29.8 show that under certain conditions, the photons emitted or absorbed by an atom have only certain discrete wavelengths or frequencies. The implication is that the energy of the atom itself can have only certain discrete values. That is, the energy of an atom is **quantized**. In the language of quantum theory, we say that the energy of an absorbed or emitted photon is equal to the difference in energy between two discrete atomic energy levels (Fig. 29.9). Through Equation 29.7, the frequencies of the emission and absorption lines give the spacing between atomic energy levels. Hence, the study of atomic spectra gives a direct window into the atom's structure.

EXAMPLE 29.2	Energy of Photons Emitted by the Hydrogen Atom

Studies of the spectral lines of hydrogen show that the highest frequency of electromagnetic radiation emitted by a hydrogen atom is about 3.28×10^{15} Hz. **(a)** What is the energy of one of the corresponding photons? **(b)** Does this light fall in the infrared, visible, or ultraviolet part of the electromagnetic spectrum? *Hint*: The approximate short wavelength end of the visible range is at $\lambda \approx 400$ nm. **(c)** How does this photon energy compare with the ionization energy of the hydrogen atom, 13.6 eV?

RECOGNIZE THE PRINCIPLE

For part (a), this problem asks for a calculation of photon energy given the frequency, which can be done with Equation 29.7. For part (b), if we are given either the frequency or wavelength, we can use results from Chapter 23 and Figure 23.8 to find where this radiation falls within the electromagnetic spectrum. For part (c), the comparison will be apparent once we know the photon energy and convert it to electron-volts.

SKETCH THE PROBLEM

A spectral line is connected with photon emission or absorption between two discrete energy levels. Figure 29.10 shows this schematically for a hydrogen atom.

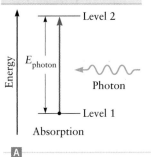

Energy is absorbed by the atom, leaving the atom in a state of higher energy.

A

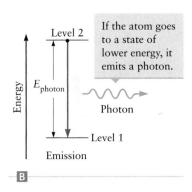

If the atom goes to a state of lower energy, it emits a photon.

B

Figure 29.9 The discrete absorption and emission lines of an atom are due to discrete atomic energy levels. An atom A absorbs or B emits a photon when it makes a transition between energy levels.

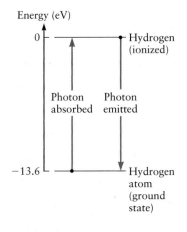

Figure 29.10 Example 29.2.

IDENTIFY THE RELATIONSHIPS AND SOLVE

(a) This problem begins with the relation $E_{photon} = hf$ (Eq. 29.7). Inserting the given value of f leads to

$$E_{photon} = hf = (6.63 \times 10^{-34}\ \text{J} \cdot \text{s})(3.28 \times 10^{15}\ \text{Hz}) = \boxed{2.17 \times 10^{-18}\ \text{J}} \quad (1)$$

(Here we keep three significant figures because they will be useful below.)

(b) To determine where this light falls in the electromagnetic spectrum, let's calculate its wavelength. The speed of light is $c = 3.00 \times 10^8$ m/s, so we have

$$\lambda = \frac{c}{f} = \frac{3.00 \times 10^8\ \text{m/s}}{3.28 \times 10^{15}\ \text{Hz}} = 91.5\ \text{nm}$$

According to the hint (and also Fig. 23.8), the visible range ends at about $\lambda \approx 400$ nm (corresponding to violet light), so this photon falls well into the $\boxed{\text{ultraviolet}}$.

(c) To compare the energy of the photon in Equation (1) with a typical atomic energy, it is useful to convert this energy to units of electron-volts. We get

$$E_{photon} = (2.17 \times 10^{-18}\ \text{J})\left(\frac{1\ \text{eV}}{1.60 \times 10^{-19}\ \text{J}}\right) = 13.6\ \text{eV}$$

Hence, this photon energy $\boxed{\text{is equal to}}$ the ionization energy.

What does it mean?

We have quoted the value of E_{photon} to three significant figures because it is a very special energy value in chemistry and physics. This result for E_{photon} is precisely equal to the ionization energy of hydrogen, the energy required to remove the electron from a hydrogen atom and thus "break" the atom apart. Our result for E_{photon} confirms the close connection between the way an electron is bound in an atom and the way the atom emits light. These processes involve the same two energy levels as sketched in Figure 29.10. The levels in hydrogen differ in energy by 13.6 eV, and by convention the upper level, corresponding to an ionized atom, has an energy $E = 0$.

29.3 | BOHR'S MODEL OF THE ATOM

The experimental facts described in the previous two sections showed that Rutherford's planetary model of the atom—and indeed *any* model based on Newton's mechanics—is a failure. The correct quantum theory of the atom did not "appear" instantaneously; there were first some intermediate proposals that eventually led to a complete quantum theory. One of these intermediate proposals is known as the **Bohr model**, invented by Danish physicist Niels Bohr (1885–1962). (Bohr later engaged Albert Einstein in famous debates over the philosophical implications of quantum theory.) Although Bohr's model has some flaws, it also contains some of the most important and most revolutionary ideas of the correct quantum theory and is an important conceptual step along the way.

Quantum States and Energy-Level Diagrams

Bohr based his thinking on the planetary model. For simplicity, he first assumed the electron orbits are circular, which allowed him to use the results for the electron kinetic and potential energies that we derived in Section 29.1 for a hypothetical hydrogen atom (consisting of just one electron and one proton). However, to explain the discrete spectral lines, Bohr needed several changes to the standard results based on Newton's laws. First, Bohr postulated that only certain electron orbits are allowed. That is, only certain specific (and special) values of the orbital radius r are permitted. If only certain discrete values of r are allowed, the electron's

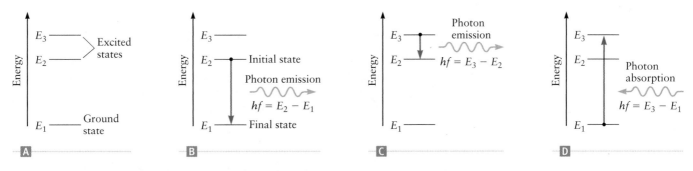

Figure 29.11 ⒜ Hypothetical energy-level diagram for an atom with three quantum states. ⒝ An atom that is initially in state 2 can undergo a transition to a state with lower energy (state 1) and emit a photon. ⒞ Photon emission due to a transition between excited states. ⒟ An atom can absorb a photon and undergo a transition to a state of higher energy.

kinetic and potential energies can have only certain values, so the total energy of the electron—the total energy of the atom—can have only certain discrete values. In the modern language of quantum theory, we say that the energy of the atom is *quantized*, and we refer to each of these allowed orbits as a *quantum state* of the atom. One way to view these energies and states is in terms of an *energy-level diagram*. Figure 29.11A shows a hypothetical diagram for an atom that can be in one of three different states. In the language of the Bohr model, there are three different allowed values of the orbital radius, and the corresponding total energy of an electron in these allowed orbits is E_1, E_2, or E_3. In this energy-level diagram, energy is plotted on the vertical axis and each state is represented by a horizontal line.

With this postulate of discrete energy levels, Bohr could explain the discrete spectral emission and absorption lines described in Section 29.2. Let's assume an atom is initially in the quantum state with energy E_2 in Figure 29.11B, corresponding to a particular orbital radius r_2. This electron may "jump" to a different orbit with energy E_1 as shown on the energy-level diagram by a vertical arrow that connects the initial and final states. After doing so, the atom has decreased in energy by an amount $\Delta E = E_2 - E_1$. According to the principle of conservation of energy, this energy cannot just vanish; total energy is conserved as the atom emits a photon with precisely this energy ΔE. The discrete set of atomic energy levels produces the discrete frequencies found in the emission spectrum of an atom (Figs. 29.7 and 29.8). We say that the atom described in Figure 29.11B was initially in an *excited state* and made a transition to the *ground state*, the state of lowest possible energy for this atom. The photon emitted in the transition has a wavelength matching one of the lines in the atom's emission spectrum. It is also possible for an atom to emit a photon and undergo a transition between two excited states such as the states with energies E_3 and E_2 in Figure 29.11C.

Transitions between the discrete energy levels in Bohr's picture also explain the absorption spectrum of an atom. The energy-level diagram in Figure 29.11D shows an atom that starts in the ground state with energy E_1 and then absorbs a photon as it undergoes a transition to an excited state with energy E_3. To conserve energy, the absorbed photon's energy must be $\Delta E = hf = E_3 - E_1$, the energy gained by the atom. This picture explains why the frequencies of the emission lines are precisely the same as the frequencies of the absorption lines. Both are determined by the difference in energy between two atomic states.

Quantized atomic states are shown on an energy-level diagram.

CONCEPT CHECK 29.1 | Spectral Lines and Energy Levels

The spectra for Na in Figure 29.8 show two closely spaced lines in both absorption and emission. Consider now a different (and hypothetical) atom. Is the *minimum* number of atomic energy levels required to give spectral lines at two or more different frequencies (a) two, (b) three, or (c) four?

Quantization of Angular Momentum Leads to Quantized States with the Correct Energies

Central to the Bohr model is the postulate that electrons can orbit at only certain allowed values of the radius r, but what determines these values? Here Bohr made a complete break with Newton's mechanics. Bohr proposed that the orbital angular momentum L of the electron could only have certain values given by the relation

$$L = n \frac{h}{2\pi} \tag{29.8}$$

where $n = 1, 2, 3, \ldots$ is an integer and h is Planck's constant. According to Equation 29.8, the allowed values of angular momentum are quantized in units of $h/(2\pi)$. What motivated Bohr to make this guess? One motivation is that this guess about L leads to the correct results for the frequencies of the emission and absorption lines for hydrogen (Figs. 29.6 and 29.7). Let's now see how Bohr's guess about the angular momentum leads to quantized energy levels.

From Chapter 7, the angular momentum of a particle of mass m traveling with speed v in a circular orbit of radius r is

$$L = mvr \tag{29.9}$$

Inserting this into Bohr's angular momentum relation, Equation 29.8, gives

$$L = mvr = n \frac{h}{2\pi}$$

Rearranging to solve for the electron's orbital speed v leads to

$$v = \frac{nh}{2\pi mr} \tag{29.10}$$

Bohr kept enough of Rutherford's planetary model to use the relationship between speed and radius for an electron moving in a circular orbit that we found in Section 29.1 (Eq. 29.3),

$$v^2 = \frac{ke^2}{mr}$$

Inserting the result for v from Equation 29.10, we obtain

$$v^2 = \left(\frac{nh}{2\pi mr} \right)^2 = \frac{ke^2}{mr}$$

We can now solve for the orbital radius r:

$$\frac{n^2 h^2}{4\pi^2 m^2 r^2} = \frac{ke^2}{mr}$$

$$r = n^2 \left(\frac{h^2}{4\pi^2 mke^2} \right) \tag{29.11}$$

The quantity in parentheses in Equation 29.11 is a combination of fundamental constants—Planck's constant, the mass and charge of an electron, and the constant k from Coulomb's law—so this term is also a constant. The variable n is an integer, so the factor n^2 on the right side of Equation 29.11 can have the values 1, $2^2 = 4$, $3^2 = 9$, and so on. In words, Equation 29.11 says that the orbital radius of an electron in a hydrogen atom can have only these particular quantized values. The smallest value of this radius is found when $n = 1$. This value is called the *Bohr radius* of the hydrogen atom and is the smallest orbit allowed in the Bohr model. We'll consider its value in Example 29.3.

Bohr radius of the hydrogen atom

Quantized Energies of the Bohr Atom

We can now continue with Bohr's approach and calculate the corresponding atomic energies using results from Section 29.1 (and Example 29.1). The kinetic energy of an orbiting electron is $KE = \frac{1}{2}mv^2$. Inserting the result for v^2 from Equation 29.3, we get

$$KE = \frac{1}{2}mv^2 = \frac{1}{2}m\frac{ke^2}{mr} = \frac{1}{2}\frac{ke^2}{r} \qquad (29.12)$$

The potential energy of this electron–proton pair is due to the electric force and is

$$PE_{elec} = -\frac{ke^2}{r} \qquad (29.13)$$

The total energy is thus

$$E_{tot} = KE + PE_{elec} = \frac{1}{2}\frac{ke^2}{r} - \frac{ke^2}{r} = -\frac{1}{2}\frac{ke^2}{r} \qquad (29.14)$$

We have already obtained the orbital radius r in Equation 29.11. Inserting that result, we find

$$E_{tot} = -\frac{1}{2}ke^2\left(\frac{1}{r}\right) = -\frac{1}{2}ke^2\left(\frac{4\pi^2mke^2}{n^2h^2}\right)$$

After some rearranging, we get

$$E_{tot} = -\left(\frac{2\pi^2k^2e^4m}{h^2}\right)\frac{1}{n^2} \qquad (29.15)$$

The factor in parentheses is a constant (because it is a combination of fundamental constants). The only variable is the integer n, which can have values $n = 1, 2, 3, \ldots$. Hence, the energy of the hydrogen atom in the Bohr model can have only certain *quantized* values corresponding to $n = 1, 2, 3, \ldots$ in Equation 29.15. In this way, Bohr's postulate about quantized angular momentum in Equation 29.8 leads to quantized energy levels for the atom.

The values of the energy levels predicted from Equation 29.15 can be used to derive the frequencies of the absorption and emission lines of hydrogen. To see how that works, we first find the values of E_{tot} predicted by Equation 29.15. Inserting the values of the various fundamental constants leads to

$$E_{tot} = -\left(\frac{2\pi^2k^2e^4m}{h^2}\right)\frac{1}{n^2}$$

$$E_{tot} = -\left[\frac{2\pi^2(8.99 \times 10^9 \text{ N} \cdot \text{m}^2/\text{C}^2)^2(1.60 \times 10^{-19} \text{ C})^4(9.11 \times 10^{-31} \text{ kg})}{(6.63 \times 10^{-34} \text{ J} \cdot \text{s})^2}\right]\frac{1}{n^2}$$

$$E_{tot} = -\frac{2.17 \times 10^{-18} \text{ J}}{n^2}$$

To compare with our previous calculations, we again convert to units of electron-volts and find

$$E_{tot} = -\frac{13.6 \text{ eV}}{n^2} \qquad (29.16)$$

Allowed energies of hydrogen in the Bohr model

We have given the value of E_{tot} to three significant figures so that we can more accurately compare it with the measured ionization energy of hydrogen.

The first few values in Equation 29.16 are plotted on the vertical axis in the energy-level diagram in Figure 29.12. The horizontal lines show levels corresponding to $n = 1, 2, 3, \ldots$. The level with $n = 1$ has the lowest energy and is the ground state of Bohr's hydrogen atom. Inserting $n = 1$ into Equation 29.16 gives $E_1 = -13.6$ eV. As n increases to $2, 3, \ldots$, the energies increase, with $E_2 = -3.4$ eV, $E_3 = -1.5$ eV, and so on, and at $n = \infty$, we have $E_\infty = 0$. All the levels except the highest one thus have negative energies. What do these negative values mean? How can an orbiting electron have a negative energy? The energy in Equation 29.16 is the total mechanical energy of Bohr's atom, the sum of the kinetic and potential energies. The kinetic energy is positive, but the potential energy is negative since the Coulomb force exerted between the proton and electron is attractive, and our convention for

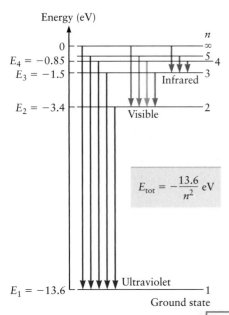

Figure 29.12 Energy-level diagram for the Bohr model of hydrogen.

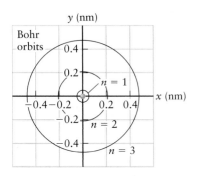

Figure 29.13 Example 29.3. Bohr orbits for the $n = 1$, 2, and 3 states of hydrogen. Note that the nucleus is not drawn to scale.

electric potential energy has $PE_{elec} = 0$ when the electron is infinitely far from the proton. The level with $E_{tot} = 0$ in Figure 29.12 and Equation 29.16 corresponds to completely removing the electron from the proton, or ionization of the atom. The energy required to take an electron from the ground state E_1 and remove it from the atom is the ionization energy. Bohr's theory thus predicts an ionization energy of 13.6 eV, in excellent agreement with the measured value.

The arrows in Figure 29.12 show some possible atomic transitions leading to emission of a photon. The energies of these transitions and the energies of the associated photons are equal to the difference in the energies of the levels at the start and end of each transition. Equation 29.16 predicts that there are an infinite number of energy levels (because $n = 1, 2, \ldots, \infty$), giving an infinite number of spectral emission lines. All these emission frequencies agree with the experimentally observed values, measured long before Bohr's work, adding convincing evidence that Bohr was indeed on the right track. More work was needed, but Bohr's ideas took physics much of the way toward the correct quantum theory of the atom.

EXAMPLE 29.3 | **The Bohr Model and the Size of a Hydrogen Atom**

In Section 29.1, we used an approximate orbital radius $r = 1.0 \times 10^{-10}$ m for the hydrogen atom as estimated from measurements of the atomic spacing in molecules and solids. Calculate r from the Bohr model for the lowest three energy levels, that is, the levels corresponding to $n = 1$, 2, and 3.

RECOGNIZE THE PRINCIPLE

The orbital radius r in the Bohr model depends on the value of n. According to the Bohr model (and Eq. 29.11), the radius is smallest for the quantum state with $n = 1$ and increases as n increases.

SKETCH THE PROBLEM

No sketch is needed to start the problem, but a plot of the results for the orbits is given in Figure 29.13.

IDENTIFY THE RELATIONSHIPS AND SOLVE

The result of Bohr's theory for the orbital radius is (Eq. 29.11)

$$r = n^2 \left(\frac{h^2}{4\pi^2 mke^2} \right)$$

Inserting the values of the constants in parentheses, we find

$$r = n^2 \left[\frac{(6.63 \times 10^{-34} \text{ J} \cdot \text{s})^2}{4\pi^2 (9.11 \times 10^{-31} \text{ kg})(8.99 \times 10^9 \text{ N} \cdot \text{m}^2/\text{C}^2)(1.60 \times 10^{-19} \text{ C})^2} \right]$$

$$r = n^2 \times 5.3 \times 10^{-11} \text{ m}$$

Evaluating for the given values of n leads to

$$n = 1 \text{ (ground state)}: r = 5.3 \times 10^{-11} \text{ m} = \boxed{0.053 \text{ nm}} \qquad (1)$$

$$n = 2: r = 2.1 \times 10^{-10} \text{ m} = \boxed{0.21 \text{ nm}}$$

$$n = 3: r = 4.8 \times 10^{-10} \text{ m} = \boxed{0.48 \text{ nm}}$$

What does it mean?

The value of r for the ground state given in Equation (1) differs by only a factor of two from the value we used in our earlier calculations with the planetary

---

model (Example 29.1), so our earlier estimates were fairly accurate. The value of r in Equation (1) is the value of the Bohr radius of the hydrogen atom and is used extensively in many estimates in atomic theory (by both physicists and chemists). Our results also show that the orbital radius increases rapidly as n increases, so electrons in these different Bohr orbits are well separated in space (Fig. 29.13).

| EXAMPLE 29.4 | Bohr Theory and the Frequency of an Emission Line |

Consider the transition in the Bohr model of a hydrogen atom from the $n = 2$ level to the $n = 1$ state. What is the wavelength of the photon emitted when the atom makes this transition?

RECOGNIZE THE PRINCIPLE

The wavelength of an emission line is related to its frequency, which is proportional to the difference in energy of the two levels. We thus need to find the energy separation of the $n = 2$ and $n = 1$ levels, which we can get from the result for the Bohr model energy in Figure 29.12 or by using Equation 29.16.

SKETCH THE PROBLEM

Figure 29.12 shows the Bohr model energy levels.

IDENTIFY THE RELATIONSHIPS

The energy of the transition is $\Delta E = E_2 - E_1$, and that equals the photon energy hf. We thus have

$$hf = \Delta E = E_2 - E_1$$

$$f = \frac{\Delta E}{h}$$

The wavelength is then

$$\lambda = \frac{c}{f} = \frac{hc}{\Delta E}$$

SOLVE

We can read the values of E_1 and E_2 from Figure 29.12 and find

$$\Delta E = E_2 - E_1 = (-3.4 \text{ eV}) - (-13.6 \text{ eV}) = 10.2 \text{ eV}$$

Converting from electron-volts to joules gives

$$\Delta E = 10.2 \text{ eV} \times \frac{1.60 \times 10^{-19} \text{ J}}{1 \text{ eV}} = 1.63 \times 10^{-18} \text{ J}$$

The wavelength is thus

$$\lambda = \frac{(6.63 \times 10^{-34} \text{ J} \cdot \text{s})(3.00 \times 10^8 \text{ m/s})}{1.63 \times 10^{-18} \text{ J}} = 1.22 \times 10^{-7} \text{ m} = \boxed{122 \text{ nm}}$$

What does it mean?

This wavelength lies in the ultraviolet; recall that the visible range ends at a wavelength of about 400 nm. Hence, this radiation would not be visible to the human eye.

Generation of X-rays by Atoms

The ionization energy of the hydrogen atom is 13.6 eV. We have seen that this is the energy of the hydrogen emission line with the highest photon energy, corresponding

Figure 29.14 ◾ For heavy atoms, the emission lines can lie in the X-ray part of the electromagnetic spectrum. ◾ The X-rays used by a dentist are photons generated by atomic transitions of atoms such as (typically) molybdenum and tungsten.

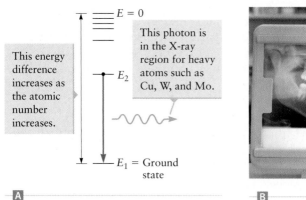

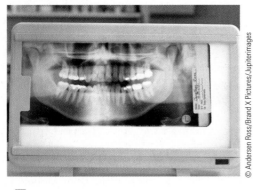

This energy difference increases as the atomic number increases.

This photon is in the X-ray region for heavy atoms such as Cu, W, and Mo.

$E = 0$

E_2

E_1 = Ground state

A

B

to a transition from the state with $E = 0$ to E_1 in Figure 29.12. The wavelength of this photon is $\lambda = 91.5$ nm (Example 29.2), which lies in the ultraviolet part of the electromagnetic spectrum. Although it is the highest photon energy that can be emitted by a hydrogen atom, other atoms can emit much more energetic photons. As the charge on the nucleus is increased (by going from hydrogen to helium and to heavier atoms), the magnitude of the electric potential energy of an electron also increases. As a result, for helium and heavier atoms the energy required to remove an electron from the $n = 1$ Bohr orbit (i.e., a tightly bound, "inner-shell" electron) is much larger than for hydrogen. The corresponding photon energies then increase into the X-ray region. Many applications employ X-ray photons generated when an electron undergoes a transition from the E_2 state to the E_1 state (Fig. 29.14A). These photons are denoted as "K X-rays," and their energies for a few elements are listed in Table 29.1. The X-ray photons used by your dentist or doctor are produced by these atomic transitions, usually involving atoms of tungsten or molybdenum (Fig. 29.14B).

CONCEPT CHECK 29.2 | *K* X-rays and the Energy Levels of an Atom

The energy of a *K* X-ray of copper (Cu) is 8.0 keV (Table 29.1). According to the Bohr model, is the energy required to remove an electron from the $n = 1$ Bohr orbit of a copper atom (and ionize the atom) (a) equal to, (b) less than, or (c) greater than 8.0 keV?

Continuous Spectra

The existence of discrete spectral lines is an essential part of quantum theory. In certain situations, however, the absorption may not involve discrete photon frequencies. Figure 29.15 shows a hydrogen atom that is initially in its ground state and then absorbs a photon. This photon has an energy greater than 13.6 eV, so it has more than enough energy to remove (eject) the electron from the atom. The extra energy goes into the kinetic energy of the ejected electron. Because the ejected electron's final kinetic energy can have a continuous range of values, the absorbed photon can also have a continuous range of energies. Hence, when an atom is ionized, the absorbed energy can have a range of values, producing what is called a continuous spectrum. Not everything is quantized in the Bohr model or in quantum mechanics.

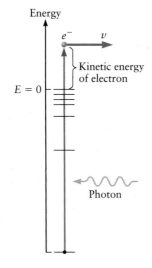

Energy

e^- v

Kinetic energy of electron

$E = 0$

Photon

Figure 29.15 When an atom absorbs a photon and is ionized, the final kinetic energy of the electron is not quantized. Hence, the photon can have a range of energies, and the absorption energy in this case is *not* quantized.

Table 29.1 Energy of *K* X-rays Produced by a Few Elements

ELEMENT	ATOMIC NUMBER (Z)	*K* X-RAY ENERGY (keV)
Ne (neon)	10	0.85
Al (aluminum)	13	1.6
Cu (copper)	29	8.0
Mo (molybdenum)	42	17
W (tungsten)	74	58

Why Is Angular Momentum Quantized?

Bohr's theory provided the first successful calculation of the frequencies of the spectral lines of any atom, but at first no one knew *why* it worked. What is the Bohr model telling us about the quantum world? Several of Bohr's assumptions can be traced to Einstein's theory of the photon and the notion that energy is conserved in atomic transitions, but Bohr's suggestion that the angular momentum of the electron is quantized (Eq. 29.8) was completely new. This assumption was needed to give quantized atomic levels, without which the Bohr model would have many of the same problems as Newton's mechanics and the planetary model. Bohr's assumption about angular momentum can be understood using the de Broglie theory of particle-waves. (But note that the de Broglie theory came about 10 years after Bohr's work.) De Broglie proposed (Chapter 28) that electrons and all other particles have a wave character, with a wavelength λ given by (Eq. 28.8)

$$\lambda = \frac{h}{p} \qquad (29.17)$$

where p is the momentum of the particle. Figure 29.16 shows such a particle-wave orbiting a nucleus, corresponding to an electron moving in a circular orbit. The allowed electron orbits in the Bohr model correspond to standing waves that fit precisely into the orbital circumference. To form a standing wave, the circumference of the orbit must be an integer multiple of the electron wavelength. Figure 29.16A shows a hypothetical case in which the electron wavelength does not match the orbit; for this value of r and λ, it is not possible to form a standing wave, and Bohr's condition for the angular momentum (Eq. 29.8) cannot be satisfied. This orbital radius is not allowed in the Bohr model. Figure 29.16B shows two cases in which the radius and wavelength match so that an integral number of wavelengths fit into one circumference and a standing wave is formed. These standing waves are two of the allowed quantized states of the Bohr model.

Figure 29.16 explains the qualitative origin of the discrete states in the Bohr model. We can also use this picture to derive Bohr's angular momentum relation in Equation 29.8 by setting the circumference of the circular orbit equal to an integer number of wavelengths. The circumference of an orbit is $2\pi r$, so if n is an integer, we require

$$2\pi r = n\lambda$$

The wavelength is given by de Broglie as $\lambda = h/p$ (Eq. 29.17), so

$$2\pi r = n\lambda = n\frac{h}{p} \qquad (29.18)$$

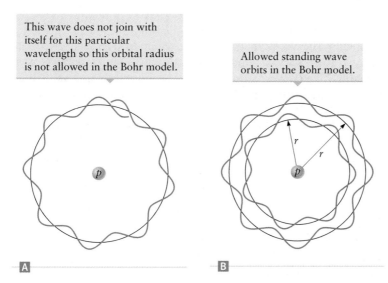

This wave does not join with itself for this particular wavelength so this orbital radius is not allowed in the Bohr model.

Allowed standing wave orbits in the Bohr model.

Figure 29.16 Bohr's quantization relation for angular momentum can be justified using the particle-wave nature of the electron, with the electron forming a standing wave in its orbit about the nucleus. The Bohr theory requires that an integer multiple of electron wavelengths equal the orbital circumference. In **A**, this condition is not met, so this orbital radius is not allowed in the Bohr theory for the wavelength shown. **B** Two allowed orbits. Note that the protons are not drawn to scale.

According to Newton's mechanics, the momentum of a particle is $p = mv$, which when inserted into Equation 29.18 gives

$$2\pi r = n\frac{h}{p} = n\frac{h}{mv}$$

Rearranging and noting that the angular momentum is $L = mvr$, we arrive at

$$2\pi r = n\frac{h}{mv}$$

$$L = mvr = n\frac{h}{2\pi} \tag{29.19}$$

which is precisely Bohr's condition for the angular momentum, Equation 29.8.

Problems with the Bohr Model: Where Do We Go Next?

The success of the Bohr model of the hydrogen atom inspired Bohr and others to apply the same ideas to other atoms, and the Bohr theory was found to work well when only a single electron is present, as in the ions He^+ and Li^{2+}. This model does not, however, correctly explain the properties of atoms or ions that contain two or more electrons. Physicists eventually concluded that the Bohr model is not the correct quantum theory, but rather a "transition theory" that helped pave the way from Newton's mechanics to modern quantum mechanics. The Bohr model gives very useful insights into why an atom has quantized states, but the correct quantum theory contains even more radical ideas.

29.4 | WAVE MECHANICS AND THE HYDROGEN ATOM

Bohr based his work on Newton's mechanics, to which he added a few new assumptions about the nature of electron motion, but his approach still relied on classical ideas based on electrons moving in mechanical orbits. The quantum theory developed by Schrödinger, Heisenberg, and others in the 1920s provided a much more radical break with Newton. Modern quantum mechanics, also called *wave mechanics*, is based not on mechanical variables such as position and velocity, but on wave functions and probability densities as explained in Section 28.5. Although we will not go into the details here, one solves a problem in quantum mechanics using *Schrödinger's equation*. The solution of this equation gives the wave function, including its dependence on position and time. In this section, we describe the wave function and some features of the quantized states of the hydrogen atom. In Section 29.5, we'll apply these ideas to multielectron atoms to explain the structure of the periodic table of the elements.

   In Bohr's theory, the electron levels are described by a single integer with allowed values $n = 1, 2, 3, \ldots$ corresponding to different quantized electron states. The integer n is called a *quantum number*. In Schrödinger's quantum theory, a full description of the quantum states of electrons in an atom requires *four* quantum numbers. These quantum numbers are collected in Table 29.2, which gives the standard notation and name for each. Each allowed electron energy level is specified by a set of values for all four quantum numbers and corresponds to one of the quantum states.

The Four Quantum Numbers for Electron States in Atoms

Quantum numbers for electron states in an atom

The four quantum numbers for electron states in atoms are n, ℓ, m, and s and are defined as follows.

n is the *principal quantum number*. It can have the values $n = 1, 2, 3, \ldots$. This quantum number is roughly similar to Bohr's quantum number. As n increases,

Table 29.2 | Quantum Numbers for Allowed Electron States in an Atom

QUANTUM NUMBER	NAME	POSSIBLE VALUES
n	Principal quantum number	$n = 1, 2, 3, \ldots$
ℓ	Orbital quantum number	$\ell = 0, 1, 2, \ldots, n - 1$
m	Orbital magnetic quantum number	$m = -\ell, -\ell + 1, \ldots, 0, \ldots, \ell - 1, \ell$
s	Spin quantum number	$s = -\frac{1}{2}$ or $+\frac{1}{2}$

the average distance from the electron to the nucleus increases. (See also Example 29.3.) The states with a particular value of n are referred to as a "shell." For example, all the states with $n = 2$ make up the "$n = 2$ shell."

ℓ is the *orbital quantum number*, with allowed values $\ell = 0, 1, \ldots, n - 1$. The angular momentum of the electron is proportional to ℓ. An electron state with $\ell = 0$ is called an "s state," while states with $\ell = 1$ are called "p states." The shorthand letters for other states are (in order) "d" ($\ell = 2$) and "f" ($\ell = 3$) as listed in Table 29.3. States with $\ell = 0$ have zero angular momentum.

m is the *orbital magnetic quantum number*, with allowed values $m = -\ell$, $-\ell + 1, \ldots, -1, 0, +1, \ldots, +\ell$. Intuitively, you can think of m as giving the direction of the angular momentum of the electron in a particular state.

s is the *spin quantum number*. While the other quantum numbers all have integer values, the spin quantum number of an electron is $s = +\frac{1}{2}$ or $-\frac{1}{2}$, often referred to as "spin up" and "spin down." This quantum number gives the direction of the electron's spin angular momentum as discussed in Section 28.4.

Warning! Do not confuse the spin quantum number s with the notion of an "s state" ($\ell = 0$); they refer to completely different quantum numbers. This notation came about for historical reasons, and we are now (unfortunately) stuck with it.

Electron Shells and Probability Distributions

A particular quantized electron state is specified by all four of the quantum numbers n, ℓ, m, and s. Solution of the Schrödinger equation also gives the wave function of each quantum state, and from the wave function we can calculate the probability for finding the electron at different locations around the nucleus. The electron probability distributions for a few states of the hydrogen atom are shown in Figure 29.17; these probability plots are often called "electron clouds."

The ground state of hydrogen is specified by $n = 1$. According to the rules in Table 29.2, the only allowed values of the orbital and orbital magnetic quantum numbers for $n = 1$ are $\ell = 0$ (an "s state") and $m = 0$, but the electron can be in either the spin-up ($s = +\frac{1}{2}$) or spin-down ($s = -\frac{1}{2}$) state. The probability of finding the electron at a particular location does not depend on the value of s, so the spin-up and spin-down probabilities are the same. A plot of the electron probability for the ground state of hydrogen is shown at the far left of Figure 29.17A. The electron probability distribution forms a spherical "cloud" around the nucleus. For this state, the probability is largest at the nucleus (where the cloud is "darkest" in Fig. 29.17), so the electron has the highest probability of being found at or near the nucleus. In this ground state, the electron occupies the level with the lowest possible energy. There are two such states, with $n = 1$, $\ell = 0$, $m = 0$, and $s = +\frac{1}{2}$ or $-\frac{1}{2}$. Together they make up the $n = 1$ "shell" and are called the $1s$ states of the atom.

Figure 29.17 also shows the electron probability for the states with principal quantum number $n = 2$, called the $n = 2$ shell. The probability for the state with $n = 2$, $\ell = 0$, and $m = 0$ has a spherical component (a spherical "shell" where the

Table 29.3 | Decoding the Configuration Shorthand for States with Various Orbital Quantum Numbers

ORBITAL QUANTUM NUMBER	CONFIGURATION LETTER
$\ell = 0$	s
$\ell = 1$	p
$\ell = 2$	d
$\ell = 3$	f
$\ell = 4$	g
$\ell = 5$	h

Figure 29.17 **A** Electron probability distributions for the 1s, 2s, and 2p states of hydrogen, calculated using the Schrödinger equation. The 2p states can be pictured in two different ways. Here in part A, the 2p electron probability clouds for $m = \pm 1$ are shown with "doughnut" shapes. In **B**, the same 2p states are redrawn in an equivalent form popular in chemistry, called p_x, p_y, and p_z states.

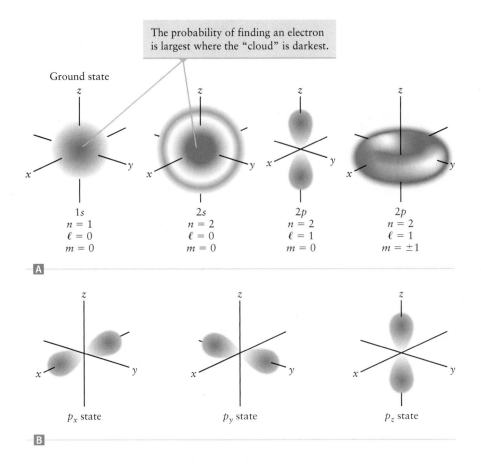

The probability of finding an electron is largest where the "cloud" is darkest.

Ground state

1s	2s	2p	2p
$n = 1$	$n = 2$	$n = 2$	$n = 2$
$\ell = 0$	$\ell = 0$	$\ell = 1$	$\ell = 1$
$m = 0$	$m = 0$	$m = 0$	$m = \pm 1$

A

p_x state　　　p_y state　　　p_z state

B

probability is large) near the nucleus and a second "layer" (another spherical shell where the probability is high) farther away. The spin quantum number for electrons in this state can again have the values $s = +\frac{1}{2}$ and $-\frac{1}{2}$, forming two 2s states of the atom. The $n = 2$ shell also contains states with $\ell = 1$, which are called the 2p states. The value of the orbital quantum number ℓ specifies the "subshell," so the 2s and 2p states are said to be in the same shell (because they have the same value of n) but in different subshells. According to the quantum number rules in Table 29.2, there are 2p states with $m = +1$, 0, and -1. There are two ways to display the probability clouds for the 2p states with $m = \pm 1$. Many physicists prefer to show them as a doughnut-like electron cloud (on the far right in Fig. 29.17A). However, the two states with $m = \pm 1$ can also be combined to make what chemists call p_x and p_y states (Fig. 29.17B). While it is not obvious, both plots show the *same* quantum states. The chemists' way of displaying these electron probabilities is useful for understanding the directionality of chemical bonding.

The electron probability distributions for all these states are independent of the value of the spin quantum number. For the hydrogen atom, the electron energy depends only on the value of n and is independent of the values of ℓ, m, and s (except for some tiny effects we will not discuss in this book). That is not the case for atoms containing more than one electron, as we'll see in the next section.

29.5 | MULTIELECTRON ATOMS

The quantum states we have discussed for hydrogen can be used to describe the states of atoms containing more than one electron; they are called **multielectron atoms**. The electron energy levels of these atoms follow the same pattern as found for hydrogen, with the same quantum numbers listed in Tables 29.2 and 29.3. The electron probability distributions are also similar. There are two main quantitative differences between the electron states in a multielectron atom and in hydrogen. First, the values of the electron energies are different for different atoms. For example, the energy of the 1s state in helium is different from the energy of the 1s state in hydrogen. Second, the spatial extent of the electron probability clouds varies from element to element. For example, the 1s electron probability "cloud" for helium is closer on average to the nucleus than it is for hydrogen.

While the electron levels of all atoms are thus similar to those of hydrogen, one crucial feature is important for atoms with more than one electron: *each quantum state can be occupied by only one electron.* That is, each electron in an atom must occupy its own quantum state, different from the states of all other electrons. This is called the **Pauli exclusion principle**. Since each quantum state is characterized by a unique set of quantum numbers, each electron is described by a unique set of quantum numbers.

Pauli exclusion principle

Figure 29.18 shows how electrons are distributed among the possible energy levels in several different cases. The diagram in Figure 29.18A shows the levels of an "empty" atom, that is, an atom before we have added any electrons. For an "empty" atom and for hydrogen, all quantum states for a given value of Z and with a particular value of the principal quantum number n have almost exactly the same energy, so we only show a single horizontal line for each value of n.

In Figure 29.18B, we apply the energy-level diagram to hydrogen, in which there is only one electron to consider. We show this electron in the ground state (occupying the lowest energy level) as an arrow, with the direction of the arrow denoting the value of the electron's spin. (Here we have arbitrarily shown the electron as spin up, but it could just as well be spin down.) In words, the diagram in Figure 29.18B indicates that for a hydrogen atom in the ground state, the electron occupies the $n = 1$, $\ell = 0$, $m = 0$, $s = +\frac{1}{2}$ energy level. This is also called a "1s^1" *electron configuration*. This notation is a useful shorthand and is explained in Equation 29.20.

$$\boxed{n = 1} \qquad \boxed{\text{one electron}}$$

$$\text{electron configuration for H: } 1s^1 \qquad\qquad (29.20)$$

$$\boxed{\ell = 0}$$

The first number in the configuration shorthand indicates the value of the principal quantum number n, while the letter indicates the value of ℓ using the code in Table 29.3. In Equation 29.20, the letter "s" indicates that $\ell = 0$. The superscript indicates that one electron is in the energy level with these values of n and ℓ. The value of the electron spin is not explicitly given in this notation.

Parts C through E of Figure 29.18 give energy-level diagrams for several other atoms. Helium (He) has two electrons, occupying levels with the quantum numbers $n = 1$, $\ell = 0$, $m = 0$, and $s = \pm\frac{1}{2}$, which satisfies the Pauli exclusion principle because

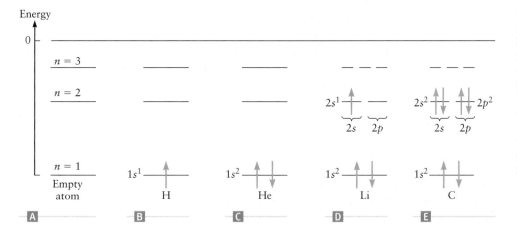

Figure 29.18 Filling the energy levels of an atom. **A** Empty levels. **B** Electron configuration of a hydrogen atom in its ground state. Configurations for **C** helium (He), **D** lithium (Li), and **E** carbon (C). These diagrams show the *order* of energy levels for each type of atom (each value of Z). The energies of these levels (e.g., the ground state and the excited states) are different for different atoms.

the electrons have different values of their spin quantum number. The electrons are indicated by the two arrows in Figure 29.18C, one pointing up (spin up, denoting $s = +\frac{1}{2}$) and the other pointing down (spin down, denoting $s = -\frac{1}{2}$). In the configuration shorthand, we have

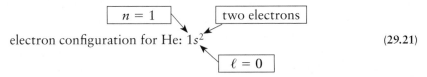

$$\text{electron configuration for He: } 1s^2 \qquad\qquad (29.21)$$

This configuration differs from that of hydrogen (Eq. 29.20) only in the superscript; here the superscript 2 indicates that in helium there are two electrons in the energy levels with $n = 1$ and $\ell = 0$.

The energy-level diagram for the ground state of lithium (Li) is shown in Figure 29.18D; a lithium atom has three electrons, indicated by the three arrows. As with helium, two of these electrons occupy the levels with $n = 1$, $\ell = 0$, $m = 0$, and $s = \pm\frac{1}{2}$. According to the rules in Table 29.2, there are no other possible energy levels with $n = 1$. Since each electron must have its own unique set of quantum numbers (according to the Pauli exclusion principle), the third electron in lithium must occupy a higher energy level. This third electron is indicated by the arrow at the $n = 2$ level in Figure 29.18D. The lowest energy level in a shell has $\ell = 0$, so Figure 29.18D shows this third electron in the $2s$ state. The full set of quantum numbers for this electron is thus $n = 2$, $\ell = 0$, $m = 0$, and $s = +\frac{1}{2}$. The corresponding configuration for lithium is then

$$\text{electron configuration for Li: } 1s^2 2s^1$$

Notice that the configuration shorthand does not indicate the value of m for any of the electrons.

Carbon (C) contains six electrons. The third, fourth, fifth, and sixth electrons in a carbon atom occupy states with $n = 2$ as indicated in Figure 29.18E. The corresponding configuration is

$$\text{electron configuration for C: } 1s^2 2s^2 2p^2$$

In words, there are two electrons (spin up and spin down) in levels with $n = 1$ and $\ell = 0$; two electrons (spin up and spin down) in levels with $n = 2$, $\ell = 0$ (the $2s$ levels); and two electrons (spin up and spin down) in levels with $n = 2$ and $\ell = 1$ (the $2p$ levels).

The energies of levels in the diagrams in Figure 29.18 depend mainly on the value of n, with the energy increasing as n becomes larger. This is approximately true at small values of n, but for atoms containing many electrons, the order of energy levels is more complicated as n increases, and for shells higher than $n = 2$, the energies of subshells from different shells begin to overlap. In general, the energy levels fill with electrons in the following order:

$$1s \quad 2s \quad 2p \quad 3s \quad 3p \quad 4s \quad 3d \quad 4p \quad 5s \quad 4d \quad 5p \quad 6s \quad 4f \qquad (29.22)$$

The $1s$ levels are filled first, followed by the $2s$ levels, then the $2p$ levels, and so on, with the overall order of the levels shown in Figure 29.19. This filling pattern is followed in Figure 29.18 and can also be used to get the results in Table 29.4 (page 1008), which lists the electron configurations for the ground states of all elements up to sodium. However, when one reaches the $3d$ levels the order of levels becomes even more complicated and can vary as electrons are added in going from one value of Z to the next.

CONCEPT CHECK 29.3 | What Element Is It?

The ground state of an atom has the configuration $1s^2 2s^2 2p^6 3s^2 3p^5$. What element is it? *Hint:* You may wish to consult the periodic table facing the last page of the index in the back of this book.

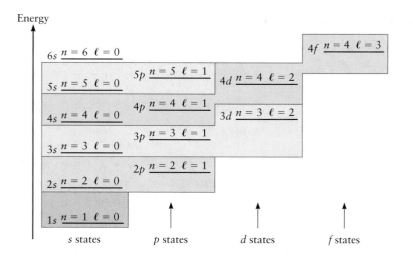

Figure 29.19 Order of energy levels in an atom.

EXAMPLE 29.5 | Electron Configuration and Energy-Level Diagram for Silicon

A silicon atom contains 14 electrons. Determine its electron configuration and construct the corresponding diagram of occupied levels.

RECOGNIZE THE PRINCIPLE

According to the Pauli exclusion principle, no two electrons can occupy the same quantum state, that is, have the same set of quantum numbers. To find the configuration of silicon, we assign the 14 electrons to the 14 quantum states that have the lowest energy.

SKETCH THE PROBLEM

We start with Figure 29.19 and add electrons to the lowest energy levels. Adding 14 electrons gives the diagram of filled energy levels in Figure 29.20.

IDENTIFY THE RELATIONSHIPS

We fill the lowest 14 energy levels in Figure 29.19 according to the rules in Table 29.2. The $1s$, $2s$, and $3s$ levels can each hold two electrons (one electron with spin up and another with spin down), and the $2p$ and $3p$ states can each hold six electrons.

SOLVE

The 14 electrons in silicon will fill completely the $1s$ (two electrons), $2s$ (two electrons), $2p$ (six electrons), and $3s$ (two electrons) energy levels, leaving two electrons for the $3p$ state. The resulting configuration is $\boxed{1s^2 2s^2 2p^6 3s^2 3p^2}$.

What does it mean?
The energy-level diagram in Figure 29.19 is the key to understanding the electron configurations of all the elements. The filled electron states of a particular atom are found by simply adding electrons to the lowest available energy levels.

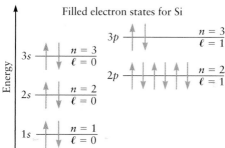

Figure 29.20 Example 29.5.

29.6 | CHEMICAL PROPERTIES OF THE ELEMENTS AND THE PERIODIC TABLE

It is probably fair to say that the periodic table of the elements is the key to all chemistry. One important triumph of quantum theory is that it explains why the periodic table has its structure. This table was first assembled by Dmitry Mendeleyev in the late 1860s. At that time, there were about 60 known elements, and Mendeleyev (along

Table 29.4 Electron Configurations for Several Elements

Element	Number of Electrons	Electron Configuration	Quantum Numbers of Occupied Electron Levels
H (hydrogen)	1	$1s^1$	$n = 1, \ell = 0, m = 0, s = +\frac{1}{2}$
He (helium)	2	$1s^2$	$n = 1, \ell = 0, m = 0, s = +\frac{1}{2}$
			$n = 1, \ell = 0, m = 0, s = -\frac{1}{2}$
Li (lithium)	3	$1s^2 2s^1$	Same as He plus
			$n = 2, \ell = 0, m = 0, s = +\frac{1}{2}$
Be (beryllium)	4	$1s^2 2s^2$	Same as He plus
			$n = 2, \ell = 0, m = 0, s = +\frac{1}{2}$
			$n = 2, \ell = 0, m = 0, s = -\frac{1}{2}$
B (boron)	5	$1s^2 2s^2 2p^1$	Same as Be plus
			$n = 2, \ell = 1, m = 0, s = +\frac{1}{2}$
C (carbon)	6	$1s^2 2s^2 2p^2$	Same as Be plus
			$n = 2, \ell = 1, m = 0, s = +\frac{1}{2}$
			$n = 2, \ell = 1, m = 0, s = -\frac{1}{2}$
N (nitrogen)	7	$1s^2 2s^2 2p^3$	Same as C plus
			$n = 2, \ell = 1, m = 1, s = +\frac{1}{2}$
O (oxygen)	8	$1s^2 2s^2 2p^4$	Same as C plus
			$n = 2, \ell = 1, m = 1, s = +\frac{1}{2}$
			$n = 2, \ell = 1, m = 1, s = -\frac{1}{2}$
F (fluorine)	9	$1s^2 2s^2 2p^5$	Same as O plus
			$n = 2, \ell = 1, m = -1, s = +\frac{1}{2}$
Ne (neon)	10	$1s^2 2s^2 2p^6$	Same as O plus
			$n = 2, \ell = 1, m = -1, s = +\frac{1}{2}$
			$n = 2, \ell = 1, m = -1, s = -\frac{1}{2}$
Na (sodium)	11	$1s^2 2s^2 2p^6 3s^1$	Same as Ne plus
			$n = 3, \ell = 0, m = 0, s = +\frac{1}{2}$

with other chemists) had noticed that many elements could be grouped according to their chemical properties. For example, the elements lithium, sodium, and potassium form very similar chemical compounds and undergo similar reactions. Mendeleyev organized his table by grouping such related elements in the same column. He also placed the lightest (i.e., low atomic weight) elements at the top of each column and the heaviest ones at the bottom. His table was essentially identical to the modern version of the periodic table given in Figure 29.21 and facing the last page of the index in this book. The main difference was that Mendeleyev's table had a number of "holes" (i.e., openings) because many elements had not been discovered yet.

While Mendeleyev discovered an intriguing regularity in the elements and their properties, he could not explain why the periodic table has the form it does. That is, why does it contain just two elements in the first row, eight in the second and third rows, and even more elements in the lower rows? The answer is contained in the organization of electron energy levels given in Table 29.4 and Figure 29.19. To appreciate this answer, we must first understand how the electron energy levels and the electron configuration of an atom are responsible for its chemical properties.

When an atom participates in a chemical reaction, some of its electrons combine with electrons from other atoms to form chemical bonds. The bonding electrons are those that are most easily removed from an atom; hence, they are the electrons occupying the highest energy levels. For example, the ground state of lithium has

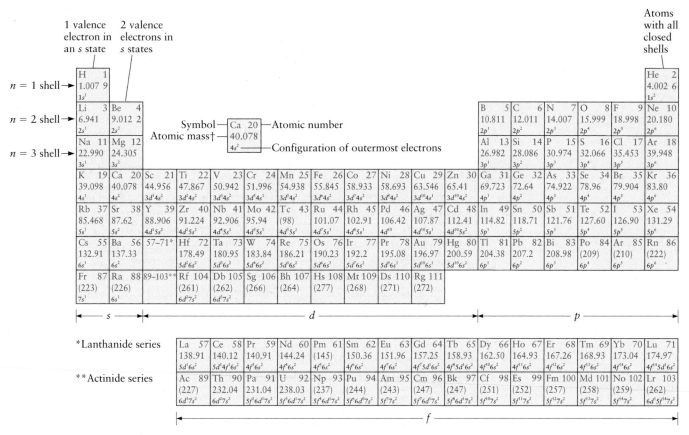

Figure 29.21 Periodic table of the elements.

the configuration $1s^2 2s^1$ (Fig. 29.18D and Table 29.4). The $2s$ electron occupies the state with the highest energy, so that electron is more weakly bound than the two electrons in the $1s$ state. This $2s$ electron forms bonds with other atoms and is called a ***valence*** electron. In contrast, the two electrons in the $1s$ state form a ***closed shell***, a shell with all possible states filled. These $1s$ electrons have a much lower energy than the $2s$ electron. The $1s$ electrons are much more difficult to remove from the atom and do not participate in bonding (in chemical reactions).

The same ideas apply to other atoms. For example, the bonding of a sodium atom involves the $3s$ electron (Table 29.4). The remaining electrons form a closed $1s$ shell, a closed $2s$ subshell, and a closed $2p$ subshell. The configuration of the closed shells is thus $1s^2 2s^2 2p^6$. Likewise, potassium also has a single bonding electron, occupying the $4s$ state outside a $1s^2 2s^2 2p^6 3s^2 3p^6$ configuration. The elements lithium, sodium, and potassium thus have the same number of valence electrons (one), all in s states. As a result, these elements have similar chemical bonding properties, which is why Mendeleyev placed them in the same column in the periodic table (Fig. 29.21).

Closed Shells and the Periodic Table

The last column in the periodic table includes helium, neon, and argon (He, Ne, and Ar). Atoms of these elements all contain completely filled shells. Since they have no valence electrons, it is very difficult for them to form chemical bonds. Such elements are largely inert, almost never participating in chemical reactions. Atoms of these elements also interact very weakly with each other, resulting in low condensation and freezing temperatures. For this reason, they are called *inert gases* or *noble gases*. In Mendeleyev's time, elements were discovered through their chemical reactions and compounds, so very little was known about these elements when Mendeleyev did his work in the 1860s. In fact, most inert gases were discovered after Mendeleyev's time, so his version of the periodic table did not even contain that column.

Structure of the Periodic Table

In his periodic table, Mendeleyev grouped elements into *columns* according to their common bonding properties and chemical reactions. These properties rely on the valence electrons and can thus be traced to the electron configurations in Table 29.4. What, though, determines the number of elements in each *row* of the table, and why do different rows have different numbers of elements? Each row in the periodic table corresponds to a particular value of the principal quantum number n. The row containing hydrogen and helium corresponds to $n = 1$, the row beginning with lithium has $n = 2$, and so forth. The number of elements in a given row is equal to the number of electrons needed to fill completely a particular shell. Because the $n = 1$ shell can hold only two electrons, this row contains just two elements. The $n = 2$ shell can hold eight electrons (two in the $2s$ states and six more in the $2p$ states), so this row contains eight elements. The number of elements in all the other rows can be found using the rules for allowed quantum numbers in Table 29.2.

CONCEPT CHECK 29.4 | Valence Electrons for Barium

How many valence electrons does barium have, and what energy levels do they occupy?
   (a) One valence electron in the $6s$ state
   (b) Two valence electrons in $6s$ states
   (c) No valence electrons

CONCEPT CHECK 29.5 | Valence Electrons of an Atom

Which atoms from this group—(a) Sr, (b) F, (c) Mg, and (d) K—have the same number of valence electrons?

29.7 | APPLICATIONS

The existence of discrete spectral lines played a major role in the development of quantum theory and also makes possible a number of important applications.

Atomic Clocks and the Definition of the Second

Atomic clocks are used as global (and U.S.) time standards (Fig. 29.22). These clocks are based on the accurate measurement of certain spectral line frequencies.

Figure 29.22 A Early atomic clock. This clock dates from the 1950s. B Modern atomic clocks are much more compact and fit on the corner of a desk.

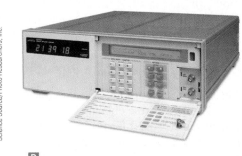

Since all atoms of a given element are identical, they all have precisely the same set of spectral lines at precisely the same frequencies.

Atomic clocks have been constructed using several different types of atoms, with cesium (Cs) being a popular choice. Two of the cesium levels differ by an energy of approximately 3.8×10^{-5} eV ($\approx 6.1 \times 10^{-24}$ J). These two levels are much closer in energy than the typical spacing of a few electron-volts seen for the levels in hydrogen. The spectral emission line produced when a cesium atom undergoes a transition between these two levels is very strong; that is, cesium atoms emit a relatively large intensity at this frequency. The frequency of this spectral line can also be measured with *extremely* high accuracy.

To make an atomic clock, a collection of cesium atoms in the gas phase is placed in a container. The atoms are excited by heating or other means so that many of them are in the higher of the two energy levels mentioned above. These atoms then undergo transitions to the lower state, all emitting photons with the same frequency. Separate calibration experiments have shown that this frequency is

$$f_{\text{Cs clock}} = 9{,}192{,}631{,}770 \text{ Hz} \tag{29.23}$$

This frequency is used as the "tick rate" of the atomic clock. A cesium clock thus ticks 9,192,631,770 times per second, forming a very accurate clock because the rate of ticking is extremely fast and is known very accurately. What's more, all cesium clocks have the same ticking rate, making them ideal for performing and comparing scientific measurements made in different laboratories at different times. In fact, in 1960 the cesium clock with Equation 29.23 was adopted as the SI definition of the second. One second is now *defined* to be the time it takes a cesium clock to complete precisely 9,192,631,770 ticks.

Atomic clocks are widely used in situations where very accurate time measurements are required. Applications such as GPS (Global Positioning System), which involves precisely timed radio signals from a set of satellites orbiting the Earth, are made possible by such clocks.

Fluorescent Lights

Not all applications of quantum and atomic theory are as exotic as atomic clocks. You probably see and use one of the applications every day: "neon" and fluorescent lighting.

There are two basic types of lightbulbs. The **incandescent bulb**, developed by Thomas Edison, contains a very thin wire filament that carries a large electric current. The electrical energy dissipated in the filament heats it to a high temperature, and the filament then acts as a blackbody and emits radiation. The other type of lightbulb uses a gas of atoms in a glass container. In common terms, bulbs filled with a gas are often called either "neon" bulbs or "fluorescent" bulbs, but the spectra emitted by neon and fluorescent bulbs are somewhat different. A neon bulb contains a gas of Ne atoms. An electric current passes through the gas, producing ions and high-energy electrons as shown schematically in Figure 29.23A. The electrons, ions, and neutral atoms in the gas undergo many collisions, causing many of the Ne atoms to be in excited states. These atoms then decay back to their ground state, emitting light in the process as discussed in Section 29.2. Neon lamps thus emit light only at certain discrete wavelengths, that is, only with certain colors. By using different types of atoms, including neon and sodium, these lightbulbs can be used to make brightly colored signs (Fig. 29.23B).

Fluorescent lightbulbs (Fig. 29.23C) also contain a gas of atoms, often mercury (Hg), inside a glass bulb, but the inside of the bulb is coated with a fluorescent material. The Hg atoms are excited and emit photons in the same way as the Ne atoms in a neon bulb emit light. Mercury, however, emits strongly in the ultraviolet, and those photons are not detected by the eye, so Hg atoms alone would not make a useful lightbulb. This problem is overcome by the fluorescent material; the photons emitted by Hg are absorbed by the fluorescent coating, exciting those atoms and

Insight 29.3
DEFINING THE "SECOND"
The second is now defined in terms of the frequency of a particular atomic transition in Cs. Before 1960, the second was defined in terms of the length of a solar day, and 1 second was 1/86,400 of the length of what astronomers call a "mean solar day." A mean solar day is the time it takes (on average) for the Earth to complete one rotation about its axis. A major difficulty with this definition is that the Earth's rotation rate is gradually slowing. For this reason, the modern definition is much more useful in scientific work.

Figure 29.23 🅐 Neon bulbs contain a gas of Ne atoms in a glass tube. The atoms are excited by an electric current, and light is emitted when the Ne atoms undergo transitions from an excited state back to the ground state. 🅑 Neon bulbs emit red light and are used in many advertisements. The other colors seen in a "neon" sign come from atoms of other elements. 🅒 In a fluorescent lamp, the ultraviolet photons from an atom (here Hg) are absorbed by a coating on the inside of the tube and then reemitted as photons with a wide range of wavelengths in the visible range.

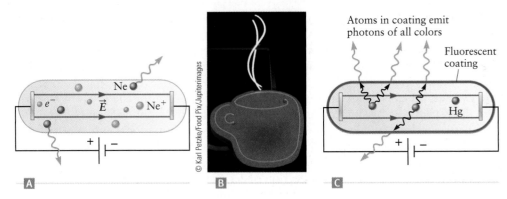

molecules to higher energy levels. When these atoms and molecules undergo transitions to lower energy states, they emit new photons. The fluorescent material is designed to emit photons throughout the visible spectrum, so a fluorescent lamp produces "white" light, that is, light of virtually all colors. Fluorescent bulbs were invented in the late nineteenth century, but their use has expanded greatly since about 2000 due to advances in the design of compact fluorescent bulbs (involving circuitry built into the base of the bulb).

EXAMPLE 29.6 | Photons Emitted by a Sodium Lamp

Sodium (Na) lamps emit light when the atoms undergo a transition from an excited $3p$ level to a $3s$ level. There are actually two levels in the $3p$ group with different values of the quantum numbers m and s and with slightly different energies, explaining why sodium emits two spectral lines that are very close in frequency (Fig. 29.8). The spectral line with a wavelength of $\lambda = 589.0$ nm is the stronger (more intense) of the two lines. What is the energy of these photons? Express your answer in units of electron-volts and compare it with the ionization energy of a hydrogen atom.

RECOGNIZE THE PRINCIPLE

This problem involves the relation between photon wavelength and energy. (See also Example 29.2.) We use the given photon wavelength λ to find the frequency f. The energy of the photon can then be found using $E_{\text{photon}} = hf$ (Eq. 29.7).

SKETCH THE PROBLEM

Figure 29.24 shows the transitions that produce the two yellow Na spectral lines.

IDENTIFY THE RELATIONSHIPS

The frequency of a photon produced by a Na lamp is

$$f = \frac{c}{\lambda}$$

where the wavelength is given above. Combining this equation with Equation 29.7, we find that the energy, frequency, and wavelength of a photon are related through

$$E = hf = \frac{hc}{\lambda}$$

SOLVE

Inserting the given value of the wavelength yields

$$E = \frac{hc}{\lambda} = \frac{(6.63 \times 10^{-34}\ \text{J} \cdot \text{s})(3.00 \times 10^{8}\ \text{m/s})}{5.890 \times 10^{-7}\ \text{m}} = 3.4 \times 10^{-19}\ \text{J}$$

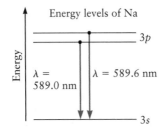

Energy levels of Na

$\lambda = 589.0$ nm $\lambda = 589.6$ nm

$3p$

$3s$

Energy

Figure 29.24 Example 29.6. These three energy levels are involved in the yellow spectral lines emitted by a sodium atom.

which expressed in units of electron-volts is

$$E = (3.4 \times 10^{-19} \text{ J})\left(\frac{1 \text{ eV}}{1.60 \times 10^{-19} \text{ J}}\right) = \boxed{2.1 \text{ eV}} \qquad (1)$$

The ionization of energy of H is 13.6 eV, so the energy of this spectral line in Na is much smaller.

What does it mean?

The energy of this photon emitted by Na lies in the yellow part of the visible spectrum, making it useful for lightbulbs. So-called high-pressure sodium bulbs are used in applications such as street lights, where very high intensities are required. The high pressures used in these bulbs cause them to emit light over a range of wavelengths (not just yellow).

Lasers

Lasers are another application of quantum and atomic theory and are commonplace in early-twenty-first-century life. Lasers depend on the coherent emission of light by many atoms, all at the same frequency.

In an ordinary fluorescent lightbulb, the transitions from the excited atomic states to lower energy levels occur in a random fashion. That is, once an atom is put into an excited state, it is impossible to predict when it will emit a photon, and the emitted photons are radiated randomly in all directions. In this *spontaneous emission* process, each atom emits photons independently of the other atoms. In a laser, an atom undergoes a transition and emits a photon in the presence of many other photons that have energies equal to the atom's transition energy (Fig. 29.25A). A process known as *stimulated emission* causes the light emitted by this atom to propagate in the same direction and with the same phase as surrounding light waves. Such a light source is called a laser, an acronym for *l*ight *a*mplification by *s*timulated *e*mission of *r*adiation. The light from a laser is thus a coherent source (Chapter 25) and is very useful for experiments involving the interference of light.

The design of many lasers is similar to that of a neon lightbulb, but there are mirrors at the ends of the bulb (called the laser "tube"). Light emitted by the gas atoms is reflected by the mirrors and travels back and forth along the tube (Fig. 29.25B). One of these mirrors is designed to let a small amount (typically a few percent) of the light pass through; that is the light you see from the laser. Just as in the neon bulb, atoms in the laser tube are in excited energy levels (produced typically by passing current through the gas). The excited atoms emit photons that stimulate the emission of more photons with the same energy, and all these photons are reflected repeatedly inside the tube. Photons resulting from stimulated emission are precisely in phase with the stimulating radiation. In this way, a laser can produce a very intense beam of light that is highly directional.

Lasers can be made with a variety of different atoms. A popular design uses a mixture of He and Ne gas and is called a helium–neon (He–Ne) laser. It was one of the first laser types invented (in 1960), making use of the energy levels shown in Figure 29.26. Helium atoms are excited by an electric current to an energy level

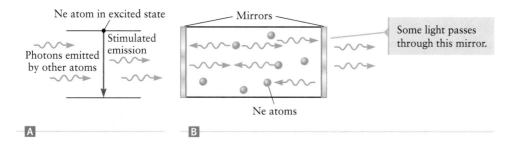

A Ne atom in excited state — Stimulated emission — Photons emitted by other atoms — **B** Mirrors — Some light passes through this mirror. — Ne atoms

Figure 29.25 **A** The emission of photons by an atom can be stimulated by the presence of other photons with the same frequency. **B** A laser uses mirrors to reflect light back and forth within the laser tube. A small amount of light is allowed to escape through one of the mirrors; that is the light you see coming from a laser.

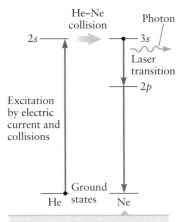

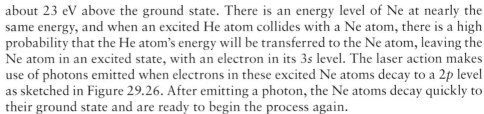

He–Ne collision

Photon

Excitation by electric current and collisions

Laser transition

Ground states

He   Ne

After emitting a photon, the Ne atom undergoes a transition back to its ground state.

Figure 29.26 Energy-level diagram for a helium–neon laser.

A

B

Figure 29.27 Some applications of lasers. A Some flatscreen televisions use laser light sources. B A barcode reader.

about 23 eV above the ground state. There is an energy level of Ne at nearly the same energy, and when an excited He atom collides with a Ne atom, there is a high probability that the He atom's energy will be transferred to the Ne atom, leaving the Ne atom in an excited state, with an electron in its $3s$ level. The laser action makes use of photons emitted when electrons in these excited Ne atoms decay to a $2p$ level as sketched in Figure 29.26. After emitting a photon, the Ne atoms decay quickly to their ground state and are ready to begin the process again.

The photons emitted by a helium–neon laser have a wavelength of about 633 nm, which is in the red part of the visible spectrum. These lasers are used in many physics laboratories, often in lecture demonstrations. Another common type of laser is based on light produced by light-emitting diodes (called LEDs). These lasers typically have a wavelength near 650 nm (which is also in the red part of the spectrum) and are used in optical barcode scanners at many stores. Figure 29.27 shows some contemporary devices that use lasers.

CONCEPT CHECK 29.6 | Operation of a CO_2 Laser

The helium–neon laser described in Figure 29.26 emits visible light, but some lasers emit photons in the infrared. A popular type of infrared laser uses photons emitted by CO_2, which have a wavelength of 10.6 μm = 1.06×10^{-5} m. This wavelength is longer than that of the Na photon in Example 29.6 by a factor of 18. Which of the following statements is correct?

(a) The energy of a photon from a CO_2 laser is greater than that of a photon from a Na lamp by a factor of 18.

(b) The energy of a photon from a CO_2 laser is less than that of a photon from a Na lamp by a factor of 18.

The Force between Two Atoms ®

How much force does it take to pull two atoms apart? This question is important because physicists have recently developed ways to manipulate individual atoms, and one application of this work is to construct (or deconstruct) molecules one atom at a time. This question is also relevant for understanding the atomic force microscope (AFM) in which atoms from a sharp tip are scanned near atoms on the surface of another object. The AFM uses the force between atoms to produce an image of the surface (Chapter 11). An accurate quantum mechanical calculation of the force between two atoms would be very complex, but we can estimate it using what we know about the relation between potential energy and force, and using typical values of the ionization energy of an atom.

Consider the two hypothetical atoms in Figure 29.28A and assume they are bound together to form a molecule. The binding energy of a molecule is the energy required to break the chemical bond between the two atoms. This energy comes from the "sharing" of valence electrons, and we can estimate this energy from the spacing between energy levels in one of the atoms. Qualitatively, the sharing of valence electrons lowers the energy of one or more of the electrons by an amount that is approximately equal to the spacing between energy levels in an atom. For a hydrogen atom, this spacing is about 1 eV to 10 eV, and it will be greater for heavier atoms, so a typical bond energy is about 10 eV. That is the change in potential energy when comparing the molecule to two unbound atoms, so we have $\Delta PE \approx 10$ eV.

If this atom is now pulled apart by separating the atoms a distance Δx, there is an associated force F between the atoms. From the relation between potential energy and force (Sections 6.3 and 6.8), the magnitude of the force F is

$$F = \left| \frac{\Delta PE}{\Delta x} \right|$$

The radius of a hydrogen atom is about 0.05 nm (Example 29.3), so separating the atoms in Figure 29.28B a distance $\Delta x \approx 1$ nm should be large enough to break the chemical bond. Our estimate for the force between two atoms is thus

$$F = \left| \frac{\Delta PE}{\Delta x} \right| \approx \frac{10 \text{ eV}}{1 \text{ nm}}$$

An energy of 10 eV is equal to $(10 \text{ eV})(1.60 \times 10^{-19} \text{ J/eV}) = 1.6 \times 10^{-18}$ J. The force is thus

$$F \approx \frac{1.6 \times 10^{-18} \text{ J}}{1 \text{ nm}} = 1.6 \times 10^{-9} \text{ N} \qquad (29.24)$$

This value is only an estimate for the force, but atomic force microscope experiments show that the force between two atoms is indeed close to that found in Equation 29.24. The keys to our calculation were the connection between force and potential energy, and our knowledge of the potential energy associated with a chemical bond.

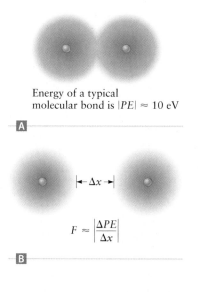

Energy of a typical molecular bond is $|PE| \approx 10$ eV

A

$$F \approx \left| \frac{\Delta PE}{\Delta x} \right|$$

B

Figure 29.28 Estimating the force between two atoms.

29.8 | QUANTUM MECHANICS AND NEWTON'S MECHANICS: SOME PHILOSOPHICAL ISSUES

Newton's mechanics fails when applied to objects such as electrons and atoms. In that regime, quantum theory is required. Does that mean we should discard Newton's laws and redo all the physics in this book using quantum theory? To answer this question, we first note that Newton's laws work extremely well in the classical regime, that is, when applied to macroscopic objects such as baseballs and planets. Quantum theory can also be applied to such macroscopic objects, giving results that are virtually identical to Newton's mechanics. How can quantum theory, which is based on a particle-wave description of all objects, be consistent with Newton's particle theory for objects such as baseballs?

The answer depends on the wavelength of the particle-waves. The wavelength λ and momentum p of a particle-wave are connected by de Broglie's relation,

$$\lambda = \frac{h}{p} \qquad (29.25)$$

An electron is a very light object (i.e., has a small mass), so its momentum is small. Equation 29.25 therefore tells us that λ for an electron is large. It is more accurate to say that the electron's wavelength is *relatively* large because we have seen that the electron wavelength in an atom is typically 0.1 nm (1×10^{-10} m). For an object such as a baseball, p is much larger than for an electron, so the wavelength is *much* smaller. For example, a baseball has a mass $m = 0.14$ kg and a speed of typically 50 m/s (about 100 mi/h). Inserting these values in Equation 29.25, the de Broglie wavelength of a baseball is

$$\lambda = \frac{h}{p} = \frac{h}{mv} = \frac{6.63 \times 10^{-34} \text{ J} \cdot \text{s}}{(0.14 \text{ kg})(50 \text{ m/s})} = 9.4 \times 10^{-35} \text{ m} \qquad (29.26)$$

which is an *extremely* short wavelength. When a wave such as light has a very short wavelength, it moves in a straight line described by a simple ray (Chapter 24). The same is true of a baseball. It will move as a simple ray; that is, it will move as a simple particle. The wave properties of the baseball are only apparent on length scales comparable to the wavelength in Equation 29.26. Hence, for all practical purposes a baseball behaves as a classical particle, and its wave properties will never be observable. That is why Newton's mechanics works so well for classical objects. Such objects always have extremely short wavelengths, making the quantum theory description in terms of particle-waves unnecessary.

Where Quantum Theory Meets Newton

If Newton's mechanics works well for macroscopic objects while quantum theory is needed to describe behavior at the atomic scale, does anything interesting happen where these two regimes meet? Physicists are now actively studying this problem. One interesting question concerns the quantum behavior of a living entity. Objects such as viruses are both very small and "alive." Do the quantized energy levels of a living thing lead to any new behavior? What if the living thing is conscious? Can we understand the wave function of biological objects such as a brain? These intriguing questions are but a few of the ones physicists and biologists are now studying. The answers to some of these questions will likely emerge in the coming years. Stay tuned.

SUMMARY | Chapter 29

KEY CONCEPTS AND PRINCIPLES

The nuclear atom
Atoms contain *electrons*, *protons*, and *neutrons*. The positive charge in an atom is located in the *nucleus*, which contains protons and neutrons. The *Rutherford planetary model* pictures the atom as a miniature solar system in which electrons orbit the nucleus.

Atomic spectra
Atoms emit and absorb light (and other electromagnetic radiation) at certain discrete frequencies. These *spectral lines* result from transitions between discrete energy levels of an atom. When an electron undergoes a transition from a quantum state with energy E_2 to another with energy E_1, a photon of energy

$$hf = E_2 - E_1$$

is emitted. Here f is the frequency of the photon.

Bohr model
The *Bohr model* of the atom was an early attempt at incorporating quantum ideas into an atomic theory. A key assumption of the Bohr model is that the orbital angular momentum of an electron is quantized with

$$L = n\frac{h}{2\pi} \qquad \text{(29.8) (page 996)}$$

where n is an integer ($n = 1, 2, 3, \ldots$) and h is Planck's constant. This assumption implies that electrons form standing waves as they orbit the nucleus, with wavelengths given by the de Broglie theory of particle-waves ($\lambda = h/p$, where p is the momentum; Eq. 28.8). The Bohr model leads to discrete atomic energy levels with energies

$$E_{\text{tot}} = -\left(\frac{2\pi^2 k^2 e^4 m}{h^2}\right)\frac{1}{n^2} \qquad \text{(29.15) (page 997)}$$

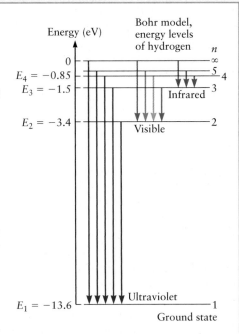

Allowed standing wave orbits in the Bohr model

The Bohr model predicts the correct values of the energy levels of hydrogen, but does not correctly describe atoms with two or more electrons.

(Continued)

The four quantum numbers of the atom

The correct quantum theory is based on the Schrödinger equation. Each energy level in an atom is described by four *quantum numbers*.

1. n is the *principal quantum number*. Possible values are $n = 1, 2, 3, \ldots$.
2. ℓ is the *orbital quantum number*. Allowed values are $\ell = 0, 1, \ldots, n - 1$.
3. m is the *orbital magnetic quantum number*. Allowed values are $m = -\ell$, $-\ell + 1, \ldots, -1, 0, +1, \ldots, \ell$.
4. s is the *spin quantum number*. Allowed values are $s = +\frac{1}{2}$ ("spin up") and $s = -\frac{1}{2}$ ("spin down").

The Pauli exclusion principle

Each electron in an atom occupies a separate energy level and thus has a unique set of quantum numbers.

APPLICATIONS

Atomic clocks and lasers use transitions between discrete energy levels to produce photons with precisely defined frequencies. They are used in many applications, including the Global Positioning System, bar code readers, and CD and DVD players. The force between two atoms is also large enough to be measured in an atomic force microscope (AFM), which uses this force to form an image of atoms on the surface of an object.

QUESTIONS

SSM = answer in Student Companion & Problem-Solving Guide   X = life science application

1. Explain why He, Ne, and Ar have similar chemical properties.

2. Explain why some periodic tables list H twice, once above Li and again above the column containing F.

3. Consider the He$^+$ ion. Like a hydrogen atom, He$^+$ has one electron, but it has two protons, so the charge of the nucleus is $+2e$. Do you expect the ionization energy of He$^+$ to be greater than or less than that of H? Explain.

4. Is there an upper limit to the wavelength of light that a hydrogen atom can emit? Explain why or why not.

5. SSM Of the following configurations, which ones are *not* allowed for the outermost shell of an atom by quantum theory, (a) $4s^4$, (b) $3d^7$, (c) $4f^9$, or (d) $2d^3$? Explain why not.

6. Explain why the plum-pudding model of the atom is inconsistent with experimental results.

7. Explain why Rutherford's planetary model of the atom is inconsistent with classical physics and what we know about atoms.

8. Explain why the total energy of an electron in the Bohr model is negative.

9. Show that $4\ell + 2$ electrons can occupy a subshell with orbital quantum number ℓ.

10. Explain why F, Cl, and Br have similar chemical properties.

11. Explain why He, Ne, Ar, and Kr rarely form stable molecules.

12. Explain why the Pauli exclusion principle is not needed to understand the quantum states of a hydrogen atom.

13. Derive a general expression for the de Broglie wavelength of an electron for the Bohr model in state n.

14. Explain how different "neon" bulbs can emit light with different colors. Use the Internet to look up what kinds of atoms are used in bulbs that emit red, green, blue, and yellow light.

15. Explain how the elements Cr through Cu in the periodic table can have fairly similar chemical properties, even though they are not arranged in the same column of the periodic table.

16. Explain why it is difficult to chemically separate the elements Ce (cerium) through Lu (lutitium) from each other.

17. SSM What would the spectrum of hydrogen look like if the electrons could move between orbits of any radius in Bohr's model?

18. What experiment would you do to determine if there is oxygen in a distant star?

19. What experiment would you do to determine if there is oxygen in the atmosphere of a planet going around a distant star?

20. When illuminated with ultraviolet light, many minerals glow in the visible part of the electromagnetic spectrum. Why?

29.1 STRUCTURE OF THE ATOM: WHAT'S INSIDE?

1. The nucleus of a hydrogen atom is a single proton. The radius of a proton is approximately 1 fm (1×10^{-15} m). (a) What is the approximate volume of a proton? (b) A hydrogen atom has an approximate radius of 5×10^{-11} m (called the Bohr radius). What is the approximate volume of a hydrogen atom? (c) What fraction of a hydrogen atom is occupied by the nucleus?

2. SSM ☆ (X) Suppose a hydrogen atom was somehow enlarged so that the nucleus is the size of a baseball (diameter approximately 8 cm). What is the approximate diameter of the volume occupied by the orbit of the electron in the planetary model? How does it compare with the diameter of the Earth?

3. ✪ Consider a Rutherford scattering experiment in which an alpha particle is fired at a sheet of atoms only one atomic layer thick (Fig. P29.3). The radius of an alpha particle is about 1 fm (1×10^{-15} m) and assume the radius of the nuclei in Figure P29.3 has the same value. Also assume the spacing between atoms in the sheet is 0.3 nm. (a) If alpha particles are fired at random at the sheet, what is the probability that one will strike a nucleus? (b) If one trillion (10^{12}) alpha particles are fired, about how many will hit a nucleus?

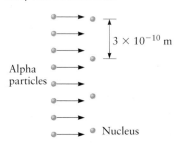

Alpha particles 3×10^{-10} m

Nucleus

Figure P29.3 Not to scale.

29.2 ATOMIC SPECTRA

4. An atom emits photons with an energy of 2.5 eV. (a) What is the spacing, in joules, between the energy levels involved in this atomic transition? (b) Where are these photons in the electromagnetic spectrum (i.e., infrared, visible, ultraviolet)?

5. A hypothetical atom contains energy levels as sketched in Figure P29.5. What are the energies of all possible emission lines involving these levels?

6. SSM ☆ A hypothetical atom absorbs photons at the following energies (all given in electron-volts): 2.0, 4.0, 6.0, 8.0, and 10.0. Construct an energy-level diagram that explains this spectrum.

7. What is the minimum frequency a photon can have so as to ionize a hydrogen atom? Assume the atom begins in its ground state.

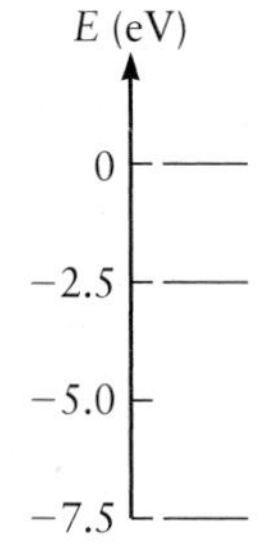

E (eV)

0 ——
−2.5 ——
−5.0 ——
−7.5 ——

Figure P29.5

29.3 BOHR'S MODEL OF THE ATOM

8. According to the Bohr model, what is the orbital radius of an electron in the $n = 10$ state of hydrogen?

9. Find the energy of an electron in the $n = 3$ state of hydrogen. If this electron undergoes a transition to the ground state, what is the energy, in electron-volts, of the photon that is emitted?

10. What is the smallest energy photon that can be absorbed by a hydrogen atom in its ground state?

11. How much energy is needed to ionize a hydrogen atom that is initially in its $n = 4$ state?

12. ☆ In the Bohr model of a hydrogen atom, what is the electric force exerted between the proton and the electron when (a) the electron is in the $n = 1$ state and (b) the electron is in the $n = 5$ state?

13. ☆ A hydrogen atom in the ground state absorbs a photon with an energy of approximately 12.8 eV. What is the value of n for the final state of the atom?

14. What is the energy needed to cause an electron to undergo a transition from the $n = 3$ state to the $n = 5$ state in hydrogen?

15. ☆ A hydrogen atom emits a photon of energy about 2.55 eV. What are the values of n for the initial and final states involved in this transition?

16. ☆ Find the de Broglie wavelength for an electron in the $n = 1$ state of hydrogen.

17. ☆ Calculate the de Broglie wavelength of an electron in the $n = 3$ state of hydrogen. Is this wavelength larger or smaller than for the $n = 1$ state?

18. ✪ In the Bohr model, the energy of the different electron states depends on the value of n (Eq. 29.16). (a) Compute the energies of the states with $n = 1, 2, 3$, and 4. (b) Plot your results from part (a) in an energy-level diagram like the one in Figure 29.12. (c) Compute the energies of all the possible emission lines involving these energy levels.

19. The radius of an electron orbit in the Bohr model depends on the value of n. (a) Find the value of n for which $r \approx 1 \ \mu$m. These atoms are called Rydberg atoms. This orbit is relatively large; objects of this size are visible with an optical microscope. (b) Find the energy (according to the Bohr model) for the electron in part (a).

20. SSM ☆ (a) What is the energy of the $n = 5$ state of hydrogen? (b) A hydrogen atom that is initially in some unknown state undergoes a transition and ends up in the $n = 5$ state. What is the highest energy photon that can possibly be *emitted*? (c) Does this photon fall in the visible part of the spectrum?

21. ☆ A hydrogen atom in its ground state absorbs a photon with an energy of 21.0 eV, and the electron is ejected. What is the kinetic energy of the ejected electron?

29.4 WAVE MECHANICS AND THE HYDROGEN ATOM

22. How many electron states in the hydrogen atom have the quantum numbers (a) $n = 2$ and $\ell = 1$? (b) $n = 3$ and $\ell = 2$?

23. What is the total number of electron states in the $n = 3$ shell?

24. What are the allowed values of ℓ for electrons in the $n = 5$ shell?

25. SSM ☆ An electron is in a subshell that contains 18 states. What is the value of ℓ?

29.5 MULTIELECTRON ATOMS

26. An atom has the electron configuration $1s^2 2s^2 2p^6 3s^2 3p^3$. What element is it?

27. Identify the following atoms from their electron configurations.
   (a) $1s^2 2s^2 2p^6 3s^2 3p^6 3d^{10} 4s^1$
   (b) $1s^2 2s^2 2p^6 3s^2 3p^6 3d^{10} 4s^2 4p^5$
   (c) $1s^2 2s^2 2p^6 3s^2 3p^6 3d^{10} 4s^2 4p^6 5s^2$

28. How many electrons are in (a) the $4p$ subshell of Br, (b) the $n = 4$ shell of Rb, (c) the $4f$ subshell of U, and (d) the $3d$ subshell of Cs?

29. SSM Give the electron configurations of the following atoms in their ground states: (a) Ge, (b) Ar, (c) Mg, (d) Si, and (e) Br.

30. ✪ Table 29.4 lists the electron configurations of various atoms in their ground (lowest energy) states. The configuration notation can also describe atoms in their excited states. Which of the following configurations describe an excited state? Name the atom in each case.
(a) $1s^2 2s^2 2p^6 3s^1 3p^2$
(b) $1s^2 2s^2 2p^6 3s^2 3p^6 3d^3 4s^1$
(c) $1s^2 2s^2 2p^6 3s^2 3p^6 3d^5 4s^1$

31. How many electrons can occupy (a) the $3d$ subshell, (b) the $5s$ subshell, (c) the $n = 4$ shell, (d) the $4f$ subshell, (e) the $5g$ subshell, and (f) any h subshell?

29.6 CHEMICAL PROPERTIES OF THE ELEMENTS AND THE PERIODIC TABLE

32. SSM List all the known elements whose chemical properties are similar to those of fluorine (F).

33. Which elements have chemical properties similar to calcium (Ca)?

29.7 APPLICATIONS

34. Most helium–neon lasers emit visible light with a wavelength of 633 nm. These lasers can also emit photons in the infrared with a wavelength of 1.15 μm. What is the frequency of these photons?

35. SSM ✪ A ruby laser uses the photons emitted by Cr^{+3} ions that are embedded in a crystal of Al_2O_3 (instead of in a gas, as with a helium–neon laser). The light emitted by a ruby laser has a wavelength of approximately 694 nm. (a) What is the frequency of this light? (b) What is the spacing, in electron-volts, between the energy levels in Cr^{+3} that are responsible for these photons?

ADDITIONAL PROBLEMS

36. ✪ We applied the Bohr model to hydrogen, but it can be extended to treat ions with one electron such as He^+ and Li^{+2}. Follow the derivation in Equations 29.10 through 29.15, but now let the nucleus have a charge $+Ze$ (for a nucleus with atomic number Z containing Z protons) instead of just $+e$ (a single proton). Derive an expression for the energy of the ground state of the ion. Compare your result for the ionization energy of He^+ with the measured value of approximately 54 eV.

37. ✪ When an atom undergoes a transition from one level to another, the energy needed is usually provided by a photon, but it may also come from the kinetic energy of a collision. Consider the collision of two hydrogen atoms, one with an initial speed v and another initially at rest. What value of v is needed to give the moving hydrogen atom a kinetic energy of 13.6 eV? If all this energy can be transferred to one of the electrons, this collision could cause one of the atoms to ionize.

38. ✪ Consider an electron in the $n = 1$ state hydrogen. (a) Find the speed v of this electron and the radius r of its orbit in the Bohr model. (b) What is the time needed for the electron to complete one orbit? (c) This orbiting electron acts as a current loop. What is the value of the current? (d) What is the magnetic field produced by this current loop at the nucleus?

39. ✪ Scientists are able to study atoms much larger than a hydrogen atom in its ground state. Consider a hydrogen atom with $n = 200$. (a) Determine the diameter of this atom (i.e., the diameter of this Bohr orbit). (b) Compare the diameter of this atom to that of a dust particle (about 10^{-6} m).

40. Determine the dissociation energy for hydrogen's electron for the states with the following principal quantum number: (a) $n = 2$, (b) $n = 10$, (c) $n = 100$.

41. ✪ Consider a hydrogen atom in its $n = 5$ state. Suppose it decays to the $n = 2$ state before proceeding to its ground state. (a) Determine the wavelengths of the emitted photons. (b) To what region of the electromagnetic spectrum do these photons correspond?

42. ✪ Using the Bohr model of the hydrogen atom, show that the speed of an electron in an orbit is given by $v_n = (2.2 \times 10^6)/n$ m/s.

43. ✪ (a) How many different possible sets of quantum numbers are there for the $n = 3$ shell? (b) Write the explicit values of all the quantum numbers (n, ℓ, m, s) for electrons in this shell.

44. ✪ How many different possible sets of quantum numbers are there for the subshells with (a) $\ell = 0$, (b) $\ell = 2$, and (c) $\ell = 3$?

45. For each of the following ground-state electron configurations, identify the atom.
(a) $1s^2 2s^2 2p^6$
(b) $1s^2 2s^2 2p^6 3s^2 3p^4$
(c) $1s^2 2s^2$

46. ✪ Imagine that electron spin did not exist. (a) How many electrons would occupy the $1s$ state? Why? (b) If there were no electron spin, what would be the first two noble gases in the periodic table?

47. ✪ Suppose the electron spin had three possible values instead of two. For this situation, determine the electron configuration of the ground state of a lithium atom.

48. ✪ Suppose a hydrogen atom is in the $6h$ state. Determine (a) the principal quantum number, (b) the energy of the state, (c) the orbital angular momentum and its quantum number ℓ, and (d) the possible values for the magnetic quantum number.

49. ✪ The energy required to remove the "outermost" electron from a boron atom is 8.26 eV. As an approximation, you can model the quantum state of this electron as that of a single electron in a hydrogen atom, but with an "effective" charge on the nucleus that is different from the charge of a single proton $(+e)$ as found for a real hydrogen atom. (a) Use the Bohr model to estimate the effective charge seen by this electron. (b) Estimate the average orbital radius for this electron.

50. ✪ Use the Bohr model to estimate the wavelength and frequency of the photon emitted during the $n = 3$ to $n = 1$ transition in molybdenum. The measured wavelength is 0.063 nm. Why do we not expect agreement?

51. ✪ The Bohr model can be accurately applied to a singly ionized helium atom. This atom has two protons in the nucleus ($Z = 2$) and a single electron orbiting the nucleus. When the electron makes a transition from a higher energy level to the $n = 4$ level, some of these transitions produce light in the visible part of the electromagnetic spectrum (about 380 nm to 750 nm). Determine which of these transitions correspond to these visible lines.

52. ✪ ✪ Consider laser eye surgery, in which laser light with a wavelength of 514 nm at a power output of 1.5 W is used to reattach a detached retina. Suppose the laser is pulsed for 0.050-s time intervals. (a) During this time, how many photons are emitted by the laser? (b) What is the difference in energy between the two energy levels involved in this laser transition?

53. ✪ In a muonic atom, the electron is replaced by a negatively charged particle called a muon. The muon has the same charge as an electron, but its mass is 207 times the mass of an electron.

Apply the Bohr model to a muonic hydrogen atom. (a) Calculate the radius of the orbit of the muon in the ground state, first excited state, and second exited state. (b) Calculate the energy of these states. (c) Determine the wavelength of the photon emitted when the muon undergoes a transition from the $n = 3$ to $n = 1$ state.

54. ✪ ® Suppose two hydrogen atoms, both initially in their ground state, are accelerated up to high speed and then collide head on. (a) Derive an expression for the speed necessary to raise both atoms to the same excited state, n. Express your answer as a function of n. (b) Evaluate this expression for $n = 2$. (c) To what temperature would you need to heat hydrogen gas to make this collision likely?

55. [SSM] ✖ LASIK, an acronym for *laser-assisted in situ ker-atomileusis*, is a form of laser eye surgery performed by ophthalmologists to correct myopia, hyperopia, and astigmatism. The procedure uses an excimer laser, which produces photons with a wavelength of 193 nm to remodel the corneal stroma. The photons vaporize thin layers of tissue less than 1 μm in thickness. Typically, the output power of the laser is around 5.0 mW, and it is pulsed for 10 nanoseconds. (a) During one of these pulses, how much energy is transferred to the cornea? (b) How many photons are contained in one pulse of the laser? (c) How many photons from a helium–neon laser would be required to deposit the same amount of energy? (d) What is the difference in energy between the two energy levels involved in this laser transition for the excimer laser? (e) Could you see the light from the LASIK laser?

56. ⬡ The wavelength of the light from a sodium-vapor lamp is 589 nm. Assume a 100-W sodium-vapor lamp radiates its energy uniformly in all directions. (a) At what rate are photons emitted from the lamp? (b) At what distance from the lamp will the average flux of photons be 1 photon/(cm² · s)? (c) What is the difference in energy between the two energy levels involved in the atomic transition that produces these photons?

57. ✪ ® **Microprobe.** Energy dispersive spectroscopy (EDS) is a technique for identifying and quantifying the elemental composition of a sample. Volumes as small as a few cubic micrometers can be probed using EDS. The characteristic X-rays of the elements in a sample are produced when the sample is bombarded with electrons in an electron beam instrument such as a scanning electron microscope. For elements heavier than hydrogen, we can approximate the energy of an electron in the $n = 1$ state (called a *K*-shell electron) using

$$E_K = -\frac{(13.6 \text{ eV})(Z - 1)^2}{n^2}$$

with $n = 1$. Here Z is the atomic number of the element, and the $(Z - 1)$ term corrects for the partial cancellation of the charge

the electron sees as a result of the other *K*-shell electron. Suppose we want to detect the presence of aluminum, copper, and tungsten in a sample. (a) Estimate the minimum potential difference through which the bombarding electron beam must be accelerated to produce the characteristic X-rays of each of these elements. (b) From the data provided in Table 29.1, estimate the energy of an electron in the $n = 2$ state for each of these elements.

58. ✪ Suppose a certain ionized atom produces an emission line in accordance with the Bohr model, but the number of protons in the nucleus Z is not known. A group of lines in the spectrum forms a series in which the shortest wavelength is 22.8 nm and the longest is 41.0 nm. Using a modified version of Equation 29.15, find the second longest wavelength in this series of lines (See Problem 57 for a hint.).

59. ✪ A quantum system (a molecule) has the energy levels shown in Figure P29.59. A large number of these molecules in their ground states make up a gas. (a) Suppose we want to excite these molecules to the state labeled E_2 by shining light on the gas. What wavelength of light should we use? (b) If we expose the molecules to a continuous spectrum of light, photons of particular wavelengths might be absorbed. What wavelength photons would you expect to be absorbed? Do not assume that the molecules all start in their ground state. Explain why that assumption makes sense.

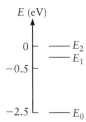

Figure P29.59

60. ✪ ✖ ® **Bioluminescence.** The process by which fireflies glow, called bioluminescence, is defined as the emission of light from living organisms. In fireflies, bioluminescence occurs in cells called photocytes that are located in the last two segments of the abdomen. Fireflies produce light through a two-step chemical process. During the first step, ATP (adenosine triphosphate), luciferin, and luciferase react to form luciferyl adenylate-luciferase and pyrophosphate. During the second step, luciferyl adenylate-luciferase decomposes in the presence of oxygen to form oxyluciferin and luciferase, releasing energy in the form of photons (visible light). These photons are in the yellow-green part of the spectrum with wavelengths around 550 nm. Typically, fireflies produce light in pulses up to a few seconds in length and sequence the pulses in a pattern designed to attract a mate. (a) Comparing the brightness of a firefly with other light sources with which you are familiar, estimate the power output of the firefly's light pulse. (b) About how much energy is released by the firefly in the form of light during a 1-s pulse? (c) Approximately how many photons are released during one pulse?

Nuclear Physics

Marie Curie was a pioneer in the study of radioactivity. She received two Nobel Prizes for her work: one in physics (along with her husband, Pierre, and French physicist Henri Becquerel) in 1903 and a second prize in chemistry in 1911. (© Photo Researchers/Alamy)

In Chapters 28 and 29, we described the development of quantum theory and discussed the quantum properties of electrons and photons. Rutherford's discovery of the atomic nucleus raised the obvious question, "How is the nucleus itself put together?" In Rutherford's time, approximately 85 different elements had been identified. Did every element have a completely different kind of nucleus? We now know that things are not that complicated: every nucleus is composed of just two different building blocks, particles called **protons** and **neutrons**. Different elements have different numbers of these two particles in their nuclei.

The protons and neutrons in most nuclei are bound together very tightly; that is, the potential energy of binding is large. Because a great deal of energy is needed to break apart a nucleus, the nucleus can usually be treated as a simple, unbreakable "particle" when dealing with phenomena such as chemical reactions and atomic spectra. In **nuclear reactions**, however, a nucleus breaks apart or is assembled when smaller nuclei and particles combine. Nuclear reactions can release very large amounts of energy that can be applied for great societal benefit, although they can also be used destructively. In this chapter, we describe the structure and properties of the nucleus and explain some applications of nuclear physics.

30.1 | STRUCTURE OF THE NUCLEUS: WHAT'S INSIDE?

The discovery of the nucleus is usually credited to Rutherford as a result of his scattering experiments described in Chapter 29. The experiments of Rutherford and others showed that all nuclei are composed of two particles, protons and neutrons. These two types of particles, called **nucleons**, are both much more massive than an electron. Their masses are

The nucleus is composed of nucleons (protons and neutrons).

$$m_p = 1.673 \times 10^{-27} \text{ kg} \quad \text{(mass of a proton)} \tag{30.1}$$

$$m_n = 1.675 \times 10^{-27} \text{ kg} \quad \text{(mass of a neutron)} \tag{30.2}$$

The values of m_p and m_n are thus very similar, and both are about 1800 times more massive than an electron. Perhaps the most important difference between protons and neutrons is that the neutron is electrically neutral, whereas the proton carries a positive charge $+e$. The charge on an electron is $-e$, so the total charge carried by one electron plus one proton is precisely zero.

Definition of atomic number

Every nucleus can be specified by the numbers of protons and neutrons it contains. The number of protons in a nucleus is called the **atomic number**, Z. Because each element has a particular and unique number of protons in its nucleus, each has a characteristic value of Z. For example, the atomic number of He (helium) is $Z = 2$, while the atomic number of O (oxygen) is $Z = 8$. Since atoms are electrically neutral, Z is also equal to the number of electrons in an atom.

The number of neutrons in a nucleus is usually denoted by the symbol N. The value of N for a particular element can vary. For example, He nuclei can have $N = 1$, $N = 2$, or $N = 4$, and other values are also possible. The number of neutrons in atomic nuclei generally increases as the atomic number increases.

Definition of mass number

The **mass number**, A, of a nucleus is the sum of the number of protons and neutrons, so

$$A = Z + N \tag{30.3}$$

All this information is described in the following shorthand notation:

$$\boxed{\text{mass number}} \searrow \atop \boxed{\text{atomic number}} \nearrow \; {}^A_Z X \; \nwarrow \boxed{\text{symbol for element}} \tag{30.4}$$

The letter "X" in Equation 30.4 denotes an element whose chemical symbol should be substituted from the periodic table. As an example, a He nucleus containing $N = 2$ neutrons is denoted

$$\boxed{A = 4 \; (2 \text{ protons} + 2 \text{ neutrons})} \searrow \atop \boxed{Z = 2 \; (2 \text{ protons})} \nearrow \; {}^4_2 He \; \nwarrow \boxed{\begin{array}{c}\text{symbol for element}\\\text{(helium)}\end{array}}$$

The various helium nuclei having $N = 1$, 2, or 4 are written as

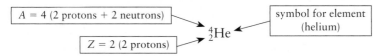

$${}^3_2 He \qquad {}^4_2 He \qquad {}^6_2 He$$

and are called different *isotopes* of He. All isotopes of a particular element have the same numbers of protons and electrons. Chemical properties are determined mainly by the bonding of electrons, so the chemical properties of different isotopes of a given element are essentially identical.

Isotopes have the same Z but different values of N.

Isotopes and Atomic Mass

The *atomic mass* of an atom is the mass of the nucleus plus Z electrons. Since different isotopes of an element have different numbers of neutrons, their atomic masses will differ. The periodic table contains an average value of the atomic mass for each element based on the *natural abundance* of each isotope. For example, several different isotopes of carbon are found in nature, with $A = 11, 12, 13,$ and 14, and the isotope with $A = 12$ has the greatest abundance. The average atomic mass for a natural sample of pure carbon (based on the abundance of different isotopes) is listed in the periodic table as the atomic mass of the element. The listed value is the mass in grams of 1 mole of carbon atoms.

| **CONCEPT CHECK 30.1** | Nucleus of Carbon

Use the notation described in Equation 30.4 to write the symbol for the nucleus of carbon that contains 8 neutrons.

| **EXAMPLE 30.1** | Identifying a Nucleus

A particular nucleus contains $N = 13$ neutrons and has mass number $A = 25$. What element is it? Write the chemical symbol for this nucleus.

RECOGNIZE THE PRINCIPLE

The chemical identity of an element is determined by its atomic number Z, the number of protons in its nucleus.

SKETCH THE PROBLEM

We follow the shorthand description of a nucleus in Equation 30.4.

IDENTIFY THE RELATIONSHIPS

The mass number A is the sum of Z and N (Eq. 30.3), so

$$A = Z + N$$

and we have been given the values of A and N.

SOLVE

The number of protons in the nucleus is thus

$$Z = A - N = 25 - 13$$
$$Z = 12$$

Using the periodic table (facing the last page of the index), the element with $Z = 12$ is magnesium (Mg) . The symbol for this nucleus is $^{25}_{12}\text{Mg}$.

What does it mean?

This element is not the only isotope of magnesium found in nature. The isotopes $^{24}_{12}\text{Mg}$ and $^{26}_{12}\text{Mg}$ are also stable. In fact, approximately 79% of the magnesium nuclei found in nature are $^{24}_{12}\text{Mg}$. (See Table A.4.)

Size of the Nucleus

Rutherford discovered the nucleus by firing alpha particles at atoms and observing how the alpha particles were deflected (Fig. 30.1). At the time, Rutherford did not

Insight 30.1

RUTHERFORD'S SCATTERING EXPERIMENT AND CONSERVATION OF MOMENTUM

Experiments like those of Rutherford, in which a projectile particle is fired at a target particle, are called *scattering experiments*. They are a very powerful way to probe a nucleus or other subatomic particle and have been used to study the internal structure of protons and neutrons (Chapter 31). Rutherford's observation that some alpha particles are "backscattered"—that is, scattered back in the direction from which they came—implies a very massive nucleus, which can be understood from the conservation of momentum and energy in an elastic collision. When a heavy object (such as a bowling ball) collides with a very light one (a marble), the bowling ball is not deflected significantly from its original path. When a marble scatters from a bowling ball, however, backscattering is possible; the marble may reverse its direction. Rutherford's results thus mean that the nucleus of an element such as gold (which he used as a target) is much more massive than an alpha particle.

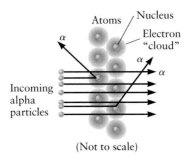

Atoms / Nucleus

α

Electron "cloud"

α

α

Incoming alpha particles

(Not to scale)

Figure 30.1 The nucleus of an atom is much smaller than the radius of the atom (i.e., the volume occupied by the electrons). When alpha particles are fired at an atom, there is only a small chance that they will collide with the nucleus; most alpha particles pass straight through. Because the mass of a nucleus is comparable to or larger than the mass of an alpha particle, the alpha particle can be scattered over a wide range of directions.

Radius of a nucleus

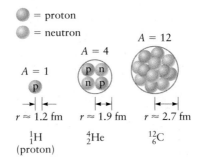

= proton

= neutron

$A = 1$

p

$r \approx 1.2$ fm

$^1_1$H
(proton)

$A = 4$

p n
n p

$r \approx 1.9$ fm

$^4_2$He

$A = 12$

$r \approx 2.7$ fm

$^{12}_6$C

Figure 30.2 The nuclear radius r increases as the number of protons and neutrons increases, implying that protons and neutrons are closely packed within the nucleus.

know exactly what an alpha particle was; we now know that it is the nucleus $^4_2$He. (We'll explain how Rutherford obtained alpha particles in Section 30.2.) As far as Rutherford was concerned, alpha particles were simply very small particles with a mass approximately equal to the mass of a helium atom. His experiments used them as tiny "billiard balls" that could bounce off atoms just as real billiard balls are used in collisions in a game of pool. Rutherford found that most alpha particles in his experiment passed right through the atoms in his target. Only a very tiny fraction of them were deflected. The implications of this result can be appreciated from Figure 30.1. The probability for an alpha particle to collide with the nucleus depends on the diameters of the alpha particle and nucleus. Since the probability of a collision is very small (as established by Rutherford's experiment), both the alpha particles and the nucleus must be much smaller than the overall size of the atom. By measuring the collision probability and assuming alpha particles are about as small as the nucleus, Rutherford could deduce the size of the nucleus.

Rutherford's collision experiments showed that the diameter of a typical nucleus is a few femtometers (1 femtometer = 1 fm = 1×10^{-15} m). Recall that the size of an atom as measured by the Bohr radius (Chapter 29) is about 5×10^{-11} m. The nucleus is thus smaller than the atom by a factor of about $(1 \times 10^{-15})/(5 \times 10^{-11})$ $= 2 \times 10^{-5} = 1/50,000$. In other words, the nucleus is about 50,000 times smaller than an atom!

Scattering experiments like those of Rutherford along with other studies have shown that most nuclei have an approximately spherical shape, with a radius r that depends on the number of nucleons A it contains. It is found that

$$r = r_0 A^{1/3} \qquad (30.5)$$

The constant $r_0 \approx 1.2$ fm $= 1.2 \times 10^{-15}$ m. We can rearrange Equation 30.5 to read

$$A = \left(\frac{r}{r_0}\right)^3$$

Since the volume of a sphere is $\frac{4}{3}\pi r^3$, the number of nucleons A is proportional to the volume of the nucleus, in agreement with the picture of nucleons packed together like tiny hard spheres in Figure 30.2. As A increases, the radius of this nuclear "sphere" increases according to Equation 30.5.

Electric Potential Energy of a Nucleus

Each proton in a nucleus has a charge of $+e$. The diameter of a typical nucleus is a few femtometers, so the protons in a nucleus are very close together. Such closely spaced positive charges have a large electric potential energy. For two point charges q_1 and q_2 separated by a distance r, the electric potential energy is (Eq. 18.6)

$$PE_{\text{elec}} = \frac{k q_1 q_2}{r}$$

For two protons in a He nucleus, we have $q_1 = q_2 = +e$, and estimating $r = 1$ fm, we get

$$PE_{\text{elec}} = \frac{k(+e)(+e)}{r} = \frac{(8.99 \times 10^9 \text{ N} \cdot \text{m}^2/\text{C}^2)(1.60 \times 10^{-19} \text{ C})^2}{1 \times 10^{-15} \text{ m}}$$

$$PE_{\text{elec}} = 2 \times 10^{-13} \text{ J} \qquad (30.6)$$

To see if this energy is small or large on an atomic scale, let's express PE_{elec} in units of electron-volts (eV). (Recall from Chapter 18 that 1 eV = 1.60×10^{-19} J.) We find

$$PE_{\text{elec}} = (2 \times 10^{-13} \text{ J})\left(\frac{1 \text{ eV}}{1.60 \times 10^{-19} \text{ J}}\right) = 1 \times 10^6 \text{ eV} = 1 \text{ MeV} \qquad (30.7)$$

In Chapter 29, we found that a typical atomic-scale energy is around 10 eV. (The ionization energy of the hydrogen atom is 13.6 eV.) Equation 30.7 indicates that a typical nuclear-scale energy is larger than that by a factor of approximately 100,000! The potential energy associated with a nucleus is thus *much* larger than the energies associated with electrons, molecules, and chemical reactions.

Forces in the Nucleus

When Rutherford first did the calculation that leads to Equation 30.7, he must have been somewhat puzzled. The electric potential energy in Equation 30.7 is positive since all protons are positively charged. The protons in a nucleus thus repel one another with a very large force. To make a nucleus stable, there must be another even larger force—what physicists call the ***strong force***—at work in the nucleus that is capable of producing an enormous attractive force between nucleons, strong enough to overcome the electrical repulsion. The strong force attracts any two nucleons to each other. Attraction due to the strong force occurs between a pair of protons, a pair of neutrons, and a proton and a neutron (Fig. 30.3). The strong force has a very large and approximately constant magnitude when the two nucleons are about 1 fm apart, but it is negligible when they are separated any farther (Fig. 30.4).

The strong force gives a large attractive force between nucleons and holds the nucleus together, but it does not act on electrons. So, when an electron approaches a nucleus, it experiences essentially just an electric force (Coulomb's law) due to the protons in the nucleus.[1] The strong force does not play any role in the binding of electrons in an atom.

The strong force is an attractive force between nucleons.

Stability of the Nucleus

The simplest nucleus is $^1_1$H (hydrogen), consisting of just a single proton. All other nuclei contain both protons and neutrons. For example, there are several nuclei of helium, including $^3_2$He and $^4_2$He, all containing neutrons. Why is there not a nucleus of helium that contains just two protons? The reason is that the strong force between two protons is not large enough to overcome the repulsive electric force in that case, so a nucleus containing just two protons is not stable (Fig. 30.3C). The addition of one or more neutrons provides the extra attractive force needed to overcome the Coulomb repulsion of the protons. If a neutron is positioned between two protons (Fig. 30.5), the strong force attracts the neutron to each of the protons. The presence of this neutron also increases the separation between the protons, reducing the repulsive Coulomb force. Neutrons are thus essential for the stability of the nucleus. The number of neutrons generally increases as the atomic number increases; as more and more protons are added to the nucleus, more and more neutrons are needed to keep all the protons sufficiently separated (at a "safe distance") so that the Coulomb force does not cause the nucleus to fly apart.

The picture in Figure 30.5 indicates that neutrons are an essential "glue" that holds the nucleus together. Based on this picture, you might also think that adding more glue (neutrons) should always result in a more stable nucleus, but adding more and more neutrons will eventually lead to an unstable nucleus. This result can be explained by the quantum theory of the nucleus called the ***nuclear shell model***, which gives the energy levels of nucleons inside a nucleus, similar to the energy levels of electrons within an atom (Chapter 29).

Mass and Energy Scales in Nuclear Physics

We have seen that the electric potential energy of protons in a nucleus is extremely large, on the order of MeV (Eq. 30.7). The strong force in the nucleus must overcome this positive potential energy, so its associated potential energy is negative and is also on the order of MeV.

THE STRONG FORCE ACTS
BETWEEN ANY TWO NUCLEONS

Two neutrons are attracted to each other by the strong force.

A

A neutron and a proton are attracted by the strong force.

B

The total force between two protons is the sum of the electric and strong forces.

C

Figure 30.3 The strong nuclear force leads to an attraction between any pair of nucleons. The neutron has a net electric charge of zero, so the dominant force between **A** a pair of neutrons or **B** a neutron and a proton is the strong force. **C** Since protons are charged, there is a repulsive electric force between two protons along with the attractive strong force. If only two protons are involved, the electric force is larger in magnitude than the strong force, so a nucleus containing just two protons is not stable.

---

[1]An electron also experiences another, much weaker force, called the "weak force," which we'll discuss in Chapter 31.

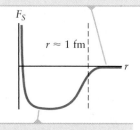

The strong force between nucleons is negligible when the separation is larger than about 1 fm.

F_S

$r \approx 1$ fm

r

The strong force is attractive for separations in this range.

Figure 30.4 The strong force is important only when the two nucleons are closer than about 1 fm. At larger separations, the strong force between two nucleons is negligible. At very small separations, the strong force is repulsive, so two nucleons can't get too close together.

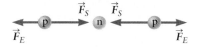

The strong force attracts these protons to a neutron located between them.

$\vec{F}_S$ $\vec{F}_S$

p   n   p

$\vec{F}_E$ $\vec{F}_E$

Figure 30.5 A neutron located between two protons acts as a sort of "glue" to bind them all together. Each proton is attracted to the neutron, and since the protons are now farther apart than they would be if the neutron were not present, the repulsive electric force is smaller than in Figure 30.3C. As a result, a nucleus with two protons and one neutron is stable. The nucleus shown here is $^3_2$He.

Another important energy scale in nuclear physics is the rest energy associated with the proton and neutron. According to the theory of relativity (Chapter 27), the rest energy of a particle of mass m is

$$E = mc^2 \qquad (30.8)$$

where c is the speed of light. For a proton, we find

$$E_p = m_p c^2 = (1.67 \times 10^{-27} \text{ kg})(3.00 \times 10^8 \text{ m/s})^2 = 1.50 \times 10^{-10} \text{ J}$$

$$E_p = 938 \text{ MeV} \qquad (30.9)$$

The mass and rest energy of any particle are related though Equation 30.8, which can be rearranged to read $m = E/c^2$. Nuclear physicists often specify the mass of a particle in units of (energy)/$c^2 = $ MeV/c^2. In these units, the mass of a proton is (using Eq. 30.9)

$$m_p = 938 \text{ MeV}/c^2 \qquad (30.10)$$

The masses of the proton, neutron, and electron are collected in Table 30.1 in units of kilograms and MeV/c^2 and also in terms of the **atomic mass unit**, denoted by "u." This unit is defined as 1/12th of the mass of a carbon atom with the nucleus $^{12}_6$C; this nucleus contains 6 neutrons and 6 protons. An atomic mass unit is related to other units of mass by

$$1 \text{ u} = 1.66 \times 10^{-27} \text{ kg} = 931.5 \text{ MeV}/c^2 \qquad (30.11)$$

Atomic mass units are handy in nuclear physics because 1 u is approximately equal to the mass of a nucleon (either a proton or a neutron).

The values in Table 30.1 include more significant figures than we have commonly used in this book because (as we will soon see) small changes in mass in a nuclear reaction correspond to large amounts of released energy, which can be calculated using the relation between mass and energy in Equation 30.8. When working problems in nuclear physics, we need to keep track of small differences and changes in mass.

30.2 | NUCLEAR REACTIONS: SPONTANEOUS DECAY OF A NUCLEUS

There are two types of nuclear reactions: *spontaneous* and *induced*. In a nuclear reaction, one or more nuclei and nucleons "react" to form new nuclei, perhaps emitting nucleons and releasing energy. For example, a collision with an alpha particle can sometimes cause a nucleus to break into fragments. This reaction is called an induced nuclear reaction because the alpha particle caused the decay. We'll discuss several important examples of induced reactions when we describe the processes of nuclear fission and fusion in Section 30.3, but let's first discuss reactions that occur spontaneously.

Radioactivity

In addition to Rutherford, many famous scientists, including the husband-and-wife team of Marie Curie and Pierre Curie, played important roles in the development

Table 30.1 Properties of the Proton, Neutron, and Electron

PARTICLE	CHARGE	MASS (kg)	MASS (atomic mass units[a] = u)	MASS (MeV/c^2)
Proton	$+e$	1.6726×10^{-27}	1.0073	938.27
Neutron	0	1.6749×10^{-27}	1.0087	939.57
Electron	$-e$	9.109×10^{-31}	5.486×10^{-4}	0.5110

[a]1 u corresponds to 931.5 MeV/c^2.

of nuclear physics. The Curies and others found that some nuclei are unstable; that is, certain nuclei decay spontaneously into two or more particles. This process is called *radioactive decay*, and the term *radioactivity* refers to the process in which a nucleus spontaneously emits either particles or radiation.

When a nucleus decays, it can emit particles with mass such as electrons or neutrons, or it can emit photons (electromagnetic radiation). At first, physicists did not know the identities of the particles emitted in radioactive decay, so these unknown decay products were initially given the names *alpha*, *beta*, and *gamma*. Eventually, the nature of these particles was deduced from experiments, but these somewhat mysterious names are still widely used. Even though their identity was initially unknown, these three types of decay products could be distinguished according to their properties, including their electric charge and their ability to penetrate matter.

Loosely speaking, we sometimes refer to alpha, beta, and gamma radiation emitted in radioactive decay. The term *radiation* is reminiscent of electromagnetic radiation (Chapter 23). We now know that only gamma radiation, also called "gamma rays," is actually a form of electromagnetic radiation, while alpha and beta "radiation" consists of particles with mass.

Alpha particles. An alpha particle is composed of two protons and two neutrons. It is thus the nucleus of a He atom and is also denoted $^4_2$He. An alpha particle produced by nuclear decay usually does not have any bound electrons, so alpha particles almost always have a charge of $+2e$ (the charge of two protons). A typical radioactive decay involving an alpha particle is that of the radium nucleus $^{226}_{88}$Ra. This process is written in a form very similar to a chemical reaction:

$$\underset{\text{parent nucleus}}{^{226}_{88}\text{Ra}} \quad \rightarrow \quad \underset{\text{daughter nucleus}}{^{222}_{86}\text{Rn}} \quad + \quad \underset{\text{alpha particle}}{^4_2\text{He}} \tag{30.12}$$

Alpha decay: emission of an alpha particle from a nucleus

This decay reaction (Fig. 30.6A) begins with a *parent nucleus* of $^{226}_{88}$Ra containing 88 protons and $N = A - Z = 226 - 88 = 138$ neutrons. This nucleus spontaneously decays into two particles called *decay products*. One decay product is an isotope of radon, $^{222}_{86}$Rn, which contains $N = A - Z = 222 - 86 = 136$ neutrons. In this reaction, $^{222}_{86}$Rn is called the *daughter nucleus*. The other decay product in Equation 30.12 is an alpha particle, $^4_2$He. The total number of nucleons is *conserved* (does not change) in the reaction. Since the reaction products contain 222 nucleons ($^{222}_{86}$Rn) plus 4 nucleons ($^4_2$He), there are a total of 226 nucleons (protons plus neutrons) at the start of the reaction and the same number at the end. This particular reaction also conserves the numbers of protons and neutrons separately. That is, there are 88 protons at the start (in $^{226}_{88}$Ra) and 88 at the end (86 in the radon daughter nucleus and 2 in the alpha particle). You can view this reaction as simply ejecting two protons and two neutrons (the alpha particle) from the original radium nucleus. In his scattering experiments, Rutherford obtained his alpha particles from the alpha decay of nuclei such as $^{214}_{84}$Po.

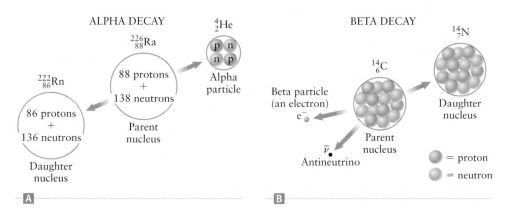

Figure 30.6 Ⓐ Example of alpha decay: the original parent nucleus ($^{226}_{88}$Ra) splits to form $^{222}_{86}$Rn and an alpha particle ($^4_2$He). Ⓑ Example of beta decay: the original $^{14}_6$C nucleus produces $^{14}_7$N plus an electron (a beta particle) and an antineutrino ($\bar{\nu}$).

The particular alpha decay reaction in Equation 30.12 has important medical consequences. It contributes to the (ill) health of miners and can also lead to a health hazard in homes. For more details, see Section 30.4.

Beta particles. There are two varieties of beta particles. One variety is negatively charged and is simply an electron. The other type is identical to an electron except it has a positive charge of precisely $+e$. This positive beta particle is called a *positron* (denoted e^+, with the superscript indicating its positive charge) and is an example of *antimatter* (Chapter 31).[2] Electrons and positrons have the same mass and are both point charges.

A typical radioactive decay producing a beta particle is

$$\underset{\substack{\text{parent}\\\text{nucleus}}}{^{14}_{6}\text{C}} \rightarrow \underset{\substack{\text{daughter}\\\text{nucleus}}}{^{14}_{7}\text{N}} + \underset{\substack{\text{beta particle}\\\text{(electron)}}}{e^-} + \underset{\substack{\text{antineutrino}}}{\bar{\nu}} \tag{30.13}$$

In this reaction (shown schematically in Figure 30.6B), the emitted beta particle is an electron denoted by e^-; the superscript negative sign indicates that the particle is negatively charged. This reaction also produces a particle called an *antineutrino*, denoted by $\bar{\nu}$. The antineutrino is another example of antimatter and is thus a "cousin" of the positron. We'll discuss neutrinos and antimatter further in Chapter 31.

As with alpha decay, beta decay also conserves the total number of nucleons. It is useful to rewrite Equation 30.13 to indicate explicitly both the number of protons and neutrons and the electric charge at the start and end of the reaction:

$$\text{number of protons: } 6 \rightarrow 7$$
$$\text{number of neutrons: } 8 \rightarrow 7$$
$$^{14}_{6}\text{C} \rightarrow {}^{14}_{7}\text{N} + e^- + \bar{\nu}$$
$$\text{charge:} \qquad +6e \rightarrow +7e - e + 0 \tag{30.14}$$

Hence, this reaction converts a neutron in the parent $^{14}_{6}\text{C}$ nucleus into a proton, conserving the total number of nucleons (neutrons plus protons). At the same time, electric charge is conserved since the total charge is $+6e$ at the beginning and at the end of the reaction. The positive charge of the new proton is balanced by the creation of an electron.

The beta decay reaction in Equation 30.13 is widely used in archeology in the technique called *carbon dating*. We'll discuss this application of radioactivity in Section 30.5.

Gamma decay. Gamma decay produces gamma "rays," which are photons, quanta of electromagnetic radiation denoted by the Greek letter γ. One example of a nuclear decay that produces gamma rays is

$$\underset{\substack{\text{parent nucleus}\\\text{(in excited state)}}}{^{14}_{7}\text{N*}} \rightarrow \underset{\substack{\text{daughter}\\\text{nucleus}}}{^{14}_{7}\text{N}} + \underset{\substack{\text{gamma}\\\text{ray}}}{\gamma} \tag{30.15}$$

The asterisk on the left in this equation denotes that the nucleus is in an excited state. In Chapter 29, we saw that an atom in excited state can undergo a transition to a state of lower energy and emit a photon in the process. In the same way, a nucleus that is initially in an excited state can undergo a transition to a state of lower energy and emit a gamma ray photon. A particular nucleus is often placed in an excited state as a result of alpha or beta decay. For example, a $^{14}_{7}\text{N*}$ nucleus in an excited state (as on the left in Eq. 30.15) can be produced by beta decay (although we did not indicate it in Eq. 30.13). A highly simplified energy-level diagram for the reaction in Equation 30.15 is shown in Figure 30.7. This diagram shows only the initial and final states in Equation 30.15. The decay from the excited to the ground

---

[2]Here and throughout this chapter we follow the standard notation of nuclear physics and denote particles such as an electron by a "non-italic" letter.

Beta decay: emission of a beta particle (an electron or positron) from a nucleus, along with an antineutrino or a neutrino

Gamma decay: emission of a gamma ray photon from an excited nucleus

state in Figure 30.7 produces a gamma ray with an energy equal to the difference in the energies of the two levels.

A more realistic energy-level diagram, this time for the excited states of $^{60}_{28}$Ni, is shown in Figure 30.8. A $^{60}_{28}$Ni* nucleus can emit gamma rays with many different energies, depending on which excited and final states are involved. For the states shown here, the emitted gamma rays have energies of a few MeV.

Gamma ray energies are much higher than those of visible light or X-ray photons. The energy of a gamma ray photon depends on the particular nuclear decay that produces it, but typical gamma rays have energies from 10 keV (10^4 eV) to 100 MeV (10^8 eV) or even higher. Recall that the spectral emission of atoms (Chapter 29) involved photons with energies of typically 10 eV. This comparison shows again that the overall energy scale of nuclear decay—and of nuclear processes in general—is much greater than a typical atomic scale energy.

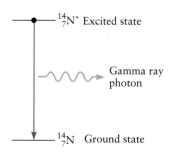

Figure 30.7 A nucleus in an excited state ($^{14}_{7}$N*) can decay to its ground state ($^{14}_{7}$N) by emitting a gamma ray photon.

CONCEPT CHECK 30.2 | Wavelength of a Gamma Ray Photon

According to the energy-level diagram for $^{60}_{28}$Ni in Figure 30.8, this nucleus can emit a gamma ray photon with an energy of 1.33 MeV. What is the wavelength of this photon?
   (a) 9.3×10^{-13} m $= 9.3 \times 10^{-4}$ nm
   (b) 2.7×10^{-31} m $= 2.7 \times 10^{-23}$ nm
   (c) 2.3×10^2 m
How does the wavelength of the gamma ray compare with the wavelength of visible light?

Conservation Rules in Nuclear Reactions

The radioactive processes of alpha, beta, and gamma decay can be understood in terms of the following conservation principles.

Conservation of mass–energy. The principle of conservation of energy applies to *all* physical processes, including nuclear reactions. The total energy at the start of a reaction such as the gamma decay in Equation 30.15 must equal the total energy at the end. Indeed, this principle allows us to use an energy-level diagram (as in Figs. 30.7 and 30.8) to calculate the energy of the gamma ray produced in such a reaction. In addition, we must account for the results of special relativity and allow for the conversion of mass to energy and vice versa. We'll discuss some specific examples later in this section.

Conservation of momentum. All nuclear reactions must conserve momentum. The application of momentum conservation in nuclear physics is similar to our work on collisions in Chapter 7.

Conservation of electric charge. All nuclear reactions conserve electric charge, but the total number of charged particles may change in a reaction. For example, the beta decay in Equation 30.14 produces a proton and an electron. The total charge produced is thus zero (so the total charge does not change), but the number of charged particles increases.

Conservation of nucleon number. The number of nucleons—that is, the number of protons plus neutrons—does not change. For example, in the beta decay reaction in Equation 30.14, a neutron is converted into a proton, but the total number of neutrons plus protons is unchanged by the reaction.

Radioactive Decay Series

When a nucleus undergoes radioactive decay through the emission of an alpha particle or a beta particle, it is converted into another type of nucleus. An important example is the alpha decay of $^{238}_{92}$U:

$$^{238}_{92}\text{U} \rightarrow ^{234}_{90}\text{Th} + ^{4}_{2}\text{He} \qquad (30.16)$$

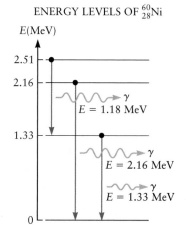

ENERGY LEVELS OF $^{60}_{28}$Ni

Figure 30.8 Energy-level diagram showing a few of the energy levels of the nucleus $^{60}_{28}$Ni. This nucleus can emit gamma ray photons with several different energies, depending on the initial and final states.

Insight 30.2

GAMMA RAYS AND X-RAYS
The terms *gamma ray* and *X-ray* both refer to portions of the electromagnetic spectrum. Typical gamma ray energies can be as low as 10 keV and as large as 100 MeV or even higher. X-ray energies are typically in the range of 100 eV to a few hundred keV. Hence, the gamma and X-ray energy ranges overlap. By convention, the distinction between them is made by the origin of the radiation. Gamma rays are produced in nuclear reactions (such as radioactive decay), while X-rays are generated by process involving atomic electrons as described in Chapter 29.

Figure 30.9 Sequence of nuclear decays starting with the parent nucleus $^{238}_{92}$U. The horizontal arrows denote beta decay, and the diagonal arrows show alpha decay.

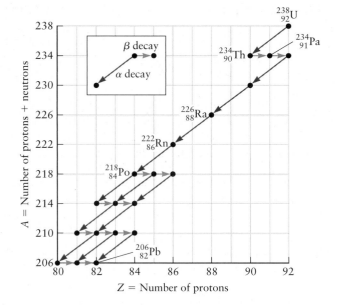

This decay converts uranium (U) into thorium (Th). The ability to convert one element into another was a goal of many chemists and alchemists throughout the Middle Ages. This conversion is precisely what happens in alpha and beta decay, but it is not an economical approach for producing any element on a commercial scale (and certainly not for producing the gold so eagerly sought by medieval alchemists).

Although Equation 30.16 describes a complete nuclear reaction, the decay reaction often continues. That is, after $^{238}_{92}$U decays into $^{234}_{90}$Th, this thorium nucleus decays further. Figure 30.9 shows a sequence of nuclear decays starting from $^{238}_{92}$U, with each decay indicated by an arrow. A horizontal arrow pointing to the right denotes a beta decay similar to the one in Equation 30.13, in which the number of protons, Z, increases by one. For example, the second decay in Figure 30.9 is the reaction

$$^{234}_{90}\text{Th} \rightarrow ^{234}_{91}\text{Pa} + e^- + \bar{\nu} \qquad (30.17)$$

in which thorium (Th) is converted into protactinium (Pa) plus an electron and an antineutrino. Each beta decay (horizontal arrow) in Figure 30.9 follows the same pattern. The diagonal arrows in the figure denote alpha decay; for example, the diagonal arrow at the upper right links $^{238}_{92}$U to $^{234}_{90}$Th in the reaction in Equation 30.16. Each alpha decay reduces the proton number by two and the neutron number by two. Some nuclei, such as $^{218}_{84}$Po, can undergo either beta decay or alpha decay, so both a horizontal arrow and a diagonal arrow originate at these nuclei.

In the radioactive decay series in Figure 30.9, $^{238}_{92}$U is the original parent nucleus while many unstable daughter nuclei are produced, and the decay proceeds until reaching the nucleus $^{206}_{82}$Pb. This nucleus is a stable final product of the decay of $^{238}_{92}$U. This particular decay series is especially important; among the daughter nuclei are $^{226}_{88}$Ra (radium), which is one of the radioactive nuclei discovered by Pierre and Marie Curie, and $^{222}_{86}$Rn (radon), which is a significant health hazard. We'll investigate other radioactive nuclei produced by other decay series in the end-of-chapter questions.

CONCEPT CHECK 30.3 Alpha Decay of Radium

Which of the following reaction equations correctly describes the alpha decay of $^{223}_{88}$Ra?

(a) $^{223}_{88}\text{Ra} \rightarrow ^{227}_{90}\text{Th} + ^4_2\text{He}$

(b) $^{227}_{90}\text{Th} \rightarrow ^{223}_{88}\text{Ra} + ^4_2\text{He}$

(c) $^{223}_{88}\text{Ra} \rightarrow ^{219}_{86}\text{Rn} + ^4_2\text{He}$

CONCEPT CHECK 30.4 Beta Decay

Consider a nucleus that undergoes beta decay and emits an electron. Which of the following statements is true?

(a) The atomic number Z decreases by one.
(b) The atomic number Z decreases by two.
(c) The atomic number Z increases by one.
(d) The atomic number Z increases by two.
(e) Z and A are unchanged.

Calculating the Binding Energy of a Nucleus

We have mentioned several times that the mass–energy relation of special relativity is important for understanding nuclear binding energies. Recall from Chapter 27 that the rest energy of a particle (such as a nucleus) with rest mass m is

$$E = mc^2 \qquad (30.18)$$

Let's now consider carefully how this relation applies to the nucleus $^4_2$He, an alpha particle. With an understanding of the mass and binding energy in this case, we will be ready to discuss the energy released in nuclear reactions and how to calculate it.

An alpha particle consists of two protons and two neutrons, and from Table 30.1 we can add up the masses of two isolated protons plus two isolated neutrons to get

$$m(2 \text{ protons} + 2 \text{ neutrons}) = 2(1.0073 \text{ u}) + 2(1.0087 \text{ u}) = 4.0320 \text{ u} \qquad (30.19)$$

Here we have included many more significant figures than usual because small mass differences will be extremely important in our final result. The result in Equation 30.19 is slightly larger than the mass of an alpha particle, which is 4.0015 u. (This is the mass of just a helium nucleus, a helium atom without the two electrons.) The difference is due to the binding energy of the alpha particle, E_{binding}. The mass difference is

$$\Delta m = 4.0015 - 4.0320 \text{ u} = -0.0305 \text{ u} \qquad (30.20)$$

Applying the relation between mass and energy from special relativity (Eq. 30.18), the mass difference in Equation 30.20 corresponds to an energy

$$E_{\text{binding}} = (\Delta m)c^2$$

To evaluate E_{binding}, we note that 1 atomic mass unit (1 u) corresponds to 931.5 MeV/c^2 (Eq. 30.11). Using this conversion factor with the value of Δm in Equation 30.20 gives

$$E_{\text{binding}} = (-0.0305 \text{ u})\left(\frac{931.5 \text{ MeV/}c^2}{\text{u}}\right)c^2 = -28.4 \text{ MeV} \qquad (30.21)$$

This binding energy is negative, indicating that an alpha particle has a *lower* energy than a collection of separated protons and neutrons. An alpha particle is therefore more stable than a collection of separate protons and neutrons and so will not decay spontaneously into protons and neutrons.

Finding the Energy Released in a Nuclear Reaction

Let's now apply our understanding of nuclear binding energy as illustrated for an alpha particle in Equations 30.19 through 30.21 to calculate the energy released in a nuclear decay process. The alpha decay reaction of radium was studied by Marie and Pierre Curie and plays an important role in many decay processes like the one diagramed in Figure 30.9:

$$^{226}_{88}\text{Ra} \quad \rightarrow \quad ^{222}_{86}\text{Rn} \quad + \quad ^4_2\text{He} \qquad (30.22)$$

$$m = 226.0254 \text{ u} \rightarrow 222.0176 \text{ u} + 4.0026 \text{ u}$$

$$\text{total mass} = 226.0254 \text{ u} \rightarrow \qquad 226.0202 \text{ u}$$

(Note that the masses here and in other examples below include the electrons for each atom, unlike Eq. 30.20 where the mass value for the alpha particle was for the

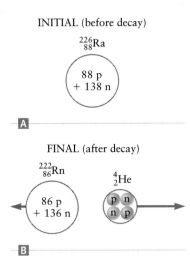

INITIAL (before decay)

$^{226}_{88}$Ra

88 p
+ 138 n

A

FINAL (after decay)

$^{222}_{86}$Rn

$^{4}_{2}$He

86 p
+ 136 n

p n
n p

B

Figure 30.10 When a $^{226}_{88}$Ra nucleus undergoes alpha decay, it emits an alpha particle ($^{4}_{2}$He) and a $^{222}_{86}$Rn nucleus. This decay leaves these two particles with some kinetic energy. To conserve momentum, the reaction products are emitted in opposite directions.

nucleus only.) For simplicity, Figure 30.10A shows the parent nucleus $^{226}_{88}$Ra as being at rest. Will the reaction products $^{222}_{86}$Rn and $^{4}_{2}$He also be at rest after the reaction? The answer is no; a nucleus such as $^{226}_{88}$Ra will decay spontaneously only if the total rest energy of the final products is less than the rest energy of the original parent nucleus. Total energy must be conserved, so the reaction in Equation 30.22 will release energy, usually in the form of kinetic energy of the decay products. (A portion of the energy is sometimes also used to put a daughter nucleus into an excited state, similar to the excited $^{14}_{7}$N* nucleus produced by beta decay; Eqs. 30.13 and 30.15.) That is why Figure 30.10B shows the $^{222}_{86}$Rn and $^{4}_{2}$He nuclei traveling away after the reaction.

To calculate the energy released in the decay of $^{226}_{88}$Ra, we must consider the masses of the nuclei in Equation 30.22. These masses are listed in Table A.4 and are given in Equation 30.22. The mass at the start of the reaction is $m_{\text{start}} = m(^{226}_{88}\text{Ra})$ = 226.0254 u, whereas the total mass at the end is $m_{\text{end}} = m(^{222}_{86}\text{Rn}) + m(^{4}_{2}\text{He}) =$ 226.0202 u. The change in mass in the reaction is thus

$$\Delta m = m_{\text{end}} - m_{\text{start}} = 226.0202 - 226.0254 \text{ u} = -0.0052 \text{ u} \quad (30.23)$$

which corresponds (from Eq. 30.18) to an energy $E_{\text{reaction}} = \Delta mc^2$. Using the definition of the mass unit u, we get

$$E_{\text{reaction}} = \Delta mc^2 = -(0.0052 \text{ u})\left(\frac{931.5 \text{ MeV}/c^2}{1 \text{ u}}\right)c^2 = -4.8 \text{ MeV} \quad (30.24)$$

This energy is negative, indicating that, as expected, the nuclei of the reaction products have a lower rest energy than the parent nucleus. This reaction releases 4.8 MeV, typically as the kinetic energy of the products. We'll explore this issue in the next example.

EXAMPLE 30.2	Speed of an Alpha Particle

The alpha decay of $^{226}_{88}$Ra releases 4.8 MeV (Eq. 30.24). If one-tenth of this energy goes into kinetic energy of the emitted alpha particle, what is the speed of the alpha particle? Use the classical relation for the kinetic energy ($KE = \frac{1}{2}mv^2$).

RECOGNIZE THE PRINCIPLE

The kinetic energy of the alpha particle is related to its speed by $KE = \frac{1}{2}mv^2$, so given the kinetic energy we can find v.

SKETCH THE PROBLEM

Figure 30.10B describes the problem.

IDENTIFY THE RELATIONSHIPS

We are given that the alpha particle has a kinetic energy of

$$KE = \frac{1}{2}m_\alpha v^2 = \frac{1}{10}(4.8 \text{ MeV}) = 4.8 \times 10^5 \text{ eV} \quad (1)$$

We must convert this energy to SI units (joules) so that we can obtain v in meters per second. The mass of the alpha particle is given in atomic mass units, and we must also convert this to SI units (kilograms) using Equation 30.11.

SOLVE

Rearranging Equation (1) to solve for v gives

$$v = \sqrt{\frac{2KE}{m_\alpha}} \quad (2)$$

Converting KE to joules gives

$$KE = (4.8 \times 10^5 \text{ eV})\left(\frac{1.60 \times 10^{-19} \text{ J}}{1 \text{ eV}}\right) = 7.7 \times 10^{-14} \text{ J}$$

Converting m_α to kilograms, we have

$$m_\alpha = (4.00 \text{ u})\left(\frac{1.66 \times 10^{-27} \text{ kg}}{1 \text{ u}}\right) = 6.6 \times 10^{-27} \text{ kg}$$

Substituting these values in Equation (2), we find

$$v = \sqrt{\frac{2KE}{m_\alpha}} = \sqrt{\frac{2(7.7 \times 10^{-14} \text{ J})}{6.6 \times 10^{-27} \text{ kg}}} = \boxed{4.8 \times 10^6 \text{ m/s}}$$

What does it mean?
While this speed is only about 2% of the speed of light, the alpha particle still carries a lot of energy compared to the binding energy of electrons in an atom or molecule. For this reason, nuclear decay products can do significant damage when they travel through human tissue (Section 30.4).

Half-life

An important aspect of spontaneous nuclear decay is the decay time; that is, if you are given a $^{226}_{88}\text{Ra}$ nucleus, how long will you have to wait for it to decay by the process in Equation 30.22? Experiments show that individual nuclei decay one at a time, at *random* times. It is not possible to predict when a particular $^{226}_{88}\text{Ra}$ nucleus will decay. This randomness is a feature of quantum theory; there is no classical analogy.

   Although one cannot predict when a particular nucleus will decay, one can give the *probability* for decay. This probability is specified in terms of the **half-life** of the nucleus, $T_{1/2}$. Suppose a total of N_0 nuclei are present at time $t = 0$. Half of these nuclei will decay during a time equal to one half-life $t = T_{1/2}$. Hence, half of the original nuclei will remain at $t = T_{1/2}$ as shown graphically in Figure 30.11. If we then wait until another half-life has passed—that is, until $t = 2T_{1/2}$—the number of original, undecayed, nuclei will fall to $N_0/4$. Half of the nuclei decay during each time interval of length $\Delta t = T_{1/2}$.

   The decay curve in Figure 30.11 is described by the exponential function. We define the **decay constant** λ in such a way that

$$N = N_0 e^{-\lambda t} \tag{30.25}$$

The value of the decay constant for a particular isotope can be determined through experimental measurements of N as a function of time and comparing the results to the exponential function in Equation 30.25. The decay constant λ is closely related to the half-life. From the definition of $T_{1/2}$, the number of nuclei is $N = N_0/2$ at $t = T_{1/2}$. Inserting these values into Equation 30.25 leads to

$$\frac{N_0}{2} = N_0 e^{-\lambda T_{1/2}}$$

$$e^{-\lambda T_{1/2}} = \tfrac{1}{2} \tag{30.26}$$

Taking the natural logarithm of both sides and doing some rearranging leads to

$$\ln(e^{-\lambda T_{1/2}}) = -\lambda T_{1/2} = \ln(1/2) = -\ln 2$$

$$\lambda T_{1/2} = \ln 2$$

$$T_{1/2} = \frac{\ln 2}{\lambda} \approx \frac{0.693}{\lambda} \tag{30.27}$$

using $\ln 2 \approx 0.693$. Equation 30.27 thus relates the value of the decay constant to the half-life. This relation is an inverse one: a large decay constant means a short half-life and vice versa.

   Values of $T_{1/2}$ vary widely; a half-life can be as short as a tiny fraction of a second or longer than the age of the Earth. Values for a few important nuclei are given in Table

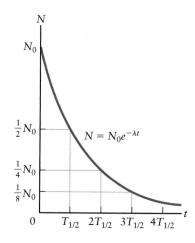

Figure 30.11 Radioactive decay is described by an exponential decay curve. The curve shows the number N of nuclei remaining after a time t. If there are N_0 nuclei at $t = 0$, only half are left at time $t = T_{1/2}$.

Table 30.2 Half-lives of Nucleons and Some Important Nuclei

Nucleus	Half-life
n (a free neutron)	10.4 min
^1_1H (a proton)	Stable
^2_1H (deuterium)	Stable
^3_1H (tritium)	12.33 yr
$^{14}_6C$	5730 yr
$^{60}_{27}Co$	5.27 yr
$^{123}_{53}I$	13.2 h
$^{222}_{86}Rn$	3.82 days
$^{226}_{88}Ra$	1600 yr
$^{235}_{92}U$	7.04×10^8 yr
$^{238}_{92}U$	4.47×10^9 yr

30.2, and a more complete listing is found in Table A.4. The value of $T_{1/2}$ is important in many effects and applications of radioactivity (Sections 30.4 and 30.5).

CONCEPT CHECK 30.5 | ⊗ PET Scans and the Half-Life of $^{20}_9F$

The isotope $^{20}_9F$ is used in a medical procedure called positron emission tomography (PET). The half-life of $^{20}_9F$ is approximately 110 minutes. If your doctor has a sample with 16 g of pure $^{20}_9F$ at $t = 0$, how much $^{20}_9F$ will he have 330 minutes later, (a) 12 g, (b) 8.0 g, (c) 4.0 g, (d) 2.0 g, or (e) 1.0 g?

EXAMPLE 30.3 | Half-life and the Radioactive Decay of Tritium

Tritium is an isotope of hydrogen containing two neutrons and is denoted by ^3_1H. It is a component in some fusion bombs (i.e., "hydrogen bombs") as we'll discuss in the next section. Tritium is radioactive and decays into an isotope of helium (^3_2He) with a half-life of $T_{1/2} = 12.33$ yr. Suppose a fusion bomb contains 10 kg of tritium when it is first assembled. How much tritium (in kilograms) will it contain after 30 years on the shelf?

RECOGNIZE THE PRINCIPLE

If there are N_0 tritium nuclei at $t = 0$, the number remaining after a time t is

$$N = N_0 e^{-\lambda t} \qquad (1)$$

We weren't given the value of λ for tritium, but we can find it from the given half-life using the relation between λ and $T_{1/2}$ in Equation 30.27.

SKETCH THE PROBLEM

Figure 30.11 describes the problem. After a time $T_{1/2}$ passes, half of the radioactive nuclei have decayed.

IDENTIFY THE PRINCIPLES

We first rearrange Equation 30.27 to find the decay constant for tritium in terms of the half-life:

$$\lambda = \frac{\ln 2}{T_{1/2}} = \frac{0.693}{12.33 \text{ yr}} = 0.056 \text{ yr}^{-1}$$

SOLVE

Using this value of λ in Equation (1), we find that the fraction N/N_0 of tritium nuclei that remain after $t = 30$ yr is

$$\frac{N}{N_0} = e^{-\lambda t} = e^{-(0.056 \text{ yr}^{-1})(30 \text{ yr})} = 0.19$$

Hence, only 19% of the original tritium nuclei will remain. The initial mass was 10 kg, so the final mass of tritium is 0.19×10 kg = $\boxed{1.9 \text{ kg}}$.

What does it mean?
Because ^3_2He is not useful for making bombs, the decay of tritium will cause a fusion bomb to stop working after a certain period of time. As a result, fusion bombs containing tritium must be periodically "reloaded."

EXAMPLE 30.4 | Isotope Abundance and $^{235}_{92}U$

The isotope $^{235}_{92}U$ has a half-life for spontaneous fission of 7.0×10^8 yr. In deposits of naturally occurring uranium, this isotope is only about 0.72% of the total fraction of uranium nuclei. Hence, if you had a sample containing 100 g of pure uranium, only 0.72 g would be $^{235}_{92}$U. The Earth is believed to be about 4.5 billion = 4.5×10^9 years old, or about $(4.5 \times 10^9)/(7.0 \times 10^8) \approx 6$ half-lives of $^{235}_{92}$U. How many grams of $^{235}_{92}$U were in your sample when the Earth was formed 4.5×10^9 yr ago?

RECOGNIZE THE PRINCIPLE

As we go back in time, the amount of $^{235}_{92}$U increases by a factor of two for every half-life. Hence, if we go back in time by one half-life, the amount of $^{235}_{92}$U is doubled; if we go back in time another half-life, the amount of $^{235}_{92}$U doubles again, and so on.

SKETCH THE PROBLEM

Figure 30.12 shows how the amount of $^{235}_{92}$U in our sample grows as we go back in time by six half-lives.

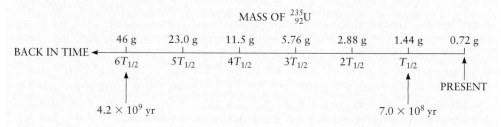

Figure 30.12 Example 30.4. As we go back in time, the amount of $^{235}_{92}$U doubles every $T_{1/2} = 7.0 \times 10^8$ yr.

IDENTIFY THE RELATIONSHIPS AND SOLVE

According to Figure 30.12, the total mass of $^{235}_{92}$U after going back in time six half-lives is approximately $\boxed{46 \text{ g}}$.

We can be a bit more mathematical by noting that after one half-life, the mass increases by a factor of two, after two half-lives it increases by $2^2 = 4$, and after six half-lives it increases by $2^6 = 64$, which again gives

$$m_{start} = 2^6 m_{end} = 2^6(0.72 \text{ g}) = \boxed{46 \text{ g}}$$

What does it mean?
Geophysicists continue to investigate the composition of the Earth when it first formed. At that time, the uranium was more equally apportioned between $^{238}_{92}$U and $^{235}_{92}$U. At present (about six half-lives later), most of the original $^{235}_{92}$U has decayed. Notice that the half-life of $^{238}_{92}$U is about 4.5×10^9 yr, so only about half of the original $^{238}_{92}$U has decayed since the formation of the Earth.

How Do We "Measure" Radioactivity?

The half-life of a nucleus tells how quickly (on average) the nucleus will undergo radioactive decay. Each decay can produce potentially harmful products such as the alpha particle produced in the decay of $^{226}_{88}$Ra (Eq. 30.22). A nucleus with a short half-life is more likely to decay, so it is potentially more dangerous than a nucleus with a long half-life. Since a sample with more radioactive nuclei will be more dangerous than one with a small number of such nuclei, to fully assess such dangers we must also know how many nuclei are present in a sample. For this reason, the "strength" of a radioactive sample is measured using a property called the *activity*, which can be measured with a Geiger counter (Fig. 30.13). The activity of a sample is proportional to the number of nuclei that decay in 1 second. If all other factors are similar, a sample with a high activity is more dangerous than a sample with a low activity. We often say that a sample with a high activity level is "hot."

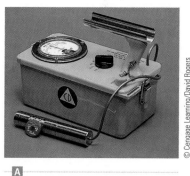

Figure 30.13 A Activity can be measured with a Geiger counter. B Charged particles passing through the detector chamber ionize gas molecules, causing a short pulse of current to flow. Some Geiger counters give an audible "click" to indicate each current pulse.

One common unit of activity is the curie (Ci), which is defined as

$$1 \text{ Ci} = 3.7 \times 10^{10} \text{ decays/s} \tag{30.28}$$

Hence, a sample of radioactive material has an activity of 1 Ci if it exhibits 3.7×10^{10} decays each second; 1 g of naturally occurring radium has an activity of approximately 1 Ci. In practice, a sample with an activity of 1 Ci is extremely dangerous due to the large number of alpha and beta particles and gamma rays that it emits. Most studies or medical procedures with radioactive substances involve samples with activities of millicuries (mCi = 10^{-3} Ci) or microcuries (μCi = 10^{-6} Ci).

The official SI unit of activity is the becquerel (Bq), named in honor of Henri Becquerel (1852–1908), who shared the Nobel Prize in Physics with Marie and Pierre Curie for his discovery of spontaneous radioactivity. The becquerel is defined by

$$1 \text{ Bq} = 1 \text{ decay/s} \tag{30.29}$$

As time passes and nuclei in a sample decay, the number of remaining unstable nuclei drops. Hence, the activity of a sample decreases with time. We thus say (loosely speaking) that a "hot" radioactive sample (such as the nuclear reactor at Chernobyl in Ukraine) becomes "cooler" with time and therefore safer to deal with or clean up.

While a sample with a high activity level is generally more dangerous than one with low activity, there are complicating factors due to the variable amount of energy carried by decay products as we'll discuss in Section 30.4.

CONCEPT CHECK 30.6 | How Does the Activity Level of a Radioactive Sample Change with Time?

Consider a radioactive material that decays with a half-life of $T_{1/2}$. How long does it take for the activity of the sample to decrease by a factor of four, (a) $T_{1/2}$, (b) $2T_{1/2}$, (c) $3T_{1/2}$, or (d) $4T_{1/2}$?

EXAMPLE 30.5 | Activity of Tritium

You are given a sample containing exactly 1 mole of tritium ($^3_1$H). The tritium decays with a half-life of 12.33 yr. What is the activity of the sample? Express your answer in units of decays per second. *Hint:* Inserting $t = 1$ s in the exponential decay law $N = N_0 e^{-\lambda t}$ gives the number of nuclei remaining after 1 s. The activity is the number of nuclei that *decay* in the same 1-s time period.

RECOGNIZE THE PRINCIPLE

If a sample initially contains N_0 nuclei, the number remaining after $t = 1$ s is

$$N_{\text{remaining}} = N_0 e^{-\lambda t}$$

Hence, the number that decay during this time is

$$N_{\text{decays}} = N_0 - N_0 e^{-\lambda t} = N_0(1 - e^{-\lambda t}) \tag{1}$$

The activity equals the number of nuclei that decay per second.

SKETCH THE PROBLEM

Figure 30.14 shows how the number of nuclei remaining decreases with time as well as the number of nuclei that decay in $t = 1$ s.

IDENTIFY THE RELATIONSHIPS

To evaluate Equation (1) for our given tritium sample, we must find both N_0 and the decay constant λ. The sample originally contains exactly 1 mole of tritium, so $N_0 = 6.02 \times 10^{23}$ (Avogadro's number). From Equation 30.27, the decay constant is related to the half-life through $\lambda = 0.693/T_{1/2}$.

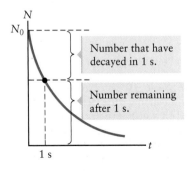

Figure 30.14 Example 30.5. Not to scale.

Inserting the known half-life of tritium (12.33 yr) gives

$$\lambda = \frac{0.693}{T_{1/2}} = \frac{0.693}{12.33 \text{ yr}} = \frac{0.693}{3.9 \times 10^8 \text{ s}} = 1.8 \times 10^{-9} \text{ s}^{-1}$$

SOLVE

Inserting our values of N_0 and λ as well as $t = 1$ s into Equation (1) leads to

$$N_{\text{decays}} = N_0(1 - e^{-\lambda t}) = (6.02 \times 10^{23})[1 - e^{-(1.8 \times 10^{-9} \text{ s}^{-1})(1 \text{ s})}]$$

The exponential factor is very close to one, so we have to keep extra significant figures in performing this calculation. We find

$$\text{activity} = N_{\text{decays}} = 1.1 \times 10^{15} \text{ per second} = \boxed{1.1 \times 10^{15} \text{ Bq}}$$

What does it mean?
Nuclei with a shorter half-life decay more rapidly (and have a larger value of λ) than those with a long half-life. Hence, when comparing two samples, the one with a short half-life will have a higher activity (and hence be more dangerous) than the one with a long half-life, assuming the number of nuclei in the two samples is the same.[3]

30.3 | STABILITY OF THE NUCLEUS: FISSION AND FUSION

To be stable, a nucleus containing two or more protons must contain neutrons. Neutrons are thus necessary for the stability of essentially all matter. Figure 30.15 shows a plot of the neutron number N as a function of the number of protons Z (i.e., the atomic number) for all known stable nuclei. The dashed line shows the function $N = Z$; nuclei that lie on this line have equal numbers of neutrons and protons, while nuclei above this line contain an excess of neutrons. Low-mass nuclei such as He and C tend to have equal numbers of protons and neutrons and are thus on or near the dashed line corresponding to $N = Z$ in Figure 30.15. Heavy nuclei, however, have many more neutrons than protons; for example, Pb (lead) nuclei contain around 40 more neutrons than protons.

The reason these additional neutrons are needed can be seen by considering the forces between protons and neutrons. Figure 30.16A shows a highly schematic nucleus with two protons and two neutrons, representing $_2^4\text{He}$. This nucleus is extremely stable, so this arrangement of protons and neutrons is particularly favored. We now add another proton to make lithium (Li). To bind this additional proton and achieve a stable arrangement of protons and neutrons similar to $_2^4\text{He}$, we need two more neutrons as shown in Figure 30.16B. This highly simplified picture leaves out the possibility that the protons in Li might rearrange to better "share" the attraction of the neutrons, just as two atoms share electrons when they form a chemical bond. In reality, $_3^6\text{Li}$ and $_3^7\text{Li}$ are both stable isotopes. However, this simple picture does explain why nuclei with large numbers of protons (high Z) require greater numbers of neutrons to be stable, accounting for the trend that $N > Z$ for heavy nuclei in Figure 30.15. This trend is essential for understanding nuclear fission.

Nuclear Binding Energy

We have discussed how the mass of a nucleus is related to its binding energy, and as an example we calculated the binding energy of an alpha particle (Eq. 30.21). We

[3]Notice also that our calculation gives the *average* activity during a 1-s interval, but because the half-life here is much greater than 1 s, the *instantaneous* activity will have almost exactly the same value.

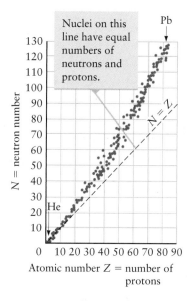

Figure 30.15 Stable nuclei usually have a greater number of neutrons than protons, so $N > Z$ for all but the lightest nuclei. This trend is especially noticeable at high Z.

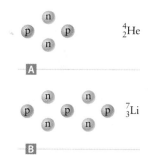

Figure 30.16 Ⓐ The isotope $_2^4\text{He}$ contains two protons and two neutrons. This nucleus is stable because the neutrons can (roughly speaking) sit between the protons, holding the nucleus together. (Compare with Fig. 30.5.) Ⓑ If a third proton is added, more neutrons are needed to keep the protons apart and provide the needed binding due to the strong force. As the number of protons grows, extra neutrons are needed. For this reason, $_3^7\text{Li}$ (pictured here) is more stable than $_3^6\text{Li}$.

found that the mass of the nucleus $^4_2$He, which contains two protons and two neutrons, is less than the total mass of two isolated protons plus two isolated neutrons. The same is true for other stable nuclei; the mass of a nucleus containing N neutrons and Z protons is *less* than the total mass of N isolated neutrons plus Z isolated protons. This mass difference is connected to the binding energy of the nucleus through the relativistic relation between mass and energy ($E = mc^2$). By measuring the mass of a nucleus and then subtracting the mass of the same numbers of isolated protons and neutrons, we can find the binding energy of the nucleus. This approach has been used to find the binding energy of every nucleus in Figure 30.15, and the results for the binding energy per nucleon are shown in Figure 30.17. In the figure, the horizontal axis is mass number A. Different isotopes (nuclei having the same Z but different A) can have different binding energies; here we plot an approximate average value for each element, that is, for each value of A. Values of the binding energy are plotted as negative because each of these nuclei has a lower energy than a collection of separated protons and neutrons. The magnitude of the binding energy is largest for nuclei close to $A = 60$, which is near Fe and Ni in the periodic table, meaning that these nuclei are the *most stable* ones. This fact and the relationship between binding energy and Z are the basis for two important nuclear phenomena: fission and fusion.

In our other discussions of binding energies in previous chapters, including the binding energy of an atom, the energy of a stable bound system has also been negative. For example, the gravitational potential energy of two masses is negative (Eq. 6.22), and the energy of a bound atom such as hydrogen is also negative (Fig. 29.12). When an attractive force exists between two or more particles, you must add energy to separate them. Hence, binding lowers the potential energy relative to the state in which the particles are at infinite separation.

Nuclear Fission

Figure 30.9 shows how a parent nucleus (in that case, $^{238}_{92}$U) can undergo spontaneous radioactive decay to form a series of daughter nuclei. Radioactive decay can also be *induced* by a collision between two nuclei or by the collision of a neutron with a nucleus. A particularly important example occurs when $^{235}_{92}$U collides with a neutron. A typical reaction of this kind is

> one neutron induces
> fission of $^{235}_{92}$U

> three neutrons
> are produced

$$^1_0n + {}^{235}_{92}U \rightarrow {}^{139}_{56}Ba + {}^{94}_{36}Kr + 3{}^1_0n \qquad (30.30)$$

Nuclear fission

Here a neutron is denoted by $^1_0$n since it "contains" no protons ($Z = 0$) and has a total of one nucleon ($A = 1$). The process in Equation 30.30 is called **nuclear fission**, a reaction whose main products are two nuclei both roughly half the size of

Figure 30.17 Binding energy per nucleon for nuclei with different mass numbers. A binding energy of high magnitude corresponds to a tightly bound nucleon, so nuclei falling in the shaded region of the graph are the most stable. The binding energy is negative because a nucleus is more stable and has a lower energy than a collection of separated protons plus neutrons.

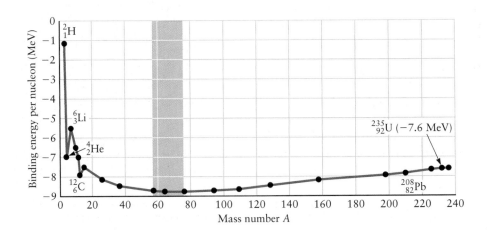

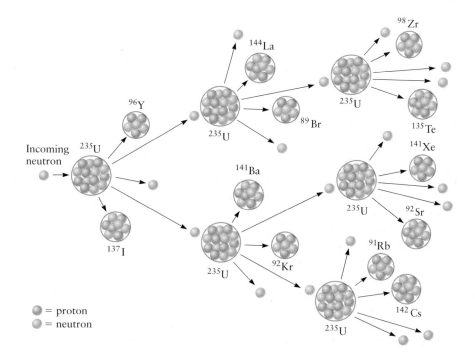

Figure 30.18 The fission of $^{235}_{92}U$ can be initiated by a collision with a neutron. Here the original $^{235}_{92}U$ nucleus splits into two fission fragments, $^{96}_{39}Y$ and $^{137}_{53}I$, along with three neutrons. These neutrons then collide with other $^{235}_{92}U$ nuclei and cause them to fission. This is an example of a chain reaction.

the original parent nucleus. (By contrast, in alpha and beta decay one of the decay products is very close in mass to the original parent nucleus.)

Fission reactions have several important features. First, there can be many different possible reaction products. For example, when $^{235}_{92}U$ collides with a neutron, it may produce $^{139}_{56}Ba$ and $^{94}_{36}Kr$ as in Equation 30.30, but it may also produce a different pair of nuclei. Another possible fission reaction involving $^{235}_{92}U$ is

$$^{1}_{0}n + {}^{235}_{92}U \rightarrow {}^{95}_{38}Sr + {}^{138}_{54}Xe + 3{}^{1}_{0}n$$

As in the reaction in Equation 30.30, the collision with a neutron causes $^{235}_{92}U$ to split into two smaller nuclei and also emit several neutrons. There are approximately 90 different fission reactions involving $^{235}_{92}U$.

A second key feature of fission is that it produces more free neutrons than it uses. For example, the reaction in Equation 30.30 begins with one neutron, but there are three neutrons at the end. Some fission reactions with $^{235}_{92}U$ emit three neutrons, whereas others emit two or four. On average, the fission of a single $^{235}_{92}U$ nucleus caused by the collision with a neutron results in about 2.5 neutrons at the end of the reaction. These neutrons can then induce the fission of other nearby $^{235}_{92}U$ nuclei, leading to a ***chain reaction*** as sketched in Figure 30.18. In this way, the fission of a single $^{235}_{92}U$ nucleus can induce the fission of an enormous number of additional $^{235}_{92}U$ nuclei. Each fission event produces additional free neutrons because the ratio of neutrons to protons in a heavy nucleus such as $^{235}_{92}U$ is greater than in either of the two nuclei that are produced (e.g., $^{139}_{56}Ba$ and $^{94}_{36}Kr$ in Eq. 30.30). This fact can be traced to the shape of the stability curve in Figure 30.15.

A third important feature of fission reactions is that they release energy as can be seen from the binding energy curve in Figure 30.17. The magnitude of the binding energy for $^{235}_{92}U$ is about 7.6 MeV/nucleon, while the magnitudes of the binding energies of $^{139}_{56}Ba$ and $^{94}_{36}Kr$ (the fission products in Eq. 30.30) are above 8.0 MeV/nucleon. This fission reaction releases the extra binding energy in the form of gamma rays and through the kinetic energies of the particles produced (Fig. 30.19). The shape of the binding energy curve guarantees that the magnitude of the binding energy per nucleon of $^{235}_{92}U$ is always smaller in magnitude than the binding energies of its fission products. The same is true for other nuclei that undergo fission such as $^{239}_{94}Pu$ (plutonium) . That is why fission produces large amounts of energy.

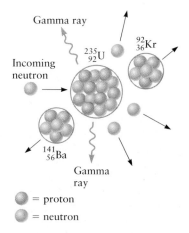

Figure 30.19 When a nucleus undergoes fission, some gamma ray photons are usually emitted along with the two fission fragments and neutrons. Typically, these particles all have large kinetic energies.

CONCEPT CHECK 30.7 | Fission Reaction of $^{235}_{92}\text{U}$

Consider the fission reaction

$$^{235}_{92}\text{U} + ^{1}_{0}\text{n} \rightarrow ^{144}_{56}\text{Ba} + ^{90}_{36}\text{Kr} + ?^{1}_{0}\text{n}$$

How many neutrons does it produce?

Energy from Fission

Nuclear fission is the basis for two modern technologies: fission weapons (bombs) and nuclear reactors for producing electrical power. Both are based on nuclear chain reactions like the one in Figure 30.18. In a nuclear fission bomb, the chain of fission reactions is allowed to run out of control, rapidly releasing extremely large amounts of energy.

Figure 30.18 implies that it's easy to set off an uncontrolled chain reaction: just fire one neutron at a sample of $^{235}_{92}\text{U}$ and the 2 to 4 neutrons (2.5 on average) produced in the initial fission event go on to split other $^{235}_{92}\text{U}$ nuclei, releasing enough energy to cause a huge explosion. Fortunately, it is not that easy. First of all, natural uranium is more than 99% $^{238}_{92}\text{U}$, an isotope that does not readily fission. The uranium used in weapons is "enriched" to increase the concentration of scarce $^{235}_{92}\text{U}$, and enrichment is a very difficult engineering process. Second, if the piece of uranium is small, many of the nuclei are close enough to the surface that many of the released neutrons escape from the sample before inducing another fission event. In such a case, the reaction may stop before a significant number of $^{235}_{92}\text{U}$ nuclei undergo fission. Increasing the size (and hence also the mass) of a uranium sample lowers the surface-to-volume ratio and lowers the probability that a free neutron will reach the surface; it is then more likely that the neutron will encounter a $^{235}_{92}\text{U}$ nucleus and induce fission. The minimum amount of nuclear fuel material needed to sustain a chain reaction is called the *critical mass*. For uranium, the critical mass is typically about 50 kg (corresponding to a uranium sphere with a radius of about 9 cm), but depending on how the sample is shaped and how the surface is prepared it can be as small as 15 kg.

When $^{235}_{92}\text{U}$ undergoes fission, it produces two nuclei of roughly equal mass and several neutrons as described in Equation 30.30. We can estimate the amount of energy released using the binding energy curve in Figure 30.17. The binding energy per nucleon of $^{235}_{92}\text{U}$ is about -7.6 MeV, whereas for many of the nuclei produced by fission it is about -8.5 MeV per nucleon. The difference is $(-8.5 + 7.6)$ MeV $= -0.9$ MeV per nucleon. Since there are 235 nucleons in one $^{235}_{92}\text{U}$ nucleus, the total amount of energy released by the fission of one such nucleus is (0.9 MeV) $\times$ 235 $\approx$ 210 MeV. To get a feeling for the scale of this value, let's calculate how much energy we can expect from the fission of 1 g of pure $^{235}_{92}\text{U}$ (about the mass of a paper clip).

The atomic mass of pure $^{235}_{92}\text{U}$ is 235 g (from the definition of the atomic mass unit), so 235 g of $^{235}_{92}\text{U}$ contains 1 mole of nuclei. Our 1-g sample thus contains

$$N_{\text{U-235}} = (1 \text{ g}) \times \frac{6.02 \times 10^{23}}{235 \text{ g}} = 2.6 \times 10^{21} \text{ nuclei}$$

Each nucleus releases about 210 MeV when it undergoes fission, so the total energy released by the fission of 1 g of $^{235}_{92}\text{U}$ is

$$E_{\text{fission}} = (2.6 \times 10^{21})(210 \text{ MeV})\left(\frac{10^6 \text{ eV}}{1 \text{ MeV}}\right)\left(\frac{1.60 \times 10^{-19} \text{ J}}{1 \text{ eV}}\right)$$

$$E_{\text{fission}} \approx 9 \times 10^{10} \text{ J per gram of } ^{235}_{92}\text{U} \tag{30.31}$$

This is an *enormous* amount of energy. By comparison, 1 g of the chemical explosive TNT releases about 4200 J. The energy released by 1 g of $^{235}_{92}\text{U}$ is more than 10 million times larger!

The amount of energy released in a fission bomb is usually measured in terms of the mass of TNT that would release the same amount of energy. The fission bombs that were exploded in World War II each released an amount of energy equivalent to more than 10,000 *tons* of TNT.

| EXAMPLE 30.6 | Energy Released in a Fission Bomb |

The fission bomb that was dropped on the Japanese city of Hiroshima in 1945 contained approximately 64 kg of $^{235}_{92}U$ and released an amount of energy equivalent to the explosion of approximately 15,000 tons (1.5×10^{10} g) of TNT (called a "15-kiloton bomb"). Did all this $^{235}_{92}U$ undergo fission? If not, what is the mass of $^{235}_{92}U$ that did?

RECOGNIZE THE PRINCIPLE

In our analysis of fission, we found that the energy released by the fission of 1 g of $^{235}_{92}U$ is about 9×10^{10} J (Eq. 30.31). We also noted that 1 g of TNT releases about 4200 J. We can use these values to compute what mass of $^{235}_{92}U$ is needed to give the energy released by 15,000 tons of TNT.

SKETCH THE PROBLEM

Fission releases energy in the form of the kinetic energy of the reaction products plus the energy of gamma rays that are emitted (Fig. 30.19).

IDENTIFY THE RELATIONSHIPS

Let's first find the energy released by 15,000 tons = 1.5×10^{10} g of TNT. From the given information, we have

$$E = (1.5 \times 10^{10} \text{ g})\left(\frac{4200 \text{ J}}{\text{g}}\right) = 6.3 \times 10^{13} \text{ J}$$

SOLVE

Using the value of $E_{\text{fission}} = 9 \times 10^{10}$ J/g of $^{235}_{92}U$ from Equation 30.31, this energy corresponds to

$$\text{mass of } ^{235}_{92}U = (6.3 \times 10^{13} \text{ J})\left(\frac{1 \text{ g of } ^{235}_{92}U}{9 \times 10^{10} \text{ J}}\right) = \boxed{700 \text{ g}}$$

What does it mean?

Only about 1% of the $^{235}_{92}U$ in the Hiroshima weapon actually underwent fission. Even so, an enormous amount of energy was released. So-called modern nuclear weapons yield much greater amounts of energy.

Nuclear Power Plants

Nuclear fission is used in a productive and controlled way in nuclear power plants. In a power reactor, fission takes place in a region called the reactor core, shown schematically in Figure 30.20A. The important components of the core are fuel rods, control rods, and moderators.

Fuel rods contain $^{235}_{92}U$ that can be added or removed from the core while the reactor is operating. The fuel rods are removed when the $^{235}_{92}U$ is exhausted or when operators need to reduce the fission rate. *Control rods* contain materials (e.g., $^{113}_{48}Cd$) that absorb neutrons. They can be inserted or removed from the reactor core, allowing the reactor operators to control the number of free neutrons and thus adjust the fission rate. The *moderator* (often water) circulates through the reactor core. One of the moderator's functions is to slow the neutrons down, which increases their probability of inducing a fission event. (Roughly speaking, a slow neutron passing near a uranium nucleus has more time to induce fission.) The water also carries away heat

Insight 30.4

SAFETY AT NUCLEAR FISSION REACTORS

Safety is always a major concern at a nuclear fission reactor, particularly those used as power reactors to generate electricity. All reactors are designed with extensive safety features to prevent a "meltdown" of the reactor core or any similar hazard. The latest reactor designs contain many "passive" safety features that do not require operator action or even electronic feedback to deal with emergency situations. For example, these safety features work even in the event of a complete electrical failure, making these reactors extremely safe.

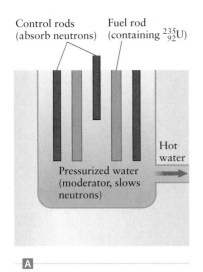

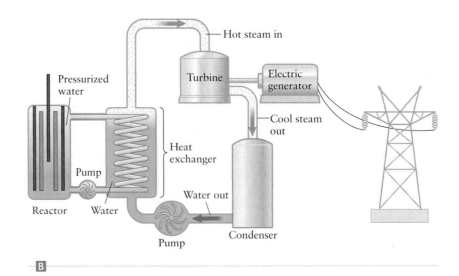

Control rods (absorb neutrons)
Fuel rod (containing $^{235}_{92}$U)

Hot water

Pressurized water (moderator, slows neutrons)

A

Pressurized water

Hot steam in

Turbine

Electric generator

Cool steam out

Heat exchanger

Pump

Reactor   Water

Water out

Pump

Condenser

B

Figure 30.20 ◭ Schematic of the core of a nuclear reactor. The fuel rods contain $^{235}_{92}$U (or some other fissionable nuclei). Neutrons emitted during fission collide with nuclei in the moderator (here, water), slowing the neutrons. Control rods containing nuclei that readily absorb neutrons are used to control the rate of fission. ◪ Heat generated in the moderator (water) is used to power a steam turbine and thereby generate electric power.

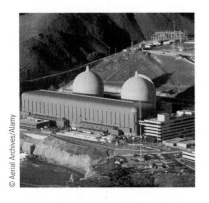

Figure 30.21 The steam engines in a nuclear power reactor must cool the steam as part of the thermodynamic cycle (Chapter 16). This cooling is accomplished with large cooling towers or with a source of cool water from an ocean or river. This photo shows the Diablo Canyon Nuclear power plant in California. The two large domes each house a reactor, and the cooling water is taken from the ocean.

to a separate steam engine (Fig. 30.20B), where the energy generated by the reactor is converted to electrical energy using a generator (Figs. 21.7 and 22.3). This heat engine operates according to the principles of thermodynamics, so it requires a cold reservoir. Nuclear power plants are often located near bodies of water such as lakes that provide this cold reservoir (Fig. 30.21).

As mentioned previously, naturally occurring uranium contains a mixture of the isotopes $^{235}_{92}$U (0.7%) and $^{238}_{92}$U (99.3%). The fission reactions in a reactor involve only $^{235}_{92}$U, so the uranium in fuel rods is usually enriched to increase the concentration of $^{235}_{92}$U. However, this concentration is always kept low enough that the fuel rod does not exceed the critical mass, so it is not possible for the core of a nuclear reactor to undergo an uncontrollable chain reaction and explode like a nuclear fission bomb. It is possible, though, to have an accident called a "meltdown" in which the fuel rods become so hot that they and the walls of the reactor melt and the water in the core rapidly turns to steam. The result can be a conventional explosion (due to the high-pressure steam), as occurred at Chernobyl in Ukraine in 1986. This accident allowed the escape of significant amounts of radioactive material into the environment and still affects the safety and health of nearby residents.

Nuclear Fusion

In **nuclear fusion**, two nuclei join together to produce one new particle. Consider two $^6_3$Li nuclei; they have $Z = 3$ and are on the left side of the binding energy curve in Figure 30.17. If they fuse, the new nucleus has $Z = 6$ and contains a total of $A = 12$ nucleons, which is $^{12}_6$C. According to Figure 30.17, $^{12}_6$C has a greater binding energy per nucleon (in magnitude) than $^6_3$Li and is thus more stable. Hence, this fusion reaction releases energy:

Nuclear fusion

$$^6_3\text{Li} + {}^6_3\text{Li} \rightarrow {}^{12}_6\text{C} + \text{energy} \qquad (30.32)$$

This energy is in the form of gamma rays plus the kinetic energy of the $^{12}_6$C nucleus. We can use Figure 30.17 to estimate how much energy this reaction will release. The binding energy per nucleon is about -5.4 MeV for $^6_3$Li and about -7.9 MeV for $^{12}_6$C. Hence, the energy *released* in this fusion reaction is approximately $7.9 - 5.4 = 2.5$ MeV per nucleon. The fission of $^{235}_{92}$U releases approximately 0.9 MeV per nucleon, so this fusion reaction releases even more energy (per nucleon) that fission, which is one reason (unfortunately) bombs employing fusion are capable of even more explosive power than fission weapons. (We'll explore this topic more in Question 10 at the end of this chapter.)

The Sun's Power Comes from Fusion

Nuclear fusion is the process that powers stars such as our Sun. When first formed, a typical star consists mainly of hydrogen—that is, protons and electrons—with a small amount of ^4_2He and much smaller amounts of heavier elements. Through a series of reactions, these protons fuse to produce ^4_2He (alpha particles), which in turn fuse to form heavier elements. Our own Sun is hot enough to produce elements up to about carbon in this way. All the heavier elements in the universe were created in either hotter stars or in supernova explosions, which can trigger even more energetic fusion reactions.

Fusion reactions can take place in a star due to the very high temperatures and pressures in its core. These conditions cause nuclei to come into very close contact when they collide, which is necessary for fusion to occur. To create a fusion reaction on Earth, we must approach or duplicate stellar conditions, giving the initial nuclei very high speeds and keeping them very close together. In a fusion bomb, this is accomplished using a fission bomb as a trigger. That is, a bomb that uses the fission of $^{235}_{92}\text{U}$ or some other nucleus is exploded in such a way that the fusion ingredients are compressed very rapidly, initiating a fusion reaction.

Fusion has one potential useful application on the Earth: as the basis for a power plant. Current (but so far hypothetical) designs for a fusion power reactor use ^2_1H (deuterium) or ^3_1H (tritium) as the starting nuclei in a fusion reaction that releases very large amounts of energy to produce electrical power. The fuel for this reactor could be obtained from water since naturally occurring water has a small but significant concentration of deuterium. The fuel for a fusion reactor would thus be very inexpensive and virtually limitless. Moreover, current designs for these reactors would produce virtually no harmful radiation or by-products. Although a practical fusion power plant has not yet been built, several exploratory engineering projects have been. The latest is the ITER (International Tokomak Engineering Reactor), which uses what is called "tokomak" geometry for confining the nuclei while they undergo fusion. The ITER will become operational in 2016 and should produce 500 MW (500,000,000 W) of power during 10-minute fusion periods, during which it consumes about 0.5 g of a deuterium–tritium mixture. The ITER will still be an experimental project, but the long-term hope is that it will lead to the construction of practical fusion-based power plants that would provide abundant amounts of cheap and clean electrical energy.

EXAMPLE 30.7 | Energy of a Fusion Reaction

A fusion reaction that might be used in a power reactor involves deuterium, ^2_1H:

$$^2_1\text{H} + {}^2_1\text{H} \rightarrow {}^3_1\text{H} + {}^1_1\text{H} \tag{1}$$

In this reaction, two deuterium nuclei fuse to form ^3_1H plus a proton, ^1_1H. Use the masses of these nuclei from Table A.4 to calculate the energy released in this reaction. Express the answer in MeV.

RECOGNIZE THE PRINCIPLE

This reaction converts some of the mass of the original nuclei into energy. The amount of energy can be found from the mass–energy relation of special relativity $E = \Delta mc^2$, where Δm is the difference in the mass of the masses on the left- and right-hand sides of Equation (1).

SKETCH THE PROBLEM

No sketch is necessary, but it is useful to rewrite Equation (1) to also show the rest masses of the nuclei; see Equation (2) below.

IDENTIFY THE RELATIONSHIPS

According to Table A.4, the masses in Equation (1) are

$$m(^1_1\text{H}) = 1.007825 \text{ u}$$

$$m(^2_1\text{H}) = 2.014102 \text{ u}$$

$$m(^3_1\text{H}) = 3.016049 \text{ u}$$

We include extra significant figures because the change in mass in Equation (1) will be small, but will still lead to a substantial energy.

SOLVE

Let's rewrite Equation (1) to show the masses before and after the reaction:

$$^2_1\text{H} \quad + \quad ^2_1\text{H} \quad \rightarrow \quad ^3_1\text{H} \quad + \quad ^1_1\text{H}$$

$$2.014102 \text{ u} + 2.014102 \text{ u} \rightarrow 3.016049 \text{ u} + 1.007825 \text{ u}$$

$$4.028204 \text{ u} \rightarrow 4.023874 \text{ u} \qquad (2)$$

The mass difference is thus

$$\Delta m = m(\text{right side}) - m(\text{left side}) = -0.004330 \text{ u} \qquad (3)$$

which corresponds to an energy of

$$E = \Delta m c^2 \qquad (4)$$

From Equation 30.11, atomic mass units can be converted to units of MeV/c^2 using the relation 1 u = 931.5 MeV/c^2. From Equations (2) and (3), we get

$$E = (-0.004330 \text{ u})\left(\frac{931.5 \text{ MeV}/c^2}{\text{u}}\right)c^2 = \boxed{-4.033 \text{ MeV}}$$

The negative value means that this is the energy *released* by the reaction.

What does it mean?
Since deuterium occurs naturally in ordinary water, this source of energy is potentially cheap. We will explore this topic in the end-of-chapter problems.

30.4 | ⊗ BIOLOGICAL EFFECTS OF RADIOACTIVITY

The biological effects of radioactivity result from the way decay products or reaction products interact with atoms and molecules. From Chapter 29, the typical binding energy of an electron in an atom is on the order of 10 eV. The energy released in a nuclear reaction is much larger (typically several or even hundreds of MeV), often appearing as the kinetic energy of an emitted alpha or beta particle. If one of these particles collides with an atomic electron or with a molecule, there is ample energy to eject the electron from the atom or to break a chemical bond in molecules such as DNA, making these molecules biologically inactive or even harmful. That is how radiation damages living tissue.

The amount of damage that a particular particle is capable of doing is a complicated issue because alpha, beta, and gamma radiation have different masses and charges and therefore interact with tissue in different ways. We defined the curie (Ci) and becquerel (Bq) as units of activity that measure the number of radioactive decays per unit time in Equations 30.28 and 30.29. Some particles produced in radioactive decay carry a great deal of kinetic energy, while others carry little. To account for this difference, we introduce a unit called the ***rad***. By definition, 1 rad is the amount of radiation that deposits 10^{-2} J of energy into 1 kg of absorbing material. Hence, this unit accounts for both the amount of energy carried by the particle and the efficiency with which this energy is absorbed.

The name "rad" denotes *r*adiation *a*bsorbed *d*ose. The definition of the rad and the term *dose* both indicate that the total amount of radiation received is important when assessing biological effects, but even the rad unit does not capture the full complexity of radiation damage. Different types of particles can do different amounts of biological damage even if they deposit the same amount of energy and hence have the same strength as measured in rads. For this reason, another unit called the *relative biological effectiveness* factor is also very useful. This factor, denoted ***RBE***, measures how efficiently a particular type of particle damages tissue. Some typical *RBE* factors for various kinds of radiation from nuclear reactions are given in Table 30.3. Notice that the *RBE* value tends to increase as the mass of the particle increases. For example, the *RBE* of an alpha particle (a helium nucleus) is greater than the *RBE* of a beta particle (an electron or positron).

Another widely used measure of radiation experienced by biological organisms is the unit called the ***rem***, which is defined as[4]

$$\text{dose in rem} = (\text{dose in rads}) \times (RBE) \qquad (30.33)$$

Since the rem includes the total amount of energy delivered by the radiation (measured in rads) and the efficiency with which this radiation damages real tissue, it combines the usefulness of both the rad and the *RBE* factor. The rem is the most widely used unit in discussions of radiation exposure and damage.

Table 30.3 Relative Biological Effectiveness (*RBE*) for Different Types of Radiation

RADIATION	RBE
Alpha particles	10–20
Beta particles	1.0–1.7
Gamma rays and X-rays	1.0
Slow neutrons	4–5
Protons	5
Fast neutrons	10
Heavy ions	20

Definition of the rem

EXAMPLE 30.8 ⊗ Comparing the Dose from Different Types of Radiation

A person is exposed to a radiation dose of 0.25 mrad (millirad) in a chest X-ray. What dose of fast neutrons would produce the same amount of tissue damage?

RECOGNIZE THE PRINCIPLE

The amount of damage done by radiation is determined by the product of the dose (which can be measured in rads) and the *RBE* (relative biological effectiveness):

$$\text{radiation damage} = \text{dose} \times RBE \qquad (1)$$

Hence, to compare the damage done by X-rays and fast neutrons, we must take their *RBE* values into account.

SKETCH THE PROBLEM

No sketch is necessary.

IDENTIFY THE RELATIONSHIPS

From Table 30.3, the *RBE* for X-rays is 1.0, while for fast neutrons we find *RBE* = 10. Hence, to achieve the same damage with these two types of radiation, the fast neutron dose must be 10 times smaller than the X-ray dose.

SOLVE

If a dose of fast neutrons produces the same amount of damage as a 0.25-mrad chest X-ray, we can use Equation (1) to write

$$\text{dose(X-rays)} \times RBE(\text{X-rays}) = \text{dose(fast neutrons)} \times RBE(\text{fast neutrons})$$

---

[4]The name "rem" stands for *r*öntgen *e*quivalent in *m*an. Wilhelm Röntgen was the discoverer of X-rays.

which leads to

$$\text{dose(fast neutrons)} = \frac{RBE(\text{X-rays})}{RBE(\text{fast neutrons})} \times \text{dose(X-rays)}$$

Inserting RBE values along with the given value of the X-ray dose, we find

$$\text{dose(fast neutrons)} = \frac{1.0}{10} \times (0.25 \text{ mrad}) = \boxed{0.025 \text{ mrad}}$$

What does it mean?

When comparing the radiation damage from different sources, one must always take the RBE of the radiation into account.

Some Facts about Radiation Damage

The physiology and cell biology of radiation damage is a complex and interesting subject. Here we can only mention some of the important facts.

1. When the radiation dose is low, cells are sometimes able to repair the damage, especially if the dose is absorbed over a long period of time. Hence, small amounts of radiation generally do not cause significant harm to living cells.
2. If the radiation dose is very large, cells can be completely destroyed (cell death).
3. At intermediate radiation doses, cells survive but often malfunction as a result of the damage or in their attempt to repair the damage. A typical result is that the affected cells reproduce in an uncontrolled fashion, leading to cancer.
4. Radiation damage is usually most severe for quickly dividing cells. Many types of blood and bone marrow cells fall into this category, along with cells in an embryo or fetus. Cancerous cells are also quickly dividing, so radiation can be used as a tool to selectively destroy cancer cells (Section 30.5).
5. The amount of damage can depend strongly on where the radiation source is located. For example, alpha particles have a very short range and can be stopped by a sheet of paper. So, if a source of alpha particles is outside the body, the particles will be stopped in the outer layer of skin and do relatively little damage. On the other hand, if a person somehow ingests an alpha source, the source may become lodged in his or her lungs or some other place, and we say that the person is **contaminated**. In this case, the alpha particles can do a great deal of damage to nearby cells, damaging their DNA and so forth.

Sadly, many of the early scientists who studied radioactivity suffered from their exposure to hazardous amounts of radiation. Röntgen (the discoverer of X-rays) died from bone cancer, and Marie Curie died from leukemia that was probably caused by her exposure to many radioactive substances. Her husband, Pierre, also suffered from radiation-induced illnesses (although he died after being run over by a horse-drawn carriage; go figure). In fact, the Curies' scientific papers are reportedly still quite radioactive.

People who work with radiation are normally required to use radiation monitors to record their exposure. These monitors typically contain a piece of photographic film (Fig. 30.22). Exposure to radiation exposes the film, and the degree of film exposure measures the radiation dose. In the event of an accidental overdose, the worker can then quickly get the appropriate medical treatment.

Sources of Radioactivity in Everyday Life

Table 30.4 lists the radiation doses of several common procedures and shows that we are all exposed to radiation during the course of everyday life. Even so, the

Figure 30.22 The radiation and particles emitted by radioactive substances cause photographic film to be exposed. To monitor their exposure to harmful radiation, people who work with radioactive materials wear small "film badges."

Table 30.4 Radiation Doses in Perspective (Typical Values)

ACTIVITY	DOSE (mrem)
Airline flight from New York City to Los Angeles	3
Dental X-ray	10
Chest X-ray	5–10
Mammogram	50–100
Approximate annual dose from natural background sources	300
Recommended maximum annual dose in addition to background dose	500
Apollo XVI astronauts	500
1-in-30 risk of cancer	10,000
Radiation sickness possible	60,000

benefits from, for instance, a chest X-ray, dental X-ray, or mammogram usually far outweigh possible risks.

There are several natural sources of radiation. Cosmic rays—a collection of many different types of particles that come to the Earth from outer space—are a significant source, especially at high altitudes. They are responsible for the increased radiation dose absorbed by astronauts and by persons at altitude (e.g., during an airplane flight or in a city such as Denver).

One of the largest natural sources of radiation exposure involves radon gas produced by the radioactive decay of $^{238}_{92}U$. There are very small concentrations of $^{238}_{92}U$ in the soil and rocks at virtually all places on this planet. These nuclei decay with a half-life of 4.5×10^9 yr, initiating the series of decays in Figure 30.9. One of these daughter nuclei is $^{222}_{86}Rn$ (radon), and being a gas, it diffuses through pores in the surrounding soil or rock. Radon gas thus seeps into mines and into the lungs of miners. These $^{222}_{86}Rn$ nuclei then decay further, producing nuclei that are absorbed in the lining of the lungs of anyone who breathes the gas, leading to alpha and beta radiation that can ultimately cause lung cancer. Radon gas is a major health hazard for miners, and in some homes. It can seep into basements and crawl spaces, where it can collect to unsafe levels, just as in mines. Fortunately, there are simple ways to remediate this hazard, typically by improving the ventilation in the mine or home (Fig. 30.23).

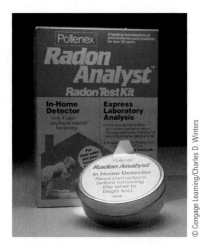

Figure 30.23 Simple radon detector for home use. This detector contains a small canister of activated charcoal, which absorbs radon gas (along with other gases). After being exposed to the atmosphere for a specified period of time, the canister is sent to a laboratory for analysis.

EXAMPLE 30.9 ⊗ Radiation Exposure from a CT Scan

In Chapter 23, we described how a CT (or CAT) scan uses X-rays to form an image of bones and soft tissue within the body. The X-ray photons in a CT scan produce radiation damage similar to that produced by gamma rays. The radiation dose for a CT scan depends on many factors, but is typically about 150 mrem for a scan of the head. How many CT scans can your doctor prescribe before the radiation damage of the scan itself becomes a potential health hazard?

RECOGNIZE THE PRINCIPLE

The "normal" dose from radiation sources in the environment (cosmic rays, the small amount of radioactive material in the soil, etc.) is about 300 mrem per year (Table 30.4). To be safe, it is recommended that the dose from other sources such as a CT scan should be less than about 500 mrem.

SKETCH THE PROBLEM

No sketch is necessary, but the data in Table 30.4 are very useful.

We want the total dose from all the CT scans in a year to be no larger than the recommended maximum dose of 500 mrem. The safe number of CT scans is therefore

$$\text{number of scans} = \frac{500 \text{ mrem}}{\text{CT dose}} = \frac{500 \text{ mrem}}{150 \text{ mrem}} = \boxed{3}$$

What does it mean?

You might now conclude that it is unsafe to have more than three CT scans per year. Our analysis is a bit simplistic, however, since a doctor must carefully weigh the potential benefits (e.g., from locating a suspected tumor) in assessing whether or not to prescribe a CT scan.

30.5 | ⊗ APPLICATIONS OF NUCLEAR PHYSICS IN MEDICINE AND OTHER FIELDS

Although radiation can damage living tissue, it has many positive uses in medicine and other fields. We'll now describe a few of these applications.

Radioactive Tracers

A *radioactive tracer* is a chemical that contains radioactive nuclei. The movement of tracer nuclei through the body can be monitored by observing the radiation they emit through radioactive decay. Tracers are widely used in medicine and many other fields.

For example, in a method known as nuclear stress testing, a small amount of a radioactive liquid containing (typically) $^{201}_{81}\text{Th}$ or $^{99}_{43}\text{Tc}$ is injected into a vein near the heart. These isotopes emit gamma rays, which are detected by special cameras placed over the heart. Gamma rays from blood containing the tracer molecules cause the blood to "light up" in the images. These images thus show how blood flows (or does not flow) in and around the heart and are used to diagnose heart disease. In this and other applications of radioactive tracers, the dose of radioactivity is kept low so that it does no significant harm. The tracer isotopes usually have short half-lives, so their activity decays to a very low level after a few hours or days.

The method of positron emission tomography (called PET scanning) works in a similar way. A glucose solution containing typically $^{18}_{9}\text{F}$ (the isotopes $^{11}_{6}\text{C}$, $^{15}_{8}\text{O}$, and $^{13}_{7}\text{N}$ are also used) is injected into a patient's blood. These nuclei decay by the emission of a positron (a type of beta decay), and these positrons in turn produce gamma rays when they encounter an electron. (We'll discuss how and why in Chapter 31.) These gamma rays are observed by a collection of detectors positioned around the patient (Fig. 30.24). By combining the images from many different detectors (the "tomography" part of the process), medical professionals construct a three-dimensional image of the positions of the radioactive nuclei. These images are used to monitor blood flow near and inside the heart and brain. Some types of cancer cells absorb large amounts of glucose, so PET scans can also be used to find these cancers.

Another use of tracers is to fight thyroid cancer. Iodine tends to accumulate in the thyroid gland, so by using a tracer containing the radioactive isotopes $^{123}_{53}\text{I}$ or $^{131}_{53}\text{I}$ and detecting the radiation they emit, one can measure how effectively the thyroid absorbs iodine. Cancerous thyroid cells can be killed using the electrons emitted in beta decay as well as by the gamma radiation emitted by these iodine nuclei.

Radioactive tracers thus have many applications. The best choice of tracer nucleus depends on the application, but all tracers should have a short half-life so that their activity decreases to negligible levels soon after use. Also, tracer nuclei cannot be stored for future use, but must be made just before they are used. Tracer nuclei are usually made by nuclear reactions induced in a nuclear reactor core.

Gamma ray detectors

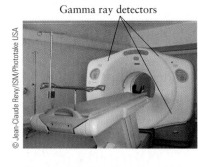

Figure 30.24 Positron emission tomography (PET scanning) is a widely used medical procedure. After ingesting a radioactive tracer, the patient enters an apparatus like the one shown here. This device measures the gamma rays emitted when positrons annihilate with electrons and determines the sites from which they are emitted.

© Jean-Claude Revy/ISM/Phototake USA

Using Radiation to Fight Cancer

Because quickly dividing cells are generally the ones most damaged by radiation, radiation treatments are effective in treating many types of cancers. For example, radiation is widely used to kill breast cancer cells. This therapy uses X-rays, so strictly speaking it does not involve radioactivity. Breast cancer is sometimes also treated by implanting radioactive nuclei, a treatment called brachytherapy. These nuclei (usually $^{90}_{38}Sr$) emit high-energy electrons (via beta decay), gamma rays, or both, killing nearby cancer cells. A similar treatment is used in dealing with prostate cancer in men. Radioactive "seeds" containing typically $^{125}_{53}I$ or $^{103}_{46}Pd$ are implanted in the prostate gland, and the gamma rays they emit kill nearby cancerous cells. These isotopes have relatively short half-lives, so their activity is very small after a few months.

Carbon Dating

One of the most important uses of radioactivity and tracers outside of medicine is the technique of *carbon dating* developed in 1947 by Willard Libby, who received the Nobel Prize in Chemistry for this work. Of the carbon nuclei occurring naturally on the Earth, approximately 99% are the isotope $^{12}_6C$ and about 1% are $^{13}_6C$. In addition to these two stable isotopes, the radioactive isotope $^{14}_6C$ is produced in the Earth's atmosphere when cosmic rays collide with $^{14}_7N$. The half-life of $^{14}_6C$ is $T_{1/2} = 5730$ yr as it undergoes beta decay, producing $^{14}_7N$ (Eq. 30.13). This decay is balanced by the cosmic-ray–induced production of $^{14}_6C$, resulting in a carbon isotopic ratio in the atmosphere[5] equal to $^{14}_6C/^{12}_6C \approx 1.3 \times 10^{-12}$.

Carbon-containing molecules including proteins and DNA are the building blocks of life. When an animal is living, it accumulates carbon from the environment as it eats and breathes, while plants take in atmospheric CO_2 during photosynthesis and absorb carbon-containing nutrients from the water and soil. Since there is a small amount of $^{14}_6C$ in the atmosphere and in their nutrients, organisms absorb $^{14}_6C$ until the percentage of $^{14}_6C$ in the organism equals the $^{14}_6C$ percentage in the atmosphere. When the organism dies, it stops accumulating new $^{14}_6C$, and the $^{14}_6C$ that it contained at death undergoes radioactive decay. Hence, after one half-life (5730 yr), the amount of $^{14}_6C$ in the organism is equal to half of what it was at the time the organism died. Likewise, after a time equal to $2T_{1/2}$, the amount of $^{14}_6C$ is a fourth of the original amount. The behavior of the $^{14}_6C$ concentration as a function of time is given by the exponential relation in Equation 30.25 and is plotted in Figure 30.25. Since the ratio $^{14}_6C/^{12}_6C$ at the time of death is known, the ratio measured today in a sample can be used to determine how much time has passed since death. In this way, the age of an animal or plant specimen can be accurately determined. This technique is now an essential part of experimental archaeology.

Carbon dating is done by first measuring the isotopic ratio of $^{14}_6C/^{12}_6C$ in a specimen, often using a mass spectrometer (Fig. 20.26) to measure the separate amounts of $^{14}_6C$ and $^{12}_6C$ that are present. Since $^{12}_6C$ is a stable isotope, the $^{14}_6C/^{12}_6C$ ratio decreases with time according to the decay of $^{14}_6C$ alone, and the isotopic ratio follows the decay curve in Figure 30.25. Given the measured isotopic ratio, the age of the sample can then be read off the curve in Figure 30.25.

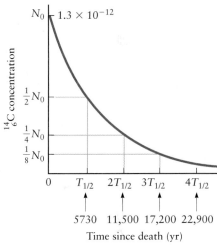

Figure 30.25 Carbon dating uses the concentration of $^{14}_6C$ (the ratio $^{14}_6C/^{12}_6C$) in a sample from a formerly living organism to deduce the time at which the organism died. After death, the concentration of $^{14}_6C$ falls with time due to its radioactive decay, with a half-life of 5730 yr.

EXAMPLE 30.10 ⊗ Carbon Dating

Some ancient plant material is found to have a fraction of $^{14}_6C$ given by $^{14}_6C/^{12}_6C = 6.0 \times 10^{-14}$. How old is it? The $^{14}_6C/^{12}_6C$ ratio at the time of death was equal to 1.3×10^{-12} (as explained above).

---

[5]Nuclear weapons testing in the 1950s and 1960s nearly doubled the $^{14}_6C/^{12}_6C$ ratio in the atmosphere. This ratio is now returning to its pre-1950s level.

RECOGNIZE THE PRINCIPLE

The radioactive decay of $^{14}_{6}C$ is described by

$$N = N_0 e^{-\lambda t} \tag{1}$$

with a half-life of 5730 yr, whereas $^{12}_{6}C$ is stable and does not decay. The $^{14}_{6}C/^{12}_{6}C$ thus decreases with time according to

$$\frac{^{14}_{6}C}{^{12}_{6}C} \text{ at time } t = \left(\frac{^{14}_{6}C}{^{12}_{6}C} \text{ at time of death} \right) \times e^{-\lambda t} \tag{2}$$

SKETCH THE PROBLEM

The decay behavior in Equations (1) and (2) is described graphically in Figure 30.25.

IDENTIFY THE RELATIONSHIPS

The given $^{14}_{6}C/^{12}_{6}C$ ratio is smaller than the ratio at the time of death by

$$\frac{^{14}_{6}C/^{12}_{6}C \text{ (at present)}}{^{14}_{6}C/^{12}_{6}C \text{ (at death)}} = \frac{6.0 \times 10^{-14}}{1.3 \times 10^{-12}} = 0.046 = 4.6\% \tag{3}$$

We can now approach this problem in two ways. One would be to use the graph in Figure 30.25 and read off the time at which the $^{14}_{6}C/^{12}_{6}C$ ratio has dropped to 4.6% of its original value, but this small value is hard to locate accurately on the vertical scale in Figure 30.25. Instead, let's use the exponential relation in Equation (1). At an unknown time t, only 4.6% of the original $^{14}_{6}C$ remains, so if N is the number of $^{14}_{6}C$ we have

$$N(\text{at present}) = N(\text{at death})e^{-\lambda t} \tag{4}$$

where λ is the decay constant for $^{14}_{6}C$. The decay constant is related to the half-life by (Eq. 30.27)

$$T_{1/2} = \frac{0.693}{\lambda}$$

SOLVE

Solving for λ, we get

$$\lambda = \frac{0.693}{T_{1/2}} = \frac{0.693}{5730 \text{ yr}} = 1.2 \times 10^{-4} \text{ yr}^{-1}$$

Rearranging Equation (4) and using the $^{14}_{6}C/^{12}_{6}C$ ratio in Equation (3) leads to

$$\frac{N(\text{at present})}{N(\text{at death})} = 0.046 = e^{-\lambda t}$$

We now take the logarithm of both sides and use the definition of logarithms $\ln(e^x) = x$. We find

$$\ln(0.046) = -\lambda t$$

$$t = -\frac{\ln(0.046)}{\lambda} = -\frac{\ln(0.046)}{1.2 \times 10^{-4} \text{ yr}^{-1}} = \boxed{26,000 \text{ yr}}$$

What does it mean?

Carbon dating is a precise way to date many kinds of material. However, one limitation on this technique is set by the half-life of $^{14}_{6}C$. If the material is very much older than the half-life, there will be very little $^{14}_{6}C$ left to measure. In practice, carbon dating is useful for material that is no older than about 50,000 years.

Magnetic Resonance Imaging

Magnetic resonance imaging (MRI) makes use of the magnetic properties of nuclei. Hydrogen has the simplest nucleus—a single proton—and a proton has an intrinsic spin angular momentum similar to the spin angular momentum of the electron discussed in Chapter 28. As with electrons, a proton has just two spin states, "spin up" and "spin down," and in a magnetic field these two states have different energies. The presence of a proton can be detected by observing the absorption of a photon that induces a transition from the lower- to the higher-energy spin state (Fig. 30.26A). This is just like the absorption of a photon by an atom as it undergoes a transition between energy levels (Chapter 29).

The absorption of a photon only occur if the separation between energy levels matches the photon energy, and this energy separation depends on the magnetic field. An MRI magnet is designed so that these energies match only at a particular spot within the body, so the MRI signal gives the density of protons (and other properties) at just that spot. By then scanning this spot around the body, a three-dimensional image can be constructed, giving a unique picture of the body's internal structure.

Most applications of MRI employ a magnet with coils in a unit large enough for a person to fit inside (Fig. 30.26B). After only a few minutes, an image with remarkable resolution can be obtained (Fig. 30.26C). MRI images are now widely used to diagnose many different types of medical conditions (including the author's knee injuries). This method does not employ radioactive tracers, but uses the magnetic properties of naturally occurring nuclei (usually hydrogen nuclei) in the patient and radio frequency photons (which have very low energies). As a result, there do not appear to be any health risks associated with MRI, making this technique a very safe and effective medical tool.

30.6 | QUESTIONS ABOUT THE NUCLEUS

In this chapter, we have described nuclear properties and nuclear reactions. We now ask *why* the nucleus has these properties. As usual, this "why" question is much more difficult than discovering the actual properties, but it also leads us to fundamental issues.

Conservation rules (Section 30.2) are fundamental principles of physics that apply to a wide range of situations. Nuclear physics reveals a new conservation rule, the conservation of nucleon number. There are just two types of nucleons—protons and neutrons—and *all* known nuclear reactions conserve the number of nucleons. The *total* number of nucleons at the start of a reaction is equal to the total number of nucleons at the end, which you can confirm for yourself by reviewing the examples of alpha, beta, and gamma decay described in Section 30.2. Physicists have spent much time and effort studying this problem, and as far as we can tell, nucleon number is always conserved. Why? We don't know. All we can say is that nucleon number conservation is a property of all known processes.

Such number conservation rules do not apply to all particles. For example, the number of electrons is *not* conserved in all reactions. Beta decay (e.g., Eq. 30.13) creates an electron where one did not previously exist, so electrons are different from nucleons in some fundamental way. We'll explore this topic further in Chapter 31.

We can also ask about the structure of a nucleon. Is there anything inside a proton or a neutron? Physicists have spent years working on this question. Scattering experiments similar (in principle) to those of Rutherford have shown that nucleons are composed of particles called *quarks*. There are several different types of quarks, all which carry an electric charge of $\pm e/3$ or $\pm 2e/3$, where $-e$ is the charge carried by an electron. Protons and neutrons are each composed of three quarks. In a proton, the quark charges add up to $+e$; in a neutron, they add up to zero. So, although a neutron is electrically neutral, it contains charged particles! In Chapter 31, we'll

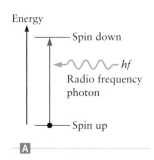

Energy — Spin down
— hf
Radio frequency photon
— Spin up

A

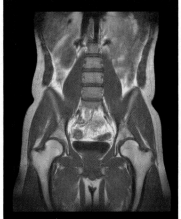

B

© Royalty-Free/Corbis/Jupiterimages

C

© Simon Fraser/Photo Researchers, Inc.

Figure 30.26 **A** When a proton (a hydrogen nucleus) is placed in a magnetic field, the spin-up and spin-down states have different energies. A proton can absorb a radio frequency photon and undergo a transition from one level to the other. **B** Magnetic resonance imaging (MRI) uses the spin magnetism of hydrogen nuclei to measure the density and other properties of tissue within the body. This procedure requires that the patient be placed inside a large magnet. **C** MRI image of a person's abdomen.

discuss the behavior of quarks and how they combine to make protons and neutrons, but we still don't know *why* nature is put together this way.

SUMMARY | Chapter 30

Nuclear structure

Nuclei contain **protons** and **neutrons**, also called **nucleons**. The number of protons in a nucleus is the **atomic number** Z, and the total number of protons plus neutrons is the **mass number** A. The number of neutrons N is thus related to Z and A by

$$A = Z + N \qquad \textbf{(30.3)} \text{ (page 1022)}$$

Different **isotopes** of a particular element contain the same number of protons, but different numbers of neutrons.

All this information is described in the following shorthand notation:

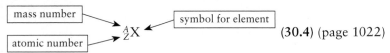

$$\textbf{(30.4)} \text{ (page 1022)}$$

Nuclear size

Nuclei have a radius r of a few femtometers (1 fm = 10^{-15} m). The radius depends on the number of nucleons A in the nucleus through the relation

$$r = r_0 A^{1/3} \qquad \textbf{(30.5)} \text{ (page 1024)}$$

with $r_0 \approx 1.2$ fm = 1.2×10^{-15} m.

The strong force

Nuclei are held together by the **strong force**, an attractive force that acts between pairs of nucleons. The strong force does not act on electrons, and it is responsible for the stability of the nucleus.

Radioactive decay

Some nuclei are unstable and decay spontaneously, a process called **radioactive decay**. In **alpha decay**, an alpha particle (a helium nucleus, $^{4}_{2}\text{He}$) is emitted. In **beta decay**, an electron or a positron is emitted. In **gamma decay**, a gamma ray (a high-energy photon) is produced. The **half-life** is the time it takes for half of the nuclei in a sample to decay. Half-lives can be less than 1 second or longer than the age of the Earth.

(Continued)

Nuclear fission

In nuclear **fission**, a nucleus splits into two nuclei that are each about half the size of the original nucleus and several neutrons. This reaction releases energy; it is the basis for nuclear weapons and nuclear power plants.

NUCLEAR FISSION (a heavy nucleus splits)

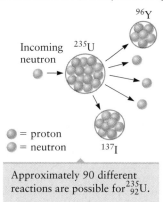

Approximately 90 different reactions are possible for $^{235}_{92}$U.

Nuclear fusion

In nuclear **fusion**, two low-mass nuclei combine to form one heavier nucleus. Fusion reactions also release large amounts of energy and are responsible for the energy output of the Sun.

NUCLEAR FUSION

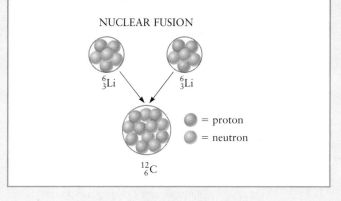

APPLICATIONS

Energy in nuclear reactions

The energy released in a typical nuclear reaction is typically on the order of MeV. It is much larger than a typical atomic or molecular energy (about 10 eV), and the particles emitted by a nuclear reaction can damage living tissue.

Units of radioactivity

Several units are used to measure the amount of radioactivity in a sample.

The **activity** of a sample is proportional to the number of nuclei in the sample that undergo radioactive decay in 1 second. One unit of activity, the curie (Ci), is defined by

$$1 \text{ Ci} = 3.7 \times 10^{10} \text{ decays/s} \qquad \textbf{(30.28)} \text{ (page 1036)}$$

Activity is also measured with the SI unit becquerel (Bq), with

$$1 \text{ Bq} = 1 \text{ decay/s} \qquad \textbf{(30.29)} \text{ (page 1036)}$$

Biological effects

The amount of damage caused by radiation is determined in part by the energy deposited by the radiation in tissue. A radiation dose of 1 **rad** deposits 10^{-2} J in 1 kg of material. Different types of radiation can do more or less damage. The **RBE** (relative biological effectiveness) factor measures this difference. The unit called the **rem** takes all these factors into account:

$$\text{dose in rem} = (\text{dose in rads}) \times (RBE) \qquad \textbf{(30.33)} \text{ (page 1045)}$$

Benefits of nuclear radiation

Radiation presents many health hazards, but it also has useful applications, including carbon dating and the treatment of many types of cancer.

1. Why do different isotopes of a given element have, for most practical purposes, the same chemical properties? That is, why do different isotopes form the same molecules and compounds even though they have different numbers of neutrons and different masses?

2. Different isotopes of an element essentially share chemical properties. How, then, is it possible to separate different isotopes?

3. What is heavy water? Would an equal number of moles of heavy water actually weigh more than plain water?

4. Was evidence for the nucleus was first obtained through (a) carbon dating, (b) the spectral lines of hydrogen, (c) alpha particle scattering, (d) cosmic rays, or (e) electromagnetic radiation?

5. What is (a) an alpha particle? (b) a beta particle? (c) a gamma ray? (d) an X-ray? (e) a daughter nucleus?

6. Is the mass of a nucleus larger or smaller than the masses of the nucleons of which it is made? Explain.

7. SSM In the mid 1800s, the elements of the periodic table were being identified and discovered, yet for many decades element 43, between molybdenum and ruthenium (Fig. Q30.7), remained unfound. Technetium (Tc) was finally discovered almost 100 years later, in 1936. (a) What is special about this element, and why was it so hard to find? (b) Where can technetium be found today?

Cr 24	Mn 25	Fe 26
51.996	54.938	55.845
$3d^54s^1$	$3d^54s^2$	$3d^64s^2$
Mo 42	43	Ru 44
95.94	?	101.07
$4d^55s^1$		$4d^75s^1$
W 74	Re 75	Os 76
183.84	186.21	190.23
$5d^46s^2$	$5d^56s^2$	$5d^66s^2$

Figure Q30.7 Element 43.

8. Why is the element plutonium not found in significant amounts in nature (on the Earth)?

9. Compare how a fission bomb works with how a fusion bomb works. Be sure to discuss what each would use for fuel and the waste products produced by each.

10. There is a critical mass for the creation of a spontaneous chain reaction involving $^{235}_{92}$U. Why would it be difficult to make a bomb that is much larger than the critical mass? Why does this limit not exist for a fusion bomb?

11. ⊗ Explain why carbon dating can in general only be used to measure the age of material that was once alive.

12. Do you expect that $^{56}_{26}$Fe would undergo fusion? Explain why or why not. *Hint*: Use the binding energy curve in Figure 30.17.

13. Consider the radioactive decay of $^{235}_{92}$U. Construct a series of alpha and beta decays (similar to Fig. 30.9) that include the daughter

nuclei $^{223}_{88}$Ra, $^{219}_{86}$Rn, and $^{211}_{83}$Bi, and terminates with $^{207}_{82}$Pb. Why can this decay series not terminate with $^{206}_{82}$Pb or $^{208}_{82}$Pb?

14. ⊗ In the treatment of cancer using radioactive "seeds" (brachytherapy), the radioactive material must be placed very close to the cancer cells. (a) Explain why cells farther away from the radioactive seeds are not significantly harmed by the emitted radiation. (b) Explain why placing the radioactive seeds outside the body is not as effective as placing them inside or next to a cancerous tumor.

15. A nucleus of $^{245}_{96}$Cm that is initially at rest undergoes alpha decay. What is the daughter nucleus? Which of the reaction products has the greater kinetic energy, the alpha particle or the daughter nucleus?

16. Explain why the neutrons emitted in a nuclear reaction can penetrate very thick sheets of lead, but the alpha and beta particles cannot.

17. Which decays change the parent nucleus into a nucleus of a different element: alpha decay, beta decay, or gamma decay?

18. ⊗ An archaeologist friend proposes to determine the age of a dinosaur bone using carbon dating. Why does your friend need to take a physics course? That is, what is wrong with this proposal?

19. SSM ⊗ Scientists in areas such as geology, paleontology, and archaeology use "radiodating" techniques (such as carbon dating) in the study of the history of the Earth and its inhabitants. There are several radiodating techniques, each using different nuclei: (i) carbon 14, (ii) potassium–argon 40, (iii) rubidium–strontium, and (iv) uranium–lead. (a) Approximately how old an object can be dated with each of these techniques? (b) Conversely, what are the youngest objects that can be reliably dated with each technique? (c) Which techniques are best for archaeology? For paleontology? *Note*: You may have to look up the half-lives of the relevant parent isotopes for the elements used in each method.

20. In a volume of soil that measures 1 mi² and 1 ft deep there are roughly 2 g of radium, predominantly ^{236}Ra. But ^{236}Ra has a half-life of just 1600 yr, while the Earth is billions of years old. Where did this radium come from?

21. The nuclei $^{235}_{92}$U, $^{253}_{100}$Fm, and $^{14}_{6}$C have very different half-lives. Explain how samples of these materials can still have the same activity.

22. How might the $^{239}_{94}$Pu used in nuclear weapons be created? *Hint*: Consider reactions in which a nucleus captures a neutron.

23. The sketch in Figure 30.10 shows a nuclear decay in which two product nuclei are produced. Why do they travel away in opposite directions?

24. ⊗ Which of the following objects could be dated with carbon dating?
(a) The logs used to make George Washington's house
(b) The bricks used to make George Washington's house
(c) Animal bones found in George Washington's barn
(d) The blade of George Washington's pocket knife

30.1 STRUCTURE OF THE NUCLEUS: WHAT'S INSIDE?

1. Is the atomic number (a) the number of neutrons in the nucleus, (b) the number of protons in the nucleus, (c) the number of elec-

trons in the nucleus, or (d) the sum of the number of neutrons and protons in the nucleus?

2. Is the mass number (a) the number of neutrons in the nucleus, (b) the number of protons in the nucleus, (c) the number of electrons in the nucleus, or (d) the sum of the number of neutrons and protons in the nucleus?

3. Do different isotopes of the same element have the same number of (a) protons, (b) neutrons, (c) protons plus neutrons, or (d) electrons (assuming a neutral atom)? (More than one answer may be correct.)

4. Name three different nuclei that contain 120 neutrons.

5. What are Z, N, and A for the nuclei (a) $^{12}_{6}C$, (b) $^{34}_{16}S$, (c) $^{208}_{82}Pb$, and (d) $^{56}_{26}Fe$?

6. (a) A nucleus has $Z = 17$ and $N = 18$. What is the symbol for this nucleus? (b) If four neutrons are added to $^{208}_{82}Pb$, what element is created? Give the symbol for this nucleus.

7. If two protons are removed from $^{206}_{82}Pb$, what element is created? Give the symbol for this nucleus.

8. SSM ✪ Ⓡ Estimate the electric potential energy of all the protons in $^{6}_{3}Li$. Assume the distance between each pair of protons is 1 fm.

9. Use Equation 30.5 to estimate the radius of (a) an alpha particle, (b) $^{34}_{16}S$, and (c) $^{56}_{26}Fe$.

10. What is the approximate density of (a) an alpha particle and (b) $^{56}_{26}Fe$?

11. What is the ratio of the radius of $^{206}_{82}Pb$ to the radius of an alpha particle?

12. A nucleus has a radius of 4.5×10^{-15} m. What nucleus might it be?

13. An alpha particle has a kinetic energy of 1.2 MeV. What is its speed?

14. A beta particle emitted in nuclear decay has a kinetic energy of 50,000 eV. What is its speed?

15. What is the speed of a gamma ray?

16. Which two of the following forces are most important in a nucleus?
(a) electric force
(b) magnetic force
(c) the strong force
(d) friction
(e) gravity

17. Which of the following forces is (most) crucial for holding a nucleus together?
(a) electric force
(b) magnetic force
(c) the strong force
(d) friction
(e) gravity

18. Name the dominant force for (a) two protons far apart, (b) a neutron and a proton in a nucleus, and (c) a proton and an electron in a nucleus.

19. An electric potential difference ΔV is used to give an alpha particle a kinetic energy of 1.5 MeV. What is the value of ΔV?

20. SSM ✪ A proton is accelerated through a potential difference ΔV_p, and an alpha particle is accelerated through a potential difference ΔV_α. If the two particles have the same kinetic energy, is the ratio $\Delta V_p / \Delta V_\alpha$ (a) 4, (b) 2, (c) 1, (d) 1/2, or (e) 1/4?

30.2 NUCLEAR REACTIONS: SPONTANEOUS DECAY OF A NUCLEUS

21. After beta decay, the daughter nucleus differs from the parent nucleus in which of the following ways?
(a) The daughter nucleus has one more proton.
(b) The daughter nucleus has one less proton.
(c) The daughter nucleus has one more neutron.

22. After alpha decay, what is the difference in mass number between the daughter and the parent nuclei?
(a) The mass number of the daughter nucleus is smaller by four.

(b) The mass number of the daughter nucleus is larger by four.
(c) The mass number of the daughter nucleus is smaller by two.

23. If $^{242}_{94}Pu$ decays via beta decay, what nucleus is produced? If it decays by alpha decay, what nucleus is produced?

24. If $^{210}_{82}Pu$ is produced by alpha decay, what was the parent nucleus?

25. SSM ✪ Complete (i.e., balance) the following reactions. Also label each reaction as alpha decay, beta decay, or gamma decay.
(a) $^{234}_{90}Th \rightarrow ^{230}_{88}Ra + ?$
(b) $^{234}_{90}Th \rightarrow ^{234}_{91}Pa + ?$
(c) $^{234}_{90}Th^* \rightarrow ^{234}_{90}Th + ?$

26. Write the reaction for the alpha decay of $^{214}_{84}Po$.

27. Write the reaction for the beta decay of $^{60}_{27}Co$.

28. Write the reaction for the gamma decay of $^{210}_{82}Pb^*$.

29. If 16 g of radon gas with a half-life of 4 days is present in a balloon at $t = 0$, what is the mass of radon gas present after (a) 4 days, (b) 16 days, and (c) 25 days?

30. The half-life of $^{224}_{88}Ra$ is 3.7 days. What is its decay constant? Express your answer in s^{-1}.

31. The decay constant of a nucleus is 25 s^{-1}. What is its half-life?

32. ✪ After 25 years, 75% of a radioactive material decays. What is the half-life?

33. ✪ The isotope $^{19}_{10}Ne$ has a half-life of 17.2 s. If you have 45 g of it at $t = 0$, how much is left after 45 s?

34. SSM ✪ A particular nucleus has a half-life of 36 minutes. How long does it take for 95% of this material to decay?

35. ✪ At $t = 0$, 950 g of a radioactive material is present. Twenty-five years later it is found that only 23 g of the material remains. What is its half-life?

36. ❂ At present, the concentration of $^{235}_{92}U$ in naturally occurring uranium deposits is approximately 0.72%. What will the concentration be one billion years from now?

37. ✪ Balance the following nuclear reactions.
(a) $^{1}_{1}H + ^{6}_{3}Li \rightarrow ^{4}_{2}He + ?$
(b) $^{1}_{0}n + ^{235}_{92}U \rightarrow ^{133}_{51}Sb + ? + 4^{1}_{0}n$
(c) $^{14}_{6}C \rightarrow ? + e^- + \bar{\nu}$

38. Use the mass data in Table A.4 to calculate the binding energy of (a) $^{12}_{6}C$, (b) $^{56}_{26}Fe$, (c) $^{235}_{92}U$, (d) $^{238}_{92}U$, and (e) $^{3}_{1}H$.

39. What is the magnitude of the binding energy per nucleon of $^{208}_{82}Pb$? What is the magnitude of the *total* binding energy of this nucleus? Give your answers in MeV.

40. What is the approximate energy needed to remove one neutron from $^{55}_{26}Fe$?

41. The isotope $^{89}_{36}Kr$ decays by beta emission. What nucleus is produced?

42. A piece of radium has an activity of 4.5 mCi. (a) What is the mass of the radium? (b) What is its activity in Bq?

43. A sample has an activity of 30 mCi (millicuries). If the sample is divided into 12 equal pieces, what is the activity of each piece?

30.3 STABILITY OF THE NUCLEUS: FISSION AND FUSION

44. How many neutrons are released when $^{235}_{92}U$ undergoes fission to produce $^{132}_{50}Sn$ and $^{100}_{42}Mo$? Do not count the neutron that initiated the fission.

45. ❂ In Example 30.7, we considered the energy produced by the fusion of two deuterium nuclei ($^{2}_{1}H$). (a) In the example, we found the energy in MeV. Convert this energy to joules. (b) In seawater, deuterium has a concentration of 0.0003% relative to that of normal hydrogen ($^{1}_{1}H$). If all the deuterium in 1.0 g of seawater undergoes fusion (as in Example 30.7), how much energy would be released?

46. Balance the fission reaction $^{235}_{92}U + ^{1}_{0}n \rightarrow ^{142}_{55}Cs + ^{90}_{37}Rb + ?^{1}_{0}n$.

47. ✪ The fission bomb that exploded at Hiroshima, Japan, in 1945 released an energy of about 6×10^{13} J. (The value is known only approximately.) Consider a fusion bomb that releases energy by the fusion of two deuterium nuclei (Example 30.7). What mass of deuterium nuclei is required to release the energy that was released at Hiroshima?

48. SSM ✫ The largest fusion bomb ever detonated (called the Tsar Bomba) was tested by the Soviet Union in 1961. The explosion from the device shown in Figure P30.48 released an energy of approximately 2.4×10^{17} J. (a) How many grams of matter were converted to energy? (b) If this bomb used the deuterium–deuterium fusion reaction (Example 30.7), how many grams of deuterium were required?

Figure P30.48 The Tsar Bomba weighed 2.7×10^4 kg and had a design yield equivalent to 50 megatons of TNT.

49. In the fission reaction $^{235}_{92}\text{U} + ^{1}_{0}\text{n} \rightarrow ^{138}_{54}\text{Xe} + ^{95}_{38}\text{Sr} + 3^{1}_{0}\text{n}$, the isotope $^{235}_{92}\text{U}$ has a mass of 235.04392 u, the mass of $^{138}_{54}\text{Xe}$ is 137.91395 u, and the mass of $^{95}_{38}\text{Sr}$ is 94.91936 u. How much energy does this reaction release?

50. ✪ ℝ **Energy production in the Sun.** The Sun produces energy through a series of fusion reactions, starting with protons ($^{1}_{1}\text{H}$) and ending with alpha particles (helium nuclei, $^{4}_{2}\text{He}$). The first reaction produces deuterium ($^{2}_{1}\text{H}$) plus a positron and a neutrino,

$$^{1}_{1}\text{H} + ^{1}_{1}\text{H} \rightarrow ^{2}_{1}\text{H} + e^+ + \nu \qquad (1)$$

and the second reaction fuses a proton with deuterium to produce $^{3}_{2}\text{He}$ (a stable isotope of helium),

$$^{1}_{1}\text{H} + ^{2}_{1}\text{H} \rightarrow ^{3}_{2}\text{He} + \gamma \qquad (2)$$

These helium nuclei are then converted to alpha particles through either of the following reactions:

$$^{1}_{1}\text{H} + ^{3}_{2}\text{He} \rightarrow ^{4}_{2}\text{He} + e^+ + \nu \qquad (3)$$

$$^{3}_{2}\text{He} + ^{3}_{2}\text{He} \rightarrow ^{4}_{2}\text{He} + ^{1}_{1}\text{H} + ^{1}_{1}\text{H} \qquad (4)$$

(a) Use the masses given in Table A.4 to find the energy released in each of the reactions (1) through (4). (Do not include any subsequent reactions involving the positron.) (b) What is the total energy released when one alpha particle is produced using the process in reactions (1), (2), and (3)? (c) What is the total energy released when one alpha particle is produced using the process in reactions (1), (2), and (4)? (d) The Sun is currently converting approximately 600,000,000 tons of hydrogen into alpha particles each second. How many years will it take to consume all the Sun's fuel? Assume all the Sun's current mass ($m_{\text{Sun}} = 2.0 \times 10^{30}$ kg) is made up of protons.

30.4 BIOLOGICAL EFFECTS OF RADIOACTIVITY

51. ⊗ A person absorbs a dose of 15 rads of protons. What dose of alpha particles causes the same amount of damage? Assume that the *RBE* for these protons is 15.

52. ⊗ Use the *RBE* factors in Table 30.3 to compare the damage produced by alpha particles, gamma rays, and fast neutrons.

For the same dose as measured in rads, rank these three types of radiation in order of least damage to most damage.

53. ⊗ Alpha particles with an *RBE* of 15 deliver 25 mrem of radiation to a patient. What dose does the patient receive? Express your answer in rad.

54. What is the activity of 5.0 g of $^{235}_{92}\text{U}$? Express your answer in decays/s and Ci.

55. ✫ ⊗ A person with a mass of 85 kg is exposed to a dose of 30 rad. How many joules of energy are deposited in the person's body?

56. SSM ✫ ⊗ A 125-rad dose of radiation is administered to a cancer patient (mass = 80 kg). If all this radiation is absorbed by the tumor, how much energy is absorbed?

57. ✫ ⊗ ℝ A typical exposure from a dental X-ray is 10 mrem. (See Table 30.4.) (a) How much energy is deposited to your head when you get a dental X-ray? (b) A weight of 1 N falls on your head. From how high above your head would the weight need to be dropped to impart the same energy as the X-ray?

58. A radioactive material has an activity of 2.5 Bq. What is its activity in Ci?

59. A radioactive sample has an initial activity of 4.5 mCi (millicuries). Its activity 15 hours later is 1.8 mCi. What is its half-life?

60. SSM ✫ A radioactive sample has an initial activity of 800 decays/s. Its activity 24 hours later is 200 decays/s. What is its half-life?

61. ⊗ An average annual dose from natural radiation sources is about 500 mrem, of which about 40 mrem comes from ^{40}K, an isotope found all around us and in our food. What is the maximum energy ($RBE = 1.7$) deposited in a 65-kg individual from the emitted beta rays of potassium each year?

62. ✪ ⊗ ℝ Approximate the activity (in units of Bq and Ci) in one banana from its ^{40}K content (Fig. P30.62). An average banana has about 600 mg of potassium. *Hint:* You will need to find the natural abundance and half-life of the ^{40}K isotope.

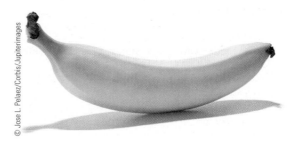

Figure P30.62 Bananas contain some ^{40}K, making them slightly radioactive.

30.5 APPLICATIONS OF NUCLEAR PHYSICS IN MEDICINE AND OTHER FIELDS

63. ⊗ A patient with thyroid problems has radioactive iodine ($^{123}_{53}\text{I}$) deposited in his thyroid gland. If the $^{123}_{53}\text{I}$ has an initial activity of 2.5 MBq, what is its activity after 20 days?

64. ✫ ⊗ A sample of material is found to be 25,000 years old using carbon dating. What fraction of the original $^{14}_{6}\text{C}$ remains?

65. ⊗ An archaeologist uses carbon dating to find the age of some ancient plant material. If the concentration of $^{14}_{6}\text{C}$ is 2.8×10^{-14} smaller than the concentration of $^{12}_{6}\text{C}$, how old is the material?

66. ⊗ A piece of very old wood is found to have a $^{14}_{6}\text{C}/^{12}_{6}\text{C}$ concentration ratio that is 0.10 times smaller than the concentration found in a tree that died in 1900. How old is the older piece of wood?

67. SSM ✫ ⊗ In 1991, hikers in the Alps discovered the remains of a man who had been trapped in a glacier (Fig. P30.67). Carbon dating revealed the $^{14}_{6}$C/$^{12}_{6}$C ratio in the remains to be 6.8×10^{-13}. How old are the remains?

Figure P30.67

68. The tracer $^{123}_{53}$I is used to treat thyroid problems. This tracer nucleus is produced by bombarding xenon with protons in the reaction $^{1}_{1}$H + $^{124}_{54}$Xe = $^{123}_{53}$I + ? What else does this reaction produce?

69. Some smoke detectors (Fig. P30.69) use the radiation produced by the decay of $^{241}_{95}$Am (half-life 432 yr) to ionize air molecules, which in turn produces a steady current across two electrodes. If smoke enters the region of ionization, the alpha particles are absorbed by the smoke particulates, reducing the current accross the electrodes. The electronic circuitry then sounds the alarm when it senses any current reduction. How long does it take for the activity of $^{241}_{95}$Am to drop to 1% of its initial value? Would it pose a problem if the smoke detector is discarded in a landfill?

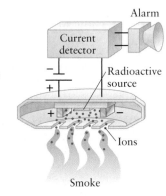

Figure P30.69

70. In positron emission tomography (PET), a positron is created by the decay of a nucleus attached to a glucose molecule. This positron then annihilates an electron (because the positron is antimatter; that is, it is an anti-electron). The annihilation reaction produces two gamma rays, traveling off in opposite directions. When the gamma rays are detected, their origin in the body reveals how and where glucose is absorbed. (a) Find the energy, in MeV, of one of these gamma rays. (b) Find the frequency of the gamma ray in part (a). (c) Explain why there are two gamma rays emitted in opposite directions.

71. A common nucleus used in PET scans is $^{18}_{9}$F, which has a half-life of 110 min. What fraction of the original activity remains in the body after 3 h?

ADDITIONAL PROBLEMS

72. **Heavy water.** The compound D_2O is called heavy water. It is essentially a water molecule in which each of the hydrogen atoms has a deuterium nucleus. (a) How much heavier, by percentage, is heavy water from regular water? (b) Estimate the density of heavy water.

73. A nuclear power plant with a 1000-MW output is 40% efficient in converting the energy released from fission into electricity. How much $^{235}_{92}$U is consumed by the power plant in one day? Express your answer in kg and in atoms.

74. An MX "Peacekeeper" missile carries 10 warheads, each rated at 20 megatons. (a) How many times greater is the energy release from one of these warheads than that of the Hiroshima explosion? (b) What quantity of mass is used to produce this energy?

75. SSM The half-life of $^{223}_{88}$Ra is 11.4 days. (a) Calculate its decay constant. (b) If the activity of a piece of $^{223}_{88}$Ra is 0.75 Ci today, how long will it take for its activity to fall to 0.10 Ci?

76. Ordinary soil contains typically 1 part per million (ppm) of uranium by mass. (a) How many uranium nuclei are in the top 10 m of soil under a typical house (20 m × 20 m)? (b) Only 0.72% of the nuclei in part (a) are $^{235}_{92}$U; the rest are $^{238}_{92}$U, which has a much longer half-life. How many $^{235}_{92}$U nuclei are in the soil under this house? How many $^{238}_{92}$U nuclei are in the soil under this house? (c) How many of the $^{235}_{92}$U nuclei in part (b) undergo radioactive decay over a period of one day? (d) How many of the $^{238}_{92}$U nuclei in part (b) undergo radioactive decay over a period of one day? (e) The decay of $^{238}_{92}$U leads to radon gas ($^{222}_{86}$Rn; see Fig. 30.9). If all the $^{238}_{92}$U decays to make $^{222}_{86}$Rn, how many $^{222}_{86}$Rn nuclei are produced under the house in one month? (f) If all the $^{222}_{86}$Rn nuclei in part (e) seep into the house above, what is the activity in the house?

77. The binding energy of a typical nucleus is comparable to the electric potential energy $PE_{elec} \approx 2 \times 10^{-13}$ J found in Equation 30.7. If that is the magnitude of the binding energy of a single helium nucleus, what is the approximate magnitude of the total binding energy of 1 g of helium?

78. One way to date ancient wood is with carbon dating. An instrument dealer comes to you with what he says is a violin made by the master violin maker Antonio Stradivari (1644–1736). You have a very small piece of wood from the violin analyzed, and the ratio of $^{14}_{6}$C/$^{12}_{6}$C is found to be 96.4% of the ratio found in trees that were cut down in 1950. Could this violin have been made by Stradivari?

79. Tritium is used in nuclear weapons and has a half-life of 12.3 yr. Suppose a nuclear weapon will cease to function if the amount of tritium is less than 20% of the amount present when the weapon is first built. What is the maximum age such a weapon can have and still be dangerous?

80. A neutron star is a star that has collapsed so that it is essentially one big nucleus, and its radius is related to its "mass number" by Equation 30.5. If our Sun collapsed to make a neutron star, what would its radius be?

81. Two radioactive samples are composed of two different types of nuclei. The samples have the same number of radioactive nuclei, but the activity of sample 1 is four times greater than the activity of sample 2. What is the ratio of the half-lives of the two different types of nuclei?

82. A Rutherford scattering experiment is performed in which alpha particles are directed at a flat sheet of carbon. This sheet contains a single layer of carbon atoms arranged on a square grid (Fig. P30.82), with an atomic spacing of 0.25 nm. Estimate the probability that an alpha particle will collide with a carbon nucleus. Assume the carbon nuclei have a diameter of 2.0 fm and that the center of an alpha particle must overlap with a carbon nucleus in order for a collision to occur (since the strong force has a very short range).

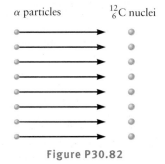

Figure P30.82

83. ✪ ℝ Consider again the Rutherford scattering experiment in Figure P30.82, but now assume there are many sheets (i.e., many layers) of carbon atoms. Assume the distance between sheets is 0.25 nm, the total thickness of the sheets is 1.0 cm, and make the same assumptions about collisions between an alpha particle and a carbon nucleus as described in Problem 82. Estimate the probability that an alpha particle will collide with a carbon nucleus as it travels through all the layers.

84. ✪ ℝ According to the Heisenberg uncertainty principle (Chapter 28), the uncertainties in the position Δx and momentum Δp of a particle are related by

$$\Delta x \Delta p \geq \frac{h}{4\pi}$$

This relation can be used to estimate an upper bound on the kinetic energy and velocity of a proton or neutron inside a nucleus by taking Δx to be the diameter of the nucleus and Δp the neutron's total momentum. (a) Use the uncertainty principle to estimate the kinetic energy of a proton inside an alpha particle. (b) What is the speed of the proton?

85. ✪ The critical mass of pure $^{235}_{92}$U can be as small as about 15 kg, but a much smaller amount has a significant activity. (a) What is the activity of a sample of $^{235}_{92}$U with a mass of 1.0 mg? (b) A Geiger counter with a detector area of 5.0 cm^2 is placed 20 cm from the $^{235}_{92}$U sample in part (a). How many decays will the Geiger counter measure?

86. ✪ **Chicago Pile One.** In 1942, the world's first nuclear fission reactor was built and operated by team of physicists headed by Enrico Fermi under the bleachers of an abandoned stadium (Fig. P30.86) at the University of Chicago. The reactor was simply a pile of graphite bricks and uranium metal, measuring 20 ft high by 25 ft wide, with cadmium rods as a moderator. The reaction lasted 33 min and produced 0.5 W of power, confirming that a nuclear chain reaction had been achieved. (a) How much energy was released during this first fission reaction? (b) How many neutrons were produced and left the pile during the reaction?

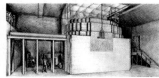

Courtesy of Department of Energy (both)

Figure P30.86 (Left) The Stagg Field bleachers. (Right) The Chicago Pile One fission reactor stack of graphite bricks and uranium within the bleacher enclosure.

87. ✪ ℝ **Supernova 1987A.** At a relatively close 51.4 kiloparsecs away from the Earth, the supernova of 1987 was the closest nova to occur since telescopes were invented, and it allowed a unique opportunity to confirm that radioactive decay rates are constant over time. Theoretical considerations of the fusion processes in the extreme environment of a supernova blast predict that many isotopes will be created, including a large amount of the short-lived isotope ^{56}Co with a half-life of 77.1 days. Figure P30.87 shows the intensity of the light measured from SN1987A as a function of time. A few weeks after the peak intensity, the light from the explosion dims logarithmically as a function of time, as one would expect if the light were being generated due to radioactive decay of the created ^{56}Co. (a) How long ago did the explosion of the supernova take place? (b) Using the light

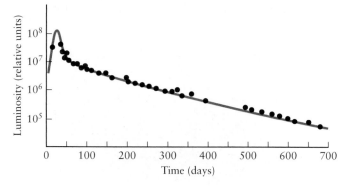

Figure P30.87 The light curve for Supernova 1987A.

curve in Figure P30.87, estimate the half-life of this ^{56}Co and compare it with the known half-life. What does this estimate tell you about the decay rate of ^{56}Co?

88. ✪ **Geochronology and potassium-argon dating.** One method of dating rocks uses the decay of ^{40}K to ^{40}Ar with a half-life of 1.26×10^9 yr. Most of the time, ^{40}K decays to ^{40}Ca, and 11% of the time it decays to ^{40}Ar. Initially, when the minerals in the rock form, there is very little argon included because argon is an inert noble gas. Potassium is abundant in nature, and most rocks have trace amounts of ^{40}K. To find the age of a rock, a sample is crushed into a fine powder and the released argon is carefully measured. The amount of ^{40}K is then extracted from the same sample and precisely measured. (a) A rock sample taken from atop the canyon wall is found to have one argon atom for every 52 atoms of ^{40}K. Approximately how old is the rock from the top strata? (b) Another sample of rock is taken from the bottom of a canyon at the base of this cliff wall. This sample is found to have one argon atom to every ten ^{40}K atoms. How much older is the rock at the canyon bottom than the rock at the top?

89. ✪ Consider conservation of energy and show that a ^{12}C nucleus cannot spontaneously decay into three alpha particles. *Hint:* Compare the masses of the ^{12}C nucleus and the three alpha particles.

90. ✪ **Space probe power supply.** The sunlight that reaches the outer solar system is too feeble to power spacecraft with solar panels. Space probes like *Cassini*, currently exploring Saturn and its moons, use a radioisotope thermoelectric generator (RTG), which converts the heat generated from radioactive decay to electricity. The RTG on *Cassini* (Fig. P30.90) generates 13 kW of heat using 23.4 kg of ^{238}Pu, which emits an alpha particle with 5.6 MeV of kinetic energy upon each decay. (a) Estimate the activity in units of Bq and Ci contained in the RTG. (b) *Cassini* was designed for an 11-yr mission. If the half-life of ^{238}Pu is 87.7 yr, how much power will be available from the RTG at the very end of the mission?

© NASA/JPL

Figure P30.90 The *Cassini* probe currently in orbit around Saturn. The RTG power supply is mounted on a long boom extending from the spacecraft.

Physics in the 21st Century

The LIGO experiment (laser interferometer gravitational wave observatory) employs a Michelson interferometer (Example 25.2) to measure extremely small changes in the distance between mirrors placed several kilometers apart. Physicists hope to use several observatories like this one, placed around the world, to detect gravitational waves generated in distant galaxies. (Courtesy of LIGO Laboratory)

Toward the end of the 19th century, some physicists believed that all the laws of physics had been discovered and that the main job left for physicists was to measure quantities like the fundamental constants and the orbital parameters of the Moon more accurately. These physicists were proven wrong, with the discoveries of relativity and quantum theory leading to a century of fantastic progress in physics. Advances in physics during the 20th century led to many important applications, including computers, space travel, cell phones, nuclear power, lasers, CD and DVD players, and magnetic resonance imaging. What's more, physics continues to be an active research area, with no end in sight. In this chapter, we describe several topics of current research in physics. We start with elementary particle physics and consider how protons, neutrons, and other subatomic particles are put together. This field began with the study of cosmic rays in the early part of the 1900s, so we'll briefly describe that history. We then discuss some topics in the field of astrophysics, which is concerned with the origin and fate of the universe. In the last section of this chapter, we briefly describe how physicists are teaming up with biologists, chemists, and engineers to address interdisciplinary questions, including the new field of nanoscience.

31.1 | COSMIC RAYS

Electrons, protons, and neutrons were discovered and identified as atomic building blocks in the early 1900s. It was also found that these and other particles bombard the Earth from outer space. Particles that arrive from space are called *cosmic rays*, a name chosen when physicists mistakenly believed that cosmic rays were a type of electromagnetic radiation. Cosmic rays are actually a mix of different particles, including protons (about 89%) and alpha particles (10%); the rest are the nuclei of elements more massive than He and other types of particles. Cosmic rays are often studied in experiments in which balloons carry particle detectors to high altitudes (Fig. 31.1). Other experiments examine the particles created when cosmic rays collide with atoms in the Earth's atmosphere. In fact, several types of particles, including positrons and muons, were first discovered in cosmic-ray experiments.

A key question for physicists is, "How are cosmic rays generated?" It is believed that some cosmic rays are created when stars collide. These violent collisions can produce cosmic-ray particles with extremely high kinetic energy. Cosmic rays are also created when a massive star uses up its fuel. The star then collapses (due to gravity) and triggers what is called a supernova explosion (Fig. 31.2). These explosions are also thought to be the main source of elements heavier than oxygen in the universe.

NASA

Figure 31.1 Some experiments use balloons to carry particle detectors to high altitudes and detect cosmic rays as they enter the atmosphere from outer space. Here NASA launches a balloon that will carry a package of instruments to the upper atmosphere.

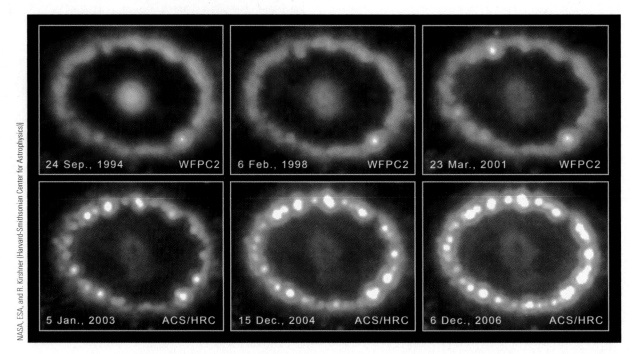

NASA, ESA, and R. Kirshner [Harvard-Smithsonian Center for Astrophysics]]

24 Sep., 1994 — WFPC2
6 Feb., 1998 — WFPC2
23 Mar., 2001 — WFPC2
5 Jan., 2003 — ACS/HRC
15 Dec., 2004 — ACS/HRC
6 Dec., 2006 — ACS/HRC

Figure 31.2 Physicists believe that cosmic-ray particles are generated in supernova explosions. These photographs of Supernova 1987A were obtained with the Hubble Space Telescope shortly after the supernova exploded in 1987. The bright ring around the star is due to a ring of gas that formed long before the explosion. When the supernova exploded, particles were ejected from the star; light from these particles is visible as violet in these photos. In the first photo (upper left), these particles were near the star (at the center of the ring). As time passed, the particles moved outward, and when they reached the gas ring, they produced the bright white spots. The bright spots at the upper-left and lower-right parts of the ring (seen most clearly in the image at the upper right) are due to stars between 1987A and the Earth.

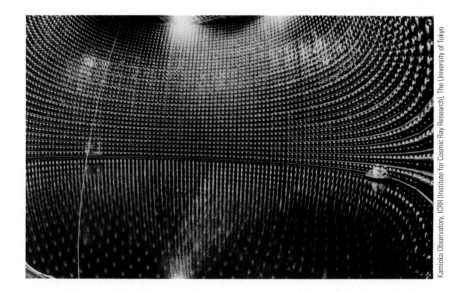

Kamioka Observatory, ICRR (Institute for Cosmic Ray Research), The University of Tokyo

Figure 31.3 This detector is called Super-Kamiokande. Located in Japan, it is designed to detect neutrinos that arrive at the Earth from space. This detector is a large container of water with walls covered with photodetectors that detect light generated when incoming particles interact with the water. This photo of the detector was taken during maintenance, when some of the water was drained out. Notice the people in a small boat (doing repair work).

Typical cosmic-ray energies are 100 MeV (1×10^8 eV), but some particles with energies above 10^{20} eV have been detected. This is an incredibly high energy—approximately the kinetic energy of a baseball moving at 100 mi/h—and is carried by a particle with about the mass of a proton! It is still not completely understood how particles are accelerated to such high energies.

New, more sensitive detectors are continually being built, with the goal of better understanding cosmic rays and other particles that arrive from outer space. Figure 31.3 shows the Super-Kamiokande detector located in Japan, built as part of a joint experiment with physicists from the United States and other countries. Super-Kamiokande consists of a large volume of water (approximately 64,000 m³) in a container whose walls are lined with photodetectors. The photodetectors detect light produced when cosmic rays and other particles from space collide with water molecules. The Super-Kamiokande experiment is designed to study neutrinos, which are generated as a result of fusion reactions in our Sun (Chapter 30) and in supernova explosions. These neutrinos give physicists a "glimpse" of the kinds of particle reactions that take place in these systems.

31.2 | MATTER AND ANTIMATTER

One of the developers of quantum physics was P. A. M. Dirac; around 1930, he formulated a theory of quantum mechanics that combined the quantum theory of Schrödinger and Heisenberg with the postulates of special relativity. Dirac's theory gave two important results: (1) it predicted the phenomenon of electron spin, and in Chapter 29 we saw how electron spin is essential for explaining the periodic table of the elements, and (2) it predicted the existence of a completely new particle having the same mass as the electron and an electric charge of the same magnitude as an electron but with opposite sign. Several years later, this particle, named the *positron*, was discovered in cosmic rays. The positron is denoted by the symbol e^+ to distinguish it from the ordinary electron, which is denoted by e^-. The superscripts indicate that the electron carries a negative charge, while the charge on the positron is positive. The positron is an example of *antimatter*, a particle of matter having the same mass and opposite charge as the corresponding particle of ordinary matter. Other examples of antimatter particles (*antiparticles*) are the antiproton and the antineutron.

Particles such as electrons, positrons, and protons undergo reactions, much like the reactions involving nuclei. When describing these reactions, we denote each of these particles by a symbol, similar to the symbols used in Chapter 30 to denote

Table 31.1 | Some Properties of Electrons, Protons, Neutrons, and Their Antiparticles

PARTICLE	SYMBOL	MASS (kg)	MASS (MeV/c^2)	CHARGE
Electron	e^-	9.109×10^{-31}	0.511	$-e$
Positron	e^+	9.109×10^{-31}	0.511	$+e$
Proton	p	1.673×10^{-27}	938	$+e$
Antiproton	$\overline{\text{p}}$	1.673×10^{-27}	938	$-e$
Neutron	n	1.675×10^{-27}	940	0
Antineutron	$\overline{\text{n}}$	1.675×10^{-27}	940	0

different nuclei. Protons are denoted p and neutrons as n; the corresponding anti-particles are denoted with an overbar, with $\overline{\text{p}}$ for the antiproton (pronounced "p bar") and $\overline{\text{n}}$ for the antineutron. Most other particles and antiparticles are denoted in a similar way, except for the positron, which is denoted by e^+ without a bar on top. For convenience, Table 31.1 lists the mass and electric charge of the electron, proton, and neutron along with the values for the corresponding antiparticles.

The annihilation and creation of antiparticles give clear examples of special relativity at work. For example, when an electron encounters its antiparticle (a positron), the two can undergo a reaction in which both particles are annihilated. This reaction is written as

$$e^- + e^+ \rightarrow \gamma + \gamma \tag{31.1}$$

In words, this reaction states that an electron plus a positron react to form two gamma rays, that is, two photons. This reaction must satisfy the conservation of energy, so the total energy before the reaction (the energies of the original electron and positron) is equal to the final energy (of the two gamma ray photons). The total initial energy includes the kinetic energies of the electron and positron plus their rest energies. According to special relativity (Chapter 27), the rest energy of an electron (rest mass m_e) is $m_e c^2$. The positron also has a rest mass m_e, so it has the same rest energy. By measuring the energies of the two gamma rays emitted in this reaction, one can determine the rest energies of the electron and positron and thus check the predictions of special relativity. The analysis is carried out in Example 31.1.

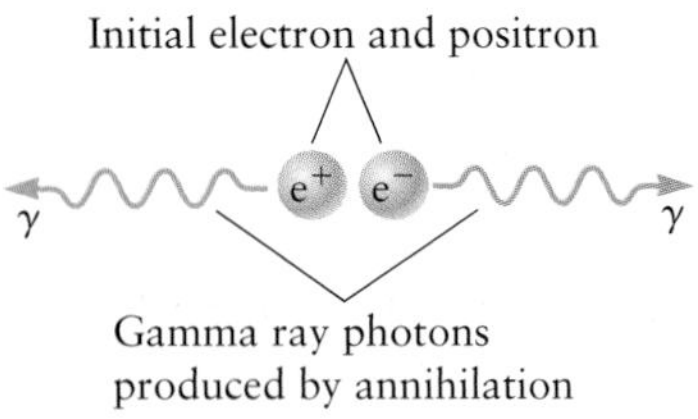

EXAMPLE 31.1 | **Electron–Positron Annihilation**

An electron and positron initially at rest annihilate each other, producing two gamma rays. Assume the gamma rays have the same energies and are emitted in opposite directions as sketched in Figure 31.4. Find the energy of one of the gamma ray photons.

RECOGNIZE THE PRINCIPLE

The annihilation reaction must conserve energy, so

$$\text{initial energy } (e^- + e^+) = \text{final energy } (\gamma \text{ rays})$$

The energy of the electron–positron pair consists of their kinetic energies (which are zero) and their rest energies. A gamma ray photon's energy is related to its frequency, with $E_{\text{photon}} = hf$.

SKETCH THE PROBLEM

See Figure 31.4. The electron and positron are at rest, so the initial momentum is zero. Since the final momentum must also be zero, the outgoing gamma ray photons are

Initial electron and positron

Gamma ray photons produced by annihilation

Figure 31.4 Example 31.1.

traveling in opposite directions. The two photons must also have the same frequency. This is the only way two photons can have a total momentum of zero.

FIND THE RELATIONSHIPS

We have

$$\text{initial energy } (e^- + e^+) = KE + \text{rest energy}$$

and the initial kinetic energy is zero because the electron and positron are initially at rest. The initial rest energy is $m_e c^2 + m_e c^2 = 2m_e c^2$ since the electron and positron have the same mass m_e. We are left with

$$\text{initial energy } (e^- + e^+) = 2m_e c^2$$

SOLVE

The energy of each of the gamma ray photons must therefore be

$$E_{\text{photon}} = m_e c^2$$

From Table 31.1, we have

$$\boxed{E_{\text{photon}} = 0.511 \text{ MeV}}$$

What does it mean?
This energy is *much* greater than that produced by atomic transitions. For example, in Chapter 29 we saw that the highest energy photon a hydrogen atom can emit is 13.6 eV.

Conservation Rules for Particle Reactions

The electron–positron reaction in Equation 31.1 obeys the principles of conservation of energy and conservation of momentum (Example 31.1). It also obeys another conservation law, the conservation of electric charge, as the total electric charge is the same (zero) before and after the reaction. All other particle reactions also conserve energy, momentum, and charge. These conservation laws are nothing new—we learned about them in our work on classical physics—but we'll soon encounter several other important conservation laws that go far beyond our classical laws of physics.

Figure 31.5 Concept Check 31.1.

CONCEPT CHECK 31.1 | Electron–Positron Creation?

Your friend says that he has observed an electron–positron reaction in which a single photon forms an electron and positron as shown in Figure 31.5. He also claims that the two particles are at rest after the reaction and that the energy of the photon is equal to the sum of the rest energies of the two particles. Is this reaction (a) possible or (b) not possible? Explain why.

The Stability of Antimatter

A positron is completely stable if it is kept away from electrons. Similarly, an antiproton is stable if it is kept away from protons. It is even possible to bring together a positron and an antiproton to form an atom of antihydrogen. Such atoms of antimatter have been studied in physics experiments, but it is very difficult to create large amounts of antimatter because it is difficult to contain it sufficiently far from regular matter to prevent annihilation.

The amount of energy released by the annihilation of an electron is very large (Example 31.1), and since a proton has a much larger mass than an electron, the

Figure 31.6 Richard Feynman (1918–1988) was one of the developers of quantum electrodynamics (QED). He also made important contributions to the Manhattan project in World War II, to our understanding of liquid helium, and to many other areas of physics. Feynman was a very entertaining and enthusiastic teacher of science to the general public.

energy released in the annihilation of a hydrogen atom with antihydrogen is much greater than calculated in Example 31.1. For this reason, antimatter is a popular fuel in science fiction stories. Unfortunately, it is not a practical energy source. All known methods for creating antiparticles require an amount of energy much greater than the rest energies of the particles that are created. The problem of making a suitable container hasn't been solved yet either.

The Earth and our galaxy are composed almost entirely of matter. One might expect to find an equal amount of antimatter in the universe, perhaps in the form of antimatter galaxies, but as far as astrophysicists can tell, the amount of matter in the universe is much greater than the amount of antimatter. Physicists are now trying to understand this puzzle.

31.3 | QUANTUM ELECTRODYNAMICS

The theory of the photon as a quantum of electromagnetic energy (i.e., a quantum of electromagnetic radiation) was first discussed by Einstein in 1905 and was an important step in the development of quantum theory (Chapter 28). Einstein's discovery of the photon was only part of a complete quantum theory of light, however. Electromagnetic waves (radiation) have an associated electric and magnetic field. Since the photon is a quantum of electromagnetic radiation, photons also "carry" electric and magnetic fields as described by Einstein's theory. Einstein's ideas were developed further by other physicists, including Richard Feynman (Fig. 31.6), Julian Schwinger, and Shinichiro Tomonaga, who developed a theory called *quantum electrodynamics* (QED), which combines electromagnetism and Maxwell's equations with quantum mechanics. According to QED, photons are the elementary quanta of electric and magnetic fields, and they play a role not just in electromagnetic radiation, but in all electric and magnetic forces.

Consider the interaction of two charged particles through Coulomb's law. A classical picture of this interaction is sketched in Figure 31.7A, showing the electric field produced by the charge on the left (q_1). The electric force on the right-hand charge (q_2) is due to this electric field. The QED picture of this force involves photons as sketched in Figure 31.7B; here the interaction between two charged particles is due to the exchange of photons, the elementary quanta of the electromagnetic field. We say that the electric force is "mediated" by photons. In this way, the photon is responsible for all electromagnetic forces.

Figure 31.7C gives a classical picture of how the exchange of particles can lead to a force between two objects. Here the "objects" are two children playing catch with a ball. The person throwing the ball experiences a recoil force during the throw

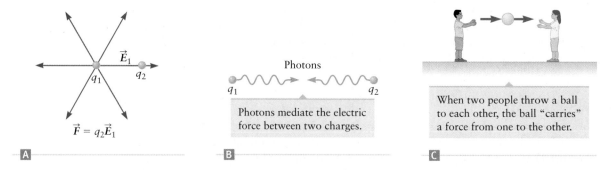

Figure 31.7 Force between two charged particles. **A** In the classical picture in Chapter 17, the force on one charge q_2 is due to the electric field produced by the charge q_1. **B** In the quantum mechanical picture of QED, the electric force is mediated (or "carried") from one charge to another by photons. **C** The force carried by photons is similar to the case of two people playing catch with a ball. The ball "carries" a force between the two people.

while the person catching the ball experiences a recoil force when she catches it. This ball plays the role of a photon in the electromagnetic force as both the ball and the photon mediate (or "carry") a force between two objects.[1] It is believed that all the fundamental forces in nature are mediated by elementary particles.

The photon is an example of an elementary particle, but it differs in two ways from the other particles we have discussed so far. First, the photon has zero rest mass. Second, the photon is its own antiparticle, so there is no "antiphoton."

31.4 | ELEMENTARY PARTICLE PHYSICS: THE STANDARD MODEL

Studies of nuclear and particle physics began in the early 1900s, when Rutherford and other physicists conducted many experiments in which various particles were fired at atoms and nuclei (Chapters 29 and 30), revealing that nuclei are composed of protons and neutrons. For a brief period, it was thought that electrons, protons, and neutrons are the fundamental particles from which all matter is composed, but that simple picture did not last long. Many other particles were discovered in cosmic-ray studies, in collision experiments, and in nuclear decay processes. It is now clear that protons and neutrons are themselves composed of particles called *quarks*, truly fundamental point particles whose charges are multiples of $\pm e/3$, one-third of the electron charge. Particles that are composed of quarks—such as protons and neutrons and their antiparticles—form a family of particles called *hadrons*. Not all particles are members of the hadron family, however. There is a second class of particles called *leptons*, which include electrons and positrons. The behavior of hadrons and leptons is described by what is called the *standard model* of elementary particles. Let's now consider the properties of hadrons and leptons.

The standard model describes the properties of fundamental particles and the interactions between them.

Quarks Bind Together to Form Hadrons

There are six different kinds of quarks, named "up" (denoted by the symbol u), "down" (d), "charm" (c), "strange" (s), "top" (t), and "bottom" (b). These names do not refer to any physical properties of the quarks; rather, they are just whimsical names coined by physicists. Each of these quarks has a corresponding antiquark; denoted $\bar{u}$, $\bar{d}$, and so on. Some properties of quarks are listed in Table 31.2.

Quarks were first discovered in collision experiments involving protons. When a high-energy electron collides with a proton, the way the electron scatters (i.e., its outgoing direction and energy) gives information about how mass and charge are distributed inside the proton. This is similar to Rutherford's experiment (Chapter 29), in which the way alpha particles are scattered from an atom indicate that a massive nucleus is located at the atom's center. Collision experiments with protons show that there are three pointlike particles inside. Table 31.3 gives the quark composition of the proton: two up quarks and one down quark. The total charge on the proton is the sum of the charges of the constituent quarks, so the proton's charge is

$$\text{proton charge} = 2 \times (\text{charge of u quark}) + (\text{charge of d quark})$$
$$= 2\left(\frac{+2e}{3}\right) + \left(\frac{-e}{3}\right)$$
$$= +e$$

as expected.

---

[1]This analogy is a very qualitative way to think about how photon exchange leads to forces and has its limitations. In particular, it is hard to give a classical explanation of how the exchange of particles can lead to an *attractive* force. The full theory of QED correctly describes both repulsive and attractive electromagnetic forces.

Table 31.2   Quarks and Their Properties

QUARK	SYMBOL	CHARGE	MASS (MeV/c^2)	ANTIPARTICLE
Up	u	$+2e/3$	4	$\bar{u}$
Down	d	$-e/3$	8	$\bar{d}$
Charm	c	$+2e/3$	1,500	$\bar{c}$
Strange	s	$-e/3$	150	$\bar{s}$
Top	t	$+2e/3$	176,000	$\bar{t}$
Bottom	b	$-e/3$	4,700	$\bar{b}$

Quarks can combine to form a hadron in two ways. Hadrons composed of three quarks are called **baryons** (Fig. 31.8A). Protons and neutrons are baryons; Table 31.3 lists a few other baryons, and dozens of other baryons have been observed and their component quarks identified. It is also possible for a quark and an antiquark to combine to form a particle. Hadronic particles composed of just two quarks are called **mesons** (Fig. 31.8B), and a few are listed in Table 31.4.

All hadrons are composed of quarks, so the interactions between quarks determine the properties of hadrons and how they interact with one another. The two most important hadrons are the proton and neutron, so the behavior of quarks also determines the properties of nuclei. Quarks are charged, so they interact via the electric (Coulomb) force. They also interact via the strong force mentioned in Chapter 30. The strong force holds quarks together to form protons and neutrons (nucleons) and is also responsible for holding protons and neutrons together to make nuclei.

Table 31.3   Properties of Some Baryons

PARTICLE	SYMBOL	CONSTITUENT QUARKS	LIFETIME (S)	MASS (MeV/c^2)
Proton	p	uud	Stable	938
Neutron	n	udd	890	940
Lambda zero	Λ^0	sud	2.6×10^{-10}	1116
Sigma plus	Σ^+	uus	0.8×10^{-10}	1189
Sigma zero	Σ^0	sud	6.0×10^{-20}	1193
Sigma minus	Σ^-	dds	1.5×10^{-10}	1197
Hyperion	Ξ^-	ssd	1.6×10^{-10}	1321

Note: There are many other baryons, which are composed of other combinations of three quarks and antiquarks.

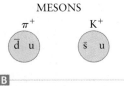

Figure 31.8 **A** Baryons and antibaryons are composed of three quarks. The three particles represented here are the proton (p), the antiproton ($\bar{p}$), and the neutron (n). **B** Mesons are composed of one quark and one antiquark. These mesons are called pions and kaons.

Table 31.4   Properties of Some Mesons

PARTICLE	S SYMBOL	CONSTITUENT QUARKS	LIFETIME (S)	MASS (MeV/c^2)
Pion (pi plus)	π^+	$u\bar{d}$	2.6×10^{-8}	140
Pi zero*	π^0	$d\bar{d}/u\bar{u}$	8.4×10^{-17}	135
Kaon (K plus)	K^+	$u\bar{s}$	1.2×10^{-8}	494
Kaon (K minus)	K^-	$\bar{u}s$	1.2×10^{-8}	494
Phi	φ	$s\bar{s}$	1.6×10^{-22}	1020

Note: There are many other mesons, which are composed of other combinations of a quark and an antiquark.

*The π^0 is a quantum mechanical combination of the $d\bar{d}$ and $u\bar{u}$ quark states.

Rules for "Making" Baryons and Mesons, and Their Reactions

The standard model places restrictions on the ways quarks can combine to form baryons and mesons and on the kinds of particle reactions that are possible. One of these restrictions has to do with the allowed electric charges of baryons and mesons. The charge of any particular baryon or meson can be found by simply adding up the charges of its constituent quarks. The quark charges are listed in Table 31.2, and the corresponding antiquarks have the opposite charges (so the $\bar{d}$ has a charge of $+e/3$, which is opposite to that of the d quark, etc.). Working out the charges of the baryons and mesons in Tables 31.3 and 31.4, you'll find that they all have electric charges of $+e$, $-e$, or zero; some other baryons (not listed here) have charges $+2e$ and $-2e$. The standard model guarantees that an isolated particle cannot have a fractional charge, that is, a charge of $\pm e/3$ or $\pm 2e/3$. Since quarks themselves have fractional charge, the implication is that individual quarks cannot be isolated. We'll explain why when we consider the strong force in more detail below.

In addition to this restriction on the charges of baryons and mesons, the standard model also predicts that *baryon number* must be conserved in all particle reactions. Each quark has a baryon number of $+\frac{1}{3}$, so the baryon number of a proton (which is composed of three quarks) is +1. Antiquarks have a baryon number of $-\frac{1}{3}$, so all mesons have a baryon number of zero. We'll discuss some examples of baryon number conservation below, after we describe leptons.

CONCEPT CHECK 31.2 | Which of these hadrons are allowed?

Which quark combination is *not* an allowed meson or baryon, (a) $c\bar{c}$, (b) bbs, (c) uud, (d) uu$\bar{d}$, or (e) $\overline{uu}\bar{d}$?

EXAMPLE 31.2 | Rest Mass of the Proton

The proton is composed of two up quarks and a down quark (p = uud). Compare the total rest mass of these three quarks with the rest mass of the proton. Are they equal?

RECOGNIZE THE PRINCIPLE

The rest masses of the proton and the u and d quarks are listed in Tables 31.2 and 31.3. We can simply add the appropriate masses and compare the results.

SKETCH THE PROBLEM

No sketch is needed.

FIND THE RELATIONSHIPS

The u quark has a mass of about 4 MeV/c^2 and the d quark mass is around 8 MeV/c^2, so the total mass of the quarks that make up a proton is

$$\text{mass of u + u + d} = (4 + 4 + 8) \text{ MeV}/c^2 = 16 \text{ MeV}/c^2 \qquad (1)$$

From Table 31.3, the rest mass of the proton is

$$\text{rest mass of proton} = 938 \text{ MeV}/c^2 \qquad (2)$$

SOLVE

Comparing Equations (1) and (2), the rest mass of the proton is *much* greater than the combined masses of its constituent quarks, so the answer is no .

What does it mean?

The difference between the masses in Equations (1) and (2) is due to the quark kinetic energies within the proton and from potential energy resulting from the

Insight 31.2
MEASURING THE MASS OF A QUARK
One way to measure the mass of a quark is through Newton's second law, $\Sigma F = ma$ (or the corresponding result of quantum mechanics). However, quarks are always confined in baryons or mesons, so it is impossible to study how a free quark would respond to an applied force. The best we can do is probe the motion or properties of quarks inside a baryon or meson and then deduce the quark mass using the theory of the strong force. Such an analysis must allow for the contributions of quark kinetic energy and the potential energy of the strong force, and these contributions are difficult to calculate with high accuracy. For this reason, the masses of the quarks are not known with the same level of precision as the masses of particles such as the proton, neutron, and electron.

strong interactions between quarks, together with the relativistic relation between energy and mass.

Leptons

We have just discussed the hadron family of particles, which are all assembled from quarks. Let's now consider a second family of particles called leptons. There are six fundamental leptons (Table 31.5) plus their corresponding antiparticles. These particles group naturally into three pairs: the electron (e^-) and the electron neutrino (ν_e); the muon (μ^-) and the muon neutrino (ν_μ); and the tau (τ^-) and the tau neutrino (ν_τ). The muon (lifetime = 2.2×10^{-6} s) and the tau (lifetime 2.9×10^{-13} s) are not stable. Electrons are stable, while the behavior of neutrinos is more complicated. Recent experiments indicate that as they travel through space, neutrinos change from one to another of the three types of neutrinos listed in Table 31.5, an effect called "neutrino oscillations." Another interesting property of neutrinos is that they have very small masses; the best experiments to date give only an approximate mass for the electron neutrino between 0.05 and 2 eV/c^2. That is more than 100,000 times less than the mass of an electron!

Leptons play important roles in certain reactions involving hadrons. For example, an "isolated" neutron (a neutron outside a nucleus) decays to form a proton, an electron, and an anti-electron neutrino:

$$n \rightarrow p + e^- + \bar{\nu}_e \tag{31.2}$$

Leptons are also produced in many nuclear decay reactions. In fact, neutrinos were first observed in the 1950s in studies of the particles emitted by a nuclear power reactor. A very large source of the neutrinos observed on the Earth are the nuclear fusion processes that power the Sun. Studying these solar neutrinos provides one of the best ways to understand the nuclear reactions that take place inside the Sun.

Reactions involving leptons, such as neutron decay (Eq. 31.2), must satisfy conservation of **lepton number**. Each lepton has a lepton number of +1, whereas each antilepton has a lepton number of −1. Since neutron decay involves both baryons and leptons, this process must conserve both baryon number and lepton number. Notice that all baryons have a lepton number of zero and all leptons have a baryon number of zero. For Equation 31.2, we have

$$
\begin{aligned}
n \quad &\rightarrow \quad p \; + \; e^- \; + \; \bar{\nu}_e \\
\text{baryon number:} \quad +1 &\rightarrow +1 + \; 0 \;\; + \; 0 \; = +1 \\
\text{lepton number:} \quad 0 \;\; &\rightarrow \;\; 0 \; + \; 1 \; - \; 1 \; = 0
\end{aligned}
\tag{31.3}
$$

Hence, although this reaction produces a lepton and an antilepton, it still conserves both baryon number and lepton number.

Table 31.5 Leptons and Their Properties

PARTICLE	SYMBOL	CHARGE	MASS/c^2	ANTIPARTICLE
Electron	e^-	$-e$	0.511 MeV	e^+
Electron neutrino	ν_e	0	$0.05 \text{ eV} < m < 2 \text{ eV}$	$\bar{\nu}_e$
Muon	μ^-	$-e$	106 MeV	μ^+
Muon neutrino	ν_μ	0	< 0.19 MeV	$\bar{\nu}_\mu$
Tau	τ^-	$-e$	1780 MeV	τ^+
Tau neutrino	ν_τ	0	< 18 MeV	$\bar{\nu}_\tau$

EXAMPLE 31.3 | Review Problem: Studying the Decay of a Tau Lepton

Elementary particles are produced and studied in collision experiments at laboratories like the one shown in Figure 31.9. Among the particles studied in this way is the tau lepton. Tau leptons have a lifetime of approximately $\Delta t_0 = 2.9 \times 10^{-13}$ s measured in the tau's reference frame (i.e., measured by an observer at rest relative to the tau). Suppose a certain collision experiment creates a tau particle with a speed $v = 0.99c$ relative to a physics laboratory. How far will this particle travel relative to the laboratory before it decays? Give the distance as measured in (a) the reference frame of the tau and (b) the reference frame of a physicist in the laboratory.

RECOGNIZE THE PRINCIPLE

The tau's speed is sufficiently high that we must consider the effects of special relativity. This example is a time-dilation problem very similar to Examples 27.2 and 27.4, in which a high-speed muon decayed in the Earth's atmosphere. According to an observer traveling with the tau (i.e., in the tau's reference frame), the lab moves at a speed of $0.99c$ for a time Δt_0. The physicist sees the tau moving by at $0.99c$, but observes a dilated lifetime. According to Equation 27.7,

$$\Delta t = \frac{\Delta t_0}{\sqrt{1 - v^2/c^2}} \qquad (1)$$

SKETCH THE PROBLEM

Figure 31.10 shows the tau lepton along with two observers. One observer travels with the tau and measures a lifetime Δt_0. The other observer (the physicist at rest in the lab) measures a dilated lifetime Δt. (Compare with Fig. 27.9.)

FIND THE RELATIONSHIPS

(a) The value Δt_0 of the tau lifetime given above is the proper time because it is measured in the tau's reference frame. The distance d_0 traveled by the tau as measured in its own reference frame is

$$d_0 = v\,\Delta t_0 \qquad (2)$$

where $v = 0.99c$.

(b) A physicist studying the moving tau measures a dilated lifetime Δt given by Equation (1). The distance traveled by the tau in the physicist's frame of reference is thus

$$d = v\,\Delta t = \frac{v\,\Delta t_0}{\sqrt{1 - v^2/c^2}} \qquad (3)$$

SOLVE

(a) Inserting the given values of v and Δt_0 from Equation (2) leads to

$$d_0 = (0.99c)\Delta t_0 = (0.99)(3.00 \times 10^8 \text{ m/s})(2.9 \times 10^{-13} \text{ s})$$

$$d_0 = 8.6 \times 10^{-5} \text{ m} = \boxed{0.086 \text{ mm}}$$

(b) Inserting the values for v and Δt_0 again from Equation (3), the distance traveled in the laboratory reference frame is

$$d = \frac{(0.99c)\Delta t_0}{\sqrt{1 - (0.99c)^2/c^2}} = \frac{(0.99)(3.00 \times 10^8 \text{ m/s})(2.9 \times 10^{-13} \text{ s})}{\sqrt{1 - (0.99)^2}}$$

$$d = 6.1 \times 10^{-4} \text{ m} = \boxed{0.61 \text{ mm}}$$

Courtesy of Fermilab

Figure 31.9 This large accelerator is at Fermi National Accelerator Laboratory, located near Chicago. These two large circular rings, each with a diameter of about 2 km, are part of a chain of accelerators in which protons and antiprotons travel with energies near 1 TeV (10^{12} eV). Collisions between protons and antiprotons are used to create and study quarks and other particles.

Observer traveling with τ^- measures proper time Δt_0.

$v = 0.99c$

τ^-
Tau lepton

Physicist in lab measures dilated time

$$\Delta t = \frac{\Delta t_0}{\sqrt{(1 - v^2/c^2)}}.$$

Figure 31.10 Example 31.3.

31.5 | THE FUNDAMENTAL FORCES OF NATURE

In Section 31.4, we described a few of the properties of hadrons and leptons such
as their charge and mass and some of the reactions they can undergo. To get a complete understanding, we must also consider the forces that act between them. There
are four fundamental forces, as listed and compared in Table 31.6.

The Strong Nuclear Force Only Acts on Quarks

Quarks carry electric charge, so they experience electric and magnetic forces, but
the largest force between two quarks is the strong force. The strong force actually
has two main "roles": it binds quarks together to form particles such as protons
and neutrons, and it holds protons and neutrons together to form nuclei. After the
quark model was proposed, physicists searched unsuccessfully for free particles
with charge $\pm e/3$ or $\pm 2e/3$. Despite much effort, a single isolated quark has never
been observed, which can be explained by the theory of the strong force and the
standard model. One striking prediction of these theories is that the energy required
to separate two quarks grows larger as the separation increases; eventually, at large
separations, this energy is large enough to produce new quark–antiquark pairs that
then bind to form baryons and mesons. This phenomenon is called *confinement*
and is why an individual quark is never observed in isolation.

A proton is composed of two up quarks and a down quark, a combination
denoted by uud. Quarks are quantum particles obeying the Pauli exclusion principle
(Chapter 29), so these two quarks cannot occupy the same quantum state. Quarks
have been found to possess a new quantum property called *color* that helps satisfy
the Pauli exclusion principle in the case of a proton. All quarks carry a *color quantum number* with the possible values red, green, and blue, like the spin quantum
number of an electron. From Section 29.5, two electrons can occupy the 1s state of
a helium atom provided that they have different spin values. In the same way, the
two up quarks in a proton must have different color values.

Table 31.6 The Fundamental Forces in Nature

FORCE	FORCE-CARRYING PARTICLE(S)	ACTS ON	RESPONSIBLE FOR	COMMENTS
Electromagnetism (QED)	Photon	All particles that carry electric charge	Electric and magnetic forces	Unified with the weak force as the electroweak force
Weak force	W^+, W^-, Z	Leptons and quarks	Beta decay and other nuclear decay processes	
Strong force (QCD)	Gluons	Quarks	Binding of quarks to form baryons and mesons	Also responsible for binding of protons and neutrons in the nucleus
Gravitation	Graviton (not yet observed)	All particles[a]	Gravity	

[a]From general relativity, gravity acts on photons.

The Strong Force Is Mediated by Gluons

In the language of quantum electrodynamics introduced in Section 31.3, photons are said to "mediate" electromagnetic interactions because they "carry" the electromagnetic force acting on charged particles. In the same way, the strong force is mediated or carried by particles called *gluons*, and there are eight (!) different types of gluons. The force exerted between two quarks depends on the type of each quark.

The theory of the strong force is called *quantum chromodynamics*, or QCD. One prediction of QCD is that baryon number is always conserved in a reaction as we mentioned in connection with the decay of a neutron (Eq. 31.3). All known reactions obey the conservation of baryon number, now believed to be a fundamental conservation principle.

Forces Acting on Leptons: Electromagnetism and the Weak Force

Three of the fundamental leptons—the electron, muon, and tau—are charged, so they experience electric and magnetic forces. These forces are described by the theory of QED. The strong force that is so important for quarks and hadrons does not act on leptons. All leptons, however, do experience what is known as the *weak force*, which is carried by three different particles called the W^+, the W^-, and the Z. The weak force acts on both leptons and quarks (and hence on all hadrons). This force has an extremely short range; two particles must be within about 10^{-18} m $=$ 0.001 fm to experience this force. This distance is about one one-thousandth the diameter of a proton! Two particles rarely come this close together; hence, the name "weak" for this force. Even so, the weak force is responsible for some important processes such as the decay of the neutron (Eq. 31.2):

$$n \rightarrow p + e^- + \bar{\nu}_e$$

The neutron and proton are composed of quarks, so we can also write this reaction as

$$udd \rightarrow uud + e^- + \bar{\nu}_e \tag{31.4}$$

Hence, the decay of the neutron converts one of the down quarks (d) in the neutron into an up quark (u). Processes in which one type of quark is converted into another type always involve the weak force.

EXAMPLE 31.4 | Particle Decay and Conservation Rules

The negatively charged pion π^- is a meson composed of two quarks $\bar{u}d$. The π^- meson can decay into a muon and an antimuon neutrino:

$$\pi^- \rightarrow \mu^- + \bar{\nu}_\mu \tag{1}$$

Show that this decay conserves both charge and lepton number.

RECOGNIZE THE PRINCIPLE

To determine if this decay conserves charge, we need to compare the total charge before and after the decay. We analyze the conservation of lepton number in the same way.

SKETCH THE PROBLEM

No sketch is necessary.

IDENTIFY THE RELATIONSHIPS

The π^- meson has a charge of $-e$. Mesons are hadrons, so the lepton number of the π^- is zero. The μ^- muon has a charge of $-e$ and a lepton number of $+1$. The antimuon neutrino is neutral, and since it is an antiparticle, its lepton number is -1.

SOLVE

We can rewrite the decay in Equation (1) as

$$\pi^- \;\rightarrow\; \mu^- \;+\; \bar{\nu}_\mu$$

$$\text{charge} \;=\; -e \quad -e \quad 0$$

$$\text{lepton count} \;=\; 0 \quad +1 \quad -1$$

Hence, both charge and lepton number are indeed $\boxed{\text{conserved}}$.

What does it mean?

We have also seen that particle decays must conserve baryon number. In Equation (1), the final baryon number is zero because the muon and antimuon neutrino are both leptons. The π^- meson contains a $\bar{u}$ and a d quark; since the $\bar{u}$ is an antiparticle, the total baryon number of the π^- meson is zero, so baryon number is also conserved in this decay.

Insight 31.3

WHAT DOES IT MEAN WHEN TWO FORCES ARE "UNIFIED"?

Two forces are "unified" when it is shown that they are two aspects of the same "underlying" force. For example, in Section 27.11 we showed that the electromagnetic force on a moving charge can be viewed as either an electric force or a magnetic force, depending on the relative motion of the charge and an observer. Hence, we say that these are two aspects of the *electromagnetic* force. Maxwell's theory gives a unified theory of electromagnetism.

Unification of the Weak Force and Electromagnetism

In the 1970s, it was discovered that the weak force and electromagnetism are actually two different aspects of the same force, similar to the relation between electricity and magnetism. Prior to Maxwell, physicists thought that electricity and magnetism were two completely different types of forces, but Maxwell showed that they are really two aspects of the same phenomenon (hence the name "electromagnetism"). Maxwell thus provided a unified theory of electricity and magnetism. In the same way, Sheldon Glashow, Abdus Salam, and Steven Weinberg developed a theory that unifies electromagnetism and the weak force. This combined theory of the *electroweak force* explains phenomena that cannot be accounted for by either separate theory. This theory predicted the existence of new particles that carry the electroweak force; these particles—the W^+, the W^-, and the Z (Table 31.6)—were then found in subsequent experiments.

Gravitation

Newton's theory of gravitation (Chapter 5) was extended by Einstein in his work on general relativity (Chapter 27). Gravitation is another fundamental force in nature. It plays an important role in our everyday terrestrial lives and is also responsible for the motion of the planets and stars in the universe. Even so, in some ways gravity is much weaker than the other fundamental forces of nature. For example, the electric force between two electrons is much larger in magnitude than the gravitational force between them.

The particles that mediate the electromagnetic, strong, and weak forces have all been observed and their properties measured (Table 31.6). It is believed that gravity is also carried by a particle; although this particle has not yet been directly observed, physicists believe it must exist and have already given it a name: the *graviton*.

EXAMPLE 31.5	Comparing the Electromagnetic and Gravitational Forces

The two up quarks (u) in a proton are separated by a distance $r \approx 1$ fm ($= 1 \times 10^{-15}$ m). These quarks each have an electric charge $+2e/3$, so they repel each other through an electric force obeying Coulomb's law. They also attract each other through a gravitational force. What is the ratio of the magnitudes of the electric and gravitational forces between these two quarks?

RECOGNIZE THE PRINCIPLE

The gravitational force between two point particles of mass m separated by a distance r is

$$F_{\text{grav}} = \frac{Gmm}{r^2}$$

while the electric force between two point charges q is

$$F_{\text{elec}} = \frac{kqq}{r^2}$$

We wish to calculate $F_{\text{grav}}/F_{\text{elec}}$ for two quarks in a proton.

SKETCH THE PROBLEM

Figure 31.11 shows a classical picture of two u quarks separated by a distance r. The electric force and the gravitational force between them are both proportional to $1/r^2$.

IDENTIFY THE RELATIONSHIPS

Let m_u be the mass of an up quark. The force of gravity between two up quarks is then

$$F_{\text{grav}} = \frac{Gm_u m_u}{r^2}$$

The charge on an up quark is $q = +2e/3$ (Table 31.2), so the electric force between two up quarks is

$$F_{\text{elec}} = \frac{kqq}{r^2} = \frac{k(2e/3)^2}{r^2} = \frac{4ke^2}{9r^2}$$

The ratio of these two forces is

$$\frac{F_{\text{grav}}}{F_{\text{elec}}} = \frac{Gm_u^2/r^2}{4ke^2/(9r^2)} = \frac{9Gm_u^2}{4ke^2} \tag{1}$$

To evaluate Equation (1), we need the mass of the u quark, m_u. From Table 31.2, the mass of the up quark is $m_u = 4\ \text{MeV}/c^2$. Converting this value to kilograms, we get

$$m_u = \frac{4\ \text{MeV}}{c^2} = \left[\frac{4 \times 10^6\ \text{eV}}{(3.00 \times 10^8\ \text{m/s})^2}\right]\left(\frac{1.60 \times 10^{-19}\ \text{J}}{1\ \text{eV}}\right) = 7 \times 10^{-30}\ \text{kg}$$

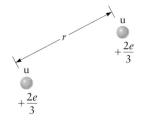

Figure 31.11 Example 31.5.

SOLVE

Inserting our value of m_u into Equation (1) along with values of the other factors leads to

$$\frac{F_{\text{grav}}}{F_{\text{elec}}} = \frac{9Gm_u^2}{4ke^2} = \frac{9(6.67 \times 10^{-11}\ \text{N}\cdot\text{m}^2/\text{kg}^2)(7 \times 10^{-30}\ \text{kg})^2}{4(8.99 \times 10^9\ \text{N}\cdot\text{m}^2/\text{C}^2)(1.60 \times 10^{-19}\ \text{C})^2}$$

$$\boxed{\frac{F_{\text{grav}}}{F_{\text{elec}}} = 3 \times 10^{-41}}$$

What does it mean?
The value of r canceled when we computed the ratio $F_{\text{grav}}/F_{\text{elec}}$, so the ratio has this value for all values of r. The gravitational force between two up quarks is thus always *much* smaller than the electric force. That is why gravity is considered the weakest of the fundamental forces.

The Four Forces in Nature and Their Unification

We have referred to the four fundamental forces in nature, but as indicated in Table 31.6 the electromagnetic and weak forces are two aspects of the same electroweak force. So, one could also say that there are only three fundamental forces in nature.

In addition, many physicists are attempting to unify the electroweak and strong forces; such a **grand unified theory** does not yet exist and is the subject of much current research. Going one step further, some physicists suspect that all forces in nature can be unified into one "theory of everything." A candidate for such a theory is string theory in which all particles are stringlike objects in a high-dimensional space. The details of such unified theories are still very uncertain. Einstein was working on this problem up until his death in 1955 and many others are working on it now, so progress has been slow.

31.6 | ELEMENTARY PARTICLE PHYSICS: IS THIS THE FINAL ANSWER?

Throughout this book, we have emphasized the simplicity of the laws of physics. In Newton's mechanics, all motion could be understood using just three laws (Newton's laws). Prior to quantum mechanics, all electromagnetic phenomena could be explained using Maxwell's four equations. At an intuitive level, scientists believe that nature is inherently "simple" and that the fundamental forces and particles should reflect this simplicity. However, adding up the number of quarks and leptons in Tables 31.2 and 31.5 along with the photon and other force-carrying particles in Table 31.6 brings the total number of fundamental particles to more than two dozen. Some physicists (including the author) suspect that a simple universe should not contain this many "fundamental" particles. For this reason, many physicists believe that our current understanding of fundamental particles and forces is not complete. The search for the ultimate laws of physics is thus still in progress. While debate over the correct laws of physics continues, it should not overshadow the theories we *do* have. These theories very accurately describe the quantum properties of atoms and molecules, the behavior of nuclei, the production of energy in the Sun, the behavior of the laser in your CD player, the motion of your car when you apply the brakes, and the propagation of the electromagnetic waves used by your cell phone. Our current theories may turn out to be only approximate, just as Newton's mechanics is an approximate theory, but our present theories do work *extremely* well.

31.7 | ASTROPHYSICS AND THE UNIVERSE

The Big Bang and the Expansion of the Universe

How did the universe begin, and how will it end? Did it even have a beginning? Will it ever end? Questions about the origin and fate of the universe have deep implications for philosophy and many other fields. Such questions came within reach of physics in the early 1900s when astronomers observed that the spectrum of light from distant galaxies can be used to determine the motion of those galaxies relative to the Earth. In Chapter 13, we discussed the Doppler effect, which causes the frequency of a wave to shift to either higher or lower values when the source and observer are in relative motion. As a result, the frequency of light or other electromagnetic radiation is shifted upward when the source and observer are moving toward each other and is shifted to a lower frequency when they are moving apart. (See also Chapter 23 and Example 23.7.)

In the early 1900s, astronomers discovered that light from all galaxies visible from the Earth is shifted to lower frequencies compared with what would be observed if their sources were at rest. This effect is called a **red shift** because red light is at the low frequency end of the visible spectrum. Using this measured frequency shift together with the Doppler effect formula, one can calculate the *apparent* velocity of a particular galaxy relative to the Earth. (We'll explain why we use the term *apparent* shortly.) Astronomer Edwin Hubble collected such data for

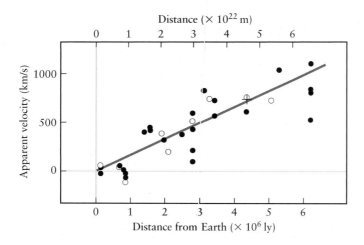

Distance ($\times 10^{22}$ m)

Apparent velocity (km/s)

Distance from Earth ($\times 10^6$ ly)

Figure 31.12 Edwin Hubble's original data showing how the apparent velocities of different galaxies increase with distance from the Earth. Galaxies farthest from the Earth have the largest velocities relative to the Earth. The bottom scale shows distance in units of light-years (ly); 1 ly is the distance light travels in 1 year.

many galaxies and plotted their apparent velocities as a function of their distances from the Earth, obtaining the result in Figure 31.12. His data showed that galaxies farthest away have the largest apparent velocities, and all galaxies are moving away from the Earth.

Hubble's results in Figure 31.12 (and much data gathered more recently) follow a linear behavior, with the apparent velocity v of a galaxy proportional to its distance d from the Earth:

$$v = H_0 d \qquad (31.5)$$

Hubble's law

This is called **Hubble's law**, and H_0 is called the Hubble constant. The best current estimate for the Hubble constant is $H_0 = 21$ (km/s)/(10^6 ly), where 1 ly = 1 light-year is the distance light travels in 1 year. So, a galaxy that is 10^6 ly (one million light-years) from the Earth is moving away at an apparent velocity of 21 km/s.

Hubble's law is evidence for the **Big Bang** model of the universe, which states that the universe originated with an explosion. At the instant of this explosion all the universe was an infinitesimal point, and since this "bang," the universe has expanded as sketched in Figure 31.13. All galaxies appear to be moving away from us, which would seem to suggest that the Earth is at the center of it all. Figure 31.13, however, shows that an observer at any location in an expanding universe will find that all galaxies are moving away from him. Galaxies with the largest apparent velocities relative to the Earth are farthest away from the Earth. Since light takes a certain amount of time to travel from these galaxies to the Earth, this light was emitted when the universe was younger than at present. In this way, studying the most distant galaxies gives a picture of the universe shortly after it was formed.

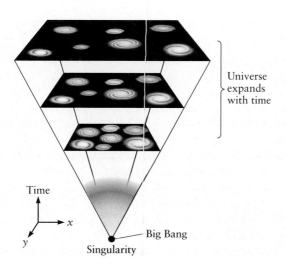

Universe expands with time

Time

x

y

Big Bang

Singularity

Figure 31.13 Schematic of the Big Bang. As time passes, the universe is expanding, carrying galaxies farther apart.

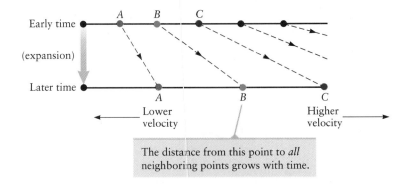

Figure 31.14 Schematic of how the relative distance between any two points in an expanding universe increases with time. As time increases (going from the top line to the bottom line), the distance between any pair of points increases. For an observer at any point, all other points are moving farther away, and each observer perceives that he or she is at the "center" of the expansion.

The distance from this point to *all* neighboring points grows with time.

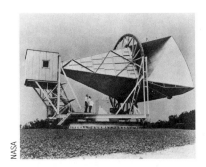

Figure 31.15 Arno Penzias and Robert Wilson were using this antenna to study how radio and microwaves propagate through the atmosphere. They discovered blackbody radiation arriving at the Earth from outer space, with an apparent blackbody temperature of approximately 3 K.

Now let's explain why we refer to the quantity v in Hubble's law as an *apparent* velocity. In our normal work in mechanics, we define and measure the velocity of an object in terms of its position relative to a fixed coordinate system, that is, relative to "markers" in space such as a coordinate axis. With an expanding universe (Fig. 31.14), these markers are themselves expanding along with the universe. That is, the "fabric" of space itself expands with time, carrying us and all other galaxies along with it. Although in some ways we can think of the expanding universe as involving "motion," it is a very different type of motion than in normal mechanics, which is why astronomers usually refer to the astronomical red shift as giving an "apparent" velocity. This picture also gives a more accurate way to understand the astronomical red shift. Light emitted by a distant galaxy has a certain wavelength. As that light travels to the Earth, the universe expands, "stretching" space and thereby increasing the wavelength of the light. This increase is similar to a Doppler red shift, but now the shift to a longer wavelength (lower frequency) is due to the expansion of space rather than the simple motion of the source.

One of the most important pieces of evidence in support of the Big Bang theory was discovered by two scientists, Arno Penzias and Robert Wilson, who were studying satellite communications systems at Bell Laboratories in the 1960s. They found that their antenna (Fig. 31.15) was detecting a small amount of radiation even when aimed at "empty" space. The intensity of this radiation was the same as that produced by a blackbody at a temperature of approximately 3 K, and this radiation from space is now known as the cosmic microwave background radiation. The existence of this radiation and its blackbody temperature are key features supporting the Big Bang model. According to the model, this radiation was emitted when the universe was only a few hundred thousand years old and then red-shifted due to subsequent expansion so that it now has an effective blackbody temperature

Figure 31.16 Recent experiments have studied the cosmic background radiation in great detail. This plot shows how the radiation intensity varies slightly from point to point in the sky, with the red regions being hottest and the blue regions coolest. The temperature variations are quite small, with the hottest regions only about 10^{-5} K higher in temperature than the average. These variations were produced by slight variations of the density of matter throughout space shortly after the Big Bang. These density variations eventually became stars and galaxies.

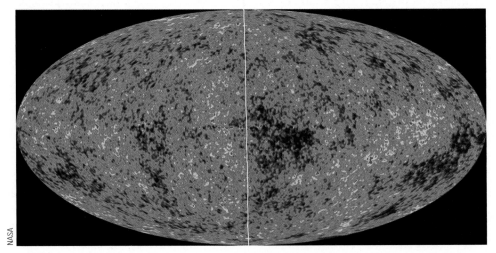

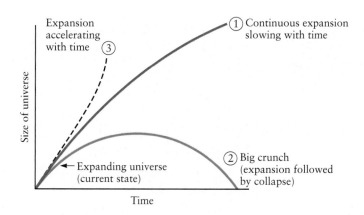

Figure 31.17 Different possible scenarios for the fate of the universe. Curve 1 shows a universe that expands forever, while curve 2 shows a universe that collapses back to a point (the big crunch). Our universe is now on the far left side of this plot and is expanding. Curve 3 shows a universe that is expanding at an ever-increasing rate ("accelerating"). Recent observations suggest that our universe is following a curve like curve 3.

of 3 K. In more recent experiments, physicists have measured tiny variations in the intensity of the cosmic microwave background radiation across the sky (Fig. 31.16). These variations give information on the state of the early universe when stars and galaxies were first forming.

Fate of the Universe

Hubble's law together with the cosmic microwave background radiation and other observations are strong evidence that the Big Bang occurred approximately 14 billion years ago. Physicists are also interested in the ultimate fate of the universe. For many years, it was believed that there were just two possibilities, labeled curves 1 and 2 in Figure 31.17. Curve 1 shows a universe that continues to expand forever, but with the rate of expansion slowing due to the gravitation attraction between stars. On the other hand, the gravitational attraction may be large enough so that the entire universe collapses back to an infinitesimal point; this model, called the "big crunch," is illustrated by curve 2 in Figure 31.17.

These "options" for the ultimate fate of the universe depend how much matter it contains; if the amount of mass is large enough, the mutual gravitational attraction will lead to the big crunch. This issue has prompted astronomers to do careful measurements of the total mass contained in the universe, and the results show that the universe contains a considerable amount of **dark matter**, matter that does not emit or absorb enough radiation to be observed directly. While dark matter cannot be "seen" with an ordinary telescope, its presence can still be measured by studying how it exerts a gravitational force on nearby galaxies. In fact, the best current measurements suggest that as much as 80% to 90% of the mass in the universe is dark matter! The nature of dark matter is not yet known. Speculations are that it could be composed of hitherto undiscovered fundamental particles, but that is not yet proven.

Another big surprise in recent years has been the discovery that the universe is actually described by curve 3 in Figure 31.17, which shows that the expansion of the universe is *accelerating*. According to the standard Big Bang model, all matter was given an initial velocity by the Big Bang. If there were no forces on a particular galaxy, it would continue forever with a constant velocity and the universe would expand forever at a constant rate. The effects of gravity should eventually cause the relative speed of a galaxy to decrease with time (as in curves 1 and 2 in Fig. 31.17), but the most recent observations indicate that galaxies are instead accelerating.

The explanation of this puzzle is not yet known. Some physicists have suggested that the acceleration is due to a phenomenon called **dark energy**. Currently, there is no agreed-upon theory of dark energy or even a precise definition of what it is. The rough idea is that an entirely new form of energy is present in "empty" space and that this energy is imperceptible to most observations ("dark"). The density of this energy is very low and is normally not noticed. In the very large volume of the

Figure 31.18 Image of the surface of Fe atoms on the surface of Cu, obtained using a scanning tunneling microscope (STM), one of the new tools of nanoscience. Each pyramid-like "bump" is an individual Fe atom. The STM is able to form images of atomic scale structure and to move atoms from place to place, producing the atomic arrangement shown here.

universe, however, dark energy produces significant effects and is responsible for repelling matter and for the accelerating expansion of the universe. The existence of dark energy and its properties are now being debated, so the picture of fundamental forces and elementary particles given in this chapter may soon need to be revised.

31.8 | PHYSICS AND INTERDISCIPLINARY SCIENCE

Physicists are actively studying many areas besides elementary particle physics and astrophysics. In fact, there are so many interesting research topics in physics that there is no way we can even list them all. We'll mention just a few of these topics and give only very brief descriptions.

Nanoscience is the study of objects with sizes in the range of roughly 1 nm to 1000 nm. It spans the regime of atoms and molecules up to objects such as cells, which can be seen with an optical microscope. There are several things that make this "nanoworld" interesting. First, quantum effects are important, so some of the "mysteries" of quantum theory play a key role in the properties of nanoscale objects. Second, nanoscientists have developed ways to manipulate matter atom by atom (Fig. 31.18), thus making things not found in nature, including materials that are both stronger and lighter than steel. Third, the new properties of nanoscale objects are beginning to lead to entirely new kinds of applications. For example, computers that store information using the quantum properties of individual electrons are now being developed; such computers, if successful, could be much faster and more powerful that those we have today.

Another promising area is *biophysics*, in which physics principles and methods are applied to problems in biology. This work is giving important new insights into a wide range of problems, including photosynthesis, molecular motors, how the nose detects odors, and how the brain works. Some of these advances involve new techniques from nanoscience. For example, nanoscientists may soon be able to probe and assemble DNA and other biomolecules atom by atom (Fig. 31.19), which could lead to the design of custom drugs, breakthrough medical therapies, new species of plants with higher yields than previously possible, and much more.

Physics in the 21st century thus has much to offer humanity and promises to be a very exciting journey.

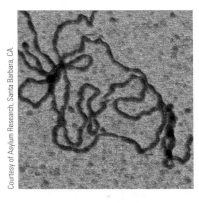

Figure 31.19 Image of strand of DNA obtained using an atomic force microscope. The "bumps" on the DNA strand are individual base pairs.

Fundamental particles

There are two families of *fundamental particles*:

There are six types of *leptons*.
electron (e^-)
electron neutrino (ν_e)
muon (μ^-)
muon neutrino (ν_μ)
tau (τ^-)
tau neutrino (ν_τ)

All leptons are point particles.

All *hadrons* are composed of *quarks*.
There are six types of quarks:

up (u)	strange (s)
down (d)	top (t)
charm (c)	bottom (b)

Quarks are point particles.
Hadrons composed of three quarks are called *baryons*. They include protons (denoted p and composed of two up quarks and one down quark, uud), and neutrons (n, quark composition udd).
Hadrons composed of two quarks are called *mesons*.

Matter and antimatter

For each fundamental particle, there is a corresponding antiparticle that has the same mass but opposite electric charge. For example, the antiparticle of the electron is the positron (denoted e^+). The electron has charge $-e$, whereas the positron has charge $+e$.

Fundamental forces

There are four fundamental forces in nature, and it is believed that each is carried or "mediated" by one or more particles.

1. *Electromagnetism* is described by Maxwell's equations. The quantum theory of this force is called *quantum electrodynamics* (QED). The electromagnetic force is carried by the photon.
2. The *strong force* is an attractive force between quarks and holds quarks together to form baryons (including protons and neutrons) and mesons. The strong force also attracts protons and neutrons and holds them together in the nucleus. The strong force is carried by particles called *gluons* and is described by *quantum chromodynamics* (QCD).
3. The *weak force* acts on all leptons and hadrons. The weak force is much smaller in magnitude than the strong force, but is responsible for processes such as beta decay of a nucleus. The weak force is carried by the W^+, W^-, and Z particles.
4. *Gravity* is the weakest of the fundamental forces. There is no quantum theory of gravity yet, but it is believed that the gravitational force is transmitted by a particle called the *graviton*. Gravitons have not yet been observed.

The *standard model* is a quantum mechanical theory for the behavior of the electromagnetic, strong, and weak forces and the interactions between leptons and hadrons.

(Continued)

Unification of forces

Electromagnetism is an example of a "unified" theory because it reveals electricity and magnetism as two aspects of a single type of force. The quantum theory of electromagnetism (QED) has also been unified with the weak force. These combined forces are called the ***electroweak*** force. Many physicists suspect that there is a single unified theory that includes all the fundamental forces, but this theory has not yet been discovered.

Measurements of the ***red shift*** of light from distant galaxies show that the universe is now expanding and provides evidence for the ***Big Bang theory*** of the universe.

QUESTIONS

SSM = answer in Student Companion & Problem-Solving Guide

⊗ = life science application

1. In Figure 31.4, an electron and a positron annihilate and produce electromagnetic radiation in the form of two gamma ray photons, each of energy 0.511 MeV. Explain why it is not possible for this reaction to give a single gamma ray photon with energy 2 × (0.511 MeV) = 1.022 MeV.

2. Explain why there are no doubly charged mesons, that is, mesons with an electric charge of $+2e$ or $-2e$.

3. How many different hadrons are stable outside the nucleus? *Hint*: They are all listed in Table 31.3.

4. Consider the hypothetical reaction in Figure Q31.4 in which an electron (e^-) annihilates a positron (e^+), with the positron initially at rest. Can this reaction result in a single emitted photon? Explain why or why not.

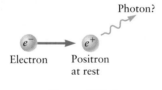

Electron   Positron at rest

Photon?

Figure Q31.4

5. The Σ^+ is composed of two up quarks (u) and one strange quark (s). The mass of an up quark is about 4 MeV/c^2, while the mass of a strange quark is about 150 MeV/c^2. Does the mass of the Σ^+ come mainly from the masses of these quarks or from the potential energy associated with binding? Explain.

6. Design an experiment that could distinguish a neutron from an antineutron.

7. Are mesons composed of (a) a total of three quarks and antiquarks, (b) three leptons, or (c) a quark and an antiquark?

8. Are baryons composed of (a) three quarks, (b) three leptons, or (c) two quarks?

9. How many quarks and antiquarks are present in an antimeson?

10. Why is it that we never see a single electron decay to a pair of photons? Could an electron decay to a pair of neutrinos? Why or why not? Name the explicit conservation rule each of these supposed decay schemes would violate.

11. SSM A particle with nonzero velocity enters a region of space that has a uniform magnetic field directed into the page as depicted in Figure Q31.11. The resulting path (trajectory) followed by the particle will depend on the particle's charge and mass. Sketch qualitatively the trajectory of an electron, a positron, a proton, and a neutron. Assume all have the same initial velocity.

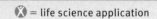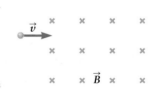

Figure Q31.11

12. At sea level, a significant number of cosmic-ray particles are muons with relativistic velocities. If the muons have a mean lifetime of only 2.2 microseconds, where do they come from and how are they produced?

13. Describe some naturally occurring common processes that produce antimatter. How common are anti-electrons (positrons) on the Earth? (Consider radioactivity.) How common are antimuons on the Earth?

14. It has been suggested that Newton's law of gravity brought about the first unification of physics principles. In what way does his law of gravity unify the motion of heavenly bodies with the motion of earthly objects?

15. Unification of physical laws often leads to new concepts and novel insights. Describe some of the consequences from the unification of the physics of electricity with that of magnetism. How was the concept of light altered?

16. Does the universe have a center? Does the universe have edges? If not, explain how that could be possible.

17. Consider the science of some hypothetical intelligent life-forms inhabiting a planet in a galaxy many millions of light-years away from the Earth. Would they measure the same Hubble constant as we do?

18. Given the current age of the universe and a telescope powerful enough, how far away is the most distant object we could observe? What kind of object is it?

19. SSM Due to the cosmological red shift, the photons that make up the cosmic microwave background are, as the name implies, in the microwave region of the electromagnetic spectrum. In what region of the electromagnetic spectrum were these photons when they were first released?

31.1 COSMIC RAYS

1. A cosmic-ray proton has a kinetic energy of 10^{19} eV. What is its speed? *Hint*: You will have to use the relativistic expression for the kinetic energy from Chapter 27.

2. The collisions of cosmic-ray particles with the Earth's atmosphere often produce muons that subsequently decay into an electron and two neutrinos according to the reaction

$$\mu^- \;\rightarrow\; e^- + \nu_\mu + \bar{\nu}_e$$

What is the maximum possible kinetic energy of the electron? Assume the muon is at rest just before it decays and the masses of the neutrinos are negligible.

3. A cosmic-ray proton has a de Broglie wavelength of 0.020 fm. What is its kinetic energy? *Hint*: Use the relativistic relations for momentum and energy along with de Broglie's relation for the wavelength of a particle-wave in Equation 28.9.

4. SSM ★ At sea level, some hundreds of muons go though your body each second. A muon has a mean lifetime of 2.2 microseconds, so why do we find so many near the Earth's surface? Consider a muon with a kinetic energy of 150 MeV that is created in the upper atmosphere when a cosmic ray collides with a nitrogen atom. (a) Determine the velocity of the muon. (b) Neglect relativity and calculate how far the muon can move from its creation assuming it decays within 2.2 microseconds. (c) Now take into consideration time dilation and calculate the distance the muon will travel.

31.2 MATTER AND ANTIMATTER

5. A proton and antiproton, both initially at rest, annihilate each other, producing two gamma ray photons. What is the frequency of one of the photons?

6. An electron–positron pair is produced by a photon with an energy of 11 MeV. (a) Name two quantities this process must conserve. (b) What is the total kinetic energy of the electron and positron combined? (c) What is the total momentum of the electron and positron combined?

7. An electron with speed $0.80c$ collides head-on with a positron with the same speed, producing two photons. What is the energy of each photon?

8. An unknown particle collides with its antiparticle and the two are annihilated, emitting two photons, each with energy 3.0 MeV. What is the rest mass of the particle, in kilograms? Assume the unknown particles are moving very slowly before annihilating.

9. SSM (a) Calculate the energy released when a proton annihilates an antiproton. Express your answer here and in the rest of this problem in MeV. (b) What is the approximate energy released in the fission of one $^{235}_{92}$U nucleus? (c) From your answers to parts (a) and (b), calculate the energy per unit mass of (1) a collection of equal numbers of protons and antiprotons and (2) a sample of $^{235}_{92}$U.

31.4 ELEMENTARY PARTICLE PHYSICS: THE STANDARD MODEL

10. At present, the most powerful accelerator employs protons and antiprotons at energies of about 1 TeV. (a) What is the kinetic energy of such a proton, in joules? (b) What is its de Broglie wavelength?

11. A pi meson (π^0) decays into two photons. What is the energy of each photon?

12. SSM ★ Describe the quark content of a particle that has a charge of $-e$ and contains one strange quark. Is this particle a baryon or a meson? Note: There is more than just one answer (but you only need to give one).

13. ★ Determine the quark content of a particle that is composed of 3 or fewer quarks and has a charge of $-2e$. Is this particle matter or antimatter? A baryon or a meson? Note: There is more than just one answer (but you only need to give one).

14. The following reactions are forbidden because they violate one or more conservation laws. Which conservation law(s) do they violate? The π^+ particle is a meson.
    (a) $p \rightarrow e^+ + \nu_e$
    (b) $p \rightarrow e^+ + \gamma$
    (c) $n \rightarrow p + e^- + \nu_\mu$
    (d) $p \rightarrow e^- + \pi^+$
    (e) $\pi^+ \rightarrow \pi^+ + n$

15. The pi mesons π^+ and π^- are each composed of two quarks. The quark composition of the π^+ is u$\bar{\text{d}}$, while the composition of the π^- is $\bar{\text{u}}$d. (a) What is the antiparticle of the π^+? (b) The π^- can decay into a muon and an antimuon neutrino according to

$$\pi^- \rightarrow \mu^- + \bar{\nu}_\mu$$

What are the final quarks involved in this decay? (c) Does this decay conserve lepton number? If so, what is the final lepton number?

16. Find the charge on each of these baryons or antibaryons and identify them: (a) uud, (b) $\overline{\text{uu}}$d, and (c) $\overline{\text{ud}}$d.

31.5 THE FUNDAMENTAL FORCES OF NATURE

17. ℝ The strong force holds the quarks in a proton together, overcoming the repulsion between the positively charged u quarks. For simplicity, ignore the effect of the d quark and estimate the magnitude of the (attractive) strong force between two u quarks needed to keep them bound in a proton.

18. SSM ★ Complete the following reactions.
    (a) $K^0 \rightarrow \pi^+ + ?$
    (b) $p + \bar{p} \rightarrow \gamma + ?$
    (c) $? + n \rightarrow p + e^-$

31.7 ASTROPHYSICS AND THE UNIVERSE

19. ★ A red shift measurement shows that the speed of a distant galaxy relative to an astronomer on the Earth is $v = 0.25c$, where c is the speed of light. (a) How far away is the galaxy? (b) If this galaxy emits blue light with $\lambda = 450$ nm (as viewed by an observer at rest in the galaxy), what is the wavelength observed on the Earth? *Hint*: Consider the Doppler shift formulas for light in Chapter 23.

20. (a) At what frequency f_{max} does the cosmic microwave background radiation from the Big Bang have its greatest intensity? (b) Our galaxy moves at a speed of about 300 km/s relative to the source of the cosmic microwave background radiation, shifting f_{max} relative to the value found for an observer at rest relative to the source. What are the magnitude and sign of this frequency shift?

21. SSM Based on his data in Figure 31.12, Hubble proposed that the velocity v of a galaxy relative to the Earth is related to its distance d from the Earth by what is now known as Hubble's law (Eq. 31.5). The red shift of light from the most distant galaxies known gives a relative velocity of about $v = 0.96c$. What is their distance from the Earth?

22. A galaxy is found to be 2.0×10^6 ly away from the Earth. Assume this galaxy does not have any motion relative to us other than that due to the expansion of space. (a) What is the galaxy's recession speed due to the expansion of space? (b) How much farther away from us will this galaxy be by next year?

23. SSM Ⓡ How much energy would be released if a baseball were annihilated with an antibaseball? State your answer in joules and in megatons of TNT. (See Example 30.6.)

24. A Σ^+ particle moving with a kinetic energy of 500 MeV decays to a pion and neutron, $\Sigma^+ \rightarrow n + \pi^+$. What is the total kinetic energy of the decay products?

25. ✪ The probability for an individual particle to have *not yet* decayed after a time t since its creation is given by $\text{prob}(t) = e^{-t/\tau}$, where τ is the mean lifetime of the particle. (a) What is the probability that a π^+ at rest will still exist after 0.10 μs? (b) What is the probability that a π^+ of momentum 500 MeV/c can travel a distance of 20 m before decaying? *Hint:* Consider time dilation. (c) What is the probability that a φ particle with the same momentum could travel the same distance? (See Table 31.4.)

26. Consider the decay of an excited hyperion described in Figure P31.26. This figure shows a "cascade" of decay reactions, producing many particles along the way. (a) How many particles, including photons, are there in the final outcome? (b) Is charge conserved in the decay? That is, do the sum of the charges in the final particles add up to the initial charge? (c) Is baryon number conserved overall?

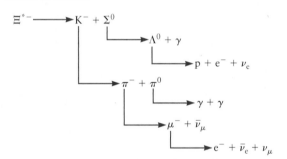

Figure P31.26 Decay scheme of an excited hyperion.

27. ★ Shortly after the discovery of the expansion of space, physicists made many attempts to measure Hubble's constant. These measurements are difficult, and values of H_0 ranging from 12 to 28 (km/s)/Mly were found. (a) Find the age of the universe for this minimum and maximum value of H_0. (b) The best current value of H_0 comes from the WMAP probe of the cosmic background radiation, giving a value of 21 (km/s)/(Mly) with an uncertainty of about 4%. What age of the universe does this measured value suggest?

28. ✪ Consider a hypothetical cluster of galaxies distributed in space as shown in the top part of Figure P31.28, where each galaxy is positioned on a grid with a spacing of two million light-years. Four billion years later, the universe has expanded such that the grid spacing is now three million light-years as seen in the bottom part of Figure P31.28. (a) Assume your planet is in galaxy A. Calculate the *change in distance* to galaxies B, C, D, and E as seen from A over the 4.0-billion-year period. (b) Which galaxy moves the most with respect to galaxy A? (c) Now assume your planet is in galaxy D. Find the change in distance to galaxies C, B, A, and F. (d) Which galaxy moves the most with respect to galaxy D over this period of time? (e) Find the velocities of galaxies B, C, D, and E that would be measured if your planet were in galaxy A. (f) Find the velocities of galaxies C, B, A, and F as seen from a planet in galaxy D. (g) Finally, calculate the Hubble constant as measured from a planet in galaxy A and compare it to what an observer in galaxy D would measure.

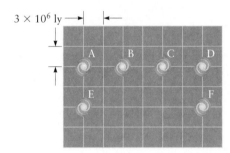

Figure P31.28

29. ✪ **The Planck time.** It is believed that there is a period of time so short that the theory of gravity (or any of the known interactions) cannot make meaningful predictions. Interestingly, there exists an algebraical arrangement of three fundamental constants that gives units of time. The constants are G (Newton's gravitational constant), h (Planck's constant), and c (the speed of light). (a) Use dimensional analysis to find this period of time (called the Planck time). Our current understanding suggests that for the first instant of the Big Bang, from $t = 0$ to the Plank time, the known laws of physics did not apply, and this epoch remains a mystery. (b) The shortest time interval that can be currently measured is about 10^{-18} s. How many Planck time intervals would occur during this minimum measurement time?

Reference Tables

Table A.1 Some Useful Physical Constants and Data

QUANTITY	SYMBOL	APPROXIMATE VALUE
Universal gravitational constant	G	6.674×10^{-11} N $\cdot$ m^2/kg^2
Acceleration due to gravity near the Earth's surface	g	9.8 m/s^2
Speed of light in vacuum	c	$2.997\ 924\ 58 \times 10^8$ m/s
Mass of the electron	m_e	$9.109\ 382 \times 10^{-31}$ kg
Mass of the proton	m_p	$1.672\ 621\ 6 \times 10^{-27}$ kg
Mass of the neutron	m_n	$1.674\ 927\ 2 \times 10^{-27}$ kg
Magnitude of the electron charge	e	$1.602\ 176\ 5 \times 10^{-19}$ C
Permittivity of free space	ε_0	$8.854\ 187\ 817 \times 10^{-12}$ C^2/(N $\cdot$ m^2)
Coulomb constant	$k = \dfrac{1}{4\pi\varepsilon_0}$	$8.987\ 551\ 788 \times 10^9$ N $\cdot$ m^2/C^2
Permeability of free space	μ_0	$4\pi \times 10^{-7}$ T $\cdot$ m/A
Boltzmann's constant	k_B	$1.380\ 650 \times 10^{-23}$ J/K
Planck's constant	h	$6.626\ 069 \times 10^{-34}$ J$\cdot$s
Avogadro's number	N_A	$6.022\ 142 \times 10^{23}$ mol^{-1}
Universal gas constant	R	$8.314\ 472$ J/(mol $\cdot$ K)
Stefan−Boltzmann constant	σ	$5.670\ 40 \times 10^{-8}$ W/(m^2 $\cdot$ K^4)
Atomic mass unit	u	$1.660\ 538\ 8 \times 10^{-27}$ kg

Note: 2006 CODATA recommended values.

Table A.2   SI Base Units

Quantity	Unit	Symbol	Quantity	Unit	Symbol
Length	meter	m	Electric current	ampere	A
Mass	kilogram	kg	Temperature	kelvin	K
Time	second	s	Luminous intensity	candela	cd

Table A.3   Solar System Data

Body	Radius at equator (km)	Mass (kg)	Average density (kg/m³)	Orbital semimajor axis (km)	Orbital period (years)	Orbital eccentricity
Mercury	2,440	3.30×10^{23}	5430	5.79×10^{7}	0.241	0.205
Venus	6,050	4.87×10^{24}	5240	1.08×10^{8}	0.615	0.007
Earth	6,370	5.97×10^{24}	5510	1.50×10^{8}	1.000	0.017
Mars	3,390	6.42×10^{23}	3930	2.28×10^{8}	1.881	0.094
Jupiter	71,500	1.90×10^{27}	1330	7.78×10^{8}	11.86	0.049
Saturn	60,300	5.68×10^{26}	690	1.43×10^{9}	29.46	0.057
Uranus	25,600	8.68×10^{25}	1270	2.87×10^{9}	84.01	0.046
Neptune	24,800	1.02×10^{26}	1640	4.50×10^{9}	164.8	0.011
Pluto	1,200	1.3×10^{22}	1750	5.91×10^{9}	247.7	0.244
Sun	6.95×10^{5}	1.99×10^{30}	1410			
Moon	1,740	7.35×10^{22}	3340	3.84×10^{5}	27.3 days	0.055

Note: From NASA Planetary Fact Sheet (2006).

Atomic Number Z	Element	Chemical Symbol	Atomic Mass (u)	Mass Number A	Atomic Mass (u)	Percent Abundance	Half-Life (if radio-active) $T_{1/2}$	Primary Decay Modes*
0	(Neutron)	n		1	1.008 665		10.4 min	β^-
1	Hydrogen	H	1.007 94	1	1.007 825	99.988 5		
	Deuterium	D		2	2.014 102	0.011 5		
	Tritium	T		3	3.016 049		12.33 years	β^-
2	Helium	He	4.002 602	3	3.016 029	0.000 137		
				4	4.002 603	99.999 863		
3	Lithium	Li	6.941	6	6.015 122	7.5		
				7	7.016 004	92.5		
4	Beryllium	Be	9.012 182	7	7.016 929		53.1 days	EC
				9	9.012 182	100		
5	Boron	B	10.811	10	10.012 937	19.9		
				11	11.009 306	80.1		
6	Carbon	C	12.010 7	10	10.016 853		19.3 s	EC, β^+
				11	11.011 434		20.4 min	EC, β^+
				12	12.000 000	98.93		
				13	13.003 355	1.07		
				14	14.003 242		5 730 years	β^-
7	Nitrogen	N	14.006 7	13	13.005 739		9.96 min	EC, β^+, β^-
				14	14.003 074	99.632		
				15	15.000 109	0.368		
8	Oxygen	O	15.999 4	15	15.003 065		122 s	EC, β^+
				16	15.994 915	99.757		
				18	17.999 160	0.205		
9	Fluorine	F	18.998 403 2	19	18.998 403	100		
10	Neon	Ne	20.179 7	20	19.992 440	90.48		
				22	21.991 385	9.25		
11	Sodium	Na	22.989 77	22	21.994 437		2.60 years	EC, β^+
				23	22.989 770	100		
				24	23.990 963		14.96 h	β^-
12	Magnesium	Mg	24.305 0	24	23.985 042	78.99		
				25	24.985 837	10.00		
				26	25.982 593	11.01		
13	Aluminum	Al	26.981 538	27	26.981 539	100		
14	Silicon	Si	28.085 5	28	27.976 926	92.229 7		
15	Phosphorus	P	30.973 761	31	30.973 762	100		
				32	31.973 907		14.26 days	β^-

*EC denotes decay by electron capture, SF denotes spontaneous fission.

Atomic Number Z	Element	Chemical Symbol	Atomic Mass (u)	Mass Number A	Atomic Mass (u)	Percent Abundance	Half-Life (if radioactive) $T_{1/2}$	Primary Decay Modes*
16	Sulfur	S	32.066	32	31.972 071	94.93		
				35	34.969 032		87.3 days	β^-
17	Chlorine	Cl	35.452 7	35	34.968 853	75.78		
				37	36.965 903	24.22		
18	Argon	Ar	39.948	40	39.962 383	99.600 3		
19	Potassium	K	39.098 3	39	38.963 707	93.258 1		
				40	39.963 999	0.011 7	1.28×10^9 years	EC, β^+, β^-
20	Calcium	Ca	40.078	40	39.962 591	96.941		
21	Scandium	Sc	44.955 910	45	44.955 910	100		
22	Titanium	Ti	47.867	48	47.947 947	73.72		
23	Vanadium	V	50.941 5	51	50.943 964	99.750		
24	Chromium	Cr	51.996 1	52	51.940 512	83.789		
25	Manganese	Mn	54.938 049	55	54.938 050	100		
26	Iron	Fe	55.845	56	55.934 942	91.754		
27	Cobalt	Co	58.933 200	59	58.933 200	100		
				60	59.933 822		5.27 years	β^-
28	Nickel	Ni	58.693 4	58	57.935 348	68.076 9		
				60	59.930 790	26.223 1		
29	Copper	Cu	63.546	63	62.929 601	69.17		
				65	64.927 794	30.83		
30	Zinc	Zn	65.39	64	63.929 147	48.63		
				66	65.926 037	27.90		
				68	67.924 848	18.75		
31	Gallium	Ga	69.723	69	68.925 581	60.108		
				71	70.924 705	39.892		
32	Germanium	Ge	72.61	70	69.924 250	20.84		
				72	71.922 076	27.54		
				74	73.921 178	36.28		
33	Arsenic	As	74.921 60	75	74.921 596	100		
34	Selenium	Se	78.96	78	77.917 310	23.77		
				80	79.916 522	49.61		
35	Bromine	Br	79.904	79	78.918 338	50.69		
				81	80.916 291	49.31		
36	Krypton	Kr	83.80	82	81.913 485	11.58		
				83	82.914 136	11.49		
				84	83.911 507	57.00		
				86	85.910 610	17.30		

*EC denotes decay by electron capture, SF denotes spontaneous fission.

Atomic Number Z	Element	Chemical Symbol	Atomic Mass (u)	Mass Number A	Atomic Mass (u)	Percent Abundance	Half-Life (if radio-active) $T_{1/2}$	Primary Decay Modes*
37	Rubidium	Rb	85.467 8	85	84.911 789	72.17		
				87	86.909 184	27.83	4.75×10^{10} years	β^-
38	Strontium	Sr	87.62	86	85.909 262	9.86		
				88	87.905 614	82.58		
				90	89.907 738		158 s	β^-
39	Yttrium	Y	88.905 85	89	88.905 848	100		
40	Zirconium	Zr	91.224	90	89.904 704	51.45		
				91	90.905 645	11.22		
				92	91.905 040	17.15		
				94	93.906 316	17.38		
41	Niobium	Nb	92.906 38	93	92.906 378	100		
42	Molybdenum	Mo	95.94	92	91.906 810	14.84		
				95	94.905 842	15.92		
				96	95.904 679	16.68		
				98	97.905 408	24.13		
43	Technetium	Tc		98	97.907 216		4.2×10^6 years	β^-
				99	98.906 255		2.1×10^5 years	β^-
44	Ruthenium	Ru	101.07	99	98.905 939	12.76		
				100	99.904 220	12.60		
				101	100.905 582	17.06		
				102	101.904 350	31.55		
				104	103.905 430	18.62		
45	Rhodium	Rh	102.905 50	103	102.905 504	100		
46	Palladium	Pd	106.42	104	103.904 035	11.14		
				105	104.905 084	22.33		
				106	105.903 483	27.33		
				108	107.903 894	26.46		
				110	109.905 152	11.72		
47	Silver	Ag	107.868 2	107	106.905 093	51.839		
				109	108.904 756	48.161		
48	Cadmium	Cd	112.411	110	109.903 006	12.49		
				111	110.904 182	12.80		
				112	111.902 757	24.13		
				113	112.904 401	12.22	7.7×10^{15} years	β^-
				114	113.903 358	28.73		
49	Indium	In	114.818	115	114.903 878	95.71	4.4×10^{14} years	β^-

*EC denotes decay by electron capture, SF denotes spontaneous fission.

Atomic Number Z	Element	Chemical Symbol	Atomic Mass (u)	Mass Number A	Atomic Mass (u)	Percent Abundance	Half-Life (If Radioactive) $T_{1/2}$	Primary Decay Modes*
50	Tin	Sn	118.710	116	115.901 744	14.54		
				118	117.901 606	24.22		
				120	119.902 197	32.58		
51	Antimony	Sb	121.760	121	120.903 818	57.21		
				123	122.904 216	42.79		
52	Tellurium	Te	127.60	126	125.903 306	18.84		
				128	127.904 461	31.74	2.2×10^{24} years	β^-
				130	129.906 223	34.08	7.9×10^{20} years	β^-
53	Iodine	I	126.904 47	127	126.904 468	100		
				129	128.904 988		1.6×10^7 years	β^-
54	Xenon	Xe	131.29	129	128.904 780	26.44		
				131	130.905 082	21.18		
				132	131.904 145	26.89		
				134	133.905 394	10.44		
				136	135.907 220	8.87	$\geq 2.36 \times 10^{21}$ years	β^-
55	Cesium	Cs	132.905 45	133	132.905 447	100		
56	Barium	Ba	137.327	137	136.905 821	11.232		
				138	137.905 241	71.698		
57	Lanthanum	La	138.905 5	139	138.906 349	99.910		
58	Cerium	Ce	140.116	140	139.905 434	88.450		
				142	141.909 240	11.114	$> 2.6 \times 10^{17}$ years	β^-
59	Praseodymium	Pr	140.907 65	141	140.907 648	100		
60	Neodymium	Nd	144.24	142	141.907 719	27.2		
				144	143.910 083	23.8	2.3×10^{15} years	α
				146	145.913 112	17.2		
61	Promethium	Pm		145	144.912 744		17.7 years	EC, α
62	Samarium	Sm	150.36	147	146.914 893	14.99	1.06×10^{11} years	α
				149	148.917 180	13.82		
				152	151.919 728	26.75		
				154	153.922 205	22.75		
63	Europium	Eu	151.964	151	150.919 846	47.81		
				153	152.921 226	52.19		
64	Gadolinium	Gd	157.25	156	155.922 120	20.47		
				158	157.924 100	24.84		
				160	159.927 051	21.86		
65	Terbium	Tb	158.925 34	159	158.925 343	100		

*EC denotes decay by electron capture, SF denotes spontaneous fission.

Atomic Number Z	Element	Chemical Symbol	Atomic Mass (u)	Mass Number A	Atomic Mass (u)	Percent Abundance	Half-Life (if radio-active) $T_{1/2}$	Primary Decay Modes*
66	Dysprosium	Dy	162.50	162	161.926 796	25.51		
				163	162.928 728	24.90		
				164	163.929 171	28.18		
67	Holmium	Ho	164.930 32	165	164.930 320	100		
68	Erbium	Er	167.6	166	165.930 290	33.61		
				167	166.932 045	22.93		
				168	167.932 368	26.78		
69	Thulium	Tm	168.934 21	169	168.934 211	100		
70	Ytterbium	Yb	173.04	172	171.936 378	21.83		
				173	172.938 207	16.13		
				174	173.938 858	31.83		
71	Lutecium	Lu	174.967	175	174.940 768	97.41		
72	Hafnium	Hf	178.49	177	176.943 220	18.60		
				178	177.943 698	27.28		
				179	178.945 815	13.62		
				180	179.946 549	35.08		
73	Tantalum	Ta	180.947 9	181	180.947 996	99.988		
74	Tungsten	W	183.84	182	181.948 206	26.50		
				183	182.950 224	14.31		
				184	183.950 933	30.64	$> 3 \times 10^{19}$ years	α
				186	185.954 362	28.43		
75	Rhenium	Re	186.207	185	184.952 956	37.40		
				187	186.955 751	62.60	4.1×10^{10} years	β^-
76	Osmium	Os	190.23	188	187.955 836	13.24		
				189	188.958 145	16.15		
				190	189.958 445	26.26		
				192	191.961 479	40.78		
77	Iridium	Ir	192.217	191	190.960 591	37.3		
				193	192.962 924	62.7		
78	Platinum	Pt	195.078	194	193.962 664	32.967		
				195	194.964 774	33.832		
				196	195.964 935	25.242		
79	Gold	Au	196.966 55	197	196.966 552	100		
80	Mercury	Hg	200.59	199	198.968 262	16.87		
				200	199.968 309	23.10		
				201	200.970 285	13.18		
				202	201.970 626	29.86		

*EC denotes decay by electron capture, SF denotes spontaneous fission.

Atomic Number Z	Element	Chemical Symbol	Atomic Mass (u)	Mass Number A	Atomic Mass (u)	Percent Abundance	Half-Life (if radioactive) $T_{1/2}$	Primary Decay Modes*
81	Thallium	Tl	204.383 3	203	202.972 329	29.524		
				205	204.974 412	70.476		
				208	207.982 005		3.053 min	β^-
				210	209.990 066		1.30 min	β^-
82	Lead	Pb	207.2	204	203.973 029	1.4	$\geq 1.4 \times 10^{17}$ years	α
				206	205.974 449	24.1		
				207	206.975 881	22.1		
				208	207.976 636	52.4		
				210	209.984 173		22.2 years	α, β^-
				211	210.988 732		36.1 min	β^-
				212	211.991 888		10.64 h	β^-
				214	213.999 798		26.8 min	β^-
83	Bismuth	Bi	208.980 38	209	208.980 383	100		
				211	210.987 258		2.14 min	α, β^-
84	Polonium	Po		210	209.982 857		138.38 days	α
				214	213.995 186		164 μs	α
85	Astatine	At		218	218.008 682		1.5 s	α, β^-
86	Radon	Rn		222	222.017 570		3.823 days	α, β^-
87	Francium	Fr		223	223.019 731		22 min	α, β^-
88	Radium	Ra		226	226.025 403		1 600 years	α
				228	228.031 064		5.75 years	β^-
89	Actinium	Ac		227	227.027 747		21.77 years	α, β^-
90	Thorium	Th	232.038 1					
				228	228.028 731		1.912 years	α
				232	232.038 050	100	1.40×10^{10} years	α
91	Protactinium	Pa	231.035 88	231	231.035 879		3.28×10^4 years	α
92	Uranium	U	238.028 9	232	232.037 146		69 years	α
				233	233.039 628		1.59×10^5 years	α
				235	235.043 923	0.720 0	7.04×10^8 years	α, SF
				236	236.045 562		2.34×10^7 years	α, SF
				238	238.050 783	99.274 5	4.47×10^9 years	α
93	Neptunium	Np		237	237.048 167		2.14×10^6 years	α
94	Plutonium	Pu		239	239.052 156		2.412×10^4 years	α
				242	242.058 737		3.75×10^5 years	α
				244	244.064 198		8.1×10^7 years	α, SF

*EC denotes decay by electron capture, SF denotes spontaneous fission.

Sources: Chemical atomic masses are from T. B. Coplen, "Atomic weights of the elements 1999," a technical report to the International Union of Pure and Applied Chemistry and published in *Pure and Applied Chemistry* 73(4):667–683 (2001). Atomic masses of the isotopes are from G. Audi and A. H. Wapstra, "The 1995 update to the atomic mass evaluation," *Nuclear Physics* A595(4):409–480 (December 25, 1995). Percent abundance values are from K. J. R. Rosman and P. D. P. Taylor, "Isotopic compositions of the elements 1999," a technical report to the International Union of Pure and Applied Chemistry and published in *Pure and Applied Chemistry* 70(1):217–236 (1998). Data on decay modes and half-lives are from Lawrence Berkeley Laboratory Isotopes Project Listing and the National Nuclear Data Center, Brookhaven National Laboratory.

Mathematical Review

B.1 | MATHEMATICAL SYMBOLS

The following symbols are used to express mathematical relationships.

Symbol	Meaning	Example	Explanation
$+$	Addition	$a + b = c$	a plus b equals c
$-$	Subtraction	$a - b = c$	a minus b equals c
$\times$ or $\cdot$	Multiplication	$a \times b = c$	a times b equals c
		$a \cdot b = c$	
$/$ or $\div$	Division	$a/b = c$	a divided by b equals c
		$a \div b = c$	
$>$	Greater than	$a > b$	a is greater than b
$<$	Less than	$a < b$	a is less than b
$\geq$	Greater than or equal to	$a \geq b$	a is greater than or equal to b
$\leq$	Less than or equal to	$a \leq b$	a is less than or equal to b
$\approx$	Approximately equal to	$a \approx b$	a is approximately equal to b
$\lim\limits_{x \to 0}$	Limit as x approaches 0	$y = \lim\limits_{x \to 0} f$	y equals the value of f as the variable x approaches zero

The basic rules of algebra are reviewed in Section 1.6. When solving an equation to find the value of a single variable, you should rearrange the equation to put the *unknown* variable (the variable you wish to solve for) on one side of the equation and the *known* quantities on the other side. For example, suppose the variable a in the equation

$$4a + 7 = T - 9$$

is the unknown. The goal is to isolate a on one side of the equal sign (usually the left). In this example, we can subtract 7 from both sides to get

$$4a = T - 16$$

We next divide both sides by the constant factor 4 to get the solution

$$a = \frac{T}{4} - 4$$

When dealing with two equations containing two unknowns, the usual approach is to use one equation to eliminate one of the unknowns from the other equation. For example, suppose a and T are unknown in the equations

$$3a = T - 9 \tag{B2.1}$$

$$2a = -T + 24 \tag{B2.2}$$

We can rearrange Equation B2.2 to get

$$T = -2a + 24$$

We then substitute this expression for T in Equation B2.1 to find

$$3a = T - 9 = (-2a + 24) - 9 = -2a + 15 \tag{B2.3}$$

We can now solve for a:

$$5a = 15$$

$$a = 3$$

This value can then be used in Equation B2.1 to find the value of T; we get $T = 18$.

Another approach when working with two equations and two unknowns is to simply add the two equations. For example, if we add Equations B2.1 and B2.2, we will eliminate T:

$$\begin{array}{rl} 3a = & T - 9 \\ 2a = & -T + 24 \\ \hline 5a = & -9 + 24 \end{array}$$

which is another way to get Equation B2.3.

The equations considered in Equations B2.1 through B2.3 all involve the unknowns (e.g., a and T) raised to the first power. In a **quadratic equation**, the unknown quantity is raised to the second power. If x is an unknown, an equation of this type is

$$ax^2 + bx + c = 0$$

The solution is given by the **quadratic formula,**

$$x = \frac{-b \pm \sqrt{b^2 - 4ac}}{2a} \tag{B2.4}$$

Note that there are two solutions as indicated by the appearance of the $\pm$ sign.

B.3 | SCIENTIFIC NOTATION

Scientific notation is a convenient way to express extremely large or extremely small numbers. For example, the number 640,000 is written as 6.4×10^5 in scientific notation. To express a number in scientific notation, move the decimal point in the original number to obtain a new number between 1 and 10. Count the number of places the decimal point has been moved; this number will become the exponent of 10 in scientific notation. If you started with a number greater than 10 (such as 150,000,000,000), the exponent of 10 is positive (1.5×10^{11}). If you started with a number less than 1 (such as 0.000055), the exponent is negative (5.5×10^{-5}).

B.4 | GEOMETRY AND TRIGONOMETRY

One way to define and measure angles is shown in Figure B.1. The angle θ equals the ratio of the arc length s measured along the circle divided by the radius r of the circle:

$$\theta = \frac{s}{r} \tag{B4.1}$$

This relation gives the value of θ in **radians.** To convert to degrees, we use the fact that one complete "trip" around the circle in Figure B.1 corresponds to both 2π radians and to $360°$. We thus have

$$\theta \text{ (in radians)} \times \frac{360°}{2\pi \text{ radians}} = \theta \text{ (in degrees)} \tag{B4.2}$$

Figure B.2 shows a right triangle with sides x, y, and r, where r is the hypotenuse. According to the **Pythagorean theorem,**

$$x^2 + y^2 = r^2 \tag{B4.3}$$

The trigonometric functions **sine, cosine,** and **tangent** are defined as

$$\sin \theta = y/r \tag{B4.4}$$

$$\cos \theta = x/r \tag{B4.5}$$

$$\tan \theta = y/x \tag{B4.6}$$

The Pythagorean theorem (Eq. B4.3) implies that

$$\sin^2 \theta + \cos^2 \theta = 1 \tag{B4.7}$$

for any value of the angle θ. Equations B4.3 through B4.6 also lead to a number of other trigonometric relations (called **identities**), some of which are the following

$$\sin(-\theta) = -\sin \theta \quad \cos(-\theta) = \cos \theta \quad \tan(-\theta) = -\tan \theta$$

$$\sin(\theta + 90°) = \sin(\theta + \pi/2) = \cos \theta$$

$$\cos(\theta + 90°) = \cos(\theta + \pi/2) = -\sin \theta$$

$$\sin(\alpha \pm \beta) = \sin \alpha \cos \beta \pm \cos \alpha \sin \beta$$

$$\cos(\alpha \pm \beta) = \cos \alpha \cos \beta \mp \sin \alpha \sin \beta \tag{B4.8}$$

The relations in Equation B4.8 hold for any values of the angles θ, α, and β. If $\alpha = \beta$, the relation for $\sin(\alpha + \beta)$ becomes

$$\sin(\alpha + \alpha) = \sin(2\alpha) = \sin \alpha \cos \alpha + \cos \alpha \sin \alpha$$

We can also replace the angle α with θ; rearranging then leads to

$$\sin \theta \cos \theta = \tfrac{1}{2}\sin(2\theta) \tag{B4.9}$$

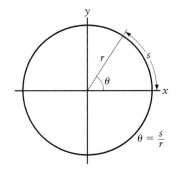

Figure B.1 The angle θ measured in radians equals the ratio of the length s measured along the circular arc divided by the radius r of the circle.

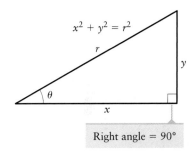

Figure B.2 The Pythagorean theorem is a relation between the lengths of the sides of a right triangle.

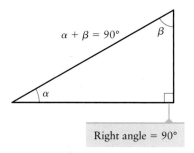

Figure B.3 The sum of the three interior angles of a right triangle is 180°. Hence, $\alpha + \beta = 90°$.

$\alpha + \beta = 90°$

β

α

Right angle = 90°

Equations B4.4 through B4.6 tell how to calculate the sine, cosine, and tangent of an angle from the lengths of the sides of the associated right triangle. We can also use this approach to find the value of the angle. For example, if we know the values of y and r, we can solve for the angle as

$$\sin \theta = y/r$$

$$\theta = \sin^{-1}(y/r) \qquad \text{(B4.10)}$$

The function $\sin^{-1}$ is the *inverse* sine function, also known as the *arcsine* (= arcsin = $\sin^{-1}$). In words, Equation B4.10 says that θ equals the angle whose sine is y/r. Scientific calculators all contain this function (as a single "button"). If the values of y and r are known, you must only take the ratio y/r and then compute the arcsine of this value to find θ in Equation B4.10. Likewise, there are also functions for the inverse cosine ($\cos^{-1}$ = arccos) and inverse tangent ($\tan^{-1}$ = arctan):

$$\theta = \cos^{-1}(x/r) \quad \text{and} \quad \theta = \tan^{-1}(y/x) \qquad \text{(B4.11)}$$

We often need to deal with triangles like the one in Figure B.3, which shows a right triangle with interior angles α and β. The sum of the three interior angles of a triangle is always 180°. For the triangle in Figure B.3, one of the angles is 90° (it is a right triangle); hence,

$$\alpha + \beta = 90° \qquad \text{(B4.12)}$$

Two angles whose sum is 90° are called complementary angles. From the definitions of the sine and cosine functions, for any pair of complementary angles α and β we have

$$\sin \alpha = \cos \beta$$

$$\cos \alpha = \sin \beta$$

Two approximate relations involving trigonometric functions are also useful. When the angle θ is small,

$$\sin \theta \approx \theta \quad \text{and} \quad \tan \theta \approx \theta \qquad \text{(B4.13)}$$

which are good approximations (accurate to a few percent or better) when θ is smaller than about 30°. Notice that the relations in Equation B4.13 apply only when the angle is measured in radians.

B.5 | VECTORS

A vector quantity has both a magnitude and a direction. There are two ways to do calculations with vectors; one is a graphical approach, which is useful for approximate or qualitative calculations, while the other involves the components of the vectors.

The graphical approach is illustrated in Figure B.4, which shows two vectors $\vec{A}$ and $\vec{B}$ along with their sum $(\vec{A} + \vec{B})$ and difference $(\vec{A} - \vec{B})$. Multiplying a vector

Figure B.4 Examples showing the effect of multiplying a vector by a scalar, adding two vectors, and subtracting one vector from another.

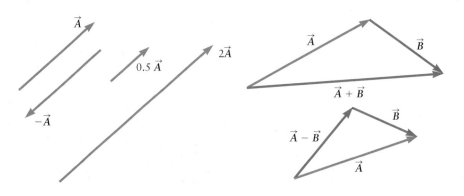

by a scalar (e.g., 0.5 or 2) changes the length but does not change the direction. Multiplying by a negative number (e.g., -1) reverses the direction. The vectors $-\vec{A}$, $0.5\vec{A}$, and $2\vec{A}$ are also shown in Figure B.4.

The components of a vector can be obtained using trigonometry. The length of a vector $\vec{A}$, also called its **magnitude,** is denoted by either A or $|\vec{A}|$. If this vector makes an angle θ with the x axis, the **components** are

$$A_x = A \cos\theta = |\vec{A}| \cos\theta \text{ and } A_y = A \sin\theta = |\vec{A}| \sin\theta \quad \text{(B5.1)}$$

The sum of two vectors can be found by adding components, as illustrated in Figure B.5. If

$$\vec{C} = \vec{A} + \vec{B} \quad \text{(B5.2)}$$

in terms of the components of these vectors we then have

$$C_x = A_x + B_x \text{ and } C_y = A_y + B_y \quad \text{(B5.3)}$$

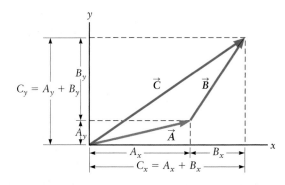

Figure B.5 To compute the sum of two vectors, we add their components. Here we add the x components of $\vec{A}$ and $\vec{B}$ to get the x component of $\vec{C}$. We follow the same procedure to find the y component of $\vec{C}$.

B.6 | EXPONENTIAL FUNCTIONS AND LOGARITHMS

Exponential functions and logarithms play a role in many areas of physics, including the behavior of fluids (Chapter 10), sound (Chapter 13), electric circuits (Chapters 19 and 22), and radioactive decay (Chapter 30).

The number $e = 2.71828\ldots$ plays an important role. Useful numerical values involving the number e are $e^{-1} \approx 0.37$ and $1 - e^{-1} \approx 0.73$.

There are two versions of the logarithm function. The **natural logarithm** function is denoted by the symbol "ln" and is related to the exponential function by

$$\ln(e^a) = a \quad \text{(B6.1)}$$

where a is a number. Other useful properties of these functions are

$$e^a \times e^b = e^{a+b}$$

$$\frac{e^a}{e^b} = e^{a-b}$$

$$\ln(a) + \ln(b) = \ln(ab) \quad \text{(B6.2)}$$

$$\ln(a) - \ln(b) = \ln(a/b)$$

$$\ln(a^b) = b \ln a$$

The **common logarithm** function (also called the base 10 logarithm) is denoted by the symbol "log"; the log function has properties similar to the natural logarithm, but it is based on the number 10 instead of e. These properties include

$$\log(10^a) = a \quad \text{(B6.3)}$$

$$10^a \times 10^b = 10^{a+b}$$

$$\frac{10^a}{10^b} = 10^{a-b}$$

$$\log(a) + \log(b) = \log(ab) \quad \text{(B6.4)}$$

$$\log(a) - \log(b) = \log(a/b)$$

$$\log(a^b) = b \log a$$

Exponential functions may also involve base numbers besides e and 10. For example, for any number y, one has

$$y^a \times y^b = y^{a+b} \quad \text{and} \quad \frac{y^a}{y^b} = y^{a-b} \tag{B6.5}$$

B.7 | SOME RELATIONS USEFUL IN SPECIAL RELATIVITY

In special relativity, factors such as v/c and v^2/c^2, where c is the speed of light and v is the speed of an object, appear often. The speed of light is very large, so the ratio v/c is usually very small. When that is the case, certain algebraic expressions can be simplified. For example, when A is a very small number, the approximation

$$\frac{1}{1 + A} \approx 1 - A \tag{B7.1}$$

is very accurate. If we replace A by v/c, we find

$$\frac{1}{1 + v/c} \approx 1 - v/c \tag{B7.2}$$

and in a similar way, we get

$$\frac{1}{1 + (v/c)^2} \approx 1 - (v/c)^2 \tag{B7.3}$$

Another expression that arises in special relativity, especially in problems involving time dilation and length contraction, is $\sqrt{1 - (v/c)^2}$. When v/c is very small, to a good approximation we have

$$\sqrt{1 - v^2/c^2} \approx 1 - \frac{v^2}{2c^2} \tag{B7.4}$$

and

$$\frac{1}{\sqrt{1 - v^2/c^2}} \approx 1 + \frac{v^2}{2c^2} \tag{B7.5}$$

The accuracy of these approximations depends on the value of v/c. If $v/c = 0.1$, the approximation in Equation B7.2 is good to about 1%, whereas Equations B7.4 and B7.5 are accurate to about 0.005%. The accuracy improves for smaller values of v/c.

CHAPTER 17

Concept Checks

17.1   (b)
17.2   (a)
17.3   (a)
17.4   (d)
17.5   (b)
17.6   (c)
17.7   (b)

Problems

1. -9.6×10^4 C

3. 5.3×10^{10}

5. -7.2×10^{-18} C

7. 3.1×10^{10}

9. 2.3×10^{-8} N

11. 9.2×10^{-8} N

13. Graph c

15. Ⓡ Equal and negative, or equal and positive, but smaller than $+Q$.

17. 0.034 m

19. (a) 5.7 N downward (b) 5.7 N at 30.4° above horizontal

21. Ⓡ $F_{grav} = 6.7 \times 10^{-17}$ N $= 2.4 \times 10^{-43}$ F_{elec}; the gravitational force is extremely small (or weak) compared with the electric force.

23. At $x = -0.27L$

25. (a) 1.0×10^{-49} N (b) 2.3×10^{-10} N (c) The ratio of the forces is the same at 1.0 m as it is at 1.0 nm.

27. 6.5×10^{16}

29. -5.1×10^{-5} C

31. $3L$

33. 1.4×10^{10} N/C

35. 2.8×10^{-5} C

37. $E_x = 1.4 \times 10^9$ N/C, $E_y = -9.0 \times 10^9$ N/C

39. 3.5×10^{14} m/s^2 in the $-x$ direction

41.

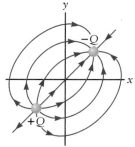

43. -5.7×10^{-3} C

45. 1/9

47. (a) 3.6×10^9 N/C downward (b) 2.4×10^9 N/C downward

49. Outside the conductor, the field is as shown. Within the conductor, the field is zero.

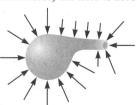

51. -1.8×10^{-8} N · m^2/C

53. 0

55. $+3Q/\varepsilon_0$

57. (a) The electric field is directed radially outward from the cylinder. (b) A cylindrical Gaussian surface of length L (c) $q_{enclosed} = \lambda L$ if the Gaussian surface has a radius larger than that of the metal cylinder; $q_{enclosed} = 0$ if the Gaussian surface has a radius smaller than that of the cylinder. (d) Outside the cylinder, $E = \lambda/2\pi R\varepsilon_0 = (1.8 \times 10^4)/R$ N/C (with R in units of meters). The electric field is zero inside the cylinder and then falls off with the inverse of the radius outside the cylinder.

59. $E_A = 5.6 \times 10^9$ N/C, $E_B = -5.6 \times 10^9$ N/C, $E_C = 5.6 \times 10^9$ N/C, $E_D = -5.6 \times 10^9$ N/C

61. $8Q/\varepsilon_0$

63. $\dfrac{kQ}{r_s^3} r$

65. 0

67. 4.4×10^{-9} C

69. $\dfrac{\rho R}{9\varepsilon_0}$

71. (a) $v_1 = 11 \times 10^{-7}$ m/s, $v_2 = 9.1 \times 10^{-7}$ m/s (b) 9.3×10^3 s (c) 1.5×10^{-3} m

73. The magnitude of the force on the quark with charge $-e/3$ is 51 N. The magnitude of the force on one of the quarks with charge $2e/3$ is 88 N. If the only force acting between the quarks is the electric force, this arrangement of quarks is not stable.

75. -8.4×10^7 C

77. $\dfrac{\lambda}{\varepsilon_0(2\pi R)}$

79. The paper will move toward the right plate.

81. (a) 3.4×10^{12} N/C downward (b) 1.7×10^{-19} C (c) 3.1×10^{-10} N

83. 3.0×10^5

85. (a) $q_{Moon} = 6.3 \times 10^{12}$ C, $q_{Earth} = 5.2 \times 10^{14}$ C (b) $n_{Moon} = 6.6 \times 10^7$ mol, $n_{Earth} = 5.4 \times 10^9$ mol

87. (a) 470 N/C (b) The bottom plate

89. 3.1×10^5 N/C

91. (a) 17 N · m²/C (b) 34 N · m²/C (c) 17 N · m²/C

93. 1.0×10^{-5} C

CHAPTER 18

Concept Checks

1. (d)

2. (c)

3. (a) and (c)

4. (a)

5. If the line carries a negative charge, the electric field lines are directed radially and point toward the line. The equipotential surfaces are unchanged (cylinders centered on the line of charge).

6. A qualitative sketch of the field lines is as shown. Electric field lines are always perpendicular to equipotential surfaces. Because the electric field is directed from regions of high potential to regions of low potential, the object at the center must have a positive charge.

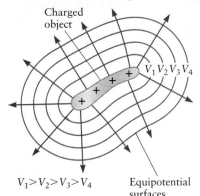

Problems

1. 0.86 m

3. 2.3×10^{-13} J

5. $+4.6 \times 10^{-18}$ J

7. -3.8×10^{-9} J

9. -7.2×10^{-3} J

11. -7.7×10^{-20} J

13. 3.5×10^{-2} J

15. 9.2×10^{-19} J

17. -1.9×10^{-17} J, -120 eV

19. 1200 V/m directed from the final position to the initial position of the electron.

21. 8.4×10^{-10} m

23. (a) The electron moves from a region of lower potential to a region of higher potential. (b) 4.8×10^{-15} J (c) 1.0×10^8 m/s

25. 1.5×10^5 V

27. -2.3×10^5 V

29. 4.1×10^{-10} m

31. (a) $-Q$ (b) $k\dfrac{Q}{r_2}$

33. $2\pi k \lambda$

35. $-1/3$

37. 2.5 V/m pointing from the 10 V equipotential surface toward the 5 V equipotential surface, perpendicular to both

39. A factor of three

41. 8.9×10^{-11} F

43. (a) 1.2×10^{-9} C (b) 7.2×10^{-9} J

45. (a) Closer together (b) 1/9

47. 3C

49. 8.3 μF

51. 15 μF

53. 5.8×10^3 V

55. (a) PE_{elec} increases by a factor of two. (b) Work must be done to pull the plates farther apart.

57. 13 V

59. 5.3×10^{-15} F

61. 2.0×10^4 V

63. The stored energy with the new dielectric is smaller by a factor of five. The capacitor does work to pull in the new dielectric, decreasing the energy stored in the capacitor.

65. Ⓡ -5.1×10^5 C; 3.2×10^{24} electrons

67. Ⓡ 170 V

69. (a) 1.0×10^3 J (b) 5.0×10^4 W

71. 12 J/m³

73. A factor of four

75. (a) 7.7×10^{-13} J (b) 3.0×10^7 m/s (c) 4.8×10^6 V

77. (a) 2.3×10^8 m² = 230 km² = 87 mi²
(b) 1.0×10^6 m² = 1.0 km² = 0.39 mi²

79. Ⓡ (a) 1.4 nF (b) 1.4 J

81. (a) 5.9 pF (b) 1.7 μF (c) 990 μC

83. (a) 5.4 V (b) 7.7×10^{11} V/m

CHAPTER 19

Concept Checks

1. (c)

2. (a)

3. (c)

4. Circuit 2

5. (b)

Problems

1. 2.6×10^{20}

3. 1.7×10^{22}

5. 2.0×10^4 A; 1.3×10^{20}

7. 3.6×10^9 s; this time is just over 114 years, so it is quite sufficient!

9. 2.2×10^{-2} Ω

11. Increases by a factor of 6.3

13. 4.7 Ω · m

15. 2×10^{-4} m

17. Increases by a factor of nine

19. 5.9×10^6

21. $15\ \Omega$

23. (a) A factor of $\frac{4}{3}$ (b) A factor of $\frac{2}{3}$ (c) A factor of $\frac{2}{3}$

25. $1.1 \times 10^3\ \Omega$

27. (a) 2.4×10^{-3} A (b) $5000\ \Omega$

29. 2.0×10^{-2} W

31. (a) 2.1 A (b) A factor of $\frac{1}{2}$

33. 1.3 A

35. Two sets of two series resistors in parallel or, equivalently, two sets in series of two parallel resistors

37. (b)

39. 3.4×10^{-4} A

41. (a) $\mathcal{E}_1 - I_1 R_1 - \mathcal{E}_2 = 0$ (b) $\mathcal{E}_2 - I_2 R_2 = 0$
(c) $I_1 = -6.3 \times 10^{-4}$ A, $I_2 = 2.1 \times 10^{-3}$ A

43. 2.6

45. 0.71

47. Bulb 1 is brightest; bulbs 3 and 4 are equally the dimmest.

49. (a) $\dfrac{\mathcal{E}}{R_1 + R_2}$ (b) 0

51. 5C

53. C/7

55. 3.7×10^{-3} A

57. 6.8×10^{-2} s

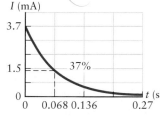

59. (a) -2.3×10^{-3} A; a negative value indicates counterclockwise current in Figure P19.58.

(b)
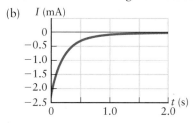

(c) 0

61. 0; 8.8×10^{-5} C

63. -6.7×10^{-5}

65. 0.0002 A

67. $0.002\mathcal{E}$

69. $\dfrac{\mathcal{E}_1}{\mathcal{E}_2}\left(\dfrac{R_3}{R_3 + R_4}\right)$

71. $7.9 \times 10^{-2}\ \Omega \cdot$ m

73. (a) 32 A (b) 2.0×10^{18}

75. (a) $660\ \Omega$ (b) The bulb's brightness will decrease as the bulb warms up. (A warm bulb draws slightly less current and is slightly less bright than a cool bulb.)

77. 2.9×10^{17}

79.
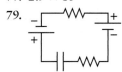

81. (a) 8.8×10^{-4} C (b) 10^{-3} s (c) 0.88 A

83. Increase

85. (a) $V_1 > V_2 > V_3 > 0$ (b) $I_1 = I_2 = I_3$

87. $80\ \mu$A

89. (a) $V_{60} = 75$ V, $V_{100} = 45$ V (b) $V_{100} = V_{60} = 120$ V

91. ℝ (a) 1.5×10^8 V (b) 0.1 C (c) 1000 A

93. ℝ $4000\ \text{m}^3/\text{s}$

95. $R_{AB} = R_{AC} = R_{CD} = R_{BD} = R_{AD} = 10\ \Omega$

97. ℝ 40 V

99. ℝ Two or three batteries

CHAPTER 20

Concept Checks

20.1 Part 1: (d), Part 2: (d)

20.2 Part 1: (b), Part 2: (d)

20.3 Part 1: (a), Part 2: (a)

20.4 (a)

20.5 (b)

20.6 (b)

20.7 (a)

20.8 (a)

20.9 (b)

Problems

1. A = out of the page, B = into the page, C = out of the page, D = into the page

3. To the left

5. Clockwise

7. (a) The force is zero. (b) Counterclockwise

9. (a) To the left (b) Counterclockwise

11. 1.2×10^{-16} N

13. 32°

15. Along the $-x$ direction

17. The force is zero.

19. Toward the top

21. 4.2×10^{-4} N

23. 2.8×10^{-5} m/s

25. 0.42 T

27. Clockwise

29. Counterclockwise helical when viewed from the $+z$ direction

31. 3.6×10^{-9} m

33. ℝ 4×10^{-21} N

35. ℝ 100 km

37. The radius would be decreased by a factor of 3; the radius would be increased by a factor of 10.

39. (a) Into the plane (b) The ratio of electron speed to proton speed is 1800:1.

41. Largest = case 1; smallest = case 3

43. Case 1 = force is into the page, case 2 = force is upward (toward the top of the page).

45. 7.4 N

47. 33°

49. Ⓡ 5×10^{-5} N; it is unlikely that you could detect the force holding the wire. The force is equal to the weight of about 5 mosquitoes.

51. 0

53. Along the $-x$ direction

55. (a) Along the $+x$ direction (b) 2.7 N

57. Case 1: The axis will rotate clockwise about the direction perpendicular to the plane of the drawing. Case 2: The axis will rotate counterclockwise about the direction perpendicular to the plane of the drawing. Case 3: There is no torque.

59. (a) 0 (b) 4.4×10^{-2} N · m

61. (a) Downward along the axis (b) Upward along the axis

63. (a) 25 V/m (b) 2.5×10^{-2} V (c) The velocity selector will work for both Li^+ and Mg^{2+} ions because the velocity selector equation has no dependence on mass. It would even work with negative charges.

65. 12 T

67. $E = vB$

69. 0.25 A

71. $\dfrac{\mu_0 I}{4\pi r}$

73. 1.9×10^{-2} m

75. 2.5 A

77. (a) 1.8×10^{-6} N toward the long, straight wire (b) 1.8×10^{-6} N toward the current loop

79. The field is zero at point B. It has the same (nonzero) magnitude at points A and C.

81. 8.6×10^{-6} T

83. Out of the page

85. 5.0×10^{-5} T; it is about equal to the Earth's magnetic field.

87. 159 A; this very high current would likely melt the wire, making this setup impractical.

89. Ⓡ 40 A

91. Ⓡ 5×10^{8} A

93. (a) 7.5×10^{-5} V (b) The top of the blood vessel (c) It would now be 1.3×10^{-5} V. (d) The narrower the artery, the smaller the induced emf that must be measured.

95. (a) 1.9 N (b) The weight of the wire is 3290 N, much heavier than the magnetic force on it.

97. 1.8 cm

99. (a) 3.5×10^{-17} J (b) 4.4×10^{7} m/s, 8.8×10^{-16} J (c) 280 MHz

CHAPTER 21

Concept Checks

21.1 (a)

21.2 (b)

21.3 (b)

21.4 (b)

21.5 (c)

21.6 (a)

21.7 (b)

Problems

1. 0.063 Wb = 0.063 T · m²

3. 6.5 m²

5. (a) 0.5 V (b) 0.3 V (c) Opposite each other

7.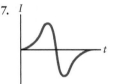

9. Ⓡ 0.05 Wb = 0.05 T · m²

11. Ⓡ 1200 V

13. (a) 0.59 T (b) The induced emf depends only on *changes* in flux (Faraday's law).

15. 2×10^{-4} V

17. (a) The magnetic flux decreases with time. (b) Clockwise when viewed looking into the plane of Figure P21.16

19. (a) Into the page (b) Decreasing (c) Into the page (d) Clockwise (e) Toward the wire

21. (b)

23. Into the plane of the drawing

25. 0.11 A; clockwise when viewed looking into the plane of Figure P21.25

27. (a) Downward (b) The force is zero.

29. (a) South pole up (b) Upward (toward $+z$) (c) The answer to part (b) does not change.

31. 1.1 mH

33. 190 mH

35. 500 Wb = 500 T · m²

37. (a) + (positive) (b) 5.0×10^{-5} V

39. At $t = 0$, the voltage is $V_L = 0$.

41. $V_R = 0$

43. $\tau = \dfrac{L}{R_1 + R_2}$

45. $V_{equiv} = \left(\dfrac{L_1 L_2}{L_1 + L_2}\right)\dfrac{\Delta I}{\Delta t}$, $L_{equiv} = \dfrac{L_1 L_2}{L_1 + L_2}$

47. (a) 17,000 (b) 51 H (c) 5.0×10^3 J (d) 32 m/s

49. $\sqrt{\varepsilon_0 \mu_0}$

51. ⓇEW 1100 loops with a radius of 2.0 cm

53. The induced emf at 800 Hz is double the induced emf at 400 Hz.

55. 3.3×10^{-12} s

57. Clockwise as viewed from above

59. (a) (1) Out of the page, (2) into the page, (3) out of the page, (4) into the page (b) (1) Counterclockwise, (2) clockwise, (3) counterclockwise, (4) clockwise

61. (a) 1.2 s (b) 0.12 s

63. (a) 0.26 ms (b) 9.6 V

65. (a) 6.0×10^{-4} H (b) 0.60 A (c) 2.1×10^{-5} s

67. Highest emf = 6.4 V; lowest emf = -3.2 V

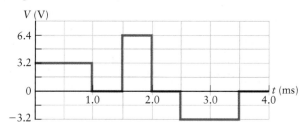

69. ⓇEW 8×10^{-4} V

CHAPTER 22

Concept Checks

22.1 (a) and (c)

22.2 (c)

22.3 (c)

22.4 (a)

22.5 (d)

22.6 (a)

22.7 (a)

22.8 (a)

Problems

1. ⓇEW (a) 240 Hz (b) 4.0 V (c) 2.8 V (d) No

3.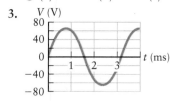

5. It will increase by a factor of four.

7. -16 V

9. (a) 0.018 A (b) 0.61 W (c) 0 (d) 0.30 W (e) 820 s

11. 2600 V

13. 0.13 A

15.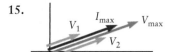

All four phasors point along the same direction because all four circuit quantities are in phase with one another. The relative lengths of these phasors depend on the values of V_{max}, R_1, and R_2.

17. 38 V

19. The 20-pF capacitor

21. $3.3 \times 10^5 \ \Omega$

23. 2.9×10^{-7} F

25. (a) 0.15 A (b) 0.11 A

27. (a) 18 μC (b) 1.1×10^{-4} J (c) 0.0033 s

29. 4200 Ω

31. Zero

33. The 80-mH inductor

35. 2100 Ω; it increases by a factor of three.

37.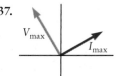

39. (a) The maximum energy occurs at about $t = 1.0$ ms, 2.2 ms, 3.2 ms, and so on when the voltage is zero; the maximum energy is 1.1×10^{-5} J.

(b) The minimum energy occurs at about $t = 0.5$ ms, 1.6 ms, 2.7 ms, and so on when current is zero; the minimum energy is zero.

41. (a) $\dfrac{1}{2\sqrt{2}} f_0$

(b) $C_{new} = \dfrac{C}{8}$ (reduce capacitance by factor of eight)

43. (a) 1.5×10^{-5} J (b) 2.8×10^{-7} C (c) 9.6×10^{-6} s (d) $I = 0$; $|q| = 2.8 \times 10^{-7}$ C

45. (a) 2.4×10^{-6} C (b) 0.10 A (c) 0.086 A

47. (a) 4.2×10^{-13} F (b) 1.0 MHz (c) Yes

49. Only (a) is possible; possible values are $L = 85$ mH and $C = 1.5$ mF.

51. (a)

53. (a) 4600 Ω (b) 13°, voltage leading

55. $4.7 \times 10^6 \ \Omega$; by the inductor; increase

57. 4.2×10^{-9} H

59. 1.6×10^7 Hz

61. (a) 0 (b) 1 (c) Low-pass filter

63. 6.7 V

65. (a) Step-down transformer (b) 1040 turns

67. (a) 0.38 A (b) 3.8 A

69. (a) 190 J; 6.8×10^5 J (b) 1.6 A

71. 4.4%

73. $\phi = \tan^{-1}(2\pi fCR)$

75. (a) 11 Hz (b) 22 ms later

77. The primary has 480 turns, and the secondary has 48 turns.

79. (a) 9.1 μH; 99 μH (b) A maximum of 75 and a minimum of 7 turns (c) 0.38 turns/mm (d) No. Changing the number of turns by one can select between stations for the low frequencies, but not for the high frequencies. One more turn = +4 kHz for low frequencies and −120 kHz for high frequencies.

81. (a) 0.71 kW (b) 1.5 μH (c) 0.71 A; yes, the thick copper wire is needed. (d) 47 nF

CHAPTER 23

Concept Checks

23.1 (a)

23.2 (c) This is much smaller than the maximum recommended delay for phone service, which is about 0.2 s. This is why you can have a pleasant conversation with a friend on the opposite coast.

23.3 (b)

23.4 Using Figure 23.8, we get the following order: (d) radio, (e) microwaves, (a) infrared, (f) red light, (c) blue light, (b) X-rays.

23.5 (c)

23.6 (c)

Problems

1. ℝ 6×10^{-9} s

3. (a) 9×10^{-11} N/C (b) 1×10^{-29} N. It is a large force on an electron because it causes the electron to accelerate at about 15 m/s^2; it is also about equal to the weight of an electron.

5. 1.0×10^{-12} kg · m/s

7. ℝ (a) 270 s, or about $4\frac{1}{2}$ min (b) 1300 s, or about 22 min

9. ℝ 5300 s, or about 88 min

11. 2.9×10^{-6} T

13. 6.7×10^{-8} T

15. ℝ 270 N/C

17. ℝ Light from the laser pointer has momentum, and conservation of momentum requires that you have momentum of equal magnitude and opposite direction; $v_{10\,\text{min}} \approx 3 \times 10^{-11}$ m/s

19. 2.0×10^{-4} s

21. ℝ 110 m

23. 600 nm

25. 5.5×10^{14} Hz

27. 3.0×10^{18} Hz

29. 0.12 m

31. 6.0×10^{6} m; no

33. ℝ (a) 8 to 10 cm (b) $\lambda \approx 20$ cm or 0.2 m; $f \approx 1.5 \times 10^{9}$ Hz

35. (a). For the best reception when using a dipole antenna, orient the receiving antenna along the same axis as the transmitting antenna.

37. (a) 37 W/m^2 (b) 170 N/C

39. 0.18 V/m

41. $I_0/8$

43. $0.72c$

45. Statement (a) is most likely, although statement (b) is possible.

47. $0.39c$ away from you

49. 0.035 V/m

51. (a) 0.13 N (b) 4.9×10^{-3} N

53. (a) 2.0×10^{5} W (b) 2.5×10^{11} W/m^2 (c) 1.4×10^{7} V/m

55. ℝ 1.2 km

57. (a) 1/8 (b) Zero (c) Removing the middle filter reduces the light intensity to zero because the first and third filters have transmission axes at right angles. In each situation, you must consider the effect of one polarizer at a time. If the first or third filter were removed instead, the final intensity would be 1/4 of the incident intensity.

59. ℝ 2.1×10^{7} m/s or about 7% of c; no man-made object has achieved this speed, and a judge would not be impressed with the argument.

61. (a) 3.0 mm (b) $I_{\text{laser}} = 7.3 \times 10^{-7}$ W/m^2, which is about one billion times less intense than sunlight ($I_{\text{sunlight}} = 1000$ W/m^2, about the same as on the Earth) and is therefore not destructive.

CHAPTER 24

Concept Checks

24.1 (a)

24.2 (b); 19°

24.3 Total internal reflection is possible in cases (a), (c), and (e) because in these cases light is traveling into the material with the larger index of refraction.

24.4 (a)

24.5 (b)

24.6 (b)

24.7 (a) positive; (b) positive; (c) positive

24.8 (a) positive; (b) negative; (c) negative

24.9 (b)

Problems

1. 8.3 m

3. 70°

5. 1; upright

7. The two images are at (70 cm, −20 cm) and (−20 cm, 80 cm).

9. 24

11. 130 m

13. 1.36

15. 1.2×10^{8} m/s

17. 4.6×10^{14} Hz

19. 65°

21. 49°

23. 1.7

25. 54°; the light starts in the glass.

27. (b) 0.70 cm

29. 21° below the horizontal

31. Ray 1:

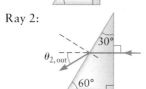

$\theta_{1,\,out} = 24°$ below the horizontal

Ray 2:

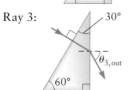

$\theta_{2,\,out} = 31°$ below the horizontal

Ray 3:

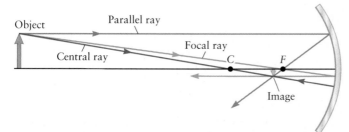

$\theta_{3,\,out} = 61°$ below the horizontal

Ray 4:

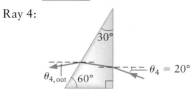

$\theta_{4,\,out} = 4°$ below the horizontal

33. No light emerges.

35. 0.49 m

37. (a)

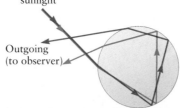

(b) $s_i = 0.090$ m = 9.0 cm; on same side of mirror as object (c) real (but inverted); 1.0 cm tall

39. Smaller

41. −0.11

43. 32 cm

45. (a)

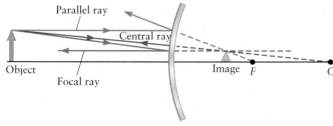

(b) $s_i = -0.083$ m = −8.3 cm; behind the mirror
(c) Virtual; 1.7 cm tall

47. Behind the mirror

49. 0.58 m = 58 cm

51. (a) +0.24 m (in front of the mirror) (b) Real (c) 0.69

53. (a) 0.24 (b) Now m is 0.45.

55. $s_i = -10$ cm (image is behind mirror)

57. (a)

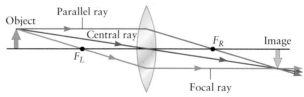

(b) $s_i = +10$ cm; $h_i = -1.5$ cm

59. (a)

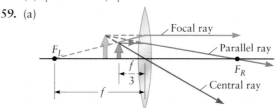

(b) $s_i = -f/2$ (a distance $f/2$ in front of the lens)
(c) Virtual

61. −70 cm

63. $s_i = -67$ cm (in front of the lens)

65. Smaller

67. The focal length increases by a factor of 2.9. The closer the index of refraction of the lens and the surrounding medium, the smaller is the refraction, so the rays travel a longer distance before crossing at the focal point.

69. (a) 14 cm from the lens (b) 6.0 cm from the lens

71. 1.3 cm

73. (a) $h_i = -0.44$ cm (inverted) (b) Smaller; it is only two-thirds as large as before.

75. (b); refraction and reflection take place in the droplets, but only refraction depends on wavelength. Different colors refract at different angles due to dispersion.

77.

Consider incoming red and blue rays as shown. Both are reflected twice and are refracted twice, but the angles of refraction for the blue rays are greater. That gives the outgoing blue ray a steeper angle than the outgoing red ray, which is opposite to the order of colors in the primary rainbow (Fig. 24.48B and C).

79. 65 cm

81. −45 cm

83. 58°

85. 2.1

87. 10 cm

89. 128°

91. (a) −9.0 cm (inside the bowl)
    (b) −6.0 cm (inside the bowl)

93. 1.6 m

95. (a) $s_{i2} = -33$ cm (behind the mirror) (b) −1.3
    (c) Virtual, inverted, enlarged

97. 1.6 mm/s toward the lens

99. (a) 45 cm (b) The transparency should be upright on the projector if the screen is at your back.

101. (a) 9.0 m (b) As the pinhole becomes smaller, the image becomes sharper but the intensity of light decreases.

103. Ⓡ $h_{Sun} = 28$ mm; $h_{Moon} = 27$ mm

CHAPTER 25

Concept Checks

25.1 It is *not* possible to observe interference with light from two *different* helium–neon lasers. Light from the two lasers is emitted by different atoms, and light emitted by independent sources is not coherent.

25.2 (a)

25.3 Wave 2 in Figure 25.12A is reflected when it reaches the air interface at the bottom of the uppermost glass plate. The index of refraction is larger for glass than for air, similar to the reflection in Figure 25.9B, so there is no phase change at this reflection. On the other hand, wave 3 reflects back into air from the air–glass interface at the top of the lower plate, with a phase change of 180°. At the far left edge of the wedge in Figure 25.12 the path length difference between the two reflected waves is zero, so the only phase difference comes from the 180° phase change of wave 3. The interference is thus destructive, and the leftmost fringe is dark.

25.4 (b)

25.5 (b)

25.6 (a)

25.7 (b)

Problems

1. Constructively

3. Constructively

5. Two possible visible wavelengths are $\lambda = 600$ nm and $\lambda = 429$ nm.

7. Constructive interference

9. 0.73 m

11. 5800

13. 1000

15. (a) 3.2×10^6 (b) The uncertainty in N (c) Yes

17. If film appears red at normal incidence, $d_{film} = 110$ nm.

19. 450 nm

21. 93 nm

23. 640 nm or 700 nm

25. (a) 3.0 cm (b) 3.0 cm

27. (a)

29. 470 nm

31. (a) 3.8 mm (b) 3.8 mm

33. 630 nm

35. Ⓡ $(3 \times 10^{-4})°$

37. 200 μm

39. 2.8°

41. 5.2×10^{-4} m

43. 0, 5.7°, and 12°

45. 180 slits/mm

47. Ⓡ 9.4 cm

49. 7.3×10^{-4} rad = 0.042°

51. Ⓡ 1300 m

53. Ⓡ (a) 1.6×10^6 m (b) 8:1

55. 96 nm

57. 46 μm

59. (a) 0.89 arc second (b) 180 m

61. (a) 0.007° (b) 0.012° (c) 0.0040° (d) No (e) 3.4 mm

63. Ⓡ $\lambda \approx 200$ nm, which is a factor of two smaller than the wavelength of blue light

CHAPTER 26

Concept Checks

26.1 (c)

26.2 (a)

26.3 A magnifying glass produces a virtual upright image as can be seen from Figure 26.6. We can predict the nature of the image from the thin-lens equation,

$$\frac{1}{s_o} + \frac{1}{s_i} = \frac{1}{f}$$

The converging lens of the magnifying glass has a positive focal length f. The object distance is positive for an object to the left of the lens in Figure 26.7C (according to our sign conventions). The object is very close to the lens, so $s_o < f$. The solution for s_i from the thin-lens equation then always gives a negative image distance ($s_i < 0$), corresponding to a virtual image. Applying the result for the magnification from Chapter 24 (Eq. 24.18) shows that this virtual image is always upright.

26.4 The image is real because light passes through the image point. Light must strike the film or CCD detector to be recorded, so all cameras must produce a real image.

26.5 (d)

26.6 (c)

26.7 (b)

Problems

1. -2.5 m
3. -8.9 m; diverging
5. 2.4 cm
7. Ⓡ (a) Diverging (b) -2.0 m (c) -0.51 diopter
9. 20 cm; it is large for a contact lens, but the assumption that $R_2 = \infty$ is not realistic.
11. 2.7 diopters
13. Diverging lens with $f = -2.0$ m
15. (a) Increase (b) 1.8 cm
17. (a) The lens with $f = 2.3$ cm (b) $m_{\theta, 1} = 14$; $m_{\theta, 2} = 5$ (c) 5.2 cm
19. 25 cm; 1.4
21. Ⓡ 7.3 μm
23. It would increase by a factor of three to a new value of 300.
25. (a) 2.5 cm (b) The objective lens produces an inverted image that serves as the object for the eyepiece.
27. Highest is 640×; lowest is 66×
29. (a) 60× (b) 76 cm
31. Ⓡ About 6 km
33. 0.72 rad = 41°; 400×
35. (a) 60 cm (b) 79 cm (c) 26 cm
37. Decrease shutter speed (increase exposure time) by factor of four.
39. Ⓡ 0.0040 s
41. (a) 2.1 (b) Reduced by factor of nine
43. (a) 3.1 μm (b) 1.5 cm (c) 1.6 cm
45. 1.6
47. Ⓡ There are 3.7× as many pits in a Blu-Ray disc. A single-layer Blu-Ray disc can store 36 times as much information as a CD, so diffraction effects do not completely explain the difference in storage capacity.
49. (a) 90° (b) 59°
51. Ⓡ Focal length of eyepiece, $f_{\text{eyepiece}} = 2.5$ cm; magnification of objective, $m_{\text{obj}} = 50×$; magnification of eyepiece, $m_{\text{eyepiece}} = 10×$
53. (a) $m_{\theta, 1} = 7.3$; $m_{\theta, 2} = 3.1$ (b) 11× (c) 3.0×; 16 cm
55. The focal length and approximate telescope length are both 19 m; yes.
57. Ⓡ (a) 261 nm (b) 65 nm; shallower (c) For a CD, 8 Mb/cm²; for a DVD, 26 Mb/cm²; for a Blu-Ray disc, 120 Mb/cm²
59. 28 m

CHAPTER 27

Concept Checks

27.1 (a) and (c)
27.2 (c)
27.3 (b)
27.4 (a)
27.5 (c)
27.6 (a)
27.7 (b). R is about 10 times the radius of the Sun.

Problems

1. (a) v in $+x$ direction (b) v in $+x$ direction (c) 0 (d) 0 (e) 0; yes. The laws of motion are the same in all inertial reference frames. Since Newton's second law is obeyed in Ted's frame, it will be obeyed in Alice's frame.
3. 2.5×10^{-5} percent
5. (a) Ted (b) 57 s
7. $0.94c$
9. 35 yr
11. 0.19 s
13. (a) No (b) The explosion at the front of the railroad car
15. $0.66c$
17. 0.37 m
19. $0.47c$
21. $0.44c$
23. 7.1 m
25. $0.75c$
27. $0.99995c$
29. 1700 s
31. 49.99999999999967 m/s
33. (a) 2.6×10^{-22} kg · m/s (b) 8.3×10^{-22} kg · m/s (c) At high speeds, the electron responds to forces and collisions as if its mass were greater than the rest mass.
35. $0.995c$
37. $0.98c$
39. $0.98c$
41. 5.1×10^5 V
43. $0.80c$
45. $(4.4 \times 10^{-23})c$
47. 1.1×10^{-35} kg
49. 5.1×10^{-11} kg
51. 5.7×10^{38} s or 1.8×10^{31} yr
53. 2.7×10^{-26} kg · m/s
55. $\sqrt{15}mc$
57. 3.0×10^9 J
59. Ⓡ (a) 4.4×10^9 kg/s (b) 1.4×10^{13} yr
61. Ⓡ (a) 1.2×10^{-29} kg (b) 1.1×10^{-12} J (c) The energy goes into kinetic energies of the emitted ^{212}Pb nucleus and the alpha particle.
63. (a) 1.6×10^{-14} J; 1.9×10^8 m/s; 1.7×10^{-22} kg · m/s (b) 1.6×10^{-14} J; 1.6×10^8 m/s; 1.8×10^{-22} kg · m/s
67. 3.54×10^{-11} eV
69. Without relativity, you die before reaching Mars in either reference frame. If relativity is included, you live.

CHAPTER 28

Concept Checks

28.1 (b)

28.2 (c)

28.3 (c)

28.4 (b)

28.5 The electron in state 2 has the higher kinetic energy.

Problems

1. Ⓡ (a) 7×10^{-26} J (b) 1.6×10^{-24} J (c) 5.7×10^{-25} J (d) 4.0×10^{-19} J

3. 3.2×10^{17}

5. 2.2×10^{-18} J

7. 1.4×10^{15} Hz

9. Ⓡ About 3×10^{21}

11. (a) 1.9×10^{-22} J (b) Microwave

13. 1.3×10^{15} Hz; ultraviolet

15. 8.9×10^{19}

17. 1.2×10^{15} Hz

19. Ⓡ (a) 5.0×10^{-16} J (b) X-ray

21. 4.1 eV

23. Ⓡ (a) 7300 K (b) 4500 K

25. (a) $f = 3.3 \times 10^{15}$ Hz, $\lambda = 91$ nm; ultraviolet; no (b) 1.4 eV ($= 2.2 \times 10^{-19}$ J), 7.0×10^5 m/s (c) $f = 1.6 \times 10^{15}$ Hz, $\lambda = 180$ nm; no (still ultraviolet)

27. 1.1 m/s

29. (a) 9.4×10^{19} (b) 1.0×10^{-7} N (c) 2.0×10^{-7} N

31. 210:1

33. (a) 9.5×10^{-25} kg · m/s (b) 570

35. (a) 5.5×10^{-13} m (b) Yes; the wavelength of the electrons is smaller for the TEM. (c) The electron wavelength is much smaller than either.

37. (a) 360 nm (b) $E_{\text{photon}}/E_{\text{electron}} = 3.0 \times 10^5$

39. (a) 2200 m/s (b) 0.18 nm (c) The wavelength is smaller than the spacing between atoms by a factor of almost two.

41. 1.3×10^{-18} J (= 8.1 eV), 0.43 nm

43. 1800:1

45. Ⓡ 5×10^{-33} m/s

47. 0.15° (= 0.0026 rad)

49. (a) 9.2×10^{-24} C · J · s/kg

51. 1.1×10^{-25} kg · m/s

53. 5×10^{-20} kg · m/s; 8×10^{-13} eV

55. (a) 6.0×10^{-22} J (b) 2.4×10^{-21} J

57. (a) 1.1×10^{-25} kg · m/s (b) 1.2×10^5 m/s (c) 4.3×10^{-15} s (d) $\Delta E = 1.2 \times 10^{-20}$ J. The kinetic energy calculated from the speed in part (b) is 0.61×10^{-20} J. It is within a factor of two.

59. Blue, 4.6×10^{-19} J; green, 3.8×10^{-19} J; red, 3.5×10^{-19} J

61. 2.0 eV

63. 3.3×10^{-28} m

65. (a) Gamma-ray photon, 1.4×10^{-16} J = 860 eV; visible photon, 3.6×10^{-19} J = 2.3 eV (b) $E_\gamma/E_{\text{visible}} = 380$; nuclear processes have higher energies.

67. (a) 820 V (b) 0.83 MV (c) $KE_{\text{proton}}/KE_{\text{electron}} = 9.9 \times 10^{-4}$

69. 2.8×10^{-25} s; the shorter the lifetime, the less well defined the mass.

71. (a) 0.0087 nm (b) 2.2 μm (c) 5.3×10^{-29} kg · m/s (d) 3.0 ns (e) 0.17 μm (f) The spread of the beam as it hits the screen as calculated by diffraction is about a factor of 4π greater than that calculated using the uncertainty principle, in agreement with the discussion in Section 28.5.

CHAPTER 29

Concept Checks

29.1 (b)

29.2 (c)

29.3 chlorine (Cl)

29.4 (b)

29.5 (a) and (c)

29.6 (b)

Problems

1. (a) 4×10^{-45} m³ (b) 5×10^{-31} m³ (c) 8×10^{-15}

3. (a) 3×10^{-10} (b) 300

5. 2.5 eV, 5.0 eV, and 7.5 eV

7. 3.28×10^{15} Hz

9. $E_3 = -1.5$ eV; $E_{\text{photon}} = 12$ eV

11. 0.85 eV

13. 4

15. $n_{\text{initial}} = 4$, $n_{\text{final}} = 2$

17. 1.0×10^{-9} m; larger

19. (a) 137 (b) -7.2×10^{-4} eV

21. 7.4 eV

23. 18

25. 4

27. (a) Cu (copper) (b) Br (bromine) (c) Sr (strontium)

29. (a) $1s^2 2s^2 2p^6 3s^2 3p^6 3d^{10} 4s^2 4p^2$ (b) $1s^2 2s^2 2p^6 3s^2 3p^6$ (c) $1s^2 2s^2 2p^6 3s^2$ (d) $1s^2 2s^2 2p^6 3s^2 3p^2$ (e) $1s^2 2s^2 2p^6 3s^2 3p^6 3d^{10} 4s^2 4p^5$

31. (a) 10 (b) 2 (c) 32 (d) 14 (e) 18 (f) 22

33. Be (beryllium), Mg (magnesium), Sr (strontium), Ba (barium), and Ra (radium)

35. (a) 4.3×10^{14} Hz (b) 1.8 eV

37. 5.1×10^4 m/s

39. (a) 4.2×10^{-6} m (b) It is about four times larger than the diameter of a dust particle.

41. (a) $\lambda_{5\to2} = 430$ nm; $\lambda_{2\to1} = 120$ nm (b) The $5 \to 2$ transition is in the visible; the $2 \to 1$ transition is in the ultraviolet.

43. (a) 18

(b)

n	ℓ	m	s
3	0	0	$\frac{1}{2}$
3	0	0	$-\frac{1}{2}$
3	1	-1	$\frac{1}{2}$
3	1	-1	$-\frac{1}{2}$
3	1	0	$\frac{1}{2}$
3	1	0	$-\frac{1}{2}$
3	1	1	$\frac{1}{2}$
3	1	1	$-\frac{1}{2}$
3	2	-2	$\frac{1}{2}$
3	2	-2	$-\frac{1}{2}$
3	2	-1	$\frac{1}{2}$
3	2	-1	$-\frac{1}{2}$
3	2	0	$\frac{1}{2}$
3	2	0	$-\frac{1}{2}$
3	2	1	$\frac{1}{2}$
3	2	1	$-\frac{1}{2}$
3	2	2	$\frac{1}{2}$
3	2	2	$-\frac{1}{2}$

45. (a) Ne (neon) (b) S (sulfur) (c) Be (beryllium)

47. $1s^3$

49. (a) 2.5×10^{-19} C = about $1.6\times$ the charge on a proton (b) 1.4×10^{-10} m

51. Any transition down to $n = 4$ starting in states $n = 6$ through $n = 21$ will result in a visible wavelength.

53. (a) Ground state $r_1 = 2.6 \times 10^{-13}$ m; first excited state $r_2 = 1.0 \times 10^{-12}$ m; second excited state $r_3 = 2.3 \times 10^{-12}$ m (b) $E_1 = -4.5 \times 10^{-16}$ J = 2800 eV; $E_2 = -1.1 \times 10^{-16}$ J = 700 eV; $E_3 = -5.0 \times 10^{-17}$ J = 310 eV (c) 0.50 nm

55. (a) 5.0×10^{-11} J (b) 4.9×10^7 (c) 1.6×10^8 (d) 6.4 eV (e) No

57. Ⓡ (a) For aluminum, 2.0 kV; for copper, 11 kV; for tungsten, 72 kV (b) E_2 (aluminum) $= -3.6 \times 10^2$ eV; E_2 (copper) $= -2.7 \times 10^3$ eV; E_2 (tungsten) $= -1.4 \times 10^4$ eV

59. (a) 500 nm and 2500 nm (b) $\lambda_{2\rightarrow1} = 2500$ nm; $\lambda_{1\rightarrow0} = 620$ nm; $\lambda_{2\rightarrow0} = 500$ nm

CHAPTER 30

Concept Checks

30.1 $_{6}^{14}$C

30.2 (a) The wavelength of blue light is approximately 400 nm, so gamma rays have a *much* shorter wavelength.

30.3 (c)

30.4 (c)

30.5 (d)

30.6 (b)

30.7 2

Problems

1. (b)

3. (a) and (d)

5. (a) $Z = 6$, $N = 6$, $A = 12$ (b) $Z = 16$, $N = 18$, $A = 34$ (c) $Z = 82$, $N = 126$, $A = 208$ (d) $Z = 26$, $N = 30$, $A = 56$

7. Mercury, $_{80}^{204}$Hg

9. (a) 1.9 fm (b) 3.9 fm (c) 4.6 fm

11. 3.7

13. 7.6×10^6 m/s

15. 3.00×10^8 m/s

17. (c)

19. 7.5×10^5 V

21. (a)

23. Beta decay produces $_{95}^{242}$Am; alpha decay produces $_{92}^{238}$U.

25. (a) $_{90}^{234}$Th $\rightarrow$ $_{88}^{230}$Ra + $_{2}^{4}$He; alpha decay

(b) $_{90}^{234}$Th $\rightarrow$ $_{91}^{234}$Pa + e^- + $\bar{\nu}$; beta decay

(c) $_{90}^{234}$Th* $\rightarrow$ $_{90}^{234}$Th + γ; gamma decay

27. $_{27}^{60}$Co $\rightarrow$ $_{28}^{60}$Ni + e^- + $\bar{\nu}$

29. (a) 8.0 g (b) 1.0 g (c) 0.21 g

31. 0.028 s

33. 7.3 g

35. 4.7 yr

37. (a) $_{1}^{1}$H + $_{3}^{6}$Li $\rightarrow$ $_{2}^{4}$He + $_{2}^{3}$He

(b) $_{0}^{1}$n + $_{92}^{235}$U $\rightarrow$ $_{51}^{133}$Sb + $_{41}^{99}$Nb + 4_{0}^{1}n

(c) $_{6}^{14}$C $\rightarrow$ $_{7}^{14}$N + e^- + $\bar{\nu}$

39. 7.9 MeV/nucleon for a total binding energy of 1600 MeV

41. $_{37}^{89}$Rb

43. 2.5 mCi

45. (a) 6.5×10^{-13} J (b) 6.5×10^4 J

47. 0.62 kg

49. 180 MeV

51. 5.0 rads

53. 1.7 mrad

55. 26 J

57. Ⓡ (a) 5×10^{-4} J (b) 0.5 mm

59. 11 h

61. 0.015 J

63. 28 μBq

65. 32,000 yr

67. 5400 yr

69. It would take 2900 yr, which would be a problem because discarded smoke detectors will keep emitting radiation for a long time.

71. 0.32

73. 2 kg or 6×10^{24} atoms

75. (a) 7.0×10^{-7} s^{-1} (b) 33 days

77. 3×10^{10} J

79. 29 yr

81. 1:4

83. Ⓡ 0.002

85. (a) 80 decays/s = 80 Bq (b) 0.080 count/s

87. Ⓡ (a) 168,000 yr ago (b) About 80 days, the same as the currently measured half-life

89. The mass of the three alpha particles is greater than the mass of ^{12}C, so the reaction cannot occur spontaneously.

CHAPTER 31

Concept Checks

31.1   (b)

31.2   (d)

Problems

1. $\approx 3.0 \times 10^8$ m/s; very nearly equal to the speed of light

3. 6.1×10^{10} eV

5. 2.3×10^{23} Hz

7. 0.85 MeV

9. (a) 1880 MeV (b) 180.0 MeV (c) For protons and antiprotons, 9×10^{16} J/kg; for $^{235}_{92}U$, 9×10^{13} J/kg

11. 67.5 MeV

13. Three up, charm, or top antiquarks; antimatter; baryon

15. (a) π^- (b) There are no quarks in the final state. (c) Yes; zero

17. Ⓡ 100 N

19. (a) 3.6×10^9 ly (b) 581 nm

21. 1.4×10^{10} ly

23. Ⓡ 2.6×10^{16} J = 6.2 megatons TNT

25. (a) 2.1% (b) 48% (c) 0

27. (a) 25 billion years, 11 billion years (b) 14 billion years

29. (a) $t_{Planck} = \sqrt{\dfrac{Gh}{c^5}} = 1.3 \times 10^{-43}$ s (b) 7.7×10^{24}

Particle-wave duality, 964–965
Particles
    as waves, 965–968
    electrons as, 955–956, 955f, 956f
    properties of, 955
    vs. waves, 841, 955
Particles, "point," 205, 240
Pascal, 310, 319, 319t
    Blaise, 310
Pascal's principle, 321–324, 322f–324f
Path length difference, 844
Pauli exclusion principle, 1005
Payload dropping, 106–107, 107f
Pendulum
    arm as, 359, 359f
    simple, 356–358, 356f–357f
Perception, of sound, 412–413, 412f–413f
Period
    measuring, 349–350, 349f
    of circular motion, 131
    of harmonic motion, 349, 349f
    of rotational motion, 244–245, 244f
Periodic table, 1007–1010, 1008t, 1009f
Periodic wave, 379
Permanent magnets, 645, 667–668
Permitivity of free space, 532
Perpendicular distance, in wave fronts, 796
Perpetual motion, 512, 512f
PET scans, 1034, 1048
Phase, 841, 849–850, 849f
Phase angle, 732
Phase changes, 439–448
Phase diagram, 440
Phases of matter, 439–448
Phasor analysis, of RC circuit, 742, 742f
Phasors, 730–731, 731f
Phi particles, 1066t
Philosophy
    quantum mechanics and, 1015–1016
    uncertainty and, 974
Photoelectric effect, 957–959, 958f
Photons, 157, 871, 957–965, 978–979, 979f, 1012
Photosynthesis, 519–520, 520f
Physics
    definition of, 2
    numbers in, 4–9
    prediction and, 2
    problem solving in, 3–4
    purpose of, 2–3
    quantities in, 9–12
Pi zero particle, 1066t
Piano string, 385–386, 386f
Pier, refraction on, 802, 802f
Pinhole camera, 904–905, 904f, 905f
Pions, 1066t
Pitch, of note, 395, 409–410
Pivot point, 247, 255–256, 256f
Planck's hypothesis, 963–964
Plane mirror, reflection from, 797–800, 798f–800f
Plane waves, 389f, 390

Plane wings, 331–332, 331f–332f
Planetary model of atom, 988–989
Planetary motion
    angular momentum and, 293–294, 293f–294f
    elliptical orbits and, 151–152
    geosynchronous, 154
    gravity and, 145, 145f
    Kepler's laws and, 150–155
    of Jupiter, 152
    of Moon, 146, 146f
    of satellites, 153–154, 154f, 182–183
Plants
    capillary pressure in, 338–339
    photosynthesis in, 519–520, 520f
Plastic deformations, 363–364, 363f
Plates, electric field between, 551–552, 551f
Platinum, 314t
Playground swing, 358
Plexiglas, 589t, 801t
Plum-pudding model of atom, 987–988, 987f
Pluto, 147t
Point charge
    electron as, 555
    flux for, 547–548, 547f
    force on, 537, 537f
    in motion, 569–570, 569f
    potential due to, 575, 575f
    potential energy of two, 566–567, 566f
Point particles, 205, 240
Poiseuille's law, 334–335, 334f
Polarization, 544, 544f, 779–783, 779f, 781f–783f
Polarizers, 779, 779f
Poles
    geographic, 670, 670f
    magnetic, 645, 645f, 670, 670f
Pollution, air, 533, 552, 552f
Polyatomic gases, 481
Polystyrene, 314t
Population, of Earth, 5
Position
    as function of time, 350f, 351, 351f, 352f
    constant nonzero acceleration and, 55–57
    final, 30
    harmonic motion and, 350f, 351
    initial, 30
    instantaneous velocity and, 32
    velocity and, 30–31
Positive charge, 530
Positive torque, 251
Positron, 1028
Positron–electron annihilation, 941, 1062–1063
Postulates of relativity, 919–921, 920f
Potential, electric
    acceleration of charged particles and, 571–573, 571f
    capacitance and, 581–582, 581f

changes in, 575
definition of, 564, 571
electric field and, 573, 573f
electric potential energy and, 565–570
equipotential surfaces and, 579–581, 579f–581f
from point charge, 575, 575f
ground and, 575
in battery, 624, 624f
in capacitors in series, 584–585, 584f
in circuits, 606–607, 607f
in one-loop circuits, 613–614, 614f
in parallel capacitors, 585–586, 586f
lightning and, 590–591, 590f, 591f
lightning rod and, 576–578, 577f
potential difference and, 572
unit of, 571
voltage and, 570–578
Potential difference, 572
Potential energy
    as scalar, 174–175
    as stored energy, 175, 566, 566f
    changes in, 181–182, 181f
    conservative forces and, 175–176
    definition of, 174
    elastic forces and, 183–185, 184f–185f
    electric
        electric field and, 570–571, 571f
        electric potential and, 564
        of atomic nucleus, 1024–1025, 1024f
        overview of, 565, 565f
        superposition and, 567–568
    electric current and, 604, 604f
    functions, 182–189
    gravitational, 182
    gravity and, 174–175, 174f
    in hydrogen atom, 567, 567f
    in muscles, 195–196, 195f
    in solar system, 182
    of two point charges, 566–567, 566f
    retrieval of, 192–193
    total, 188–189, 188f
    with multiple forces, 188–189, 188f
    work and, 174–182
    work–energy theorem and, 176
Power
    definition of, 193
    efficiency and, 195
    force and, 195
    in AC circuit, 727–728
    in capacitor circuit, 733–734, 734f
    in inductor circuit, 736, 736f
    in lightbulb, 729, 729f
    notations, 730t
    of household appliances, 193, 194t
    time and, 193
    units of, 193
    velocity and, 195
    wave, 389
Power plant, nuclear, 1041–1042, 1042f
Powers of ten, 12
Precession, 296–297, 296f
Prediction, in physics, 2

Periodic Table of the Elements

Transition elements

Legend:
- Symbol — Ca 20 — Atomic number
- Atomic mass† — 40.078
- $4s^2$ — Configuration of outermost electrons

Main Table

Group I	Group II											Group III	Group IV	Group V	Group VI	Group VII	Group 0
H 1 1.007 9 $1s^1$																	He 2 4.002 6 $1s^2$
Li 3 6.941 $2s^1$	Be 4 9.012 2 $2s^2$											B 5 10.811 $2p^1$	C 6 12.011 $2p^2$	N 7 14.007 $2p^3$	O 8 15.999 $2p^4$	F 9 18.998 $2p^5$	Ne 10 20.180 $2p^6$
Na 11 22.990 $3s^1$	Mg 12 24.305 $3s^2$											Al 13 26.982 $3p^1$	Si 14 28.086 $3p^2$	P 15 30.974 $3p^3$	S 16 32.066 $3p^4$	Cl 17 35.453 $3p^5$	Ar 18 39.948 $3p^6$
K 19 39.098 $4s^1$	Ca 20 40.078 $4s^2$	Sc 21 44.956 $3d^14s^2$	Ti 22 47.867 $3d^24s^2$	V 23 50.942 $3d^34s^2$	Cr 24 51.996 $3d^54s^1$	Mn 25 54.938 $3d^54s^2$	Fe 26 55.845 $3d^64s^2$	Co 27 58.933 $3d^74s^2$	Ni 28 58.693 $3d^84s^2$	Cu 29 63.546 $3d^{10}4s^1$	Zn 30 65.41 $3d^{10}4s^2$	Ga 31 69.723 $4p^1$	Ge 32 72.64 $4p^2$	As 33 74.922 $4p^3$	Se 34 78.96 $4p^4$	Br 35 79.904 $4p^5$	Kr 36 83.80 $4p^6$
Rb 37 85.468 $5s^1$	Sr 38 87.62 $5s^2$	Y 39 88.906 $4d^15s^2$	Zr 40 91.224 $4d^25s^2$	Nb 41 92.906 $4d^45s^1$	Mo 42 95.94 $4d^55s^1$	Tc 43 (98) $4d^55s^2$	Ru 44 101.07 $4d^75s^1$	Rh 45 102.91 $4d^85s^1$	Pd 46 106.42 $4d^{10}$	Ag 47 107.87 $4d^{10}5s^1$	Cd 48 112.41 $4d^{10}5s^2$	In 49 114.82 $5p^1$	Sn 50 118.71 $5p^2$	Sb 51 121.76 $5p^3$	Te 52 127.60 $5p^4$	I 53 126.90 $5p^5$	Xe 54 131.29 $5p^6$
Cs 55 132.91 $6s^1$	Ba 56 137.33 $6s^2$	57–71*	Hf 72 178.49 $5d^26s^2$	Ta 73 180.95 $5d^36s^2$	W 74 183.84 $5d^46s^2$	Re 75 186.21 $5d^56s^2$	Os 76 190.23 $5d^66s^2$	Ir 77 192.2 $5d^76s^2$	Pt 78 195.08 $5d^96s^1$	Au 79 196.97 $5d^{10}6s^1$	Hg 80 200.59 $5d^{10}6s^2$	Tl 81 204.38 $6p^1$	Pb 82 207.2 $6p^2$	Bi 83 208.98 $6p^3$	Po 84 (209) $6p^4$	At 85 (210) $6p^5$	Rn 86 (222) $6p^6$
Fr 87 (223) $7s^1$	Ra 88 (226) $7s^2$	89–103**	Rf 104 (261) $6d^27s^2$	Db 105 (262) $6d^37s^2$	Sg 106 (266)	Bh 107 (264)	Hs 108 (277)	Mt 109 (268)	Ds 110 (271)	Rg 111 (272)							

*Lanthanide series

La 57 138.91 $5d^16s^2$	Ce 58 140.12 $5d^14f^16s^2$	Pr 59 140.91 $4f^36s^2$	Nd 60 144.24 $4f^46s^2$	Pm 61 (145) $4f^56s^2$	Sm 62 150.36 $4f^66s^2$	Eu 63 151.96 $4f^76s^2$	Gd 64 157.25 $4f^75d^16s^2$	Tb 65 158.93 $4f^85d^16s^2$	Dy 66 162.50 $4f^{10}6s^2$	Ho 67 164.93 $4f^{11}6s^2$	Er 68 167.26 $4f^{12}6s^2$	Tm 69 168.93 $4f^{13}6s^2$	Yb 70 173.04 $4f^{14}6s^2$	Lu 71 174.97 $4f^{14}5d^16s^2$

**Actinide series

Ac 89 (227) $6d^17s^2$	Th 90 232.04 $6d^27s^2$	Pa 91 231.04 $5f^26d^17s^2$	U 92 238.03 $5f^36d^17s^2$	Np 93 (237) $5f^46d^17s^2$	Pu 94 (244) $5f^67s^2$	Am 95 (243) $5f^77s^2$	Cm 96 (247) $5f^76d^17s^2$	Bk 97 (247) $5f^86d^17s^2$	Cf 98 (251) $5f^{10}7s^2$	Es 99 (252) $5f^{11}7s^2$	Fm 100 (257) $5f^{12}7s^2$	Md 101 (258) $5f^{13}7s^2$	No 102 (259) $5f^{14}7s^2$	Lr 103 (262) $6d^15f^{14}7s^2$

Note: Atomic mass values given are averaged over isotopes in the percentages in which they exist in nature.

† For an unstable element, mass number of the most stable known isotope is given in parentheses.

†† From the International Union of Pure and Applied Chemistry, June 2007.

See *old.iupac.org/reports/periodic_table/index.html.*

Powers of 10

Prefix	Factor
atto (a)	10^{-18}
femto (f)	10^{-15}
pico (p)	10^{-12}
nano (n)	10^{-9}
micro (μ)	10^{-6}
milli (m)	10^{-3}
centi (c)	10^{-2}
deci (d)	10^{-1}
kilo (k)	10^{3}
mega (M)	10^{6}
giga (G)	10^{9}
tera (T)	10^{12}
peta (P)	10^{15}
exa (E)	10^{18}

Useful Mathematical Formulas

Circumference of a circle $= 2\pi r$

Area of a circle $= \pi r^2$

Area of a triangle $= \frac{1}{2}(\text{base} \times \text{height})$

Surface area of a sphere $= 4\pi r^2$

Volume of a sphere $= \frac{4}{3}\pi r^3$

Volume of a cylinder $= \pi r^2 L$

$$ax^2 + bx + c = 0 \Rightarrow x = \frac{-b \pm \sqrt{b^2 - 4ac}}{2a}$$

Degrees and Radians

2π rad $= 360°$

1 revolution $= 1$ rev $= 360° = 2\pi$ rad

$1° = 60$ arcmin $= 60'$

1 arcmin $= 60$ arcsec $= 60''$

$45° = \pi/4$ rad

$90° = \pi/2$ rad

Exponential Functions and Logarithms

$x = e^y \Leftrightarrow \ln x = y$

$e^{xy} = e^x e^y$

$\ln(xy) = \ln x + \ln y$

$\ln\left(\dfrac{x}{y}\right) = \ln x - \ln y$

$x = 10^y \Leftrightarrow \log x = y$

$e^0 = 1$

$\ln(1) = 0$

$10^0 = 1$

$\log(1) = 0$

$\log(10) = 1$